# General Physics
# PHY 132

## DAVID HALLIDAY • ROBERT RESNICK • JEARL WALKER

### California State Polytechnic University

— Wiley Custom Learning Solutions —

ISBN 978-1-119-31450-9
Printed and bound by LSC Communications.

10 9 8 7 6 5 4 3 2

# Contents

**13** Gravitation   354

**14** Fluids   386

**15** Oscillations   413

**16** Waves–I   444

**17** Waves–II   479

**18** Temperature, Heat, and the First Law of Thermodynamics   514

**19** The Kinetic Theory of Gases   549

**20** Entropy and the Second Law of Thermodynamics   583

**Appendices**   A-1

**Answers to Checkpoints and Odd-Numbered Questions and Problems**   AN-1

**Index**   I-1

# MATHEMATICAL FORMULAS*

### Quadratic Formula

If $ax^2 + bx + c = 0$, then $x = \dfrac{-b \pm \sqrt{b^2 - 4ac}}{2a}$

### Binomial Theorem

$$(1 + x)^n = 1 + \frac{nx}{1!} + \frac{n(n-1)x^2}{2!} + \cdots \qquad (x^2 < 1)$$

### Products of Vectors

Let $\theta$ be the smaller of the two angles between $\vec{a}$ and $\vec{b}$. Then

$$\vec{a} \cdot \vec{b} = \vec{b} \cdot \vec{a} = a_x b_x + a_y b_y + a_z b_z = ab \cos \theta$$

$$\vec{a} \times \vec{b} = -\vec{b} \times \vec{a} = \begin{vmatrix} \hat{i} & \hat{j} & \hat{k} \\ a_x & a_y & a_z \\ b_x & b_y & b_z \end{vmatrix}$$

$$= \hat{i} \begin{vmatrix} a_y & a_z \\ b_y & b_z \end{vmatrix} - \hat{j} \begin{vmatrix} a_x & a_z \\ b_x & b_z \end{vmatrix} + \hat{k} \begin{vmatrix} a_x & a_y \\ b_x & b_y \end{vmatrix}$$

$$= (a_y b_z - b_y a_z)\hat{i} + (a_z b_x - b_z a_x)\hat{j} + (a_x b_y - b_x a_y)\hat{k}$$

$$|\vec{a} \times \vec{b}| = ab \sin \theta$$

### Trigonometric Identities

$$\sin \alpha \pm \sin \beta = 2 \sin \tfrac{1}{2}(\alpha \pm \beta) \cos \tfrac{1}{2}(\alpha \mp \beta)$$

$$\cos \alpha + \cos \beta = 2 \cos \tfrac{1}{2}(\alpha + \beta) \cos \tfrac{1}{2}(\alpha - \beta)$$

*See Appendix E for a more complete list.

### Derivatives and Integrals

$$\frac{d}{dx} \sin x = \cos x \qquad \int \sin x \, dx = -\cos x$$

$$\frac{d}{dx} \cos x = -\sin x \qquad \int \cos x \, dx = \sin x$$

$$\frac{d}{dx} e^x = e^x \qquad \int e^x \, dx = e^x$$

$$\int \frac{dx}{\sqrt{x^2 + a^2}} = \ln(x + \sqrt{x^2 + a^2})$$

$$\int \frac{x \, dx}{(x^2 + a^2)^{3/2}} = -\frac{1}{(x^2 + a^2)^{1/2}}$$

$$\int \frac{dx}{(x^2 + a^2)^{3/2}} = \frac{x}{a^2(x^2 + a^2)^{1/2}}$$

### Cramer's Rule

Two simultaneous equations in unknowns $x$ and $y$,

$$a_1 x + b_1 y = c_1 \quad \text{and} \quad a_2 x + b_2 y = c_2,$$

have the solutions

$$x = \frac{\begin{vmatrix} c_1 & b_1 \\ c_2 & b_2 \end{vmatrix}}{\begin{vmatrix} a_1 & b_1 \\ a_2 & b_2 \end{vmatrix}} = \frac{c_1 b_2 - c_2 b_1}{a_1 b_2 - a_2 b_1}$$

and

$$y = \frac{\begin{vmatrix} a_1 & c_1 \\ a_2 & c_2 \end{vmatrix}}{\begin{vmatrix} a_1 & b_1 \\ a_2 & b_2 \end{vmatrix}} = \frac{a_1 c_2 - a_2 c_1}{a_1 b_2 - a_2 b_1}.$$

## SI PREFIXES*

| Factor | Prefix | Symbol | Factor | Prefix | Symbol |
|--------|--------|--------|--------|--------|--------|
| $10^{24}$ | yotta | Y | $10^{-1}$ | deci | d |
| $10^{21}$ | zetta | Z | $10^{-2}$ | centi | c |
| $10^{18}$ | exa | E | $10^{-3}$ | milli | m |
| $10^{15}$ | peta | P | $10^{-6}$ | micro | $\mu$ |
| $10^{12}$ | tera | T | $10^{-9}$ | nano | n |
| $10^{9}$ | giga | G | $10^{-12}$ | pico | p |
| $10^{6}$ | mega | M | $10^{-15}$ | femto | f |
| $10^{3}$ | kilo | k | $10^{-18}$ | atto | a |
| $10^{2}$ | hecto | h | $10^{-21}$ | zepto | z |
| $10^{1}$ | deka | da | $10^{-24}$ | yocto | y |

*In all cases, the first syllable is accented, as in ná-no-mé-ter.

# FUNDAMENTALS OF PHYSICS

## TENTH EDITION

EXTENDED

Halliday & Resnick

# FUNDAMENTALS OF PHYSICS
## TENTH EDITION

**JEARL WALKER**
CLEVELAND STATE UNIVERSITY

# WILEY

EXECUTIVE EDITOR  Stuart Johnson
SENIOR PRODUCT DESIGNER  Geraldine Osnato
CONTENT EDITOR  Alyson Rentrop
ASSOCIATE MARKETING DIRECTOR  Christine Kushner
TEXT and COVER DESIGNER Madelyn Lesure
PAGE MAKE-UP  Lee Goldstein
PHOTO EDITOR  Jennifer Atkins
COPYEDITOR  Helen Walden
PROOFREADER  Lilian Brady
SENIOR PRODUCTION EDITOR  Elizabeth Swain


COVER IMAGE © 2007 CERN


This book was set in 10/12 Times Ten by cMPreparé, CSR Francesca Monaco, and was printed and bound by Quad Graphics. The cover was printed by Quad Graphics.


This book is printed on acid free paper.

Library of Congress Cataloging-in-Publication Data

Walker, Jearl
  Fundamentals of physics / Jearl Walker, David Halliday, Robert Resnick—10th edition.
    volumes cm
  Includes index.
ISBN 978-1-118-23072-5 (Extended edition)
Binder-ready version  ISBN 978-1-118-23061-9 (Extended edition)
  1. Physics—Textbooks. I. Resnick, Robert. II. Halliday, David. III. Title.
    QC21.3.H35 2014
    530—dc23
                                                        2012035307

Printed in the United States of America
10 9 8 7 6 5 4 3 2 1

# BRIEF CONTENTS

VOLUME 1

1 Measurement

2 Motion Along a Straight Line

3 Vectors

4 Motion in Two and Three Dimensions

5 Force and Motion–I

6 Force and Motion–II

7 Kinetic Energy and Work

8 Potential Energy and Conservation of Energy

9 Center of Mass and Linear Momentum

10 Rotation

11 Rolling, Torque, and Angular Momentum

12 Equilibrium and Elasticity

13 Gravitation

14 Fluids

15 Oscillations

16 Waves–I

17 Waves–II

18 Temperature, Heat, and the First Law of Thermodynamics

19 The Kinetic Theory of Gases

20 Entropy and the Second Law of Thermodynamics

VOLUME 2

21 Coulomb's Law

22 Electric Fields

23 Gauss' Law

24 Electric Potential

25 Capacitance

26 Current and Resistance

27 Circuits

28 Magnetic Fields

29 Magnetic Fields Due to Currents

30 Induction and Inductance

31 Electromagnetic Oscillations and Alternating Current

32 Maxwell's Equations; Magnetism of Matter

33 Electromagnetic Waves

34 Images

35 Interference

36 Diffraction

37 Relativity

38 Photons and Matter Waves

39 More About Matter Waves

40 All About Atoms

41 Conduction of Electricity in Solids

42 Nuclear Physics

43 Energy from the Nucleus

44 Quarks, Leptons, and the Big Bang

Appendices / Answers to Checkpoints and Odd-Numbered Questions and Problems / Index

# C O N T E N T S

**1 Measurement** 1

**1-1 MEASURING THINGS, INCLUDING LENGTHS** 1
What Is Physics? 1
Measuring Things 1
The International System of Units 2
Changing Units 3
Length 3
Significant Figures and Decimal Places 4

**1-2 TIME** 5
Time 5

**1-3 MASS** 6
Mass 6
REVIEW & SUMMARY 8 PROBLEMS 8

**2 Motion Along a Straight Line** 13

**2-1 POSITION, DISPLACEMENT, AND AVERAGE VELOCITY** 13
What Is Physics? 13
Motion 14
Position and Displacement 14
Average Velocity and Average Speed 15

**2-2 INSTANTANEOUS VELOCITY AND SPEED** 18
Instantaneous Velocity and Speed 18

**2-3 ACCELERATION** 20
Acceleration 20

**2-4 CONSTANT ACCELERATION** 23
Constant Acceleration: A Special Case 23
Another Look at Constant Acceleration 26

**2-5 FREE-FALL ACCELERATION** 27
Free-Fall Acceleration 27

**2-6 GRAPHICAL INTEGRATION IN MOTION ANALYSIS** 29
Graphical Integration in Motion Analysis 29
REVIEW & SUMMARY 30 QUESTIONS 31 PROBLEMS 32

**3 Vectors** 40

**3-1 VECTORS AND THEIR COMPONENTS** 40
What Is Physics? 40
Vectors and Scalars 40
Adding Vectors Geometrically 41
Components of Vectors 42

**3-2 UNIT VECTORS, ADDING VECTORS BY COMPONENTS** 46
Unit Vectors 46

Adding Vectors by Components 46
Vectors and the Laws of Physics 47

**3-3 MULTIPLYING VECTORS** 50
Multiplying Vectors 50
REVIEW & SUMMARY 55 QUESTIONS 56 PROBLEMS 57

**4 Motion in Two and Three Dimensions** 62

**4-1 POSITION AND DISPLACEMENT** 62
What Is Physics? 62
Position and Displacement 63

**4-2 AVERAGE VELOCITY AND INSTANTANEOUS VELOCITY** 64
Average Velocity and Instantaneous Velocity 65

**4-3 AVERAGE ACCELERATION AND INSTANTANEOUS ACCELERATION** 67
Average Acceleration and Instantaneous Acceleration 68

**4-4 PROJECTILE MOTION** 70
Projectile Motion 70

**4-5 UNIFORM CIRCULAR MOTION** 76
Uniform Circular Motion 76

**4-6 RELATIVE MOTION IN ONE DIMENSION** 78
Relative Motion in One Dimension 78

**4-7 RELATIVE MOTION IN TWO DIMENSIONS** 80
Relative Motion in Two Dimensions 80
REVIEW & SUMMARY 81 QUESTIONS 82 PROBLEMS 84

**5 Force and Motion—I** 94

**5-1 NEWTON'S FIRST AND SECOND LAWS** 94
What Is Physics? 94
Newtonian Mechanics 95
Newton's First Law 95
Force 96
Mass 97
Newton's Second Law 98

**5-2 SOME PARTICULAR FORCES** 102
Some Particular Forces 102

**5-3 APPLYING NEWTON'S LAWS** 106
Newton's Third Law 106
Applying Newton's Laws 108
REVIEW & SUMMARY 114 QUESTIONS 114 PROBLEMS 116

CONTENTS **vii**

**6** Force and Motion—II **124**

6-1 FRICTION **124**
What Is Physics? **124**
Friction **124**
Properties of Friction **127**

6-2 THE DRAG FORCE AND TERMINAL SPEED **130**
The Drag Force and Terminal Speed **130**

6-3 UNIFORM CIRCULAR MOTION **133**
Uniform Circular Motion **133**
REVIEW & SUMMARY **138**  QUESTIONS **139**  PROBLEMS **140**

**7** Kinetic Energy and Work **149**

7-1 KINETIC ENERGY **149**
What Is Physics? **149**
What Is Energy? **149**
Kinetic Energy **150**

7-2 WORK AND KINETIC ENERGY **151**
Work **151**
Work and Kinetic Energy **152**

7-3 WORK DONE BY THE GRAVITATIONAL FORCE **155**
Work Done by the Gravitational Force **156**

7-4 WORK DONE BY A SPRING FORCE **159**
Work Done by a Spring Force **159**

7-5 WORK DONE BY A GENERAL VARIABLE FORCE **162**
Work Done by a General Variable Force **162**

7-6 POWER **166**
Power **166**
REVIEW & SUMMARY **168**  QUESTIONS **169**  PROBLEMS **170**

**8** Potential Energy and Conservation of Energy **177**

8-1 POTENTIAL ENERGY **177**
What Is Physics? **177**
Work and Potential Energy **178**
Path Independence of Conservative Forces **179**
Determining Potential Energy Values **181**

8-2 CONSERVATION OF MECHANICAL ENERGY **184**
Conservation of Mechanical Energy **184**

8-3 READING A POTENTIAL ENERGY CURVE **187**
Reading a Potential Energy Curve **187**

8-4 WORK DONE ON A SYSTEM BY AN EXTERNAL FORCE **191**
Work Done on a System by an External Force **192**

8-5 CONSERVATION OF ENERGY **195**
Conservation of Energy **195**
REVIEW & SUMMARY **199**  QUESTIONS **200**  PROBLEMS **202**

**9** Center of Mass and Linear Momentum **214**

9-1 CENTER OF MASS **214**
What Is Physics? **214**
The Center of Mass **215**

9-2 NEWTON'S SECOND LAW FOR A SYSTEM OF PARTICLES **220**
Newton's Second Law for a System of Particles **220**

9-3 LINEAR MOMENTUM **224**
Linear Momentum **224**
The Linear Momentum of a System of Particles **225**

9-4 COLLISION AND IMPULSE **226**
Collision and Impulse **226**

9-5 CONSERVATION OF LINEAR MOMENTUM **230**
Conservation of Linear Momentum **230**

9-6 MOMENTUM AND KINETIC ENERGY IN COLLISIONS **233**
Momentum and Kinetic Energy in Collisions **233**
Inelastic Collisions in One Dimension **234**

9-7 ELASTIC COLLISIONS IN ONE DIMENSION **237**
Elastic Collisions in One Dimension **237**

9-8 COLLISIONS IN TWO DIMENSIONS **240**
Collisions in Two Dimensions **240**

9-9 SYSTEMS WITH VARYING MASS: A ROCKET **241**
Systems with Varying Mass: A Rocket **241**
REVIEW & SUMMARY **243**  QUESTIONS **245**  PROBLEMS **246**

**10** Rotation **257**

10-1 ROTATIONAL VARIABLES **257**
What Is Physics? **258**
Rotational Variables **259**
Are Angular Quantities Vectors? **264**

10-2 ROTATION WITH CONSTANT ANGULAR ACCELERATION **266**
Rotation with Constant Angular Acceleration **266**

10-3 RELATING THE LINEAR AND ANGULAR VARIABLES **268**
Relating the Linear and Angular Variables **268**

**10-4 KINETIC ENERGY OF ROTATION** 271
Kinetic Energy of Rotation 271

**10-5 CALCULATING THE ROTATIONAL INERTIA** 273
Calculating the Rotational Inertia 273

**10-6 TORQUE** 277
Torque 278

**10-7 NEWTON'S SECOND LAW FOR ROTATION** 279
Newton's Second Law for Rotation 279

**10-8 WORK AND ROTATIONAL KINETIC ENERGY** 282
Work and Rotational Kinetic Energy 282
REVIEW & SUMMARY 285    QUESTIONS 286    PROBLEMS 287

**11 Rolling, Torque, and Angular Momentum** 295
**11-1 ROLLING AS TRANSLATION AND ROTATION COMBINED** 295
What Is Physics? 295
Rolling as Translation and Rotation Combined 295

**11-2 FORCES AND KINETIC ENERGY OF ROLLING** 298
The Kinetic Energy of Rolling 298
The Forces of Rolling 299

**11-3 THE YO-YO** 301
The Yo-Yo 302

**11-4 TORQUE REVISITED** 302
Torque Revisited 303

**11-5 ANGULAR MOMENTUM** 305
Angular Momentum 305

**11-6 NEWTON'S SECOND LAW IN ANGULAR FORM** 307
Newton's Second Law in Angular Form 307

**11-7 ANGULAR MOMENTUM OF A RIGID BODY** 310
The Angular Momentum of a System of Particles 310
The Angular Momentum of a Rigid Body Rotating About a Fixed Axis 311

**11-8 CONSERVATION OF ANGULAR MOMENTUM** 312
Conservation of Angular Momentum 312

**11-9 PRECESSION OF A GYROSCOPE** 317
Precession of a Gyroscope 317
REVIEW & SUMMARY 318    QUESTIONS 319    PROBLEMS 320

**12 Equilibrium and Elasticity** 327
**12-1 EQUILIBRIUM** 327
What Is Physics? 327

Equilibrium 327
The Requirements of Equilibrium 329
The Center of Gravity 330

**12-2 SOME EXAMPLES OF STATIC EQUILIBRIUM** 332
Some Examples of Static Equilibrium 332

**12-3 ELASTICITY** 338
Indeterminate Structures 338
Elasticity 339
REVIEW & SUMMARY 343    QUESTIONS 343    PROBLEMS 345

**13 Gravitation** 354
**13-1 NEWTON'S LAW OF GRAVITATION** 354
What Is Physics? 354
Newton's Law of Gravitation 355

**13-2 GRAVITATION AND THE PRINCIPLE OF SUPERPOSITION** 357
Gravitation and the Principle of Superposition 357

**13-3 GRAVITATION NEAR EARTH'S SURFACE** 359
Gravitation Near Earth's Surface 360

**13-4 GRAVITATION INSIDE EARTH** 362
Gravitation Inside Earth 363

**13-5 GRAVITATIONAL POTENTIAL ENERGY** 364
Gravitational Potential Energy 364

**13-6 PLANETS AND SATELLITES: KEPLER'S LAWS** 368
Planets and Satellites: Kepler's Laws 369

**13-7 SATELLITES: ORBITS AND ENERGY** 371
Satellites: Orbits and Energy 371

**13-8 EINSTEIN AND GRAVITATION** 374
Einstein and Gravitation 374
REVIEW & SUMMARY 376    QUESTIONS 377    PROBLEMS 378

**14 Fluids** 386
**14-1 FLUIDS, DENSITY, AND PRESSURE** 386
What Is Physics? 386
What Is a Fluid? 386
Density and Pressure 387

**14-2 FLUIDS AT REST** 388
Fluids at Rest 389

**14-3 MEASURING PRESSURE** 392
Measuring Pressure 392

14-4 PASCAL'S PRINCIPLE  393
Pascal's Principle  393

14-5 ARCHIMEDES' PRINCIPLE  394
Archimedes' Principle  395

14-6 THE EQUATION OF CONTINUITY  398
Ideal Fluids in Motion  398
The Equation of Continuity  399

14-7 BERNOULLI'S EQUATION  401
Bernoulli's Equation  401
REVIEW & SUMMARY  405    QUESTIONS  405    PROBLEMS  406

15 Oscillations  413
15-1 SIMPLE HARMONIC MOTION  413
What Is Physics?  414
Simple Harmonic Motion  414
The Force Law for Simple Harmonic Motion  419

15-2 ENERGY IN SIMPLE HARMONIC MOTION  421
Energy in Simple Harmonic Motion  421

15-3 AN ANGULAR SIMPLE HARMONIC OSCILLATOR  423
An Angular Simple Harmonic Oscillator  423

15-4 PENDULUMS, CIRCULAR MOTION  424
Pendulums  425
Simple Harmonic Motion and Uniform Circular Motion  428

15-5 DAMPED SIMPLE HARMONIC MOTION  430
Damped Simple Harmonic Motion  430

15-6 FORCED OSCILLATIONS AND RESONANCE  432
Forced Oscillations and Resonance  432
REVIEW & SUMMARY  434    QUESTIONS  434    PROBLEMS  436

16 Waves—I  444
16-1 TRANSVERSE WAVES  444
What Is Physics?  445
Types of Waves  445
Transverse and Longitudinal Waves  445
Wavelength and Frequency  446
The Speed of a Traveling Wave  449

16-2 WAVE SPEED ON A STRETCHED STRING  452
Wave Speed on a Stretched String  452

16-3 ENERGY AND POWER OF A WAVE TRAVELING ALONG A STRING  454
Energy and Power of a Wave Traveling Along a String  454

16-4 THE WAVE EQUATION  456
The Wave Equation  456

16-5 INTERFERENCE OF WAVES  458
The Principle of Superposition for Waves  458
Interference of Waves  459

16-6 PHASORS  462
Phasors  462

16-7 STANDING WAVES AND RESONANCE  465
Standing Waves  465
Standing Waves and Resonance  467
REVIEW & SUMMARY  470    QUESTIONS  471    PROBLEMS  472

17 Waves—II  479
17-1 SPEED OF SOUND  479
What Is Physics?  479
Sound Waves  479
The Speed of Sound  480

17-2 TRAVELING SOUND WAVES  482
Traveling Sound Waves  482

17-3 INTERFERENCE  485
Interference  485

17-4 INTENSITY AND SOUND LEVEL  488
Intensity and Sound Level  489

17-5 SOURCES OF MUSICAL SOUND  492
Sources of Musical Sound  493

17-6 BEATS  496
Beats  497

17-7 THE DOPPLER EFFECT  498
The Doppler Effect  499

17-8 SUPERSONIC SPEEDS, SHOCK WAVES  503
Supersonic Speeds, Shock Waves  503
REVIEW & SUMMARY  504    QUESTIONS  505    PROBLEMS  506

18 Temperature, Heat, and the First Law of Thermodynamics  514
18-1 TEMPERATURE  514
What Is Physics?  514
Temperature  515
The Zeroth Law of Thermodynamics  515
Measuring Temperature  516

18-2 THE CELSIUS AND FAHRENHEIT SCALES  518
The Celsius and Fahrenheit Scales  518

**18-3 THERMAL EXPANSION  520**
Thermal Expansion  520

**18-4 ABSORPTION OF HEAT  522**
Temperature and Heat  523
The Absorption of Heat by Solids and Liquids  524

**18-5 THE FIRST LAW OF THERMODYNAMICS  528**
A Closer Look at Heat and Work  528
The First Law of Thermodynamics  531
Some Special Cases of the First Law of
Thermodynamics  532

**18-6 HEAT TRANSFER MECHANISMS  534**
Heat Transfer Mechanisms  534
REVIEW & SUMMARY  538    QUESTIONS  540    PROBLEMS  541

**19 The Kinetic Theory of Gases  549**
**19-1 AVOGADRO'S NUMBER  549**
What Is Physics?  549
Avogadro's Number  550

**19-2 IDEAL GASES  550**
Ideal Gases  551

**19-3 PRESSURE, TEMPERATURE, AND RMS SPEED  554**
Pressure, Temperature, and RMS Speed  554

**19-4 TRANSLATIONAL KINETIC ENERGY  557**
Translational Kinetic Energy  557

**19-5 MEAN FREE PATH  558**
Mean Free Path  558

**19-6 THE DISTRIBUTION OF MOLECULAR SPEEDS  560**
The Distribution of Molecular Speeds  561

**19-7 THE MOLAR SPECIFIC HEATS OF AN IDEAL GAS  564**
The Molar Specific Heats of an Ideal Gas  564

**19-8 DEGREES OF FREEDOM AND MOLAR SPECIFIC HEATS  568**
Degrees of Freedom and Molar Specific Heats  568
A Hint of Quantum Theory  570

**19-9 THE ADIABATIC EXPANSION OF AN IDEAL GAS  571**
The Adiabatic Expansion of an Ideal Gas  571
REVIEW & SUMMARY  575    QUESTIONS  576    PROBLEMS  577

**20 Entropy and the Second Law of Thermodynamics  583**
**20-1 ENTROPY  583**
What Is Physics?  584
Irreversible Processes and Entropy  584

Change in Entropy  585
The Second Law of Thermodynamics  588

**20-2 ENTROPY IN THE REAL WORLD: ENGINES  590**
Entropy in the Real World: Engines  590

**20-3 REFRIGERATORS AND REAL ENGINES  595**
Entropy in the Real World: Refrigerators  596
The Efficiencies of Real Engines  597

**20-4 A STATISTICAL VIEW OF ENTROPY  598**
A Statistical View of Entropy  598
REVIEW & SUMMARY  602    QUESTIONS  603    PROBLEMS  604

**21 Coulomb's Law  609**
**21-1 COULOMB'S LAW  609**
What Is Physics?  610
Electric Charge  610
Conductors and Insulators  612
Coulomb's Law  613

**21-2 CHARGE IS QUANTIZED  619**
Charge Is Quantized  619

**21-3 CHARGE IS CONSERVED  621**
Charge Is Conserved  621
REVIEW & SUMMARY  622    QUESTIONS  623    PROBLEMS  624

**22 Electric Fields  630**
**22-1 THE ELECTRIC FIELD  630**
What Is Physics?  630
The Electric Field  631
Electric Field Lines  631

**22-2 THE ELECTRIC FIELD DUE TO A CHARGED PARTICLE  633**
The Electric Field Due to a Point Charge  633

**22-3 THE ELECTRIC FIELD DUE TO A DIPOLE  635**
The Electric Field Due to an Electric Dipole  636

**22-4 THE ELECTRIC FIELD DUE TO A LINE OF CHARGE  638**
The Electric Field Due to Line of Charge  638

**22-5 THE ELECTRIC FIELD DUE TO A CHARGED DISK  643**
The Electric Field Due to a Charged Disk  643

**22-6 A POINT CHARGE IN AN ELECTRIC FIELD  645**
A Point Charge in an Electric Field  645

**22-7 A DIPOLE IN AN ELECTRIC FIELD  647**
A Dipole in an Electric Field  648
REVIEW & SUMMARY  650    QUESTIONS  651    PROBLEMS  652

**23** Gauss' Law  659

23-1 ELECTRIC FLUX  659

What Is Physics  659

Electric Flux  660

23-2 GAUSS' LAW  664

Gauss' Law  664

Gauss' Law and Coulomb's Law  666

23-3 A CHARGED ISOLATED CONDUCTOR  668

A Charged Isolated Conductor  668

23-4 APPLYING GAUSS' LAW: CYLINDRICAL SYMMETRY  671

Applying Gauss' Law: Cylindrical Symmetry  671

23-5 APPLYING GAUSS' LAW: PLANAR SYMMETRY  673

Applying Gauss' Law: Planar Symmetry  673

23-6 APPLYING GAUSS' LAW: SPHERICAL SYMMETRY  675

Applying Gauss' Law: Spherical Symmetry  675

REVIEW & SUMMARY  677   QUESTIONS  677   PROBLEMS  679

**24** Electric Potential  685

24-1 ELECTRIC POTENTIAL  685

What Is Physics?  685

Electric Potential and Electric Potential Energy  686

24-2 EQUIPOTENTIAL SURFACES AND THE ELECTRIC FIELD  690

Equipotential Surfaces  690

Calculating the Potential from the Field  691

24-3 POTENTIAL DUE TO A CHARGED PARTICLE  694

Potential Due to a Charged Particle  694

Potential Due a Group of Charged Particles  695

24-4 POTENTIAL DUE TO AN ELECTRIC DIPOLE  697

Potential Due to an Electric Dipole  697

24-5 POTENTIAL DUE TO A CONTINUOUS CHARGE DISTRIBUTION  698

Potential Due to a Continuous Charge Distribution  698

24-6 CALCULATING THE FIELD FROM THE POTENTIAL  701

Calculating the Field from the Potential  701

24-7 ELECTRIC POTENTIAL ENERGY OF A SYSTEM OF CHARGED PARTICLES  703

Electric Potential Energy of a System of Charged Particles  703

24-8 POTENTIAL OF A CHARGED ISOLATED CONDUCTOR  706

Potential of Charged Isolated Conductor  706

REVIEW & SUMMARY  707   QUESTIONS  708   PROBLEMS  710

**25** Capacitance  717

25-1 CAPACITANCE  717

What Is Physics?  717

Capacitance  717

25-2 CALCULATING THE CAPACITANCE  719

Calculating the Capacitance  720

25-3 CAPACITORS IN PARALLEL AND IN SERIES  723

Capacitors in Parallel and in Series  724

25-4 ENERGY STORED IN AN ELECTRIC FIELD  728

Energy Stored in an Electric Field  728

25-5 CAPACITOR WITH A DIELECTRIC  731

Capacitor with a Dielectric  731

Dielectrics: An Atomic View  733

25-6 DIELECTRICS AND GAUSS' LAW  735

Dielectrics and Gauss' Law  735

REVIEW & SUMMARY  738   QUESTIONS  738   PROBLEMS  739

**26** Current and Resistance  745

26-1 ELECTRIC CURRENT  745

What Is Physics?  745

Electric Current  746

26-2 CURRENT DENSITY  748

Current Density  749

26-3 RESISTANCE AND RESISTIVITY  752

Resistance and Resistivity  753

26-4 OHM'S LAW  756

Ohm's Law  756

A Microscopic View of Ohm's Law  758

26-5 POWER, SEMICONDUCTORS, SUPERCONDUCTORS  760

Power in Electric Circuits  760

Semiconductors  762

Superconductors  763

REVIEW & SUMMARY  763   QUESTIONS  764   PROBLEMS  765

**27** Circuits  771

27-1 SINGLE-LOOP CIRCUITS  771

What Is Physics?  772

"Pumping" Charges  772

Work, Energy, and Emf  773

Calculating the Current in a Single-Loop Circuit  774

Other Single-Loop Circuits  776

Potential Difference Between Two Points  777

**27-2 MULTILOOP CIRCUITS** **781**

Multiloop Circuits **781**

**27-3 THE AMMETER AND THE VOLTMETER** **788**

The Ammeter and the Voltmeter **788**

**27-4 RC CIRCUITS** **788**

RC Circuits **789**

REVIEW & SUMMARY **793** QUESTIONS **793** PROBLEMS **795**

**28** Magnetic Fields **803**

**28-1 MAGNETIC FIELDS AND THE DEFINITION OF $\vec{B}$** **803**

What Is Physics? **803**

What Produces a Magnetic Field? **804**

The Definition of $\vec{B}$ **804**

**28-2 CROSSED FIELDS: DISCOVERY OF THE ELECTRON** **808**

Crossed Fields: Discovery of the Electron **809**

**28-3 CROSSED FIELDS: THE HALL EFFECT** **810**

Crossed Fields: The Hall Effect **811**

**28-4 A CIRCULATING CHARGED PARTICLE** **814**

A Circulating Charged Particle **814**

**28-5 CYCLOTRONS AND SYNCHROTRONS** **817**

Cyclotrons and Synchrotrons **818**

**28-6 MAGNETIC FORCE ON A CURRENT-CARRYING WIRE** **820**

Magnetic Force on a Current-Carrying Wire **820**

**28-7 TORQUE ON A CURRENT LOOP** **822**

Torque on a Current Loop 822

**28-8 THE MAGNETIC DIPOLE MOMENT** **824**

The Magnetic Dipole Moment **825**

REVIEW & SUMMARY **827** QUESTIONS **827** PROBLEMS **829**

**29** Magnetic Fields Due to Currents **836**

**29-1 MAGNETIC FIELD DUE TO A CURRENT** **836**

What Is Physics? **836**

Calculating the Magnetic Field Due to a Current **837**

**29-2 FORCE BETWEEN TWO PARALLEL CURRENTS** **842**

Force Between Two Parallel Currents **842**

**29-3 AMPERE'S LAW** **844**

Ampere's Law **844**

**29-4 SOLENOIDS AND TOROIDS** **848**

Solenoids and Toroids **848**

**29-5 A CURRENT-CARRYING COIL AS A MAGNETIC DIPOLE** **851**

A Current-Carrying Coil as a Magnetic Dipole **851**

REVIEW & SUMMARY **854** QUESTIONS **855** PROBLEMS **856**

**30** Induction and Inductance **864**

**30-1 FARADAY'S LAW AND LENZ'S LAW** **864**

What Is Physics **864**

Two Experiments **865**

Faraday's Law of Induction **865**

Lenz's Law **868**

**30-2 INDUCTION AND ENERGY TRANSFERS** **871**

Induction and Energy Transfers **871**

**30-3 INDUCED ELECTRIC FIELDS** **874**

Induced Electric Fields **875**

**30-4 INDUCTORS AND INDUCTANCE** **879**

Inductors and Inductance **879**

**30-5 SELF-INDUCTION** **881**

Self-Induction **881**

**30-6 *RL* CIRCUITS** **882**

*RL* Circuits **883**

**30-7 ENERGY STORED IN A MAGNETIC FIELD** **887**

Energy Stored in a Magnetic Field **887**

**30-8 ENERGY DENSITY OF A MAGNETIC FIELD** **889**

Energy Density of a Magnetic Field **889**

**30-9 MUTUAL INDUCTION** **890**

Mutual Induction **890**

REVIEW & SUMMARY **893** QUESTIONS **893** PROBLEMS **895**

**31** Electromagnetic Oscillations and Alternating Current **903**

**31-1 *LC* OSCILLATIONS** **903**

What Is Physics? **904**

*LC* Oscillations, Qualitatively **904**

The Electrical-Mechanical Analogy **906**

*LC* Oscillations, Quantitatively **907**

**31-2 DAMPED OSCILLATIONS IN AN *RLC* CIRCUIT** **910**

Damped Oscillations in an *RLC* Circuit **911**

**31-3 FORCED OSCILLATIONS OF THREE SIMPLE CIRCUITS** **912**

Alternating Current **913**

Forced Oscillations **914**

Three Simple Circuits **914**

**31-4 THE SERIES *RLC* CIRCUIT** **921**

The Series *RLC* Circuit **921**

31-5 POWER IN ALTERNATING-CURRENT CIRCUITS  **927**

Power in Alternating-Current Circuits  **927**

31-6 TRANSFORMERS  **930**

Transformers  **930**

REVIEW & SUMMARY **933**    QUESTIONS **934**    PROBLEMS **935**

**32** Maxwell's Equations; Magnetism of Matter  **941**

32-1 GAUSS' LAW FOR MAGNETIC FIELDS  **941**

What Is Physics?  **941**

Gauss' Law for Magnetic Fields  **942**

32-2 INDUCED MAGNETIC FIELDS  **943**

Induced Magnetic Fields  **943**

32-3 DISPLACEMENT CURRENT  **946**

Displacement Current  **947**

Maxwell's Equations  **949**

32-4 MAGNETS  **950**

Magnets  **950**

32-5 MAGNETISM AND ELECTRONS  **952**

Magnetism and Electrons  **953**

Magnetic Materials  **956**

32-6 DIAMAGNETISM  **957**

Diamagnetism  **957**

32-7 PARAMAGNETISM  **959**

Paramagnetism  **959**

32-8 FERROMAGNETISM  **961**

Ferromagnetism  **961**

REVIEW & SUMMARY **964**    QUESTIONS **965**    PROBLEMS **967**

**33** Electromagnetic Waves  **972**

33-1 ELECTROMAGNETIC WAVES  **972**

What Is Physics?  **972**

Maxwell's Rainbow  **973**

The Traveling Electromagnetic Wave, Qualitatively  **974**

The Traveling Electromagnetic Wave, Quantitatively  **977**

33-2 ENERGY TRANSPORT AND THE POYNTING VECTOR  **980**

Energy Transport and the Poynting Vector  **981**

33-3 RADIATION PRESSURE  **983**

Radiation Pressure  **983**

33-4 POLARIZATION  **985**

Polarization  **985**

33-5 REFLECTION AND REFRACTION  **990**

Reflection and Refraction  **991**

33-6 TOTAL INTERNAL REFLECTION  **996**

Total Internal Reflection  **996**

33-7 POLARIZATION BY REFLECTION  **997**

Polarization by Reflection  **998**

REVIEW & SUMMARY **999**    QUESTIONS **1000**    PROBLEMS **1001**

**34** Images  **1010**

34-1 IMAGES AND PLANE MIRRORS  **1010**

What Is Physics?  **1010**

Two Types of Image  **1010**

Plane Mirrors  **1012**

34-2 SPHERICAL MIRRORS  **1014**

Spherical Mirrors  **1015**

Images from Spherical Mirrors  **1016**

34-3 SPHERICAL REFRACTING SURFACES  **1020**

Spherical Refracting Surfaces  **1020**

34-4 THIN LENSES  **1023**

Thin Lenses  **1023**

34-5 OPTICAL INSTRUMENTS  **1030**

Optical Instruments  **1030**

34-6 THREE PROOFS  **1033**

REVIEW & SUMMARY **1036**    QUESTIONS **1037**    PROBLEMS **1038**

**35** Interference  **1047**

35-1 LIGHT AS A WAVE  **1047**

What Is Physics?  **1047**

Light as a Wave  **1048**

35-2 YOUNG'S INTERFERENCE EXPERIMENT  **1053**

Diffraction  **1053**

Young's Interference Experiment  **1054**

35-3 INTERFERENCE AND DOUBLE-SLIT INTENSITY  **1059**

Coherence  **1059**

Intensity in Double-Slit Interference  **1060**

35-4 INTERFERENCE FROM THIN FILMS  **1063**

Interference from Thin Films  **1064**

35-5 MICHELSON'S INTERFEROMETER  **1070**

Michelson's Interferometer  **1071**

REVIEW & SUMMARY **1072**    QUESTIONS **1072**    PROBLEMS **1074**

**36 Diffraction**   1081

**36-1 SINGLE-SLIT DIFFRACTION**   1081
What Is Physics?   1081
Diffraction and the Wave Theory of Light   1081
Diffraction by a Single Slit: Locating the Minima   1083

**36-2 INTENSITY IN SINGLE-SLIT DIFFRACTION**   1086
Intensity in Single-Slit Diffraction   1086
Intensity in Single-Slit Diffraction, Quantitatively   1088

**36-3 DIFFRACTION BY A CIRCULAR APERTURE**   1090
Diffraction by a Circular Aperture   1091

**36-4 DIFFRACTION BY A DOUBLE SLIT**   1094
Diffraction by a Double Slit   1095

**36-5 DIFFRACTION GRATINGS**   1098
Diffraction Gratings   1098

**36-6 GRATINGS: DISPERSION AND RESOLVING POWER**   1101
Gratings: Dispersion and Resolving Power   1101

**36-7 X-RAY DIFFRACTION**   1104
X-Ray Diffraction   1104
REVIEW & SUMMARY 1107   QUESTIONS 1107   PROBLEMS 1108

**37 Relativity**   1116

**37-1 SIMULTANEITY AND TIME DILATION**   1116
What Is Physics?   1116
The Postulates   1117
Measuring an Event   1118
The Relativity of Simultaneity   1120
The Relativity of Time   1121

**37-2 THE RELATIVITY OF LENGTH**   1125
The Relativity of Length   1126

**37-3 THE LORENTZ TRANSFORMATION**   1129
The Lorentz Transformation   1129
Some Consequences of the Lorentz Equations   1131

**37-4 THE RELATIVITY OF VELOCITIES**   1133
The Relativity of Velocities   1133

**37-5 DOPPLER EFFECT FOR LIGHT**   1134
Doppler Effect for Light   1135

**37-6 MOMENTUM AND ENERGY**   1137
A New Look at Momentum   1138
A New Look at Energy   1138
REVIEW & SUMMARY 1143   QUESTIONS 1144   PROBLEMS 1145

**38 Photons and Matter Waves**   1153

**38-1 THE PHOTON, THE QUANTUM OF LIGHT**   1153
What Is Physics?   1153
The Photon, the Quantum of Light   1154

**38-2 THE PHOTOELECTRIC EFFECT**   1155
The Photoelectric Effect   1156

**38-3 PHOTONS, MOMENTUM, COMPTON SCATTERING, LIGHT INTERFERENCE**   1158
Photons Have Momentum   1159
Light as a Probability Wave   1162

**38-4 THE BIRTH OF QUANTUM PHYSICS**   1164
The Birth of Quantum Physics   1165

**38-5 ELECTRONS AND MATTER WAVES**   1166
Electrons and Matter Waves   1167

**38-6 SCHRÖDINGER'S EQUATION**   1170
Schrödinger's Equation   1170

**38-7 HEISENBERG'S UNCERTAINTY PRINCIPLE**   1172
Heisenberg's Uncertainty Principle   1173

**38-8 REFLECTION FROM A POTENTIAL STEP**   1174
Reflection from a Potential Step   1174

**38-9 TUNNELING THROUGH A POTENTIAL BARRIER**   1176
Tunneling Through a Potential Barrier   1176
REVIEW & SUMMARY 1179   QUESTIONS 1180   PROBLEMS 1181

**39 More About Matter Waves**   1186

**39-1 ENERGIES OF A TRAPPED ELECTRON**   1186
What Is Physics?   1186
String Waves and Matter Waves   1187
Energies of a Trapped Electron   1187

**39-2 WAVE FUNCTIONS OF A TRAPPED ELECTRON**   1191
Wave Functions of a Trapped Electron   1192

**39-3 AN ELECTRON IN A FINITE WELL**   1195
An Electron in a Finite Well   1195

**39-4 TWO- AND THREE-DIMENSIONAL ELECTRON TRAPS**   1197
More Electron Traps   1197
Two- and Three-Dimensional Electron Traps   1200

**39-5 THE HYDROGEN ATOM**   1201
The Hydrogen Atom Is an Electron Trap   1202
The Bohr Model of Hydrogen, a Lucky Break   1203
Schrödinger's Equation and the Hydrogen Atom   1205
REVIEW & SUMMARY 1213   QUESTIONS 1213   PROBLEMS 1214

**40** All About Atoms **1219**

**40-1** PROPERTIES OF ATOMS **1219**

What Is Physics? **1220**

Some Properties of Atoms **1220**

Angular Momentum, Magnetic Dipole Moments **1222**

**40-2** THE STERN-GERLACH EXPERIMENT **1226**

The Stern-Gerlach Experiment **1226**

**40-3** MAGNETIC RESONANCE **1229**

Magnetic Resonance **1229**

**40-4** EXCLUSION PRINCIPLE AND MULTIPLE ELECTRONS IN A TRAP **1230**

The Pauli Exclusion Principle **1230**

Multiple Electrons in Rectangular Traps **1231**

**40-5** BUILDING THE PERIODIC TABLE **1234**

Building the Periodic Table **1234**

**40-6** X RAYS AND THE ORDERING OF THE ELEMENTS **1236**

X Rays and the Ordering of the Elements **1237**

**40-7** LASERS **1240**

Lasers and Laser Light **1241**

How Lasers Work **1242**

REVIEW & SUMMARY **1245**     QUESTIONS **1246**     PROBLEMS **1247**

**41** Conduction of Electricity in Solids **1252**

**41-1** THE ELECTRICAL PROPERTIES OF METALS **1252**

What Is Physics? **1252**

The Electrical Properties of Solids **1253**

Energy Levels in a Crystalline Solid **1254**

Insulators **1254**

Metals **1255**

**41-2** SEMICONDUCTORS AND DOPING **1261**

Semiconductors **1262**

Doped Semiconductors **1263**

**41-3** THE *p-n* JUNCTION AND THE TRANSISTOR **1265**

The *p-n* Junction **1266**

The Junction Rectifier **1267**

The Light-Emitting Diode (LED) **1268**

The Transistor **1270**

REVIEW & SUMMARY **1271**     QUESTIONS **1272**     PROBLEMS **1272**

**42** Nuclear Physics **1276**

**42-1** DISCOVERING THE NUCLEUS **1276**

What Is Physics? **1276**

Discovering the Nucleus **1276**

**42-2** SOME NUCLEAR PROPERTIES **1279**

Some Nuclear Properties **1280**

**42-3** RADIOACTIVE DECAY **1286**

Radioactive Decay **1286**

**42-4** ALPHA DECAY **1289**

Alpha Decay **1289**

**42-5** BETA DECAY **1292**

Beta Decay **1292**

**42-6** RADIOACTIVE DATING **1295**

Radioactive Dating **1295**

**42-7** MEASURING RADIATION DOSAGE **1296**

Measuring Radiation Dosage **1296**

**42-8** NUCLEAR MODELS **1297**

Nuclear Models **1297**

REVIEW & SUMMARY **1300**     QUESTIONS **1301**     PROBLEMS **1302**

**43** Energy from the Nucleus **1309**

**43-1** NUCLEAR FISSION **1309**

What Is Physics? **1309**

Nuclear Fission: The Basic Process **1310**

A Model for Nuclear Fission **1312**

**43-2** THE NUCLEAR REACTOR **1316**

The Nuclear Reactor **1316**

**43-3** A NATURAL NUCLEAR REACTOR **1320**

A Natural Nuclear Reactor **1320**

**43-4** THERMONUCLEAR FUSION: THE BASIC PROCESS **1322**

Thermonuclear Fusion: The Basic Process **1322**

**43-5** THERMONUCLEAR FUSION IN THE SUN AND OTHER STARS **1324**

Thermonuclear Fusion in the Sun and Other Stars **1324**

**43-6** CONTROLLED THERMONUCLEAR FUSION **1326**

Controlled Thermonuclear Fusion **1326**

REVIEW & SUMMARY **1329**     QUESTIONS **1329**     PROBLEMS **1330**

**44** Quarks, Leptons, and the Big Bang **1334**

**44-1** GENERAL PROPERTIES OF ELEMENTARY PARTICLES **1334**

What Is Physics? **1334**

Particles, Particles, Particles **1335**

An Interlude **1339**

**44-2** LEPTONS, HADRONS, AND STRANGENESS **1343**

The Leptons **1343**

The Hadrons **1345**

Still Another Conservation Law **1346**

The Eightfold Way **1347**

44-3 QUARKS AND MESSENGER PARTICLES **1349**

The Quark Model **1349**

Basic Forces and Messenger Particles **1352**

44-4 COSMOLOGY **1355**

A Pause for Reflection **1355**

The Universe Is Expanding **1356**

The Cosmic Background Radiation **1357**

Dark Matter **1358**

The Big Bang **1358**

A Summing Up **1361**

REVIEW & SUMMARY **1362**   QUESTIONS **1362**   PROBLEMS **1363**

**APPENDICES**

A The International System of Units (SI)   **A-1**

B Some Fundamental Constants of Physics   **A-3**

C Some Astronomical Data   **A-4**

D Conversion Factors   **A-5**

E Mathematical Formulas   **A-9**

F Properties of The Elements   **A-12**

G Periodic Table of The Elements   **A-15**

**ANSWERS**

to Checkpoints and Odd-Numbered Questions and Problems   **AN-1**

**INDEX** **I-1**

## WHY I WROTE THIS BOOK

Fun with a big challenge. That is how I have regarded physics since the day when Sharon, one of the students in a class I taught as a graduate student, suddenly demanded of me, "What has any of this got to do with my life?" Of course I immediately responded, "Sharon, this has everything to do with your life—this is physics."

She asked me for an example. I thought and thought but could not come up with a single one. That night I began writing the book *The Flying Circus of Physics* (John Wiley & Sons Inc., 1975) for Sharon but also for me because I realized her complaint was mine. I had spent six years slugging my way through many dozens of physics textbooks that were carefully written with the best of pedagogical plans, but there was something missing. Physics is the most interesting subject in the world because it is about how the world works, and yet the textbooks had been thoroughly wrung of any connection with the real world. The fun was missing.

I have packed a lot of real-world physics into *Fundamentals of Physics*, connecting it with the new edition of *The Flying Circus of Physics*. Much of the material comes from the introductory physics classes I teach, where I can judge from the faces and blunt comments what material and presentations work and what do not. The notes I make on my successes and failures there help form the basis of this book. My message here is the same as I had with every student I've met since Sharon so long ago: "Yes, you *can* reason from basic physics concepts all the way to valid conclusions about the real world, and that understanding of the real world is where the fun is."

I have many goals in writing this book but the overriding one is to provide instructors with tools by which they can teach students how to effectively read scientific material, identify fundamental concepts, reason through scientific questions, and solve quantitative problems. This process is not easy for either students or instructors. Indeed, the course associated with this book may be one of the most challenging of all the courses taken by a student. However, it can also be one of the most rewarding because it reveals the world's fundamental clockwork from which all scientific and engineering applications spring.

Many users of the ninth edition (both instructors and students) sent in comments and suggestions to improve the book. These improvements are now incorporated into the narrative and problems throughout the book. The publisher John Wiley & Sons and I regard the book as an ongoing project and encourage more input from users. You can send suggestions, corrections, and positive or negative comments to John Wiley & Sons or Jearl Walker (mail address: Physics Department, Cleveland State University, Cleveland, OH 44115 USA; or the blog site at www.flyingcircusofphysics.com). We may not be able to respond to all suggestions, but we keep and study each of them.

## WHAT'S NEW?

**Modules and Learning Objectives** "What was I supposed to learn from this section?" Students have asked me this question for decades, from the weakest student to the strongest. The problem is that even a thoughtful student may not feel confident that the important points were captured while reading a section. I felt the same way back when I was using the first edition of Halliday and Resnick while taking first-year physics.

To ease the problem in this edition, I restructured the chapters into concept modules based on a primary theme and begin each module with a list of the module's learning objectives. The list is an explicit statement of the skills and learning points that should be gathered in reading the module. Each list is following by a brief summary of the key ideas that should also be gathered. For example, check out the first module in Chapter 16, where a student faces a truck load of concepts and terms. Rather than depending on the student's ability to gather and sort those ideas, I now provide an explicit checklist that functions somewhat like the checklist a pilot works through before taxiing out to the runway for takeoff.

**Links Between Homework Problems and Learning Objectives**   In *WileyPLUS*, every question and problem at the end of the chapter is linked to a learning objective, to answer the (usually unspoken) questions, "Why am I working this problem? What am I supposed to learn from it?" By being explicit about a problem's purpose, I believe that a student might better transfer the learning objective to other problems with a different wording but the same key idea. Such transference would help defeat the common trouble that a student learns to work a particular problem but cannot then apply its key idea to a problem in a different setting.

**Rewritten Chapters**   My students have continued to be challenged by several key chapters and by spots in several other chapters and so, in this edition, I rewrote a lot of the material. For example, I redesigned the chapters on Gauss' law and electric potential, which have proved to be tough-going for my students. The presentations are now smoother and more direct to the key points. In the quantum chapters, I expanded the coverage of the Schrödinger equation, including reflection of matter waves from a step potential. At the request of several instructors, I decoupled the discussion of the Bohr atom from the Schrödinger solution for the hydrogen atom so that the historical account of Bohr's work can be bypassed. Also, there is now a module on Planck's blackbody radiation.

**New Sample Problems and Homework Questions and Problems**   Sixteen new sample problems have been added to the chapters, written so as to spotlight some of the difficult areas for my students. Also, about 250 problems and 50 questions have been added to the homework sections of the chapters. Some of these problems come from earlier editions of the book, as requested by several instructors.

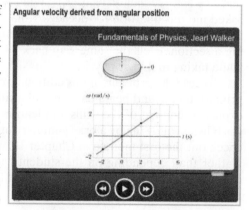

**Video Illustrations**   In the eVersion of the text available in *WileyPLUS*, David Maiullo of Rutgers University has created video versions of approximately 30 of the photographs and figures from the text. Much of physics is the study of things that move and video can often provide a better representation than a static photo or figure.

**Online Aid**   *WileyPLUS* is not just an online grading program. Rather, it is a dynamic learning center stocked with many different learning aids, including just-in-time problem-solving tutorials, embedded reading quizzes to encourage reading, animated figures, hundreds of sample problems, loads of simulations and demonstrations, and over 1500 videos ranging from math reviews to mini-lectures to examples. More of these learning aids are added every semester. For this 10th edition of HRW, some of the photos involving motion have been converted into videos so that the motion can be slowed and analyzed.

These thousands of learning aids are available 24/7 and can be repeated as many times as desired. Thus, if a student gets stuck on a homework problem at, say, 2:00 AM (which appears to be a popular time for doing physics homework), friendly and helpful resources are available at the click of a mouse.

## LEARNING TOOLS

When I learned first-year physics in the first edition of Halliday and Resnick, I caught on by repeatedly rereading a chapter. These days we better understand that students have a wide range of learning styles. So, I have produced a wide range of learning tools, both in this new edition and online in *WileyPLUS*:

**Animations** of one of the key figures in each chapter. Here in the book, those figures are flagged with the swirling icon. In the online chapter in *WileyPLUS*, a mouse click begins the animation. I have chosen the figures that are rich in information so that a student can see the physics in action and played out over a minute or two instead of just being flat on a printed page. Not only does this give life to the physics, but the animation can be repeated as many times as a student wants.

 **Videos**   I have made well over 1500 instructional videos, with more coming each semester. Students can watch me draw or type on the screen as they hear me talk about a solution, tutorial, sample problem, or review, very much as they would experience were they sitting next to me in my office while I worked out something on a notepad. An instructor's lectures and tutoring will always be the most valuable learning tools, but my videos are available 24 hours a day, 7 days a week, and can be repeated indefinitely.

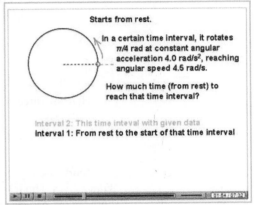

- **Video tutorials on subjects in the chapters.** I chose the subjects that challenge the students the most, the ones that my students scratch their heads about.

- **Video reviews of high school math**, such as basic algebraic manipulations, trig functions, and simultaneous equations.

- **Video introductions to math**, such as vector multiplication, that will be new to the students.

- **Video presentations of every Sample Problem** in the textbook chapters . My intent is to work out the physics, starting with the Key Ideas instead of just grabbing a formula. However, I also want to demonstrate how to read a sample problem, that is, how to read technical material to learn problem-solving procedures that can be transferred to other types of problems.

- **Video solutions to 20% of the end-of chapter problems.** The availability and timing of these solutions are controlled by the instructor. For example, they might be available after a homework deadline or a quiz. Each solution is not simply a plug-and-chug recipe. Rather I build a solution from the Key Ideas to the first step of reasoning and to a final solution. The student learns not just how to solve a particular problem but how to tackle any problem, even those that require *physics courage.*

- **Video examples of how to read data from graphs** (more than simply reading off a number with no comprehension of the physics).

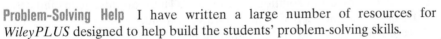 **Problem-Solving Help**   I have written a large number of resources for *WileyPLUS* designed to help build the students' problem-solving skills.

- **Every sample problem in the textbook** is available online in both reading and video formats.

- **Hundreds of additional sample problems.** These are available as standalone resources but (at the discretion of the instructor) they are also linked out of the homework problems. So, if a homework problem deals with, say, forces on a block on a ramp, a link to a related sample problem is provided. However, the sample problem is not just a replica of the homework problem and thus does not provide a solution that can be merely duplicated without comprehension.

- **GO Tutorials** for 15% of the end-of-chapter homework problems. In multiple steps, I lead a student through a homework problem, starting with the Key Ideas and giving hints when wrong answers are submitted. However, I purposely leave the last step (for the final answer) to the student so that they are responsible at the end. Some online tutorial systems trap a student when wrong answers are given, which can generate a lot of frustration. My GO Tutorials are not traps, because at any step along the way, a student can return to the main problem.

- **Hints on every end-of-chapter homework problem** are available (at the discretion of the instructor). I wrote these as true hints about the main ideas and the general procedure for a solution, not as recipes that provide an answer without any comprehension.

**Evaluation Materials**

- **Reading questions are available within each online section.** I wrote these so that they do not require analysis or any deep understanding; rather they simply test whether a student has read the section. When a student opens up a section, a randomly chosen reading question (from a bank of questions) appears at the end. The instructor can decide whether the question is part of the grading for that section or whether it is just for the benefit of the student.

- **Checkpoints are available within most sections.** I wrote these so that they require analysis and decisions about the physics in the section. *Answers to all checkpoints are in the back of the book.*

---

 **Checkpoint 1**

Here are three pairs of initial and final positions, respectively, along an $x$ axis. Which pairs give a negative displacement: (a) $-3$ m, $+5$ m; (b) $-3$ m, $-7$ m; (c) 7 m, $-3$ m?

---

- **All end-of-chapter homework Problems** in the book (and many more problems) are available in *WileyPLUS*. The instructor can construct a homework assignment and control how it is graded when the answers are submitted online. For example, the instructor controls the deadline for submission and how many attempts a student is allowed on an answer. The instructor also controls which, if any, learning aids are available with each homework problem. Such links can include hints, sample problems, in-chapter reading materials, video tutorials, video math reviews, and even video solutions (which can be made available to the students after, say, a homework deadline).

- **Symbolic notation problems** that require algebraic answers are available in every chapter.

- **All end-of-chapter homework Questions** in the book are available for assignment in *WileyPLUS*. These Questions (in a multiple choice format) are designed to evaluate the students' conceptual understanding.

**Icons for Additional Help**   When worked-out solutions are provided either in print or electronically for certain of the odd-numbered problems, the statements for those problems include an icon to alert both student and instructor as to where the solutions are located. There are also icons indicating which problems have GO Tutorial, an Interactive LearningWare, or a link to the *The Flying Circus of Physics*. An icon guide is provided here and at the beginning of each set of problems.

---

| | |
|---|---|
| **GO** Tutoring problem available (at instructor's discretion) in *WileyPLUS* and WebAssign | |
|  **SSM** Worked-out solution available in Student Solutions Manual | **WWW** Worked-out solution is at |
| • – ••• Number of dots indicates level of problem difficulty | **ILW** Interactive solution is at http://www.wiley.com/college/halliday |
| 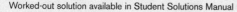 Additional information available in *The Flying Circus of Physics* and at flyingcircusofphysics.com | |

---

## VERSIONS OF THE TEXT

To accommodate the individual needs of instructors and students, the ninth edition of *Fundamentals of Physics* is available in a number of different versions.

The **Regular Edition** consists of Chapters 1 through 37 (ISBN 9781118230718).

The **Extended Edition** contains seven additional chapters on quantum physics and cosmology, Chapters 1–44 (ISBN 9781118230725).

**Volume 1** — Chapters 1–20 (Mechanics and Thermodynamics), hardcover, ISBN 9781118233764

**Volume 2** — Chapters 21–44 (E&M, Optics, and Quantum Physics), hardcover, ISBN 9781118230732

## INSTRUCTOR SUPPLEMENTS

**Instructor's Solutions Manual** by Sen-Ben Liao, Lawrence Livermore National Laboratory. This manual provides worked-out solutions for all problems found at the end of each chapter. It is available in both MSWord and PDF.

**Instructor Companion Site** http://www.wiley.com/college/halliday

• **Instructor's Manual** This resource contains lecture notes outlining the most important topics of each chapter; demonstration experiments; laboratory and computer projects; film and video sources; answers to all Questions, Exercises, Problems, and Checkpoints; and a correlation guide to the Questions, Exercises, and Problems in the previous edition. It also contains a complete list of all problems for which solutions are available to students (SSM,WWW, and ILW).

• **Lecture PowerPoint Slides** These PowerPoint slides serve as a helpful starter pack for instructors, outlining key concepts and incorporating figures and equations from the text.

• **Classroom Response Systems ("Clicker") Questions** by David Marx, Illinois State University. There are two sets of questions available: Reading Quiz questions and Interactive Lecture questions. The Reading Quiz questions are intended to be relatively straightforward for any student who reads the assigned material. The Interactive Lecture questions are intended for use in an interactive lecture setting.

• **Wiley Physics Simulations** by Andrew Duffy, Boston University and John Gastineau, Vernier Software. This is a collection of 50 interactive simulations (Java applets) that can be used for classroom demonstrations.

• **Wiley Physics Demonstrations** by David Maiullo, Rutgers University. This is a collection of digital videos of 80 standard physics demonstrations. They can be shown in class or accessed from *WileyPLUS*. There is an accompanying Instructor's Guide that includes "clicker" questions.

• **Test Bank** For the 10th edition, the Test Bank has been completely over-hauled by Suzanne Willis, Northern Illinois University. The Test Bank includes more than 2200 multiple-choice questions. These items are also available in the Computerized Test Bank which provides full editing features to help you customize tests (available in both IBM and Macintosh versions).

• **All text illustrations** suitable for both classroom projection and printing.

**Online Homework and Quizzing.**  In addition to *WileyPLUS*, *Fundamentals of Physics*, tenth edition, also supports WebAssignPLUS and LON-CAPA, which are other programs that give instructors the ability to deliver and grade homework and quizzes online. WebAssign PLUS also offers students an online version of the text.

## STUDENT SUPPLEMENTS

**Student Companion Site.** The web site http://www.wiley.com/college/halliday was developed specifically for *Fundamentals of Physics*, tenth edition, and is designed to further assist students in the study of physics. It includes solutions to selected end-of-chapter problems (which are identified with a www icon in the text); simulation exercises; tips on how to make best use of a programmable calculator; and the Interactive LearningWare tutorials that are described below.

**Student Study Guide** (ISBN 9781118230787) by Thomas Barrett of Ohio State University. The Student Study Guide consists of an overview of the chapter's important concepts, problem solving techniques and detailed examples.

**Student Solutions Manual** (ISBN 9781118230664) by Sen-Ben Liao, Lawrence Livermore National Laboratory. This manual provides students with complete worked-out solutions to 15 percent of the problems found at the end of each chapter within the text. The Student Solutions Manual for the 10th edition is written using an innovative approach called TEAL which stands for Think, Express, Analyze, and Learn. This learning strategy was originally developed at the Massachusetts Institute of Technology and has proven to be an effective learning tool for students. These problems with TEAL solutions are indicated with an SSM icon in the text.

**Interactive Learningware.**  This software guides students through solutions to 200 of the end-of-chapter problems. These problems are indicated with an ILW icon in the text. The solutions process is developed interactively, with appropriate feedback and access to error-specific help for the most common mistakes.

**Introductory Physics with Calculus as a Second Language:** (ISBN 9780471739104) *Mastering Problem Solving* by Thomas Barrett of Ohio State University. This brief paperback teaches the student how to approach problems more efficiently and effectively. The student will learn how to recognize common patterns in physics problems, break problems down into manageable steps, and apply appropriate techniques. The book takes the student step by step through the solutions to numerous examples.

# ACKNOWLEDGMENTS

A great many people have contributed to this book. Sen-Ben Liao of Lawrence Livermore National Laboratory, James Whitenton of Southern Polytechnic State University, and Jerry Shi, of Pasadena City College, performed the Herculean task of working out solutions for every one of the homework problems in the book. At John Wiley publishers, the book received support from Stuart Johnson, Geraldine Osnato and Aly Rentrop, the editors who oversaw the entire project from start to finish. We thank Elizabeth Swain, the production editor, for pulling all the pieces together during the complex production process. We also thank Maddy Lesure for her design of the text and the cover; Lee Goldstein for her page make-up; Helen Walden for her copyediting; and Lilian Brady for her proofreading. Jennifer Atkins was inspired in the search for unusual and interesting photographs. Both the publisher John Wiley & Sons, Inc. and Jearl Walker would like to thank the following for comments and ideas about the recent editions:

Jonathan Abramson, *Portland State University*; Omar Adawi, *Parkland College*; Edward Adelson, *The Ohio State University*; Steven R. Baker, *Naval Postgraduate School*; George Caplan, *Wellesley College*; Richard Kass, *The Ohio State University*; M. R. Khoshbin-e-Khoshnazar, *Research Institution for Curriculum Development & Educational Innovations (Tehran)*; Craig Kletzing, *University of Iowa*, Stuart Loucks, *American River College*; Laurence Lurio, *Northern Illinois University*; Ponn Maheswaranathan, *Winthrop University;* Joe McCullough, *Cabrillo College*; Carl E. Mungan, *U. S. Naval Academy*, Don N. Page, *University of Alberta*; Elie Riachi, *Fort Scott Community College*; Andrew G. Rinzler, *University of Florida*; Dubravka Rupnik, *Louisiana State University*; Robert Schabinger, *Rutgers University*; Ruth Schwartz, *Milwaukee School of Engineering*; Carol Strong, *University of Alabama at Huntsville*, Nora Thornber, *Raritan Valley Community College*; Frank Wang, *LaGuardia Community College*; Graham W. Wilson, *University of Kansas*; Roland Winkler, *Northern Illinois University*; William Zacharias, *Cleveland State University*; Ulrich Zurcher, *Cleveland State University*.

Finally, our external reviewers have been outstanding and we acknowledge here our debt to each member of that team.

Maris A. Abolins, *Michigan State University*

Edward Adelson, *Ohio State University*

Nural Akchurin, *Texas Tech*

Yildirim Aktas, *University of North Carolina-Charlotte*

Barbara Andereck, *Ohio Wesleyan University*

Tetyana Antimirova, *Ryerson University*

Mark Arnett, *Kirkwood Community College*

Arun Bansil, *Northeastern University*

Richard Barber, *Santa Clara University*

Neil Basecu, *Westchester Community College*

Anand Batra, *Howard University*

Kenneth Bolland, *The Ohio State University*

Richard Bone, *Florida International University*

Michael E. Browne, *University of Idaho*

Timothy J. Burns, *Leeward Community College*

Joseph Buschi, *Manhattan College*

Philip A. Casabella, *Rensselaer Polytechnic Institute*

Randall Caton, *Christopher Newport College*

Roger Clapp, *University of South Florida*

W. R. Conkie, *Queen's University*

Renate Crawford, *University of Massachusetts-Dartmouth*

Mike Crivello, *San Diego State University*

Robert N. Davie, Jr., *St. Petersburg Junior College*

Cheryl K. Dellai, *Glendale Community College*

Eric R. Dietz, *California State University at Chico*

N. John DiNardo, *Drexel University*

Eugene Dunnam, *University of Florida*

Robert Endorf, *University of Cincinnati*

F. Paul Esposito, *University of Cincinnati*

Jerry Finkelstein, *San Jose State University*

Robert H. Good, *California State University-Hayward*

Michael Gorman, *University of Houston*

Benjamin Grinstein, *University of California, San Diego*

John B. Gruber, *San Jose State University*

Ann Hanks, *American River College*

Randy Harris, *University of California-Davis*

Samuel Harris, *Purdue University*

Harold B. Hart, *Western Illinois University*

Rebecca Hartzler, *Seattle Central Community College*

John Hubisz, *North Carolina State University*

Joey Huston, *Michigan State University*

David Ingram, *Ohio University*

Shawn Jackson, *University of Tulsa*

Hector Jimenez, *University of Puerto Rico*

Sudhakar B. Joshi, *York University*

Leonard M. Kahn, *University of Rhode Island*

Sudipa Kirtley, *Rose-Hulman Institute*

Leonard Kleinman, *University of Texas at Austin*

Craig Kletzing, *University of Iowa*

Peter F. Koehler, *University of Pittsburgh*

Arthur Z. Kovacs, *Rochester Institute of Technology*

Kenneth Krane, *Oregon State University*

Hadley Lawler, *Vanderbilt University*

Priscilla Laws, *Dickinson College*

Edbertho Leal, *Polytechnic University of Puerto Rico*

Vern Lindberg, *Rochester Institute of Technology*

Peter Loly, *University of Manitoba*

James MacLaren, *Tulane University*

Andreas Mandelis, *University of Toronto*

Robert R. Marchini, *Memphis State University*

Andrea Markelz, *University at Buffalo, SUNY*

Paul Marquard, *Caspar College*

David Marx, *Illinois State University*

Dan Mazilu, *Washington and LeeUniversity*

James H. McGuire, *Tulane University*

David M. McKinstry, *Eastern Washington University*

Jordon Morelli, *Queen's University*

Eugene Mosca, *United States Naval Academy*

Eric R. Murray, *Georgia Institute of Technology, School of Physics*

James Napolitano, *Rensselaer Polytechnic Institute*

Blaine Norum, *University of Virginia*

Michael O'Shea, *Kansas State University*

Patrick Papin, *San Diego State University*

Kiumars Parvin, *San Jose State University*

Robert Pelcovits, *Brown University*

Oren P. Quist, *South Dakota State University*

Joe Redish, *University of Maryland*

Timothy M. Ritter, *University of North Carolina at Pembroke*

Dan Styer, *Oberlin College*

Frank Wang, *LaGuardia Community College*

Robert Webb, *Texas A&M University*

Suzanne Willis, *Northern Illinois University*

Shannon Willoughby, *Montana State University*

# Gravitation

## 13-1 NEWTON'S LAW OF GRAVITATION

### Learning Objectives

*After reading this module, you should be able to . . .*

**13.01** Apply Newton's law of gravitation to relate the gravitational force between two particles to their masses and their separation.

**13.02** Identify that a uniform spherical shell of matter attracts a particle that is outside the shell as if all the shell's mass were concentrated as a particle at its center.

**13.03** Draw a free-body diagram to indicate the gravitational force on a particle due to another particle or a uniform, spherical distribution of matter.

### Key Ideas

● Any particle in the universe attracts any other particle with a gravitational force whose magnitude is

$$F = G\frac{m_1 m_2}{r^2} \quad \text{(Newton's law of gravitation)},$$

where $m_1$ and $m_2$ are the masses of the particles, $r$ is their separation, and $G$ ($= 6.67 \times 10^{-11}$ N·m$^2$/kg$^2$) is the gravitational constant.

● The gravitational force between extended bodies is found by adding (integrating) the individual forces on individual particles within the bodies. However, if either of the bodies is a uniform spherical shell or a spherically symmetric solid, the net gravitational force it exerts on an *external* object may be computed as if all the mass of the shell or body were located at its center.

### What Is Physics?

One of the long-standing goals of physics is to understand the gravitational force—the force that holds you to Earth, holds the Moon in orbit around Earth, and holds Earth in orbit around the Sun. It also reaches out through the whole of our Milky Way galaxy, holding together the billions and billions of stars in the Galaxy and the countless molecules and dust particles between stars. We are located somewhat near the edge of this disk-shaped collection of stars and other matter, $2.6 \times 10^4$ light-years ($2.5 \times 10^{20}$ m) from the galactic center, around which we slowly revolve.

The gravitational force also reaches across intergalactic space, holding together the Local Group of galaxies, which includes, in addition to the Milky Way, the Andromeda Galaxy (Fig. 13-1) at a distance of $2.3 \times 10^6$ light-years away from Earth, plus several closer dwarf galaxies, such as the Large Magellanic Cloud. The Local Group is part of the Local Supercluster of galaxies that is being drawn by the gravitational force toward an exceptionally massive region of space called the Great Attractor. This region appears to be about $3.0 \times 10^8$ light-years from Earth, on the opposite side of the Milky Way. And the gravitational force is even more far-reaching because it attempts to hold together the entire universe, which is expanding.

This force is also responsible for some of the most mysterious structures in the universe: *black holes.* When a star considerably larger than our Sun burns out, the gravitational force between all its particles can cause the star to collapse in on itself and thereby to form a black hole. The gravitational force at the surface of such a collapsed star is so strong that neither particles nor light can escape from the surface (thus the term "black hole"). Any star coming too near a black hole can be ripped apart by the strong gravitational force and pulled into the hole. Enough captures like this yields a *supermassive black hole.* Such mysterious monsters appear to be common in the universe. Indeed, such a monster lurks at the center of our Milky Way galaxy—the black hole there, called Sagittarius A*, has a mass of about $3.7 \times 10^6$ solar masses. The gravitational force near this black hole is so strong that it causes orbiting stars to whip around the black hole, completing an orbit in as little as 15.2 y.

Although the gravitational force is still not fully understood, the starting point in our understanding of it lies in the *law of gravitation* of Isaac Newton.

## Newton's Law of Gravitation

Before we get to the equations, let's just think for a moment about something that we take for granted. We are held to the ground just about right, not so strongly that we have to crawl to get to school (though an occasional exam may leave you crawling home) and not so lightly that we bump our heads on the ceiling when we take a step. It is also just about right so that we are held to the ground but not to each other (that would be awkward in any classroom) or to the objects around us (the phrase "catching a bus" would then take on a new meaning). The attraction obviously depends on how much "stuff" there is in ourselves and other objects: Earth has lots of "stuff" and produces a big attraction but another person has less "stuff" and produces a smaller (even negligible) attraction. Moreover, this "stuff" always attracts other "stuff," never repelling it (or a hard sneeze could put us into orbit).

In the past people obviously knew that they were being pulled downward (especially if they tripped and fell over), but they figured that the downward force was unique to Earth and unrelated to the apparent movement of astronomical bodies across the sky. But in 1665, the 23-year-old Isaac Newton recognized that this force is responsible for holding the Moon in its orbit. Indeed he showed that every body in the universe attracts every other body. This tendency of bodies to move toward one another is called **gravitation**, and the "stuff" that is involved is the mass of each body. If the myth were true that a falling apple inspired Newton to his **law of gravitation**, then the attraction is between the mass of the apple and the mass of Earth. It is appreciable because the mass of Earth is so large, but even then it is only about 0.8 N. The attraction between two people standing near each other on a bus is (thankfully) much less (less than 1 $\mu$N) and imperceptible.

The gravitational attraction between extended objects such as two people can be difficult to calculate. Here we shall focus on Newton's force law between two *particles* (which have no size). Let the masses be $m_1$ and $m_2$ and $r$ be their separation. Then the magnitude of the gravitational force acting on each due to the presence of the other is given by

$$F = G \frac{m_1 m_2}{r^2} \quad \text{(Newton's law of gravitation).} \tag{13-1}$$

$G$ is the **gravitational constant:**

$$G = 6.67 \times 10^{-11} \, \text{N} \cdot \text{m}^2/\text{kg}^2$$
$$= 6.67 \times 10^{-11} \, \text{m}^3/\text{kg} \cdot \text{s}^2. \tag{13-2}$$

Courtesy NASA

**Figure 13-1** The Andromeda Galaxy. Located $2.3 \times 10^6$ light-years from us, and faintly visible to the naked eye, it is very similar to our home galaxy, the Milky Way.

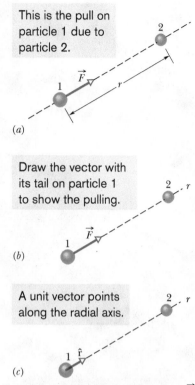

This is the pull on particle 1 due to particle 2.

*(a)*

Draw the vector with its tail on particle 1 to show the pulling.

*(b)*

A unit vector points along the radial axis.

*(c)*

**Figure 13-2** *(a)* The gravitational force $\vec{F}$ on particle 1 due to particle 2 is an attractive force because particle 1 is attracted to particle 2. *(b)* Force $\vec{F}$ is directed along a radial coordinate axis $r$ extending from particle 1 through particle 2. *(c)* $\vec{F}$ is in the direction of a unit vector $\hat{r}$ along the $r$ axis.

$\nabla F = 0.80$ N
$\Delta F = 0.80$ N

**Figure 13-3** The apple pulls up on Earth just as hard as Earth pulls down on the apple.

In Fig. 13-2*a*, $\vec{F}$ is the gravitational force acting on particle 1 (mass $m_1$) due to particle 2 (mass $m_2$). The force is directed toward particle 2 and is said to be an *attractive force* because particle 1 is attracted toward particle 2. The magnitude of the force is given by Eq. 13-1. We can describe $\vec{F}$ as being in the positive direction of an $r$ axis extending radially from particle 1 through particle 2 (Fig. 13-2*b*). We can also describe $\vec{F}$ by using a radial unit vector $\hat{r}$ (a dimensionless vector of magnitude 1) that is directed away from particle 1 along the $r$ axis (Fig. 13-2*c*). From Eq. 13-1, the force on particle 1 is then

$$\vec{F} = G \frac{m_1 m_2}{r^2} \hat{r}. \tag{13-3}$$

The gravitational force on particle 2 due to particle 1 has the same magnitude as the force on particle 1 but the opposite direction. These two forces form a third-law force pair, and we can speak of the gravitational force *between* the two particles as having a magnitude given by Eq. 13-1. This force between two particles is not altered by other objects, even if they are located between the particles. Put another way, no object can shield either particle from the gravitational force due to the other particle.

The strength of the gravitational force—that is, how strongly two particles with given masses at a given separation attract each other—depends on the value of the gravitational constant $G$. If $G$—by some miracle—were suddenly multiplied by a factor of 10, you would be crushed to the floor by Earth's attraction. If $G$ were divided by this factor, Earth's attraction would be so weak that you could jump over a building.

**Nonparticles.** Although Newton's law of gravitation applies strictly to particles, we can also apply it to real objects as long as the sizes of the objects are small relative to the distance between them. The Moon and Earth are far enough apart so that, to a good approximation, we can treat them both as particles—but what about an apple and Earth? From the point of view of the apple, the broad and level Earth, stretching out to the horizon beneath the apple, certainly does not look like a particle.

Newton solved the apple–Earth problem with the *shell theorem:*

> A uniform spherical shell of matter attracts a particle that is outside the shell as if all the shell's mass were concentrated at its center.

Earth can be thought of as a nest of such shells, one within another and each shell attracting a particle outside Earth's surface as if the mass of that shell were at the center of the shell. Thus, from the apple's point of view, Earth *does* behave like a particle, one that is located at the center of Earth and has a mass equal to that of Earth.

**Third-Law Force Pair.** Suppose that, as in Fig. 13-3, Earth pulls down on an apple with a force of magnitude 0.80 N. The apple must then pull up on Earth with a force of magnitude 0.80 N, which we take to act at the center of Earth. In the language of Chapter 5, these forces form a force pair in Newton's third law. Although they are matched in magnitude, they produce different accelerations when the apple is released. The acceleration of the apple is about 9.8 m/s², the familiar acceleration of a falling body near Earth's surface. The acceleration of Earth, however, measured in a reference frame attached to the center of mass of the apple–Earth system, is only about $1 \times 10^{-25}$ m/s².

### ✓ Checkpoint 1

A particle is to be placed, in turn, outside four objects, each of mass $m$: (1) a large uniform solid sphere, (2) a large uniform spherical shell, (3) a small uniform solid sphere, and (4) a small uniform shell. In each situation, the distance between the particle and the center of the object is $d$. Rank the objects according to the magnitude of the gravitational force they exert on the particle, greatest first.

# 13-2 GRAVITATION AND THE PRINCIPLE OF SUPERPOSITION

## Learning Objectives

*After reading this module, you should be able to . . .*

**13.04** If more than one gravitational force acts on a particle, draw a free-body diagram showing those forces, with the tails of the force vectors anchored on the particle.

**13.05** If more than one gravitational force acts on a particle, find the net force by adding the individual forces as vectors.

## Key Ideas

● Gravitational forces obey the principle of superposition; that is, if $n$ particles interact, the net force $\vec{F}_{1,\text{net}}$ on a particle labeled particle 1 is the sum of the forces on it from all the other particles taken one at a time:

$$\vec{F}_{1,\text{net}} = \sum_{i=2}^{n} \vec{F}_{1i},$$

in which the sum is a vector sum of the forces $\vec{F}_{1i}$ on particle 1 from particles 2, 3, . . . , $n$.

● The gravitational force $\vec{F}_1$ on a particle from an extended body is found by first dividing the body into units of differential mass $dm$, each of which produces a differential force $d\vec{F}$ on the particle, and then integrating over all those units to find the sum of those forces:

$$\vec{F}_1 = \int d\vec{F}.$$

## Gravitation and the Principle of Superposition

Given a group of particles, we find the net (or resultant) gravitational force on any one of them from the others by using the **principle of superposition.** This is a general principle that says a net effect is the sum of the individual effects. Here, the principle means that we first compute the individual gravitational forces that act on our selected particle due to each of the other particles. We then find the net force by adding these forces vectorially, just as we have done when adding forces in earlier chapters.

Let's look at two important points in that last (probably quickly read) sentence. (1) Forces are vectors and can be in different directions, and thus we must *add them as vectors*, taking into account their directions. (If two people pull on you in the opposite direction, their net force on you is clearly different than if they pull in the same direction.) (2) We *add* the individual forces. Think how impossible the world would be if the net force depended on some multiplying factor that varied from force to force depending on the situation, or if the presence of one force somehow amplified the magnitude of another force. No, thankfully, the world requires only simple vector addition of the forces.

For $n$ interacting particles, we can write the principle of superposition for the gravitational forces on particle 1 as

$$\vec{F}_{1,\text{net}} = \vec{F}_{12} + \vec{F}_{13} + \vec{F}_{14} + \vec{F}_{15} + \cdots + \vec{F}_{1n}. \qquad (13\text{-}4)$$

Here $\vec{F}_{1,\text{net}}$ is the net force on particle 1 due to the other particles and, for example, $\vec{F}_{13}$ is the force on particle 1 from particle 3. We can express this equation more compactly as a vector sum:

$$\vec{F}_{1,\text{net}} = \sum_{i=2}^{n} \vec{F}_{1i}. \qquad (13\text{-}5)$$

***Real Objects.*** What about the gravitational force on a particle from a real (extended) object? This force is found by dividing the object into parts small enough to treat as particles and then using Eq. 13-5 to find the vector sum of the forces on the particle from all the parts. In the limiting case, we can divide the extended object into differential parts each of mass $dm$ and each producing a differential force $d\vec{F}$

## Sample Problem 13.01    Net gravitational force, 2D, three particles

Figure 13-4a shows an arrangement of three particles, particle 1 of mass $m_1 = 6.0$ kg and particles 2 and 3 of mass $m_2 = m_3 = 4.0$ kg, and distance $a = 2.0$ cm. What is the net gravitational force $\vec{F}_{1,\text{net}}$ on particle 1 due to the other particles?

### KEY IDEAS

(1) Because we have particles, the magnitude of the gravitational force on particle 1 due to either of the other particles is given by Eq. 13-1 ($F = Gm_1m_2/r^2$). (2) The direction of either gravitational force on particle 1 is toward the particle responsible for it. (3) Because the forces are not along a single axis, we *cannot* simply add or subtract their magnitudes or their components to get the net force. Instead, we must add them as vectors.

**Calculations:** From Eq. 13-1, the magnitude of the force $\vec{F}_{12}$ on particle 1 from particle 2 is

$$F_{12} = \frac{Gm_1m_2}{a^2} \tag{13-7}$$

$$= \frac{(6.67 \times 10^{-11} \text{ m}^3/\text{kg}\cdot\text{s}^2)(6.0 \text{ kg})(4.0 \text{ kg})}{(0.020 \text{ m})^2}$$

$$= 4.00 \times 10^{-6} \text{ N}.$$

Similarly, the magnitude of force $\vec{F}_{13}$ on particle 1 from particle 3 is

$$F_{13} = \frac{Gm_1m_3}{(2a)^2} \tag{13-8}$$

$$= \frac{(6.67 \times 10^{-11} \text{ m}^3/\text{kg}\cdot\text{s}^2)(6.0 \text{ kg})(4.0 \text{ kg})}{(0.040 \text{ m})^2}$$

$$= 1.00 \times 10^{-6} \text{ N}.$$

Force $\vec{F}_{12}$ is directed in the positive direction of the y axis (Fig. 13-4b) and has only the y component $F_{12}$. Similarly, $\vec{F}_{13}$ is directed in the negative direction of the x axis and has only the x component $-F_{13}$ (Fig. 13-4c). (Note something important: We draw the force diagrams with the tail of a force vector anchored on the particle experiencing the force. Drawing them in other ways invites errors, especially on exams.)

To find the net force $\vec{F}_{1,\text{net}}$ on particle 1, we must add the two forces as vectors (Figs. 13-4d and e). We can do so on a vector-capable calculator. However, here we note that $-F_{13}$ and $F_{12}$ are actually the x and y components of $\vec{F}_{1,\text{net}}$. Therefore, we can use Eq. 3-6 to find first the magnitude and then the direction of $\vec{F}_{1,\text{net}}$. The magnitude is

$$F_{1,\text{net}} = \sqrt{(F_{12})^2 + (-F_{13})^2}$$

$$= \sqrt{(4.00 \times 10^{-6} \text{ N})^2 + (-1.00 \times 10^{-6} \text{ N})^2}$$

$$= 4.1 \times 10^{-6} \text{ N}. \tag{Answer}$$

Relative to the positive direction of the x axis, Eq. 3-6 gives the direction of $\vec{F}_{1,\text{net}}$ as

$$\theta = \tan^{-1}\frac{F_{12}}{-F_{13}} = \tan^{-1}\frac{4.00 \times 10^{-6} \text{ N}}{-1.00 \times 10^{-6} \text{ N}} = -76°.$$

Is this a reasonable direction (Fig. 13-4f)? No, because the direction of $\vec{F}_{1,\text{net}}$ must be between the directions of $\vec{F}_{12}$ and $\vec{F}_{13}$. Recall from Chapter 3 that a calculator displays only one of the two possible answers to a $\tan^{-1}$ function. We find the other answer by adding 180°:

$$-76° + 180° = 104°, \tag{Answer}$$

which *is* a reasonable direction for $\vec{F}_{1,\text{net}}$ (Fig. 13-4g).

 **PLUS** Additional examples, video, and practice available at *WileyPLUS*

on the particle. In this limit, the sum of Eq. 13-5 becomes an integral and we have

$$\vec{F}_1 = \int d\vec{F}, \tag{13-6}$$

in which the integral is taken over the entire extended object and we drop the subscript "net." If the extended object is a uniform sphere or a spherical shell, we can avoid the integration of Eq. 13-6 by assuming that the object's mass is concentrated at the object's center and using Eq. 13-1.

 **Checkpoint 2**

The figure shows four arrangements of three particles of equal masses. (a) Rank the arrangements according to the magnitude of the net gravitational force on the particle labeled $m$, greatest first. (b) In arrangement 2, is the direction of the net force closer to the line of length $d$ or to the line of length $D$?

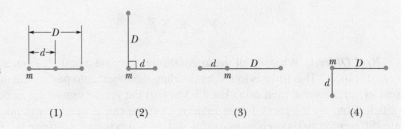

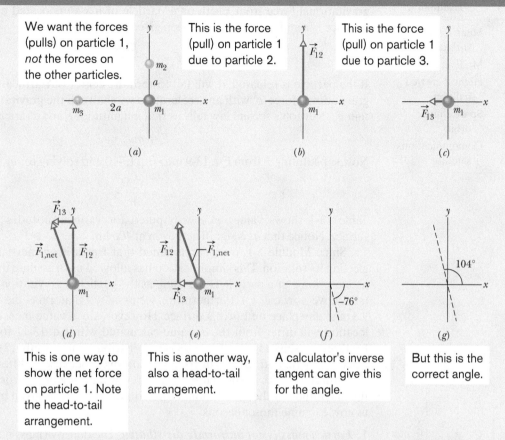

**Figure 13-4** (a) An arrangement of three particles. The force on particle 1 due to (b) particle 2 and (c) particle 3. (d)–(g) Ways to combine the forces to get the net force magnitude and orientation. In *WileyPLUS*, this figure is available as an animation with voiceover.

# 13-3 GRAVITATION NEAR EARTH'S SURFACE

## Learning Objectives

*After reading this module, you should be able to . . .*

**13.06** Distinguish between the free-fall acceleration and the gravitational acceleration.

**13.07** Calculate the gravitational acceleration near but outside a uniform, spherical astronomical body.

**13.08** Distinguish between measured weight and the magnitude of the gravitational force.

## Key Ideas

● The gravitational acceleration $a_g$ of a particle (of mass $m$) is due solely to the gravitational force acting on it. When the particle is at distance $r$ from the center of a uniform, spherical body of mass $M$, the magnitude $F$ of the gravitational force on the particle is given by Eq. 13-1. Thus, by Newton's second law,

$$F = ma_g,$$

which gives

$$a_g = \frac{GM}{r^2}.$$

● Because Earth's mass is not distributed uniformly, because the planet is not perfectly spherical, and because it rotates, the actual free-fall acceleration $\vec{g}$ of a particle near Earth differs slightly from the gravitational acceleration $\vec{a}_g$, and the particle's weight (equal to $mg$) differs from the magnitude of the gravitational force on it.

**Table 13-1** Variation of $a_g$ with Altitude

| Altitude (km) | $a_g$ (m/s²) | Altitude Example |
|---|---|---|
| 0 | 9.83 | Mean Earth surface |
| 8.8 | 9.80 | Mt. Everest |
| 36.6 | 9.71 | Highest crewed balloon |
| 400 | 8.70 | Space shuttle orbit |
| 35 700 | 0.225 | Communications satellite |

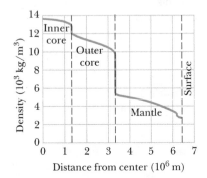

**Figure 13-5** The density of Earth as a function of distance from the center. The limits of the solid inner core, the largely liquid outer core, and the solid mantle are shown, but the crust of Earth is too thin to show clearly on this plot.

# Gravitation Near Earth's Surface

Let us assume that Earth is a uniform sphere of mass $M$. The magnitude of the gravitational force from Earth on a particle of mass $m$, located outside Earth a distance $r$ from Earth's center, is then given by Eq. 13-1 as

$$F = G\frac{Mm}{r^2}. \qquad (13\text{-}9)$$

If the particle is released, it will fall toward the center of Earth, as a result of the gravitational force $\vec{F}$, with an acceleration we shall call the **gravitational acceleration** $\vec{a}_g$. Newton's second law tells us that magnitudes $F$ and $a_g$ are related by

$$F = ma_g. \qquad (13\text{-}10)$$

Now, substituting $F$ from Eq. 13-9 into Eq. 13-10 and solving for $a_g$, we find

$$a_g = \frac{GM}{r^2}. \qquad (13\text{-}11)$$

Table 13-1 shows values of $a_g$ computed for various altitudes above Earth's surface. Notice that $a_g$ is significant even at 400 km.

Since Module 5-1, we have assumed that Earth is an inertial frame by neglecting its rotation. This simplification has allowed us to assume that the free-fall acceleration $g$ of a particle is the same as the particle's gravitational acceleration (which we now call $a_g$). Furthermore, we assumed that $g$ has the constant value 9.8 m/s² any place on Earth's surface. However, any $g$ value measured at a given location will differ from the $a_g$ value calculated with Eq. 13-11 for that location for three reasons: (1) Earth's mass is not distributed uniformly, (2) Earth is not a perfect sphere, and (3) Earth rotates. Moreover, because $g$ differs from $a_g$, the same three reasons mean that the measured weight $mg$ of a particle differs from the magnitude of the gravitational force on the particle as given by Eq. 13-9. Let us now examine those reasons.

1. **Earth's mass is not uniformly distributed.** The density (mass per unit volume) of Earth varies radially as shown in Fig. 13-5, and the density of the crust (outer section) varies from region to region over Earth's surface. Thus, $g$ varies from region to region over the surface.

2. **Earth is not a sphere.** Earth is approximately an ellipsoid, flattened at the poles and bulging at the equator. Its equatorial radius (from its center point out to the equator) is greater than its polar radius (from its center point out to either north or south pole) by 21 km. Thus, a point at the poles is closer to the dense core of Earth than is a point on the equator. This is one reason the free-fall acceleration $g$ increases if you were to measure it while moving at sea level from the equator toward the north or south pole. As you move, you are actually getting closer to the center of Earth and thus, by Newton's law of gravitation, $g$ increases.

3. **Earth is rotating.** The rotation axis runs through the north and south poles of Earth. An object located on Earth's surface anywhere except at those poles must rotate in a circle about the rotation axis and thus must have a centripetal acceleration directed toward the center of the circle. This centripetal acceleration requires a centripetal net force that is also directed toward that center.

To see how Earth's rotation causes $g$ to differ from $a_g$, let us analyze a simple situation in which a crate of mass $m$ is on a scale at the equator. Figure 13-6a shows this situation as viewed from a point in space above the north pole.

Figure 13-6b, a free-body diagram for the crate, shows the two forces on the crate, both acting along a radial $r$ axis that extends from Earth's center. The normal force $\vec{F}_N$ on the crate from the scale is directed outward, in the positive direction of the $r$ axis. The gravitational force, represented with its equivalent $m\vec{a}_g$, is directed inward. Because it travels in a circle about the center of Earth

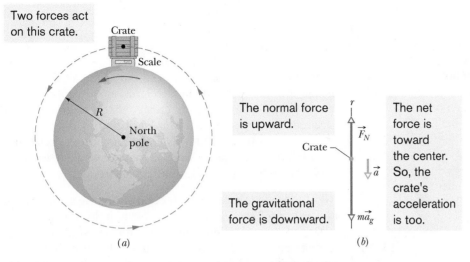

(a)    (b)

**Figure 13-6** (a) A crate sitting on a scale at Earth's equator, as seen by an observer positioned on Earth's rotation axis at some point above the north pole. (b) A free-body diagram for the crate, with a radial $r$ axis extending from Earth's center. The gravitational force on the crate is represented with its equivalent $m\vec{a}_g$. The normal force on the crate from the scale is $\vec{F}_N$. Because of Earth's rotation, the crate has a centripetal acceleration $\vec{a}$ that is directed toward Earth's center.

as Earth turns, the crate has a centripetal acceleration $\vec{a}$ directed toward Earth's center. From Eq. 10-23 ($a_r = \omega^2 r$), we know this acceleration is equal to $\omega^2 R$, where $\omega$ is Earth's angular speed and $R$ is the circle's radius (approximately Earth's radius). Thus, we can write Newton's second law for forces along the $r$ axis ($F_{\text{net},r} = ma_r$) as

$$F_N - ma_g = m(-\omega^2 R). \tag{13-12}$$

The magnitude $F_N$ of the normal force is equal to the weight $mg$ read on the scale. With $mg$ substituted for $F_N$, Eq. 13-12 gives us

$$mg = ma_g - m(\omega^2 R), \tag{13-13}$$

which says

$$\left(\begin{array}{c}\text{measured}\\\text{weight}\end{array}\right) = \left(\begin{array}{c}\text{magnitude of}\\\text{gravitational force}\end{array}\right) - \left(\begin{array}{c}\text{mass times}\\\text{centripetal acceleration}\end{array}\right).$$

Thus, the measured weight is less than the magnitude of the gravitational force on the crate, because of Earth's rotation.

***Acceleration Difference.*** To find a corresponding expression for $g$ and $a_g$, we cancel $m$ from Eq. 13-13 to write

$$g = a_g - \omega^2 R, \tag{13-14}$$

which says

$$\left(\begin{array}{c}\text{free-fall}\\\text{acceleration}\end{array}\right) = \left(\begin{array}{c}\text{gravitational}\\\text{acceleration}\end{array}\right) - \left(\begin{array}{c}\text{centripetal}\\\text{acceleration}\end{array}\right).$$

Thus, the measured free-fall acceleration is less than the gravitational acceleration because of Earth's rotation.

***Equator.*** The difference between accelerations $g$ and $a_g$ is equal to $\omega^2 R$ and is greatest on the equator (for one reason, the radius of the circle traveled by the crate is greatest there). To find the difference, we can use Eq. 10-5 ($\omega = \Delta\theta/\Delta t$) and Earth's radius $R = 6.37 \times 10^6$ m. For one rotation of Earth, $\theta$ is $2\pi$ rad and the time period $\Delta t$ is about 24 h. Using these values (and converting hours to seconds), we find that $g$ is less than $a_g$ by only about 0.034 m/s$^2$ (small compared to 9.8 m/s$^2$). Therefore, neglecting the difference in accelerations $g$ and $a_g$ is often justified. Similarly, neglecting the difference between weight and the magnitude of the gravitational force is also often justified.

**Sample Problem 13.02** **Difference in acceleration at head and feet**

(a) An astronaut whose height $h$ is 1.70 m floats "feet down" in an orbiting space shuttle at distance $r = 6.77 \times 10^6$ m away from the center of Earth. What is the difference between the gravitational acceleration at her feet and at her head?

### KEY IDEAS

We can approximate Earth as a uniform sphere of mass $M_E$. Then, from Eq. 13-11, the gravitational acceleration at any distance $r$ from the center of Earth is

$$a_g = \frac{GM_E}{r^2}. \qquad (13\text{-}15)$$

We might simply apply this equation twice, first with $r = 6.77 \times 10^6$ m for the location of the feet and then with $r = 6.77 \times 10^6$ m + 1.70 m for the location of the head. However, a calculator may give us the same value for $a_g$ twice, and thus a difference of zero, because $h$ is so much smaller than $r$. Here's a more promising approach: Because we have a differential change $dr$ in $r$ between the astronaut's feet and head, we should differentiate Eq. 13-15 with respect to $r$.

*Calculations:* The differentiation gives us

$$da_g = -2\frac{GM_E}{r^3}\,dr, \qquad (13\text{-}16)$$

where $da_g$ is the differential change in the gravitational acceleration due to the differential change $dr$ in $r$. For the astronaut, $dr = h$ and $r = 6.77 \times 10^6$ m. Substituting data into Eq. 13-16, we find

$$da_g = -2\frac{(6.67 \times 10^{-11}\ \text{m}^3/\text{kg}\cdot\text{s}^2)(5.98 \times 10^{24}\ \text{kg})}{(6.77 \times 10^6\ \text{m})^3}\,(1.70\ \text{m})$$

$$= -4.37 \times 10^{-6}\ \text{m/s}^2, \qquad \text{(Answer)}$$

where the $M_E$ value is taken from Appendix C. This result means that the gravitational acceleration of the astronaut's feet toward Earth is slightly greater than the gravitational acceleration of her head toward Earth. This difference in acceleration (often called a *tidal effect*) tends to stretch her body, but the difference is so small that she would never even sense the stretching, much less suffer pain from it.

(b) If the astronaut is now "feet down" at the same orbital radius $r = 6.77 \times 10^6$ m about a black hole of mass $M_h = 1.99 \times 10^{31}$ kg (10 times our Sun's mass), what is the difference between the gravitational acceleration at her feet and at her head? The black hole has a mathematical surface (*event horizon*) of radius $R_h = 2.95 \times 10^4$ m. Nothing, not even light, can escape from that surface or anywhere inside it. Note that the astronaut is well outside the surface (at $r = 229R_h$).

*Calculations:* We again have a differential change $dr$ in $r$ between the astronaut's feet and head, so we can again use Eq. 13-16. However, now we substitute $M_h = 1.99 \times 10^{31}$ kg for $M_E$. We find

$$da_g = -2\frac{(6.67 \times 10^{-11}\ \text{m}^3/\text{kg}\cdot\text{s}^2)(1.99 \times 10^{31}\ \text{kg})}{(6.77 \times 10^6\ \text{m})^3}\,(1.70\ \text{m})$$

$$= -14.5\ \text{m/s}^2. \qquad \text{(Answer)}$$

This means that the gravitational acceleration of the astronaut's feet toward the black hole is noticeably larger than that of her head. The resulting tendency to stretch her body would be bearable but quite painful. If she drifted closer to the black hole, the stretching tendency would increase drastically.

**WILEY** **PLUS** Additional examples, video, and practice available at *WileyPLUS*

# 13-4 GRAVITATION INSIDE EARTH

## Learning Objectives

*After reading this module, you should be able to . . .*

**13.09** Identify that a uniform shell of matter exerts no net gravitational force on a particle located inside it.

**13.10** Calculate the gravitational force that is exerted on a particle at a given radius inside a nonrotating uniform sphere of matter.

## Key Ideas

● A uniform shell of matter exerts no *net* gravitational force on a particle located inside it.

● The gravitational force $\vec{F}$ on a particle inside a uniform solid sphere, at a distance $r$ from the center, is due only to mass $M_{\text{ins}}$ in an "inside sphere" with that radius $r$:

$$M_{\text{ins}} = \tfrac{4}{3}\pi r^3 \rho = \frac{M}{R^3}r^3,$$

where $\rho$ is the solid sphere's density, $R$ is its radius, and $M$ is its mass. We can assign this inside mass to be that of a particle at the center of the solid sphere and then apply Newton's law of gravitation for particles. We find that the magnitude of the force acting on mass $m$ is

$$F = \frac{GmM}{R^3}r.$$

## Gravitation Inside Earth

Newton's shell theorem can also be applied to a situation in which a particle is located *inside* a uniform shell, to show the following:

> A uniform shell of matter exerts no net gravitational force on a particle located inside it.

*Caution:* This statement does *not* mean that the gravitational forces on the particle from the various elements of the shell magically disappear. Rather, it means that the *sum* of the force vectors on the particle from all the elements is zero.

If Earth's mass were uniformly distributed, the gravitational force acting on a particle would be a maximum at Earth's surface and would decrease as the particle moved outward, away from the planet. If the particle were to move inward, perhaps down a deep mine shaft, the gravitational force would change for two reasons. (1) It would tend to increase because the particle would be moving closer to the center of Earth. (2) It would tend to decrease because the thickening shell of material lying outside the particle's radial position would not exert any net force on the particle.

To find an expression for the gravitational force inside a uniform Earth, let's use the plot in *Pole to Pole*, an early science fiction story by George Griffith. Three explorers attempt to travel by capsule through a naturally formed (and, of course, fictional) tunnel directly from the south pole to the north pole. Figure 13-7 shows the capsule (mass $m$) when it has fallen to a distance $r$ from Earth's center. At that moment, the *net* gravitational force on the capsule is due to the mass $M_{ins}$ inside the sphere with radius $r$ (the mass enclosed by the dashed outline), not the mass in the outer spherical shell (outside the dashed outline). Moreover, we can assume that the inside mass $M_{ins}$ is concentrated as a particle at Earth's center. Thus, we can write Eq. 13-1, for the magnitude of the gravitational force on the capsule, as

$$F = \frac{GmM_{ins}}{r^2}. \tag{13-17}$$

Because we assume a uniform density $\rho$, we can write this inside mass in terms of Earth's total mass $M$ and its radius $R$:

$$\text{density} = \frac{\text{inside mass}}{\text{inside volume}} = \frac{\text{total mass}}{\text{total volume}},$$

$$\rho = \frac{M_{ins}}{\frac{4}{3}\pi r^3} = \frac{M}{\frac{4}{3}\pi R^3}.$$

Solving for $M_{ins}$ we find

$$M_{ins} = \tfrac{4}{3}\pi r^3 \rho = \frac{M}{R^3}\, r^3. \tag{13-18}$$

Substituting the second expression for $M_{ins}$ into Eq. 13-17 gives us the magnitude of the gravitational force on the capsule as a function of the capsule's distance $r$ from Earth's center:

$$F = \frac{GmM}{R^3}r. \tag{13-19}$$

According to Griffith's story, as the capsule approaches Earth's center, the gravitational force on the explorers becomes alarmingly large and, exactly at the center, it suddenly but only momentarily disappears. From Eq. 13-19 we see that, in fact, the force magnitude decreases linearly as the capsule approaches the center, until it is zero at the center. At least Griffith got the zero-at-the-center detail correct.

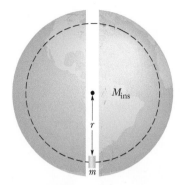

**Figure 13-7** A capsule of mass $m$ falls from rest through a tunnel that connects Earth's south and north poles. When the capsule is at distance $r$ from Earth's center, the portion of Earth's mass that is contained in a sphere of that radius is $M_{ins}$.

Equation 13-19 can also be written in terms of the force vector $\vec{F}$ and the capsule's position vector $\vec{r}$ along a radial axis extending from Earth's center. Letting $K$ represent the collection of constants in Eq. 13-19, we can rewrite the force in vector form as

$$\vec{F} = -K\vec{r}, \qquad (13\text{-}20)$$

in which we have inserted a minus sign to indicate that $\vec{F}$ and $\vec{r}$ have opposite directions. Equation 13-20 has the form of Hooke's law (Eq. 7-20, $\vec{F} = -k\vec{d}$). Thus, under the idealized conditions of the story, the capsule would oscillate like a block on a spring, with the center of the oscillation at Earth's center. After the capsule had fallen from the south pole to Earth's center, it would travel from the center to the north pole (as Griffith said) and then back again, repeating the cycle forever.

For the real Earth, which certainly has a nonuniform distribution of mass (Fig. 13-5), the force on the capsule would initially *increase* as the capsule descends. The force would then reach a maximum at a certain depth, and only then would it begin to decrease as the capsule further descends.

# 13-5 GRAVITATIONAL POTENTIAL ENERGY

## Learning Objectives

*After reading this module, you should be able to . . .*

**13.11** Calculate the gravitational potential energy of a system of particles (or uniform spheres that can be treated as particles).

**13.12** Identify that if a particle moves from an initial point to a final point while experiencing a gravitational force, the work done by that force (and thus the change in gravitational potential energy) is independent of which path is taken.

**13.13** Using the gravitational force on a particle near an astronomical body (or some second body that is fixed in

place), calculate the work done by the force when the body moves.

**13.14** Apply the conservation of mechanical energy (including gravitational potential energy) to a particle moving relative to an astronomical body (or some second body that is fixed in place).

**13.15** Explain the energy requirements for a particle to escape from an astronomical body (usually assumed to be a uniform sphere).

**13.16** Calculate the escape speed of a particle in leaving an astronomical body.

## Key Ideas

● The gravitational potential energy $U(r)$ of a system of two particles, with masses $M$ and $m$ and separated by a distance $r$, is the negative of the work that would be done by the gravitational force of either particle acting on the other if the separation between the particles were changed from infinite (very large) to $r$. This energy is

$$U = -\frac{GMm}{r} \quad \text{(gravitational potential energy).}$$

● If a system contains more than two particles, its total gravitational potential energy $U$ is the sum of the terms rep-

resenting the potential energies of all the pairs. As an example, for three particles, of masses $m_1$, $m_2$, and $m_3$,

$$U = -\left( \frac{Gm_1m_2}{r_{12}} + \frac{Gm_1m_3}{r_{13}} + \frac{Gm_2m_3}{r_{23}} \right).$$

● An object will escape the gravitational pull of an astronomical body of mass $M$ and radius $R$ (that is, it will reach an infinite distance) if the object's speed near the body's surface is at least equal to the escape speed, given by

$$v = \sqrt{\frac{2GM}{R}}.$$

## Gravitational Potential Energy

In Module 8-1, we discussed the gravitational potential energy of a particle–Earth system. We were careful to keep the particle near Earth's surface, so that we could regard the gravitational force as constant. We then chose some reference configuration of the system as having a gravitational potential energy of zero. Often, in this configuration the particle was on Earth's surface. For particles not on Earth's surface, the gravitational potential energy decreased when the separation between the particle and Earth decreased.

Here, we broaden our view and consider the gravitational potential energy $U$ of two particles, of masses $m$ and $M$, separated by a distance $r$. We again choose a reference configuration with $U$ equal to zero. However, to simplify the equations, the separation distance $r$ in the reference configuration is now large enough to be approximated as *infinite*. As before, the gravitational potential energy decreases when the separation decreases. Since $U = 0$ for $r = \infty$, the potential energy is negative for any finite separation and becomes progressively more negative as the particles move closer together.

With these facts in mind and as we shall justify next, we take the gravitational potential energy of the two-particle system to be

$$U = -\frac{GMm}{r} \quad \text{(gravitational potential energy).} \quad (13\text{-}21)$$

Note that $U(r)$ approaches zero as $r$ approaches infinity and that for any finite value of $r$, the value of $U(r)$ is negative.

***Language.*** The potential energy given by Eq. 13-21 is a property of the system of two particles rather than of either particle alone. There is no way to divide this energy and say that so much belongs to one particle and so much to the other. However, if $M \gg m$, as is true for Earth (mass $M$) and a baseball (mass $m$), we often speak of "the potential energy of the baseball." We can get away with this because, when a baseball moves in the vicinity of Earth, changes in the potential energy of the baseball–Earth system appear almost entirely as changes in the kinetic energy of the baseball, since changes in the kinetic energy of Earth are too small to be measured. Similarly, in Module 13-7 we shall speak of "the potential energy of an artificial satellite" orbiting Earth, because the satellite's mass is so much smaller than Earth's mass. When we speak of the potential energy of bodies of comparable mass, however, we have to be careful to treat them as a system.

***Multiple Particles.*** If our system contains more than two particles, we consider each pair of particles in turn, calculate the gravitational potential energy of that pair with Eq. 13-21 as if the other particles were not there, and then algebraically sum the results. Applying Eq. 13-21 to each of the three pairs of Fig. 13-8, for example, gives the potential energy of the system as

$$U = -\left( \frac{Gm_1m_2}{r_{12}} + \frac{Gm_1m_3}{r_{13}} + \frac{Gm_2m_3}{r_{23}} \right). \quad (13\text{-}22)$$

## Proof of Equation 13-21

Let us shoot a baseball directly away from Earth along the path in Fig. 13-9. We want to find an expression for the gravitational potential energy $U$ of the ball at point $P$ along its path, at radial distance $R$ from Earth's center. To do so, we first find the work $W$ done on the ball by the gravitational force as the ball travels from point $P$ to a great (infinite) distance from Earth. Because the gravitational force $\vec{F}(r)$ is a variable force (its magnitude depends on $r$), we must use the techniques of Module 7-5 to find the work. In vector notation, we can write

$$W = \int_R^\infty \vec{F}(r) \cdot d\vec{r}. \quad (13\text{-}23)$$

The integral contains the scalar (or dot) product of the force $\vec{F}(r)$ and the differential displacement vector $d\vec{r}$ along the ball's path. We can expand that product as

$$\vec{F}(r) \cdot d\vec{r} = F(r)\, dr \cos \phi, \quad (13\text{-}24)$$

where $\phi$ is the angle between the directions of $\vec{F}(r)$ and $d\vec{r}$. When we substitute

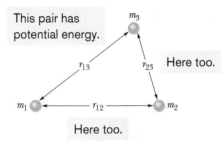

This pair has potential energy.

Here too.

Here too.

**Figure 13-8** A system consisting of three particles. The gravitational potential energy *of the system* is the sum of the gravitational potential energies of all three pairs of particles.

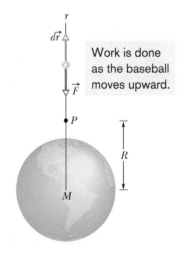

Work is done as the baseball moves upward.

**Figure 13-9** A baseball is shot directly away from Earth, through point $P$ at radial distance $R$ from Earth's center. The gravitational force $\vec{F}$ on the ball and a differential displacement vector $d\vec{r}$ are shown, both directed along a radial $r$ axis.

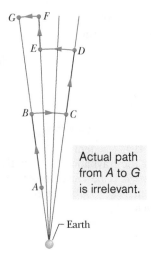

**Figure 13-10** Near Earth, a baseball is moved from point $A$ to point $G$ along a path consisting of radial lengths and circular arcs.

180° for $\phi$ and Eq. 13-1 for $F(r)$, Eq. 13-24 becomes

$$\vec{F}(r) \cdot d\vec{r} = -\frac{GMm}{r^2}\, dr,$$

where $M$ is Earth's mass and $m$ is the mass of the ball.

Substituting this into Eq. 13-23 and integrating give us

$$W = -GMm \int_R^\infty \frac{1}{r^2}\, dr = \left[\frac{GMm}{r}\right]_R^\infty$$

$$= 0 - \frac{GMm}{R} = -\frac{GMm}{R}, \tag{13-25}$$

where $W$ is the work required to move the ball from point $P$ (at distance $R$) to infinity. Equation 8-1 ($\Delta U = -W$) tells us that we can also write that work in terms of potential energies as

$$U_\infty - U = -W.$$

Because the potential energy $U_\infty$ at infinity is zero, $U$ is the potential energy at $P$, and $W$ is given by Eq. 13-25, this equation becomes

$$U = W = -\frac{GMm}{R}.$$

Switching $R$ to $r$ gives us Eq. 13-21, which we set out to prove.

### Path Independence

In Fig. 13-10, we move a baseball from point $A$ to point $G$ along a path consisting of three radial lengths and three circular arcs (centered on Earth). We are interested in the total work $W$ done by Earth's gravitational force $\vec{F}$ on the ball as it moves from $A$ to $G$. The work done along each circular arc is zero, because the direction of $\vec{F}$ is perpendicular to the arc at every point. Thus, $W$ is the sum of only the works done by $\vec{F}$ along the three radial lengths.

Now, suppose we mentally shrink the arcs to zero. We would then be moving the ball directly from $A$ to $G$ along a single radial length. Does that change $W$? No. Because no work was done along the arcs, eliminating them does not change the work. The path taken from $A$ to $G$ now is clearly different, but the work done by $\vec{F}$ is the same.

We discussed such a result in a general way in Module 8-1. Here is the point: The gravitational force is a conservative force. Thus, the work done by the gravitational force on a particle moving from an initial point $i$ to a final point $f$ is independent of the path taken between the points. From Eq. 8-1, the change $\Delta U$ in the gravitational potential energy from point $i$ to point $f$ is given by

$$\Delta U = U_f - U_i = -W. \tag{13-26}$$

Since the work $W$ done by a conservative force is independent of the actual path taken, the change $\Delta U$ in gravitational potential energy is *also independent* of the path taken.

### Potential Energy and Force

In the proof of Eq. 13-21, we derived the potential energy function $U(r)$ from the force function $\vec{F}(r)$. We should be able to go the other way—that is, to start from the potential energy function and derive the force function. Guided by Eq. 8-22 ($F(x) = -dU(x)/dx$), we can write

$$F = -\frac{dU}{dr} = -\frac{d}{dr}\left(-\frac{GMm}{r}\right)$$

$$= -\frac{GMm}{r^2}. \tag{13-27}$$

This is Newton's law of gravitation (Eq. 13-1). The minus sign indicates that the force on mass $m$ points radially inward, toward mass $M$.

### Escape Speed

If you fire a projectile upward, usually it will slow, stop momentarily, and return to Earth. There is, however, a certain minimum initial speed that will cause it to move upward forever, theoretically coming to rest only at infinity. This minimum initial speed is called the (Earth) **escape speed.**

Consider a projectile of mass $m$, leaving the surface of a planet (or some other astronomical body or system) with escape speed $v$. The projectile has a kinetic energy $K$ given by $\frac{1}{2}mv^2$ and a potential energy $U$ given by Eq. 13-21:

$$U = -\frac{GMm}{R},$$

in which $M$ is the mass of the planet and $R$ is its radius.

When the projectile reaches infinity, it stops and thus has no kinetic energy. It also has no potential energy because an infinite separation between two bodies is our zero-potential-energy configuration. Its total energy at infinity is therefore zero. From the principle of conservation of energy, its total energy at the planet's surface must also have been zero, and so

$$K + U = \tfrac{1}{2}mv^2 + \left(-\frac{GMm}{R}\right) = 0.$$

This yields

$$v = \sqrt{\frac{2GM}{R}}. \tag{13-28}$$

Note that $v$ does not depend on the direction in which a projectile is fired from a planet. However, attaining that speed is easier if the projectile is fired in the direction the launch site is moving as the planet rotates about its axis. For example, rockets are launched eastward at Cape Canaveral to take advantage of the Cape's eastward speed of 1500 km/h due to Earth's rotation.

Equation 13-28 can be applied to find the escape speed of a projectile from any astronomical body, provided we substitute the mass of the body for $M$ and the radius of the body for $R$. Table 13-2 shows some escape speeds.

Table 13-2 **Some Escape Speeds**

| Body | Mass (kg) | Radius (m) | Escape Speed (km/s) |
|---|---|---|---|
| Ceres[a] | $1.17 \times 10^{21}$ | $3.8 \times 10^{5}$ | 0.64 |
| Earth's moon[a] | $7.36 \times 10^{22}$ | $1.74 \times 10^{6}$ | 2.38 |
| Earth | $5.98 \times 10^{24}$ | $6.37 \times 10^{6}$ | 11.2 |
| Jupiter | $1.90 \times 10^{27}$ | $7.15 \times 10^{7}$ | 59.5 |
| Sun | $1.99 \times 10^{30}$ | $6.96 \times 10^{8}$ | 618 |
| Sirius B[b] | $2 \times 10^{30}$ | $1 \times 10^{7}$ | 5200 |
| Neutron star[c] | $2 \times 10^{30}$ | $1 \times 10^{4}$ | $2 \times 10^{5}$ |

[a]The most massive of the asteroids.
[b]A *white dwarf* (a star in a final stage of evolution) that is a companion of the bright star Sirius.
[c]The collapsed core of a star that remains after that star has exploded in a *supernova* event.

 **Checkpoint 3**

You move a ball of mass $m$ away from a sphere of mass $M$. (a) Does the gravitational potential energy of the system of ball and sphere increase or decrease? (b) Is positive work or negative work done by the gravitational force between the ball and the sphere?

**Sample Problem 13.03** **Asteroid falling from space, mechanical energy**

An asteroid, headed directly toward Earth, has a speed of 12 km/s relative to the planet when the asteroid is 10 Earth radii from Earth's center. Neglecting the effects of Earth's atmosphere on the asteroid, find the asteroid's speed $v_f$ when it reaches Earth's surface.

**KEY IDEAS**

Because we are to neglect the effects of the atmosphere on the asteroid, the mechanical energy of the asteroid–Earth system is conserved during the fall. Thus, the final mechanical energy (when the asteroid reaches Earth's surface) is equal to the initial mechanical energy. With kinetic energy $K$ and gravitational potential energy $U$, we can write this as

$$K_f + U_f = K_i + U_i. \qquad (13\text{-}29)$$

Also, if we assume the system is isolated, the system's linear momentum must be conserved during the fall. Therefore, the momentum change of the asteroid and that of Earth must be equal in magnitude and opposite in sign. However, because Earth's mass is so much greater than the asteroid's mass, the change in Earth's speed is negligible relative to the change in the asteroid's speed. So, the change in Earth's kinetic energy is also negligible. Thus, we can assume that the kinetic energies in Eq. 13-29 are those of the asteroid alone.

*Calculations:* Let $m$ represent the asteroid's mass and $M$ represent Earth's mass ($5.98 \times 10^{24}$ kg). The asteroid is ini-

tially at distance $10R_E$ and finally at distance $R_E$, where $R_E$ is Earth's radius ($6.37 \times 10^6$ m). Substituting Eq. 13-21 for $U$ and $\frac{1}{2}mv^2$ for $K$, we rewrite Eq. 13-29 as

$$\tfrac{1}{2}mv_f^2 - \frac{GMm}{R_E} = \tfrac{1}{2}mv_i^2 - \frac{GMm}{10R_E}.$$

Rearranging and substituting known values, we find

$$v_f^2 = v_i^2 + \frac{2GM}{R_E}\left(1 - \frac{1}{10}\right)$$

$$= (12 \times 10^3 \text{ m/s})^2$$

$$+ \frac{2(6.67 \times 10^{-11} \text{ m}^3/\text{kg} \cdot \text{s}^2)(5.98 \times 10^{24} \text{ kg})}{6.37 \times 10^6 \text{ m}} 0.9$$

$$= 2.567 \times 10^8 \text{ m}^2/\text{s}^2,$$

and $\qquad v_f = 1.60 \times 10^4 \text{ m/s} = 16 \text{ km/s}.$ (Answer)

At this speed, the asteroid would not have to be particularly large to do considerable damage at impact. If it were only 5 m across, the impact could release about as much energy as the nuclear explosion at Hiroshima. Alarmingly, about 500 million asteroids of this size are near Earth's orbit, and in 1994 one of them apparently penetrated Earth's atmosphere and exploded 20 km above the South Pacific (setting off nuclear-explosion warnings on six military satellites).

 Additional examples, video, and practice available at *WileyPLUS*

# 13-6 PLANETS AND SATELLITES: KEPLER'S LAWS

## Learning Objectives

*After reading this module, you should be able to . . .*

**13.17** Identify Kepler's three laws.

**13.18** Identify which of Kepler's laws is equivalent to the law of conservation of angular momentum.

**13.19** On a sketch of an elliptical orbit, identify the semimajor axis, the eccentricity, the perihelion, the aphelion, and the focal points.

**13.20** For an elliptical orbit, apply the relationships between the semimajor axis, the eccentricity, the perihelion, and the aphelion.

**13.21** For an orbiting natural or artificial satellite, apply Kepler's relationship between the orbital period and radius and the mass of the astronomical body being orbited.

## Key Ideas

● The motion of satellites, both natural and artificial, is governed by Kepler's laws:

1. *The law of orbits.* All planets move in elliptical orbits with the Sun at one focus.

2. *The law of areas.* A line joining any planet to the Sun sweeps out equal areas in equal time intervals. (This statement is equivalent to conservation of angular momentum.)

3. *The law of periods.* The square of the period $T$ of any planet is proportional to the cube of the semimajor axis $a$ of its orbit. For circular orbits with radius $r$,

$$T^2 = \left(\frac{4\pi^2}{GM}\right)r^3 \qquad \text{(law of periods)},$$

where $M$ is the mass of the attracting body—the Sun in the case of the solar system. For elliptical planetary orbits, the semimajor axis $a$ is substituted for $r$.

## Planets and Satellites: Kepler's Laws

The motions of the planets, as they seemingly wander against the background of the stars, have been a puzzle since the dawn of history. The "loop-the-loop" motion of Mars, shown in Fig. 13-11, was particularly baffling. Johannes Kepler (1571–1630), after a lifetime of study, worked out the empirical laws that govern these motions. Tycho Brahe (1546–1601), the last of the great astronomers to make observations without the help of a telescope, compiled the extensive data from which Kepler was able to derive the three laws of planetary motion that now bear Kepler's name. Later, Newton (1642–1727) showed that his law of gravitation leads to Kepler's laws.

In this section we discuss each of Kepler's three laws. Although here we apply the laws to planets orbiting the Sun, they hold equally well for satellites, either natural or artificial, orbiting Earth or any other massive central body.

    **1. THE LAW OF ORBITS:** All planets move in elliptical orbits, with the Sun at one focus.

Figure 13-12 shows a planet of mass $m$ moving in such an orbit around the Sun, whose mass is $M$. We assume that $M \gg m$, so that the center of mass of the planet–Sun system is approximately at the center of the Sun.

The orbit in Fig. 13-12 is described by giving its **semimajor axis** $a$ and its **eccentricity** $e$, the latter defined so that $ea$ is the distance from the center of the ellipse to either focus $F$ or $F'$. *An eccentricity of zero corresponds to a circle*, in which the two foci merge to a single central point. The eccentricities of the planetary orbits are not large; so if the orbits are drawn to scale, they look circular. The eccentricity of the ellipse of Fig. 13-12, which has been exaggerated for clarity, is 0.74. The eccentricity of Earth's orbit is only 0.0167.

    **2. THE LAW OF AREAS:** A line that connects a planet to the Sun sweeps out equal areas in the plane of the planet's orbit in equal time intervals; that is, the rate $dA/dt$ at which it sweeps out area $A$ is constant.

Qualitatively, this second law tells us that the planet will move most slowly when it is farthest from the Sun and most rapidly when it is nearest to the Sun. As it turns out, Kepler's second law is totally equivalent to the law of conservation of angular momentum. Let us prove it.

The area of the shaded wedge in Fig. 13-13a closely approximates the area swept out in time $\Delta t$ by a line connecting the Sun and the planet, which are separated by distance $r$. The area $\Delta A$ of the wedge is approximately the area of

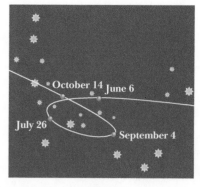

**Figure 13-11** The path seen from Earth for the planet Mars as it moved against a background of the constellation Capricorn during 1971. The planet's position on four days is marked. Both Mars and Earth are moving in orbits around the Sun so that we see the position of Mars relative to us; this relative motion sometimes results in an apparent loop in the path of Mars.

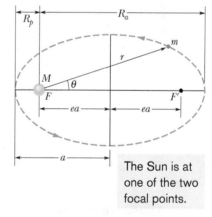

The Sun is at one of the two focal points.

**Figure 13-12** A planet of mass $m$ moving in an elliptical orbit around the Sun. The Sun, of mass $M$, is at one focus $F$ of the ellipse. The other focus is $F'$, which is located in empty space. The semimajor axis $a$ of the ellipse, the perihelion (nearest the Sun) distance $R_p$, and the aphelion (farthest from the Sun) distance $R_a$ are also shown.

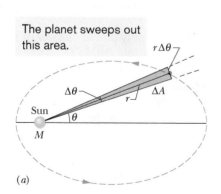

The planet sweeps out this area.

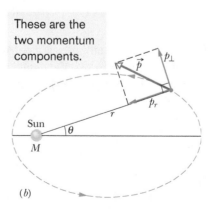

These are the two momentum components.

(a)                (b)

**Figure 13-13** (a) In time $\Delta t$, the line $r$ connecting the planet to the Sun moves through an angle $\Delta\theta$, sweeping out an area $\Delta A$ (shaded). (b) The linear momentum $\vec{p}$ of the planet and the components of $\vec{p}$.

a triangle with base $r\,\Delta\theta$ and height $r$. Since the area of a triangle is one-half of the base times the height, $\Delta A \approx \frac{1}{2}r^2\Delta\theta$. This expression for $\Delta A$ becomes more exact as $\Delta t$ (hence $\Delta\theta$) approaches zero. The instantaneous rate at which area is being swept out is then

$$\frac{dA}{dt} = \tfrac{1}{2}r^2\frac{d\theta}{dt} = \tfrac{1}{2}r^2\omega, \tag{13-30}$$

in which $\omega$ is the angular speed of the line connecting Sun and planet, as the line rotates around the Sun.

Figure 13-13b shows the linear momentum $\vec{p}$ of the planet, along with the radial and perpendicular components of $\vec{p}$. From Eq. 11-20 ($L = rp_\perp$), the magnitude of the angular momentum $\vec{L}$ of the planet about the Sun is given by the product of $r$ and $p_\perp$, the component of $\vec{p}$ perpendicular to $r$. Here, for a planet of mass $m$,

$$L = rp_\perp = (r)(mv_\perp) = (r)(m\omega r)$$
$$= mr^2\omega, \tag{13-31}$$

where we have replaced $v_\perp$ with its equivalent $\omega r$ (Eq. 10-18). Eliminating $r^2\omega$ between Eqs. 13-30 and 13-31 leads to

$$\frac{dA}{dt} = \frac{L}{2m}. \tag{13-32}$$

If $dA/dt$ is constant, as Kepler said it is, then Eq. 13-32 means that $L$ must also be constant — angular momentum is conserved. Kepler's second law is indeed equivalent to the law of conservation of angular momentum.

 **3. THE LAW OF PERIODS:** The square of the period of any planet is proportional to the cube of the semimajor axis of its orbit.

To see this, consider the circular orbit of Fig. 13-14, with radius $r$ (the radius of a circle is equivalent to the semimajor axis of an ellipse). Applying Newton's second law ($F = ma$) to the orbiting planet in Fig. 13-14 yields

$$\frac{GMm}{r^2} = (m)(\omega^2 r). \tag{13-33}$$

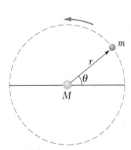

**Figure 13-14** A planet of mass $m$ moving around the Sun in a circular orbit of radius $r$.

Here we have substituted from Eq. 13-1 for the force magnitude $F$ and used Eq. 10-23 to substitute $\omega^2 r$ for the centripetal acceleration. If we now use Eq. 10-20 to replace $\omega$ with $2\pi/T$, where $T$ is the period of the motion, we obtain Kepler's third law:

$$T^2 = \left(\frac{4\pi^2}{GM}\right)r^3 \quad \text{(law of periods).} \tag{13-34}$$

The quantity in parentheses is a constant that depends only on the mass $M$ of the central body about which the planet orbits.

Equation 13-34 holds also for elliptical orbits, provided we replace $r$ with $a$, the semimajor axis of the ellipse. This law predicts that the ratio $T^2/a^3$ has essentially the same value for every planetary orbit around a given massive body. Table 13-3 shows how well it holds for the orbits of the planets of the solar system.

**Table 13-3** Kepler's Law of Periods for the Solar System

| Planet | Semimajor Axis $a$ ($10^{10}$ m) | Period $T$ (y) | $T^2/a^3$ ($10^{-34}$ y²/m³) |
|---|---|---|---|
| Mercury | 5.79 | 0.241 | 2.99 |
| Venus | 10.8 | 0.615 | 3.00 |
| Earth | 15.0 | 1.00 | 2.96 |
| Mars | 22.8 | 1.88 | 2.98 |
| Jupiter | 77.8 | 11.9 | 3.01 |
| Saturn | 143 | 29.5 | 2.98 |
| Uranus | 287 | 84.0 | 2.98 |
| Neptune | 450 | 165 | 2.99 |
| Pluto | 590 | 248 | 2.99 |

✓ **Checkpoint 4**

Satellite 1 is in a certain circular orbit around a planet, while satellite 2 is in a larger circular orbit. Which satellite has (a) the longer period and (b) the greater speed?

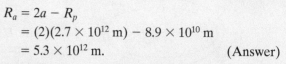

## Sample Problem 13.04    Kepler's law of periods, Comet Halley

Comet Halley orbits the Sun with a period of 76 years and, in 1986, had a distance of closest approach to the Sun, its *perihelion distance* $R_p$, of $8.9 \times 10^{10}$ m. Table 13-3 shows that this is between the orbits of Mercury and Venus.

(a) What is the comet's farthest distance from the Sun, which is called its *aphelion distance* $R_a$?

### KEY IDEAS

From Fig. 13-12, we see that $R_a + R_p = 2a$, where $a$ is the semimajor axis of the orbit. Thus, we can find $R_a$ if we first find $a$. We can relate $a$ to the given period via the law of periods (Eq. 13-34) if we simply substitute the semimajor axis $a$ for $r$.

*Calculations:* Making that substitution and then solving for $a$, we have

$$a = \left( \frac{GMT^2}{4\pi^2} \right)^{1/3}. \qquad (13\text{-}35)$$

If we substitute the mass $M$ of the Sun, $1.99 \times 10^{30}$ kg, and the period $T$ of the comet, 76 years or $2.4 \times 10^9$ s, into Eq. 13-35, we find that $a = 2.7 \times 10^{12}$ m. Now we have

$$R_a = 2a - R_p$$
$$= (2)(2.7 \times 10^{12} \text{ m}) - 8.9 \times 10^{10} \text{ m}$$
$$= 5.3 \times 10^{12} \text{ m}. \qquad \text{(Answer)}$$

Table 13-3 shows that this is a little less than the semimajor axis of the orbit of Pluto. Thus, the comet does not get farther from the Sun than Pluto.

(b) What is the eccentricity $e$ of the orbit of comet Halley?

### KEY IDEA

We can relate $e$, $a$, and $R_p$ via Fig. 13-12, in which we see that $ea = a - R_p$.

*Calculation:* We have

$$e = \frac{a - R_p}{a} = 1 - \frac{R_p}{a} \qquad (13\text{-}36)$$
$$= 1 - \frac{8.9 \times 10^{10} \text{ m}}{2.7 \times 10^{12} \text{ m}} = 0.97. \qquad \text{(Answer)}$$

This tells us that, with an eccentricity approaching unity, this orbit must be a long thin ellipse.

 **PLUS** Additional examples, video, and practice available at *WileyPLUS*

---

# 13-7 SATELLITES: ORBITS AND ENERGY

## Learning Objectives

*After reading this module, you should be able to . . .*

**13.22** For a satellite in a circular orbit around an astronomical body, calculate the gravitational potential energy, the kinetic energy, and the total energy.

**13.23** For a satellite in an elliptical orbit, calculate the total energy.

## Key Ideas

● When a planet or satellite with mass $m$ moves in a circular orbit with radius $r$, its potential energy $U$ and kinetic energy $K$ are given by

$$U = -\frac{GMm}{r} \quad \text{and} \quad K = \frac{GMm}{2r}.$$

The mechanical energy $E = K + U$ is then

$$E = -\frac{GMm}{2r}.$$

For an elliptical orbit of semimajor axis $a$,

$$E = -\frac{GMm}{2a}.$$

---

## Satellites: Orbits and Energy

As a satellite orbits Earth in an elliptical path, both its speed, which fixes its kinetic energy $K$, and its distance from the center of Earth, which fixes its gravitational potential energy $U$, fluctuate with fixed periods. However, the mechanical energy $E$ of the satellite remains constant. (Since the satellite's mass is so much smaller than Earth's mass, we assign $U$ and $E$ for the Earth–satellite system to the satellite alone.)

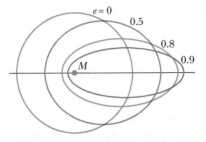

**Figure 13-15** Four orbits with different eccentricities $e$ about an object of mass $M$. All four orbits have the same semimajor axis $a$ and thus correspond to the same total mechanical energy $E$.

This is a plot of a satellite's energies versus orbit radius.

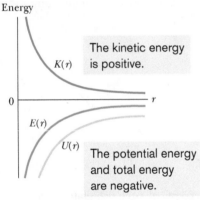

The kinetic energy is positive.

The potential energy and total energy are negative.

**Figure 13-16** The variation of kinetic energy $K$, potential energy $U$, and total energy $E$ with radius $r$ for a satellite in a circular orbit. For any value of $r$, the values of $U$ and $E$ are negative, the value of $K$ is positive, and $E = -K$. As $r \to \infty$, all three energy curves approach a value of zero.

The potential energy of the system is given by Eq. 13-21:

$$U = -\frac{GMm}{r}$$

(with $U = 0$ for infinite separation). Here $r$ is the radius of the satellite's orbit, assumed for the time being to be circular, and $M$ and $m$ are the masses of Earth and the satellite, respectively.

To find the kinetic energy of a satellite in a circular orbit, we write Newton's second law ($F = ma$) as

$$\frac{GMm}{r^2} = m\,\frac{v^2}{r}, \tag{13-37}$$

where $v^2/r$ is the centripetal acceleration of the satellite. Then, from Eq. 13-37, the kinetic energy is

$$K = \tfrac{1}{2}mv^2 = \frac{GMm}{2r}, \tag{13-38}$$

which shows us that for a satellite in a circular orbit,

$$K = -\frac{U}{2} \quad \text{(circular orbit)}. \tag{13-39}$$

The total mechanical energy of the orbiting satellite is

$$E = K + U = \frac{GMm}{2r} - \frac{GMm}{r}$$

or

$$E = -\frac{GMm}{2r} \quad \text{(circular orbit)}. \tag{13-40}$$

This tells us that for a satellite in a circular orbit, the total energy $E$ is the negative of the kinetic energy $K$:

$$E = -K \quad \text{(circular orbit)}. \tag{13-41}$$

For a satellite in an elliptical orbit of semimajor axis $a$, we can substitute $a$ for $r$ in Eq. 13-40 to find the mechanical energy:

$$E = -\frac{GMm}{2a} \quad \text{(elliptical orbit)}. \tag{13-42}$$

Equation 13-42 tells us that the total energy of an orbiting satellite depends only on the semimajor axis of its orbit and not on its eccentricity $e$. For example, four orbits with the same semimajor axis are shown in Fig. 13-15; the same satellite would have the same total mechanical energy $E$ in all four orbits. Figure 13-16 shows the variation of $K$, $U$, and $E$ with $r$ for a satellite moving in a circular orbit about a massive central body. Note that as $r$ is increased, the kinetic energy (and thus also the orbital speed) decreases.

 **Checkpoint 5**

In the figure here, a space shuttle is initially in a circular orbit of radius $r$ about Earth. At point $P$, the pilot briefly fires a forward-pointing thruster to decrease the shuttle's kinetic energy $K$ and mechanical energy $E$. (a) Which of the dashed elliptical orbits shown in the figure will the shuttle then take? (b) Is the orbital period $T$ of the shuttle (the time to return to $P$) then greater than, less than, or the same as in the circular orbit?

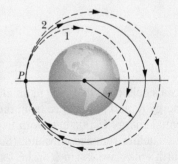

### Sample Problem 13.05  Mechanical energy of orbiting bowling ball

A playful astronaut releases a bowling ball, of mass $m = 7.20$ kg, into circular orbit about Earth at an altitude $h$ of 350 km.

**(a)** What is the mechanical energy $E$ of the ball in its orbit?

#### KEY IDEA

We can get $E$ from the orbital energy, given by Eq. 13-40 ($E = -GMm/2r$), if we first find the orbital radius $r$. (It is *not* simply the given altitude.)

*Calculations:* The orbital radius must be

$$r = R + h = 6370 \text{ km} + 350 \text{ km} = 6.72 \times 10^6 \text{ m},$$

in which $R$ is the radius of Earth. Then, from Eq. 13-40 with Earth mass $M = 5.98 \times 10^{24}$ kg, the mechanical energy is

$$E = -\frac{GMm}{2r}$$
$$= -\frac{(6.67 \times 10^{-11} \text{ N} \cdot \text{m}^2/\text{kg}^2)(5.98 \times 10^{24} \text{ kg})(7.20 \text{ kg})}{(2)(6.72 \times 10^6 \text{ m})}$$
$$= -2.14 \times 10^8 \text{ J} = -214 \text{ MJ}. \qquad \text{(Answer)}$$

**(b)** What is the mechanical energy $E_0$ of the ball on the launchpad at the Kennedy Space Center (before launch)? From there to the orbit, what is the change $\Delta E$ in the ball's mechanical energy?

#### KEY IDEA

On the launchpad, the ball is *not* in orbit and thus Eq. 13-40 does *not* apply. Instead, we must find $E_0 = K_0 + U_0$, where $K_0$ is the ball's kinetic energy and $U_0$ is the gravitational potential energy of the ball–Earth system.

*Calculations:* To find $U_0$, we use Eq. 13-21 to write

$$U_0 = -\frac{GMm}{R}$$
$$= -\frac{(6.67 \times 10^{-11} \text{ N} \cdot \text{m}^2/\text{kg}^2)(5.98 \times 10^{24} \text{ kg})(7.20 \text{ kg})}{6.37 \times 10^6 \text{ m}}$$
$$= -4.51 \times 10^8 \text{ J} = -451 \text{ MJ}.$$

The kinetic energy $K_0$ of the ball is due to the ball's motion with Earth's rotation. You can show that $K_0$ is less than 1 MJ, which is negligible relative to $U_0$. Thus, the mechanical energy of the ball on the launchpad is

$$E_0 = K_0 + U_0 \approx 0 - 451 \text{ MJ} = -451 \text{ MJ}. \qquad \text{(Answer)}$$

The *increase* in the mechanical energy of the ball from launchpad to orbit is

$$\Delta E = E - E_0 = (-214 \text{ MJ}) - (-451 \text{ MJ})$$
$$= 237 \text{ MJ}. \qquad \text{(Answer)}$$

This is worth a few dollars at your utility company. Obviously the high cost of placing objects into orbit is not due to their required mechanical energy.

### Sample Problem 13.06  Transforming a circular orbit into an elliptical orbit

A spaceship of mass $m = 4.50 \times 10^3$ kg is in a circular Earth orbit of radius $r = 8.00 \times 10^6$ m and period $T_0 = 118.6$ min $= 7.119 \times 10^3$ s when a thruster is fired in the forward direction to decrease the speed to 96.0% of the original speed. What is the period $T$ of the resulting elliptical orbit (Fig. 13-17)?

#### KEY IDEAS

(1) An elliptical orbit period is related to the semimajor axis $a$ by Kepler's third law, written as Eq. 13-34 ($T^2 = 4\pi^2 r^3/GM$) but with $a$ replacing $r$. (2) The semimajor axis $a$ is related to the total mechanical energy $E$ of the ship by Eq. 13-42 ($E = -GMm/2a$), in which Earth's mass is $M = 5.98 \times 10^{24}$ kg. (3) The potential energy of the ship at radius $r$ from Earth's center is given by Eq. 13-21 ($U = -GMm/r$).

*Calculations:* Looking over the Key Ideas, we see that we need to calculate the total energy $E$ to find the semimajor axis $a$, so that we can then determine the period of the elliptical orbit. Let's start with the kinetic energy, calculating it just after the thruster is fired. The speed $v$ just then is 96% of the initial speed $v_0$, which was equal to the ratio of the circumfer-

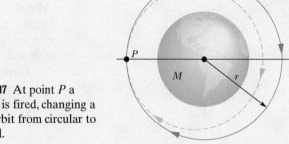

**Figure 13-17** At point $P$ a thruster is fired, changing a ship's orbit from circular to elliptical.

ence of the initial circular orbit to the initial period of the orbit. Thus, just after the thruster is fired, the kinetic energy is

$$K = \tfrac{1}{2}mv^2 = \tfrac{1}{2}m(0.96v_0)^2 = \tfrac{1}{2}m(0.96)^2\left(\frac{2\pi r}{T_0}\right)^2$$
$$= \tfrac{1}{2}(4.50 \times 10^3 \text{ kg})(0.96)^2\left(\frac{2\pi(8.00 \times 10^6 \text{ m})}{7.119 \times 10^3 \text{ s}}\right)^2$$
$$= 1.0338 \times 10^{11} \text{ J}.$$

Just after the thruster is fired, the ship is still at orbital radius $r$, and thus its gravitational potential energy is

$$U = -\frac{GMm}{r}$$

$$= -\frac{(6.67 \times 10^{-11}\,\mathrm{N \cdot m^2/kg^2})(5.98 \times 10^{24}\,\mathrm{kg})(4.50 \times 10^3\,\mathrm{kg})}{8.00 \times 10^6\,\mathrm{m}}$$

$$= -2.2436 \times 10^{11}\,\mathrm{J}.$$

We can now find the semimajor axis by rearranging Eq. 13-42, substituting $a$ for $r$, and then substituting in our energy results:

$$a = -\frac{GMm}{2E} = -\frac{GMm}{2(K+U)}$$

$$= -\frac{(6.67 \times 10^{-11}\,\mathrm{N \cdot m^2/kg^2})(5.98 \times 10^{24}\,\mathrm{kg})(4.50 \times 10^3\,\mathrm{kg})}{2(1.0338 \times 10^{11}\,\mathrm{J} - 2.2436 \times 10^{11}\,\mathrm{J})}$$

$$= 7.418 \times 10^6\,\mathrm{m}.$$

OK, one more step to go. We substitute $a$ for $r$ in Eq. 13-34 and then solve for the period $T$, substituting our result for $a$:

$$T = \left(\frac{4\pi^2 a^3}{GM}\right)^{1/2}$$

$$= \left(\frac{4\pi^2(7.418 \times 10^6\,\mathrm{m})^3}{(6.67 \times 10^{-11}\,\mathrm{N \cdot m^2/kg^2})(5.98 \times 10^{24}\,\mathrm{kg})}\right)^{1/2}$$

$$= 6.356 \times 10^3\,\mathrm{s} = 106\,\mathrm{min}. \qquad \text{(Answer)}$$

This is the period of the elliptical orbit that the ship takes after the thruster is fired. It is less than the period $T_0$ for the circular orbit for two reasons. (1) The orbital path length is now less. (2) The elliptical path takes the ship closer to Earth everywhere except at the point of firing (Fig. 13-17). The resulting decrease in gravitational potential energy increases the kinetic energy and thus also the speed of the ship.

 **PLUS** Additional examples, video, and practice available at *WileyPLUS*

# 13-8 EINSTEIN AND GRAVITATION

## Learning Objectives

*After reading this module, you should be able to . . .*

13.24 Explain Einstein's principle of equivalence.

13.25 Identify Einstein's model for gravitation as being due to the curvature of spacetime.

## Key Idea

● Einstein pointed out that gravitation and acceleration are equivalent. This principle of equivalence led him to a theory of gravitation (the general theory of relativity) that explains gravitational effects in terms of a curvature of space.

## Einstein and Gravitation

### Principle of Equivalence

Albert Einstein once said: "I was . . . in the patent office at Bern when all of a sudden a thought occurred to me: 'If a person falls freely, he will not feel his own weight.' I was startled. This simple thought made a deep impression on me. It impelled me toward a theory of gravitation."

Thus Einstein tells us how he began to form his **general theory of relativity.** The fundamental postulate of this theory about gravitation (the gravitating of objects toward each other) is called the **principle of equivalence,** which says that gravitation and acceleration are equivalent. If a physicist were locked up in a small box as in Fig. 13-18, he would not be able to tell whether the box was at

**Figure 13-18** (*a*) A physicist in a box resting on Earth sees a cantaloupe falling with acceleration $a = 9.8\,\mathrm{m/s^2}$. (*b*) If he and the box accelerate in deep space at $9.8\,\mathrm{m/s^2}$, the cantaloupe has the same acceleration relative to him. It is not possible, by doing experiments within the box, for the physicist to tell which situation he is in. For example, the platform scale on which he stands reads the same weight in both situations.

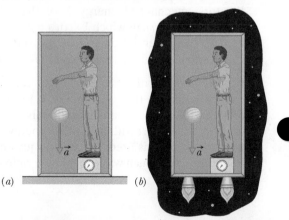

(*a*)      (*b*)

rest on Earth (and subject only to Earth's gravitational force), as in Fig. 13-18*a*, or accelerating through interstellar space at 9.8 m/s$^2$ (and subject only to the force producing that acceleration), as in Fig. 13-18*b*. In both situations he would feel the same and would read the same value for his weight on a scale. Moreover, if he watched an object fall past him, the object would have the same acceleration relative to him in both situations.

### Curvature of Space

We have thus far explained gravitation as due to a force between masses. Einstein showed that, instead, gravitation is due to a curvature of space that is caused by the masses. (As is discussed later in this book, space and time are entangled, so the curvature of which Einstein spoke is really a curvature of *spacetime,* the combined four dimensions of our universe.)

Picturing how space (such as vacuum) can have curvature is difficult. An analogy might help: Suppose that from orbit we watch a race in which two boats begin on Earth's equator with a separation of 20 km and head due south (Fig. 13-19*a*). To the sailors, the boats travel along flat, parallel paths. However, with time the boats draw together until, nearer the south pole, they touch. The sailors in the boats can interpret this drawing together in terms of a force acting on the boats. Looking on from space, however, we can see that the boats draw together simply because of the curvature of Earth's surface. We can see this because we are viewing the race from "outside" that surface.

Figure 13-19*b* shows a similar race: Two horizontally separated apples are dropped from the same height above Earth. Although the apples may appear to travel along parallel paths, they actually move toward each other because they both fall toward Earth's center. We can interpret the motion of the apples in terms of the gravitational force on the apples from Earth. We can also interpret the motion in terms of a curvature of the space near Earth, a curvature due to the presence of Earth's mass. This time we cannot see the curvature because we cannot get "outside" the curved space, as we got "outside" the curved Earth in the boat example. However, we can depict the curvature with a drawing like Fig. 13-19*c*; there the apples would move along a surface that curves toward Earth because of Earth's mass.

When light passes near Earth, the path of the light bends slightly because of the curvature of space there, an effect called *gravitational lensing*. When light passes a more massive structure, like a galaxy or a black hole having large mass, its path can be bent more. If such a massive structure is between us and a quasar (an extremely bright, extremely distant source of light), the light from the quasar

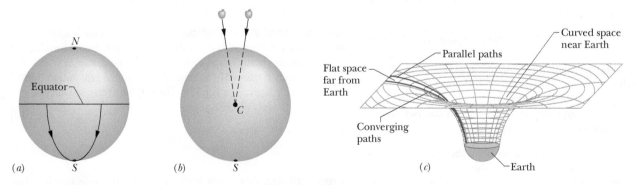

**Figure 13-19** (*a*) Two objects moving along lines of longitude toward the south pole converge because of the curvature of Earth's surface. (*b*) Two objects falling freely near Earth move along lines that converge toward the center of Earth because of the curvature of space near Earth. (*c*) Far from Earth (and other masses), space is flat and parallel paths remain parallel. Close to Earth, the parallel paths begin to converge because space is curved by Earth's mass.

**Figure 13-20** (*a*) Light from a distant quasar follows curved paths around a galaxy or a large black hole because the mass of the galaxy or black hole has curved the adjacent space. If the light is detected, it appears to have originated along the backward extensions of the final paths (dashed lines). (*b*) The Einstein ring known as MG1131+0456 on the computer screen of a telescope. The source of the light (actually, radio waves, which are a form of invisible light) is far behind the large, unseen galaxy that produces the ring; a portion of the source appears as the two bright spots seen along the ring.

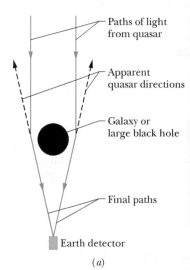

Paths of light from quasar

Apparent quasar directions

Galaxy or large black hole

Final paths

Earth detector

(*a*)

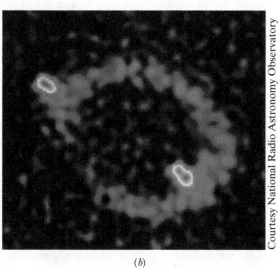

*Courtesy National Radio Astronomy Observatory*

(*b*)

can bend around the massive structure and toward us (Fig. 13-20*a*). Then, because the light seems to be coming to us from a number of slightly different directions in the sky, we see the same quasar in all those different directions. In some situations, the quasars we see blend together to form a giant luminous arc, which is called an *Einstein ring* (Fig. 13-20*b*).

Should we attribute gravitation to the curvature of spacetime due to the presence of masses or to a force between masses? Or should we attribute it to the actions of a type of fundamental particle called a *graviton*, as conjectured in some modern physics theories? Although our theories about gravitation have been enormously successful in describing everything from falling apples to planetary and stellar motions, we still do not fully understand it on either the cosmological scale or the quantum physics scale.

## Review & Summary

**The Law of Gravitation** Any particle in the universe attracts any other particle with a **gravitational force** whose magnitude is

$$F = G \frac{m_1 m_2}{r^2} \quad \text{(Newton's law of gravitation)}, \quad (13\text{-}1)$$

where $m_1$ and $m_2$ are the masses of the particles, $r$ is their separation, and $G \,(= 6.67 \times 10^{-11} \, \text{N} \cdot \text{m}^2/\text{kg}^2)$ is the *gravitational constant*.

**Gravitational Behavior of Uniform Spherical Shells** The gravitational force between extended bodies is found by adding (integrating) the individual forces on individual particles within the bodies. However, if either of the bodies is a uniform spherical shell or a spherically symmetric solid, the net gravitational force it exerts on an *external* object may be computed as if all the mass of the shell or body were located at its center.

**Superposition** Gravitational forces obey the **principle of superposition;** that is, if $n$ particles interact, the net force $\vec{F}_{1,\text{net}}$ on a particle labeled particle 1 is the sum of the forces on it from all the other particles taken one at a time:

$$\vec{F}_{1,\text{net}} = \sum_{i=2}^{n} \vec{F}_{1i}, \quad (13\text{-}5)$$

in which the sum is a vector sum of the forces $\vec{F}_{1i}$ on particle 1 from particles 2, 3, . . . , $n$. The gravitational force $\vec{F}_1$ on a particle from an extended body is found by dividing the body into units of differential mass $dm$, each of which produces a differential force $d\vec{F}$ on the particle, and then integrating to find the sum of those forces:

$$\vec{F}_1 = \int d\vec{F}. \quad (13\text{-}6)$$

**Gravitational Acceleration** The *gravitational acceleration* $a_g$ of a particle (of mass $m$) is due solely to the gravitational force acting on it. When the particle is at distance $r$ from the center of a uniform, spherical body of mass $M$, the magnitude $F$ of the gravitational force on the particle is given by Eq. 13-1. Thus, by Newton's second law,

$$F = ma_g, \quad (13\text{-}10)$$

which gives

$$a_g = \frac{GM}{r^2}. \quad (13\text{-}11)$$

**Free-Fall Acceleration and Weight** Because Earth's mass is not distributed uniformly, because the planet is not perfectly spherical, and because it rotates, the actual free-fall acceleration $\vec{g}$ of a particle near Earth differs slightly from the gravitational acceleration $\vec{a}_g$, and the particle's weight (equal to $mg$) differs from the magnitude of the gravitational force on it as calculated by Newton's law of gravitation (Eq. 13-1).

**Gravitation Within a Spherical Shell**  A uniform shell of matter exerts no net gravitational force on a particle located inside it. This means that if a particle is located inside a uniform solid sphere at distance $r$ from its center, the gravitational force exerted on the particle is due only to the mass that lies inside a sphere of radius $r$ (the *inside sphere*). The force magnitude is given by

$$F = \frac{GmM}{R^3}\, r,  \qquad (13\text{-}19)$$

where $M$ is the sphere's mass and $R$ is its radius.

**Gravitational Potential Energy**  The gravitational potential energy $U(r)$ of a system of two particles, with masses $M$ and $m$ and separated by a distance $r$, is the negative of the work that would be done by the gravitational force of either particle acting on the other if the separation between the particles were changed from infinite (very large) to $r$. This energy is

$$U = -\frac{GMm}{r}  \qquad \text{(gravitational potential energy).}  \qquad (13\text{-}21)$$

**Potential Energy of a System**  If a system contains more than two particles, its total gravitational potential energy $U$ is the sum of the terms representing the potential energies of all the pairs. As an example, for three particles, of masses $m_1, m_2,$ and $m_3$,

$$U = -\left( \frac{Gm_1 m_2}{r_{12}} + \frac{Gm_1 m_3}{r_{13}} + \frac{Gm_2 m_3}{r_{23}} \right).  \qquad (13\text{-}22)$$

**Escape Speed**  An object will escape the gravitational pull of an astronomical body of mass $M$ and radius $R$ (that is, it will reach an infinite distance) if the object's speed near the body's surface is at least equal to the **escape speed,** given by

$$v = \sqrt{\frac{2GM}{R}}.  \qquad (13\text{-}28)$$

**Kepler's Laws**  The motion of satellites, both natural and artificial, is governed by these laws:

1. *The law of orbits.* All planets move in elliptical orbits with the Sun at one focus.

2. *The law of areas.* A line joining any planet to the Sun sweeps out equal areas in equal time intervals. (This statement is equivalent to conservation of angular momentum.)

3. *The law of periods.* The square of the period $T$ of any planet is proportional to the cube of the semimajor axis $a$ of its orbit. For circular orbits with radius $r$,

$$T^2 = \left( \frac{4\pi^2}{GM} \right) r^3  \quad \text{(law of periods),}  \qquad (13\text{-}34)$$

where $M$ is the mass of the attracting body—the Sun in the case of the solar system. For elliptical planetary orbits, the semimajor axis $a$ is substituted for $r$.

**Energy in Planetary Motion**  When a planet or satellite with mass $m$ moves in a circular orbit with radius $r$, its potential energy $U$ and kinetic energy $K$ are given by

$$U = -\frac{GMm}{r} \quad \text{and} \quad K = \frac{GMm}{2r}.  \qquad (13\text{-}21, 13\text{-}38)$$

The mechanical energy $E = K + U$ is then

$$E = -\frac{GMm}{2r}.  \qquad (13\text{-}40)$$

For an elliptical orbit of semimajor axis $a$,

$$E = -\frac{GMm}{2a}.  \qquad (13\text{-}42)$$

**Einstein's View of Gravitation**  Einstein pointed out that gravitation and acceleration are equivalent. This **principle of equivalence** led him to a theory of gravitation (the **general theory of relativity**) that explains gravitational effects in terms of a curvature of space.

## Questions

**1**  In Fig. 13-21, a central particle of mass $M$ is surrounded by a square array of other particles, separated by either distance $d$ or distance $d/2$ along the perimeter of the square. What are the magnitude and direction of the net gravitational force on the central particle due to the other particles?

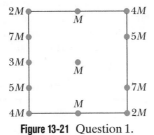

**Figure 13-21**  Question 1.

**2**  Figure 13-22 shows three arrangements of the same identical particles, with three of them placed on a circle of radius 0.20 m and the fourth one placed at the center of the circle. (a) Rank the arrangements according to the magnitude of the net gravitational force on the central particle due to the other three particles, greatest first. (b) Rank them according to the gravitational potential energy of the four-particle system, least negative first.

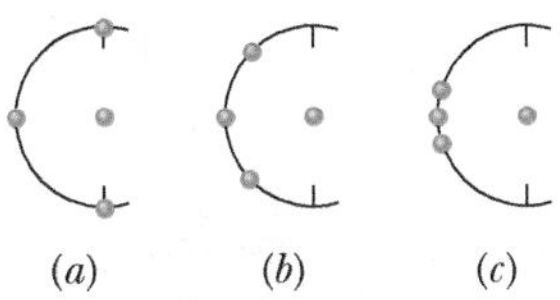

(a)  (b)  (c)

**Figure 13-22**  Question 2.

**3**  In Fig. 13-23, a central particle is surrounded by two circular rings of particles, at radii $r$ and $R$, with $R > r$. All the particles have mass $m$. What are the magnitude and direction of the net gravitational force on the central particle due to the particles in the rings?

**Figure 13-23**  Question 3.

**4**  In Fig. 13-24, two particles, of masses $m$ and $2m$, are fixed in place on an axis. (a) Where on the axis can a third particle of mass $3m$ be placed (other than at infinity) so that the net gravitational force on it from the first two particles is zero: to the left of the first two particles, to their right, between them but closer to the more massive particle, or between them but closer to the less massive particle? (b) Does the answer change if the third particle has, instead, a mass of $16m$? (c) Is there a point off the axis (other than infinity) at which the net force on the third particle would be zero?

**Figure 13-24**  Question 4.

**5** Figure 13-25 shows three situations involving a point particle $P$ with mass $m$ and a spherical shell with a uniformly distributed mass $M$. The radii of the shells are given. Rank the situations according to the magnitude of the gravitational force on particle $P$ due to the shell, greatest first.

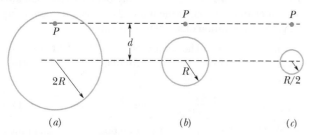

(a)    (b)    (c)

**Figure 13-25** Question 5.

**6** In Fig. 13-26, three particles are fixed in place. The mass of $B$ is greater than the mass of $C$. Can a fourth particle (particle $D$) be placed somewhere so that the net gravitational force on particle $A$ from particles $B$, $C$, and $D$ is zero? If so, in which quadrant should it be placed and which axis should it be near?

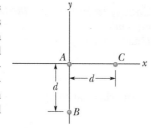

**Figure 13-26** Question 6.

**7** Rank the four systems of equal-mass particles shown in Checkpoint 2 according to the absolute value of the gravitational potential energy of the system, greatest first.

**8** Figure 13-27 gives the gravitational acceleration $a_g$ for four planets as a function of the radial distance $r$ from the center of the planet, starting at the surface of the planet (at radius $R_1, R_2, R_3,$ or $R_4$). Plots 1 and 2 coincide for $r \geq R_2$; plots 3 and 4 coincide for $r \geq R_4$. Rank the four planets according to (a) mass and (b) mass per unit volume, greatest first.

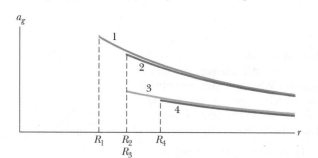

**Figure 13-27** Question 8.

**9** Figure 13-28 shows three particles initially fixed in place, with $B$ and $C$ identical and positioned symmetrically about the $y$ axis, at distance $d$ from $A$. (a) In what direction is the net gravitational force $\vec{F}_{net}$ on $A$? (b) If we move $C$ directly away from the origin, does $\vec{F}_{net}$ change in direction? If so, how and what is the limit of the change?

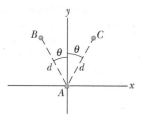

**Figure 13-28** Question 9.

**10** Figure 13-29 shows six paths by which a rocket orbiting a moon might move from point $a$ to point $b$. Rank the paths according to (a) the corresponding change in the gravitational potential energy of the rocket–moon system and (b) the net work done on the rocket by the gravitational force from the moon, greatest first.

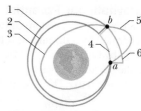

**Figure 13-29** Question 10.

**11** Figure 13-30 shows three uniform spherical planets that are identical in size and mass. The periods of rotation $T$ for the planets are given, and six lettered points are indicated—three points are on the equators of the planets and three points are on the north poles. Rank the points according to the value of the free-fall acceleration $g$ at them, greatest first.

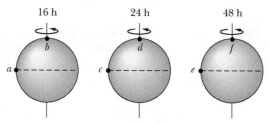

**Figure 13-30** Question 11.

**12** In Fig. 13-31, a particle of mass $m$ (which is not shown) is to be moved from an infinite distance to one of the three possible locations $a$, $b$, and $c$. Two other particles, of masses $m$ and $2m$, are already fixed in place on the axis, as shown. Rank the three possible locations according to the work done by the net gravitational force on the moving particle due to the fixed particles, greatest first.

**Figure 13-31** Question 12.

# Problems

GO Tutoring problem available (at instructor's discretion) in *WileyPLUS* and WebAssign
SSM Worked-out solution available in Student Solutions Manual
• – ••• Number of dots indicates level of problem difficulty
WWW Worked-out solution is at
ILW Interactive solution is at
http://www.wiley.com/college/halliday
Additional information available in *The Flying Circus of Physics* and at flyingcircusofphysics.com

## Module 13-1 Newton's Law of Gravitation

•**1** ILW A mass $M$ is split into two parts, $m$ and $M - m$, which are then separated by a certain distance. What ratio $m/M$ maximizes the magnitude of the gravitational force between the parts?

•**2** *Moon effect.* Some people believe that the Moon controls their activities. If the Moon moves from being directly on the opposite side of Earth from you to being directly overhead, by what percent does (a) the Moon's gravitational pull on you

9687

9

379

increase and (b) your weight (as measured on a scale) decrease? Assume that the Earth–Moon (center-to-center) distance is $3.82 \times 10^8$ m and Earth's radius is $6.37 \times 10^6$ m.

•3 SSM What must the separation be between a 5.2 kg particle and a 2.4 kg particle for their gravitational attraction to have a magnitude of $2.3 \times 10^{-12}$ N?

•4 The Sun and Earth each exert a gravitational force on the Moon. What is the ratio $F_{Sun}/F_{Earth}$ of these two forces? (The average Sun–Moon distance is equal to the Sun–Earth distance.)

•5 *Miniature black holes.* Left over from the big-bang beginning of the universe, tiny black holes might still wander through the universe. If one with a mass of $1 \times 10^{11}$ kg (and a radius of only $1 \times 10^{-16}$ m) reached Earth, at what distance from your head would its gravitational pull on you match that of Earth's?

## Module 13-2  Gravitation and the Principle of Superposition

•6 GO In Fig. 13-32, a square of edge length 20.0 cm is formed by four spheres of masses $m_1 = 5.00$ g, $m_2 = 3.00$ g, $m_3 = 1.00$ g, and $m_4 = 5.00$ g. In unit-vector notation, what is the net gravitational force from them on a central sphere with mass $m_5 = 2.50$ g?

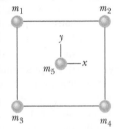

**Figure 13-32**
Problem 6.

•7 *One dimension.* In Fig. 13-33, two point particles are fixed on an x axis separated by distance d. Particle A has mass $m_A$ and particle B has mass $3.00m_A$. A third particle C, of mass $75.0m_A$, is to be placed on the x axis and near particles A and B. In terms of distance d, at what x coordinate should C be placed so that the net gravitational force on particle A from particles B and C is zero?

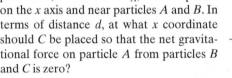

**Figure 13-33**
Problem 7.

•8 In Fig. 13-34, three 5.00 kg spheres are located at distances $d_1 = 0.300$ m and $d_2 = 0.400$ m. What are the (a) magnitude and (b) direction (relative to the positive direction of the x axis) of the net gravitational force on sphere B due to spheres A and C?

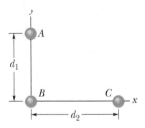

**Figure 13-34** Problem 8.

•9 SSM WWW We want to position a space probe along a line that extends directly toward the Sun in order to monitor solar flares. How far from Earth's center is the point on the line where the Sun's gravitational pull on the probe balances Earth's pull?

••10 GO *Two dimensions.* In Fig. 13-35, three point particles are fixed in place in an xy plane. Particle A has mass $m_A$, particle B has mass $2.00m_A$, and particle C has mass $3.00m_A$. A fourth particle D, with mass $4.00m_A$, is to be placed near the other three particles. In terms of dis-

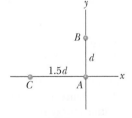

**Figure 13-35** Problem 10.

tance d, at what (a) x coordinate and (b) y coordinate should particle D be placed so that the net gravitational force on particle A from particles B, C, and D is zero?

••11 As seen in Fig. 13-36, two spheres of mass m and a third sphere of mass M form an equilateral triangle, and a fourth sphere of mass $m_4$ is at the center of the triangle. The net gravitational force on that central sphere from the three other spheres is zero. (a) What is M in terms of m? (b) If we double the value of $m_4$, what then is the magnitude of the net gravitational force on the central sphere?

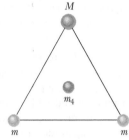

**Figure 13-36**
Problem 11.

••12 GO In Fig. 13-37a, particle A is fixed in place at $x = -0.20$ m on the x axis and particle B, with a mass of 1.0 kg, is fixed in place at the origin. Particle C (not shown) can be moved along the x axis, between particle B and $x = \infty$. Figure 13-37b shows the x component $F_{net,x}$ of the net gravitational force on particle B due to particles A and C, as a function of position x of particle C. The plot actually extends to the right, approaching an asymptote of $-4.17 \times 10^{-10}$ N as $x \to \infty$. What are the masses of (a) particle A and (b) particle C?

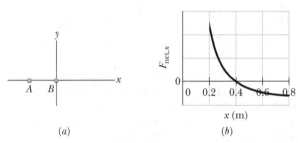

(a)              (b)

**Figure 13-37** Problem 12.

••13 Figure 13-38 shows a spherical hollow inside a lead sphere of radius $R = 4.00$ cm; the surface of the hollow passes through the center of the sphere and "touches" the right side of the sphere. The mass of the sphere before hollowing was $M = 2.95$ kg. With what gravitational force does the hollowed-out lead sphere attract a small sphere of mass $m = 0.431$ kg that lies at a distance $d = 9.00$ cm from the center of the lead sphere, on the straight line connecting the centers of the spheres and of the hollow?

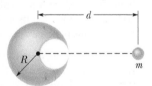

**Figure 13-38** Problem 13.

••14 GO Three point particles are fixed in position in an xy plane. Two of them, particle A of mass 6.00 g and particle B of mass 12.0 g, are shown in Fig. 13-39, with a separation of $d_{AB} = 0.500$ m at angle $\theta = 30°$. Particle C, with mass 8.00 g, is not shown. The net gravitational force acting on particle A due to particles B and C is $2.77 \times 10^{-14}$ N at an angle of $-163.8°$ from the positive direction of the x axis. What are (a) the x coordinate and (b) the y coordinate of particle C?

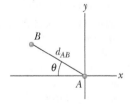

**Figure 13-39** Problem 14.

•••15 GO *Three dimensions.* Three point particles are fixed in place in an xyz coordinate system. Particle A, at the origin, has mass $m_A$.

Particle $B$, at $xyz$ coordinates $(2.00d, 1.00d, 2.00d)$, has mass $2.00m_A$, and particle $C$, at coordinates $(-1.00d, 2.00d, -3.00d)$, has mass $3.00m_A$. A fourth particle $D$, with mass $4.00m_A$, is to be placed near the other particles. In terms of distance $d$, at what (a) $x$, (b) $y$, and (c) $z$ coordinate should $D$ be placed so that the net gravitational force on $A$ from $B$, $C$, and $D$ is zero?

•••16 ⓖⓒ In Fig. 13-40, a particle of mass $m_1 = 0.67$ kg is a distance $d = 23$ cm from one end of a uniform rod with length $L = 3.0$ m and mass $M = 5.0$ kg. What is the magnitude of the gravitational force $\vec{F}$ on the particle from the rod?

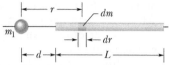

**Figure 13-40** Problem 16.

## Module 13-3 Gravitation Near Earth's Surface

•17 (a) What will an object weigh on the Moon's surface if it weighs 100 N on Earth's surface? (b) How many Earth radii must this same object be from the center of Earth if it is to weigh the same as it does on the Moon?

•18 *Mountain pull.* A large mountain can slightly affect the direction of "down" as determined by a plumb line. Assume that we can model a mountain as a sphere of radius $R = 2.00$ km and density (mass per unit volume) $2.6 \times 10^3$ kg/m³. Assume also that we hang a 0.50 m plumb line at a distance of $3R$ from the sphere's center and such that the sphere pulls horizontally on the lower end. How far would the lower end move toward the sphere?

•19 ssm At what altitude above Earth's surface would the gravitational acceleration be 4.9 m/s²?

•20 *Mile-high building.* In 1956, Frank Lloyd Wright proposed the construction of a mile-high building in Chicago. Suppose the building had been constructed. Ignoring Earth's rotation, find the change in your weight if you were to ride an elevator from the street level, where you weigh 600 N, to the top of the building.

•••21 ilw Certain neutron stars (extremely dense stars) are believed to be rotating at about 1 rev/s. If such a star has a radius of 20 km, what must be its minimum mass so that material on its surface remains in place during the rapid rotation?

•••22 The radius $R_h$ and mass $M_h$ of a black hole are related by $R_h = 2GM_h/c^2$, where $c$ is the speed of light. Assume that the gravitational acceleration $a_g$ of an object at a distance $r_o = 1.001R_h$ from the center of a black hole is given by Eq. 13-11 (it is, for large black holes). (a) In terms of $M_h$, find $a_g$ at $r_o$. (b) Does $a_g$ at $r_o$ increase or decrease as $M_h$ increases? (c) What is $a_g$ at $r_o$ for a very large black hole whose mass is $1.55 \times 10^{12}$ times the solar mass of $1.99 \times 10^{30}$ kg? (d) If an astronaut of height 1.70 m is at $r_o$ with her feet down, what is the difference in gravitational acceleration between her head and feet? (e) Is the tendency to stretch the astronaut severe?

•••23 One model for a certain planet has a core of radius $R$ and mass $M$ surrounded by an outer shell of inner radius $R$, outer radius $2R$, and mass $4M$. If $M = 4.1 \times 10^{24}$ kg and $R = 6.0 \times 10^6$ m, what is the gravitational acceleration of a particle at points (a) $R$ and (b) $3R$ from the center of the planet?

## Module 13-4 Gravitation Inside Earth

•24 Two concentric spherical shells with uniformly distributed masses $M_1$ and $M_2$ are situated as shown in Fig. 13-41. Find the magnitude of the net gravitational force on a particle of mass $m$, due to the

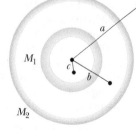

**Figure 13-41** Problem 24.

shells, when the particle is located at radial distance (a) $a$, (b) $b$, and (c) $c$.

•••25 A solid sphere has a uniformly distributed mass of $1.0 \times 10^4$ kg and a radius of 1.0 m. What is the magnitude of the gravitational force due to the sphere on a particle of mass $m$ when the particle is located at a distance of (a) 1.5 m and (b) 0.50 m from the center of the sphere? (c) Write a general expression for the magnitude of the gravitational force on the particle at a distance $r \le 1.0$ m from the center of the sphere.

•••26 A uniform solid sphere of radius $R$ produces a gravitational acceleration of $a_g$ on its surface. At what distance from the sphere's center are there points (a) inside and (b) outside the sphere where the gravitational acceleration is $a_g/3$?

•••27 Figure 13-42 shows, not to scale, a cross section through the interior of Earth. Rather than being uniform throughout, Earth is divided into three zones: an outer *crust*, a *mantle*, and an inner *core*. The dimensions of these zones and the masses contained within them are shown on the figure. Earth has a total mass of $5.98 \times 10^{24}$ kg and a radius of 6370 km. Ignore rotation and assume that Earth is spherical. (a) Calculate $a_g$ at the surface. (b) Suppose that a bore hole (the *Mohole*) is driven to the crust–mantle interface at a depth of 25.0 km; what would be the value of $a_g$ at the bottom of the hole? (c) Suppose that Earth were a uniform sphere with the same total mass and size. What would be the value of $a_g$ at a depth of 25.0 km? (Precise measurements of $a_g$ are sensitive probes of the interior structure of Earth, although results can be clouded by local variations in mass distribution.)

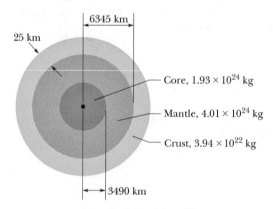

**Figure 13-42** Problem 27.

•••28 ⓖⓒ Assume a planet is a uniform sphere of radius $R$ that (somehow) has a narrow radial tunnel through its center (Fig. 13-7). Also assume we can position an apple anywhere along the tunnel or outside the sphere. Let $F_R$ be the magnitude of the gravitational force on the apple when it is located at the planet's surface. How far from the surface is there a point where the magnitude is $\frac{1}{2}F_R$ if we move the apple (a) away from the planet and (b) into the tunnel?

## Module 13-5 Gravitational Potential Energy

•29 Figure 13-43 gives the potential energy function $U(r)$ of a projectile, plotted outward from

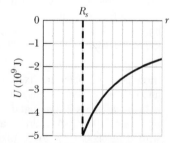

**Figure 13-43** Problems 29 and 34.

the surface of a planet of radius $R_s$. What least kinetic energy is required of a projectile launched at the surface if the projectile is to "escape" the planet?

•30   In Problem 1, what ratio $m/M$ gives the least gravitational potential energy for the system?

•31 SSM   The mean diameters of Mars and Earth are $6.9 \times 10^3$ km and $1.3 \times 10^4$ km, respectively. The mass of Mars is 0.11 times Earth's mass. (a) What is the ratio of the mean density (mass per unit volume) of Mars to that of Earth? (b) What is the value of the gravitational acceleration on Mars? (c) What is the escape speed on Mars?

•32   (a) What is the gravitational potential energy of the two-particle system in Problem 3? If you triple the separation between the particles, how much work is done (b) by the gravitational force between the particles and (c) by you?

•33   What multiple of the energy needed to escape from Earth gives the energy needed to escape from (a) the Moon and (b) Jupiter?

•34   Figure 13-43 gives the potential energy function $U(r)$ of a projectile, plotted outward from the surface of a planet of radius $R_s$. If the projectile is launched radially outward from the surface with a mechanical energy of $-2.0 \times 10^9$ J, what are (a) its kinetic energy at radius $r = 1.25R_s$ and (b) its *turning point* (see Module 8-3) in terms of $R_s$?

••35 GO   Figure 13-44 shows four particles, each of mass 20.0 g, that form a square with an edge length of $d = 0.600$ m. If $d$ is reduced to 0.200 m, what is the change in the gravitational potential energy of the four-particle system?

**Figure 13-44**
Problem 35.

••36 GO   Zero, a hypothetical planet, has a mass of $5.0 \times 10^{23}$ kg, a radius of $3.0 \times 10^6$ m, and no atmosphere. A 10 kg space probe is to be launched vertically from its surface. (a) If the probe is launched with an initial energy of $5.0 \times 10^7$ J, what will be its kinetic energy when it is $4.0 \times 10^6$ m from the center of Zero? (b) If the probe is to achieve a maximum distance of $8.0 \times 10^6$ m from the center of Zero, with what initial kinetic energy must it be launched from the surface of Zero?

••37 GO   The three spheres in Fig. 13-45, with masses $m_A = 80$ g, $m_B = 10$ g, and $m_C = 20$ g, have their centers on a common line, with $L = 12$ cm and $d = 4.0$ cm. You move sphere $B$ along the line until its center-to-center separation from $C$ is $d = 4.0$ cm. How much work is done on sphere $B$ (a) by you and (b) by the net gravitational force on $B$ due to spheres $A$ and $C$?

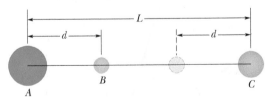

**Figure 13-45**  Problem 37.

••38   In deep space, sphere $A$ of mass 20 kg is located at the origin of an $x$ axis and sphere $B$ of mass 10 kg is located on the axis at $x = 0.80$ m. Sphere $B$ is released from rest while sphere $A$ is held at the origin. (a) What is the gravitational potential energy of the two-sphere system just as $B$ is released? (b) What is the kinetic energy of $B$ when it has moved 0.20 m toward $A$?

••39 SSM   (a) What is the escape speed on a spherical asteroid whose radius is 500 km and whose gravitational acceleration at the surface is 3.0 m/s²? (b) How far from the surface will a particle go if it leaves the asteroid's surface with a radial speed of 1000 m/s? (c) With what speed will an object hit the asteroid if it is dropped from 1000 km above the surface?

••40   A projectile is shot directly away from Earth's surface. Neglect the rotation of Earth. What multiple of Earth's radius $R_E$ gives the radial distance a projectile reaches if (a) its initial speed is 0.500 of the escape speed from Earth and (b) its initial kinetic energy is 0.500 of the kinetic energy required to escape Earth? (c) What is the least initial mechanical energy required at launch if the projectile is to escape Earth?

••41 SSM   Two neutron stars are separated by a distance of $1.0 \times 10^{10}$ m. They each have a mass of $1.0 \times 10^{30}$ kg and a radius of $1.0 \times 10^5$ m. They are initially at rest with respect to each other. As measured from that rest frame, how fast are they moving when (a) their separation has decreased to one-half its initial value and (b) they are about to collide?

••42 GO   Figure 13-46a shows a particle $A$ that can be moved along a $y$ axis from an infinite distance to the origin. That origin lies at the midpoint between particles $B$ and $C$, which have identical masses, and the $y$ axis is a perpendicular bisector between them. Distance $D$ is 0.3057 m. Figure 13-46b shows the potential energy $U$ of the three-particle system as a function of the position of particle $A$ along the $y$ axis. The curve actually extends rightward and approaches an asymptote of $-2.7 \times 10^{-11}$ J as $y \to \infty$. What are the masses of (a) particles $B$ and $C$ and (b) particle $A$?

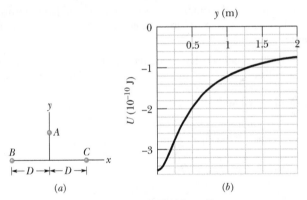

**Figure 13-46**  Problem 42.

## Module 13-6   Planets and Satellites: Kepler's Laws

•43   (a) What linear speed must an Earth satellite have to be in a circular orbit at an altitude of 160 km above Earth's surface? (b) What is the period of revolution?

•44   A satellite is put in a circular orbit about Earth with a radius equal to one-half the radius of the Moon's orbit. What is its period of revolution in lunar months? (A lunar month is the period of revolution of the Moon.)

•45   The Martian satellite Phobos travels in an approximately circular orbit of radius $9.4 \times 10^6$ m with a period of 7 h 39 min. Calculate the mass of Mars from this information.

•46   The first known collision between space debris and a functioning satellite occurred in 1996: At an altitude of 700 km, a year-old French spy satellite was hit by a piece of an Ariane rocket. A stabilizing boom on the satellite was demolished, and the satellite was sent spinning out of control. Just before the collision and in kilometers per hour, what was the speed of the rocket piece relative to the satellite if both were in circular orbits and the collision was (a) head-on and (b) along perpendicular paths?

•47 SSM WWW The Sun, which is $2.2 \times 10^{20}$ m from the center of the Milky Way galaxy, revolves around that center once every $2.5 \times 10^8$ years. Assuming each star in the Galaxy has a mass equal to the Sun's mass of $2.0 \times 10^{30}$ kg, the stars are distributed uniformly in a sphere about the galactic center, and the Sun is at the edge of that sphere, estimate the number of stars in the Galaxy.

•48 The mean distance of Mars from the Sun is 1.52 times that of Earth from the Sun. From Kepler's law of periods, calculate the number of years required for Mars to make one revolution around the Sun; compare your answer with the value given in Appendix C.

•49 A comet that was seen in April 574 by Chinese astronomers on a day known by them as the Woo Woo day was spotted again in May 1994. Assume the time between observations is the period of the Woo Woo day comet and its eccentricity is 0.9932. What are (a) the semimajor axis of the comet's orbit and (b) its greatest distance from the Sun in terms of the mean orbital radius $R_P$ of Pluto?

•50 An orbiting satellite stays over a certain spot on the equator of (rotating) Earth. What is the altitude of the orbit (called a *geosynchronous orbit*)?

•51 SSM A satellite, moving in an elliptical orbit, is 360 km above Earth's surface at its farthest point and 180 km above at its closest point. Calculate (a) the semimajor axis and (b) the eccentricity of the orbit.

•52 The Sun's center is at one focus of Earth's orbit. How far from this focus is the other focus, (a) in meters and (b) in terms of the solar radius, $6.96 \times 10^8$ m? The eccentricity is 0.0167, and the semimajor axis is $1.50 \times 10^{11}$ m.

••53 A 20 kg satellite has a circular orbit with a period of 2.4 h and a radius of $8.0 \times 10^6$ m around a planet of unknown mass. If the magnitude of the gravitational acceleration on the surface of the planet is 8.0 m/s², what is the radius of the planet?

••54 GO *Hunting a black hole.* Observations of the light from a certain star indicate that it is part of a binary (two-star) system. This visible star has orbital speed $v = 270$ km/s, orbital period $T = 1.70$ days, and approximate mass $m_1 = 6M_s$, where $M_s$ is the Sun's mass, $1.99 \times 10^{30}$ kg. Assume that the visible star and its companion star, which is dark and unseen, are both in circular orbits (Fig. 13-47). What integer multiple of $M_s$ gives the *approximate* mass $m_2$ of the dark star?

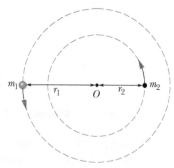

**Figure 13-47** Problem 54.

••55 In 1610, Galileo used his telescope to discover four moons around Jupiter, with these mean orbital radii $a$ and periods $T$:

| Name | $a$ ($10^8$ m) | $T$ (days) |
| --- | --- | --- |
| Io | 4.22 | 1.77 |
| Europa | 6.71 | 3.55 |
| Ganymede | 10.7 | 7.16 |
| Callisto | 18.8 | 16.7 |

(a) Plot log $a$ ($y$ axis) against log $T$ ($x$ axis) and show that you get a straight line. (b) Measure the slope of the line and compare it with the value that you expect from Kepler's third law. (c) Find the mass of Jupiter from the intercept of this line with the $y$ axis.

••56 In 1993 the spacecraft *Galileo* sent an image (Fig. 13-48) of asteroid 243 Ida and a tiny orbiting moon (now known as Dactyl), the first confirmed example of an asteroid–moon system. In the image, the moon, which is 1.5 km wide, is 100 km from the center of the asteroid, which is 55 km long. Assume the moon's orbit is circular with a period of 27 h. (a) What is the mass of the asteroid? (b) The volume of the asteroid, measured from the *Galileo* images, is 14 100 km³. What is the density (mass per unit volume) of the asteroid?

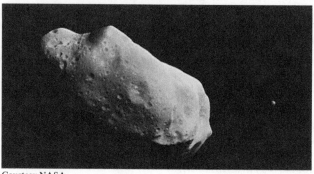

Courtesy NASA

**Figure 13-48** Problem 56. A tiny moon (at right) orbits asteroid 243 Ida.

••57 ILW In a certain binary-star system, each star has the same mass as our Sun, and they revolve about their center of mass. The distance between them is the same as the distance between Earth and the Sun. What is their period of revolution in years?

•••58 GO The presence of an unseen planet orbiting a distant star can sometimes be inferred from the motion of the star as we see it. As the star and planet orbit the center of mass of the star–planet system, the star moves toward and away from us with what is called the *line of sight velocity*, a motion that can be detected. Figure 13-49 shows a graph of the line of sight velocity versus time for the star 14 Herculis. The star's mass is believed to be 0.90 of the mass of our Sun. Assume that only one planet orbits the star and that our view is along the plane of the orbit. Then approximate (a) the planet's mass in terms of Jupiter's mass $m_J$ and (b) the planet's orbital radius in terms of Earth's orbital radius $r_E$.

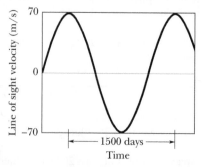

**Figure 13-49** Problem 58.

•••59 Three identical stars of mass $M$ form an equilateral triangle that rotates around the triangle's center as the stars move in a common circle about that center. The triangle has edge length $L$. What is the speed of the stars?

## Module 13-7 Satellites: Orbits and Energy

•60 In Fig. 13-50, two satellites, $A$ and $B$, both of mass $m = 125$ kg, move in the same circular orbit of radius $r = 7.87 \times 10^6$ m around Earth but in opposite senses of rotation and therefore on a collision course. (a) Find the total mechanical energy $E_A + E_B$ of the *two satellites + Earth* system before the collision. (b) If the collision is completely inelastic so that the wreckage remains as one piece of tangled material (mass $= 2m$), find the total mechanical energy immediately after the collision. (c) Just after the collision, is the wreckage falling directly toward Earth's center or orbiting around Earth?

**Figure 13-50**
Problem 60.

•61 (a) At what height above Earth's surface is the energy required to lift a satellite to that height equal to the kinetic energy required for the satellite to be in orbit at that height? (b) For greater heights, which is greater, the energy for lifting or the kinetic energy for orbiting?

•62 Two Earth satellites, $A$ and $B$, each of mass $m$, are to be launched into circular orbits about Earth's center. Satellite $A$ is to orbit at an altitude of 6370 km. Satellite $B$ is to orbit at an altitude of 19 110 km. The radius of Earth $R_E$ is 6370 km. (a) What is the ratio of the potential energy of satellite $B$ to that of satellite $A$, in orbit? (b) What is the ratio of the kinetic energy of satellite $B$ to that of satellite $A$, in orbit? (c) Which satellite has the greater total energy if each has a mass of 14.6 kg? (d) By how much?

•63 SSM WWW An asteroid, whose mass is $2.0 \times 10^{-4}$ times the mass of Earth, revolves in a circular orbit around the Sun at a distance that is twice Earth's distance from the Sun. (a) Calculate the period of revolution of the asteroid in years. (b) What is the ratio of the kinetic energy of the asteroid to the kinetic energy of Earth?

•64 A satellite orbits a planet of unknown mass in a circle of radius $2.0 \times 10^7$ m. The magnitude of the gravitational force on the satellite from the planet is $F = 80$ N. (a) What is the kinetic energy of the satellite in this orbit? (b) What would $F$ be if the orbit radius were increased to $3.0 \times 10^7$ m?

••65 A satellite is in a circular Earth orbit of radius $r$. The area $A$ enclosed by the orbit depends on $r^2$ because $A = \pi r^2$. Determine how the following properties of the satellite depend on $r$: (a) period, (b) kinetic energy, (c) angular momentum, and (d) speed.

••66 One way to attack a satellite in Earth orbit is to launch a swarm of pellets in the same orbit as the satellite but in the opposite direction. Suppose a satellite in a circular orbit 500 km above Earth's surface collides with a pellet having mass 4.0 g. (a) What is the kinetic energy of the pellet in the reference frame of the satellite just before the collision? (b) What is the ratio of this kinetic energy to the kinetic energy of a 4.0 g bullet from a modern army rifle with a muzzle speed of 950 m/s?

•••67 What are (a) the speed and (b) the period of a 220 kg satellite in an approximately circular orbit 640 km above the surface of Earth? Suppose the satellite loses mechanical energy at the average rate of $1.4 \times 10^5$ J per orbital revolution. Adopting the reasonable approximation that the satellite's orbit becomes a "circle of slowly diminishing radius," determine the satellite's (c) altitude, (d) speed, and (e) period at the end of its 1500th revolution. (f) What

is the magnitude of the average retarding force on the satellite? Is angular momentum around Earth's center conserved for (g) the satellite and (h) the satellite–Earth system (assuming that system is isolated)?

•••68 GO Two small spaceships, each with mass $m = 2000$ kg, are in the circular Earth orbit of Fig. 13-51, at an altitude $h$ of 400 km. Igor, the commander of one of the ships, arrives at any fixed point in the orbit 90 s ahead of Picard, the commander of the other ship. What are the (a) period $T_0$ and (b) speed $v_0$ of the ships? At point $P$ in Fig. 13-51, Picard fires an instantaneous burst in the forward direction, *reducing* his ship's speed by 1.00%. After this burst, he follows the elliptical orbit shown dashed in the figure. What are the (c) kinetic energy and (d) potential energy of his ship immediately after the burst?

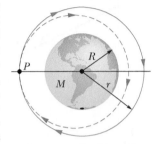

**Figure 13-51** Problem 68.

In Picard's new elliptical orbit, what are (e) the total energy $E$, (f) the semimajor axis $a$, and (g) the orbital period $T$? (h) How much earlier than Igor will Picard return to $P$?

## Module 13-8 Einstein and Gravitation

•69 In Fig. 13-18b, the scale on which the 60 kg physicist stands reads 220 N. How long will the cantaloupe take to reach the floor if the physicist drops it (from rest relative to himself) at a height of 2.1 m above the floor?

### Additional Problems

70 GO The radius $R_h$ of a black hole is the radius of a mathematical sphere, called the event horizon, that is centered on the black hole. Information from events inside the event horizon cannot reach the outside world. According to Einstein's general theory of relativity, $R_h = 2GM/c^2$, where $M$ is the mass of the black hole and $c$ is the speed of light.

Suppose that you wish to study a black hole near it, at a radial distance of $50R_h$. However, you do not want the difference in gravitational acceleration between your feet and your head to exceed 10 m/s² when you are feet down (or head down) toward the black hole. (a) As a multiple of our Sun's mass $M_S$, approximately what is the limit to the mass of the black hole you can tolerate at the given radial distance? (You need to estimate your height.) (b) Is the limit an upper limit (you can tolerate smaller masses) or a lower limit (you can tolerate larger masses)?

71 Several planets (Jupiter, Saturn, Uranus) are encircled by rings, perhaps composed of material that failed to form a satellite. In addition, many galaxies contain ring-like structures. Consider a homogeneous thin ring of mass $M$ and outer radius $R$ (Fig. 13-52). (a) What gravitational attraction does it exert on a particle of mass $m$ located on the ring's central axis a distance $x$ from the

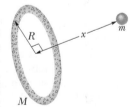

**Figure 13-52**
Problem 71.

ring center? (b) Suppose the particle falls from rest as a result of the attraction of the ring of matter. What is the speed with which it passes through the center of the ring?

72 A typical neutron star may have a mass equal to that of the Sun but a radius of only 10 km. (a) What is the gravitational acceleration at the surface of such a star? (b) How fast would an object be

moving if it fell from rest through a distance of 1.0 m on such a star? (Assume the star does not rotate.)

**73** Figure 13-53 is a graph of the kinetic energy $K$ of an asteroid versus its distance $r$ from Earth's center, as the asteroid falls directly in toward that center. (a) What is the (approximate) mass of the asteroid? (b) What is its speed at $r = 1.945 \times 10^7$ m?

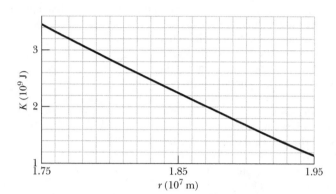

**Figure 13-53** Problem 73.

**74** ✈ The mysterious visitor that appears in the enchanting story *The Little Prince* was said to come from a planet that "was scarcely any larger than a house!" Assume that the mass per unit volume of the planet is about that of Earth and that the planet does not appreciably spin. Approximate (a) the free-fall acceleration on the planet's surface and (b) the escape speed from the planet.

**75** ILW The masses and coordinates of three spheres are as follows: 20 kg, $x = 0.50$ m, $y = 1.0$ m; 40 kg, $x = -1.0$ m, $y = -1.0$ m; 60 kg, $x = 0$ m, $y = -0.50$ m. What is the magnitude of the gravitational force on a 20 kg sphere located at the origin due to these three spheres?

**76** SSM A very early, simple satellite consisted of an inflated spherical aluminum balloon 30 m in diameter and of mass 20 kg. Suppose a meteor having a mass of 7.0 kg passes within 3.0 m of the surface of the satellite. What is the magnitude of the gravitational force on the meteor from the satellite at the closest approach?

**77** GO Four uniform spheres, with masses $m_A = 40$ kg, $m_B = 35$ kg, $m_C = 200$ kg, and $m_D = 50$ kg, have $(x, y)$ coordinates of $(0, 50$ cm), $(0, 0)$, $(-80$ cm, $0)$, and $(40$ cm, $0)$, respectively. In unit-vector notation, what is the net gravitational force on sphere $B$ due to the other spheres?

**78** (a) In Problem 77, remove sphere $A$ and calculate the gravitational potential energy of the remaining three-particle system. (b) If $A$ is then put back in place, is the potential energy of the four-particle system more or less than that of the system in (a)? (c) In (a), is the work done by you to remove $A$ positive or negative? (d) In (b), is the work done by you to replace $A$ positive or negative?

**79** SSM A certain triple-star system consists of two stars, each of mass $m$, revolving in the same circular orbit of radius $r$ around a central star of mass $M$ (Fig. 13-54). The two orbiting stars are always at opposite ends of a diameter of the orbit. Derive an expression for the period of revolution of the stars.

**Figure 13-54**
Problem 79.

**80** The fastest possible rate of rotation of a planet is that for which the gravitational force on material at the equator just barely provides the centripetal force needed for the rotation. (Why?) (a) Show that the corresponding shortest period of rotation is

$$T = \sqrt{\frac{3\pi}{G\rho}},$$

where $\rho$ is the uniform density (mass per unit volume) of the spherical planet. (b) Calculate the rotation period assuming a density of 3.0 g/cm³, typical of many planets, satellites, and asteroids. No astronomical object has ever been found to be spinning with a period shorter than that determined by this analysis.

**81** SSM In a double-star system, two stars of mass $3.0 \times 10^{30}$ kg each rotate about the system's center of mass at radius $1.0 \times 10^{11}$ m. (a) What is their common angular speed? (b) If a meteoroid passes through the system's center of mass perpendicular to their orbital plane, what minimum speed must it have at the center of mass if it is to escape to "infinity" from the two-star system?

**82** A satellite is in elliptical orbit with a period of $8.00 \times 10^4$ s about a planet of mass $7.00 \times 10^{24}$ kg. At aphelion, at radius $4.5 \times 10^7$ m, the satellite's angular speed is $7.158 \times 10^{-5}$ rad/s. What is its angular speed at perihelion?

**83** SSM In a shuttle craft of mass $m = 3000$ kg, Captain Janeway orbits a planet of mass $M = 9.50 \times 10^{25}$ kg, in a circular orbit of radius $r = 4.20 \times 10^7$ m. What are (a) the period of the orbit and (b) the speed of the shuttle craft? Janeway briefly fires a forward-pointing thruster, reducing her speed by 2.00%. Just then, what are (c) the speed, (d) the kinetic energy, (e) the gravitational potential energy, and (f) the mechanical energy of the shuttle craft? (g) What is the semimajor axis of the elliptical orbit now taken by the craft? (h) What is the difference between the period of the original circular orbit and that of the new elliptical orbit? (i) Which orbit has the smaller period?

**84** Consider a pulsar, a collapsed star of extremely high density, with a mass $M$ equal to that of the Sun ($1.98 \times 10^{30}$ kg), a radius $R$ of only 12 km, and a rotational period $T$ of 0.041 s. By what percentage does the free-fall acceleration $g$ differ from the gravitational acceleration $a_g$ at the equator of this spherical star?

**85** ILW A projectile is fired vertically from Earth's surface with an initial speed of 10 km/s. Neglecting air drag, how far above the surface of Earth will it go?

**86** An object lying on Earth's equator is accelerated (a) toward the center of Earth because Earth rotates, (b) toward the Sun because Earth revolves around the Sun in an almost circular orbit, and (c) toward the center of our galaxy because the Sun moves around the galactic center. For the latter, the period is $2.5 \times 10^8$ y and the radius is $2.2 \times 10^{20}$ m. Calculate these three accelerations as multiples of $g = 9.8$ m/s².

**87** (a) If the legendary apple of Newton could be released from rest at a height of 2 m from the surface of a neutron star with a mass 1.5 times that of our Sun and a radius of 20 km, what would be the apple's speed when it reached the surface of the star? (b) If the apple could rest on the surface of the star, what would be the approximate difference between the gravitational acceleration at the top and at the bottom of the apple? (Choose a reasonable size for an apple; the answer indicates that an apple would never survive near a neutron star.)

**88** With what speed would mail pass through the center of Earth if falling in a tunnel through the center?

**89** **SSM** The orbit of Earth around the Sun is *almost* circular: The closest and farthest distances are $1.47 \times 10^8$ km and $1.52 \times 10^8$ km respectively. Determine the corresponding variations in (a) total energy, (b) gravitational potential energy, (c) kinetic energy, and (d) orbital speed. (*Hint:* Use conservation of energy and conservation of angular momentum.)

**90** A 50 kg satellite circles planet Cruton every 6.0 h. The magnitude of the gravitational force exerted on the satellite by Cruton is 80 N. (a) What is the radius of the orbit? (b) What is the kinetic energy of the satellite? (c) What is the mass of planet Cruton?

**91** We watch two identical astronomical bodies $A$ and $B$, each of mass $m$, fall toward each other from rest because of the gravitational force on each from the other. Their initial center-to-center separation is $R_i$. Assume that we are in an inertial reference frame that is stationary with respect to the center of mass of this two-body system. Use the principle of conservation of mechanical energy ($K_f + U_f = K_i + U_i$) to find the following when the center-to-center separation is $0.5R_i$: (a) the total kinetic energy of the system, (b) the kinetic energy of each body, (c) the speed of each body relative to us, and (d) the speed of body $B$ relative to body $A$.

Next assume that we are in a reference frame attached to body $A$ (we ride on the body). Now we see body $B$ fall from rest toward us. From this reference frame, again use $K_f + U_f = K_i + U_i$ to find the following when the center-to-center separation is $0.5R_i$: (e) the kinetic energy of body $B$ and (f) the speed of body $B$ relative to body $A$. (g) Why are the answers to (d) and (f) different? Which answer is correct?

**92** A 150.0 kg rocket moving radially outward from Earth has a speed of 3.70 km/s when its engine shuts off 200 km above Earth's surface. (a) Assuming negligible air drag acts on the rocket, find the rocket's kinetic energy when the rocket is 1000 km above Earth's surface. (b) What maximum height above the surface is reached by the rocket?

**93** Planet Roton, with a mass of $7.0 \times 10^{24}$ kg and a radius of 1600 km, gravitationally attracts a meteorite that is initially at rest relative to the planet, at a distance great enough to take as infinite. The meteorite falls toward the planet. Assuming the planet is airless, find the speed of the meteorite when it reaches the planet's surface.

**94** Two 20 kg spheres are fixed in place on a $y$ axis, one at $y = 0.40$ m and the other at $y = -0.40$ m. A 10 kg ball is then released from rest at a point on the $x$ axis that is at a great distance (effectively infinite) from the spheres. If the only forces acting on the ball are the gravitational forces from the spheres, then when the ball reaches the $(x, y)$ point (0.30 m, 0), what are (a) its kinetic energy and (b) the net force on it from the spheres, in unit-vector notation?

**95** Sphere $A$ with mass 80 kg is located at the origin of an $xy$ coordinate system; sphere $B$ with mass 60 kg is located at coordinates (0.25 m, 0); sphere $C$ with mass 0.20 kg is located in the first quadrant 0.20 m from $A$ and 0.15 m from $B$. In unit-vector notation, what is the gravitational force on $C$ due to $A$ and $B$?

**96** In his 1865 science fiction novel *From the Earth to the Moon,* Jules Verne described how three astronauts are shot to the Moon by means of a huge gun. According to Verne, the aluminum capsule containing the astronauts is accelerated by ignition of nitrocellulose to a speed of 11 km/s along the gun barrel's length of 220 m. (a) In $g$ units, what is the average acceleration of the capsule and astronauts in the gun barrel? (b) Is that acceleration tolerable or deadly to the astronauts?

A modern version of such gun-launched spacecraft (although without passengers) has been proposed. In this modern version, called the SHARP (Super High Altitude Research Project) gun, ignition of methane and air shoves a piston along the gun's tube, compressing hydrogen gas that then launches a rocket. During this launch, the rocket moves 3.5 km and reaches a speed of 7.0 km/s. Once launched, the rocket can be fired to gain additional speed. (c) In $g$ units, what would be the average acceleration of the rocket within the launcher? (d) How much additional speed is needed (via the rocket engine) if the rocket is to orbit Earth at an altitude of 700 km?

**97** An object of mass $m$ is initially held in place at radial distance $r = 3R_E$ from the center of Earth, where $R_E$ is the radius of Earth. Let $M_E$ be the mass of Earth. A force is applied to the object to move it to a radial distance $r = 4R_E$, where it again is held in place. Calculate the work done by the applied force during the move by integrating the force magnitude.

**98** To alleviate the traffic congestion between two cities such as Boston and Washington, D.C., engineers have proposed building a rail tunnel along a chord line connecting the cities (Fig. 13-55). A train, unpropelled by any engine and starting from rest, would fall through the first half of the tunnel and then move up the second half. Assuming Earth is a uniform sphere and ignoring air drag and friction, find the city-to-city travel time.

**Figure 13-55** Problem 98.

**99** A thin rod with mass $M = 5.00$ kg is bent in a semicircle of radius $R = 0.650$ m (Fig. 13-56). (a) What is its gravitational force (both magnitude and direction) on a particle with mass $m = 3.0 \times 10^{-3}$ kg at $P$, the center of curvature? (b) What would be the force on the particle if the rod were a complete circle?

**Figure 13-56** Problem 99.

**100** In Fig. 13-57, identical blocks with identical masses $m = 2.00$ kg hang from strings of different lengths on a balance at Earth's surface. The strings have negligible mass and differ in length by $h = 5.00$ cm. Assume Earth is spherical with a uniform density $\rho = 5.50$ g/cm³. What is the difference in the weight of the blocks due to one being closer to Earth than the other?

**Figure 13-57** Problem 100.

**101** A spaceship is on a straight-line path between Earth and the Moon. At what distance from Earth is the net gravitational force on the spaceship zero?

# Fluids

## 14-1 FLUIDS, DENSITY, AND PRESSURE

### Learning Objectives

*After reading this module, you should be able to . . .*

14.01 Distinguish fluids from solids.

14.02 When mass is uniformly distributed, relate density to mass and volume.

14.03 Apply the relationship between hydrostatic pressure, force, and the surface area over which that force acts.

### Key Ideas

● The density $\rho$ of any material is defined as the material's mass per unit volume:

$$\rho = \frac{\Delta m}{\Delta V}.$$

Usually, where a material sample is much larger than atomic dimensions, we can write this as

$$\rho = \frac{m}{V}.$$

● A fluid is a substance that can flow; it conforms to the boundaries of its container because it cannot withstand shearing stress. It can, however, exert a force perpendicular to its surface. That force is described in terms of pressure $p$:

$$p = \frac{\Delta F}{\Delta A},$$

in which $\Delta F$ is the force acting on a surface element of area $\Delta A$. If the force is uniform over a flat area, this can be written as

$$p = \frac{F}{A}.$$

● The force resulting from fluid pressure at a particular point in a fluid has the same magnitude in all directions.

## What Is Physics?

The physics of fluids is the basis of hydraulic engineering, a branch of engineering that is applied in a great many fields. A nuclear engineer might study the fluid flow in the hydraulic system of an aging nuclear reactor, while a medical engineer might study the blood flow in the arteries of an aging patient. An environmental engineer might be concerned about the drainage from waste sites or the efficient irrigation of farmlands. A naval engineer might be concerned with the dangers faced by a deep-sea diver or with the possibility of a crew escaping from a downed submarine. An aeronautical engineer might design the hydraulic systems controlling the wing flaps that allow a jet airplane to land. Hydraulic engineering is also applied in many Broadway and Las Vegas shows, where huge sets are quickly put up and brought down by hydraulic systems.

Before we can study any such application of the physics of fluids, we must first answer the question "What is a fluid?"

## What Is a Fluid?

A **fluid,** in contrast to a solid, is a substance that can flow. Fluids conform to the boundaries of any container in which we put them. They do so because a fluid cannot sustain a force that is tangential to its surface. (In the more formal language of Module 12-3, a fluid is a substance that flows because it cannot withstand a shearing stress. It can, however, exert a force in the direction perpendicular to its surface.) Some materials, such as pitch, take a long time to conform to the boundaries of a container, but they do so eventually; thus, we classify even those materials as fluids.

You may wonder why we lump liquids and gases together and call them fluids. After all (you may say), liquid water is as different from steam as it is from ice. Actually, it is not. Ice, like other crystalline solids, has its constituent atoms organized in a fairly rigid three-dimensional array called a crystalline lattice. In neither steam nor liquid water, however, is there any such orderly long-range arrangement.

## Density and Pressure

When we discuss rigid bodies, we are concerned with particular lumps of matter, such as wooden blocks, baseballs, or metal rods. Physical quantities that we find useful, and in whose terms we express Newton's laws, are mass and force. We might speak, for example, of a 3.6 kg block acted on by a 25 N force.

With fluids, we are more interested in the extended substance and in properties that can vary from point to point in that substance. It is more useful to speak of **density** and **pressure** than of mass and force.

### Density

To find the density $\rho$ of a fluid at any point, we isolate a small volume element $\Delta V$ around that point and measure the mass $\Delta m$ of the fluid contained within that element. The **density** is then

$$\rho = \frac{\Delta m}{\Delta V}. \tag{14-1}$$

In theory, the density at any point in a fluid is the limit of this ratio as the volume element $\Delta V$ at that point is made smaller and smaller. In practice, we assume that a fluid sample is large relative to atomic dimensions and thus is "smooth" (with uniform density), rather than "lumpy" with atoms. This assumption allows us to write the density in terms of the mass $m$ and volume $V$ of the sample:

$$\rho = \frac{m}{V} \quad \text{(uniform density)}. \tag{14-2}$$

Density is a scalar property; its SI unit is the kilogram per cubic meter. Table 14-1 shows the densities of some substances and the average densities of some objects. Note that the density of a gas (see Air in the table) varies considerably with pressure, but the density of a liquid (see Water) does not; that is, gases are readily *compressible* but liquids are not.

### Pressure

Let a small pressure-sensing device be suspended inside a fluid-filled vessel, as in Fig. 14-1a. The sensor (Fig. 14-1b) consists of a piston of surface area $\Delta A$ riding in a close-fitting cylinder and resting against a spring. A readout arrangement allows us to record the amount by which the (calibrated) spring is compressed by the surrounding fluid, thus indicating the magnitude $\Delta F$ of the force that acts normal to the piston. We define the **pressure** on the piston as

$$p = \frac{\Delta F}{\Delta A}. \tag{14-3}$$

In theory, the pressure at any point in the fluid is the limit of this ratio as the surface area $\Delta A$ of the piston, centered on that point, is made smaller and smaller. However, if the force is uniform over a flat area $A$ (it is evenly distributed over every point of

**Table 14-1** Some Densities

| Material or Object | Density (kg/m³) |
|---|---|
| Interstellar space | $10^{-20}$ |
| Best laboratory vacuum | $10^{-17}$ |
| Air: 20°C and 1 atm pressure | 1.21 |
| 20°C and 50 atm | 60.5 |
| Styrofoam | $1 \times 10^2$ |
| Ice | $0.917 \times 10^3$ |
| Water: 20°C and 1 atm | $0.998 \times 10^3$ |
| 20°C and 50 atm | $1.000 \times 10^3$ |
| Seawater: 20°C and 1 atm | $1.024 \times 10^3$ |
| Whole blood | $1.060 \times 10^3$ |
| Iron | $7.9 \times 10^3$ |
| Mercury (the metal, not the planet) | $13.6 \times 10^3$ |
| Earth: average | $5.5 \times 10^3$ |
| core | $9.5 \times 10^3$ |
| crust | $2.8 \times 10^3$ |
| Sun: average | $1.4 \times 10^3$ |
| core | $1.6 \times 10^5$ |
| White dwarf star (core) | $10^{10}$ |
| Uranium nucleus | $3 \times 10^{17}$ |
| Neutron star (core) | $10^{18}$ |

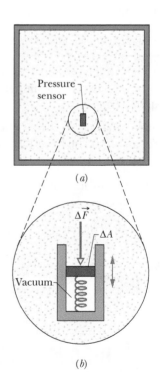

**Figure 14-1** (a) A fluid-filled vessel containing a small pressure sensor, shown in (b). The pressure is measured by the relative position of the movable piston in the sensor.

**Table 14-2** Some Pressures

|  | Pressure (Pa) |
|---|---|
| Center of the Sun | $2 \times 10^{16}$ |
| Center of Earth | $4 \times 10^{11}$ |
| Highest sustained laboratory pressure | $1.5 \times 10^{10}$ |
| Deepest ocean trench (bottom) | $1.1 \times 10^{8}$ |
| Spike heels on a dance floor | $10^{6}$ |
| Automobile tire[a] | $2 \times 10^{5}$ |
| Atmosphere at sea level | $1.0 \times 10^{5}$ |
| Normal blood systolic pressure[a,b] | $1.6 \times 10^{4}$ |
| Best laboratory vacuum | $10^{-12}$ |

[a]Pressure in excess of atmospheric pressure.
[b]Equivalent to 120 torr on the physician's pressure gauge.

the area), we can write Eq. 14-3 as

$$p = \frac{F}{A} \quad \text{(pressure of uniform force on flat area),} \quad (14\text{-}4)$$

where $F$ is the magnitude of the normal force on area $A$.

We find by experiment that at a given point in a fluid at rest, the pressure $p$ defined by Eq. 14-4 has the same value no matter how the pressure sensor is oriented. Pressure is a scalar, having no directional properties. It is true that the force acting on the piston of our pressure sensor is a vector quantity, but Eq. 14-4 involves only the *magnitude* of that force, a scalar quantity.

The SI unit of pressure is the newton per square meter, which is given a special name, the **pascal** (Pa). In metric countries, tire pressure gauges are calibrated in kilopascals. The pascal is related to some other common (non-SI) pressure units as follows:

$$1 \text{ atm} = 1.01 \times 10^{5} \text{ Pa} = 760 \text{ torr} = 14.7 \text{ lb/in.}^{2}.$$

The *atmosphere* (atm) is, as the name suggests, the approximate average pressure of the atmosphere at sea level. The *torr* (named for Evangelista Torricelli, who invented the mercury barometer in 1674) was formerly called the *millimeter of mercury* (mm Hg). The pound per square inch is often abbreviated psi. Table 14-2 shows some pressures.

### Sample Problem 14.01 Atmospheric pressure and force

A living room has floor dimensions of 3.5 m and 4.2 m and a height of 2.4 m.

(a) What does the air in the room weigh when the air pressure is 1.0 atm?

#### KEY IDEAS

(1) The air's weight is equal to $mg$, where $m$ is its mass.
(2) Mass $m$ is related to the air density $\rho$ and the air volume $V$ by Eq. 14-2 ($\rho = m/V$).

*Calculation:* Putting the two ideas together and taking the density of air at 1.0 atm from Table 14-1, we find

$$mg = (\rho V)g$$
$$= (1.21 \text{ kg/m}^{3})(3.5 \text{ m} \times 4.2 \text{ m} \times 2.4 \text{ m})(9.8 \text{ m/s}^{2})$$
$$= 418 \text{ N} \approx 420 \text{ N.} \quad \text{(Answer)}$$

This is the weight of about 110 cans of Pepsi.

(b) What is the magnitude of the atmosphere's downward force on the top of your head, which we take to have an area of 0.040 m²?

#### KEY IDEA

When the fluid pressure $p$ on a surface of area $A$ is uniform, the fluid force on the surface can be obtained from Eq. 14-4 ($p = F/A$).

*Calculation:* Although air pressure varies daily, we can approximate that $p = 1.0$ atm. Then Eq. 14-4 gives

$$F = pA = (1.0 \text{ atm})\left(\frac{1.01 \times 10^{5} \text{ N/m}^{2}}{1.0 \text{ atm}}\right)(0.040 \text{ m}^{2})$$

$$= 4.0 \times 10^{3} \text{ N.} \quad \text{(Answer)}$$

This large force is equal to the weight of the air column from the top of your head to the top of the atmosphere.

**WILEY PLUS** Additional examples, video, and practice available at *WileyPLUS*

# 14-2 FLUIDS AT REST

## Learning Objectives

*After reading this module, you should be able to . . .*

**14.04** Apply the relationship between the hydrostatic pressure, fluid density, and the height above or below a reference level.

**14.05** Distinguish between total pressure (absolute pressure) and gauge pressure.

## Key Ideas

- Pressure in a fluid at rest varies with vertical position $y$. For $y$ measured positive upward,

$$p_2 = p_1 + \rho g(y_1 - y_2).$$

If $h$ is the *depth* of a fluid sample *below* some reference level at which the pressure is $p_0$, this equation becomes

$$p = p_0 + \rho g h,$$

where $p$ is the pressure in the sample.

- The pressure in a fluid is the same for all points at the same level.

- Gauge pressure is the difference between the actual pressure (or absolute pressure) at a point and the atmospheric pressure.

## Fluids at Rest

Figure 14-2a shows a tank of water—or other liquid—open to the atmosphere. As every diver knows, the pressure *increases* with depth below the air–water interface. The diver's depth gauge, in fact, is a pressure sensor much like that of Fig. 14-1b. As every mountaineer knows, the pressure *decreases* with altitude as one ascends into the atmosphere. The pressures encountered by the diver and the mountaineer are usually called *hydrostatic pressures*, because they are due to fluids that are static (at rest). Here we want to find an expression for hydrostatic pressure as a function of depth or altitude.

Let us look first at the increase in pressure with depth below the water's surface. We set up a vertical $y$ axis in the tank, with its origin at the air–water interface and the positive direction upward. We next consider a water sample con-

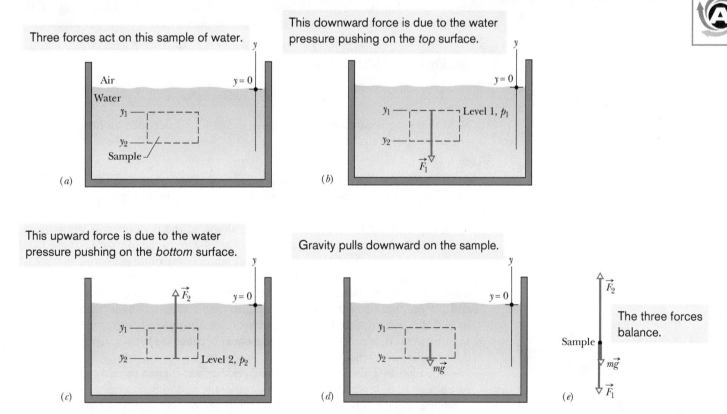

**Figure 14-2** (a) A tank of water in which a sample of water is contained in an imaginary cylinder of horizontal base area $A$. (b)–(d) Force $\vec{F}_1$ acts at the top surface of the cylinder; force $\vec{F}_2$ acts at the bottom surface of the cylinder; the gravitational force on the water in the cylinder is represented by $m\vec{g}$. (e) A free-body diagram of the water sample. In *WileyPLUS*, this figure is available as an animation with voiceover.

tained in an imaginary right circular cylinder of horizontal base (or face) area $A$, such that $y_1$ and $y_2$ (both of which are *negative* numbers) are the depths below the surface of the upper and lower cylinder faces, respectively.

Figure 14-2e is a free-body diagram for the water in the cylinder. The water is in *static equilibrium;* that is, it is stationary and the forces on it balance. Three forces act on it vertically: Force $\vec{F}_1$ acts at the top surface of the cylinder and is due to the water above the cylinder (Fig. 14-2b). Force $\vec{F}_2$ acts at the bottom surface of the cylinder and is due to the water just below the cylinder (Fig. 14-2c). The gravitational force on the water is $m\vec{g}$, where $m$ is the mass of the water in the cylinder (Fig. 14-2d). The balance of these forces is written as

$$F_2 = F_1 + mg. \tag{14-5}$$

To involve pressures, we use Eq. 14-4 to write

$$F_1 = p_1 A \quad \text{and} \quad F_2 = p_2 A. \tag{14-6}$$

The mass $m$ of the water in the cylinder is, from Eq. 14-2, $m = \rho V$, where the cylinder's volume $V$ is the product of its face area $A$ and its height $y_1 - y_2$. Thus, $m$ is equal to $\rho A(y_1 - y_2)$. Substituting this and Eq. 14-6 into Eq. 14-5, we find

$$p_2 A = p_1 A + \rho A g(y_1 - y_2)$$

or

$$p_2 = p_1 + \rho g(y_1 - y_2). \tag{14-7}$$

This equation can be used to find pressure both in a liquid (as a function of depth) and in the atmosphere (as a function of altitude or height). For the former, suppose we seek the pressure $p$ at a depth $h$ below the liquid surface. Then we choose level 1 to be the surface, level 2 to be a distance $h$ below it (as in Fig. 14-3), and $p_0$ to represent the atmospheric pressure on the surface. We then substitute

$$y_1 = 0, \quad p_1 = p_0 \quad \text{and} \quad y_2 = -h, \quad p_2 = p$$

into Eq. 14-7, which becomes

$$p = p_0 + \rho g h \quad \text{(pressure at depth } h\text{).} \tag{14-8}$$

Note that the pressure at a given depth in the liquid depends on that depth but not on any horizontal dimension.

The pressure at a point in a fluid in static equilibrium depends on the depth of that point but not on any horizontal dimension of the fluid or its container.

Thus, Eq. 14-8 holds no matter what the shape of the container. If the bottom surface of the container is at depth $h$, then Eq. 14-8 gives the pressure $p$ there.

In Eq. 14-8, $p$ is said to be the total pressure, or **absolute pressure,** at level 2. To see why, note in Fig. 14-3 that the pressure $p$ at level 2 consists of two contributions: (1) $p_0$, the pressure due to the atmosphere, which bears down on the liquid, and (2) $\rho g h$, the pressure due to the liquid above level 2, which bears down on level 2. In general, the difference between an absolute pressure and an atmospheric pressure is called the **gauge pressure** (because we use a gauge to measure this pressure difference). For Fig. 14-3, the gauge pressure is $\rho g h$.

Equation 14-7 also holds above the liquid surface: It gives the atmospheric pressure at a given distance above level 1 in terms of the atmospheric pressure $p_1$ at level 1 (*assuming* that the atmospheric density is uniform over that distance). For example, to find the atmospheric pressure at a distance $d$ above level 1 in Fig. 14-3, we substitute

$$y_1 = 0, \quad p_1 = p_0 \quad \text{and} \quad y_2 = d, \quad p_2 = p.$$

Then with $\rho = \rho_{\text{air}}$, we obtain

$$p = p_0 - \rho_{\text{air}} g d.$$

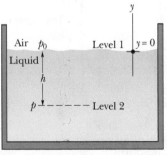

**Figure 14-3** The pressure $p$ increases with depth $h$ below the liquid surface according to Eq. 14-8.

## Checkpoint 1

The figure shows four containers of olive oil. Rank them according to the pressure at depth $h$, greatest first.

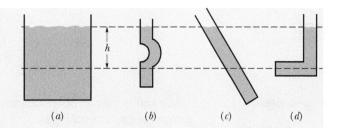

(a)          (b)          (c)          (d)

---

### Sample Problem 14.02    Gauge pressure on a scuba diver

A novice scuba diver practicing in a swimming pool takes enough air from his tank to fully expand his lungs before abandoning the tank at depth $L$ and swimming to the surface, failing to exhale during his ascent. At the surface, the difference $\Delta p$ between the external pressure on him and the air pressure in his lungs is 9.3 kPa. From what depth does he start? What potentially lethal danger does he face?

### KEY IDEA

The pressure at depth $h$ in a liquid of density $\rho$ is given by Eq. 14-8 ($p = p_0 + \rho g h$), where the gauge pressure $\rho g h$ is added to the atmospheric pressure $p_0$.

*Calculations:* Here, when the diver fills his lungs at depth $L$, the external pressure on him (and thus the air pressure within his lungs) is greater than normal and given by Eq. 14-8 as

$$p = p_0 + \rho g L,$$

where $\rho$ is the water's density (998 kg/m³, Table 14-1). As he ascends, the external pressure on him decreases, until it is atmospheric pressure $p_0$ at the surface. His blood pressure also decreases, until it is normal. However, because he does not exhale, the air pressure in his lungs remains at the value it had at depth $L$. At the surface, the pressure difference $\Delta p$ is

$$\Delta p = p - p_0 = \rho g L,$$

so

$$L = \frac{\Delta p}{\rho g} = \frac{9300 \text{ Pa}}{(998 \text{ kg/m}^3)(9.8 \text{ m/s}^2)}$$

$$= 0.95 \text{ m.} \qquad \text{(Answer)}$$

This is not deep! Yet, the pressure difference of 9.3 kPa (about 9% of atmospheric pressure) is sufficient to rupture the diver's lungs and force air from them into the depressurized blood, which then carries the air to the heart, killing the diver. If the diver follows instructions and gradually exhales as he ascends, he allows the pressure in his lungs to equalize with the external pressure, and then there is no danger.

---

### Sample Problem 14.03    Balancing of pressure in a U-tube

The U-tube in Fig. 14-4 contains two liquids in static equilibrium: Water of density $\rho_w$ (= 998 kg/m³) is in the right arm, and oil of unknown density $\rho_x$ is in the left. Measurement gives $l$ = 135 mm and $d$ = 12.3 mm. What is the density of the oil?

### KEY IDEAS

(1) The pressure $p_{int}$ at the level of the oil–water interface in the left arm depends on the density $\rho_x$ and height of the oil above the interface. (2) The water in the right arm *at the same level* must be at the same pressure $p_{int}$. The reason is that, because the water is in static equilibrium, pressures at points in the water at the same level must be the same.

*Calculations:* In the right arm, the interface is a distance $l$ below the free surface of the *water,* and we have, from Eq. 14-8,

$$p_{int} = p_0 + \rho_w g l \quad \text{(right arm)}.$$

In the left arm, the interface is a distance $l + d$ below the free surface of the *oil,* and we have, again from Eq. 14-8,

$$p_{int} = p_0 + \rho_x g (l + d) \quad \text{(left arm)}.$$

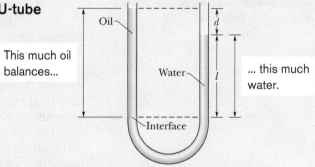

**Figure 14-4** The oil in the left arm stands higher than the water.

Equating these two expressions and solving for the unknown density yield

$$\rho_x = \rho_w \frac{l}{l + d} = (998 \text{ kg/m}^3) \frac{135 \text{ mm}}{135 \text{ mm} + 12.3 \text{ mm}}$$

$$= 915 \text{ kg/m}^3. \qquad \text{(Answer)}$$

Note that the answer does not depend on the atmospheric pressure $p_0$ or the free-fall acceleration $g$.

# 14-3 MEASURING PRESSURE

## Learning Objectives

*After reading this module, you should be able to . . .*

**14.06** Describe how a barometer can measure atmospheric pressure.

**14.07** Describe how an open-tube manometer can measure the gauge pressure of a gas.

## Key Ideas

● A mercury barometer can be used to measure atmospheric pressure.

● An open-tube manometer can be used to measure the gauge pressure of a confined gas.

## Measuring Pressure

### The Mercury Barometer

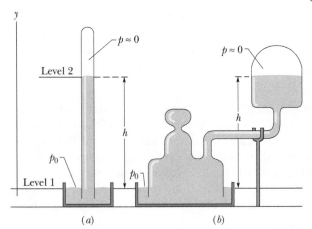

**Figure 14-5** (*a*) A mercury barometer. (*b*) Another mercury barometer. The distance *h* is the same in both cases.

Figure 14-5*a* shows a very basic *mercury barometer,* a device used to measure the pressure of the atmosphere. The long glass tube is filled with mercury and inverted with its open end in a dish of mercury, as the figure shows. The space above the mercury column contains only mercury vapor, whose pressure is so small at ordinary temperatures that it can be neglected.

We can use Eq. 14-7 to find the atmospheric pressure $p_0$ in terms of the height *h* of the mercury column. We choose level 1 of Fig. 14-2 to be that of the air–mercury interface and level 2 to be that of the top of the mercury column, as labeled in Fig. 14-5*a*. We then substitute

$$y_1 = 0, \quad p_1 = p_0 \quad \text{and} \quad y_2 = h, \quad p_2 = 0$$

into Eq. 14-7, finding that

$$p_0 = \rho g h, \tag{14-9}$$

where $\rho$ is the density of the mercury.

For a given pressure, the height *h* of the mercury column does not depend on the cross-sectional area of the vertical tube. The fanciful mercury barometer of Fig. 14-5*b* gives the same reading as that of Fig. 14-5*a*; all that counts is the vertical distance *h* between the mercury levels.

Equation 14-9 shows that, for a given pressure, the height of the column of mercury depends on the value of *g* at the location of the barometer and on the density of mercury, which varies with temperature. The height of the column (in millimeters) is numerically equal to the pressure (in torr) *only* if the barometer is at a place where *g* has its accepted standard value of 9.80665 m/s² *and* the temperature of the mercury is 0°C. If these conditions do not prevail (and they rarely do), small corrections must be made before the height of the mercury column can be transformed into a pressure.

### The Open-Tube Manometer

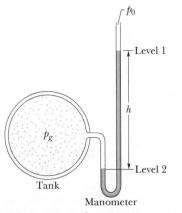

**Figure 14-6** An open-tube manometer, connected to measure the gauge pressure of the gas in the tank on the left. The right arm of the **U**-tube is open to the atmosphere.

An *open-tube manometer* (Fig. 14-6) measures the gauge pressure $p_g$ of a gas. It consists of a **U**-tube containing a liquid, with one end of the tube connected to the vessel whose gauge pressure we wish to measure and the other end open to the atmosphere. We can use Eq. 14-7 to find the gauge pressure in terms of the height *h* shown in Fig. 14-6. Let us choose levels 1 and 2 as shown in Fig. 14-6. With

$$y_1 = 0, \quad p_1 = p_0 \quad \text{and} \quad y_2 = -h, \quad p_2 = p$$

substituted into Eq. 14-7, we find that

$$p_g = p - p_0 = \rho g h, \tag{14-10}$$

where $\rho$ is the liquid's density. The gauge pressure $p_g$ is directly proportional to *h*.

The gauge pressure can be positive or negative, depending on whether $p > p_0$ or $p < p_0$. In inflated tires or the human circulatory system, the (absolute) pressure is greater than atmospheric pressure, so the gauge pressure is a positive quantity, sometimes called the *overpressure*. If you suck on a straw to pull fluid up the straw, the (absolute) pressure in your lungs is actually less than atmospheric pressure. The gauge pressure in your lungs is then a negative quantity.

# 14-4 PASCAL'S PRINCIPLE

## Learning Objectives

*After reading this module, you should be able to . . .*

**14.08** Identify Pascal's principle.

**14.09** For a hydraulic lift, apply the relationship between the input area and displacement and the output area and displacement.

## Key Idea

● Pascal's principle states that a change in the pressure applied to an enclosed fluid is transmitted undiminished to every portion of the fluid and to the walls of the containing vessel.

## Pascal's Principle

When you squeeze one end of a tube to get toothpaste out the other end, you are watching **Pascal's principle** in action. This principle is also the basis for the Heimlich maneuver, in which a sharp pressure increase properly applied to the abdomen is transmitted to the throat, forcefully ejecting food lodged there. The principle was first stated clearly in 1652 by Blaise Pascal (for whom the unit of pressure is named):

> A change in the pressure applied to an enclosed incompressible fluid is transmitted undiminished to every portion of the fluid and to the walls of its container.

### Demonstrating Pascal's Principle

Consider the case in which the incompressible fluid is a liquid contained in a tall cylinder, as in Fig. 14-7. The cylinder is fitted with a piston on which a container of lead shot rests. The atmosphere, container, and shot exert pressure $p_{ext}$ on the piston and thus on the liquid. The pressure $p$ at any point $P$ in the liquid is then

$$p = p_{ext} + \rho g h. \qquad (14\text{-}11)$$

Let us add a little more lead shot to the container to increase $p_{ext}$ by an amount $\Delta p_{ext}$. The quantities $\rho$, $g$, and $h$ in Eq. 14-11 are unchanged, so the pressure change at $P$ is

$$\Delta p = \Delta p_{ext}. \qquad (14\text{-}12)$$

This pressure change is independent of $h$, so it must hold for all points within the liquid, as Pascal's principle states.

### Pascal's Principle and the Hydraulic Lever

Figure 14-8 shows how Pascal's principle can be made the basis of a hydraulic lever. In operation, let an external force of magnitude $F_i$ be directed downward on the left-hand (or input) piston, whose surface area is $A_i$. An incompressible liquid in the device then produces an upward force of magnitude $F_o$ on the right-hand (or output) piston, whose surface area is $A_o$. To keep the system in equilibrium, there must be a downward force of magnitude $F_o$ on the output piston from an external load (not

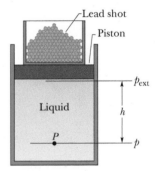

**Figure 14-7** Lead shot (small balls of lead) loaded onto the piston create a pressure $p_{ext}$ at the top of the enclosed (incompressible) liquid. If $p_{ext}$ is increased, by adding more lead shot, the pressure increases by the same amount at all points within the liquid.

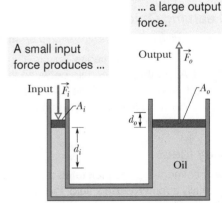

**Figure 14-8** A hydraulic arrangement that can be used to magnify a force $\vec{F}_i$. The work done is, however, not magnified and is the same for both the input and output forces.

shown). The force $\vec{F_i}$ applied on the left and the downward force $\vec{F_o}$ from the load on the right produce a change $\Delta p$ in the pressure of the liquid that is given by

$$\Delta p = \frac{F_i}{A_i} = \frac{F_o}{A_o},$$

so
$$F_o = F_i \frac{A_o}{A_i}. \qquad (14\text{-}13)$$

Equation 14-13 shows that the output force $F_o$ on the load must be greater than the input force $F_i$ if $A_o > A_i$, as is the case in Fig. 14-8.

If we move the input piston downward a distance $d_i$, the output piston moves upward a distance $d_o$, such that the same volume $V$ of the incompressible liquid is displaced at both pistons. Then

$$V = A_i d_i = A_o d_o,$$

which we can write as

$$d_o = d_i \frac{A_i}{A_o}. \qquad (14\text{-}14)$$

This shows that, if $A_o > A_i$ (as in Fig. 14-8), the output piston moves a smaller distance than the input piston moves.

From Eqs. 14-13 and 14-14 we can write the output work as

$$W = F_o d_o = \left( F_i \frac{A_o}{A_i} \right)\left( d_i \frac{A_i}{A_o} \right) = F_i d_i, \qquad (14\text{-}15)$$

which shows that the work $W$ done *on* the input piston by the applied force is equal to the work $W$ done *by* the output piston in lifting the load placed on it.

The advantage of a hydraulic lever is this:

With a hydraulic lever, a given force applied over a given distance can be transformed to a greater force applied over a smaller distance.

The product of force and distance remains unchanged so that the same work is done. However, there is often tremendous advantage in being able to exert the larger force. Most of us, for example, cannot lift an automobile directly but can with a hydraulic jack, even though we have to pump the handle farther than the automobile rises and in a series of small strokes.

# 14-5 ARCHIMEDES' PRINCIPLE

## Learning Objectives

*After reading this module, you should be able to . . .*

14.10 Describe Archimedes' principle.
14.11 Apply the relationship between the buoyant force on a body and the mass of the fluid displaced by the body.
14.12 For a floating body, relate the buoyant force to the gravitational force.

14.13 For a floating body, relate the gravitational force to the mass of the fluid displaced by the body.
14.14 Distinguish between apparent weight and actual weight.
14.15 Calculate the apparent weight of a body that is fully or partially submerged.

## Key Ideas

● Archimedes' principle states that when a body is fully or partially submerged in a fluid, the fluid pushes upward with a buoyant force with magnitude

$$F_b = m_f g,$$

where $m_f$ is the mass of the fluid that has been pushed out of the way by the body.

● When a body floats in a fluid, the magnitude $F_b$ of the (upward) buoyant force on the body is equal to the magnitude $F_g$ of the (downward) gravitational force on the body.

● The apparent weight of a body on which a buoyant force acts is related to its actual weight by

$$\text{weight}_{app} = \text{weight} - F_b.$$

## Archimedes' Principle

Figure 14-9 shows a student in a swimming pool, manipulating a very thin plastic sack (of negligible mass) that is filled with water. She finds that the sack and its contained water are in static equilibrium, tending neither to rise nor to sink. The downward gravitational force $\vec{F_g}$ on the contained water must be balanced by a net upward force from the water surrounding the sack.

This net upward force is a **buoyant force** $\vec{F_b}$. It exists because the pressure in the surrounding water increases with depth below the surface. Thus, the pressure near the bottom of the sack is greater than the pressure near the top, which means the forces on the sack due to this pressure are greater in magnitude near the bottom of the sack than near the top. Some of the forces are represented in Fig. 14-10a, where the space occupied by the sack has been left empty. Note that the force vectors drawn near the bottom of that space (with upward components) have longer lengths than those drawn near the top of the sack (with downward components). If we vectorially add all the forces on the sack from the water, the horizontal components cancel and the vertical components add to yield the upward buoyant force $\vec{F_b}$ on the sack. (Force $\vec{F_b}$ is shown to the right of the pool in Fig. 14-10a.)

Because the sack of water is in static equilibrium, the magnitude of $\vec{F_b}$ is equal to the magnitude $m_f g$ of the gravitational force $\vec{F_g}$ on the sack of water: $F_b = m_f g$. (Subscript $f$ refers to *fluid,* here the water.) In words, the magnitude of the buoyant force is equal to the weight of the water in the sack.

In Fig. 14-10b, we have replaced the sack of water with a stone that exactly fills the hole in Fig. 14-10a. The stone is said to *displace* the water, meaning that the stone occupies space that would otherwise be occupied by water. We have changed nothing about the shape of the hole, so the forces at the hole's surface must be the same as when the water-filled sack was in place. Thus, the same upward buoyant force that acted on the water-filled sack now acts on the stone; that is, the magnitude $F_b$ of the buoyant force is equal to $m_f g$, the weight of the water displaced by the stone.

Unlike the water-filled sack, the stone is not in static equilibrium. The downward gravitational force $\vec{F_g}$ on the stone is greater in magnitude than the upward buoyant force (Fig. 14-10b). The stone thus accelerates downward, sinking.

Let us next exactly fill the hole in Fig. 14-10a with a block of lightweight wood, as in Fig. 14-10c. Again, nothing has changed about the forces at the hole's surface, so the magnitude $F_b$ of the buoyant force is still equal to $m_f g$, the weight

The upward buoyant force on this sack of water equals the weight of the water.

**Figure 14-9** A thin-walled plastic sack of water is in static equilibrium in the pool. The gravitational force on the sack must be balanced by a net upward force on it from the surrounding water.

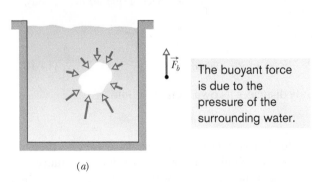

The buoyant force is due to the pressure of the surrounding water.

(a)

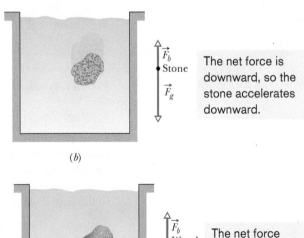

The net force is downward, so the stone accelerates downward.

(b)

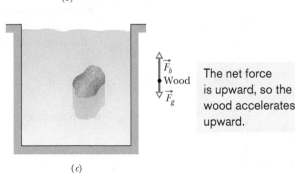

The net force is upward, so the wood accelerates upward.

(c)

**Figure 14-10** (a) The water surrounding the hole in the water produces a net upward buoyant force on whatever fills the hole. (b) For a stone of the same volume as the hole, the gravitational force exceeds the buoyant force in magnitude. (c) For a lump of wood of the same volume, the gravitational force is less than the buoyant force in magnitude.

of the displaced water. Like the stone, the block is not in static equilibrium. However, this time the gravitational force $\vec{F}_g$ is lesser in magnitude than the buoyant force (as shown to the right of the pool), and so the block accelerates upward, rising to the top surface of the water.

Our results with the sack, stone, and block apply to all fluids and are summarized in **Archimedes' principle:**

When a body is fully or partially submerged in a fluid, a buoyant force $\vec{F}_b$ from the surrounding fluid acts on the body. The force is directed upward and has a magnitude equal to the weight $m_f g$ of the fluid that has been displaced by the body.

The buoyant force on a body in a fluid has the magnitude

$$F_b = m_f g \quad \text{(buoyant force),} \tag{14-16}$$

where $m_f$ is the mass of the fluid that is displaced by the body.

## Floating

When we release a block of lightweight wood just above the water in a pool, the block moves into the water because the gravitational force on it pulls it downward. As the block displaces more and more water, the magnitude $F_b$ of the upward buoyant force acting on it increases. Eventually, $F_b$ is large enough to equal the magnitude $F_g$ of the downward gravitational force on the block, and the block comes to rest. The block is then in static equilibrium and is said to be *floating* in the water. In general,

When a body floats in a fluid, the magnitude $F_b$ of the buoyant force on the body is equal to the magnitude $F_g$ of the gravitational force on the body.

We can write this statement as

$$F_b = F_g \quad \text{(floating).} \tag{14-17}$$

From Eq. 14-16, we know that $F_b = m_f g$. Thus,

When a body floats in a fluid, the magnitude $F_g$ of the gravitational force on the body is equal to the weight $m_f g$ of the fluid that has been displaced by the body.

We can write this statement as

$$F_g = m_f g \quad \text{(floating).} \tag{14-18}$$

In other words, a floating body displaces its own weight of fluid.

## Apparent Weight in a Fluid

If we place a stone on a scale that is calibrated to measure weight, then the reading on the scale is the stone's weight. However, if we do this underwater, the upward buoyant force on the stone from the water decreases the reading. That reading is then an apparent weight. In general, an **apparent weight** is related to the actual weight of a body and the buoyant force on the body by

$$\begin{pmatrix} \text{apparent} \\ \text{weight} \end{pmatrix} = \begin{pmatrix} \text{actual} \\ \text{weight} \end{pmatrix} - \begin{pmatrix} \text{magnitude of} \\ \text{buoyant force} \end{pmatrix},$$

which we can write as

$$\text{weight}_{\text{app}} = \text{weight} - F_b \quad \text{(apparent weight).} \tag{14-19}$$

If, in some test of strength, you had to lift a heavy stone, you could do it more easily with the stone underwater. Then your applied force would need to exceed only the stone's apparent weight, not its larger actual weight.

The magnitude of the buoyant force on a floating body is equal to the body's weight. Equation 14-19 thus tells us that a floating body has an apparent weight of zero—the body would produce a reading of zero on a scale. For example, when astronauts prepare to perform a complex task in space, they practice the task floating underwater, where their suits are adjusted to give them an apparent weight of zero.

## ✓ Checkpoint 2

A penguin floats first in a fluid of density $\rho_0$, then in a fluid of density $0.95\rho_0$, and then in a fluid of density $1.1\rho_0$. (a) Rank the densities according to the magnitude of the buoyant force on the penguin, greatest first. (b) Rank the densities according to the amount of fluid displaced by the penguin, greatest first.

### Sample Problem 14.04  Floating, buoyancy, and density

In Fig. 14-11, a block of density $\rho = 800$ kg/m³ floats face down in a fluid of density $\rho_f = 1200$ kg/m³. The block has height $H = 6.0$ cm.

(a) By what depth $h$ is the block submerged?

#### KEY IDEAS

(1) Floating requires that the upward buoyant force on the block match the downward gravitational force on the block. (2) The buoyant force is equal to the weight $m_f g$ of the fluid displaced by the submerged portion of the block.

*Calculations:* From Eq. 14-16, we know that the buoyant force has the magnitude $F_b = m_f g$, where $m_f$ is the mass of the fluid displaced by the block's submerged volume $V_f$. From Eq. 14-2 ($\rho = m/V$), we know that the mass of the displaced fluid is $m_f = \rho_f V_f$. We don't know $V_f$ but if we symbolize the block's face length as $L$ and its width as $W$, then from Fig. 14-11 we see that the submerged volume must be $V_f = LWh$. If we now combine our three expressions, we find that the upward buoyant force has magnitude

$$F_b = m_f g = \rho_f V_f g = \rho_f LWhg. \qquad (14\text{-}20)$$

Similarly, we can write the magnitude $F_g$ of the gravitational force on the block, first in terms of the block's mass $m$, then in terms of the block's density $\rho$ and (full) volume $V$, and then in terms of the block's dimensions $L$, $W$, and $H$ (the full height):

$$F_g = mg = \rho Vg = \rho_f LWHg. \qquad (14\text{-}21)$$

The floating block is stationary. Thus, writing Newton's second law for components along a vertical $y$ axis with the positive direction upward ($F_{net,y} = ma_y$), we have

$$F_b - F_g = m(0),$$

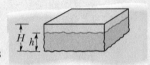
*Floating* means that the buoyant force matches the gravitational force.

**Figure 14-11** Block of height $H$ floats in a fluid, to a depth of $h$.

or from Eqs. 14-20 and 14-21,

$$\rho_f LWhg - \rho LWHg = 0,$$

which gives us

$$h = \frac{\rho}{\rho_f} H = \frac{800 \text{ kg/m}^3}{1200 \text{ kg/m}^3} (6.0 \text{ cm})$$

$$= 4.0 \text{ cm}. \qquad \text{(Answer)}$$

(b) If the block is held fully submerged and then released, what is the magnitude of its acceleration?

*Calculations:* The gravitational force on the block is the same but now, with the block fully submerged, the volume of the displaced water is $V = LWH$. (The full height of the block is used.) This means that the value of $F_b$ is now larger, and the block will no longer be stationary but will accelerate upward. Now Newton's second law yields

$$F_b - F_g = ma,$$

or

$$\rho_f LWHg - \rho LWHg = \rho LWHa,$$

where we inserted $\rho LWH$ for the mass $m$ of the block. Solving for $a$ leads to

$$a = \left(\frac{\rho_f}{\rho} - 1\right)g = \left(\frac{1200 \text{ kg/m}^3}{800 \text{ kg/m}^3} - 1\right)(9.8 \text{ m/s}^2)$$

$$= 4.9 \text{ m/s}^2. \qquad \text{(Answer)}$$

 Additional examples, video, and practice available at *WileyPLUS*

# 14-6 THE EQUATION OF CONTINUITY

## Learning Objectives

*After reading this module, you should be able to . . .*

**14.16** Describe steady flow, incompressible flow, nonviscous flow, and irrotational flow.

**14.17** Explain the term streamline.

**14.18** Apply the equation of continuity to relate the cross-sectional area and flow speed at one point in a tube to those quantities at a different point.

**14.19** Identify and calculate volume flow rate.

**14.20** Identify and calculate mass flow rate.

## Key Ideas

● An ideal fluid is incompressible and lacks viscosity, and its flow is steady and irrotational.

● A *streamline* is the path followed by an individual fluid particle.

● A *tube of flow* is a bundle of streamlines.

● The flow within any tube of flow obeys the equation of continuity:

$$R_V = Av = \text{a constant,}$$

in which $R_V$ is the volume flow rate, $A$ is the cross-sectional area of the tube of flow at any point, and $v$ is the speed of the fluid at that point.

● The mass flow rate $R_m$ is

$$R_m = \rho R_V = \rho Av = \text{a constant.}$$

---

Will McIntyre/Photo Researchers, Inc.

**Figure 14-12** At a certain point, the rising flow of smoke and heated gas changes from steady to turbulent.

## Ideal Fluids in Motion

The motion of *real fluids* is very complicated and not yet fully understood. Instead, we shall discuss the motion of an **ideal fluid,** which is simpler to handle mathematically and yet provides useful results. Here are four assumptions that we make about our ideal fluid; they all are concerned with *flow:*

1. **Steady flow** In *steady* (or *laminar*) *flow,* the velocity of the moving fluid at any fixed point does not change with time. The gentle flow of water near the center of a quiet stream is steady; the flow in a chain of rapids is not. Figure 14-12 shows a transition from steady flow to *nonsteady* (or *nonlaminar* or *turbulent*) *flow* for a rising stream of smoke. The speed of the smoke particles increases as they rise and, at a certain critical speed, the flow changes from steady to nonsteady.

2. **Incompressible flow** We assume, as for fluids at rest, that our ideal fluid is incompressible; that is, its density has a constant, uniform value.

3. **Nonviscous flow** Roughly speaking, the viscosity of a fluid is a measure of how resistive the fluid is to flow. For example, thick honey is more resistive to flow than water, and so honey is said to be more viscous than water. Viscosity is the fluid analog of friction between solids; both are mechanisms by which the kinetic energy of moving objects can be transferred to thermal energy. In the absence of friction, a block could glide at constant speed along a horizontal surface. In the same way, an object moving through a nonviscous fluid would experience no *viscous drag force*—that is, no resistive force due to viscosity; it could move at constant speed through the fluid. The British scientist Lord Rayleigh noted that in an ideal fluid a ship's propeller would not work, but, on the other hand, in an ideal fluid a ship (once set into motion) would not need a propeller!

4. **Irrotational flow** Although it need not concern us further, we also assume that the flow is *irrotational.* To test for this property, let a tiny grain of dust move with the fluid. Although this test body may (or may not) move in a circular path, in irrotational flow the test body will not rotate about an axis through its own center of mass. For a loose analogy, the motion of a Ferris wheel is rotational; that of its passengers is irrotational.

We can make the flow of a fluid visible by adding a *tracer.* This might be a dye injected into many points across a liquid stream (Fig. 14-13) or smoke

**Figure 14-13** The steady flow of a fluid around a cylinder, as revealed by a dye tracer that was injected into the fluid upstream of the cylinder.

Courtesy D. H. Peregrine, University of Bristol

particles added to a gas flow (Fig. 14-12). Each bit of a tracer follows a *streamline*, which is the path that a tiny element of the fluid would take as the fluid flows. Recall from Chapter 4 that the velocity of a particle is always tangent to the path taken by the particle. Here the particle is the fluid element, and its velocity $\vec{v}$ is always tangent to a streamline (Fig. 14-14). For this reason, two streamlines can never intersect; if they did, then an element arriving at their intersection would have two different velocities simultaneously — an impossibility.

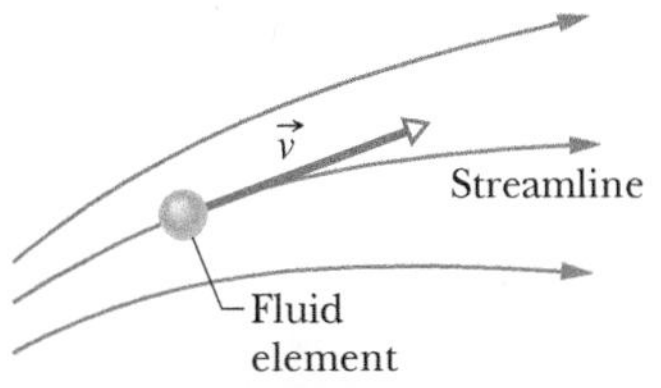

**Figure 14-14** A fluid element traces out a streamline as it moves. The velocity vector of the element is tangent to the streamline at every point.

## The Equation of Continuity

You may have noticed that you can increase the speed of the water emerging from a garden hose by partially closing the hose opening with your thumb. Apparently the speed $v$ of the water depends on the cross-sectional area $A$ through which the water flows.

Here we wish to derive an expression that relates $v$ and $A$ for the steady flow of an ideal fluid through a tube with varying cross section, like that in Fig. 14-15. The flow there is toward the right, and the tube segment shown (part of a longer tube) has length $L$. The fluid has speeds $v_1$ at the left end of the segment and $v_2$ at the right end. The tube has cross-sectional areas $A_1$ at the left end and $A_2$ at the right end. Suppose that in a time interval $\Delta t$ a volume $\Delta V$ of fluid enters the tube segment at its left end (that volume is colored purple in Fig. 14-15). Then, because the fluid is incompressible, an identical volume $\Delta V$ must emerge from the right end of the segment (it is colored green in Fig. 14-15).

The volume flow per second here must match ...

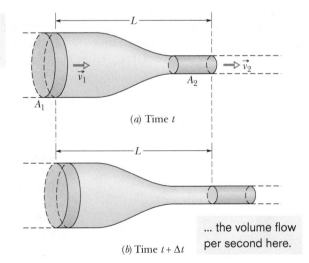

... the volume flow per second here.

**Figure 14-15** Fluid flows from left to right at a steady rate through a tube segment of length $L$. The fluid's speed is $v_1$ at the left side and $v_2$ at the right side. The tube's cross-sectional area is $A_1$ at the left side and $A_2$ at the right side. From time $t$ in (*a*) to time $t + \Delta t$ in (*b*), the amount of fluid shown in purple enters at the left side and the equal amount of fluid shown in green emerges at the right side.

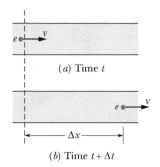

(a) Time $t$

(b) Time $t + \Delta t$

**Figure 14-16** Fluid flows at a constant speed $v$ through a tube. (a) At time $t$, fluid element $e$ is about to pass the dashed line. (b) At time $t + \Delta t$, element $e$ is a distance $\Delta x = v\,\Delta t$ from the dashed line.

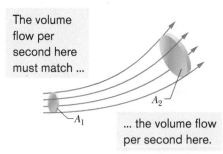

The volume flow per second here must match ...

... the volume flow per second here.

**Figure 14-17** A tube of flow is defined by the streamlines that form the boundary of the tube. The volume flow rate must be the same for all cross sections of the tube of flow.

We can use this common volume $\Delta V$ to relate the speeds and areas. To do so, we first consider Fig. 14-16, which shows a side view of a tube of *uniform* cross-sectional area $A$. In Fig. 14-16a, a fluid element $e$ is about to pass through the dashed line drawn across the tube width. The element's speed is $v$, so during a time interval $\Delta t$, the element moves along the tube a distance $\Delta x = v\,\Delta t$. The volume $\Delta V$ of fluid that has passed through the dashed line in that time interval $\Delta t$ is

$$\Delta V = A\,\Delta x = Av\,\Delta t. \tag{14-22}$$

Applying Eq. 14-22 to both the left and right ends of the tube segment in Fig. 14-15, we have

$$\Delta V = A_1 v_1\,\Delta t = A_2 v_2\,\Delta t$$

or $\qquad\qquad A_1 v_1 = A_2 v_2 \quad$ (equation of continuity). $\tag{14-23}$

This relation between speed and cross-sectional area is called the **equation of continuity** for the flow of an ideal fluid. It tells us that the flow speed increases when we decrease the cross-sectional area through which the fluid flows.

Equation 14-23 applies not only to an actual tube but also to any so-called *tube of flow,* or imaginary tube whose boundary consists of streamlines. Such a tube acts like a real tube because no fluid element can cross a streamline; thus, all the fluid within a tube of flow must remain within its boundary. Figure 14-17 shows a tube of flow in which the cross-sectional area increases from area $A_1$ to area $A_2$ along the flow direction. From Eq. 14-23 we know that, with the increase in area, the speed must decrease, as is indicated by the greater spacing between streamlines at the right in Fig. 14-17. Similarly, you can see that in Fig. 14-13 the speed of the flow is greatest just above and just below the cylinder.

We can rewrite Eq. 14-23 as

$$R_V = Av = \text{a constant} \quad \text{(volume flow rate, equation of continuity)}, \tag{14-24}$$

in which $R_V$ is the **volume flow rate** of the fluid (volume past a given point per unit time). Its SI unit is the cubic meter per second (m³/s). If the density $\rho$ of the fluid is uniform, we can multiply Eq. 14-24 by that density to get the **mass flow rate** $R_m$ (mass per unit time):

$$R_m = \rho R_V = \rho Av = \text{a constant} \quad \text{(mass flow rate)}. \tag{14-25}$$

The SI unit of mass flow rate is the kilogram per second (kg/s). Equation 14-25 says that the mass that flows into the tube segment of Fig. 14-15 each second must be equal to the mass that flows out of that segment each second.

✓ **Checkpoint 3**

The figure shows a pipe and gives the volume flow rate (in cm³/s) and the direction of flow for all but one section. What are the volume flow rate and the direction of flow for that section?

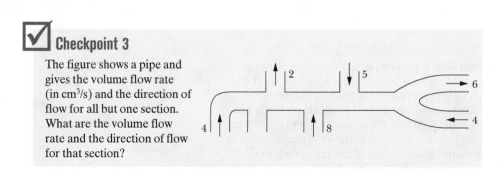

## Sample Problem 14.05   A water stream narrows as it falls

Figure 14-18 shows how the stream of water emerging from a faucet "necks down" as it falls. This change in the horizontal cross-sectional area is characteristic of any laminar (non-turbulent) falling stream because the gravitational force increases the speed of the stream. Here the indicated cross-sectional areas are $A_0 = 1.2$ cm$^2$ and $A = 0.35$ cm$^2$. The two levels are separated by a vertical distance $h = 45$ mm. What is the volume flow rate from the tap?

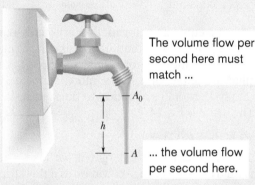

The volume flow per second here must match ...

... the volume flow per second here.

**Figure 14-18** As water falls from a tap, its speed increases. Because the volume flow rate must be the same at all horizontal cross sections of the stream, the stream must "neck down" (narrow).

**KEY IDEA**

The volume flow rate through the higher cross section must be the same as that through the lower cross section.

*Calculations:* From Eq. 14-24, we have

$$A_0 v_0 = Av, \qquad (14\text{-}26)$$

where $v_0$ and $v$ are the water speeds at the levels corresponding to $A_0$ and $A$. From Eq. 2-16 we can also write, because the water is falling freely with acceleration $g$,

$$v^2 = v_0^2 + 2gh. \qquad (14\text{-}27)$$

Eliminating $v$ between Eqs. 14-26 and 14-27 and solving for $v_0$, we obtain

$$v_0 = \sqrt{\frac{2ghA^2}{A_0^2 - A^2}}$$

$$= \sqrt{\frac{(2)(9.8 \text{ m/s}^2)(0.045 \text{ m})(0.35 \text{ cm}^2)^2}{(1.2 \text{ cm}^2)^2 - (0.35 \text{ cm}^2)^2}}$$

$$= 0.286 \text{ m/s} = 28.6 \text{ cm/s}.$$

From Eq. 14-24, the volume flow rate $R_V$ is then
$$R_V = A_0 v_0 = (1.2 \text{ cm}^2)(28.6 \text{ cm/s})$$
$$= 34 \text{ cm}^3/\text{s}. \qquad \text{(Answer)}$$

 Additional examples, video, and practice available at *WileyPLUS*

# 14-7 BERNOULLI'S EQUATION

## Learning Objectives

*After reading this module, you should be able to . . .*

**14.21** Calculate the kinetic energy density in terms of a fluid's density and flow speed.
**14.22** Identify the fluid pressure as being a type of energy density.
**14.23** Calculate the gravitational potential energy density.

**14.24** Apply Bernoulli's equation to relate the total energy density at one point on a streamline to the value at another point.
**14.25** Identify that Bernoulli's equation is a statement of the conservation of energy.

## Key Idea

● Applying the principle of conservation of mechanical energy to the flow of an ideal fluid leads to Bernoulli's equation:
$$p + \tfrac{1}{2}\rho v^2 + \rho g y = \text{a constant}$$
along any tube of flow.

## Bernoulli's Equation

Figure 14-19 represents a tube through which an ideal fluid is flowing at a steady rate. In a time interval $\Delta t$, suppose that a volume of fluid $\Delta V$, colored purple in Fig. 14-19, enters the tube at the left (or input) end and an identical volume,

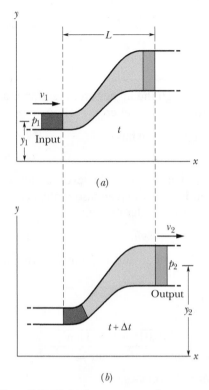

(a)

(b)

**Figure 14-19** Fluid flows at a steady rate through a length $L$ of a tube, from the input end at the left to the output end at the right. From time $t$ in (a) to time $t + \Delta t$ in (b), the amount of fluid shown in purple enters the input end and the equal amount shown in green emerges from the output end.

colored green in Fig. 14-19, emerges at the right (or output) end. The emerging volume must be the same as the entering volume because the fluid is incompressible, with an assumed constant density $\rho$.

Let $y_1$, $v_1$, and $p_1$ be the elevation, speed, and pressure of the fluid entering at the left, and $y_2$, $v_2$, and $p_2$ be the corresponding quantities for the fluid emerging at the right. By applying the principle of conservation of energy to the fluid, we shall show that these quantities are related by

$$p_1 + \tfrac{1}{2}\rho v_1^2 + \rho g y_1 = p_2 + \tfrac{1}{2}\rho v_2^2 + \rho g y_2. \qquad (14\text{-}28)$$

In general, the term $\tfrac{1}{2}\rho v^2$ is called the fluid's **kinetic energy density** (kinetic energy per unit volume). We can also write Eq. 14-28 as

$$p + \tfrac{1}{2}\rho v^2 + \rho g y = \text{a constant} \qquad \text{(Bernoulli's equation).} \qquad (14\text{-}29)$$

Equations 14-28 and 14-29 are equivalent forms of **Bernoulli's equation,** after Daniel Bernoulli, who studied fluid flow in the 1700s.* Like the equation of continuity (Eq. 14-24), Bernoulli's equation is not a new principle but simply the reformulation of a familiar principle in a form more suitable to fluid mechanics. As a check, let us apply Bernoulli's equation to fluids at rest, by putting $v_1 = v_2 = 0$ in Eq. 14-28. The result is Eq. 14-7:

$$p_2 = p_1 + \rho g (y_1 - y_2).$$

A major prediction of Bernoulli's equation emerges if we take $y$ to be a constant ($y = 0$, say) so that the fluid does not change elevation as it flows. Equation 14-28 then becomes

$$p_1 + \tfrac{1}{2}\rho v_1^2 = p_2 + \tfrac{1}{2}\rho v_2^2, \qquad (14\text{-}30)$$

which tells us that:

If the speed of a fluid element increases as the element travels along a horizontal streamline, the pressure of the fluid must decrease, and conversely.

Put another way, where the streamlines are relatively close together (where the velocity is relatively great), the pressure is relatively low, and conversely.

The link between a change in speed and a change in pressure makes sense if you consider a fluid element that travels through a tube of various widths. Recall that the element's speed in the narrower regions is fast and its speed in the wider regions is slow. By Newton's second law, forces (or pressures) must cause the changes in speed (the accelerations). When the element nears a narrow region, the higher pressure behind it accelerates it so that it then has a greater speed in the narrow region. When it nears a wide region, the higher pressure ahead of it decelerates it so that it then has a lesser speed in the wide region.

Bernoulli's equation is strictly valid only to the extent that the fluid is ideal. If viscous forces are present, thermal energy will be involved, which here we neglect.

### Proof of Bernoulli's Equation

Let us take as our system the entire volume of the (ideal) fluid shown in Fig. 14-19. We shall apply the principle of conservation of energy to this system as it moves from its initial state (Fig. 14-19a) to its final state (Fig. 14-19b). The fluid lying between the two vertical planes separated by a distance $L$ in Fig. 14-19 does not change its properties during this process; we need be concerned only with changes that take place at the input and output ends.

---

*For irrotational flow (which we assume), the constant in Eq. 14-29 has the same value for all points within the tube of flow; the points do not have to lie along the same streamline. Similarly, the points 1 and 2 in Eq. 14-28 can lie anywhere within the tube of flow.

First, we apply energy conservation in the form of the work–kinetic energy theorem,

$$W = \Delta K, \tag{14-31}$$

which tells us that the change in the kinetic energy of our system must equal the net work done on the system. The change in kinetic energy results from the change in speed between the ends of the tube and is

$$\Delta K = \tfrac{1}{2}\Delta m\, v_2^2 - \tfrac{1}{2}\Delta m\, v_1^2$$
$$= \tfrac{1}{2}\rho\, \Delta V(v_2^2 - v_1^2), \tag{14-32}$$

in which $\Delta m\ (= \rho\, \Delta V)$ is the mass of the fluid that enters at the input end and leaves at the output end during a small time interval $\Delta t$.

The work done on the system arises from two sources. The work $W_g$ done by the gravitational force ($\Delta m\, \vec{g}$) on the fluid of mass $\Delta m$ during the vertical lift of the mass from the input level to the output level is

$$W_g = -\Delta m\, g(y_2 - y_1)$$
$$= -\rho g\, \Delta V(y_2 - y_1). \tag{14-33}$$

This work is negative because the upward displacement and the downward gravitational force have opposite directions.

Work must also be done *on* the system (at the input end) to push the entering fluid into the tube and *by* the system (at the output end) to push forward the fluid that is located ahead of the emerging fluid. In general, the work done by a force of magnitude $F$, acting on a fluid sample contained in a tube of area $A$ to move the fluid through a distance $\Delta x$, is

$$F\,\Delta x = (pA)(\Delta x) = p(A\,\Delta x) = p\,\Delta V.$$

The work done on the system is then $p_1\,\Delta V$, and the work done by the system is $-p_2\,\Delta V$. Their sum $W_p$ is

$$W_p = -p_2\,\Delta V + p_1\,\Delta V$$
$$= -(p_2 - p_1)\,\Delta V. \tag{14-34}$$

The work–kinetic energy theorem of Eq. 14-31 now becomes

$$W = W_g + W_p = \Delta K.$$

Substituting from Eqs. 14-32, 14-33, and 14-34 yields

$$-\rho g\, \Delta V(y_2 - y_1) - \Delta V(p_2 - p_1) = \tfrac{1}{2}\rho\, \Delta V(v_2^2 - v_1^2).$$

This, after a slight rearrangement, matches Eq. 14-28, which we set out to prove.

## ✓ Checkpoint 4

Water flows smoothly through the pipe shown in the figure, descending in the process. Rank the four numbered sections of pipe according to (a) the volume flow rate $R_V$ through them, (b) the flow speed $v$ through them, and (c) the water pressure $p$ within them, greatest first.

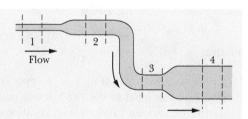

---

### Sample Problem 14.06   Bernoulli principle of fluid through a narrowing pipe

Ethanol of density $\rho = 791$ kg/m³ flows smoothly through a horizontal pipe that tapers (as in Fig. 14-15) in cross-sectional area from $A_1 = 1.20 \times 10^{-3}$ m² to $A_2 = A_1/2$. The pressure difference between the wide and narrow sections of pipe is 4120 Pa. What is the volume flow rate $R_V$ of the ethanol?

## KEY IDEAS

(1) Because the fluid flowing through the wide section of pipe must entirely pass through the narrow section, the volume flow rate $R_V$ must be the same in the two sections. Thus, from Eq. 14-24,

$$R_V = v_1A_1 = v_2A_2. \qquad (14\text{-}35)$$

However, with two unknown speeds, we cannot evaluate this equation for $R_V$. (2) Because the flow is smooth, we can apply Bernoulli's equation. From Eq. 14-28, we can write

$$p_1 + \tfrac{1}{2}\rho v_1^2 + \rho g y = p_2 + \tfrac{1}{2}\rho v_2^2 + \rho g y, \qquad (14\text{-}36)$$

where subscripts 1 and 2 refer to the wide and narrow sections of pipe, respectively, and $y$ is their common elevation. This equation hardly seems to help because it does not contain the desired $R_V$ and it contains the unknown speeds $v_1$ and $v_2$.

*Calculations:* There is a neat way to make Eq. 14-36 work for us: First, we can use Eq. 14-35 and the fact that $A_2 = A_1/2$ to write

$$v_1 = \frac{R_V}{A_1} \quad \text{and} \quad v_2 = \frac{R_V}{A_2} = \frac{2R_V}{A_1}. \qquad (14\text{-}37)$$

Then we can substitute these expressions into Eq. 14-36 to eliminate the unknown speeds and introduce the desired volume flow rate. Doing this and solving for $R_V$ yield

$$R_V = A_1 \sqrt{\frac{2(p_1 - p_2)}{3\rho}}. \qquad (14\text{-}38)$$

We still have a decision to make: We know that the pressure difference between the two sections is 4120 Pa, but does that mean that $p_1 - p_2$ is 4120 Pa or $-4120$ Pa? We could guess the former is true, or otherwise the square root in Eq. 14-38 would give us an imaginary number. However, let's try some reasoning. From Eq. 14-35 we see that speed $v_2$ in the narrow section (small $A_2$) must be greater than speed $v_1$ in the wider section (larger $A_1$). Recall that if the speed of a fluid increases as the fluid travels along a horizontal path (as here), the pressure of the fluid must decrease. Thus, $p_1$ is greater than $p_2$, and $p_1 - p_2 = 4120$ Pa. Inserting this and known data into Eq. 14-38 gives

$$R_V = 1.20 \times 10^{-3}\,\text{m}^2 \sqrt{\frac{(2)(4120\,\text{Pa})}{(3)(791\,\text{kg/m}^3)}}$$

$$= 2.24 \times 10^{-3}\,\text{m}^3/\text{s}. \qquad (\text{Answer})$$

### Sample Problem 14.07   Bernoulli principle for a leaky water tank

In the old West, a desperado fires a bullet into an open water tank (Fig. 14-20), creating a hole a distance $h$ below the water surface. What is the speed $v$ of the water exiting the tank?

## KEY IDEAS

(1) This situation is essentially that of water moving (downward) with speed $v_0$ through a wide pipe (the tank) of cross-sectional area $A$ and then moving (horizontally) with speed $v$ through a narrow pipe (the hole) of cross-sectional area $a$. (2) Because the water flowing through the wide pipe must entirely pass through the narrow pipe, the volume flow rate $R_V$ must be the same in the two "pipes." (3) We can also relate $v$ to $v_0$ (and to $h$) through Bernoulli's equation (Eq. 14-28).

*Calculations:* From Eq. 14-24,

$$R_V = av = Av_0$$

and thus

$$v_0 = \frac{a}{A}v.$$

Because $a \ll A$, we see that $v_0 \ll v$. To apply Bernoulli's equation, we take the level of the hole as our reference level for measuring elevations (and thus gravitational potential energy). Noting that the pressure at the top of the tank and at the bullet hole is the atmospheric pressure $p_0$ (because both places are exposed to the atmosphere), we write Eq. 14-28 as

$$p_0 + \tfrac{1}{2}\rho v_0^2 + \rho g h = p_0 + \tfrac{1}{2}\rho v^2 + \rho g(0). \qquad (14\text{-}39)$$

**Figure 14-20** Water pours through a hole in a water tank, at a distance $h$ below the water surface. The pressure at the water surface and at the hole is atmospheric pressure $p_0$.

(Here the top of the tank is represented by the left side of the equation and the hole by the right side. The zero on the right indicates that the hole is at our reference level.) Before we solve Eq. 14-39 for $v$, we can use our result that $v_0 \ll v$ to simplify it: We assume that $v_0^2$, and thus the term $\tfrac{1}{2}\rho v_0^2$ in Eq. 14-39, is negligible relative to the other terms, and we drop it. Solving the remaining equation for $v$ then yields

$$v = \sqrt{2gh}. \qquad (\text{Answer})$$

This is the same speed that an object would have when falling a height $h$ from rest.

 **PLUS** Additional examples, video, and practice available at *WileyPLUS*

# Review & Summary

**Density** The **density** $\rho$ of any material is defined as the material's mass per unit volume:

$$\rho = \frac{\Delta m}{\Delta V}. \tag{14-1}$$

Usually, where a material sample is much larger than atomic dimensions, we can write Eq. 14-1 as

$$\rho = \frac{m}{V}. \tag{14-2}$$

**Fluid Pressure** A **fluid** is a substance that can flow; it conforms to the boundaries of its container because it cannot withstand shearing stress. It can, however, exert a force perpendicular to its surface. That force is described in terms of **pressure** $p$:

$$p = \frac{\Delta F}{\Delta A}, \tag{14-3}$$

in which $\Delta F$ is the force acting on a surface element of area $\Delta A$. If the force is uniform over a flat area, Eq. 14-3 can be written as

$$p = \frac{F}{A}. \tag{14-4}$$

The force resulting from fluid pressure at a particular point in a fluid has the same magnitude in all directions. **Gauge pressure** is the difference between the actual pressure (or *absolute pressure*) at a point and the atmospheric pressure.

**Pressure Variation with Height and Depth** Pressure in a fluid at rest varies with vertical position $y$. For $y$ measured positive upward,

$$p_2 = p_1 + \rho g(y_1 - y_2). \tag{14-7}$$

The pressure in a fluid is the same for all points at the same level. If $h$ is the *depth* of a fluid sample below some reference level at which the pressure is $p_0$, then the pressure in the sample is

$$p = p_0 + \rho g h. \tag{14-8}$$

**Pascal's Principle** A change in the pressure applied to an enclosed fluid is transmitted undiminished to every portion of the fluid and to the walls of the containing vessel.

**Archimedes' Principle** When a body is fully or partially submerged in a fluid, a buoyant force $\vec{F}_b$ from the surrounding fluid acts on the body. The force is directed upward and has a magnitude given by

$$F_b = m_f g, \tag{14-16}$$

where $m_f$ is the mass of the fluid that has been displaced by the body (that is, the fluid that has been pushed out of the way by the body).

When a body floats in a fluid, the magnitude $F_b$ of the (upward) buoyant force on the body is equal to the magnitude $F_g$ of the (downward) gravitational force on the body. The **apparent weight** of a body on which a buoyant force acts is related to its actual weight by

$$\text{weight}_{\text{app}} = \text{weight} - F_b. \tag{14-19}$$

**Flow of Ideal Fluids** An **ideal fluid** is incompressible and lacks viscosity, and its flow is steady and irrotational. A *streamline* is the path followed by an individual fluid particle. A *tube of flow* is a bundle of streamlines. The flow within any tube of flow obeys the **equation of continuity**:

$$R_V = Av = \text{a constant}, \tag{14-24}$$

in which $R_V$ is the **volume flow rate,** $A$ is the cross-sectional area of the tube of flow at any point, and $v$ is the speed of the fluid at that point. The **mass flow rate** $R_m$ is

$$R_m = \rho R_V = \rho Av = \text{a constant}. \tag{14-25}$$

**Bernoulli's Equation** Applying the principle of conservation of mechanical energy to the flow of an ideal fluid leads to **Bernoulli's equation** along any tube of flow:

$$p + \tfrac{1}{2}\rho v^2 + \rho g y = \text{a constant}. \tag{14-29}$$

# Questions

**1** We fully submerge an irregular 3 kg lump of material in a certain fluid. The fluid that would have been in the space now occupied by the lump has a mass of 2 kg. (a) When we release the lump, does it move upward, move downward, or remain in place? (b) If we next fully submerge the lump in a less dense fluid and again release it, what does it do?

**2** Figure 14-21 shows four situations in which a red liquid and a gray liquid are in a **U**-tube. In one situation the liquids cannot be in static equilibrium. (a) Which situation is that? (b) For the other three situations, assume static equilibrium. For each of them, is the density of the red liquid greater than, less than, or equal to the density of the gray liquid?

**3** A boat with an anchor on board floats in a swimming pool that is somewhat wider than the boat. Does the pool water level move up, move down, or remain the same if the anchor is (a) dropped into the water or (b) thrown onto the surrounding ground? (c) Does the water level in the pool move upward, move downward, or remain the same if, instead, a cork is dropped from the boat into the water, where it floats?

**4** Figure 14-22 shows a tank filled with water. Five horizontal floors and ceilings are indicated; all have the same area and are located at distances $L$, $2L$, or $3L$ below the top of the tank. Rank them according to the force on them due to the water, greatest first.

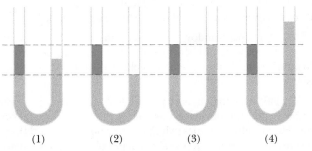

**Figure 14-21** Question 2.

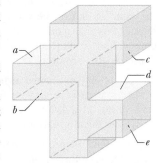

**Figure 14-22** Question 4.

**5**  *The teapot effect.* Water poured slowly from a teapot spout can double back under the spout for a considerable distance (held there by atmospheric pressure) before detaching and falling. In Fig. 14-23, the four points are at the top or bottom of the water layers, inside or outside. Rank those four points according to the gauge pressure in the water there, most positive first.

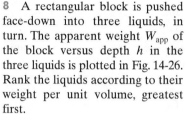

**Figure 14-23** Question 5.

**6** Figure 14-24 shows three identical open-top containers filled to the brim with water; toy ducks float in two of them. Rank the containers and contents according to their weight, greatest first.

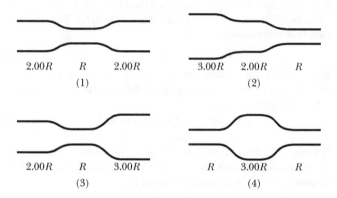

**Figure 14-24** Question 6.

**7** Figure 14-25 shows four arrangements of pipes through which

2.00R    R    2.00R
(1)

3.00R    2.00R    R
(2)

2.00R    R    3.00R
(3)

R    3.00R    R
(4)

**Figure 14-25** Question 7.

water flows smoothly toward the right. The radii of the pipe sections are indicated. In which arrangements is the net work done on a unit volume of water moving from the leftmost section to the rightmost section (a) zero, (b) positive, and (c) negative?

**8** A rectangular block is pushed face-down into three liquids, in turn. The apparent weight $W_{app}$ of the block versus depth $h$ in the three liquids is plotted in Fig. 14-26. Rank the liquids according to their weight per unit volume, greatest first.

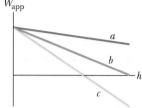

**Figure 14-26** Question 8.

**9** Water flows smoothly in a horizontal pipe. Figure 14-27 shows the kinetic energy $K$ of a water element as it moves along an $x$ axis that runs along the pipe. Rank the three lettered sections of the pipe according to the pipe radius, greatest first.

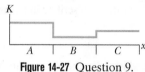

**Figure 14-27** Question 9.

**10** We have three containers with different liquids. The gauge pressure $p_g$ versus depth $h$ is plotted in Fig. 14-28 for the liquids. In each container, we will fully submerge a rigid plastic bead. Rank the plots according to the magnitude of the buoyant force on the bead, greatest first.

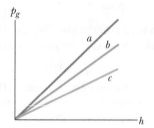

**Figure 14-28** Question 10.

## Problems

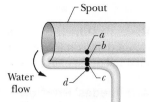

### Module 14-1 Fluids, Density, and Pressure

**•1** **ILW** A fish maintains its depth in fresh water by adjusting the air content of porous bone or air sacs to make its average density the same as that of the water. Suppose that with its air sacs collapsed, a fish has a density of 1.08 g/cm³. To what fraction of its expanded body volume must the fish inflate the air sacs to reduce its density to that of water?

**•2** A partially evacuated airtight container has a tight-fitting lid of surface area 77 in² and negligible mass. If the force required to remove the lid is 480 N and the atmospheric pressure is $1.0 \times 10^5$ Pa, what is the internal air pressure?

**•3** **SSM** Find the pressure increase in the fluid in a syringe when a nurse applies a force of 42 N to the syringe's circular piston, which has a radius of 1.1 cm.

**•4** Three liquids that will not mix are poured into a cylindrical container. The volumes and densities of the liquids are 0.50 L, 2.6 g/cm³; 0.25 L, 1.0 g/cm³; and 0.40 L, 0.80 g/cm³. What is the force on the bottom of the container due to these liquids? One liter = 1 L = 1000 cm³. (Ignore the contribution due to the atmosphere.)

**•5** **SSM** An office window has dimensions 3.4 m by 2.1 m. As a result of the passage of a storm, the outside air pressure drops to 0.96 atm, but inside the pressure is held at 1.0 atm. What net force pushes out on the window?

**•6** You inflate the front tires on your car to 28 psi. Later, you measure your blood pressure, obtaining a reading of 120/80, the readings being in mm Hg. In metric countries (which is to say, most of the world), these pressures are customarily reported in kilopascals (kPa). In kilopascals, what are (a) your tire pressure and (b) your blood pressure?

**••7** In 1654 Otto von Guericke, inventor of the air pump, gave a demonstration before the noblemen of the Holy Roman Empire in which two teams of eight horses could not pull apart two evacuated brass hemispheres. (a) Assuming the hemispheres have (strong) thin

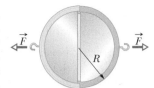

**Figure 14-29** Problem 7.

walls, so that R in Fig. 14-29 may be considered both the inside and outside radius, show that the force $\vec{F}$ required to pull apart the hemispheres has magnitude $F = \pi R^2 \Delta p$, where $\Delta p$ is the difference between the pressures outside and inside the sphere. (b) Taking R as 30 cm, the inside pressure as 0.10 atm, and the outside pressure as 1.00 atm, find the force magnitude the teams of horses would have had to exert to pull apart the hemispheres. (c) Explain why one team of horses could have proved the point just as well if the hemispheres were attached to a sturdy wall.

## Module 14-2 Fluids at Rest

**•8** _The bends during flight._ Anyone who scuba dives is advised not to fly within the next 24 h because the air mixture for diving can introduce nitrogen to the bloodstream. Without allowing the nitrogen to come out of solution slowly, any sudden air-pressure reduction (such as during airplane ascent) can result in the nitrogen forming bubbles in the blood, creating the _bends_, which can be painful and even fatal. Military special operation forces are especially at risk. What is the change in pressure on such a special-op soldier who must scuba dive at a depth of 20 m in seawater one day and parachute at an altitude of 7.6 km the next day? Assume that the average air density within the altitude range is 0.87 kg/m³.

**•9** _Blood pressure in Argentinosaurus._ (a) If this long-necked, gigantic sauropod had a head height of 21 m and a heart height of 9.0 m, what (hydrostatic) gauge pressure in its blood was required at the heart such that the blood pressure at the brain was 80 torr (just enough to perfuse the brain with blood)? Assume the blood had a density of $1.06 \times 10^3$ kg/m³. (b) What was the blood pressure (in torr or mm Hg) at the feet?

**•10** The plastic tube in Fig. 14-30 has a cross-sectional area of 5.00 cm². The tube is filled with water until the short arm (of length d = 0.800 m) is full. Then the short arm is sealed and more water is gradually poured into the long arm. If the seal will pop off when the force on it exceeds 9.80 N, what total height of water in the long arm will put the seal on the verge of popping?

**Figure 14-30**
Problems 10 and 81.

**•11** _Giraffe bending to drink._ In a giraffe with its head 2.0 m above its heart, and its heart 2.0 m above its feet, the (hydrostatic) gauge pressure in the blood at its heart is 250 torr. Assume that the giraffe stands upright and the blood density is $1.06 \times 10^3$ kg/m³. In torr (or mm Hg), find the (gauge) blood pressure (a) at the brain (the pressure is enough to perfuse the brain with blood, to keep the giraffe from fainting) and (b) at the feet (the pressure must be countered by tight-fitting skin acting like a pressure stocking). (c) If the giraffe were to lower its head to drink from a pond without splaying its legs and moving slowly, what would be the increase in the blood pressure in the brain? (Such action would probably be lethal.)

**•12** The maximum depth $d_{max}$ that a diver can snorkel is set by the density of the water and the fact that human lungs can func-

tion against a maximum pressure difference (between inside and outside the chest cavity) of 0.050 atm. What is the difference in $d_{max}$ for fresh water and the water of the Dead Sea (the saltiest natural water in the world, with a density of $1.5 \times 10^3$ kg/m³)?

**•13** At a depth of 10.9 km, the Challenger Deep in the Marianas Trench of the Pacific Ocean is the deepest site in any ocean. Yet, in 1960, Donald Walsh and Jacques Piccard reached the Challenger Deep in the bathyscaph _Trieste_. Assuming that seawater has a uniform density of 1024 kg/m³, approximate the hydrostatic pressure (in atmospheres) that the _Trieste_ had to withstand. (Even a slight defect in the _Trieste_ structure would have been disastrous.)

**•14** Calculate the hydrostatic difference in blood pressure between the brain and the foot in a person of height 1.83 m. The density of blood is $1.06 \times 10^3$ kg/m³.

**•15** What gauge pressure must a machine produce in order to suck mud of density 1800 kg/m³ up a tube by a height of 1.5 m?

**•16** _Snorkeling by humans and elephants._ When a person snorkels, the lungs are connected directly to the atmosphere through the snorkel tube and thus are at atmospheric pressure. In atmospheres, what is the difference $\Delta p$ between this internal air pressure and the water pressure against the body if the length of the snorkel

**Figure 14-31** Problem 16.

tube is (a) 20 cm (standard situation) and (b) 4.0 m (probably lethal situation)? In the latter, the pressure difference causes blood vessels on the walls of the lungs to rupture, releasing blood into the lungs. As depicted in Fig. 14-31, an elephant can safely snorkel through its trunk while swimming with its lungs 4.0 m below the water surface because the membrane around its lungs contains connective tissue that holds and protects the blood vessels, preventing rupturing.

**•17 SSM** Crew members attempt to escape from a damaged submarine 100 m below the surface. What force must be applied to a pop-out hatch, which is 1.2 m by 0.60 m, to push it out at that depth? Assume that the density of the ocean water is 1024 kg/m³ and the internal air pressure is at 1.00 atm.

**•18** In Fig. 14-32, an open tube of length L = 1.8 m and cross-sectional area A = 4.6 cm² is fixed to the top of a cylindrical barrel of diameter D = 1.2 m and height H = 1.8 m. The barrel and tube are filled with water (to the top of the tube). Calculate the ratio of the hydrostatic force on the bottom of the barrel to the gravitational force on the water contained in the barrel. Why is that ratio not equal to 1.0? (You need not consider the atmospheric pressure.)

**••19 GO** A large aquarium of height 5.00 m is filled with fresh water to a depth of 2.00 m. One wall of the aquarium consists of thick plastic 8.00 m wide. By how much does the total force on that wall increase if the aquarium is next filled to a depth of 4.00 m?

**Figure 14-32**
Problem 18.

**••20** The **L**-shaped fish tank shown in Fig. 14-33 is filled with water and is open at the top. If $d = 5.0$ m, what is the (total) force exerted by the water (a) on face $A$ and (b) on face $B$?

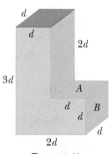

**Figure 14-33**
Problem 20.

**••21** SSM Two identical cylindrical vessels with their bases at the same level each contain a liquid of density $1.30 \times 10^3$ kg/m³. The area of each base is 4.00 cm², but in one vessel the liquid height is 0.854 m and in the other it is 1.560 m. Find the work done by the gravitational force in equalizing the levels when the two vessels are connected.

**••22** *g-LOC in dogfights.* When a pilot takes a tight turn at high speed in a modern fighter airplane, the blood pressure at the brain level decreases, blood no longer perfuses the brain, and the blood in the brain drains. If the heart maintains the (hydrostatic) gauge pressure in the aorta at 120 torr (or mm Hg) when the pilot undergoes a horizontal centripetal acceleration of 4g, what is the blood pressure (in torr) at the brain, 30 cm radially inward from the heart? The perfusion in the brain is small enough that the vision switches to black and white and narrows to "tunnel vision" and the pilot can undergo g-LOC ("g-induced loss of consciousness"). Blood density is $1.06 \times 10^3$ kg/m³.

**••23** GO In analyzing certain geological features, it is often appropriate to assume that the pressure at some horizontal *level of compensation,* deep inside Earth, is the same over a large region and is equal to the pressure due to the gravitational force on the overlying material. Thus, the pressure on the level of compensation is given by the fluid pressure formula. This model requires, for one thing, that mountains have *roots* of continental rock extending into the denser

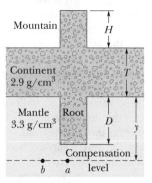

**Figure 14-34** Problem 23.

mantle (Fig. 14-34). Consider a mountain of height $H = 6.0$ km on a continent of thickness $T = 32$ km. The continental rock has a density of 2.9 g/cm³, and beneath this rock the mantle has a density of 3.3 g/cm³. Calculate the depth $D$ of the root. (*Hint:* Set the pressure at points $a$ and $b$ equal; the depth $y$ of the level of compensation will cancel out.)

**•••24** GO In Fig. 14-35, water stands at depth $D = 35.0$ m behind the vertical upstream face of a dam of width $W = 314$ m. Find (a) the net horizontal force on the dam from the gauge pressure of the water and (b) the net torque due to that force about a horizontal line through $O$

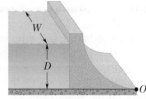

**Figure 14-35** Problem 24.

parallel to the (long) width of the dam. This torque tends to rotate the dam around that line, which would cause the dam to fail. (c) Find the moment arm of the torque.

## Module 14-3  Measuring Pressure

**•25** In one observation, the column in a mercury barometer (as is shown in Fig. 14-5a) has a measured height $h$ of 740.35 mm. The temperature is $-5.0°C$, at which temperature the density of mercury $\rho$ is $1.3608 \times 10^4$ kg/m³. The free-fall acceleration g at the site of the barom-

eter is 9.7835 m/s². What is the atmospheric pressure at that site in pascals and in torr (which is the common unit for barometer readings)?

**•26** To suck lemonade of density 1000 kg/m³ up a straw to a maximum height of 4.0 cm, what minimum gauge pressure (in atmospheres) must you produce in your lungs?

**••27** SSM What would be the height of the atmosphere if the air density (a) were uniform and (b) decreased linearly to zero with height? Assume that at sea level the air pressure is 1.0 atm and the air density is 1.3 kg/m³.

## Module 14-4  Pascal's Principle

**•28** A piston of cross-sectional area $a$ is used in a hydraulic press to exert a small force of magnitude $f$ on the enclosed liquid. A connecting pipe leads to a larger piston of cross-sectional area $A$ (Fig. 14-36). (a) What force magnitude $F$ will the larger piston sustain without moving? (b) If the piston diameters are 3.80 cm and

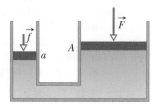

**Figure 14-36** Problem 28.

53.0 cm, what force magnitude on the small piston will balance a 20.0 kN force on the large piston?

**••29** In Fig. 14-37, a spring of spring constant $3.00 \times 10^4$ N/m is between a rigid beam and the output piston of a hydraulic lever. An empty container with negligible mass sits on the input piston. The input piston has area $A_i$, and the output piston has area $18.0A_i$. Initially the spring is at its rest length. How many kilograms of sand must be

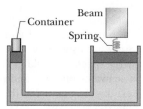

**Figure 14-37** Problem 29.

(slowly) poured into the container to compress the spring by 5.00 cm?

## Module 14-5  Archimedes' Principle

**•30** A 5.00 kg object is released from rest while fully submerged in a liquid. The liquid displaced by the submerged object has a mass of 3.00 kg. How far and in what direction does the object move in 0.200 s, assuming that it moves freely and that the drag force on it from the liquid is negligible?

**•31** SSM A block of wood floats in fresh water with two-thirds of its volume $V$ submerged and in oil with 0.90$V$ submerged. Find the density of (a) the wood and (b) the oil.

**•32** In Fig. 14-38, a cube of edge length $L = 0.600$ m and mass 450 kg is suspended by a rope in an open tank of liquid of density 1030 kg/m³. Find (a) the magnitude of the total downward force on the top of the cube from the liquid and the atmosphere, assuming atmospheric pressure is 1.00 atm, (b) the magnitude of the total upward force on the bottom of the cube, and (c) the tension

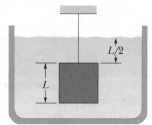

**Figure 14-38** Problem 32.

in the rope. (d) Calculate the magnitude of the buoyant force on the cube using Archimedes' principle. What relation exists among all these quantities?

**•33** SSM An iron anchor of density 7870 kg/m³ appears 200 N lighter in water than in air. (a) What is the volume of the anchor? (b) How much does it weigh in air?

**•34** A boat floating in fresh water displaces water weighing

35.6 kN. (a) What is the weight of the water this boat displaces when floating in salt water of density $1.10 \times 10^3$ kg/m³? (b) What is the difference between the volume of fresh water displaced and the volume of salt water displaced?

•35 Three children, each of weight 356 N, make a log raft by lashing together logs of diameter 0.30 m and length 1.80 m. How many logs will be needed to keep them afloat in fresh water? Take the density of the logs to be 800 kg/m³.

••36 GO In Fig. 14-39a, a rectangular block is gradually pushed face-down into a liquid. The block has height $d$; on the bottom and top the face area is $A = 5.67$ cm². Figure 14-39b gives the apparent weight $W_{app}$ of the block as a function of the depth $h$ of its lower face. The scale on the vertical axis is set by $W_s = 0.20$ N. What is the density of the liquid?

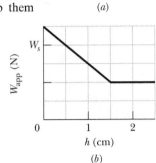
(a)

(b)

**Figure 14-39** Problem 36.

••37 ILW A hollow spherical iron shell floats almost completely submerged in water. The outer diameter is 60.0 cm, and the density of iron is 7.87 g/cm³. Find the inner diameter.

••38 GO A small solid ball is released from rest while fully submerged in a liquid and then its kinetic energy is measured when it has moved 4.0 cm in the liquid. Figure 14-40 gives the results after many liquids are used: The kinetic energy $K$ is plotted versus the liquid density $\rho_{liq}$, and $K_s = 1.60$ J sets the scale on the vertical axis. What are (a) the density and (b) the volume of the ball?

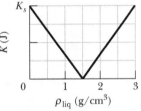

**Figure 14-40** Problem 38.

••39 SSM WWW A hollow sphere of inner radius 8.0 cm and outer radius 9.0 cm floats half-submerged in a liquid of density 800 kg/m³. (a) What is the mass of the sphere? (b) Calculate the density of the material of which the sphere is made.

••40 *Lurking alligators.* An alligator waits for prey by floating with only the top of its head exposed, so that the prey cannot easily see it. One way it can adjust the extent of sinking is by controlling the size of its lungs. Another way may be by swallowing stones (*gastrolithes*) that then reside in the stomach. Figure 14-41 shows a highly simplified model (a "rhombohedron gater") of mass 130 kg that roams with its head partially exposed. The top head surface has area 0.20 m². If the alligator were to swallow stones with a total mass of 1.0% of its body mass (a typical amount), how far would it sink?

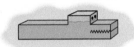

**Figure 14-41** Problem 40.

••41 What fraction of the volume of an iceberg (density 917 kg/m³) would be visible if the iceberg floats (a) in the ocean (salt water, density 1024 kg/m³) and (b) in a river (fresh water, density 1000 kg/m³)? (When salt water freezes to form ice, the salt is excluded. So, an iceberg could provide fresh water to a community.)

••42 A flotation device is in the shape of a right cylinder, with a height of 0.500 m and a face area of 4.00 m² on top and bottom, and its density is 0.400 times that of fresh water. It is initially held fully submerged in fresh water, with its top face at the water surface. Then

it is allowed to ascend gradually until it begins to float. How much work does the buoyant force do on the device during the ascent?

••43 When researchers find a reasonably complete fossil of a dinosaur, they can determine the mass and weight of the living dinosaur with a scale model sculpted from plastic and based on the dimensions of the fossil bones. The scale of the model is 1/20; that is, lengths are 1/20 actual length, areas are $(1/20)^2$ actual areas, and volumes are $(1/20)^3$ actual volumes. First, the model is suspended from one arm of a balance and weights are added to the other arm until equilibrium is reached. Then the model is fully submerged in water and enough weights are removed from the second arm to reestablish equilibrium (Fig. 14-42). For a model of a particular *T. rex* fossil, 637.76 g had to be removed to reestablish equilibrium. What was the volume of (a) the model and (b) the actual *T. rex*? (c) If the density of *T. rex* was approximately the density of water, what was its mass?

**Figure 14-42** Problem 43.

••44 A wood block (mass 3.67 kg, density 600 kg/m³) is fitted with lead (density $1.14 \times 10^4$ kg/m³) so that it floats in water with 0.900 of its volume submerged. Find the lead mass if the lead is fitted to the block's (a) top and (b) bottom.

••45 GO An iron casting containing a number of cavities weighs 6000 N in air and 4000 N in water. What is the total cavity volume in the casting? The density of solid iron is 7.87 g/cm³.

••46 GO Suppose that you release a small ball from rest at a depth of 0.600 m below the surface in a pool of water. If the density of the ball is 0.300 that of water and if the drag force on the ball from the water is negligible, how high above the water surface will the ball shoot as it emerges from the water? (Neglect any transfer of energy to the splashing and waves produced by the emerging ball.)

••47 The volume of air space in the passenger compartment of an 1800 kg car is 5.00 m³. The volume of the motor and front wheels is 0.750 m³, and the volume of the rear wheels, gas tank, and trunk is 0.800 m³; water cannot enter these two regions. The car rolls into a lake. (a) At first, no water enters the passenger compartment. How much of the car, in cubic meters, is below the water surface with the car floating (Fig. 14-43)? (b) As water slowly enters the car sinks. How many cubic meters of water are in the car as it disappears below the water surface? (The car, with a heavy load in the trunk, remains horizontal.)

**Figure 14-43** Problem 47.

•••48 GO Figure 14-44 shows an iron ball suspended by thread of negligible mass from an upright cylinder that floats partially submerged in water. The cylinder has a height of 6.00 cm, a face area of 12.0 cm² on the top and bottom, and a density of 0.30 g/cm³, and 2.00 cm of its height is above the water surface. What is the radius of the iron ball?

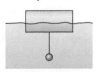

**Figure 14-44** Problem 48.

## Module 14-6 The Equation of Continuity

•49 [image] *Canal effect.* Figure 14-45 shows an anchored barge that extends across a canal by distance $d = 30$ m and into the water by distance $b = 12$ m. The canal has a width $D = 55$ m, a water depth $H = 14$ m, and a uniform water-flow speed $v_i = 1.5$ m/s. Assume that the flow around the barge is uniform. As the water passes the bow, the water level undergoes a dramatic dip known as the canal effect. If the dip

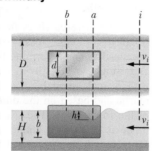

**Figure 14-45** Problem 49.

has depth $h = 0.80$ m, what is the water speed alongside the boat through the vertical cross sections at (a) point $a$ and (b) point $b$? The erosion due to the speed increase is a common concern to hydraulic engineers.

•50 Figure 14-46 shows two sections of an old pipe system that runs through a hill, with distances $d_A = d_B = 30$ m and $D = 110$ m. On each side of the hill, the pipe radius is

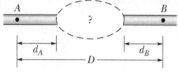

**Figure 14-46** Problem 50.

2.00 cm. However, the radius of the pipe inside the hill is no longer known. To determine it, hydraulic engineers first establish that water flows through the left and right sections at 2.50 m/s. Then they release a dye in the water at point $A$ and find that it takes 88.8 s to reach point $B$. What is the average radius of the pipe within the hill?

•51 SSM A garden hose with an internal diameter of 1.9 cm is connected to a (stationary) lawn sprinkler that consists merely of a container with 24 holes, each 0.13 cm in diameter. If the water in the hose has a speed of 0.91 m/s, at what speed does it leave the sprinkler holes?

•52 Two streams merge to form a river. One stream has a width of 8.2 m, depth of 3.4 m, and current speed of 2.3 m/s. The other stream is 6.8 m wide and 3.2 m deep, and flows at 2.6 m/s. If the river has width 10.5 m and speed 2.9 m/s, what is its depth?

••53 SSM Water is pumped steadily out of a flooded basement at 5.0 m/s through a hose of radius 1.0 cm, passing through a window 3.0 m above the waterline. What is the pump's power?

••54 [GO] The water flowing through a 1.9 cm (inside diameter) pipe flows out through three 1.3 cm pipes. (a) If the flow rates in the three smaller pipes are 26, 19, and 11 L/min, what is the flow rate in the 1.9 cm pipe? (b) What is the ratio of the speed in the 1.9 cm pipe to that in the pipe carrying 26 L/min?

## Module 14-7 Bernoulli's Equation

•55 How much work is done by pressure in forcing 1.4 m³ of water through a pipe having an internal diameter of 13 mm if the difference in pressure at the two ends of the pipe is 1.0 atm?

•56 Suppose that two tanks, 1 and 2, each with a large opening at the top, contain different liquids. A small hole is made in the side of each tank at the same depth $h$ below the liquid surface, but the hole in tank 1 has half the cross-sectional area of the hole in tank 2. (a) What is the ratio $\rho_1/\rho_2$ of the densities of the liquids if the mass flow rate is the same for the two holes? (b) What is the ratio $R_{V1}/R_{V2}$ of the volume flow rates from the two tanks? (c) At one instant, the liquid in tank 1 is 12.0 cm above the hole. If the tanks are to have *equal* volume flow rates, what height above the hole must the liquid in tank 2 be just then?

•57 SSM A cylindrical tank with a large diameter is filled with water to a depth $D = 0.30$ m. A hole of cross-sectional area $A = 6.5$ cm² in the bottom of the tank allows water to drain out. (a) What is the drainage rate in cubic meters per second? (b) At what distance below the bottom of the tank is the cross-sectional area of the stream equal to one-half the area of the hole?

•58 The intake in Fig. 14-47 has cross-sectional area of 0.74 m² and water flow at 0.40 m/s. At the outlet, distance $D = 180$ m below the intake, the cross-sectional area is smaller than at the intake and the water flows out at 9.5 m/s into equipment. What is the pressure difference between inlet and outlet?

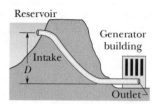

**Figure 14-47** Problem 58.

•59 SSM Water is moving with a speed of 5.0 m/s through a pipe with a cross-sectional area of 4.0 cm². The water gradually descends 10 m as the pipe cross-sectional area increases to 8.0 cm². (a) What is the speed at the lower level? (b) If the pressure at the upper level is $1.5 \times 10^5$ Pa, what is the pressure at the lower level?

•60 Models of torpedoes are sometimes tested in a horizontal pipe of flowing water, much as a wind tunnel is used to test model airplanes. Consider a circular pipe of internal diameter 25.0 cm and a torpedo model aligned along the long axis of the pipe. The model has a 5.00 cm diameter and is to be tested with water flowing past it at 2.50 m/s. (a) With what speed must the water flow in the part of the pipe that is unconstricted by the model? (b) What will the pressure difference be between the constricted and unconstricted parts of the pipe?

•61 ILW A water pipe having a 2.5 cm inside diameter carries water into the basement of a house at a speed of 0.90 m/s and a pressure of 170 kPa. If the pipe tapers to 1.2 cm and rises to the second floor 7.6 m above the input point, what are the (a) speed and (b) water pressure at the second floor?

••62 A pitot tube (Fig. 14-48) is used to determine the airspeed of an airplane. It consists of an outer tube with a number of small holes $B$ (four are shown) that allow air into the tube; that tube is connected to one arm of a U-tube. The other arm of the U-tube is connected to hole $A$ at the front end of the device, which points in the direction the plane is headed. At $A$ the air becomes stagnant so that $v_A = 0$. At $B$, however, the speed of the air presumably equals the airspeed $v$ of the plane. (a) Use Bernoulli's equation to show that

$$v = \sqrt{\frac{2\rho g h}{\rho_{air}}},$$

where $\rho$ is the density of the liquid in the U-tube and $h$ is the difference in the liquid levels in that tube. (b) Suppose that the tube contains alcohol and the level difference $h$ is 26.0 cm. What is the plane's speed relative to the air? The density of the air is 1.03 kg/m³ and that of alcohol is 810 kg/m³.

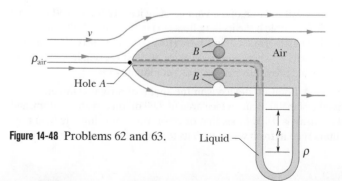

**Figure 14-48** Problems 62 and 63.

••63  A pitot tube (see Problem 62) on a high-altitude aircraft measures a differential pressure of 180 Pa. What is the aircraft's airspeed if the density of the air is 0.031 kg/m³?

••64  GO  In Fig. 14-49, water flows through a horizontal pipe and then out into the atmosphere at a speed $v_1 = 15$ m/s. The diameters of the left and right sections of the pipe are 5.0 cm and 3.0 cm. (a) What volume of water flows

**Figure 14-49** Problem 64.

into the atmosphere during a 10 min period? In the left section of the pipe, what are (b) the speed $v_2$ and (c) the gauge pressure?

••65  SSM  WWW  A *venturi meter* is used to measure the flow speed of a fluid in a pipe. The meter is connected between two sections of the pipe (Fig. 14-50); the cross-sectional area $A$ of the entrance and exit of the meter matches the pipe's cross-sectional area. Between the entrance and exit, the fluid flows from the pipe with speed $V$ and then through a narrow "throat" of cross-sectional area $a$ with speed $v$. A manometer connects the wider portion of the meter to the narrower portion. The change in the fluid's speed is accompanied by a change $\Delta p$ in the fluid's pressure, which causes a height difference $h$ of the liquid in the two arms of the manometer. (Here $\Delta p$ means pressure in the throat minus pressure in the pipe.) (a) By applying Bernoulli's equation and the equation of continuity to points 1 and 2 in Fig. 14-50, show that

$$V = \sqrt{\frac{2a^2\,\Delta p}{\rho(a^2 - A^2)}},$$

where $\rho$ is the density of the fluid. (b) Suppose that the fluid is fresh water, that the cross-sectional areas are 64 cm² in the pipe and 32 cm² in the throat, and that the pressure is 55 kPa in the pipe and 41 kPa in the throat. What is the rate of water flow in cubic meters per second?

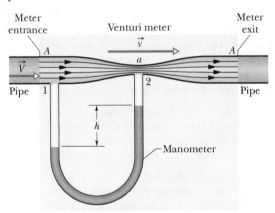

**Figure 14-50** Problems 65 and 66.

••66  Consider the venturi tube of Problem 65 and Fig. 14-50 without the manometer. Let $A$ equal $5a$. Suppose the pressure $p_1$ at $A$ is 2.0 atm. Compute the values of (a) the speed $V$ at $A$ and (b) the speed $v$ at $a$ that make the pressure $p_2$ at $a$ equal to zero. (c) Compute the corresponding volume flow rate if the diameter at $A$ is 5.0 cm. The phenomenon that occurs at $a$ when $p_2$ falls to nearly zero is known as cavitation. The water vaporizes into small bubbles.

••67  ILW  In Fig. 14-51, the fresh water behind a reservoir dam has depth $D = 15$ m. A horizontal pipe 4.0 cm in diameter passes through the dam at depth $d = 6.0$ m. A plug secures the pipe opening. (a) Find the magnitude of the frictional force between plug and pipe wall. (b) The plug is removed. What water volume exits the pipe in 3.0 h?

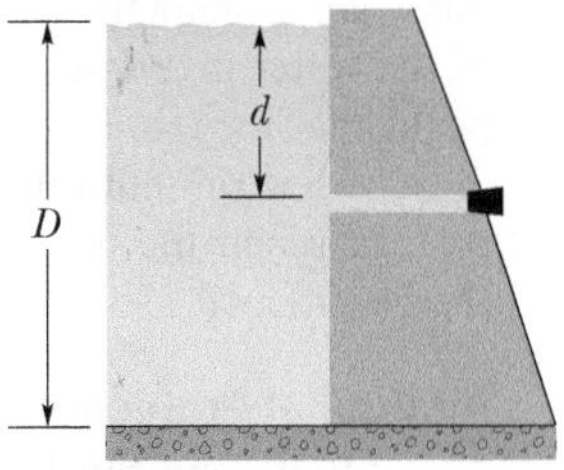

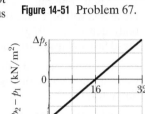

**Figure 14-51** Problem 67.

••68  GO  Fresh water flows horizontally from pipe section 1 of cross-sectional area $A_1$ into pipe section 2 of cross-sectional area $A_2$. Figure 14-52 gives a plot of the pressure difference $p_2 - p_1$ versus the inverse area squared $A_1^{-2}$ that would be expected for a volume flow rate of a certain value if the water flow were laminar under all circumstances. The scale on the vertical axis is set by $\Delta p_s = 300$ kN/m². For the conditions of the figure, what are the values of (a) $A_2$ and (b) the volume flow rate?

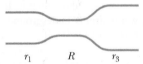

**Figure 14-52** Problem 68.

••69  A liquid of density 900 kg/m³ flows through a horizontal pipe that has a cross-sectional area of $1.90 \times 10^{-2}$ m² in region $A$ and a cross-sectional area of $9.50 \times 10^{-2}$ m² in region $B$. The pressure difference between the two regions is $7.20 \times 10^3$ Pa. What are (a) the volume flow rate and (b) the mass flow rate?

••70  GO  In Fig. 14-53, water flows steadily from the left pipe section (radius $r_1 = 2.00R$), through the middle section (radius $R$), and into the right section (radius $r_3 = 3.00R$). The speed of the water in the middle section is 0.500 m/s. What is the net work done on 0.400 m³ of the water as it moves from the left section to the right section?

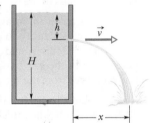

**Figure 14-53** Problem 70.

••71  Figure 14-54 shows a stream of water flowing through a hole at depth $h = 10$ cm in a tank holding water to height $H = 40$ cm. (a) At what distance $x$ does the stream strike the floor? (b) At what depth should a second hole be made to give the same value of $x$? (c) At what depth should a hole be made to maximize $x$?

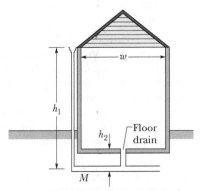

**Figure 14-54** Problem 71.

•••72  GO  A very simplified schematic of the rain drainage system for a home is shown in Fig. 14-55. Rain falling on the slanted roof runs off into gutters around the roof edge; it then drains through downspouts (only one is shown) into a main drainage pipe $M$ below the basement, which carries the water to an even larger pipe below the street. In Fig. 14-55, a floor drain in the basement is also connected to drainage pipe $M$. Suppose the following apply:

(1) the downspouts have height $h_1 = 11$ m, (2) the floor drain has height $h_2 = 1.2$ m, (3) pipe $M$ has radius 3.0 cm, (4) the house has side width $w = 30$ m and front length $L = 60$ m, (5) all

**Figure 14-55** Problem 72.

the water striking the roof goes through pipe $M$, (6) the initial speed of the water in a downspout is negligible, and (7) the wind speed is negligible (the rain falls vertically).

At what rainfall rate, in centimeters per hour, will water from pipe $M$ reach the height of the floor drain and threaten to flood the basement?

### Additional Problems

**73** About one-third of the body of a person floating in the Dead Sea will be above the waterline. Assuming that the human body density is $0.98$ g/cm$^3$, find the density of the water in the Dead Sea. (Why is it so much greater than $1.0$ g/cm$^3$?)

**74** A simple open U-tube contains mercury. When 11.2 cm of water is poured into the right arm of the tube, how high above its initial level does the mercury rise in the left arm?

**75** If a bubble in sparkling water accelerates upward at the rate of 0.225 m/s$^2$ and has a radius of 0.500 mm, what is its mass? Assume that the drag force on the bubble is negligible.

**76** Suppose that your body has a uniform density of 0.95 times that of water. (a) If you float in a swimming pool, what fraction of your body's volume is above the water surface?

Quicksand is a fluid produced when water is forced up into sand, moving the sand grains away from one another so they are no longer locked together by friction. Pools of quicksand can form when water drains underground from hills into valleys where there are sand pockets. (b) If you float in a deep pool of quicksand that has a density 1.6 times that of water, what fraction of your body's volume is above the quicksand surface? (c) Are you unable to breathe?

**77** A glass ball of radius 2.00 cm sits at the bottom of a container of milk that has a density of 1.03 g/cm$^3$. The normal force on the ball from the container's lower surface has magnitude $9.48 \times 10^{-2}$ N. What is the mass of the ball?

**78** Caught in an avalanche, a skier is fully submerged in flowing snow of density 96 kg/m$^3$. Assume that the average density of the skier, clothing, and skiing equipment is 1020 kg/m$^3$. What percentage of the gravitational force on the skier is offset by the buoyant force from the snow?

**79** An object hangs from a spring balance. The balance registers 30 N in air, 20 N when this object is immersed in water, and 24 N when the object is immersed in another liquid of unknown density. What is the density of that other liquid?

**80** In an experiment, a rectangular block with height $h$ is allowed to float in four separate liquids. In the first liquid, which is water, it floats fully submerged. In liquids $A$, $B$, and $C$, it floats with heights $h/2$, $2h/3$, and $h/4$ above the liquid surface, respectively. What are the *relative densities* (the densities relative to that of water) of (a) $A$, (b) $B$, and (c) $C$?

**81** SSM Figure 14-30 shows a modified U-tube: the right arm is shorter than the left arm. The open end of the right arm is height $d = 10.0$ cm above the laboratory bench. The radius throughout the tube is 1.50 cm. Water is gradually poured into the open end of the left arm until the water begins to flow out the open end of the right arm. Then a liquid of density 0.80 g/cm$^3$ is gradually added to the left arm until its height in that arm is 8.0 cm (it does not mix with the water). How much water flows out of the right arm?

**82** What is the acceleration of a rising hot-air balloon if the ratio of the air density outside the balloon to that inside is 1.39? Neglect the mass of the balloon fabric and the basket.

**83** Figure 14-56 shows a *siphon*, which is a device for removing liquid from a container. Tube $ABC$ must initially be filled, but once this has been done, liquid will flow through the tube until the liquid surface in the container is level with the tube opening at $A$. The liquid has density 1000 kg/m$^3$ and negligible viscosity. The distances shown are $h_1 = 25$ cm, $d = 12$ cm, and $h_2 = 40$ cm. (a) With what speed does the liquid emerge from the tube at $C$? (b) If the atmospheric pressure is $1.0 \times 10^5$ Pa, what is the pressure in the liquid at the topmost point $B$? (c) Theoretically, what is the greatest possible height $h_1$ that a siphon can lift water?

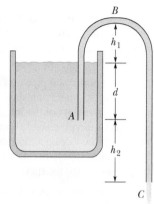

**Figure 14-56** Problem 83.

**84** When you cough, you expel air at high speed through the trachea and upper bronchi so that the air will remove excess mucus lining the pathway. You produce the high speed by this procedure: You breathe in a large amount of air, trap it by closing the glottis (the narrow opening in the larynx), increase the air pressure by contracting the lungs, partially collapse the trachea and upper bronchi to narrow the pathway, and then expel the air through the pathway by suddenly reopening the glottis. Assume that during the expulsion the volume flow rate is $7.0 \times 10^{-3}$ m$^3$/s. What multiple of 343 m/s (the speed of sound $v_s$) is the airspeed through the trachea if the trachea diameter (a) remains its normal value of 14 mm and (b) contracts to 5.2 mm?

**85** A tin can has a total volume of 1200 cm$^3$ and a mass of 130 g. How many grams of lead shot of density 11.4 g/cm$^3$ could it carry without sinking in water?

**86** The tension in a string holding a solid block below the surface of a liquid (of density greater than the block) is $T_0$ when the container (Fig. 14-57) is at rest. When the container is given an upward acceleration of $0.250g$, what multiple of $T_0$ gives the tension in the string?

**Figure 14-57** Problem 86.

**87** What is the minimum area (in square meters) of the top surface of an ice slab 0.441 m thick floating on fresh water that will hold up a 938 kg automobile? Take the densities of ice and fresh water to be 917 kg/m$^3$ and 998 kg/m$^3$, respectively.

**88** A 8.60 kg sphere of radius 6.22 cm is at a depth of 2.22 km in seawater that has an average density of 1025 kg/m$^3$. What are the (a) gauge pressure, (b) total pressure, and (c) corresponding total force compressing the sphere's surface? What are (d) the magnitude of the buoyant force on the sphere and (e) the magnitude of the sphere's acceleration if it is free to move? Take atmospheric pressure to be $1.01 \times 10^5$ Pa.

**89** (a) For seawater of density 1.03 g/cm$^3$, find the weight of water on top of a submarine at a depth of 255 m if the horizontal cross-sectional hull area is 2200.0 m$^2$. (b) In atmospheres, what water pressure would a diver experience at this depth?

**90** The sewage outlet of a house constructed on a slope is 6.59 m below street level. If the sewer is 2.16 m below street level, find the minimum pressure difference that must be created by the sewage pump to transfer waste of average density 1000.00 kg/m$^3$ from outlet to sewer.

# Oscillations

## 15-1 SIMPLE HARMONIC MOTION

### Learning Objectives

*After reading this module, you should be able to . . .*

**15.01** Distinguish simple harmonic motion from other types of periodic motion.

**15.02** For a simple harmonic oscillator, apply the relationship between position $x$ and time $t$ to calculate either if given a value for the other.

**15.03** Relate period $T$, frequency $f$, and angular frequency $\omega$.

**15.04** Identify (displacement) amplitude $x_m$, phase constant (or phase angle) $\phi$, and phase $\omega t + \phi$.

**15.05** Sketch a graph of the oscillator's position $x$ versus time $t$, identifying amplitude $x_m$ and period $T$.

**15.06** From a graph of position versus time, velocity versus time, or acceleration versus time, determine the amplitude of the plot and the value of the phase constant $\phi$.

**15.07** On a graph of position $x$ versus time $t$ describe the effects of changing period $T$, frequency $f$, amplitude $x_m$, or phase constant $\phi$.

**15.08** Identify the phase constant $\phi$ that corresponds to the starting time ($t = 0$) being set when a particle in SHM is at an extreme point or passing through the center point.

**15.09** Given an oscillator's position $x(t)$ as a function of time, find its velocity $v(t)$ as a function of time, identify the velocity amplitude $v_m$ in the result, and calculate the velocity at any given time.

**15.10** Sketch a graph of an oscillator's velocity $v$ versus time $t$, identifying the velocity amplitude $v_m$.

**15.11** Apply the relationship between velocity amplitude $v_m$, angular frequency $\omega$, and (displacement) amplitude $x_m$.

**15.12** Given an oscillator's velocity $v(t)$ as a function of time, calculate its acceleration $a(t)$ as a function of time, identify the acceleration amplitude $a_m$ in the result, and calculate the acceleration at any given time.

**15.13** Sketch a graph of an oscillator's acceleration $a$ versus time $t$, identifying the acceleration amplitude $a_m$.

**15.14** Identify that for a simple harmonic oscillator the acceleration $a$ at any instant is *always* given by the product of a negative constant and the displacement $x$ just then.

**15.15** For any given instant in an oscillation, apply the relationship between acceleration $a$, angular frequency $\omega$, and displacement $x$.

**15.16** Given data about the position $x$ and velocity $v$ at one instant, determine the phase $\omega t + \phi$ and phase constant $\phi$.

**15.17** For a spring–block oscillator, apply the relationships between spring constant $k$ and mass $m$ and either period $T$ or angular frequency $\omega$.

**15.18** Apply Hooke's law to relate the force $F$ on a simple harmonic oscillator at any instant to the displacement $x$ of the oscillator at that instant.

### Key Ideas

● The frequency $f$ of periodic, or oscillatory, motion is the number of oscillations per second. In the SI system, it is measured in hertz: $1 \text{ Hz} = 1 \text{ s}^{-1}$.

● The period $T$ is the time required for one complete oscillation, or cycle. It is related to the frequency by $T = 1/f$.

● In simple harmonic motion (SHM), the displacement $x(t)$ of a particle from its equilibrium position is described by the equation

$$x = x_m \cos(\omega t + \phi) \quad \text{(displacement)},$$

in which $x_m$ is the amplitude of the displacement, $\omega t + \phi$ is the phase of the motion, and $\phi$ is the phase constant. The angular frequency $\omega$ is related to the period and frequency of the motion by $\omega = 2\pi/T = 2\pi f$.

● Differentiating $x(t)$ leads to equations for the particle's SHM velocity and acceleration as functions of time:

$$v = -\omega x_m \sin(\omega t + \phi) \quad \text{(velocity)}$$

and
$$a = -\omega^2 x_m \cos(\omega t + \phi) \quad \text{(acceleration)}.$$

In the velocity function, the positive quantity $\omega x_m$ is the velocity amplitude $v_m$. In the acceleration function, the positive quantity $\omega^2 x_m$ is the acceleration amplitude $a_m$.

● A particle with mass $m$ that moves under the influence of a Hooke's law restoring force given by $F = -kx$ is a linear simple harmonic oscillator with

$$\omega = \sqrt{\frac{k}{m}} \quad \text{(angular frequency)}$$

and
$$T = 2\pi\sqrt{\frac{m}{k}} \quad \text{(period)}.$$

## What Is Physics?

Our world is filled with oscillations in which objects move back and forth repeatedly. Many oscillations are merely amusing or annoying, but many others are dangerous or financially important. Here are a few examples: When a bat hits a baseball, the bat may oscillate enough to sting the batter's hands or even to break apart. When wind blows past a power line, the line may oscillate ("gallop" in electrical engineering terms) so severely that it rips apart, shutting off the power supply to a community. When an airplane is in flight, the turbulence of the air flowing past the wings makes them oscillate, eventually leading to metal fatigue and even failure. When a train travels around a curve, its wheels oscillate horizontally ("hunt" in mechanical engineering terms) as they are forced to turn in new directions (you can hear the oscillations).

When an earthquake occurs near a city, buildings may be set oscillating so severely that they are shaken apart. When an arrow is shot from a bow, the feathers at the end of the arrow manage to snake around the bow staff without hitting it because the arrow oscillates. When a coin drops into a metal collection plate, the coin oscillates with such a familiar ring that the coin's denomination can be determined from the sound. When a rodeo cowboy rides a bull, the cowboy oscillates wildly as the bull jumps and turns (at least the cowboy hopes to be oscillating).

The study and control of oscillations are two of the primary goals of both physics and engineering. In this chapter we discuss a basic type of oscillation called *simple harmonic motion.*

**Heads Up.**  This material is quite challenging to most students. One reason is that there is a truckload of definitions and symbols to sort out, but the main reason is that we need to relate an object's oscillations (something that we can see or even experience) to the equations and graphs for the oscillations. Relating the real, visible motion to the abstraction of an equation or graph requires a lot of hard work.

## Simple Harmonic Motion

Figure 15-1 shows a particle that is oscillating about the origin of an $x$ axis, repeatedly going left and right by identical amounts. The **frequency** $f$ of the oscillation is the number of times per second that it completes a full oscillation (a *cycle*) and has the unit of hertz (abbreviated Hz), where

$$1 \text{ hertz} = 1 \text{ Hz} = 1 \text{ oscillation per second} = 1 \text{ s}^{-1}. \qquad (15\text{-}1)$$

The time for one full cycle is the **period** $T$ of the oscillation, which is

$$T = \frac{1}{f}. \qquad (15\text{-}2)$$

Any motion that repeats at regular intervals is called periodic motion or harmonic motion. However, here we are interested in a particular type of periodic motion called **simple harmonic motion** (SHM). Such motion is a sinusoidal function of time $t$. That is, it can be written as a sine or a cosine of time $t$. Here we arbitrarily choose the cosine function and write the displacement (or position) of the particle in Fig. 15-1 as

$$x(t) = x_m \cos(\omega t + \phi) \quad \text{(displacement)}, \qquad (15\text{-}3)$$

in which $x_m$, $\omega$, and $\phi$ are quantities that we shall define.

**Freeze-Frames.**  Let's take some freeze-frames of the motion and then arrange them one after another down the page (Fig. 15-2a). Our first freeze-frame is at $t = 0$ when the particle is at its rightmost position on the $x$ axis. We label that coordinate as $x_m$ (the subscript means *maximum*); it is the symbol in front of the cosine

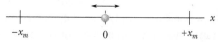

**Figure 15-1** A particle repeatedly oscillates left and right along an $x$ axis, between extreme points $x_m$ and $-x_m$.

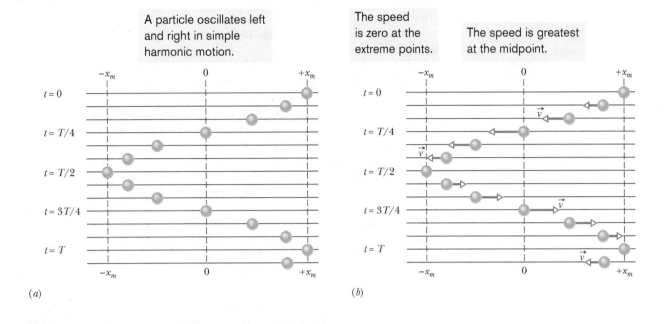

A particle oscillates left and right in simple harmonic motion.

The speed is zero at the extreme points.

The speed is greatest at the midpoint.

(a)

(b)

Rotating the figure reveals that the motion forms a cosine function.

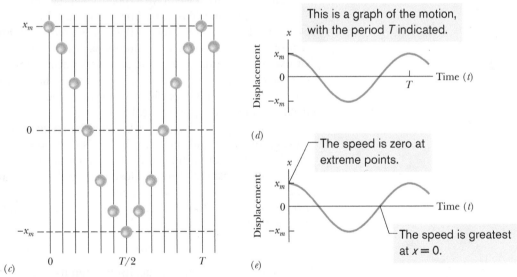

This is a graph of the motion, with the period $T$ indicated.

(d)

(c)

The speed is zero at extreme points.

The speed is greatest at $x = 0$.

(e)

**Figure 15-2** (a) A sequence of "freeze-frames" (taken at equal time intervals) showing the position of a particle as it oscillates back and forth about the origin of an $x$ axis, between the limits $+x_m$ and $-x_m$. (b) The vector arrows are scaled to indicate the speed of the particle. The speed is maximum when the particle is at the origin and zero when it is at $\pm x_m$. If the time $t$ is chosen to be zero when the particle is at $+x_m$, then the particle returns to $+x_m$ at $t = T$, where $T$ is the period of the motion. The motion is then repeated. (c) Rotating the figure reveals the motion forms a cosine function of time, as shown in (d). (e) The speed (the slope) changes.

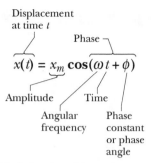

**Figure 15-3** A handy guide to the quantities in Eq. 15-3 for simple harmonic motion.

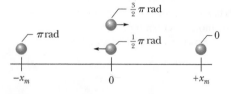

**Figure 15-4** Values of $\phi$ corresponding to the position of the particle at time $t = 0$.

function in Eq. 15-3. In the next freeze-frame, the particle is a bit to the left of $x_m$. It continues to move in the negative direction of $x$ until it reaches the leftmost position, at coordinate $-x_m$. Thereafter, as time takes us down the page through more freeze-frames, the particle moves back to $x_m$ and thereafter repeatedly oscillates between $x_m$ and $-x_m$. In Eq. 15-3, the cosine function itself oscillates between +1 and −1. The value of $x_m$ determines how far the particle moves in *its* oscillations and is called the **amplitude** of the oscillations (as labeled in the handy guide of Fig. 15-3).

Figure 15-2b indicates the velocity of the particle with respect to time, in the series of freeze-frames. We'll get to a function for the velocity soon, but for now just notice that the particle comes to a momentary stop at the extreme points and has its greatest speed (longest velocity vector) as it passes through the center point.

Mentally rotate Fig. 15-2a counterclockwise by 90°, so that the freeze-frames then progress rightward with time. We set time $t = 0$ when the particle is at $x_m$. The particle is back at $x_m$ at time $t = T$ (the period of the oscillation), when it starts the next cycle of oscillation. If we filled in lots of the intermediate freeze-frames and drew a line through the particle positions, we would have the cosine curve shown in Fig. 15-2d. What we already noted about the speed is displayed in Fig. 15-2e. What we have in the whole of Fig. 15-2 is a transformation of what we can see (the reality of an oscillating particle) into the abstraction of a graph. (In *WileyPLUS* the transformation of Fig. 15-2 is available as an animation with voiceover.) Equation 15-3 is a concise way to capture the motion in the abstraction of an equation.

*More Quantities.* The handy guide of Fig. 15-3 defines more quantities about the motion. The argument of the cosine function is called the **phase** of the motion. As it varies with time, the value of the cosine function varies. The constant $\phi$ is called the **phase angle** or **phase constant**. It is in the argument only because we want to use Eq. 15-3 to describe the motion *regardless* of where the particle is in its oscillation when we happen to set the clock time to 0. In Fig. 15-2, we set $t = 0$ when the particle is at $x_m$. For that choice, Eq. 15-3 works just fine if we also set $\phi = 0$. However, if we set $t = 0$ when the particle happens to be at some other location, we need a different value of $\phi$. A few values are indicated in Fig. 15-4. For example, suppose the particle is at its leftmost position when we happen to start the clock at $t = 0$. Then Eq. 15-3 describes the motion if $\phi = \pi$ rad. To check, substitute $t = 0$ and $\phi = \pi$ rad into Eq. 15-3. See, it gives $x = -x_m$ just then. Now check the other examples in Fig. 15-4.

The quantity $\omega$ in Eq. 15-3 is the **angular frequency** of the motion. To relate it to the frequency $f$ and the period $T$, let's first note that the position $x(t)$ of the particle must (by definition) return to its initial value at the end of a period. That is, if $x(t)$ is the position at some chosen time $t$, then the particle must return to that same position at time $t + T$. Let's use Eq. 15-3 to express this condition, but let's also just set $\phi = 0$ to get it out of the way. Returning to the same position can then be written as

$$x_m \cos \omega t = x_m \cos \omega(t + T). \tag{15-4}$$

The cosine function first repeats itself when its argument (the *phase*, remember) has increased by $2\pi$ rad. So, Eq. 15-4 tells us that

$$\omega(t + T) = \omega t + 2\pi$$

or

$$\omega T = 2\pi.$$

Thus, from Eq. 15-2 the angular frequency is

$$\omega = \frac{2\pi}{T} = 2\pi f. \tag{15-5}$$

The SI unit of angular frequency is the radian per second.

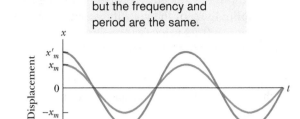

The amplitudes are different, but the frequency and period are the same.

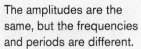

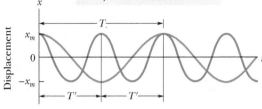

The amplitudes are the same, but the frequencies and periods are different.

(a)

(b)

**Figure 15-5** In all three cases, the blue curve is obtained from Eq. 15-3 with $\phi = 0$. (a) The red curve differs from the blue curve *only* in that the red-curve amplitude $x'_m$ is greater (the red-curve extremes of displacement are higher and lower). (b) The red curve differs from the blue curve *only* in that the red-curve period is $T' = T/2$ (the red curve is compressed horizontally). (c) The red curve differs from the blue curve *only* in that for the red curve $\phi = -\pi/4$ rad rather than zero (the negative value of $\phi$ shifts the red curve to the right).

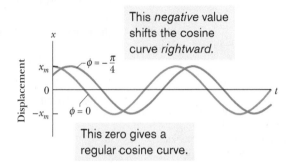

This *negative* value shifts the cosine curve *rightward*.

This zero gives a regular cosine curve.

(c)

We've had a lot of quantities here, quantities that we could experimentally change to see the effects on the particle's SHM. Figure 15-5 gives some examples. The curves in Fig. 15-5a show the effect of changing the amplitude. Both curves have the same period. (See how the "peaks" line up?) And both are for $\phi = 0$. (See how the maxima of the curves both occur at $t = 0$?) In Fig. 15-5b, the two curves have the same amplitude $x_m$ but one has twice the period as the other (and thus half the frequency as the other). Figure 15-5c is probably more difficult to understand. The curves have the same amplitude and same period but one is shifted relative to the other because of the different $\phi$ values. See how the one with $\phi = 0$ is just a regular cosine curve? The one with the negative $\phi$ is shifted rightward from it. That is a general result: negative $\phi$ values shift the regular cosine curve rightward and positive $\phi$ values shift it leftward. (Try this on a graphing calculator.)

 **Checkpoint 1**

A particle undergoing simple harmonic oscillation of period $T$ (like that in Fig. 15-2) is at $-x_m$ at time $t = 0$. Is it at $-x_m$, at $+x_m$, at 0, between $-x_m$ and 0, or between 0 and $+x_m$ when (a) $t = 2.00T$, (b) $t = 3.50T$, and (c) $t = 5.25T$?

**The Velocity of SHM**

We briefly discussed velocity as shown in Fig. 15-2b, finding that it varies in magnitude and direction as the particle moves between the extreme points (where the speed is momentarily zero) and through the central point (where the speed is maximum). To find the velocity $v(t)$ as a function of time, let's take a time derivative of the position function $x(t)$ in Eq. 15-3:

$$v(t) = \frac{dx(t)}{dt} = \frac{d}{dt}[x_m \cos(\omega t + \phi)]$$

or

$$v(t) = -\omega x_m \sin(\omega t + \phi) \quad \text{(velocity).} \tag{15-6}$$

The velocity depends on time because the sine function varies with time, between the values of $+1$ and $-1$. The quantities in front of the sine function

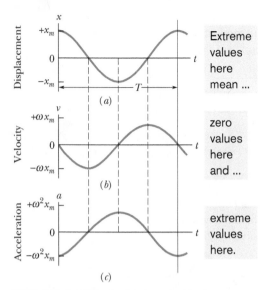

**Figure 15-6** (*a*) The displacement $x(t)$ of a particle oscillating in SHM with phase angle $\phi$ equal to zero. The period $T$ marks one complete oscillation. (*b*) The velocity $v(t)$ of the particle. (*c*) The acceleration $a(t)$ of the particle.

determine the extent of the variation in the velocity, between $+\omega x_m$ and $-\omega x_m$. We say that $\omega x_m$ is the **velocity amplitude** $v_m$ of the velocity variation. When the particle is moving rightward through $x = 0$, its velocity is positive and the magnitude is at this greatest value. When it is moving leftward through $x = 0$, its velocity is negative and the magnitude is again at this greatest value. This variation with time (a negative sine function) is displayed in the graph of Fig. 15-6*b* for a phase constant of $\phi = 0$, which corresponds to the cosine function for the displacement versus time shown in Fig. 15-6*a*.

Recall that we use a cosine function for $x(t)$ regardless of the particle's position at $t = 0$. We simply choose an appropriate value of $\phi$ so that Eq. 15-3 gives us the correct position at $t = 0$. That decision about the cosine function leads us to a negative sine function for the velocity in Eq. 15-6, and the value of $\phi$ now gives the correct velocity at $t = 0$.

### The Acceleration of SHM

Let's go one more step by differentiating the velocity function of Eq. 15-6 with respect to time to get the acceleration function of the particle in simple harmonic motion:

$$a(t) = \frac{dv(t)}{dt} = \frac{d}{dt}\left[-\omega x_m \sin(\omega t + \phi)\right]$$

or
$$a(t) = -\omega^2 x_m \cos(\omega t + \phi) \quad \text{(acceleration).} \tag{15-7}$$

We are back to a cosine function but with a minus sign out front. We know the drill by now. The acceleration varies because the cosine function varies with time, between $+1$ and $-1$. The variation in the magnitude of the acceleration is set by the **acceleration amplitude** $a_m$, which is the product $\omega^2 x_m$ that multiplies the cosine function.

Figure 15-6*c* displays Eq. 15-7 for a phase constant $\phi = 0$, consistent with Figs. 15-6*a* and 15-6*b*. Note that the acceleration magnitude is zero when the cosine is zero, which is when the particle is at $x = 0$. And the acceleration magnitude is maximum when the cosine magnitude is maximum, which is when the particle is at an extreme point, where it has been slowed to a stop so that its motion can be reversed. Indeed, comparing Eqs. 15-3 and 15-7 we see an extremely neat relationship:

$$a(t) = -\omega^2 x(t). \tag{15-8}$$

This is the hallmark of SHM: (1) The particle's acceleration is always opposite its displacement (hence the minus sign) and (2) the two quantities are always related by a constant ($\omega^2$). If you ever see such a relationship in an oscillating situation (such as with, say, the current in an electrical circuit, or the rise and fall of water in a tidal bay), you can immediately say that the motion is SHM and immediately identify the angular frequency $\omega$ of the motion. In a nutshell:

 In SHM, the acceleration $a$ is proportional to the displacement $x$ but opposite in sign, and the two quantities are related by the square of the angular frequency $\omega$.

 **Checkpoint 2**

Which of the following relationships between a particle's acceleration $a$ and its position $x$ indicates simple harmonic oscillation: (a) $a = 3x^2$, (b) $a = 5x$, (c) $a = -4x$, (d) $a = -2/x$? For the SHM, what is the angular frequency (assume the unit of rad/s)?

## The Force Law for Simple Harmonic Motion

Now that we have an expression for the acceleration in terms of the displacement in Eq. 15-8, we can apply Newton's second law to describe the force responsible for SHM:

$$F = ma = m(-\omega^2 x) = -(m\omega^2)x. \qquad (15\text{-}9)$$

The minus sign means that the direction of the force on the particle is *opposite* the direction of the displacement of the particle. That is, in SHM the force is a *restoring force* in the sense that it fights against the displacement, attempting to restore the particle to the center point at $x = 0$. We've seen the general form of Eq. 15-9 back in Chapter 8 when we discussed a block on a spring as in Fig. 15-7. There we wrote Hooke's law,

$$F = -kx, \qquad (15\text{-}10)$$

for the force acting on the block. Comparing Eqs. 15-9 and 15-10, we can now relate the spring constant $k$ (a measure of the stiffness of the spring) to the mass of the block and the resulting angular frequency of the SHM:

$$k = m\omega^2. \qquad (15\text{-}11)$$

Equation 15-10 is another way to write the hallmark equation for SHM.

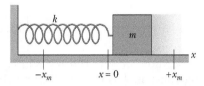

**Figure 15-7** A linear simple harmonic oscillator. The surface is frictionless. Like the particle of Fig. 15-2, the block moves in simple harmonic motion once it has been either pulled or pushed away from the $x = 0$ position and released. Its displacement is then given by Eq. 15-3.

> Simple harmonic motion is the motion of a particle when the force acting on it is proportional to the particle's displacement but in the opposite direction.

The block–spring system of Fig. 15-7 is called a **linear simple harmonic oscillator** (linear oscillator, for short), where *linear* indicates that $F$ is proportional to $x$ to the *first* power (and not to some other power).

If you ever see a situation in which the force in an oscillation is always proportional to the displacement but in the opposite direction, you can immediately say that the oscillation is SHM. You can also immediately identify the associated spring constant $k$. If you know the oscillating mass, you can then determine the angular frequency of the motion by rewriting Eq. 15-11 as

$$\omega = \sqrt{\frac{k}{m}} \quad \text{(angular frequency)}. \qquad (15\text{-}12)$$

(This is usually more important than the value of $k$.) Further, you can determine the period of the motion by combining Eqs. 15-5 and 15-12 to write

$$T = 2\pi\sqrt{\frac{m}{k}} \quad \text{(period)}. \qquad (15\text{-}13)$$

Let's make a bit of physical sense of Eqs. 15-12 and 15-13. Can you see that a stiff spring (large $k$) tends to produce a large $\omega$ (rapid oscillations) and thus a small period $T$? Can you also see that a large mass $m$ tends to result in a small $\omega$ (sluggish oscillations) and thus a large period $T$?

Every oscillating system, be it a diving board or a violin string, has some element of "springiness" and some element of "inertia" or mass. In Fig. 15-7, these elements are separated: The springiness is entirely in the spring, which we assume to be massless, and the inertia is entirely in the block, which we assume to be rigid. In a violin string, however, the two elements are both within the string.

## ✓ Checkpoint 3

Which of the following relationships between the force $F$ on a particle and the particle's position $x$ gives SHM: (a) $F = -5x$, (b) $F = -400x^2$, (c) $F = 10x$, (d) $F = 3x^2$?

## Sample Problem 15.01 Block–spring SHM, amplitude, acceleration, phase constant

A block whose mass $m$ is 680 g is fastened to a spring whose spring constant $k$ is 65 N/m. The block is pulled a distance $x = 11$ cm from its equilibrium position at $x = 0$ on a frictionless surface and released from rest at $t = 0$.

**(a)** What are the angular frequency, the frequency, and the period of the resulting motion?

### KEY IDEA

The block–spring system forms a linear simple harmonic oscillator, with the block undergoing SHM.

*Calculations:* The angular frequency is given by Eq. 15-12:

$$\omega = \sqrt{\frac{k}{m}} = \sqrt{\frac{65 \text{ N/m}}{0.68 \text{ kg}}} = 9.78 \text{ rad/s}$$

$$\approx 9.8 \text{ rad/s}. \qquad \text{(Answer)}$$

The frequency follows from Eq. 15-5, which yields

$$f = \frac{\omega}{2\pi} = \frac{9.78 \text{ rad/s}}{2\pi \text{ rad}} = 1.56 \text{ Hz} \approx 1.6 \text{ Hz}. \quad \text{(Answer)}$$

The period follows from Eq. 15-2, which yields

$$T = \frac{1}{f} = \frac{1}{1.56 \text{ Hz}} = 0.64 \text{ s} = 640 \text{ ms}. \quad \text{(Answer)}$$

**(b)** What is the amplitude of the oscillation?

### KEY IDEA

With no friction involved, the mechanical energy of the spring–block system is conserved.

*Reasoning:* The block is released from rest 11 cm from its equilibrium position, with zero kinetic energy and the elastic potential energy of the system at a maximum. Thus, the block will have zero kinetic energy whenever it is again 11 cm from its equilibrium position, which means it will never be farther than 11 cm from that position. Its maximum displacement is 11 cm:

$$x_m = 11 \text{ cm}. \qquad \text{(Answer)}$$

**(c)** What is the maximum speed $v_m$ of the oscillating block, and where is the block when it has this speed?

### KEY IDEA

The maximum speed $v_m$ is the velocity amplitude $\omega x_m$ in Eq. 15-6.

*Calculation:* Thus, we have

$$v_m = \omega x_m = (9.78 \text{ rad/s})(0.11 \text{ m})$$

$$= 1.1 \text{ m/s}. \qquad \text{(Answer)}$$

This maximum speed occurs when the oscillating block is rushing through the origin; compare Figs. 15-6a and 15-6b, where you can see that the speed is a maximum whenever $x = 0$.

**(d)** What is the magnitude $a_m$ of the maximum acceleration of the block?

### KEY IDEA

The magnitude $a_m$ of the maximum acceleration is the acceleration amplitude $\omega^2 x_m$ in Eq. 15-7.

*Calculation:* So, we have

$$a_m = \omega^2 x_m = (9.78 \text{ rad/s})^2 (0.11 \text{ m})$$

$$= 11 \text{ m/s}^2. \qquad \text{(Answer)}$$

This maximum acceleration occurs when the block is at the ends of its path, where the block has been slowed to a stop so that its motion can be reversed. At those extreme points, the force acting on the block has its maximum magnitude; compare Figs. 15-6a and 15-6c, where you can see that the magnitudes of the displacement and acceleration are maximum at the same times, when the speed is zero, as you can see in Fig. 15-6b.

**(e)** What is the phase constant $\phi$ for the motion?

*Calculations:* Equation 15-3 gives the displacement of the block as a function of time. We know that at time $t = 0$, the block is located at $x = x_m$. Substituting these *initial conditions,* as they are called, into Eq. 15-3 and canceling $x_m$ give us

$$1 = \cos \phi. \qquad (15\text{-}14)$$

Taking the inverse cosine then yields

$$\phi = 0 \text{ rad}. \qquad \text{(Answer)}$$

(Any angle that is an integer multiple of $2\pi$ rad also satisfies Eq. 15-14; we chose the smallest angle.)

**(f)** What is the displacement function $x(t)$ for the spring–block system?

*Calculation:* The function $x(t)$ is given in general form by Eq. 15-3. Substituting known quantities into that equation gives us

$$x(t) = x_m \cos(\omega t + \phi)$$

$$= (0.11 \text{ m}) \cos[(9.8 \text{ rad/s})t + 0]$$

$$= 0.11 \cos(9.8t), \qquad \text{(Answer)}$$

where $x$ is in meters and $t$ is in seconds.

 Additional examples, video, and practice available at *WileyPLUS*

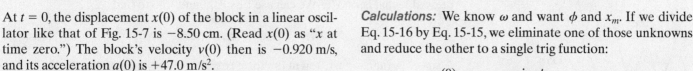

### Sample Problem 15.02 Finding SHM phase constant from displacement and velocity

At $t = 0$, the displacement $x(0)$ of the block in a linear oscillator like that of Fig. 15-7 is $-8.50$ cm. (Read $x(0)$ as "$x$ at time zero.") The block's velocity $v(0)$ then is $-0.920$ m/s, and its acceleration $a(0)$ is $+47.0$ m/s$^2$.

(a) What is the angular frequency $\omega$ of this system?

**KEY IDEA**

With the block in SHM, Eqs. 15-3, 15-6, and 15-7 give its displacement, velocity, and acceleration, respectively, and each contains $\omega$.

*Calculations:* Let's substitute $t = 0$ into each to see whether we can solve any one of them for $\omega$. We find

$$x(0) = x_m \cos \phi, \quad (15\text{-}15)$$
$$v(0) = -\omega x_m \sin \phi, \quad (15\text{-}16)$$

and
$$a(0) = -\omega^2 x_m \cos \phi. \quad (15\text{-}17)$$

In Eq. 15-15, $\omega$ has disappeared. In Eqs. 15-16 and 15-17, we know values for the left sides, but we do not know $x_m$ and $\phi$. However, if we divide Eq. 15-17 by Eq. 15-15, we neatly eliminate both $x_m$ and $\phi$ and can then solve for $\omega$ as

$$\omega = \sqrt{-\frac{a(0)}{x(0)}} = \sqrt{-\frac{47.0 \text{ m/s}^2}{-0.0850 \text{ m}}}$$
$$= 23.5 \text{ rad/s.} \quad \text{(Answer)}$$

(b) What are the phase constant $\phi$ and amplitude $x_m$?

*Calculations:* We know $\omega$ and want $\phi$ and $x_m$. If we divide Eq. 15-16 by Eq. 15-15, we eliminate one of those unknowns and reduce the other to a single trig function:

$$\frac{v(0)}{x(0)} = \frac{-\omega x_m \sin \phi}{x_m \cos \phi} = -\omega \tan \phi.$$

Solving for $\tan \phi$, we find

$$\tan \phi = -\frac{v(0)}{\omega x(0)} = -\frac{-0.920 \text{ m/s}}{(23.5 \text{ rad/s})(-0.0850 \text{ m})}$$
$$= -0.461.$$

This equation has two solutions:

$$\phi = -25° \quad \text{and} \quad \phi = 180° + (-25°) = 155°.$$

Normally only the first solution here is displayed by a calculator, but it may not be the physically possible solution. To choose the proper solution, we test them both by using them to compute values for the amplitude $x_m$. From Eq. 15-15, we find that if $\phi = -25°$, then

$$x_m = \frac{x(0)}{\cos \phi} = \frac{-0.0850 \text{ m}}{\cos(-25°)} = -0.094 \text{ m.}$$

We find similarly that if $\phi = 155°$, then $x_m = 0.094$ m. Because the amplitude of SHM must be a positive constant, the correct phase constant and amplitude here are

$$\phi = 155° \quad \text{and} \quad x_m = 0.094 \text{ m} = 9.4 \text{ cm.} \quad \text{(Answer)}$$

 Additional examples, video, and practice available at *WileyPLUS*

# 15-2 ENERGY IN SIMPLE HARMONIC MOTION

## Learning Objectives

*After reading this module, you should be able to . . .*

**15.19** For a spring–block oscillator, calculate the kinetic energy and elastic potential energy at any given time.

**15.20** Apply the conservation of energy to relate the total energy of a spring–block oscillator at one instant to the total energy at another instant.

**15.21** Sketch a graph of the kinetic energy, potential energy, and total energy of a spring–block oscillator, first as a function of time and then as a function of the oscillator's position.

**15.22** For a spring–block oscillator, determine the block's position when the total energy is entirely kinetic energy and when it is entirely potential energy.

## Key Ideas

● A particle in simple harmonic motion has, at any time, kinetic energy $K = \frac{1}{2}mv^2$ and potential energy $U = \frac{1}{2}kx^2$. If no

friction is present, the mechanical energy $E = K + U$ remains constant even though $K$ and $U$ change.

## Energy in Simple Harmonic Motion

Let's now examine the linear oscillator of Chapter 8, where we saw that the energy transfers back and forth between kinetic energy and potential energy, while the sum of the two—the mechanical energy $E$ of the oscillator—remains constant. The

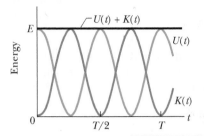

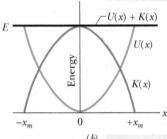

As *time* changes, the energy shifts between the two types, but the total is constant.

As *position* changes, the energy shifts between the two types, but the total is constant.

**Figure 15-8** (*a*) Potential energy $U(t)$, kinetic energy $K(t)$, and mechanical energy $E$ as functions of time $t$ for a linear harmonic oscillator. Note that all energies are positive and that the potential energy and the kinetic energy peak twice during every period. (*b*) Potential energy $U(x)$, kinetic energy $K(x)$, and mechanical energy $E$ as functions of position $x$ for a linear harmonic oscillator with amplitude $x_m$. For $x = 0$ the energy is all kinetic, and for $x = \pm x_m$ it is all potential.

potential energy of a linear oscillator like that of Fig. 15-7 is associated entirely with the spring. Its value depends on how much the spring is stretched or compressed—that is, on $x(t)$. We can use Eqs. 8-11 and 15-3 to find

$$U(t) = \tfrac{1}{2}kx^2 = \tfrac{1}{2}kx_m^2 \cos^2(\omega t + \phi). \quad (15\text{-}18)$$

*Caution:* A function written in the form $\cos^2 A$ (as here) means $(\cos A)^2$ and is *not* the same as one written $\cos A^2$, which means $\cos(A^2)$.

The kinetic energy of the system of Fig. 15-7 is associated entirely with the block. Its value depends on how fast the block is moving—that is, on $v(t)$. We can use Eq. 15-6 to find

$$K(t) = \tfrac{1}{2}mv^2 = \tfrac{1}{2}m\omega^2 x_m^2 \sin^2(\omega t + \phi). \quad (15\text{-}19)$$

If we use Eq. 15-12 to substitute $k/m$ for $\omega^2$, we can write Eq. 15-19 as

$$K(t) = \tfrac{1}{2}mv^2 = \tfrac{1}{2}kx_m^2 \sin^2(\omega t + \phi). \quad (15\text{-}20)$$

The mechanical energy follows from Eqs. 15-18 and 15-20 and is

$$E = U + K$$
$$= \tfrac{1}{2}kx_m^2 \cos^2(\omega t + \phi) + \tfrac{1}{2}kx_m^2 \sin^2(\omega t + \phi)$$
$$= \tfrac{1}{2}kx_m^2 [\cos^2(\omega t + \phi) + \sin^2(\omega t + \phi)].$$

For any angle $\alpha$,

$$\cos^2 \alpha + \sin^2 \alpha = 1.$$

Thus, the quantity in the square brackets above is unity and we have

$$E = U + K = \tfrac{1}{2}kx_m^2. \quad (15\text{-}21)$$

The mechanical energy of a linear oscillator is indeed constant and independent of time. The potential energy and kinetic energy of a linear oscillator are shown as functions of time $t$ in Fig. 15-8*a* and as functions of displacement $x$ in Fig. 15-8*b*. In any oscillating system, an element of springiness is needed to store the potential energy and an element of inertia is needed to store the kinetic energy.

 **Checkpoint 4**

In Fig. 15-7, the block has a kinetic energy of 3 J and the spring has an elastic potential energy of 2 J when the block is at $x = +2.0$ cm. (a) What is the kinetic energy when the block is at $x = 0$? What is the elastic potential energy when the block is at (b) $x = -2.0$ cm and (c) $x = -x_m$?

**Sample Problem 15.03** **SHM potential energy, kinetic energy, mass dampers**

Many tall buildings have *mass dampers*, which are anti-sway devices to prevent them from oscillating in a wind. The device might be a block oscillating at the end of a spring and on a lubricated track. If the building sways, say, eastward, the block also moves eastward but delayed enough so that when it finally moves, the building is then moving back westward. Thus, the motion of the oscillator is out of step with the motion of the building.

Suppose the block has mass $m = 2.72 \times 10^5$ kg and is designed to oscillate at frequency $f = 10.0$ Hz and with amplitude $x_m = 20.0$ cm.

(a) What is the total mechanical energy $E$ of the spring–block system?

**KEY IDEA**

The mechanical energy $E$ (the sum of the kinetic energy $K = \tfrac{1}{2}mv^2$ of the block and the potential energy $U = \tfrac{1}{2}kx^2$ of the spring) is constant throughout the motion of the oscillator. Thus, we can evaluate $E$ at any point during the motion.

*Calculations:* Because we are given amplitude $x_m$ of the oscillations, let's evaluate $E$ when the block is at position $x = x_m$, where it has velocity $v = 0$. However, to evaluate $U$ at that point, we first need to find the spring constant $k$. From Eq. 15-12 ($\omega = \sqrt{k/m}$) and Eq. 15-5 ($\omega = 2\pi f$), we find

$$k = m\omega^2 = m(2\pi f)^2$$
$$= (2.72 \times 10^5 \text{ kg})(2\pi)^2(10.0 \text{ Hz})^2$$
$$= 1.073 \times 10^9 \text{ N/m}.$$

We can now evaluate $E$ as

$$E = K + U = \tfrac{1}{2}mv^2 + \tfrac{1}{2}kx^2$$
$$= 0 + \tfrac{1}{2}(1.073 \times 10^9 \text{ N/m})(0.20 \text{ m})^2$$
$$= 2.147 \times 10^7 \text{ J} \approx 2.1 \times 10^7 \text{ J}. \qquad \text{(Answer)}$$

(b) What is the block's speed as it passes through the equilibrium point?

*Calculations:* We want the speed at $x = 0$, where the potential energy is $U = \tfrac{1}{2}kx^2 = 0$ and the mechanical energy is entirely kinetic energy. So, we can write

$$E = K + U = \tfrac{1}{2}mv^2 + \tfrac{1}{2}kx^2$$
$$2.147 \times 10^7 \text{ J} = \tfrac{1}{2}(2.72 \times 10^5 \text{ kg})v^2 + 0,$$

or $\qquad\qquad\qquad v = 12.6 \text{ m/s}.$ (Answer)

Because $E$ is entirely kinetic energy, this is the maximum speed $v_m$.

**WILEY PLUS** Additional examples, video, and practice available at *WileyPLUS*

# 15-3 AN ANGULAR SIMPLE HARMONIC OSCILLATOR

## Learning Objectives

*After reading this module, you should be able to . . .*

**15.23** Describe the motion of an angular simple harmonic oscillator.

**15.24** For an angular simple harmonic oscillator, apply the relationship between the torque $\tau$ and the angular displacement $\theta$ (from equilibrium).

**15.25** For an angular simple harmonic oscillator, apply the relationship between the period $T$ (or frequency $f$), the rotational inertia $I$, and the torsion constant $\kappa$.

**15.26** For an angular simple harmonic oscillator at any instant, apply the relationship between the angular acceleration $\alpha$, the angular frequency $\omega$, and the angular displacement $\theta$.

## Key Idea

● A torsion pendulum consists of an object suspended on a wire. When the wire is twisted and then released, the object oscillates in angular simple harmonic motion with a period given by

$$T = 2\pi\sqrt{\frac{I}{\kappa}},$$

where $I$ is the rotational inertia of the object about the axis of rotation and $\kappa$ is the torsion constant of the wire.

## An Angular Simple Harmonic Oscillator

Figure 15-9 shows an angular version of a simple harmonic oscillator; the element of springiness or elasticity is associated with the twisting of a suspension wire rather than the extension and compression of a spring as we previously had. The device is called a **torsion pendulum,** with *torsion* referring to the twisting.

If we rotate the disk in Fig. 15-9 by some angular displacement $\theta$ from its rest position (where the reference line is at $\theta = 0$) and release it, it will oscillate about that position in **angular simple harmonic motion.** Rotating the disk through an angle $\theta$ in either direction introduces a restoring torque given by

$$\tau = -\kappa\theta. \qquad (15\text{-}22)$$

Here $\kappa$ (Greek *kappa*) is a constant, called the **torsion constant,** that depends on the length, diameter, and material of the suspension wire.

Comparison of Eq. 15-22 with Eq. 15-10 leads us to suspect that Eq. 15-22 is the angular form of Hooke's law, and that we can transform Eq. 15-13, which gives the period of linear SHM, into an equation for the period of angular SHM: We replace the spring constant $k$ in Eq. 15-13 with its equivalent, the constant

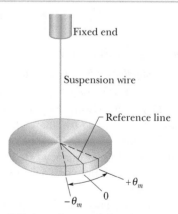

**Figure 15-9** A torsion pendulum is an angular version of a linear simple harmonic oscillator. The disk oscillates in a horizontal plane; the reference line oscillates with angular amplitude $\theta_m$. The twist in the suspension wire stores potential energy as a spring does and provides the restoring torque.

$\kappa$ of Eq. 15-22, and we replace the mass $m$ in Eq. 15-13 with *its* equivalent, the rotational inertia $I$ of the oscillating disk. These replacements lead to

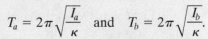

$$T = 2\pi\sqrt{\frac{I}{\kappa}} \quad \text{(torsion pendulum).} \qquad (15\text{-}23)$$

---

**Sample Problem 15.04**   **Angular simple harmonic oscillator, rotational inertia, period**

Figure 15-10$a$ shows a thin rod whose length $L$ is 12.4 cm and whose mass $m$ is 135 g, suspended at its midpoint from a long wire. Its period $T_a$ of angular SHM is measured to be 2.53 s. An irregularly shaped object, which we call object $X$, is then hung from the same wire, as in Fig. 15-10$b$, and its period $T_b$ is found to be 4.76 s. What is the rotational inertia of object $X$ about its suspension axis?

**KEY IDEA**

The rotational inertia of either the rod or object $X$ is related to the measured period by Eq. 15-23.

**Calculations:** In Table 10-2$e$, the rotational inertia of a thin rod about a perpendicular axis through its midpoint is given as $\frac{1}{12}mL^2$. Thus, we have, for the rod in Fig. 15-10$a$,

$$I_a = \tfrac{1}{12}mL^2 = (\tfrac{1}{12})(0.135 \text{ kg})(0.124 \text{ m})^2$$
$$= 1.73 \times 10^{-4} \text{ kg} \cdot \text{m}^2.$$

Now let us write Eq. 15-23 twice, once for the rod and once for object $X$:

$$T_a = 2\pi\sqrt{\frac{I_a}{\kappa}} \quad \text{and} \quad T_b = 2\pi\sqrt{\frac{I_b}{\kappa}}.$$

The constant $\kappa$, which is a property of the wire, is the same for both figures; only the periods and the rotational inertias differ.

Let us square each of these equations, divide the second by the first, and solve the resulting equation for $I_b$. The result is

$$I_b = I_a\frac{T_b^2}{T_a^2} = (1.73 \times 10^{-4} \text{ kg} \cdot \text{m}^2)\frac{(4.76 \text{ s})^2}{(2.53 \text{ s})^2}$$
$$= 6.12 \times 10^{-4} \text{ kg} \cdot \text{m}^2. \qquad \text{(Answer)}$$

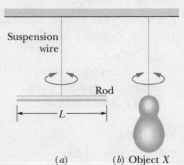

**Figure 15-10** Two torsion pendulums, consisting of ($a$) a wire and a rod and ($b$) the same wire and an irregularly shaped object.

 Additional examples, video, and practice available at *WileyPLUS*

---

# 15-4 PENDULUMS, CIRCULAR MOTION

## Learning Objectives

*After reading this module, you should be able to . . .*

**15.27** Describe the motion of an oscillating simple pendulum.

**15.28** Draw a free-body diagram of a pendulum bob with the pendulum at angle $\theta$ to the vertical.

**15.29** For small-angle oscillations of a *simple pendulum*, relate the period $T$ (or frequency $f$) to the pendulum's length $L$.

**15.30** Distinguish between a simple pendulum and a physical pendulum.

**15.31** For small-angle oscillations of a *physical pendulum*, relate the period $T$ (or frequency $f$) to the distance $h$ between the pivot and the center of mass.

**15.32** For an angular oscillating system, determine the angular frequency $\omega$ from either an equation relating torque $\tau$ and angular displacement $\theta$ or an equation relating angular acceleration $\alpha$ and angular displacement $\theta$.

**15.33** Distinguish between a pendulum's angular frequency $\omega$ (having to do with the rate at which cycles are completed) and its $d\theta/dt$ (the rate at which its angle with the vertical changes).

**15.34** Given data about the angular position $\theta$ and rate of change $d\theta/dt$ at one instant, determine the phase constant $\phi$ and amplitude $\theta_m$.

**15.35** Describe how the free-fall acceleration can be measured with a simple pendulum.

**15.36** For a given physical pendulum, determine the location of the center of oscillation and identify the meaning of that phrase in terms of a simple pendulum.

**15.37** Describe how simple harmonic motion is related to uniform circular motion.

### Key Ideas

- A simple pendulum consists of a rod of negligible mass that pivots about its upper end, with a particle (the bob) attached at its lower end. If the rod swings through only small angles, its motion is approximately simple harmonic motion with a period given by

$$T = 2\pi\sqrt{\frac{I}{mgL}} \qquad \text{(simple pendulum)},$$

where $I$ is the particle's rotational inertia about the pivot, $m$ is the particle's mass, and $L$ is the rod's length.

- A physical pendulum has a more complicated distribution of mass. For small angles of swinging, its motion is simple harmonic motion with a period given by

$$T = 2\pi\sqrt{\frac{I}{mgh}} \qquad \text{(physical pendulum)},$$

where $I$ is the pendulum's rotational inertia about the pivot, $m$ is the pendulum's mass, and $h$ is the distance between the pivot and the pendulum's center of mass.

- Simple harmonic motion corresponds to the projection of uniform circular motion onto a diameter of the circle.

## Pendulums

We turn now to a class of simple harmonic oscillators in which the springiness is associated with the gravitational force rather than with the elastic properties of a twisted wire or a compressed or stretched spring.

### The Simple Pendulum

If an apple swings on a long thread, does it have simple harmonic motion? If so, what is the period $T$? To answer, we consider a **simple pendulum,** which consists of a particle of mass $m$ (called the *bob* of the pendulum) suspended from one end of an unstretchable, massless string of length $L$ that is fixed at the other end, as in Fig. 15-11a. The bob is free to swing back and forth in the plane of the page, to the left and right of a vertical line through the pendulum's pivot point.

 ***The Restoring Torque.*** The forces acting on the bob are the force $\vec{T}$ from the string and the gravitational force $\vec{F}_g$, as shown in Fig. 15-11b, where the string makes an angle $\theta$ with the vertical. We resolve $\vec{F}_g$ into a radial component $F_g \cos\theta$ and a component $F_g \sin\theta$ that is tangent to the path taken by the bob. This tangential component produces a restoring torque about the pendulum's pivot point because the component always acts opposite the displacement of the bob so as to bring the bob back toward its central location. That location is called the *equilibrium position* ($\theta = 0$) because the pendulum would be at rest there were it not swinging.

From Eq. 10-41 ($\tau = r_\perp F$), we can write this restoring torque as

$$\tau = -L(F_g \sin\theta), \qquad (15\text{-}24)$$

where the minus sign indicates that the torque acts to reduce $\theta$ and $L$ is the moment arm of the force component $F_g \sin\theta$ about the pivot point. Substituting Eq. 15-24 into Eq. 10-44 ($\tau = I\alpha$) and then substituting $mg$ as the magnitude of $F_g$, we obtain

$$-L(mg \sin\theta) = I\alpha, \qquad (15\text{-}25)$$

where $I$ is the pendulum's rotational inertia about the pivot point and $\alpha$ is its angular acceleration about that point.

We can simplify Eq. 15-25 if we assume the angle $\theta$ is small, for then we can approximate $\sin\theta$ with $\theta$ (expressed in radian measure). (As an example, if $\theta = 5.00° = 0.0873$ rad, then $\sin\theta = 0.0872$, a difference of only about 0.1%.) With that approximation and some rearranging, we then have

$$\alpha = -\frac{mgL}{I}\theta. \qquad (15\text{-}26)$$

This equation is the angular equivalent of Eq. 15-8, the hallmark of SHM. It tells us that the angular acceleration $\alpha$ of the pendulum is proportional to the angular displacement $\theta$ but opposite in sign. Thus, as the pendulum bob moves to the right, as in Fig. 15-11a, its acceleration *to the left* increases until the bob stops and

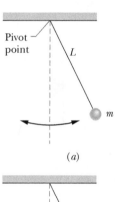

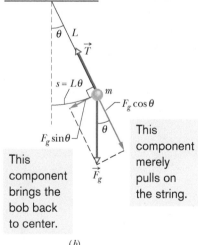

**Figure 15-11** (a) A simple pendulum. (b) The forces acting on the bob are the gravitational force $\vec{F}_g$ and the force $\vec{T}$ from the string. The tangential component $F_g \sin\theta$ of the gravitational force is a restoring force that tends to bring the pendulum back to its central position.

begins moving to the left. Then, when it is to the left of the equilibrium position, its acceleration to the right tends to return it to the right, and so on, as it swings back and forth in SHM. More precisely, the motion of a *simple pendulum swinging through only small angles* is approximately SHM. We can state this restriction to small angles another way: The **angular amplitude** $\theta_m$ of the motion (the maximum angle of swing) must be small.

**Angular Frequency.** Here is a neat trick. Because Eq. 15-26 has the same form as Eq. 15-8 for SHM, we can immediately identify the pendulum's angular frequency as being the square root of the constants in front of the displacement:

$$\omega = \sqrt{\frac{mgL}{I}}.$$

In the homework problems you might see oscillating systems that do not seem to resemble pendulums. However, if you can relate the acceleration (linear or angular) to the displacement (linear or angular), you can then immediately identify the angular frequency as we have just done here.

**Period.** Next, if we substitute this expression for $\omega$ into Eq. 15-5 ($\omega = 2\pi/T$), we see that the period of the pendulum may be written as

$$T = 2\pi\sqrt{\frac{I}{mgL}}. \tag{15-27}$$

All the mass of a simple pendulum is concentrated in the mass $m$ of the particle-like bob, which is at radius $L$ from the pivot point. Thus, we can use Eq. 10-33 ($I = mr^2$) to write $I = mL^2$ for the rotational inertia of the pendulum. Substituting this into Eq. 15-27 and simplifying then yield

$$T = 2\pi\sqrt{\frac{L}{g}} \qquad \text{(simple pendulum, small amplitude).} \tag{15-28}$$

We assume small-angle swinging in this chapter.

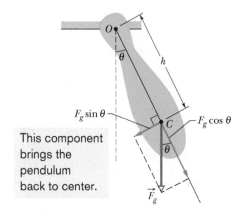

**Figure 15-12** A physical pendulum. The restoring torque is $hF_g \sin\theta$. When $\theta = 0$, center of mass $C$ hangs directly below pivot point $O$.

### The Physical Pendulum

A real pendulum, usually called a **physical pendulum,** can have a complicated distribution of mass. Does it also undergo SHM? If so, what is its period?

Figure 15-12 shows an arbitrary physical pendulum displaced to one side by angle $\theta$. The gravitational force $\vec{F}_g$ acts at its center of mass $C$, at a distance $h$ from the pivot point $O$. Comparison of Figs. 15-12 and 15-11$b$ reveals only one important difference between an arbitrary physical pendulum and a simple pendulum. For a physical pendulum the restoring component $F_g \sin\theta$ of the gravitational force has a moment arm of distance $h$ about the pivot point, rather than of string length $L$. In all other respects, an analysis of the physical pendulum would duplicate our analysis of the simple pendulum up through Eq. 15-27. Again (for small $\theta_m$), we would find that the motion is approximately SHM.

If we replace $L$ with $h$ in Eq. 15-27, we can write the period as

$$T = 2\pi\sqrt{\frac{I}{mgh}} \qquad \text{(physical pendulum, small amplitude).} \tag{15-29}$$

As with the simple pendulum, $I$ is the rotational inertia of the pendulum about $O$. However, now $I$ is not simply $mL^2$ (it depends on the shape of the physical pendulum), but it is still proportional to $m$.

A physical pendulum will not swing if it pivots at its center of mass. Formally, this corresponds to putting $h = 0$ in Eq. 15-29. That equation then predicts $T \to \infty$, which implies that such a pendulum will never complete one swing.

Corresponding to any physical pendulum that oscillates about a given pivot point $O$ with period $T$ is a simple pendulum of length $L_0$ with the same period $T$. We can find $L_0$ with Eq. 15-28. The point along the physical pendulum at distance $L_0$ from point $O$ is called the *center of oscillation* of the physical pendulum for the given suspension point.

### Measuring *g*

We can use a physical pendulum to measure the free-fall acceleration $g$ at a particular location on Earth's surface. (Countless thousands of such measurements have been made during geophysical prospecting.)

To analyze a simple case, take the pendulum to be a uniform rod of length $L$, suspended from one end. For such a pendulum, $h$ in Eq. 15-29, the distance between the pivot point and the center of mass, is $\frac{1}{2}L$. Table 10-2e tells us that the rotational inertia of this pendulum about a perpendicular axis through its center of mass is $\frac{1}{12}mL^2$. From the parallel-axis theorem of Eq. 10-36 ($I = I_{\text{com}} + Mh^2$), we then find that the rotational inertia about a perpendicular axis through one end of the rod is

$$I = I_{\text{com}} + mh^2 = \tfrac{1}{12}mL^2 + m(\tfrac{1}{2}L)^2 = \tfrac{1}{3}mL^2. \tag{15-30}$$

If we put $h = \frac{1}{2}L$ and $I = \frac{1}{3}mL^2$ in Eq. 15-29 and solve for $g$, we find

$$g = \frac{8\pi^2 L}{3T^2}. \tag{15-31}$$

Thus, by measuring $L$ and the period $T$, we can find the value of $g$ at the pendulum's location. (If precise measurements are to be made, a number of refinements are needed, such as swinging the pendulum in an evacuated chamber.)

### ✓ Checkpoint 5

Three physical pendulums, of masses $m_0$, $2m_0$, and $3m_0$, have the same shape and size and are suspended at the same point. Rank the masses according to the periods of the pendulums, greatest first.

## Sample Problem 15.05    Physical pendulum, period and length

In Fig. 15-13a, a meter stick swings about a pivot point at one end, at distance $h$ from the stick's center of mass.

(a) What is the period of oscillation $T$?

### KEY IDEA

The stick is not a simple pendulum because its mass is not concentrated in a bob at the end opposite the pivot point — so the stick is a physical pendulum.

*Calculations:* The period for a physical pendulum is given by Eq. 15-29, for which we need the rotational inertia $I$ of the stick about the pivot point. We can treat the stick as a uniform rod of length $L$ and mass $m$. Then Eq. 15-30 tells us that $I = \frac{1}{3}mL^2$, and the distance $h$ in Eq. 15-29 is $\frac{1}{2}L$. Substituting these quantities into Eq. 15-29,

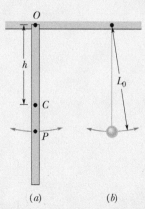

**Figure 15-13** (*a*) A meter stick suspended from one end as a physical pendulum. (*b*) A simple pendulum whose length $L_0$ is chosen so that the periods of the two pendulums are equal. Point $P$ on the pendulum of (*a*) marks the center of oscillation.

we find

$$T = 2\pi\sqrt{\frac{I}{mgh}} = 2\pi\sqrt{\frac{\frac{1}{3}mL^2}{mg(\frac{1}{2}L)}} \quad (15\text{-}32)$$

$$= 2\pi\sqrt{\frac{2L}{3g}} \quad (15\text{-}33)$$

$$= 2\pi\sqrt{\frac{(2)(1.00\ \text{m})}{(3)(9.8\ \text{m/s}^2)}} = 1.64\ \text{s}. \quad (\text{Answer})$$

Note the result is independent of the pendulum's mass $m$.

(b) What is the distance $L_0$ between the pivot point $O$ of the stick and the center of oscillation of the stick?

*Calculations:* We want the length $L_0$ of the simple pendu-

lum (drawn in Fig. 15-13$b$) that has the same period as the physical pendulum (the stick) of Fig. 15-13$a$. Setting Eqs. 15-28 and 15-33 equal yields

$$T = 2\pi\sqrt{\frac{L_0}{g}} = 2\pi\sqrt{\frac{2L}{3g}}. \quad (15\text{-}34)$$

You can see by inspection that

$$L_0 = \tfrac{2}{3}L \quad (15\text{-}35)$$

$$= (\tfrac{2}{3})(100\ \text{cm}) = 66.7\ \text{cm}. \quad (\text{Answer})$$

In Fig. 15-13$a$, point $P$ marks this distance from suspension point $O$. Thus, point $P$ is the stick's center of oscillation for the given suspension point. Point $P$ would be different for a different suspension choice.

**WILEY PLUS** Additional examples, video, and practice available at *WileyPLUS*

## Simple Harmonic Motion and Uniform Circular Motion

In 1610, Galileo, using his newly constructed telescope, discovered the four prin- cipal moons of Jupiter. Over weeks of observation, each moon seemed to him to be moving back and forth relative to the planet in what today we would call simple harmonic motion; the disk of the planet was the midpoint of the motion. The record of Galileo's observations, written in his own hand, is actually still available. A. P. French of MIT used Galileo's data to work out the position of the moon Callisto relative to Jupiter (actually, the angular distance from Jupiter as seen from Earth) and found that the data approximates the curve shown in Fig. 15-14. The curve strongly suggests Eq. 15-3, the displacement function for simple harmonic motion. A period of about 16.8 days can be measured from the plot, but it is a period of what exactly? After all, a moon cannot possibly be oscillating back and forth like a block on the end of a spring, and so why would Eq. 15-3 have anything to do with it?

*Actually,* Callisto moves with essentially constant speed in an essentially cir- cular orbit around Jupiter. Its true motion—far from being simple harmonic— is uniform circular motion along that orbit. What Galileo saw—and what you can see with a good pair of binoculars and a little patience—is the projection of this uniform circular motion on a line in the plane of the motion. We are led by Galileo's remarkable observations to the conclusion that simple harmonic

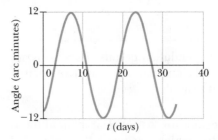

**Figure 15-14** The angle between Jupiter and its moon Callisto as seen from Earth. Galileo's 1610 measurements approximate this curve, which suggests simple harmonic motion. At Jupiter's mean distance from Earth, 10 minutes of arc corresponds to about $2 \times 10^6$ km. (Based on A. P. French, *Newtonian Mechanics,* W. W. Norton & Company, New York, 1971, p. 288.)

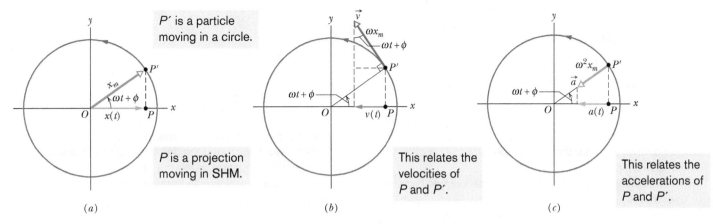

(a)    (b)    (c)

**Figure 15-15** (a) A reference particle $P'$ moving with uniform circular motion in a reference circle of radius $x_m$. Its projection $P$ on the $x$ axis executes simple harmonic motion. (b) The projection of the velocity $\vec{v}$ of the reference particle is the velocity of SHM. (c) The projection of the radial acceleration $\vec{a}$ of the reference particle is the acceleration of SHM.

motion is uniform circular motion viewed edge-on. In more formal language:

> Simple harmonic motion is the projection of uniform circular motion on a diameter of the circle in which the circular motion occurs.

Figure 15-15a gives an example. It shows a *reference particle P'* moving in uniform circular motion with (constant) angular speed $\omega$ in a *reference circle*. The radius $x_m$ of the circle is the magnitude of the particle's position vector. At any time $t$, the angular position of the particle is $\omega t + \phi$, where $\phi$ is its angular position at $t = 0$.

***Position.*** The projection of particle $P'$ onto the $x$ axis is a point $P$, which we take to be a second particle. The projection of the position vector of particle $P'$ onto the $x$ axis gives the location $x(t)$ of $P$. (Can you see the $x$ component in the triangle in Fig. 15-15a?) Thus, we find

$$x(t) = x_m \cos(\omega t + \phi), \tag{15-36}$$

which is precisely Eq. 15-3. Our conclusion is correct. If reference particle $P'$ moves in uniform circular motion, its projection particle $P$ moves in simple harmonic motion along a diameter of the circle.

***Velocity.*** Figure 15-15b shows the velocity $\vec{v}$ of the reference particle. From Eq. 10-18 ($v = \omega r$), the magnitude of the velocity vector is $\omega x_m$; its projection on the $x$ axis is

$$v(t) = -\omega x_m \sin(\omega t + \phi), \tag{15-37}$$

which is exactly Eq. 15-6. The minus sign appears because the velocity component of $P$ in Fig. 15-15b is directed to the left, in the negative direction of $x$. (The minus sign is consistent with the derivative of Eq. 15-36 with respect to time.)

***Acceleration.*** Figure 15-15c shows the radial acceleration $\vec{a}$ of the reference particle. From Eq. 10-23 ($a_r = \omega^2 r$), the magnitude of the radial acceleration vector is $\omega^2 x_m$; its projection on the $x$ axis is

$$a(t) = -\omega^2 x_m \cos(\omega t + \phi), \tag{15-38}$$

which is exactly Eq. 15-7. Thus, whether we look at the displacement, the velocity, or the acceleration, the projection of uniform circular motion is indeed simple harmonic motion.

# 15-5 DAMPED SIMPLE HARMONIC MOTION

## Learning Objectives

*After reading this module, you should be able to . . .*

**15.38** Describe the motion of a damped simple harmonic oscillator and sketch a graph of the oscillator's position as a function of time.

**15.39** For any particular time, calculate the position of a damped simple harmonic oscillator.

**15.40** Determine the amplitude of a damped simple harmonic oscillator at any given time.

**15.41** Calculate the angular frequency of a damped simple harmonic oscillator in terms of the spring constant, the damping constant, and the mass, and approximate the angular frequency when the damping constant is small.

**15.42** Apply the equation giving the (approximate) total energy of a damped simple harmonic oscillator as a function of time.

## Key Ideas

● The mechanical energy $E$ in a real oscillating system decreases during the oscillations because external forces, such as a drag force, inhibit the oscillations and transfer mechanical energy to thermal energy. The real oscillator and its motion are then said to be damped.

● If the damping force is given by $\vec{F}_d = -b\vec{v}$, where $\vec{v}$ is the velocity of the oscillator and $b$ is a damping constant, then the displacement of the oscillator is given by

$$x(t) = x_m \, e^{-bt/2m} \cos(\omega' t + \phi),$$

where $\omega'$, the angular frequency of the damped oscillator, is given by

$$\omega' = \sqrt{\frac{k}{m} - \frac{b^2}{4m^2}}.$$

● If the damping constant is small ($b \ll \sqrt{km}$), then $\omega' \approx \omega$, where $\omega$ is the angular frequency of the undamped oscillator. For small $b$, the mechanical energy $E$ of the oscillator is given by

$$E(t) \approx \tfrac{1}{2} k x_m^2 \, e^{-bt/m}.$$

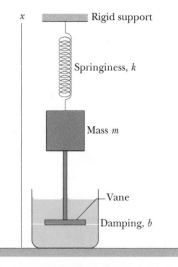

**Figure 15-16** An idealized damped simple harmonic oscillator. A vane immersed in a liquid exerts a damping force on the block as the block oscillates parallel to the *x* axis.

## Damped Simple Harmonic Motion

A pendulum will swing only briefly underwater, because the water exerts on the pendulum a drag force that quickly eliminates the motion. A pendulum swinging in air does better, but still the motion dies out eventually, because the air exerts a drag force on the pendulum (and friction acts at its support point), transferring energy from the pendulum's motion.

When the motion of an oscillator is reduced by an external force, the oscillator and its motion are said to be **damped.** An idealized example of a damped oscillator is shown in Fig. 15-16, where a block with mass $m$ oscillates vertically on a spring with spring constant $k$. From the block, a rod extends to a vane (both assumed massless) that is submerged in a liquid. As the vane moves up and down, the liquid exerts an inhibiting drag force on it and thus on the entire oscillating system. With time, the mechanical energy of the block–spring system decreases, as energy is transferred to thermal energy of the liquid and vane.

Let us assume the liquid exerts a **damping force** $\vec{F}_d$ that is proportional to the velocity $\vec{v}$ of the vane and block (an assumption that is accurate if the vane moves slowly). Then, for force and velocity components along the *x* axis in Fig. 15-16, we have

$$F_d = -bv, \tag{15-39}$$

where $b$ is a **damping constant** that depends on the characteristics of both the vane and the liquid and has the SI unit of kilogram per second. The minus sign indicates that $\vec{F}_d$ opposes the motion.

**Damped Oscillations.** The force on the block from the spring is $F_s = -kx$. Let us assume that the gravitational force on the block is negligible relative to $F_d$ and $F_s$. Then we can write Newton's second law for components along the *x* axis ($F_{net,x} = ma_x$) as

$$-bv - kx = ma. \tag{15-40}$$

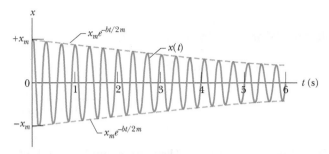

**Figure 15-17** The displacement function $x(t)$ for the damped oscillator of Fig. 15-16. The amplitude, which is $x_m\, e^{-bt/2m}$, decreases exponentially with time.

Substituting $dx/dt$ for $v$ and $d^2x/dt^2$ for $a$ and rearranging give us the differential equation

$$m\frac{d^2x}{dt^2} + b\frac{dx}{dt} + kx = 0. \qquad (15\text{-}41)$$

The solution of this equation is

$$x(t) = x_m\, e^{-bt/2m} \cos(\omega' t + \phi), \qquad (15\text{-}42)$$

where $x_m$ is the amplitude and $\omega'$ is the angular frequency of the damped oscillator. This angular frequency is given by

$$\omega' = \sqrt{\frac{k}{m} - \frac{b^2}{4m^2}}. \qquad (15\text{-}43)$$

If $b = 0$ (there is no damping), then Eq. 15-43 reduces to Eq. 15-12 ($\omega = \sqrt{k/m}$) for the angular frequency of an undamped oscillator, and Eq. 15-42 reduces to Eq. 15-3 for the displacement of an undamped oscillator. If the damping constant is small but not zero (so that $b \ll \sqrt{km}$), then $\omega' \approx \omega$.

**Damped Energy.** We can regard Eq. 15-42 as a cosine function whose amplitude, which is $x_m\, e^{-bt/2m}$, gradually decreases with time, as Fig. 15-17 suggests. For an undamped oscillator, the mechanical energy is constant and is given by Eq. 15-21 ($E = \frac{1}{2}kx_m^2$). If the oscillator is damped, the mechanical energy is not constant but decreases with time. If the damping is small, we can find $E(t)$ by replacing $x_m$ in Eq. 15-21 with $x_m\, e^{-bt/2m}$, the amplitude of the damped oscillations. By doing so, we find that

$$E(t) \approx \tfrac{1}{2}kx_m^2 e^{-bt/m}, \qquad (15\text{-}44)$$

which tells us that, like the amplitude, the mechanical energy decreases exponentially with time.

 **Checkpoint 6**

Here are three sets of values for the spring constant, damping constant, and mass for the damped oscillator of Fig. 15-16. Rank the sets according to the time required for the mechanical energy to decrease to one-fourth of its initial value, greatest first.

| | | | |
|---|---|---|---|
| Set 1 | $2k_0$ | $b_0$ | $m_0$ |
| Set 2 | $k_0$ | $6b_0$ | $4m_0$ |
| Set 3 | $3k_0$ | $3b_0$ | $m_0$ |

### Sample Problem 15.06 Damped harmonic oscillator, time to decay, energy

For the damped oscillator of Fig. 15-16, $m = 250$ g, $k = 85$ N/m, and $b = 70$ g/s.

(a) What is the period of the motion?

**KEY IDEA**

Because $b \ll \sqrt{km} = 4.6$ kg/s, the period is approximately that of the undamped oscillator.

*Calculation:* From Eq. 15-13, we then have

$$T = 2\pi\sqrt{\frac{m}{k}} = 2\pi\sqrt{\frac{0.25 \text{ kg}}{85 \text{ N/m}}} = 0.34 \text{ s.} \qquad \text{(Answer)}$$

(b) How long does it take for the amplitude of the damped oscillations to drop to half its initial value?

**KEY IDEA**

The amplitude at time $t$ is displayed in Eq. 15-42 as $x_m e^{-bt/2m}$.

*Calculations:* The amplitude has the value $x_m$ at $t = 0$. Thus, we must find the value of $t$ for which

$$x_m e^{-bt/2m} = \tfrac{1}{2}x_m.$$

Canceling $x_m$ and taking the natural logarithm of the equation that remains, we have $\ln \frac{1}{2}$ on the right side and

$$\ln(e^{-bt/2m}) = -bt/2m$$

on the left side. Thus,

$$t = \frac{-2m \ln \frac{1}{2}}{b} = \frac{-(2)(0.25 \text{ kg})(\ln \frac{1}{2})}{0.070 \text{ kg/s}}$$

$$= 5.0 \text{ s.} \qquad \text{(Answer)}$$

Because $T = 0.34$ s, this is about 15 periods of oscillation.

(c) How long does it take for the mechanical energy to drop to one-half its initial value?

**KEY IDEA**

From Eq. 15-44, the mechanical energy at time $t$ is $\tfrac{1}{2}kx_m^2 e^{-bt/m}$.

*Calculations:* The mechanical energy has the value $\tfrac{1}{2}kx_m^2$ at $t = 0$. Thus, we must find the value of $t$ for which

$$\tfrac{1}{2}kx_m^2 e^{-bt/m} = \tfrac{1}{2}(\tfrac{1}{2}kx_m^2).$$

If we divide both sides of this equation by $\tfrac{1}{2}kx_m^2$ and solve for $t$ as we did above, we find

$$t = \frac{-m \ln \frac{1}{2}}{b} = \frac{-(0.25 \text{ kg})(\ln \frac{1}{2})}{0.070 \text{ kg/s}} = 2.5 \text{ s.} \qquad \text{(Answer)}$$

This is exactly half the time we calculated in (b), or about 7.5 periods of oscillation. Figure 15-17 was drawn to illustrate this sample problem.

  Additional examples, video, and practice available at *WileyPLUS*

# 15-6 FORCED OSCILLATIONS AND RESONANCE

## Learning Objectives

*After reading this module, you should be able to . . .*

**15.43** Distinguish between natural angular frequency $\omega$ and driving angular frequency $\omega_d$.

**15.44** For a forced oscillator, sketch a graph of the oscillation amplitude versus the ratio $\omega_d/\omega$ of driving angular fre-

quency to natural angular frequency, identify the approximate location of resonance, and indicate the effect of increasing the damping constant.

**15.45** For a given natural angular frequency $\omega$, identify the approximate driving angular frequency $\omega_d$ that gives resonance.

## Key Ideas

● If an external driving force with angular frequency $\omega_d$ acts on an oscillating system with natural angular frequency $\omega$, the system oscillates with angular frequency $\omega_d$.

● The velocity amplitude $v_m$ of the system is greatest when

$$\omega_d = \omega,$$

a condition called resonance. The amplitude $x_m$ of the system is (approximately) greatest under the same condition.

## Forced Oscillations and Resonance

A person swinging in a swing without anyone pushing it is an example of *free oscillation*. However, if someone pushes the swing periodically, the swing has *forced,* or *driven, oscillations. Two* angular frequencies are associated with a system undergoing driven oscillations: (1) the *natural* angular frequency $\omega$ of the system, which is the angular frequency at which it would oscillate if it were suddenly disturbed and then left to oscillate freely, and (2) the angular frequency $\omega_d$ of the external driving force causing the driven oscillations.

We can use Fig. 15-16 to represent an idealized forced simple harmonic oscillator if we allow the structure marked "rigid support" to move up and down at a variable angular frequency $\omega_d$. Such a forced oscillator oscillates at the angular frequency $\omega_d$ of the driving force, and its displacement $x(t)$ is given by

$$x(t) = x_m \cos(\omega_d t + \phi), \tag{15-45}$$

where $x_m$ is the amplitude of the oscillations.

How large the displacement amplitude $x_m$ is depends on a complicated function of $\omega_d$ and $\omega$. The velocity amplitude $v_m$ of the oscillations is easier to describe: it is greatest when

$$\omega_d = \omega \quad \text{(resonance)}, \tag{15-46}$$

a condition called **resonance.** Equation 15-46 is also *approximately* the condition at which the displacement amplitude $x_m$ of the oscillations is greatest. Thus, if you push a swing at its natural angular frequency, the displacement and velocity amplitudes will increase to large values, a fact that children learn quickly by trial and error. If you push at other angular frequencies, either higher or lower, the displacement and velocity amplitudes will be smaller.

Figure 15-18 shows how the displacement amplitude of an oscillator depends on the angular frequency $\omega_d$ of the driving force, for three values of the damping coefficient $b$. Note that for all three the amplitude is approximately greatest when $\omega_d/\omega = 1$ (the resonance condition of Eq. 15-46). The curves of Fig. 15-18 show that less damping gives a taller and narrower *resonance peak.*

***Examples.*** All mechanical structures have one or more natural angular frequencies, and if a structure is subjected to a strong external driving force that matches one of these angular frequencies, the resulting oscillations of the structure may rupture it. Thus, for example, aircraft designers must make sure that none of the natural angular frequencies at which a wing can oscillate matches the angular frequency of the engines in flight. A wing that flaps violently at certain engine speeds would obviously be dangerous.

Resonance appears to be one reason buildings in Mexico City collapsed in September 1985 when a major earthquake (8.1 on the Richter scale) occurred on the western coast of Mexico. The seismic waves from the earthquake should have been too weak to cause extensive damage when they reached Mexico City about 400 km away. However, Mexico City is largely built on an ancient lake bed, where the soil is still soft with water. Although the amplitude of the seismic waves was small in the firmer ground en route to Mexico City, their amplitude substantially increased in the loose soil of the city. Acceleration amplitudes of the waves were as much as 0.20$g$, and the angular frequency was (surprisingly) concentrated around 3 rad/s. Not only was the ground severely oscillated, but many intermediate-height buildings had resonant angular frequencies of about 3 rad/s. Most of those buildings collapsed during the shaking (Fig. 15-19), while shorter buildings (with higher resonant angular frequencies) and taller buildings (with lower resonant angular frequencies) remained standing.

During a 1989 earthquake in the San Francisco–Oakland area, a similar resonant oscillation collapsed part of a freeway, dropping an upper deck onto a lower deck. That section of the freeway had been constructed on a loosely structured mudfill.

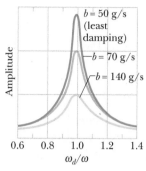

**Figure 15-18** The displacement amplitude $x_m$ of a forced oscillator varies as the angular frequency $\omega_d$ of the driving force is varied. The curves here correspond to three values of the damping constant $b$.

**Figure 15-19** In 1985, buildings of intermediate height collapsed in Mexico City as a result of an earthquake far from the city. Taller and shorter buildings remained standing.

# Review & Summary

**Frequency** The *frequency f* of periodic, or oscillatory, motion is the number of oscillations per second. In the SI system, it is measured in hertz:

$$1 \text{ hertz} = 1 \text{ Hz} = 1 \text{ oscillation per second} = 1 \text{ s}^{-1}. \quad (15\text{-}1)$$

**Period** The *period T* is the time required for one complete oscillation, or **cycle.** It is related to the frequency by

$$T = \frac{1}{f}. \quad (15\text{-}2)$$

**Simple Harmonic Motion** In *simple harmonic motion* (SHM), the displacement $x(t)$ of a particle from its equilibrium position is described by the equation

$$x = x_m \cos(\omega t + \phi) \quad \text{(displacement)}, \quad (15\text{-}3)$$

in which $x_m$ is the **amplitude** of the displacement, $\omega t + \phi$ is the **phase** of the motion, and $\phi$ is the **phase constant.** The **angular frequency** $\omega$ is related to the period and frequency of the motion by

$$\omega = \frac{2\pi}{T} = 2\pi f \quad \text{(angular frequency)}. \quad (15\text{-}5)$$

Differentiating Eq. 15-3 leads to equations for the particle's SHM velocity and acceleration as functions of time:

$$v = -\omega x_m \sin(\omega t + \phi) \quad \text{(velocity)} \quad (15\text{-}6)$$

and

$$a = -\omega^2 x_m \cos(\omega t + \phi) \quad \text{(acceleration)}. \quad (15\text{-}7)$$

In Eq. 15-6, the positive quantity $\omega x_m$ is the **velocity amplitude** $v_m$ of the motion. In Eq. 15-7, the positive quantity $\omega^2 x_m$ is the **acceleration amplitude** $a_m$ of the motion.

**The Linear Oscillator** A particle with mass $m$ that moves under the influence of a Hooke's law restoring force given by $F = -kx$ exhibits simple harmonic motion with

$$\omega = \sqrt{\frac{k}{m}} \quad \text{(angular frequency)} \quad (15\text{-}12)$$

and

$$T = 2\pi \sqrt{\frac{m}{k}} \quad \text{(period)}. \quad (15\text{-}13)$$

Such a system is called a **linear simple harmonic oscillator.**

**Energy** A particle in simple harmonic motion has, at any time, kinetic energy $K = \frac{1}{2}mv^2$ and potential energy $U = \frac{1}{2}kx^2$. If no friction is present, the mechanical energy $E = K + U$ remains constant even though $K$ and $U$ change.

**Pendulums** Examples of devices that undergo simple harmonic motion are the **torsion pendulum** of Fig. 15-9, the **simple pendulum** of Fig. 15-11, and the **physical pendulum** of Fig. 15-12. Their periods of oscillation for small oscillations are, respectively,

$$T = 2\pi \sqrt{I/\kappa} \quad \text{(torsion pendulum)}, \quad (15\text{-}23)$$

$$T = 2\pi \sqrt{L/g} \quad \text{(simple pendulum)}, \quad (15\text{-}28)$$

$$T = 2\pi \sqrt{I/mgh} \quad \text{(physical pendulum)}. \quad (15\text{-}29)$$

**Simple Harmonic Motion and Uniform Circular Motion** Simple harmonic motion is the projection of uniform circular motion onto the diameter of the circle in which the circular motion occurs. Figure 15-15 shows that all parameters of circular motion (position, velocity, and acceleration) project to the corresponding values for simple harmonic motion.

**Damped Harmonic Motion** The mechanical energy $E$ in a real oscillating system decreases during the oscillations because external forces, such as a drag force, inhibit the oscillations and transfer mechanical energy to thermal energy. The real oscillator and its motion are then said to be **damped.** If the **damping force** is given by $\vec{F}_d = -b\vec{v}$, where $\vec{v}$ is the velocity of the oscillator and $b$ is a **damping constant,** then the displacement of the oscillator is given by

$$x(t) = x_m e^{-bt/2m} \cos(\omega' t + \phi), \quad (15\text{-}42)$$

where $\omega'$, the angular frequency of the damped oscillator, is given by

$$\omega' = \sqrt{\frac{k}{m} - \frac{b^2}{4m^2}}. \quad (15\text{-}43)$$

If the damping constant is small ($b \ll \sqrt{km}$), then $\omega' \approx \omega$, where $\omega$ is the angular frequency of the undamped oscillator. For small $b$, the mechanical energy $E$ of the oscillator is given by

$$E(t) \approx \frac{1}{2}kx_m^2 e^{-bt/m}. \quad (15\text{-}44)$$

**Forced Oscillations and Resonance** If an external driving force with angular frequency $\omega_d$ acts on an oscillating system with *natural* angular frequency $\omega$, the system oscillates with angular frequency $\omega_d$. The velocity amplitude $v_m$ of the system is greatest when

$$\omega_d = \omega, \quad (15\text{-}46)$$

a condition called **resonance.** The amplitude $x_m$ of the system is (approximately) greatest under the same condition.

# Questions

1 Which of the following describe $\phi$ for the SHM of Fig. 15-20a:

(a) $-\pi < \phi < -\pi/2$,

(b) $\pi < \phi < 3\pi/2$,

(c) $-3\pi/2 < \phi < -\pi$?

2 The velocity $v(t)$ of a particle undergoing SHM is graphed in Fig. 15-20b. Is the particle momentarily stationary, headed toward $-x_m$, or headed toward $+x_m$ at (a) point $A$ on the graph and (b) point $B$? Is the particle at $-x_m$, at $+x_m$, at 0, between $-x_m$ and 0, or between 0 and $+x_m$ when its velocity is represented by (c) point $A$

and (d) point $B$? Is the speed of the particle increasing or decreasing at (e) point $A$ and (f) point $B$?

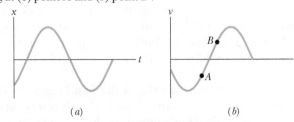

(a)           (b)

**Figure 15-20** Questions 1 and 2.

**3** The acceleration $a(t)$ of a particle undergoing SHM is graphed in Fig. 15-21. (a) Which of the labeled points corresponds to the particle at $-x_m$? (b) At point 4, is the velocity of the particle positive, negative, or zero? (c) At point 5, is the particle at $-x_m$, at $+x_m$, at 0, between $-x_m$ and 0, or between 0 and $+x_m$?

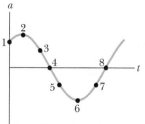

**Figure 15-21** Question 3.

**4** Which of the following relationships between the acceleration $a$ and the displacement $x$ of a particle involve SHM: (a) $a = 0.5x$, (b) $a = 400x^2$, (c) $a = -20x$, (d) $a = -3x^2$?

**5** You are to complete Fig. 15-22a so that it is a plot of velocity $v$ versus time $t$ for the spring–block oscillator that is shown in Fig. 15-22b for $t = 0$. (a) In Fig. 15-22a, at which lettered point or in what region between the points should the (vertical) $v$ axis intersect the $t$ axis? (For example, should it intersect at point $A$, or maybe in the region between points $A$ and $B$?) (b) If the block's velocity is given by $v = -v_m \sin(\omega t + \phi)$, what is the value of $\phi$? Make it positive, and if you cannot specify the value (such as $+\pi/2$ rad), then give a range of values (such as between 0 and $\pi/2$ rad).

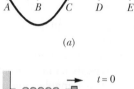

(a)

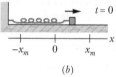

(b)

**Figure 15-22** Question 5.

**6** You are to complete Fig. 15-23a so that it is a plot of acceleration $a$ versus time $t$ for the spring–block oscillator that is shown in Fig. 15-23b for $t = 0$. (a) In Fig. 15-23a, at which lettered point or in what region between the points should the (vertical) $a$ axis intersect the $t$ axis? (For example, should it intersect at point $A$, or maybe in the region between points $A$ and $B$?) (b) If the block's acceleration is given by $a = -a_m \cos(\omega t + \phi)$, what is the value of $\phi$? Make it positive, and if you cannot specify the value (such as $+\pi/2$ rad), then give a range of values (such as between 0 and $\pi/2$).

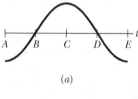

(a)

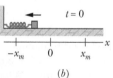

(b)

**Figure 15-23** Question 6.

**7** Figure 15-24 shows the $x(t)$ curves for three experiments involving a particular spring–box system oscillating in SHM. Rank the curves according to (a) the system's angular frequency, (b) the spring's potential energy at time $t = 0$, (c) the box's kinetic energy at $t = 0$, (d) the box's speed at $t = 0$, and (e) the box's maximum kinetic energy, greatest first.

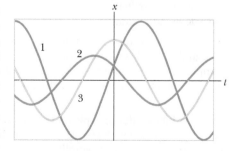

**Figure 15-24** Question 7.

**8** Figure 15-25 shows plots of the kinetic energy $K$ versus position $x$ for three harmonic oscillators that have the same mass.

Rank the plots according to (a) the corresponding spring constant and (b) the corresponding period of the oscillator, greatest first.

**9** Figure 15-26 shows three physical pendulums consisting of identical uniform spheres of the same mass that are rigidly connected by identical rods of negligible mass. Each pendulum is vertical and can pivot about suspension point $O$. Rank the pendulums according to their period of oscillation, greatest first.

**10** You are to build the oscillation transfer device shown in Fig. 15-27. It consists of two spring–block systems hanging from a flexible rod. When the spring of system 1 is stretched and then released, the resulting SHM of system 1 at frequency $f_1$ oscillates the rod. The rod then exerts a driving force on system 2, at the same frequency $f_1$. You can choose from four springs with spring constants $k$ of 1600, 1500, 1400, and 1200 N/m, and four blocks with masses $m$ of 800, 500, 400, and 200 kg. Mentally determine which spring should go with which block in each of the two systems to maximize the amplitude of oscillations in system 2.

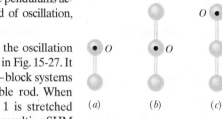

**Figure 15-25** Question 8.

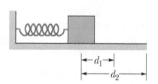

(a)          (b)          (c)

**Figure 15-26** Question 9.

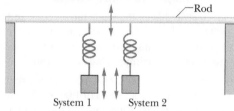

System 1          System 2

**Figure 15-27** Question 10.

**11** In Fig. 15-28, a spring–block system is put into SHM in two experiments. In the first, the block is pulled from the equilibrium position through a displacement $d_1$ and then released. In the second, it is pulled from the equilibrium position through a greater displacement $d_2$ and then released. Are the (a) amplitude, (b) period, (c) frequency, (d) maximum kinetic energy, and (e) maximum potential energy in the second experiment greater than, less than, or the same as those in the first experiment?

**Figure 15-28** Question 11.

**12** Figure 15-29 gives, for three situations, the displacements $x(t)$ of a pair of simple harmonic oscillators ($A$ and $B$) that are identical except for phase. For each pair, what phase shift (in radians and in degrees) is needed to shift the curve for $A$ to coincide with the curve for $B$? Of the many possible answers, choose the shift with the smallest absolute magnitude.

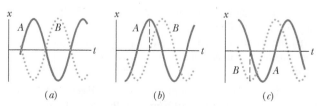

(a)          (b)          (c)

**Figure 15-29** Question 12.

# Problems

| | |
|---|---|
| **GO** Tutoring problem available (at instructor's discretion) in *WileyPLUS* and WebAssign | |
| **SSM** Worked-out solution available in Student Solutions Manual | **WWW** Worked-out solution is at |
| • – ••• Number of dots indicates level of problem difficulty | **ILW** Interactive solution is at http://www.wiley.com/college/halliday |
| Additional information available in *The Flying Circus of Physics* and at flyingcircusofphysics.com | |

## Module 15-1 Simple Harmonic Motion

•1 An object undergoing simple harmonic motion takes 0.25 s to travel from one point of zero velocity to the next such point. The distance between those points is 36 cm. Calculate the (a) period, (b) frequency, and (c) amplitude of the motion.

•2 A 0.12 kg body undergoes simple harmonic motion of amplitude 8.5 cm and period 0.20 s. (a) What is the magnitude of the maximum force acting on it? (b) If the oscillations are produced by a spring, what is the spring constant?

•3 What is the maximum acceleration of a platform that oscillates at amplitude 2.20 cm and frequency 6.60 Hz?

•4 An automobile can be considered to be mounted on four identical springs as far as vertical oscillations are concerned. The springs of a certain car are adjusted so that the oscillations have a frequency of 3.00 Hz. (a) What is the spring constant of each spring if the mass of the car is 1450 kg and the mass is evenly distributed over the springs? (b) What will be the oscillation frequency if five passengers, averaging 73.0 kg each, ride in the car with an even distribution of mass?

•5 SSM In an electric shaver, the blade moves back and forth over a distance of 2.0 mm in simple harmonic motion, with frequency 120 Hz. Find (a) the amplitude, (b) the maximum blade speed, and (c) the magnitude of the maximum blade acceleration.

•6 A particle with a mass of $1.00 \times 10^{-20}$ kg is oscillating with simple harmonic motion with a period of $1.00 \times 10^{-5}$ s and a maximum speed of $1.00 \times 10^3$ m/s. Calculate (a) the angular frequency and (b) the maximum displacement of the particle.

•7 SSM A loudspeaker produces a musical sound by means of the oscillation of a diaphragm whose amplitude is limited to 1.00 μm. (a) At what frequency is the magnitude $a$ of the diaphragm's acceleration equal to $g$? (b) For greater frequencies, is $a$ greater than or less than $g$?

•8 What is the phase constant for the harmonic oscillator with the position function $x(t)$ given in Fig. 15-30 if the position function has the form $x = x_m \cos(\omega t + \phi)$? The vertical axis scale is set by $x_s = 6.0$ cm.

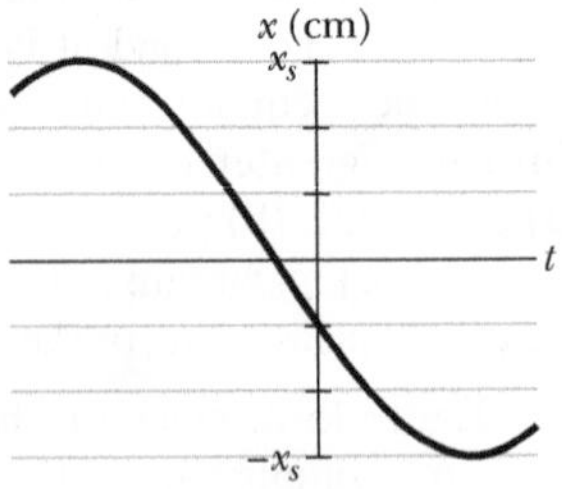

•9 The position function $x = (6.0 \text{ m}) \cos[(3\pi \text{ rad/s})t + \pi/3 \text{ rad}]$ gives the simple harmonic motion of a body. At $t = 2.0$ s, what are the (a) displacement, (b) velocity, (c) acceleration, and (d) phase of the motion? Also, what are the (e) frequency and (f) period of the motion?

**Figure 15-30** Problem 8.

•10 An oscillating block–spring system takes 0.75 s to begin repeating its motion. Find (a) the period, (b) the frequency in hertz, and (c) the angular frequency in radians per second.

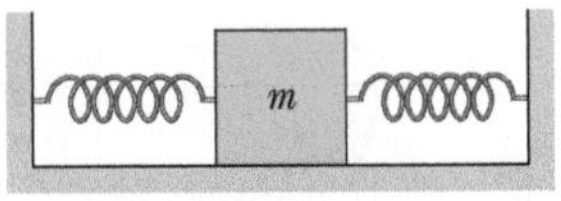

•11 In Fig. 15-31, two identical springs of spring constant 7580 N/m

**Figure 15-31** Problems 11 and 21.

are attached to a block of mass 0.245 kg. What is the frequency of oscillation on the frictionless floor?

•12 What is the phase constant for the harmonic oscillator with the velocity function $v(t)$ given in Fig. 15-32 if the position function $x(t)$ has the form $x = x_m \cos(\omega t + \phi)$? The vertical axis scale is set by $v_s = 4.0$ cm/s.

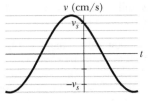

**Figure 15-32** Problem 12.

•13 SSM An oscillator consists of a block of mass 0.500 kg connected to a spring. When set into oscillation with amplitude 35.0 cm, the oscillator repeats its motion every 0.500 s. Find the (a) period, (b) frequency, (c) angular frequency, (d) spring constant, (e) maximum speed, and (f) magnitude of the maximum force on the block from the spring.

••14 A simple harmonic oscillator consists of a block of mass 2.00 kg attached to a spring of spring constant 100 N/m. When $t = 1.00$ s, the position and velocity of the block are $x = 0.129$ m and $v = 3.415$ m/s. (a) What is the amplitude of the oscillations? What were the (b) position and (c) velocity of the block at $t = 0$ s?

••15 SSM Two particles oscillate in simple harmonic motion along a common straight-line segment of length $A$. Each particle has a period of 1.5 s, but they differ in phase by $\pi/6$ rad. (a) How far apart are they (in terms of $A$) 0.50 s after the lagging particle leaves one end of the path? (b) Are they then moving in the same direction, toward each other, or away from each other?

••16 Two particles execute simple harmonic motion of the same amplitude and frequency along close parallel lines. They pass each other moving in opposite directions each time their displacement is half their amplitude. What is their phase difference?

••17 ILW An oscillator consists of a block attached to a spring ($k = 400$ N/m). At some time $t$, the position (measured from the system's equilibrium location), velocity, and acceleration of the block are $x = 0.100$ m, $v = -13.6$ m/s, and $a = -123$ m/s². Calculate (a) the frequency of oscillation, (b) the mass of the block, and (c) the amplitude of the motion.

••18 GO At a certain harbor, the tides cause the ocean surface to rise and fall a distance $d$ (from highest level to lowest level) in simple harmonic motion, with a period of 12.5 h. How long does it take for the water to fall a distance $0.250d$ from its highest level?

••19 A block rides on a piston (a squat cylindrical piece) that is moving vertically with simple harmonic motion. (a) If the SHM has period 1.0 s, at what amplitude of motion will the block and piston separate? (b) If the piston has an amplitude of 5.0 cm, what is the maximum frequency for which the block and piston will be in contact continuously?

••20 GO Figure 15-33a is a partial graph of the position function $x(t)$ for a simple harmonic oscillator with an angular frequency of

1.20 rad/s; Fig. 15-33b is a partial graph of the corresponding velocity function $v(t)$. The vertical axis scales are set by $x_s = 5.0$ cm and $v_s = 5.0$ cm/s. What is the phase constant of the SHM if the position function $x(t)$ is in the general form $x = x_m \cos(\omega t + \phi)$?

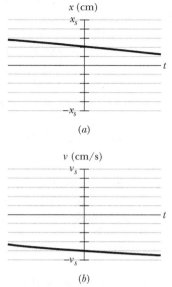

(a)

(b)

**Figure 15-33** Problem 20.

**••21 ILW** In Fig. 15-31, two springs are attached to a block that can oscillate over a frictionless floor. If the left spring is removed, the block oscillates at a frequency of 30 Hz. If, instead, the spring on the right is removed, the block oscillates at a frequency of 45 Hz. At what frequency does the block oscillate with both springs attached?

**••22 GO** Figure 15-34 shows block 1 of mass 0.200 kg sliding to the right over a frictionless elevated surface at a speed of 8.00 m/s. The block undergoes an elastic collision with stationary block 2, which is attached to a spring of spring constant 1208.5 N/m. (Assume that the spring does not affect the collision.) After the collision, block 2 oscillates in SHM with a period of 0.140 s, and block 1 slides off the opposite end of the elevated surface, landing a distance $d$ from the base of that surface after falling height $h = 4.90$ m. What is the value of $d$?

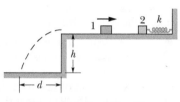

**Figure 15-34** Problem 22.

**••23 SSM WWW** A block is on a horizontal surface (a shake table) that is moving back and forth horizontally with simple harmonic motion of frequency 2.0 Hz. The coefficient of static friction between block and surface is 0.50. How great can the amplitude of the SHM be if the block is not to slip along the surface?

**•••24** In Fig. 15-35, two springs are joined and connected to a block of mass 0.245 kg that is set oscillating over a frictionless floor. The springs each have spring constant $k = 6430$ N/m. What is the frequency of the oscillations?

**Figure 15-35** Problem 24.

**•••25 GO** In Fig. 15-36, a block weighing 14.0 N, which can slide without friction on an incline at angle $\theta = 40.0°$, is connected to the top of the incline by a massless spring of unstretched length 0.450 m and spring constant 120 N/m. (a) How far from the top of the incline is the block's equilibrium point? (b) If the block is pulled slightly down the incline and released, what is the period of the resulting oscillations?

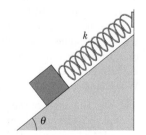

**Figure 15-36** Problem 25.

**•••26 GO** In Fig. 15-37, two blocks ($m = 1.8$ kg and $M = 10$ kg) and a spring ($k = 200$ N/m) are arranged on a horizontal, frictionless surface. The coefficient of static friction between the two blocks is 0.40. What amplitude of simple harmonic motion of the spring–blocks system puts the smaller block on the verge of slipping over the larger block?

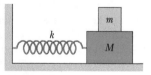

**Figure 15-37** Problem 26.

## Module 15-2    Energy in Simple Harmonic Motion

**•27 SSM** When the displacement in SHM is one-half the amplitude $x_m$, what fraction of the total energy is (a) kinetic energy and (b) potential energy? (c) At what displacement, in terms of the amplitude, is the energy of the system half kinetic energy and half potential energy?

**•28** Figure 15-38 gives the one-dimensional potential energy well for a 2.0 kg particle (the function $U(x)$ has the form $bx^2$ and the vertical axis scale is set by $U_s = 2.0$ J). (a) If the particle passes through the equilibrium position with a velocity of 85 cm/s, will it be turned back before it reaches $x = 15$ cm? (b) If yes, at what position, and if no, what is the speed of the particle at $x = 15$ cm?

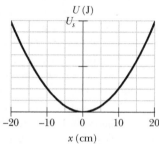

**Figure 15-38** Problem 28.

**•29 SSM** Find the mechanical energy of a block–spring system with a spring constant of 1.3 N/cm and an amplitude of 2.4 cm.

**•30** An oscillating block–spring system has a mechanical energy of 1.00 J, an amplitude of 10.0 cm, and a maximum speed of 1.20 m/s. Find (a) the spring constant, (b) the mass of the block, and (c) the frequency of oscillation.

**•31 ILW** A 5.00 kg object on a horizontal frictionless surface is attached to a spring with $k = 1000$ N/m. The object is displaced from equilibrium 50.0 cm horizontally and given an initial velocity of 10.0 m/s back toward the equilibrium position. What are (a) the motion's frequency, (b) the initial potential energy of the block–spring system, (c) the initial kinetic energy, and (d) the motion's amplitude?

**•32** Figure 15-39 shows the kinetic energy $K$ of a simple harmonic oscillator versus its position $x$. The vertical axis scale is set by $K_s = 4.0$ J. What is the spring constant?

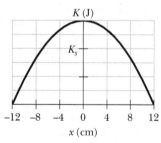

**Figure 15-39** Problem 32.

**••33 GO** A block of mass $M = 5.4$ kg, at rest on a horizontal frictionless table, is attached to a rigid support by a spring of constant $k = 6000$ N/m. A bullet of mass $m = 9.5$ g and velocity $\vec{v}$ of magnitude 630 m/s strikes and is embedded in the block (Fig. 15-40). Assuming the compression of the spring is negligible until the bullet is embedded, determine (a) the speed of the block immediately after the collision and (b) the amplitude of the resulting simple harmonic motion.

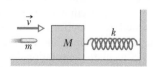

**Figure 15-40** Problem 33.

**••34** In Fig. 15-41, block 2 of mass 2.0 kg oscillates on the end of a spring in SHM with a period of 20 ms. The block's position is given by $x = (1.0\text{ cm})\cos(\omega t + \pi/2)$. Block 1 of mass 4.0 kg slides toward block 2 with a velocity of magnitude 6.0 m/s, directed along the spring's length. The two blocks undergo a completely inelastic collision at time $t = 5.0$ ms. (The duration of the collision is much less than the period of motion.) What is the amplitude of the SHM after the collision?

**Figure 15-41** Problem 34.

**••35** A 10 g particle undergoes SHM with an amplitude of 2.0 mm, a maximum acceleration of magnitude $8.0 \times 10^3$ m/s², and an unknown phase constant $\phi$. What are (a) the period of the motion, (b) the maximum speed of the particle, and (c) the total mechanical energy of the oscillator? What is the magnitude of the force on the particle when the particle is at (d) its maximum displacement and (e) half its maximum displacement?

**••36** If the phase angle for a block–spring system in SHM is $\pi/6$ rad and the block's position is given by $x = x_m \cos(\omega t + \phi)$, what is the ratio of the kinetic energy to the potential energy at time $t = 0$?

**•••37** A massless spring hangs from the ceiling with a small object attached to its lower end. The object is initially held at rest in a position $y_i$ such that the spring is at its rest length. The object is then released from $y_i$ and oscillates up and down, with its lowest position being 10 cm below $y_i$. (a) What is the frequency of the oscillation? (b) What is the speed of the object when it is 8.0 cm below the initial position? (c) An object of mass 300 g is attached to the first object, after which the system oscillates with half the original frequency. What is the mass of the first object? (d) How far below $y_i$ is the new equilibrium (rest) position with both objects attached to the spring?

### Module 15-3 An Angular Simple Harmonic Oscillator

**•38** A 95 kg solid sphere with a 15 cm radius is suspended by a vertical wire. A torque of 0.20 N·m is required to rotate the sphere through an angle of 0.85 rad and then maintain that orientation. What is the period of the oscillations that result when the sphere is then released?

**••39** SSM WWW The balance wheel of an old-fashioned watch oscillates with angular amplitude $\pi$ rad and period 0.500 s. Find (a) the maximum angular speed of the wheel, (b) the angular speed at displacement $\pi/2$ rad, and (c) the magnitude of the angular acceleration at displacement $\pi/4$ rad.

### Module 15-4 Pendulums, Circular Motion

**•40** ILW A physical pendulum consists of a meter stick that is pivoted at a small hole drilled through the stick a distance $d$ from the 50 cm mark. The period of oscillation is 2.5 s. Find $d$.

**•41** SSM In Fig. 15-42, the pendulum consists of a uniform disk with radius $r = 10.0$ cm and mass 500 g attached to a uniform rod with length $L = 500$ mm and mass 270 g. (a) Calculate the rotational inertia of the pendulum about the pivot point. (b) What is the distance between the pivot point and

**Figure 15-42** Problem 41.

the center of mass of the pendulum? (c) Calculate the period of oscillation.

**•42** Suppose that a simple pendulum consists of a small 60.0 g bob at the end of a cord of negligible mass. If the angle $\theta$ between the cord and the vertical is given by

$$\theta = (0.0800\text{ rad})\cos[(4.43\text{ rad/s})t + \phi],$$

what are (a) the pendulum's length and (b) its maximum kinetic energy?

**•43** (a) If the physical pendulum of Fig. 15-13 and the associated sample problem is inverted and suspended at point $P$, what is its period of oscillation? (b) Is the period now greater than, less than, or equal to its previous value?

**•44** A physical pendulum consists of two meter-long sticks joined together as shown in Fig. 15-43. What is the pendulum's period of oscillation about a pin inserted through point $A$ at the center of the horizontal stick?

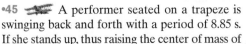

**Figure 15-43** Problem 44.

**•45** A performer seated on a trapeze is swinging back and forth with a period of 8.85 s. If she stands up, thus raising the center of mass of the *trapeze + performer* system by 35.0 cm, what will be the new period of the system? Treat *trapeze + performer* as a simple pendulum.

**•46** A physical pendulum has a center of oscillation at distance $2L/3$ from its point of suspension. Show that the distance between the point of suspension and the center of oscillation for a physical pendulum of any form is $I/mh$, where $I$ and $h$ have the meanings in Eq. 15-29 and $m$ is the mass of the pendulum.

**•47** In Fig. 15-44, a physical pendulum consists of a uniform solid disk (of radius $R = 2.35$ cm) supported in a vertical plane by a pivot located a distance $d = 1.75$ cm from the center of the disk. The disk is displaced by a small angle and released. What is the period of the resulting simple harmonic motion?

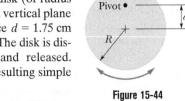

**Figure 15-44** Problem 47.

**••48** A rectangular block, with face lengths $a = 35$ cm and $b = 45$ cm, is to be suspended on a thin horizontal rod running through a narrow hole in the block. The block is then to be set swinging about the rod like a pendulum, through small angles so that it is in SHM. Figure 15-45 shows one possible position of the hole, at distance $r$ from the block's center, along a line connecting the center with a corner. (a) Plot the period versus distance $r$ along that line such that the minimum in the curve is apparent. (b) For what value of $r$ does that minimum occur? There is a line of points around the block's center for which the period of swinging has the same minimum value. (c) What shape does that line make?

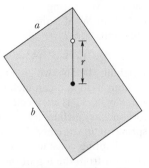

**Figure 15-45** Problem 48.

**••49** The angle of the pendulum of Fig. 15-11b is given by $\theta = \theta_m \cos[(4.44\text{ rad/s})t + \phi]$. If at $t = 0$, $\theta = 0.040$ rad and $d\theta/dt = -0.200$ rad/s, what are (a) the phase constant $\phi$ and (b) the maximum angle $\theta_m$? (*Hint:* Don't confuse the rate $d\theta/dt$ at which $\theta$ changes with the $\omega$ of the SHM.)

**••50**  A thin uniform rod (mass = 0.50 kg) swings about an axis that passes through one end of the rod and is perpendicular to the plane of the swing. The rod swings with a period of 1.5 s and an angular amplitude of 10°. (a) What is the length of the rod? (b) What is the maximum kinetic energy of the rod as it swings?

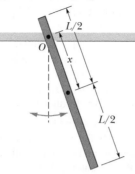

Figure 15-46  Problem 51.

**••51**  In Fig. 15-46, a stick of length $L = 1.85$ m oscillates as a physical pendulum. (a) What value of distance $x$ between the stick's center of mass and its pivot point $O$ gives the least period? (b) What is that least period?

**••52**  The 3.00 kg cube in Fig. 15-47 has edge lengths $d = 6.00$ cm and is mounted on an axle through its center. A spring ($k = 1200$ N/m) connects the cube's upper corner to a rigid wall. Initially the spring is at its rest length. If the cube is rotated 3° and released, what is the period of the resulting SHM?

Figure 15-47
Problem 52.

**••53  SSM  ILW**  In the overhead view of Fig. 15-48, a long uniform rod of mass 0.600 kg is free to rotate in a horizontal plane about a vertical axis through its center. A spring with force constant $k = 1850$ N/m is connected horizontally between one end of the rod and a fixed wall. When the rod is in equilibrium, it is parallel to the wall. What is the period of the small oscillations that result when the rod is rotated slightly and released?

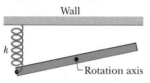

Figure 15-48  Problem 53.

**••54**  In Fig. 15-49a, a metal plate is mounted on an axle through its center of mass. A spring with $k = 2000$ N/m connects a wall with a point on the rim a distance $r = 2.5$ cm from the center of mass. Initially the spring is at its rest length. If the plate is rotated by 7° and released, it rotates about the axle in SHM, with its angular position given by Fig. 15-49b. The horizontal axis scale is set by $t_s = 20$ ms. What is the rotational inertia of the plate about its center of mass?

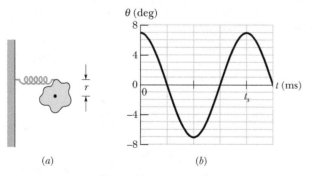

Figure 15-49  Problem 54.

**•••55**  A pendulum is formed by pivoting a long thin rod about a point on the rod. In a series of experiments, the period is measured as a function of the distance $x$ between the pivot point and the rod's center. (a) If the rod's length is $L = 2.20$ m and its mass is $m = 22.1$ g, what is the minimum period? (b) If $x$ is cho-

sen to minimize the period and then $L$ is increased, does the period increase, decrease, or remain the same? (c) If, instead, $m$ is increased without $L$ increasing, does the period increase, decrease, or remain the same?

**•••56**  In Fig. 15-50, a 2.50 kg disk of diameter $D = 42.0$ cm is supported by a rod of length $L = 76.0$ cm and negligible mass that is pivoted at its end. (a) With the massless torsion spring unconnected, what is the period of oscillation? (b) With the torsion spring connected, the rod is vertical at equilibrium. What is the torsion constant of the spring if the period of oscillation has been decreased by 0.500 s?

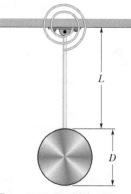

Figure 15-50  Problem 56.

## Module 15-5  Damped Simple Harmonic Motion

**•57**  The amplitude of a lightly damped oscillator decreases by 3.0% during each cycle. What percentage of the mechanical energy of the oscillator is lost in each cycle?

**•58**  For the damped oscillator system shown in Fig. 15-16, with $m = 250$ g, $k = 85$ N/m, and $b = 70$ g/s, what is the ratio of the oscillation amplitude at the end of 20 cycles to the initial oscillation amplitude?

**•59  SSM  WWW**  For the damped oscillator system shown in Fig. 15-16, the block has a mass of 1.50 kg and the spring constant is 8.00 N/m. The damping force is given by $-b(dx/dt)$, where $b = 230$ g/s. The block is pulled down 12.0 cm and released. (a) Calculate the time required for the amplitude of the resulting oscillations to fall to one-third of its initial value. (b) How many oscillations are made by the block in this time?

**••60**  The suspension system of a 2000 kg automobile "sags" 10 cm when the chassis is placed on it. Also, the oscillation amplitude decreases by 50% each cycle. Estimate the values of (a) the spring constant $k$ and (b) the damping constant $b$ for the spring and shock absorber system of one wheel, assuming each wheel supports 500 kg.

## Module 15-6  Forced Oscillations and Resonance

**•61**  For Eq. 15-45, suppose the amplitude $x_m$ is given by

$$x_m = \frac{F_m}{[m^2(\omega_d^2 - \omega^2)^2 + b^2\omega_d^2]^{1/2}},$$

where $F_m$ is the (constant) amplitude of the external oscillating force exerted on the spring by the rigid support in Fig. 15-16. At resonance, what are the (a) amplitude and (b) velocity amplitude of the oscillating object?

**•62**  Hanging from a horizontal beam are nine simple pendulums of the following lengths: (a) 0.10, (b) 0.30, (c) 0.40, (d) 0.80, (e) 1.2, (f) 2.8, (g) 3.5, (h) 5.0, and (i) 6.2 m. Suppose the beam undergoes horizontal oscillations with angular frequencies in the range from 2.00 rad/s to 4.00 rad/s. Which of the pendulums will be (strongly) set in motion?

**••63**  A 1000 kg car carrying four 82 kg people travels over a "washboard" dirt road with corrugations 4.0 m apart. The car bounces with maximum amplitude when its speed is 16 km/h. When the car stops, and the people get out, by how much does the car body rise on its suspension?

**Additional Problems**

**64** ✈ Although California is known for earthquakes, it has large regions dotted with precariously balanced rocks that would be easily toppled by even a mild earthquake. Apparently no major earthquakes have occurred in those regions. If an earthquake were to put such a rock into sinusoidal oscillation (parallel to the ground) with a frequency of 2.2 Hz, an oscillation amplitude of 1.0 cm would cause the rock to topple. What would be the magnitude of the maximum acceleration of the oscillation, in terms of $g$?

**65** A loudspeaker diaphragm is oscillating in simple harmonic motion with a frequency of 440 Hz and a maximum displacement of 0.75 mm. What are the (a) angular frequency, (b) maximum speed, and (c) magnitude of the maximum acceleration?

**66** A uniform spring with $k = 8600$ N/m is cut into pieces 1 and 2 of unstretched lengths $L_1 = 7.0$ cm and $L_2 = 10$ cm. What are (a) $k_1$ and (b) $k_2$? A block attached to the original spring as in Fig. 15-7 oscillates at 200 Hz. What is the oscillation frequency of the block attached to (c) piece 1 and (d) piece 2?

**67** ⓖⓞ In Fig. 15-51, three 10 000 kg ore cars are held at rest on a mine railway using a cable that is parallel to the rails, which are inclined at angle $\theta = 30°$. The cable stretches 15 cm just before the coupling between the two lower cars breaks, detaching the lowest car. Assuming that the cable obeys Hooke's law, find the (a) frequency and (b) amplitude of the resulting oscillations of the remaining two cars.

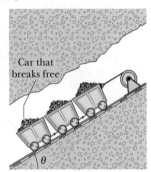

**Figure 15-51** Problem 67.

**68** A 2.00 kg block hangs from a spring. A 300 g body hung below the block stretches the spring 2.00 cm farther. (a) What is the spring constant? (b) If the 300 g body is removed and the block is set into oscillation, find the period of the motion.

**69** SSM In the engine of a locomotive, a cylindrical piece known as a piston oscillates in SHM in a cylinder head (cylindrical chamber) with an angular frequency of 180 rev/min. Its stroke (twice the amplitude) is 0.76 m. What is its maximum speed?

**70** ⓖⓞ A wheel is free to rotate about its fixed axle. A spring is attached to one of its spokes a distance $r$ from the axle, as shown in Fig. 15-52. (a) Assuming that the wheel is a hoop of mass $m$ and radius $R$, what is the angular frequency $\omega$ of small oscillations of this system in terms of $m$, $R$, $r$, and the spring constant $k$? What is $\omega$ if (b) $r = R$ and (c) $r = 0$?

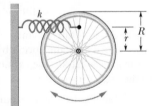

**Figure 15-52** Problem 70.

**71** A 50.0 g stone is attached to the bottom of a vertical spring and set vibrating. If the maximum speed of the stone is 15.0 cm/s and the period is 0.500 s, find the (a) spring constant of the spring, (b) amplitude of the motion, and (c) frequency of oscillation.

**72** A uniform circular disk whose radius $R$ is 12.6 cm is suspended as a physical pendulum from a point on its rim. (a) What is its period? (b) At what radial distance $r < R$ is there a pivot point that gives the same period?

**73** SSM A vertical spring stretches 9.6 cm when a 1.3 kg block is hung from its end. (a) Calculate the spring constant. This block is then displaced an additional 5.0 cm downward and released from rest. Find the (b) period, (c) frequency, (d) amplitude, and (e) maximum speed of the resulting SHM.

**74** A massless spring with spring constant 19 N/m hangs vertically. A body of mass 0.20 kg is attached to its free end and then released. Assume that the spring was unstretched before the body was released. Find (a) how far below the initial position the body descends, and the (b) frequency and (c) amplitude of the resulting SHM.

**75** A 4.00 kg block is suspended from a spring with $k = 500$ N/m. A 50.0 g bullet is fired into the block from directly below with a speed of 150 m/s and becomes embedded in the block. (a) Find the amplitude of the resulting SHM. (b) What percentage of the original kinetic energy of the bullet is transferred to mechanical energy of the oscillator?

**76** A 55.0 g block oscillates in SHM on the end of a spring with $k = 1500$ N/m according to $x = x_m \cos(\omega t + \phi)$. How long does the block take to move from position $+0.800x_m$ to (a) position $+0.600x_m$ and (b) position $-0.800x_m$?

**77** Figure 15-53 gives the position of a 20 g block oscillating in SHM on the end of a spring. The horizontal axis scale is set by $t_s = 40.0$ ms. What are (a) the maximum kinetic energy of the block and (b) the number of times per second that maximum is reached? (*Hint:* Measuring a slope will probably not be very accurate. Find another approach.)

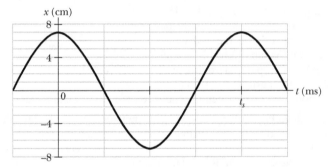

**Figure 15-53** Problems 77 and 78.

**78** Figure 15-53 gives the position $x(t)$ of a block oscillating in SHM on the end of a spring ($t_s = 40.0$ ms). What are (a) the speed and (b) the magnitude of the radial acceleration of a particle in the corresponding uniform circular motion?

**79** Figure 15-54 shows the kinetic energy $K$ of a simple pendulum versus its angle $\theta$ from the vertical. The vertical axis scale is set by $K_s = 10.0$ mJ. The pendulum bob has mass 0.200 kg. What is the length of the pendulum?

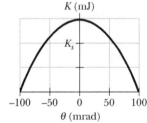

**Figure 15-54** Problem 79.

**80** A block is in SHM on the end of a spring, with position given by $x = x_m \cos(\omega t + \phi)$. If $\phi = \pi/5$ rad, then at $t = 0$ what percentage of the total mechanical energy is potential energy?

**81** A simple harmonic oscillator consists of a 0.50 kg block attached to a spring. The block slides back and forth along a straight line on a frictionless surface with equilibrium point $x = 0$. At $t = 0$ the block is at $x = 0$ and moving in the positive $x$ direction. A graph of the magnitude of the net force $\vec{F}$ on the block as a function of its position is shown in Fig. 15-55. The vertical scale is set by $F_s = 75.0$ N. What are (a) the amplitude and (b) the period of the motion, (c) the magnitude of the maximum acceleration, and (d) the maximum kinetic energy?

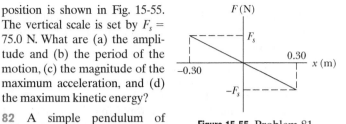

**Figure 15-55** Problem 81.

**82** A simple pendulum of length 20 cm and mass 5.0 g is suspended in a race car traveling with constant speed 70 m/s around a circle of radius 50 m. If the pendulum undergoes small oscillations in a radial direction about its equilibrium position, what is the frequency of oscillation?

**83** The scale of a spring balance that reads from 0 to 15.0 kg is 12.0 cm long. A package suspended from the balance is found to oscillate vertically with a frequency of 2.00 Hz. (a) What is the spring constant? (b) How much does the package weigh?

**84** A 0.10 kg block oscillates back and forth along a straight line on a frictionless horizontal surface. Its displacement from the origin is given by

$$x = (10 \text{ cm}) \cos[(10 \text{ rad/s})t + \pi/2 \text{ rad}].$$

(a) What is the oscillation frequency? (b) What is the maximum speed acquired by the block? (c) At what value of $x$ does this occur? (d) What is the magnitude of the maximum acceleration of the block? (e) At what value of $x$ does this occur? (f) What force, applied to the block by the spring, results in the given oscillation?

**85** The end point of a spring oscillates with a period of 2.0 s when a block with mass $m$ is attached to it. When this mass is increased by 2.0 kg, the period is found to be 3.0 s. Find $m$.

**86** The tip of one prong of a tuning fork undergoes SHM of frequency 1000 Hz and amplitude 0.40 mm. For this tip, what is the magnitude of the (a) maximum acceleration, (b) maximum velocity, (c) acceleration at tip displacement 0.20 mm, and (d) velocity at tip displacement 0.20 mm?

**87** A flat uniform circular disk has a mass of 3.00 kg and a radius of 70.0 cm. It is suspended in a horizontal plane by a vertical wire attached to its center. If the disk is rotated 2.50 rad about the wire, a torque of 0.0600 N·m is required to maintain that orientation. Calculate (a) the rotational inertia of the disk about the wire, (b) the torsion constant, and (c) the angular frequency of this torsion pendulum when it is set oscillating.

**88** A block weighing 20 N oscillates at one end of a vertical spring for which $k = 100$ N/m; the other end of the spring is attached to a ceiling. At a certain instant the spring is stretched 0.30 m beyond its relaxed length (the length when no object is attached) and the block has zero velocity. (a) What is the net force on the block at this instant? What are the (b) amplitude and (c) period of the resulting simple harmonic motion? (d) What is the maximum kinetic energy of the block as it oscillates?

**89** A 3.0 kg particle is in simple harmonic motion in one dimension and moves according to the equation

$$x = (5.0 \text{ m}) \cos[(\pi/3 \text{ rad/s})t - \pi/4 \text{ rad}],$$

with $t$ in seconds. (a) At what value of $x$ is the potential energy of the particle equal to half the total energy? (b) How long does the particle take to move to this position $x$ from the equilibrium position?

**90** A particle executes linear SHM with frequency 0.25 Hz about the point $x = 0$. At $t = 0$, it has displacement $x = 0.37$ cm and zero velocity. For the motion, determine the (a) period, (b) angular frequency, (c) amplitude, (d) displacement $x(t)$, (e) velocity $v(t)$, (f) maximum speed, (g) magnitude of the maximum acceleration, (h) displacement at $t = 3.0$ s, and (i) speed at $t = 3.0$ s.

**91** **SSM** What is the frequency of a simple pendulum 2.0 m long (a) in a room, (b) in an elevator accelerating upward at a rate of 2.0 m/s², and (c) in free fall?

**92** A grandfather clock has a pendulum that consists of a thin brass disk of radius $r = 15.00$ cm and mass 1.000 kg that is attached to a long thin rod of negligible mass. The pendulum swings freely about an axis perpendicular to the rod and through the end of the rod opposite the disk, as shown in Fig. 15-56. If the pendulum is to have a period of 2.000 s for small oscillations at a place where $g = 9.800$ m/s², what must be the rod length $L$ to the nearest tenth of a millimeter?

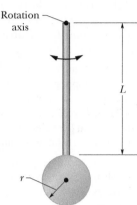

**Figure 15-56** Problem 92.

**93** A 4.00 kg block hangs from a spring, extending it 16.0 cm from its unstretched position. (a) What is the spring constant? (b) The block is removed, and a 0.500 kg body is hung from the same spring. If the spring is then stretched and released, what is its period of oscillation?

**94** What is the phase constant for SMH with $a(t)$ given in Fig. 15-57 if the position function $x(t)$ has the form $x = x_m \cos(\omega t + \phi)$ and $a_s = 4.0$ m/s²?

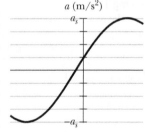

**Figure 15-57** Problem 94.

**95** An engineer has an odd-shaped 10 kg object and needs to find its rotational inertia about an axis through its center of mass. The object is supported on a wire stretched along the desired axis. The wire has a torsion constant $\kappa = 0.50$ N·m. If this torsion pendulum oscillates through 20 cycles in 50 s, what is the rotational inertia of the object?

**96** A spider can tell when its web has captured, say, a fly because the fly's thrashing causes the web threads to oscillate. A spider can even determine the size of the fly by the frequency of the oscillations. Assume that a fly oscillates on the *capture thread* on which it is caught like a block on a spring. What is the ratio of oscillation frequency for a fly with mass $m$ to a fly with mass $2.5m$?

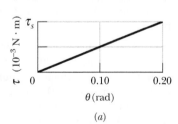

**97** A torsion pendulum consists of a metal disk with a wire running through its center and soldered in place. The wire is mounted vertically on clamps and pulled taut. Figure 15-58a gives the magnitude $\tau$ of the torque

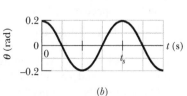

**Figure 15-58** Problem 97.

needed to rotate the disk about its center (and thus twist the wire) versus the rotation angle $\theta$. The vertical axis scale is set by $\tau_s = 4.0 \times 10^{-3}$ N·m. The disk is rotated to $\theta = 0.200$ rad and then released. Figure 15-58$b$ shows the resulting oscillation in terms of angular position $\theta$ versus time $t$. The horizontal axis scale is set by $t_s = 0.40$ s. (a) What is the rotational inertia of the disk about its center? (b) What is the maximum angular speed $d\theta/dt$ of the disk? (*Caution:* Do not confuse the (constant) angular frequency of the SHM with the (varying) angular speed of the rotating disk, even though they usually have the same symbol $\omega$. *Hint:* The potential energy $U$ of a torsion pendulum is equal to $\frac{1}{2}\kappa\theta^2$, analogous to $U = \frac{1}{2}kx^2$ for a spring.)

**98** When a 20 N can is hung from the bottom of a vertical spring, it causes the spring to stretch 20 cm. (a) What is the spring constant? (b) This spring is now placed horizontally on a frictionless table. One end of it is held fixed, and the other end is attached to a 5.0 N can. The can is then moved (stretching the spring) and released from rest. What is the period of the resulting oscillation?

**99** For a simple pendulum, find the angular amplitude $\theta_m$ at which the restoring torque required for simple harmonic motion deviates from the actual restoring torque by 1.0%. (See "Trigonometric Expansions" in Appendix E.)

**100** In Fig. 15-59, a solid cylinder attached to a horizontal spring ($k = 3.00$ N/m) rolls without slipping along a horizontal surface. If the system is released from rest when the spring is stretched by 0.250 m, find (a) the translational kinetic energy

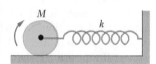

**Figure 15-59** Problem 100.

and (b) the rotational kinetic energy of the cylinder as it passes through the equilibrium position. (c) Show that under these conditions the cylinder's center of mass executes simple harmonic motion with period

$$T = 2\pi\sqrt{\frac{3M}{2k}},$$

where $M$ is the cylinder mass. (*Hint:* Find the time derivative of the total mechanical energy.)

**101** SSM A 1.2 kg block sliding on a horizontal frictionless surface is attached to a horizontal spring with $k = 480$ N/m. Let $x$ be the displacement of the block from the position at which the spring is unstretched. At $t = 0$ the block passes through $x = 0$ with a speed of 5.2 m/s in the positive $x$ direction. What are the (a) frequency and (b) amplitude of the block's motion? (c) Write an expression for $x$ as a function of time.

**102** A simple harmonic oscillator consists of an 0.80 kg block attached to a spring ($k = 200$ N/m). The block slides on a horizontal frictionless surface about the equilibrium point $x = 0$ with a total mechanical energy of 4.0 J. (a) What is the amplitude of the oscillation? (b) How many oscillations does the block complete in 10 s? (c) What is the maximum kinetic energy attained by the block? (d) What is the speed of the block at $x = 0.15$ m?

**103** A block sliding on a horizontal frictionless surface is attached to a horizontal spring with a spring constant of 600 N/m. The block executes SHM about its equilibrium position with a period of 0.40 s and an amplitude of 0.20 m. As the block slides through its equilibrium position, a 0.50 kg putty wad is dropped

vertically onto the block. If the putty wad sticks to the block, determine (a) the new period of the motion and (b) the new amplitude of the motion.

**104** A damped harmonic oscillator consists of a block ($m = 2.00$ kg), a spring ($k = 10.0$ N/m), and a damping force ($F = -bv$). Initially, it oscillates with an amplitude of 25.0 cm; because of the damping, the amplitude falls to three-fourths of this initial value at the completion of four oscillations. (a) What is the value of $b$? (b) How much energy has been "lost" during these four oscillations?

**105** A block weighing 10.0 N is attached to the lower end of a vertical spring ($k = 200.0$ N/m), the other end of which is attached to a ceiling. The block oscillates vertically and has a kinetic energy of 2.00 J as it passes through the point at which the spring is unstretched. (a) What is the period of the oscillation? (b) Use the law of conservation of energy to determine the maximum distance the block moves both above and below the point at which the spring is unstretched. (These are not necessarily the same.) (c) What is the amplitude of the oscillation? (d) What is the maximum kinetic energy of the block as it oscillates?

**106** A simple harmonic oscillator consists of a block attached to a spring with $k = 200$ N/m. The block slides on a frictionless surface, with equilibrium point $x = 0$ and amplitude 0.20 m. A graph of the block's velocity $v$ as a function of time $t$ is shown in Fig. 15-60. The horizontal scale is set by $t_s = 0.20$ s. What are (a) the period of the SHM, (b) the block's mass, (c) its displacement at $t = 0$, (d) its acceleration at $t = 0.10$ s, and (e) its maximum kinetic energy?

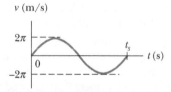

**Figure 15-60** Problem 106.

**107** The vibration frequencies of atoms in solids at normal temperatures are of the order of $10^{13}$ Hz. Imagine the atoms to be connected to one another by springs. Suppose that a single silver atom in a solid vibrates with this frequency and that all the other atoms are at rest. Compute the effective spring constant. One mole of silver (6.02 × $10^{23}$ atoms) has a mass of 108 g.

**108** Figure 15-61 shows that if we hang a block on the end of a spring with spring constant $k$, the spring is stretched by distance $h = 2.0$ cm. If we pull down on the block a short distance and then release it, it oscillates vertically with a certain frequency. What length must a simple pendulum have to swing with that frequency?

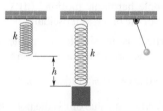

**Figure 15-61** Problem 108.

**109** The physical pendulum in Fig. 15-62 has two possible pivot points $A$ and $B$. Point $A$ has a fixed position but $B$ is adjustable along the length of the pendulum as indicated by the scaling. When suspended from $A$, the pendulum has a period of $T = 1.80$ s. The pendulum is then suspended from $B$, which is moved until the pendulum again has that period. What is the distance $L$ between $A$ and $B$?

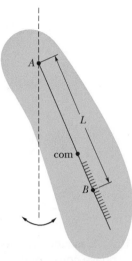

**Figure 15-62** Problem 109.

**110** A common device for entertaining a toddler is a *jump seat* that hangs from the horizontal portion of a doorframe via elastic cords (Fig. 15-63). Assume that only one cord is on each side in spite of the more realistic arrangement shown. When a child is placed in the seat, they both descend by a distance $d_s$ as the cords stretch (treat them as springs). Then the seat is pulled down an extra distance $d_m$ and released, so that the child oscillates vertically, like a block on the end of a spring. Suppose you are the safety engineer for the manufacturer of the seat. You do not want the magnitude of the child's acceleration to exceed $0.20g$ for fear of hurting the child's neck. If $d_m = 10$ cm, what value of $d_s$ corresponds to that acceleration magnitude?

**Figure 15-63** Problem 110.

**111** A 2.0 kg block executes SHM while attached to a horizontal spring of spring constant 200 N/m. The maximum speed of the block as it slides on a horizontal frictionless surface is 3.0 m/s. What are (a) the amplitude of the block's motion, (b) the magnitude of its maximum acceleration, and (c) the magnitude of its minimum acceleration? (d) How long does the block take to complete 7.0 cycles of its motion?

**112** In Fig. 15-64, a 2500 kg demolition ball swings from the end of a crane. The length of the swinging segment of cable is 17 m. (a) Find the period of the swinging, assuming that the system can be treated as a simple pendulum. (b) Does the period depend on the ball's mass?

**Figure 15-64** Problem 112.

**113** 🡒 The center of oscillation of a physical pendulum has this interesting property: If an impulse (assumed horizontal and in the plane of oscillation) acts at the center of oscillation, no oscillations are felt at the point of support. Baseball players (and players of many other sports) know that unless the ball hits the bat at this point (called the "sweet spot" by athletes), the oscillations due to the impact will sting their hands. To prove this property, let the stick in Fig. 15-13$a$ simulate a baseball bat. Suppose that a horizontal force $\vec{F}$ (due to impact with the ball) acts toward the right at $P$, the center of oscillation. The batter is assumed to hold the bat at $O$, the pivot point of the stick. (a) What acceleration does the point $O$ undergo as a result of $\vec{F}$? (b) What angular acceleration is produced by $\vec{F}$ about the center of mass of the stick? (c) As a result of the angular acceleration in (b), what linear acceleration does point $O$ undergo? (d) Considering the magnitudes and directions of the accelerations in (a) and (c), convince yourself that $P$ is indeed the "sweet spot."

**114** A (hypothetical) large slingshot is stretched 2.30 m to launch a 170 g projectile with speed sufficient to escape from Earth (11.2 km/s). Assume the elastic bands of the slingshot obey Hooke's law. (a) What is the spring constant of the device if all the elastic potential energy is converted to kinetic energy? (b) Assume that an average person can exert a force of 490 N. How many people are required to stretch the elastic bands?

**115** What is the length of a simple pendulum whose full swing from left to right and then back again takes 3.2 s?

**116** A 2.0 kg block is attached to the end of a spring with a spring constant of 350 N/m and forced to oscillate by an applied force $F = (15$ N$) \sin(\omega_d t)$, where $\omega_d = 35$ rad/s. The damping constant is $b = 15$ kg/s. At $t = 0$, the block is at rest with the spring at its rest length. (a) Use numerical integration to plot the displacement of the block for the first 1.0 s. Use the motion near the end of the 1.0 s interval to estimate the amplitude, period, and angular frequency. Repeat the calculation for (b) $\omega_d = \sqrt{k/m}$ and (c) $\omega_d = 20$ rad/s.

# Waves-I

## 16-1 TRANSVERSE WAVES

### Learning Objectives

*After reading this module, you should be able to . . .*

**16.01** Identify the three main types of waves.

**16.02** Distinguish between transverse waves and longitudinal waves.

**16.03** Given a displacement function for a traverse wave, determine amplitude $y_m$, angular wave number $k$, angular frequency $\omega$, phase constant $\phi$, and direction of travel, and calculate the phase $kx \pm \omega t + \phi$ and the displacement at any given time and position.

**16.04** Given a displacement function for a traverse wave, calculate the time between two given displacements.

**16.05** Sketch a graph of a transverse wave as a function of position, identifying amplitude $y_m$, wavelength $\lambda$, where the slope is greatest, where it is zero, and where the string elements have positive velocity, negative velocity, and zero velocity.

**16.06** Given a graph of displacement versus time for a transverse wave, determine amplitude $y_m$ and period $T$.

**16.07** Describe the effect on a transverse wave of changing phase constant $\phi$.

**16.08** Apply the relation between the wave speed $v$, the distance traveled by the wave, and the time required for that travel.

**16.09** Apply the relationships between wave speed $v$, angular frequency $\omega$, angular wave number $k$, wavelength $\lambda$, period $T$, and frequency $f$.

**16.10** Describe the motion of a string element as a transverse wave moves through its location, and identify when its transverse speed is zero and when it is maximum.

**16.11** Calculate the transverse velocity $u(t)$ of a string element as a transverse wave moves through its location.

**16.12** Calculate the transverse acceleration $a(t)$ of a string element as a transverse wave moves through its location.

**16.13** Given a graph of displacement, transverse velocity, or transverse acceleration, determine the phase constant $\phi$.

### Key Ideas

● Mechanical waves can exist only in material media and are governed by Newton's laws. Transverse mechanical waves, like those on a stretched string, are waves in which the particles of the medium oscillate perpendicular to the wave's direction of travel. Waves in which the particles of the medium oscillate parallel to the wave's direction of travel are longitudinal waves.

● A sinusoidal wave moving in the positive direction of an $x$ axis has the mathematical form

$$y(x, t) = y_m \sin(kx - \omega t),$$

where $y_m$ is the amplitude (magnitude of the maximum displacement) of the wave, $k$ is the angular wave number, $\omega$ is the angular frequency, and $kx - \omega t$ is the phase. The wavelength $\lambda$ is related to $k$ by

$$k = \frac{2\pi}{\lambda}.$$

● The period $T$ and frequency $f$ of the wave are related to $\omega$ by

$$\frac{\omega}{2\pi} = f = \frac{1}{T}.$$

● The wave speed $v$ (the speed of the wave along the string) is related to these other parameters by

$$v = \frac{\omega}{k} = \frac{\lambda}{T} = \lambda f.$$

● Any function of the form

$$y(x, t) = h(kx \pm \omega t)$$

can represent a traveling wave with a wave speed as given above and a wave shape given by the mathematical form of $h$. The plus sign denotes a wave traveling in the negative direction of the $x$ axis, and the minus sign a wave traveling in the positive direction.

## What Is Physics?

One of the primary subjects of physics is waves. To see how important waves are in the modern world, just consider the music industry. Every piece of music you hear, from some retro-punk band playing in a campus dive to the most eloquent concerto playing on the web, depends on performers producing waves and your detecting those waves. In between production and detection, the information carried by the waves might need to be transmitted (as in a live performance on the web) or recorded and then reproduced (as with CDs, DVDs, or the other devices currently being developed in engineering labs worldwide). The financial importance of controlling music waves is staggering, and the rewards to engineers who develop new control techniques can be rich.

This chapter focuses on waves traveling along a stretched string, such as on a guitar. The next chapter focuses on sound waves, such as those produced by a guitar string being played. Before we do all this, though, our first job is to classify the countless waves of the everyday world into basic types.

## Types of Waves

Waves are of three main types:

1. *Mechanical waves.* These waves are most familiar because we encounter them almost constantly; common examples include water waves, sound waves, and seismic waves. All these waves have two central features: They are governed by Newton's laws, and they can exist only within a material medium, such as water, air, and rock.

2. *Electromagnetic waves.* These waves are less familiar, but you use them constantly; common examples include visible and ultraviolet light, radio and television waves, microwaves, x rays, and radar waves. These waves require no material medium to exist. Light waves from stars, for example, travel through the vacuum of space to reach us. All electromagnetic waves travel through a vacuum at the same speed $c = 299\ 792\ 458$ m/s.

3. *Matter waves.* Although these waves are commonly used in modern technology, they are probably very unfamiliar to you. These waves are associated with electrons, protons, and other fundamental particles, and even atoms and molecules. Because we commonly think of these particles as constituting matter, such waves are called matter waves.

Much of what we discuss in this chapter applies to waves of all kinds. However, for specific examples we shall refer to mechanical waves.

## Transverse and Longitudinal Waves

A wave sent along a stretched, taut string is the simplest mechanical wave. If you give one end of a stretched string a single up-and-down jerk, a wave in the form of a single *pulse* travels along the string. This pulse and its motion can occur because the string is under tension. When you pull your end of the string upward, it begins to pull upward on the adjacent section of the string via tension between the two sections. As the adjacent section moves upward, it begins to pull the next section upward, and so on. Meanwhile, you have pulled down on your end of the string. As each section moves upward in turn, it begins to be pulled back downward by neighboring sections that are already on the way down. The net result is that a distortion in the string's shape (a pulse, as in Fig. 16-1a) moves along the string at some velocity $\vec{v}$.

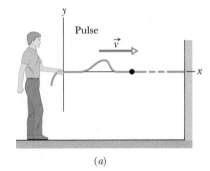

(a)

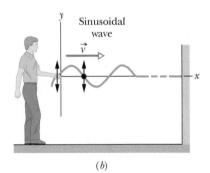

(b)

**Figure 16-1** (a) A single pulse is sent along a stretched string. A typical string element (marked with a dot) moves up once and then down as the pulse passes. The element's motion is perpendicular to the wave's direction of travel, so the pulse is a *transverse wave.* (b) A sinusoidal wave is sent along the string. A typical string element moves up and down continuously as the wave passes. This too is a transverse wave.

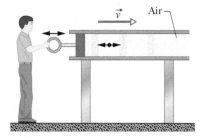

**Figure 16-2** A sound wave is set up in an air-filled pipe by moving a piston back and forth. Because the oscillations of an element of the air (represented by the dot) are parallel to the direction in which the wave travels, the wave is a *longitudinal wave.*

If you move your hand up and down in continuous simple harmonic motion, a continuous wave travels along the string at velocity $\vec{v}$. Because the motion of your hand is a sinusoidal function of time, the wave has a sinusoidal shape at any given instant, as in Fig. 16-1*b*; that is, the wave has the shape of a sine curve or a cosine curve.

We consider here only an "ideal" string, in which no friction-like forces within the string cause the wave to die out as it travels along the string. In addition, we assume that the string is so long that we need not consider a wave rebounding from the far end.

One way to study the waves of Fig. 16-1 is to monitor the **wave forms** (shapes of the waves) as they move to the right. Alternatively, we could monitor the motion of an element of the string as the element oscillates up and down while a wave passes through it. We would find that the displacement of every such oscillating string element is *perpendicular* to the direction of travel of the wave, as indicated in Fig. 16-1*b*. This motion is said to be **transverse,** and the wave is said to be a **transverse wave.**

***Longitudinal Waves.*** Figure 16-2 shows how a sound wave can be produced by a piston in a long, air-filled pipe. If you suddenly move the piston rightward and then leftward, you can send a pulse of sound along the pipe. The rightward motion of the piston moves the elements of air next to it rightward, changing the air pressure there. The increased air pressure then pushes rightward on the elements of air somewhat farther along the pipe. Moving the piston leftward reduces the air pressure next to it. As a result, first the elements nearest the piston and then farther elements move leftward. Thus, the motion of the air and the change in air pressure travel rightward along the pipe as a pulse.

If you push and pull on the piston in simple harmonic motion, as is being done in Fig. 16-2, a sinusoidal wave travels along the pipe. Because the motion of the elements of air is parallel to the direction of the wave's travel, the motion is said to be **longitudinal,** and the wave is said to be a **longitudinal wave.** In this chapter we focus on transverse waves, and string waves in particular; in Chapter 17 we focus on longitudinal waves, and sound waves in particular.

Both a transverse wave and a longitudinal wave are said to be **traveling waves** because they both travel from one point to another, as from one end of the string to the other end in Fig. 16-1 and from one end of the pipe to the other end in Fig. 16-2. Note that it is the wave that moves from end to end, not the material (string or air) through which the wave moves.

## Wavelength and Frequency

To completely describe a wave on a string (and the motion of any element along its length), we need a function that gives the shape of the wave. This means that we need a relation in the form

$$y = h(x, t), \tag{16-1}$$

in which $y$ is the transverse displacement of any string element as a function $h$ of the time $t$ and the position $x$ of the element along the string. In general, a sinusoidal shape like the wave in Fig. 16-1*b* can be described with $h$ being either a sine or cosine function; both give the same general shape for the wave. In this chapter we use the sine function.

***Sinusoidal Function.*** Imagine a sinusoidal wave like that of Fig. 16-1*b* traveling in the positive direction of an $x$ axis. As the wave sweeps through succeeding elements (that is, very short sections) of the string, the elements oscillate parallel to the $y$ axis. At time $t$, the displacement $y$ of the element located at position $x$ is given by

$$y(x, t) = y_m \sin(kx - \omega t). \tag{16-2}$$

Because this equation is written in terms of position $x$, it can be used to find the displacements of all the elements of the string as a function of time. Thus, it can tell us the shape of the wave at any given time.

The names of the quantities in Eq. 16-2 are displayed in Fig. 16-3 and defined next. Before we discuss them, however, let us examine Fig. 16-4, which shows five "snapshots" of a sinusoidal wave traveling in the positive direction of an $x$ axis. The movement of the wave is indicated by the rightward progress of the short arrow pointing to a high point of the wave. From snapshot to snapshot, the short arrow moves to the right with the wave shape, but the string moves *only* parallel to the $y$ axis. To see that, let us follow the motion of the red-dyed string element at $x = 0$. In the first snapshot (Fig. 16-4$a$), this element is at displacement $y = 0$. In the next snapshot, it is at its extreme downward displacement because a *valley* (or extreme low point) of the wave is passing through it. It then moves back up through $y = 0$. In the fourth snapshot, it is at its extreme upward displacement because a *peak* (or extreme high point) of the wave is passing through it. In the fifth snapshot, it is again at $y = 0$, having completed one full oscillation.

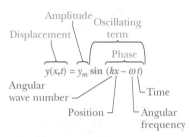

Figure 16-3 The names of the quantities in Eq. 16-2, for a transverse sinusoidal wave.

### Amplitude and Phase

The **amplitude** $y_m$ of a wave, such as that in Fig. 16-4 , is the magnitude of the maximum displacement of the elements from their equilibrium positions as the wave passes through them. (The subscript $m$ stands for maximum.) Because $y_m$ is a magnitude, it is always a positive quantity, even if it is measured downward instead of upward as drawn in Fig. 16-4$a$.

The **phase** of the wave is the *argument* $kx - \omega t$ of the sine in Eq. 16-2. As the wave sweeps through a string element at a particular position $x$, the phase changes linearly with time $t$. This means that the sine also changes, oscillating between $+1$ and $-1$. Its extreme positive value $(+1)$ corresponds to a peak of the wave moving through the element; at that instant the value of $y$ at position $x$ is $y_m$. Its extreme negative value $(-1)$ corresponds to a valley of the wave moving through the element; at that instant the value of $y$ at position $x$ is $-y_m$. Thus, the sine function and the time-dependent phase of a wave correspond to the oscillation of a string element, and the amplitude of the wave determines the extremes of the element's displacement.

*Caution:* When evaluating the phase, rounding off the numbers before you evaluate the sine function can throw of the calculation considerably.

### Wavelength and Angular Wave Number

The **wavelength** $\lambda$ of a wave is the distance (parallel to the direction of the wave's travel) between repetitions of the shape of the wave (or *wave shape*). A typical wavelength is marked in Fig. 16-4$a$, which is a snapshot of the wave at time $t = 0$. At that time, Eq. 16-2 gives, for the description of the wave shape,

$$y(x, 0) = y_m \sin kx. \tag{16-3}$$

By definition, the displacement $y$ is the same at both ends of this wavelength—that is, at $x = x_1$ and $x = x_1 + \lambda$. Thus, by Eq. 16-3,

$$y_m \sin kx_1 = y_m \sin k(x_1 + \lambda)$$
$$= y_m \sin(kx_1 + k\lambda). \tag{16-4}$$

A sine function begins to repeat itself when its angle (or argument) is increased by $2\pi$ rad, so in Eq. 16-4 we must have $k\lambda = 2\pi$, or

$$k = \frac{2\pi}{\lambda} \quad \text{(angular wave number).} \tag{16-5}$$

We call $k$ the **angular wave number** of the wave; its SI unit is the radian per meter, or the inverse meter. (Note that the symbol $k$ here does *not* represent a spring constant as previously.)

Notice that the wave in Fig. 16-4 moves to the right by $\frac{1}{4}\lambda$ from one snapshot to the next. Thus, by the fifth snapshot, it has moved to the right by $1\lambda$.

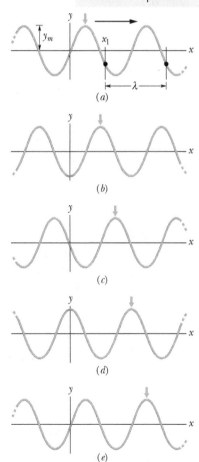

Watch this spot in this series of snapshots.

Figure 16-4 Five "snapshots" of a string wave traveling in the positive direction of an $x$ axis. The amplitude $y_m$ is indicated. A typical wavelength $\lambda$, measured from an arbitrary position $x_1$, is also indicated.

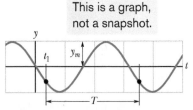

**Figure 16-5** A graph of the displacement of the string element at $x = 0$ as a function of time, as the sinusoidal wave of Fig. 16-4 passes through the element. The amplitude $y_m$ is indicated. A typical period $T$, measured from an arbitrary time $t_1$, is also indicated.

### Period, Angular Frequency, and Frequency

Figure 16-5 shows a graph of the displacement $y$ of Eq. 16-2 versus time $t$ at a certain position along the string, taken to be $x = 0$. If you were to monitor the string, you would see that the single element of the string at that position moves up and down in simple harmonic motion given by Eq. 16-2 with $x = 0$:

$$y(0, t) = y_m \sin(-\omega t)$$
$$= -y_m \sin \omega t \quad (x = 0). \quad (16\text{-}6)$$

Here we have made use of the fact that $\sin(-\alpha) = -\sin \alpha$, where $\alpha$ is any angle. Figure 16-5 is a graph of this equation, with displacement plotted versus time; it *does not* show the shape of the wave. (Figure 16-4 shows the shape and is a picture of reality; Fig. 16-5 is a graph and thus an abstraction.)

We define the **period** of oscillation $T$ of a wave to be the time any string element takes to move through one full oscillation. A typical period is marked on the graph of Fig. 16-5. Applying Eq. 16-6 to both ends of this time interval and equating the results yield

$$-y_m \sin \omega t_1 = -y_m \sin \omega(t_1 + T)$$
$$= -y_m \sin(\omega t_1 + \omega T). \quad (16\text{-}7)$$

This can be true only if $\omega T = 2\pi$, or if

$$\omega = \frac{2\pi}{T} \quad \text{(angular frequency).} \quad (16\text{-}8)$$

We call $\omega$ the **angular frequency** of the wave; its SI unit is the radian per second.

Look back at the five snapshots of a traveling wave in Fig. 16-4. The time between snapshots is $\frac{1}{4}T$. Thus, by the fifth snapshot, every string element has made one full oscillation.

The **frequency** $f$ of a wave is defined as $1/T$ and is related to the angular frequency $\omega$ by

$$f = \frac{1}{T} = \frac{\omega}{2\pi} \quad \text{(frequency).} \quad (16\text{-}9)$$

Like the frequency of simple harmonic motion in Chapter 15, this frequency $f$ is a number of oscillations per unit time—here, the number made by a string element as the wave moves through it. As in Chapter 15, $f$ is usually measured in hertz or its multiples, such as kilohertz.

### ✓ Checkpoint 1

The figure is a composite of three snapshots, each of a wave traveling along a particular string. The phases for the waves are given by (a) $2x - 4t$, (b) $4x - 8t$, and (c) $8x - 16t$. Which phase corresponds to which wave in the figure?

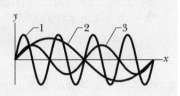

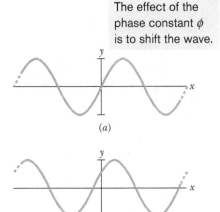

**Figure 16-6** A sinusoidal traveling wave at $t = 0$ with a phase constant $\phi$ of (a) 0 and (b) $\pi/5$ rad.

### Phase Constant

When a sinusoidal traveling wave is given by the wave function of Eq. 16-2, the wave near $x = 0$ looks like Fig. 16-6a when $t = 0$. Note that at $x = 0$, the displacement is $y = 0$ and the slope is at its maximum positive value. We can generalize Eq. 16-2 by inserting a **phase constant** $\phi$ in the wave function:

$$y = y_m \sin(kx - \omega t + \phi). \quad (16\text{-}10)$$

The value of $\phi$ can be chosen so that the function gives some other displacement and slope at $x = 0$ when $t = 0$. For example, a choice of $\phi = +\pi/5$ rad gives the displacement and slope shown in Fig. 16-6b when $t = 0$. The wave is still sinusoidal with the same values of $y_m$, $k$, and $\omega$, but it is now shifted from what you see in Fig. 16-6a (where $\phi = 0$). Note also the direction of the shift. A positive value of $\phi$ shifts the curve in the negative direction of the $x$ axis; a negative value shifts the curve in the positive direction.

## The Speed of a Traveling Wave

Figure 16-7 shows two snapshots of the wave of Eq. 16-2, taken a small time interval $\Delta t$ apart. The wave is traveling in the positive direction of $x$ (to the right in Fig. 16-7), the entire wave pattern moving a distance $\Delta x$ in that direction during the interval $\Delta t$. The ratio $\Delta x/\Delta t$ (or, in the differential limit, $dx/dt$) is the **wave speed** $v$. How can we find its value?

As the wave in Fig. 16-7 moves, each point of the moving wave form, such as point $A$ marked on a peak, retains its displacement $y$. (Points on the string do not retain their displacement, but points on the wave *form* do.) If point $A$ retains its displacement as it moves, the phase in Eq. 16-2 giving it that displacement must remain a constant:

$$kx - \omega t = \text{a constant.} \tag{16-11}$$

Note that although this argument is constant, both $x$ and $t$ are changing. In fact, as $t$ increases, $x$ must also, to keep the argument constant. This confirms that the wave pattern is moving in the positive direction of $x$.

To find the wave speed $v$, we take the derivative of Eq. 16-11, getting

$$k \frac{dx}{dt} - \omega = 0$$

or

$$\frac{dx}{dt} = v = \frac{\omega}{k}. \tag{16-12}$$

Using Eq. 16-5 ($k = 2\pi/\lambda$) and Eq. 16-8 ($\omega = 2\pi/T$), we can rewrite the wave speed as

$$v = \frac{\omega}{k} = \frac{\lambda}{T} = \lambda f \quad \text{(wave speed).} \tag{16-13}$$

The equation $v = \lambda/T$ tells us that the wave speed is one wavelength per period; the wave moves a distance of one wavelength in one period of oscillation.

Equation 16-2 describes a wave moving in the positive direction of $x$. We can find the equation of a wave traveling in the opposite direction by replacing $t$ in Eq. 16-2 with $-t$. This corresponds to the condition

$$kx + \omega t = \text{a constant,} \tag{16-14}$$

which (compare Eq. 16-11) requires that $x$ *decrease* with time. Thus, a wave traveling in the negative direction of $x$ is described by the equation

$$y(x, t) = y_m \sin(kx + \omega t). \tag{16-15}$$

If you analyze the wave of Eq. 16-15 as we have just done for the wave of Eq. 16-2, you will find for its velocity

$$\frac{dx}{dt} = -\frac{\omega}{k}. \tag{16-16}$$

The minus sign (compare Eq. 16-12) verifies that the wave is indeed moving in the negative direction of $x$ and justifies our switching the sign of the time variable.

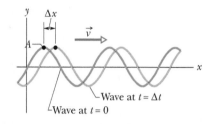

**Figure 16-7** Two snapshots of the wave of Fig. 16-4, at time $t = 0$ and then at time $t = \Delta t$. As the wave moves to the right at velocity $\vec{v}$, the entire curve shifts a distance $\Delta x$ during $\Delta t$. Point $A$ "rides" with the wave form, but the string elements move only up and down.

Consider now a wave of arbitrary shape, given by

$$y(x, t) = h(kx \pm \omega t), \tag{16-17}$$

where $h$ represents *any* function, the sine function being one possibility. Our previous analysis shows that all waves in which the variables $x$ and $t$ enter into the combination $kx \pm \omega t$ are traveling waves. Furthermore, all traveling waves *must* be of the form of Eq. 16-17. Thus, $y(x, t) = \sqrt{ax + bt}$ represents a possible (though perhaps physically a little bizarre) traveling wave. The function $y(x, t) = \sin(ax^2 - bt)$, on the other hand, does *not* represent a traveling wave.

 **Checkpoint 2**

Here are the equations of three waves:
(1) $y(x, t) = 2 \sin(4x - 2t)$, (2) $y(x, t) = \sin(3x - 4t)$, (3) $y(x, t) = 2 \sin(3x - 3t)$.
Rank the waves according to their (a) wave speed and (b) maximum speed perpendicular to the wave's direction of travel (the transverse speed), greatest first.

## Sample Problem 16.01    Determining the quantities in an equation for a transverse wave

A transverse wave traveling along an $x$ axis has the form given by

$$y = y_m \sin(kx \pm \omega t + \phi). \tag{16-18}$$

Figure 16-8a gives the displacements of string elements as a function of $x$, all at time $t = 0$. Figure 16-8b gives the displacements of the element at $x = 0$ as a function of $t$. Find the values of the quantities shown in Eq. 16-18, including the correct choice of sign.

### KEY IDEAS

(1) Figure 16-8a is effectively a snapshot of reality (something that we can see), showing us motion spread out over the $x$ axis. From it we can determine the wavelength $\lambda$ of the wave along that axis, and then we can find the angular wave number $k$ ($= 2\pi/\lambda$) in Eq. 16-18. (2) Figure 16-8b is an ab-

straction, showing us motion spread out over time. From it we can determine the period $T$ of the string element in its SHM and thus also of the wave itself. From $T$ we can then find angular frequency $\omega$ ($= 2\pi/T$) in Eq. 16-18. (3) The phase constant $\phi$ is set by the displacement of the string at $x = 0$ and $t = 0$.

*Amplitude:* From either Fig. 16-8a or 16-8b we see that the maximum displacement is 3.0 mm. Thus, the wave's amplitude $x_m = 3.0$ mm.

*Wavelength:* In Fig. 16-8a, the wavelength $\lambda$ is the distance along the $x$ axis between repetitions in the pattern. The easiest way to measure $\lambda$ is to find the distance from one crossing point to the next crossing point where the string has the same slope. Visually we can roughly measure that distance with the scale on the axis. Instead, we can lay the edge of a

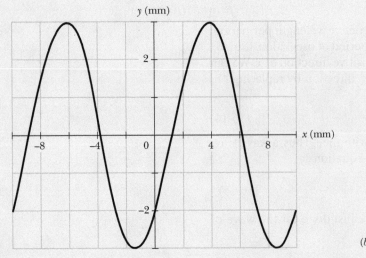

(a)

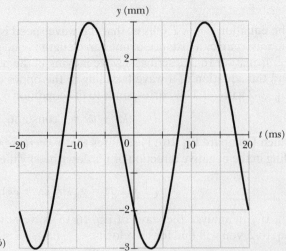

(b)

**Figure 16-8** (a) A snapshot of the displacement $y$ versus position $x$ along a string, at time $t = 0$. (b) A graph of displacement $y$ versus time $t$ for the string element at $x = 0$.

paper sheet on the graph, mark those crossing points, slide the sheet to align the left-hand mark with the origin, and then read off the location of the right-hand mark. Either way we find $\lambda = 10$ mm. From Eq. 16-5, we then have

$$k = \frac{2\pi}{\lambda} = \frac{2\pi}{0.010 \text{ m}} = 200\pi \text{ rad/m}.$$

**Period:** The period $T$ is the time interval that a string element's SHM takes to begin repeating itself. In Fig. 16-8b, $T$ is the distance along the $t$ axis from one crossing point to the next crossing point where the plot has the same slope. Measuring the distance visually or with the aid of a sheet of paper, we find $T = 20$ ms. From Eq. 16-8, we then have

$$\omega = \frac{2\pi}{T} = \frac{2\pi}{0.020 \text{ s}} = 100\pi \text{ rad/s}.$$

**Direction of travel:** To find the direction, we apply a bit of reasoning to the figures. In the snapshot at $t = 0$ given in Fig. 16-8a, note that if the wave is moving rightward, then just after the snapshot, the depth of the wave at $x = 0$ should in-

crease (mentally slide the curve slightly rightward). If, instead, the wave is moving leftward, then just after the snapshot, the depth at $x = 0$ should decrease. Now let's check the graph in Fig. 16-8b. It tells us that just after $t = 0$, the depth increases. Thus, the wave is moving rightward, in the positive direction of $x$, and we choose the minus sign in Eq. 16-18.

**Phase constant:** The value of $\phi$ is set by the conditions at $x = 0$ at the instant $t = 0$. From either figure we see that at that location and time, $y = -2.0$ mm. Substituting these three values and also $y_m = 3.0$ mm into Eq. 16-18 gives us

$$-2.0 \text{ mm} = (3.0 \text{ mm}) \sin(0 + 0 + \phi)$$

or

$$\phi = \sin^{-1}(-\tfrac{2}{3}) = -0.73 \text{ rad}.$$

Note that this is consistent with the rule that on a plot of $y$ versus $x$, a negative phase constant shifts the normal sine function rightward, which is what we see in Fig. 16-8a.

**Equation:** Now we can fill out Eq. 16-18:

$$y = (3.0 \text{ mm}) \sin(200\pi x - 100\pi t - 0.73 \text{ rad}), \quad \text{(Answer)}$$

with $x$ in meters and $t$ in seconds.

---

**Sample Problem 16.02  Transverse velocity and transverse acceleration of a string element**

A wave traveling along a string is described by

$$y(x,t) = (0.00327 \text{ m}) \sin(72.1x - 2.72t),$$

in which the numerical constants are in SI units (72.1 rad/m and 2.72 rad/s).

(a) What is the transverse velocity $u$ of the string element at $x = 22.5$ cm at time $t = 18.9$ s? (This velocity, which is associated with the transverse oscillation of a string element, is parallel to the $y$ axis. Don't confuse it with $v$, the constant velocity at which the wave form moves along the $x$ axis.)

**KEY IDEAS**

The transverse velocity $u$ is the rate at which the displacement $y$ of the element is changing. In general, that displacement is given by

$$y(x,t) = y_m \sin(kx - \omega t). \quad (16\text{-}19)$$

For an element at a certain location $x$, we find the rate of change of $y$ by taking the derivative of Eq. 16-19 with respect to $t$ while treating $x$ as a constant. A derivative taken while one (or more) of the variables is treated as a constant is called a partial derivative and is represented by a symbol such as $\partial/\partial t$ rather than $d/dt$.

**Calculations:** Here we have

$$u = \frac{\partial y}{\partial t} = -\omega y_m \cos(kx - \omega t). \quad (16\text{-}20)$$

Next, substituting numerical values but suppressing the units, which are SI, we write

$$u = (-2.72)(0.00327) \cos[(72.1)(0.225) - (2.72)(18.9)]$$
$$= 0.00720 \text{ m/s} = 7.20 \text{ mm/s}. \quad \text{(Answer)}$$

Thus, at $t = 18.9$ s our string element is moving in the positive direction of $y$ with a speed of 7.20 mm/s. (*Caution:* In evaluating the cosine function, we keep all the significant figures in the argument or the calculation can be off considerably. For example, round off the numbers to two significant figures and then see what you get for $u$.)

(b) What is the transverse acceleration $a_y$ of our string element at $t = 18.9$ s?

**KEY IDEA**

The transverse acceleration $a_y$ is the rate at which the element's transverse velocity is changing.

**Calculations:** From Eq. 16-20, again treating $x$ as a constant but allowing $t$ to vary, we find

$$a_y = \frac{\partial u}{\partial t} = -\omega^2 y_m \sin(kx - \omega t). \quad (16\text{-}21)$$

Substituting numerical values but suppressing the units, which are SI, we have

$$a_y = -(2.72)^2(0.00327) \sin[(72.1)(0.225) - (2.72)(18.9)]$$
$$= -0.0142 \text{ m/s}^2 = -14.2 \text{ mm/s}^2. \quad \text{(Answer)}$$

From part (a) we learn that at $t = 18.9$ s our string element is moving in the positive direction of $y$, and here we learn that it is slowing because its acceleration is in the opposite direction of $u$.

  Additional examples, video, and practice available at *WileyPLUS*

# 16-2 WAVE SPEED ON A STRETCHED STRING

## Learning Objectives

*After reading this module, you should be able to . . .*

16.14 Calculate the linear density $\mu$ of a uniform string in terms of the total mass and total length.

16.15 Apply the relationship between wave speed $v$, tension $\tau$, and linear density $\mu$.

## Key Ideas

● The speed of a wave on a stretched string is set by properties of the string, not properties of the wave such as frequency or amplitude.

● The speed of a wave on a string with tension $\tau$ and linear density $\mu$ is

$$v = \sqrt{\frac{\tau}{\mu}}.$$

## Wave Speed on a Stretched String

The speed of a wave is related to the wave's wavelength and frequency by Eq. 16-13, but *it is set by the properties of the medium*. If a wave is to travel through a medium such as water, air, steel, or a stretched string, it must cause the particles of that medium to oscillate as it passes, which requires both mass (for kinetic energy) and elasticity (for potential energy). Thus, the mass and elasticity determine how fast the wave can travel. Here, we find that dependency in two ways.

### Dimensional Analysis

In dimensional analysis we carefully examine the dimensions of all the physical quantities that enter into a given situation to determine the quantities they produce. In this case, we examine mass and elasticity to find a speed $v$, which has the dimension of length divided by time, or $LT^{-1}$.

For the mass, we use the mass of a string element, which is the mass $m$ of the string divided by the length $l$ of the string. We call this ratio the *linear density* $\mu$ of the string. Thus, $\mu = m/l$, its dimension being mass divided by length, $ML^{-1}$.

You cannot send a wave along a string unless the string is under tension, which means that it has been stretched and pulled taut by forces at its two ends. The tension $\tau$ in the string is equal to the common magnitude of those two forces. As a wave travels along the string, it displaces elements of the string by causing additional stretching, with adjacent sections of string pulling on each other because of the tension. Thus, we can associate the tension in the string with the stretching (elasticity) of the string. The tension and the stretching forces it produces have the dimension of a force—namely, $MLT^{-2}$ (from $F = ma$).

We need to combine $\mu$ (dimension $ML^{-1}$) and $\tau$ (dimension $MLT^{-2}$) to get $v$ (dimension $LT^{-1}$). A little juggling of various combinations suggests

$$v = C\sqrt{\frac{\tau}{\mu}}, \qquad (16\text{-}22)$$

in which $C$ is a dimensionless constant that cannot be determined with dimensional analysis. In our second approach to determining wave speed, you will see that Eq. 16-22 is indeed correct and that $C = 1$.

### Derivation from Newton's Second Law

Instead of the sinusoidal wave of Fig. 16-1*b*, let us consider a single symmetrical pulse such as that of Fig. 16-9, moving from left to right along a string with speed *v*. For convenience, we choose a reference frame in which the pulse remains stationary; that is, we run along with the pulse, keeping it constantly in view. In this frame, the string appears to move past us, from right to left in Fig. 16-9, with speed *v*.

Consider a small string element of length Δ*l* within the pulse, an element that forms an arc of a circle of radius *R* and subtending an angle 2θ at the center of that circle. A force $\vec{\tau}$ with a magnitude equal to the tension in the string pulls tangentially on this element at each end. The horizontal components of these forces cancel, but the vertical components add to form a radial restoring force $\vec{F}$. In magnitude,

$$F = 2(\tau \sin \theta) \approx \tau(2\theta) = \tau\frac{\Delta l}{R} \quad \text{(force)}, \tag{16-23}$$

where we have approximated sin θ as θ for the small angles θ in Fig. 16-9. From that figure, we have also used 2θ = Δ*l*/*R*. The mass of the element is given by

$$\Delta m = \mu \, \Delta l \quad \text{(mass)}, \tag{16-24}$$

where μ is the string's linear density.

At the moment shown in Fig. 16-9, the string element Δ*l* is moving in an arc of a circle. Thus, it has a centripetal acceleration toward the center of that circle, given by

$$a = \frac{v^2}{R} \quad \text{(acceleration)}. \tag{16-25}$$

Equations 16-23, 16-24, and 16-25 contain the elements of Newton's second law. Combining them in the form

$$\text{force} = \text{mass} \times \text{acceleration}$$

gives

$$\frac{\tau \, \Delta l}{R} = (\mu \, \Delta l)\frac{v^2}{R}.$$

Solving this equation for the speed *v* yields

$$v = \sqrt{\frac{\tau}{\mu}} \quad \text{(speed)}, \tag{16-26}$$

in exact agreement with Eq. 16-22 if the constant *C* in that equation is given the value unity. Equation 16-26 gives the speed of the pulse in Fig. 16-9 and the speed of *any* other wave on the same string under the same tension.

Equation 16-26 tells us:

> The speed of a wave along a stretched ideal string depends only on the tension and linear density of the string and not on the frequency of the wave.

The *frequency* of the wave is fixed entirely by whatever generates the wave (for example, the person in Fig. 16-1*b*). The *wavelength* of the wave is then fixed by Eq. 16-13 in the form λ = *v*/*f*.

### Checkpoint 3

You send a traveling wave along a particular string by oscillating one end. If you increase the frequency of the oscillations, do (a) the speed of the wave and (b) the wavelength of the wave increase, decrease, or remain the same? If, instead, you increase the tension in the string, do (c) the speed of the wave and (d) the wavelength of the wave increase, decrease, or remain the same?

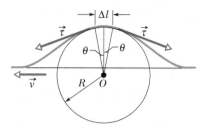

**Figure 16-9** A symmetrical pulse, viewed from a reference frame in which the pulse is stationary and the string appears to move right to left with speed *v*. We find speed *v* by applying Newton's second law to a string element of length Δ*l*, located at the top of the pulse.

# 16-3 ENERGY AND POWER OF A WAVE TRAVELING ALONG A STRING

## Learning Objective

*After reading this module, you should be able to . . .*

**16.16** Calculate the average rate at which energy is transported by a transverse wave.

## Key Idea

● The average power of, or average rate at which energy is transmitted by, a sinusoidal wave on a stretched string is given by

$$P_{\text{avg}} = \tfrac{1}{2}\mu v \omega^2 y_m^2.$$

## Energy and Power of a Wave Traveling Along a String

When we set up a wave on a stretched string, we provide energy for the motion of the string. As the wave moves away from us, it transports that energy as both kinetic energy and elastic potential energy. Let us consider each form in turn.

### Kinetic Energy

A string element of mass $dm$, oscillating transversely in simple harmonic motion as the wave passes through it, has kinetic energy associated with its transverse velocity $\vec{u}$. When the element is rushing through its $y = 0$ position (element $b$ in Fig. 16-10), its transverse velocity—and thus its kinetic energy—is a maximum. When the element is at its extreme position $y = y_m$ (as is element $a$), its transverse velocity—and thus its kinetic energy—is zero.

### Elastic Potential Energy

To send a sinusoidal wave along a previously straight string, the wave must necessarily stretch the string. As a string element of length $dx$ oscillates transversely, its length must increase and decrease in a periodic way if the string element is to fit the sinusoidal wave form. Elastic potential energy is associated with these length changes, just as for a spring.

When the string element is at its $y = y_m$ position (element $a$ in Fig. 16-10), its length has its normal undisturbed value $dx$, so its elastic potential energy is zero. However, when the element is rushing through its $y = 0$ position, it has maximum stretch and thus maximum elastic potential energy.

### Energy Transport

The oscillating string element thus has both its maximum kinetic energy and its maximum elastic potential energy at $y = 0$. In the snapshot of Fig. 16-10, the regions of the string at maximum displacement have no energy, and the regions at zero displacement have maximum energy. As the wave travels along the string, forces due to the tension in the string continuously do work to transfer energy from regions with energy to regions with no energy.

As in Fig. 16-1$b$, let's set up a wave on a string stretched along a horizontal $x$ axis such that Eq. 16-2 applies. As we oscillate one end of the string, we continuously provide energy for the motion and stretching of the string—as the string sections oscillate perpendicularly to the $x$ axis, they have kinetic energy and elastic potential energy. As the wave moves into sections that were previously at rest, energy is transferred into those new sections. Thus, we say that the wave *transports* the energy along the string.

### The Rate of Energy Transmission

The kinetic energy $dK$ associated with a string element of mass $dm$ is given by

$$dK = \tfrac{1}{2}\, dm\, u^2, \tag{16-27}$$

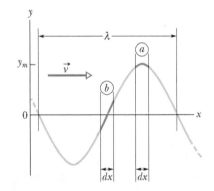

**Figure 16-10** A snapshot of a traveling wave on a string at time $t = 0$. String element $a$ is at displacement $y = y_m$, and string element $b$ is at displacement $y = 0$. The kinetic energy of the string element at each position depends on the transverse velocity of the element. The potential energy depends on the amount by which the string element is stretched as the wave passes through it.

where $u$ is the transverse speed of the oscillating string element. To find $u$, we differentiate Eq. 16-2 with respect to time while holding $x$ constant:

$$u = \frac{\partial y}{\partial t} = -\omega y_m \cos(kx - \omega t). \qquad (16\text{-}28)$$

Using this relation and putting $dm = \mu\, dx$, we rewrite Eq. 16-27 as

$$dK = \tfrac{1}{2}(\mu\, dx)(-\omega y_m)^2 \cos^2(kx - \omega t). \qquad (16\text{-}29)$$

Dividing Eq. 16-29 by $dt$ gives the rate at which kinetic energy passes through a string element, and thus the rate at which kinetic energy is carried along by the wave. The $dx/dt$ that then appears on the right of Eq. 16-29 is the wave speed $v$, so

$$\frac{dK}{dt} = \tfrac{1}{2}\mu v \omega^2 y_m^2 \cos^2(kx - \omega t). \qquad (16\text{-}30)$$

The *average* rate at which kinetic energy is transported is

$$\left(\frac{dK}{dt}\right)_{\text{avg}} = \tfrac{1}{2}\mu v \omega^2 y_m^2 \left[\cos^2(kx - \omega t)\right]_{\text{avg}}$$

$$= \tfrac{1}{4}\mu v \omega^2 y_m^2. \qquad (16\text{-}31)$$

Here we have taken the average over an integer number of wavelengths and have used the fact that the average value of the square of a cosine function over an integer number of periods is $\tfrac{1}{2}$.

Elastic potential energy is also carried along with the wave, and at the same average rate given by Eq. 16-31. Although we shall not examine the proof, you should recall that, in an oscillating system such as a pendulum or a spring–block system, the average kinetic energy and the average potential energy are equal.

The **average power,** which is the average rate at which energy of both kinds is transmitted by the wave, is then

$$P_{\text{avg}} = 2\left(\frac{dK}{dt}\right)_{\text{avg}} \qquad (16\text{-}32)$$

or, from Eq. 16-31,

$$P_{\text{avg}} = \tfrac{1}{2}\mu v \omega^2 y_m^2 \quad \text{(average power)}. \qquad (16\text{-}33)$$

The factors $\mu$ and $v$ in this equation depend on the material and tension of the string. The factors $\omega$ and $y_m$ depend on the process that generates the wave. The dependence of the average power of a wave on the square of its amplitude and also on the square of its angular frequency is a general result, true for waves of all types.

---

### Sample Problem 16.03    Average power of a transverse wave

A string has linear density $\mu = 525$ g/m and is under tension $\tau = 45$ N. We send a sinusoidal wave with frequency $f = 120$ Hz and amplitude $y_m = 8.5$ mm along the string. At what average rate does the wave transport energy?

**KEY IDEA**

The average rate of energy transport is the average power $P_{\text{avg}}$ as given by Eq. 16-33.

***Calculations:*** To use Eq. 16-33, we first must calculate

angular frequency $\omega$ and wave speed $v$. From Eq. 16-9,

$$\omega = 2\pi f = (2\pi)(120\ \text{Hz}) = 754\ \text{rad/s}.$$

From Eq. 16-26 we have

$$v = \sqrt{\frac{\tau}{\mu}} = \sqrt{\frac{45\ \text{N}}{0.525\ \text{kg/m}}} = 9.26\ \text{m/s}.$$

Equation 16-33 then yields

$$P_{\text{avg}} = \tfrac{1}{2}\mu v \omega^2 y_m^2$$

$$= (\tfrac{1}{2})(0.525\ \text{kg/m})(9.26\ \text{m/s})(754\ \text{rad/s})^2(0.0085\ \text{m})^2$$

$$\approx 100\ \text{W.} \qquad \text{(Answer)}$$

 Additional examples, video, and practice available at *WileyPLUS*

# 16-4 THE WAVE EQUATION

## Learning Objective

*After reading this module, you should be able to . . .*

**16.17** For the equation giving a string-element displacement as a function of position $x$ and time $t$, apply the relationship between the second derivative with respect to $x$ and the second derivative with respect to $t$.

## Key Idea

● The general differential equation that governs the travel of waves of all types is

$$\frac{\partial^2 y}{\partial x^2} = \frac{1}{v^2}\frac{\partial^2 y}{\partial t^2}.$$

Here the waves travel along an $x$ axis and oscillate parallel to the $y$ axis, and they move with speed $v$, in either the positive $x$ direction or the negative $x$ direction.

## The Wave Equation

As a wave passes through any element on a stretched string, the element moves perpendicularly to the wave's direction of travel (we are dealing with a transverse wave). By applying Newton's second law to the element's motion, we can derive a general differential equation, called the *wave equation*, that governs the travel of waves of any type.

Figure 16-11*a* shows a snapshot of a string element of mass $dm$ and length $\ell$ as a wave travels along a string of linear density $\mu$ that is stretched along a horizontal $x$ axis. Let us assume that the wave amplitude is small so that the element can be tilted only slightly from the $x$ axis as the wave passes. The force $\vec{F}_2$ on the right end of the element has a magnitude equal to tension $\tau$ in the string and is directed slightly upward. The force $\vec{F}_1$ on the left end of the element also has a magnitude equal to the tension $\tau$ but is directed slightly downward. Because of the slight curvature of the element, these two forces are not simply in opposite direction so that they cancel. Instead, they combine to produce a net force that causes the element to have an upward acceleration $a_y$. Newton's second law written for $y$ components ($F_{net,y} = ma_y$) gives us

$$F_{2y} - F_{1y} = dm\, a_y. \tag{16-34}$$

Let's analyze this equation in parts, first the mass $dm$, then the acceleration component $a_y$, then the individual force components $F_{2y}$ and $F_{1y}$, and then finally the net force that is on the left side of Eq. 16-34.

**Mass.** The element's mass $dm$ can be written in terms of the string's linear density $\mu$ and the element's length $\ell$ as $dm = \mu\ell$. Because the element can have only a slight tilt, $\ell \approx dx$ (Fig. 16-11*a*) and we have the approximation

$$dm = \mu\, dx. \tag{16-35}$$

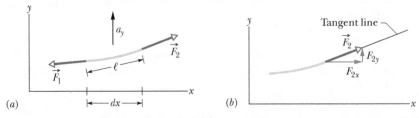

**Figure 16-11** (*a*) A string element as a sinusoidal transverse wave travels on a stretched string. Forces $\vec{F}_1$ and $\vec{F}_2$ act at the left and right ends, producing acceleration $\vec{a}$ having a vertical component $a_y$. (*b*) The force at the element's right end is directed along a tangent to the element's right side.

***Acceleration.*** The acceleration $a_y$ in Eq. 16-34 is the second derivative of the displacement $y$ with respect to time:

$$a_y = \frac{d^2y}{dt^2}.$$  (16-36)

***Forces.*** Figure 16-11$b$ shows that $\vec{F}_2$ is tangent to the string at the right end of the string element. Thus we can relate the components of the force to the string slope $S_2$ at the right end as

$$\frac{F_{2y}}{F_{2x}} = S_2.$$  (16-37)

We can also relate the components to the magnitude $F_2$ ($= \tau$) with

$$F_2 = \sqrt{F_{2x}^2 + F_{2y}^2}$$

or   $$\tau = \sqrt{F_{2x}^2 + F_{2y}^2}.$$  (16-38)

However, because we assume that the element is only slightly tilted, $F_{2y} \ll F_{2x}$ and therefore we can rewrite Eq. 16-38 as

$$\tau = F_{2x}.$$  (16-39)

Substituting this into Eq. 16-37 and solving for $F_{2y}$ yield

$$F_{2y} = \tau S_2.$$  (16-40)

Similar analysis at the left end of the string element gives us

$$F_{1y} = \tau S_1.$$  (16-41)

***Net Force.*** We can now substitute Eqs. 16-35, 16-36, 16-40, and 16-41 into Eq. 16-34 to write

$$\tau S_2 - \tau S_1 = (\mu \, dx)\frac{d^2y}{dt^2},$$

or   $$\frac{S_2 - S_1}{dx} = \frac{\mu}{\tau}\frac{d^2y}{dt^2}.$$  (16-42)

Because the string element is short, slopes $S_2$ and $S_1$ differ by only a differential amount $dS$, where $S$ is the slope at any point:

$$S = \frac{dy}{dx}.$$  (16-43)

First replacing $S_2 - S_1$ in Eq. 16-42 with $dS$ and then using Eq. 16-43 to substitute $dy/dx$ for $S$, we find

$$\frac{dS}{dx} = \frac{\mu}{\tau}\frac{d^2y}{dt^2},$$

$$\frac{d(dy/dx)}{dx} = \frac{\mu}{\tau}\frac{d^2y}{dt^2},$$

and   $$\frac{\partial^2y}{\partial x^2} = \frac{\mu}{\tau}\frac{\partial^2y}{\partial t^2}.$$  (16-44)

In the last step, we switched to the notation of partial derivatives because on the left we differentiate only with respect to $x$ and on the right we differentiate only with respect to $t$. Finally, substituting from Eq. 16-26 ($v = \sqrt{\tau/\mu}$), we find

$$\frac{\partial^2y}{\partial x^2} = \frac{1}{v^2}\frac{\partial^2y}{\partial t^2} \quad \text{(wave equation)}.$$  (16-45)

This is the general differential equation that governs the travel of waves of all types.

# 16-5 INTERFERENCE OF WAVES

## Learning Objectives

*After reading this module, you should be able to . . .*

**16.18** Apply the principle of superposition to show that two overlapping waves add algebraically to give a resultant (or net) wave.

**16.19** For two transverse waves with the same amplitude and wavelength and that travel together, find the displacement equation for the resultant wave and calculate the amplitude in terms of the individual wave amplitude and the phase difference.

**16.20** Describe how the phase difference between two transverse waves (with the same amplitude and wavelength) can result in fully constructive interference, fully destructive interference, and intermediate interference.

**16.21** With the phase difference between two interfering waves expressed in terms of wavelengths, quickly determine the type of interference the waves have.

## Key Ideas

● When two or more waves traverse the same medium, the displacement of any particle of the medium is the sum of the displacements that the individual waves would give it, an effect known as the principle of superposition for waves.

● Two sinusoidal waves on the same string exhibit interference, adding or canceling according to the principle of superposition. If the two are traveling in the same direction and have the same amplitude $y_m$ and

frequency (hence the same wavelength) but differ in phase by a phase constant $\phi$, the result is a single wave with this same frequency:

$$y'(x, t) = [2y_m \cos \tfrac{1}{2}\phi] \sin(kx - \omega t + \tfrac{1}{2}\phi).$$

If $\phi = 0$, the waves are exactly in phase and their interference is fully constructive; if $\phi = \pi$ rad, they are exactly out of phase and their interference is fully destructive.

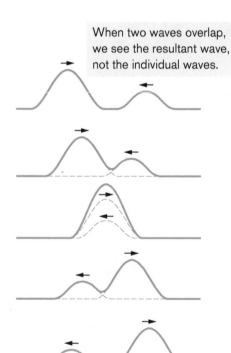

When two waves overlap, we see the resultant wave, not the individual waves.

**Figure 16-12** A series of snapshots that show two pulses traveling in opposite directions along a stretched string. The superposition principle applies as the pulses move through each other.

## The Principle of Superposition for Waves

It often happens that two or more waves pass simultaneously through the same region. When we listen to a concert, for example, sound waves from many instruments fall simultaneously on our eardrums. The electrons in the antennas of our radio and television receivers are set in motion by the net effect of many electromagnetic waves from many different broadcasting centers. The water of a lake or harbor may be churned up by waves in the wakes of many boats.

Suppose that two waves travel simultaneously along the same stretched string. Let $y_1(x, t)$ and $y_2(x, t)$ be the displacements that the string would experience if each wave traveled alone. The displacement of the string when the waves overlap is then the algebraic sum

$$y'(x, t) = y_1(x, t) + y_2(x, t). \qquad (16\text{-}46)$$

This summation of displacements along the string means that

 Overlapping waves algebraically add to produce a **resultant wave** (or **net wave**).

This is another example of the **principle of superposition,** which says that when several effects occur simultaneously, their net effect is the sum of the individual effects. (We should be thankful that only a simple sum is needed. If two effects somehow amplified each other, the resulting nonlinear world would be very difficult to manage and understand.)

Figure 16-12 shows a sequence of snapshots of two pulses traveling in opposite directions on the same stretched string. When the pulses overlap, the resultant pulse is their sum. Moreover,

 Overlapping waves do not in any way alter the travel of each other.

## Interference of Waves

Suppose we send two sinusoidal waves of the same wavelength and amplitude in the same direction along a stretched string. The superposition principle applies. What resultant wave does it predict for the string?

The resultant wave depends on the extent to which the waves are *in phase* (in step) with respect to each other—that is, how much one wave form is shifted from the other wave form. If the waves are exactly in phase (so that the peaks and valleys of one are exactly aligned with those of the other), they combine to double the displacement of either wave acting alone. If they are exactly out of phase (the peaks of one are exactly aligned with the valleys of the other), they combine to cancel everywhere, and the string remains straight. We call this phenomenon of combining waves **interference,** and the waves are said to **interfere.** (These terms refer only to the wave displacements; the travel of the waves is unaffected.)

Let one wave traveling along a stretched string be given by

$$y_1(x, t) = y_m \sin(kx - \omega t) \tag{16-47}$$

and another, shifted from the first, by

$$y_2(x, t) = y_m \sin(kx - \omega t + \phi). \tag{16-48}$$

These waves have the same angular frequency $\omega$ (and thus the same frequency $f$), the same angular wave number $k$ (and thus the same wavelength $\lambda$), and the same amplitude $y_m$. They both travel in the positive direction of the $x$ axis, with the same speed, given by Eq. 16-26. They differ only by a constant angle $\phi$, the phase constant. These waves are said to be *out of phase* by $\phi$ or to have a *phase difference* of $\phi$, or one wave is said to be *phase-shifted* from the other by $\phi$.

From the principle of superposition (Eq. 16-46), the resultant wave is the algebraic sum of the two interfering waves and has displacement

$$y'(x, t) = y_1(x, t) + y_2(x, t)$$
$$= y_m \sin(kx - \omega t) + y_m \sin(kx - \omega t + \phi). \tag{16-49}$$

In Appendix E we see that we can write the sum of the sines of two angles $\alpha$ and $\beta$ as

$$\sin \alpha + \sin \beta = 2 \sin \tfrac{1}{2}(\alpha + \beta) \cos \tfrac{1}{2}(\alpha - \beta). \tag{16-50}$$

Applying this relation to Eq. 16-49 leads to

$$y'(x, t) = [2y_m \cos \tfrac{1}{2}\phi] \sin(kx - \omega t + \tfrac{1}{2}\phi). \tag{16-51}$$

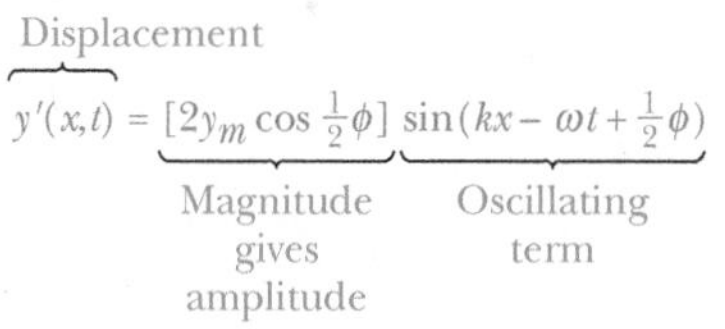

As Fig. 16-13 shows, the resultant wave is also a sinusoidal wave traveling in the direction of increasing $x$. It is the only wave you would actually see on the string (you would *not* see the two interfering waves of Eqs. 16-47 and 16-48).

If two sinusoidal waves of the same amplitude and wavelength travel in the *same* direction along a stretched string, they interfere to produce a resultant sinusoidal wave traveling in that direction.

The resultant wave differs from the interfering waves in two respects: (1) its phase constant is $\tfrac{1}{2}\phi$, and (2) its amplitude $y'_m$ is the magnitude of the quantity in the brackets in Eq. 16-51:

$$y'_m = |2y_m \cos \tfrac{1}{2}\phi| \quad \text{(amplitude).} \tag{16-52}$$

If $\phi = 0$ rad (or $0°$), the two interfering waves are exactly in phase and Eq. 16-51 reduces to

$$y'(x, t) = 2y_m \sin(kx - \omega t) \quad (\phi = 0). \tag{16-53}$$

Displacement
$$y'(x,t) = \overbrace{[2y_m \cos \tfrac{1}{2}\phi]}^{\substack{\text{Magnitude} \\ \text{gives} \\ \text{amplitude}}} \underbrace{\sin(kx - \omega t + \tfrac{1}{2}\phi)}_{\substack{\text{Oscillating} \\ \text{term}}}$$

**Figure 16-13** The resultant wave of Eq. 16-51, due to the interference of two sinusoidal transverse waves, is also a sinusoidal transverse wave, with an amplitude and an oscillating term.

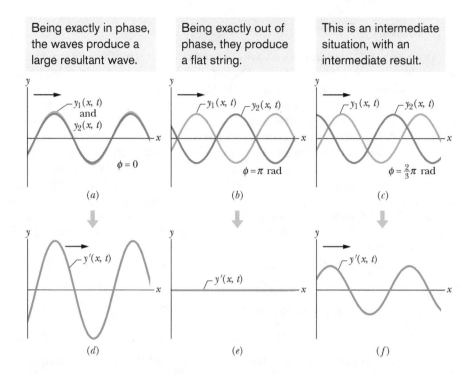

| Being exactly in phase, the waves produce a large resultant wave. | Being exactly out of phase, they produce a flat string. | This is an intermediate situation, with an intermediate result. |

**Figure 16-14** Two identical sinusoidal waves, $y_1(x, t)$ and $y_2(x, t)$, travel along a string in the positive direction of an $x$ axis. They interfere to give a resultant wave $y'(x, t)$. The resultant wave is what is actually seen on the string. The phase difference $\phi$ between the two interfering waves is (a) 0 rad or 0°, (b) $\pi$ rad or 180°, and (c) $\frac{2}{3}\pi$ rad or 120°. The corresponding resultant waves are shown in (d), (e), and (f).

The two waves are shown in Fig. 16-14$a$, and the resultant wave is plotted in Fig. 16-14$d$. Note from both that plot and Eq. 16-53 that the amplitude of the resultant wave is twice the amplitude of either interfering wave. That is the greatest amplitude the resultant wave can have, because the cosine term in Eqs. 16-51 and 16-52 has its greatest value (unity) when $\phi = 0$. Interference that produces the greatest possible amplitude is called *fully constructive interference*.

If $\phi = \pi$ rad (or 180°), the interfering waves are exactly out of phase as in Fig. 16-14$b$. Then $\cos\frac{1}{2}\phi$ becomes $\cos \pi/2 = 0$, and the amplitude of the resultant wave as given by Eq. 16-52 is zero. We then have, for all values of $x$ and $t$,

$$y'(x, t) = 0 \quad (\phi = \pi\text{rad}). \tag{16-54}$$

The resultant wave is plotted in Fig. 16-14$e$. Although we sent two waves along the string, we see no motion of the string. This type of interference is called *fully destructive interference*.

Because a sinusoidal wave repeats its shape every $2\pi$ rad, a phase difference of $\phi = 2\pi$ rad (or 360°) corresponds to a shift of one wave relative to the other wave by a distance equivalent to one wavelength. Thus, phase differences can be described in terms of wavelengths as well as angles. For example, in Fig. 16-14$b$ the waves may be said to be 0.50 wavelength out of phase. Table 16-1 shows some other examples of phase differences and the interference they produce. Note that when interference is neither fully constructive nor fully destructive, it is called *intermediate interference*. The amplitude of the resultant wave is then intermediate between 0 and $2y_m$. For example, from Table 16-1, if the interfering waves have a phase difference of 120° ($\phi = \frac{2}{3}\pi$ rad = 0.33 wavelength), then the resultant wave has an amplitude of $y_m$, the same as that of the interfering waves (see Figs. 16-14$c$ and $f$).

Two waves with the same wavelength are in phase if their phase difference is zero or any integer number of wavelengths. Thus, the integer part of any phase difference *expressed in wavelengths* may be discarded. For example, a phase difference of 0.40 wavelength (an intermediate interference, close to fully destructive interference) is equivalent in every way to one of 2.40 wavelengths,

**Table 16-1  Phase Difference and Resulting Interference Types**[a]

| Phase Difference, in | | | Amplitude of Resultant Wave | Type of Interference |
|---|---|---|---|---|
| Degrees | Radians | Wavelengths | | |
| 0 | 0 | 0 | $2y_m$ | Fully constructive |
| 120 | $\frac{2}{3}\pi$ | 0.33 | $y_m$ | Intermediate |
| 180 | $\pi$ | 0.50 | 0 | Fully destructive |
| 240 | $\frac{4}{3}\pi$ | 0.67 | $y_m$ | Intermediate |
| 360 | $2\pi$ | 1.00 | $2y_m$ | Fully constructive |
| 865 | 15.1 | 2.40 | $0.60y_m$ | Intermediate |

[a]The phase difference is between two otherwise identical waves, with amplitude $y_m$, moving in the same direction.

and so the simpler of the two numbers can be used in computations. Thus, by looking at only the decimal number and comparing it to 0, 0.5, or 1.0 wavelength, you can quickly tell what type of interference two waves have.

 **Checkpoint 4**

Here are four possible phase differences between two identical waves, expressed in wavelengths: 0.20, 0.45, 0.60, and 0.80. Rank them according to the amplitude of the resultant wave, greatest first.

## Sample Problem 16.04    Interference of two waves, same direction, same amplitude

Two identical sinusoidal waves, moving in the same direction along a stretched string, interfere with each other. The amplitude $y_m$ of each wave is 9.8 mm, and the phase difference $\phi$ between them is 100°.

(a) What is the amplitude $y'_m$ of the resultant wave due to the interference, and what is the type of this interference?

### KEY IDEA

These are identical sinusoidal waves traveling in the *same direction* along a string, so they interfere to produce a sinusoidal traveling wave.

*Calculations:* Because they are identical, the waves have the *same amplitude*. Thus, the amplitude $y'_m$ of the resultant wave is given by Eq. 16-52:

$$y'_m = |2y_m \cos \tfrac{1}{2}\phi| = |(2)(9.8 \text{ mm}) \cos(100°/2)|$$
$$= 13 \text{ mm}. \quad \text{(Answer)}$$

We can tell that the interference is *intermediate* in two ways. The phase difference is between 0 and 180°, and, correspondingly, the amplitude $y'_m$ is between 0 and $2y_m$ (= 19.6 mm).

(b) What phase difference, in radians and wavelengths, will give the resultant wave an amplitude of 4.9 mm?

*Calculations:* Now we are given $y'_m$ and seek $\phi$. From Eq. 16-52,

$$y'_m = |2y_m \cos \tfrac{1}{2}\phi|,$$

we now have

$$4.9 \text{ mm} = (2)(9.8 \text{ mm}) \cos \tfrac{1}{2}\phi,$$

which gives us (with a calculator in the radian mode)

$$\phi = 2 \cos^{-1} \frac{4.9 \text{ mm}}{(2)(9.8 \text{ mm})}$$
$$= \pm 2.636 \text{ rad} \approx \pm 2.6 \text{ rad}. \quad \text{(Answer)}$$

There are two solutions because we can obtain the same resultant wave by letting the first wave *lead* (travel ahead of) or *lag* (travel behind) the second wave by 2.6 rad. In wavelengths, the phase difference is

$$\frac{\phi}{2\pi \text{ rad/wavelength}} = \frac{\pm 2.636 \text{ rad}}{2\pi \text{ rad/wavelength}}$$
$$= \pm 0.42 \text{ wavelength}. \quad \text{(Answer)}$$

# 16-6 PHASORS

## Learning Objectives

*After reading this module, you should be able to . . .*

**16.22** Using sketches, explain how a phasor can represent the oscillations of a string element as a wave travels through its location.

**16.23** Sketch a phasor diagram for two overlapping waves traveling together on a string, indicating their amplitudes and phase difference on the sketch.

**16.24** By using phasors, find the resultant wave of two transverse waves traveling together along a string, calculating the amplitude and phase and writing out the displacement equation, and then displaying all three phasors in a phasor diagram that shows the amplitudes, the leading or lagging, and the relative phases.

## Key Idea

● A wave $y(x, t)$ can be represented with a phasor. This is a vector that has a magnitude equal to the amplitude $y_m$ of the wave and that rotates about an origin with an angular speed equal to the angular frequency $\omega$ of the wave. The projection of the rotating phasor on a vertical axis gives the displacement $y$ of a point along the wave's travel.

## Phasors

Adding two waves as discussed in the preceding module is strictly limited to waves with *identical* amplitudes. If we have such waves, that technique is easy enough to use, but we need a more general technique that can be applied to any waves, whether or not they have the same amplitudes. One neat way is to use phasors to represent the waves. Although this may seem bizarre at first, it is essentially a graphical technique that uses the vector addition rules of Chapter 3 instead of messy trig additions.

A **phasor** is a vector that rotates around its tail, which is pivoted at the origin of a coordinate system. The magnitude of the vector is equal to the amplitude $y_m$ of the wave that it represents. The angular speed of the rotation is equal to the angular frequency $\omega$ of the wave. For example, the wave

$$y_1(x, t) = y_{m1} \sin(kx - \omega t) \tag{16-55}$$

is represented by the phasor shown in Figs. 16-15*a* to *d*. The magnitude of the phasor is the amplitude $y_{m1}$ of the wave. As the phasor rotates around the origin at angular speed $\omega$, its projection $y_1$ on the vertical axis varies sinusoidally, from a maximum of $y_{m1}$ through zero to a minimum of $-y_{m1}$ and then back to $y_{m1}$. This variation corresponds to the sinusoidal variation in the displacement $y_1$ of any point along the string as the wave passes through that point. (All this is shown as an animation with voiceover in *WileyPLUS*.)

When two waves travel along the same string in the same direction, we can represent them and their resultant wave in a *phasor diagram*. The phasors in Fig. 16-15*e* represent the wave of Eq. 16-55 and a second wave given by

$$y_2(x, t) = y_{m2} \sin(kx - \omega t + \phi). \tag{16-56}$$

This second wave is phase-shifted from the first wave by phase constant $\phi$. Because the phasors rotate at the same angular speed $\omega$, the angle between the two phasors is always $\phi$. If $\phi$ is a *positive* quantity, then the phasor for wave 2 *lags* the phasor for wave 1 as they rotate, as drawn in Fig. 16-15*e*. If $\phi$ is a negative quantity, then the phasor for wave 2 *leads* the phasor for wave 1.

Because waves $y_1$ and $y_2$ have the same angular wave number $k$ and angular frequency $\omega$, we know from Eqs. 16-51 and 16-52 that their resultant is of the form

$$y'(x, t) = y'_m \sin(kx - \omega t + \beta), \tag{16-57}$$

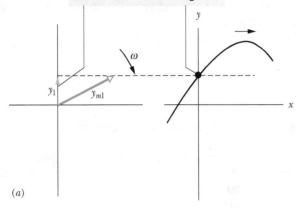

This projection matches this displacement of the dot as the wave moves through it.

(a)

Zero projection, zero displacement

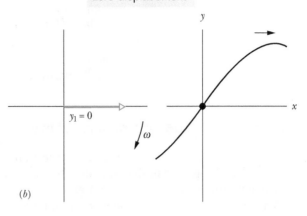

(b)

Maximum negative projection

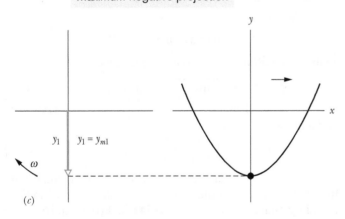

(c)

The next crest is about to move through the dot.

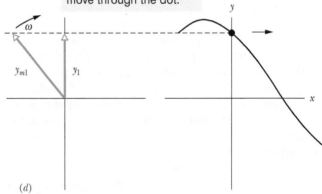

(d)

This is a snapshot of the two phasors for two waves.

These are the projections of the two phasors.

Wave 2, delayed by $\phi$ radians

Wave 1

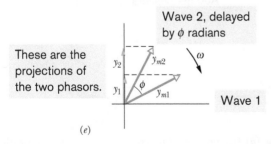

(e)

Adding the two phasors as vectors gives the resultant phasor of the resultant wave.

This is the projection of the resultant phasor.

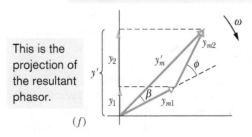

(f)

**Figure 16-15** (a)–(d) A phasor of magnitude $y_{m1}$ rotating about an origin at angular speed $\omega$ represents a sinusoidal wave. The phasor's projection $y_1$ on the vertical axis represents the displacement of a point through which the wave passes. (e) A second phasor, also of angular speed $\omega$ but of magnitude $y_{m2}$ and rotating at a constant angle $\phi$ from the first phasor, represents a second wave, with a phase constant $\phi$. (f) The resultant wave is represented by the vector sum $y'_m$ of the two phasors.

where $y'_m$ is the amplitude of the resultant wave and $\beta$ is its phase constant. To find the values of $y'_m$ and $\beta$, we would have to sum the two combining waves, as we did to obtain Eq. 16-51. To do this on a phasor diagram, we vectorially add the two phasors at any instant during their rotation, as in Fig. 16-15*f* where phasor $y_{m2}$ has been shifted to the head of phasor $y_{m1}$. The magnitude of the vector sum equals the amplitude $y'_m$ in Eq. 16-57. The angle between the vector sum and the phasor for $y_1$ equals the phase constant $\beta$ in Eq. 16-57.

Note that, in contrast to the method of Module 16-5:

 We can use phasors to combine waves *even if their amplitudes are different*.

## Sample Problem 16.05 Interference of two waves, same direction, phasors, any amplitudes

Two sinusoidal waves $y_1(x, t)$ and $y_2(x, t)$ have the same wavelength and travel together in the same direction along a string. Their amplitudes are $y_{m1} = 4.0$ mm and $y_{m2} = 3.0$ mm, and their phase constants are 0 and $\pi/3$ rad, respectively. What are the amplitude $y'_m$ and phase constant $\beta$ of the resultant wave? Write the resultant wave in the form of Eq. 16-57.

### KEY IDEAS

(1) The two waves have a number of properties in common: Because they travel along the same string, they must have the same speed $v$, as set by the tension and linear density of the string according to Eq. 16-26. With the same wavelength $\lambda$, they have the same angular wave number $k$ ($= 2\pi/\lambda$). Also, because they have the same wave number $k$ and speed $v$, they must have the same angular frequency $\omega$ ($= kv$).

(2) The waves (call them waves 1 and 2) can be represented by phasors rotating at the same angular speed $\omega$ about an origin. Because the phase constant for wave 2 is *greater* than that for wave 1 by $\pi/3$, phasor 2 must *lag* phasor 1 by $\pi/3$ rad in their clockwise rotation, as shown in Fig. 16-16*a*. The resultant wave due to the interference of waves 1 and 2 can then be represented by a phasor that is the vector sum of phasors 1 and 2.

*Calculations:* To simplify the vector summation, we drew phasors 1 and 2 in Fig. 16-16*a* at the instant when phasor 1 lies along the horizontal axis. We then drew lagging phasor 2 at positive angle $\pi/3$ rad. In Fig. 16-16*b* we shifted phasor 2 so its tail is at the head of phasor 1. Then we can draw the phasor $y'_m$ of the resultant wave from the tail of phasor 1 to the head of phasor 2. The phase constant $\beta$ is the angle phasor $y'_m$ makes with phasor 1.

To find values for $y'_m$ and $\beta$, we can sum phasors 1 and 2 as vectors on a vector-capable calculator. However, here

we shall sum them by components. (They are called horizontal and vertical components, because the symbols $x$ and $y$ are already used for the waves themselves.) For the horizontal components we have

$$y'_{mh} = y_{m1} \cos 0 + y_{m2} \cos \pi/3$$
$$= 4.0 \text{ mm} + (3.0 \text{ mm}) \cos \pi/3 = 5.50 \text{ mm}.$$

For the vertical components we have

$$y'_{mv} = y_{m1} \sin 0 + y_{m2} \sin \pi/3$$
$$= 0 + (3.0 \text{ mm}) \sin \pi/3 = 2.60 \text{ mm}.$$

Thus, the resultant wave has an amplitude of

$$y'_m = \sqrt{(5.50 \text{ mm})^2 + (2.60 \text{ mm})^2}$$
$$= 6.1 \text{ mm} \qquad \text{(Answer)}$$

and a phase constant of

$$\beta = \tan^{-1} \frac{2.60 \text{ mm}}{5.50 \text{ mm}} = 0.44 \text{ rad.} \qquad \text{(Answer)}$$

From Fig. 16-16*b*, phase constant $\beta$ is a *positive* angle relative to phasor 1. Thus, the resultant wave *lags* wave 1 in their travel by phase constant $\beta = +0.44$ rad. From Eq. 16-57, we can write the resultant wave as

$$y'(x, t) = (6.1 \text{ mm}) \sin(kx - \omega t + 0.44 \text{ rad}). \qquad \text{(Answer)}$$

Add the phasors as vectors.

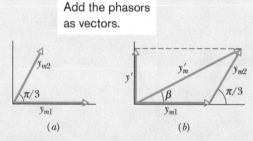

(a)          (b)

**Figure 16-16** (*a*) Two phasors of magnitudes $y_{m1}$ and $y_{m2}$ and with phase difference $\pi/3$. (*b*) Vector addition of these phasors at any instant during their rotation gives the magnitude $y'_m$ of the phasor for the resultant wave.

# 16-7 STANDING WAVES AND RESONANCE

## Learning Objectives

*After reading this module, you should be able to . . .*

**16.25** For two overlapping waves (same amplitude and wavelength) that are traveling in opposite directions, sketch snapshots of the resultant wave, indicating nodes and antinodes.

**16.26** For two overlapping waves (same amplitude and wavelength) that are traveling in opposite directions, find the displacement equation for the resultant wave and calculate the amplitude in terms of the individual wave amplitude.

**16.27** Describe the SHM of a string element at an antinode of a standing wave.

**16.28** For a string element at an antinode of a standing wave, write equations for the displacement, transverse velocity, and transverse acceleration as functions of time.

**16.29** Distinguish between "hard" and "soft" reflections of string waves at a boundary.

**16.30** Describe resonance on a string tied taut between two supports, and sketch the first several standing wave patterns, indicating nodes and antinodes.

**16.31** In terms of string length, determine the wavelengths required for the first several harmonics on a string under tension.

**16.32** For any given harmonic, apply the relationship between frequency, wave speed, and string length.

## Key Ideas

● The interference of two identical sinusoidal waves moving in opposite directions produces standing waves. For a string with fixed ends, the standing wave is given by

$$y'(x, t) = [2y_m \sin kx] \cos \omega t.$$

Standing waves are characterized by fixed locations of zero displacement called nodes and fixed locations of maximum displacement called antinodes.

● Standing waves on a string can be set up by reflection of traveling waves from the ends of the string. If an end is fixed, it must be the position of a node. This limits the frequencies at which standing waves will occur on a given string. Each possible frequency is a resonant frequency, and the corresponding standing wave pattern is an oscillation mode. For a stretched string of length $L$ with fixed ends, the resonant frequencies are

$$f = \frac{v}{\lambda} = n \frac{v}{2L}, \qquad \text{for } n = 1, 2, 3, \ldots .$$

The oscillation mode corresponding to $n = 1$ is called the *fundamental mode* or the *first harmonic*; the mode corresponding to $n = 2$ is the *second harmonic*; and so on.

## Standing Waves

In Module 16-5, we discussed two sinusoidal waves of the same wavelength and amplitude traveling *in the same direction* along a stretched string. What if they travel in opposite directions? We can again find the resultant wave by applying the superposition principle.

Figure 16-17 suggests the situation graphically. It shows the two combining waves, one traveling to the left in Fig. 16-17a, the other to the right in Fig. 16-17b. Figure 16-17c shows their sum, obtained by applying the superposition

**Figure 16-17** (a) Five snapshots of a wave traveling to the left, at the times $t$ indicated below part (c) ($T$ is the period of oscillation). (b) Five snapshots of a wave identical to that in (a) but traveling to the right, at the same times $t$. (c) Corresponding snapshots for the superposition of the two waves on the same string. At $t = 0, \frac{1}{2}T$, and $T$, fully constructive interference occurs because of the alignment of peaks with peaks and valleys with valleys. At $t = \frac{1}{4}T$ and $\frac{3}{4}T$, fully destructive interference occurs because of the alignment of peaks with valleys. Some points (the nodes, marked with dots) never oscillate; some points (the antinodes) oscillate the most.

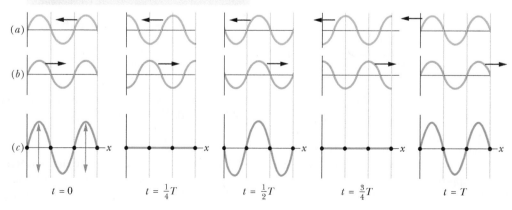

As the waves move through each other, some points never move and some move the most.

$t = 0 \qquad t = \frac{1}{4}T \qquad t = \frac{1}{2}T \qquad t = \frac{3}{4}T \qquad t = T$

principle graphically. The outstanding feature of the resultant wave is that there are places along the string, called **nodes,** where the string never moves. Four such nodes are marked by dots in Fig. 16-17c. Halfway between adjacent nodes are **antinodes,** where the amplitude of the resultant wave is a maximum. Wave patterns such as that of Fig. 16-17c are called **standing waves** because the wave patterns do not move left or right; the locations of the maxima and minima do not change.

> If two sinusoidal waves of the same amplitude and wavelength travel in *opposite* directions along a stretched string, their interference with each other produces a standing wave.

To analyze a standing wave, we represent the two waves with the equations

$$y_1(x, t) = y_m \sin(kx - \omega t) \tag{16-58}$$

and

$$y_2(x, t) = y_m \sin(kx + \omega t). \tag{16-59}$$

The principle of superposition gives, for the combined wave,

$$y'(x, t) = y_1(x, t) + y_2(x, t) = y_m \sin(kx - \omega t) + y_m \sin(kx + \omega t).$$

Applying the trigonometric relation of Eq. 16-50 leads to Fig. 16-18 and

$$y'(x, t) = [2y_m \sin kx] \cos \omega t. \tag{16-60}$$

This equation does not describe a traveling wave because it is not of the form of Eq. 16-17. Instead, it describes a standing wave.

The quantity $2y_m \sin kx$ in the brackets of Eq. 16-60 can be viewed as the amplitude of oscillation of the string element that is located at position $x$. However, since an amplitude is always positive and $\sin kx$ can be negative, we take the absolute value of the quantity $2y_m \sin kx$ to be the amplitude at $x$.

In a traveling sinusoidal wave, the amplitude of the wave is the same for all string elements. That is not true for a standing wave, in which the amplitude *varies with position*. In the standing wave of Eq. 16-60, for example, the amplitude is zero for values of $kx$ that give $\sin kx = 0$. Those values are

$$kx = n\pi, \quad \text{for } n = 0, 1, 2, \ldots . \tag{16-61}$$

Substituting $k = 2\pi/\lambda$ in this equation and rearranging, we get

$$x = n\frac{\lambda}{2}, \quad \text{for } n = 0, 1, 2, \ldots \quad \text{(nodes)}, \tag{16-62}$$

as the positions of zero amplitude—the nodes—for the standing wave of Eq. 16-60. Note that adjacent nodes are separated by $\lambda/2$, half a wavelength.

The amplitude of the standing wave of Eq. 16-60 has a maximum value of $2y_m$, which occurs for values of $kx$ that give $|\sin kx| = 1$. Those values are

$$kx = \tfrac{1}{2}\pi, \tfrac{3}{2}\pi, \tfrac{5}{2}\pi, \ldots$$
$$= (n + \tfrac{1}{2})\pi, \quad \text{for } n = 0, 1, 2, \ldots . \tag{16-63}$$

Substituting $k = 2\pi/\lambda$ in Eq. 16-63 and rearranging, we get

$$x = \left(n + \frac{1}{2}\right)\frac{\lambda}{2}, \quad \text{for } n = 0, 1, 2, \ldots \quad \text{(antinodes)}, \tag{16-64}$$

as the positions of maximum amplitude—the antinodes—of the standing wave of Eq. 16-60. Antinodes are separated by $\lambda/2$ and are halfway between nodes.

### Reflections at a Boundary

We can set up a standing wave in a stretched string by allowing a traveling wave to be reflected from the far end of the string so that the wave travels back

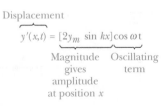

**Figure 16-18** The resultant wave of Eq. 16-60 is a standing wave and is due to the interference of two sinusoidal waves of the same amplitude and wavelength that travel in opposite directions.

through itself. The incident (original) wave and the reflected wave can then be described by Eqs. 16-58 and 16-59, respectively, and they can combine to form a pattern of standing waves.

In Fig. 16-19, we use a single pulse to show how such reflections take place. In Fig. 16-19a, the string is fixed at its left end. When the pulse arrives at that end, it exerts an upward force on the support (the wall). By Newton's third law, the support exerts an opposite force of equal magnitude on the string. This second force generates a pulse at the support, which travels back along the string in the direction opposite that of the incident pulse. In a "hard" reflection of this kind, there must be a node at the support because the string is fixed there. The reflected and incident pulses must have opposite signs, so as to cancel each other at that point.

In Fig. 16-19b, the left end of the string is fastened to a light ring that is free to slide without friction along a rod. When the incident pulse arrives, the ring moves up the rod. As the ring moves, it pulls on the string, stretching the string and producing a reflected pulse with the same sign and amplitude as the incident pulse. Thus, in such a "soft" reflection, the incident and reflected pulses reinforce each other, creating an antinode at the end of the string; the maximum displacement of the ring is twice the amplitude of either of these two pulses.

## Checkpoint 5

Two waves with the same amplitude and wavelength interfere in three different situations to produce resultant waves with the following equations:

(1) $y'(x, t) = 4 \sin(5x - 4t)$

(2) $y'(x, t) = 4 \sin(5x) \cos(4t)$

(3) $y'(x, t) = 4 \sin(5x + 4t)$

In which situation are the two combining waves traveling (a) toward positive $x$, (b) toward negative $x$, and (c) in opposite directions?

## Standing Waves and Resonance

Consider a string, such as a guitar string, that is stretched between two clamps. Suppose we send a continuous sinusoidal wave of a certain frequency along the string, say, toward the right. When the wave reaches the right end, it reflects and begins to travel back to the left. That left-going wave then overlaps the wave that is still traveling to the right. When the left-going wave reaches the left end, it reflects again and the newly reflected wave begins to travel to the right, overlapping the left-going and right-going waves. In short, we very soon have many overlapping traveling waves, which interfere with one another.

For certain frequencies, the interference produces a standing wave pattern (or **oscillation mode**) with nodes and large antinodes like those in Fig. 16-20. Such a standing wave is said to be produced at **resonance,** and the string is said to *resonate* at these certain frequencies, called **resonant frequencies.** If the string

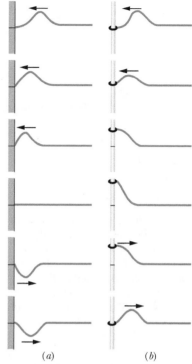

There are two ways a pulse can reflect from the end of a string.

**Figuer 16-19** (*a*) A pulse incident from the right is reflected at the left end of the string, which is tied to a wall. Note that the reflected pulse is inverted from the incident pulse. (*b*) Here the left end of the string is tied to a ring that can slide without friction up and down the rod. Now the pulse is not inverted by the reflection.

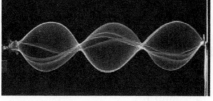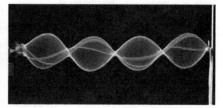

Richard Megna/Fundamental Photographs

**Figure 16-20** Stroboscopic photographs reveal (imperfect) standing wave patterns on a string being made to oscillate by an oscillator at the left end. The patterns occur at certain frequencies of oscillation.

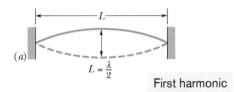

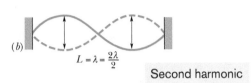

First harmonic

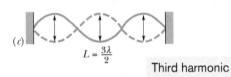

Second harmonic

Third harmonic

**Figure 16-21** A string, stretched between two clamps, is made to oscillate in standing wave patterns. (a) The simplest possible pattern consists of one *loop*, which refers to the composite shape formed by the string in its extreme displacements (the solid and dashed lines). (b) The next simplest pattern has two loops. (c) The next has three loops.

Courtesy Thomas D. Rossing, Northern Illinois University

**Figure 16-22** One of many possible standing wave patterns for a kettledrum head, made visible by dark powder sprinkled on the drumhead. As the head is set into oscillation at a single frequency by a mechanical oscillator at the upper left of the photograph, the powder collects at the nodes, which are circles and straight lines in this two-dimensional example.

is oscillated at some frequency other than a resonant frequency, a standing wave is not set up. Then the interference of the right-going and left-going traveling waves results in only small, temporary (perhaps even imperceptible) oscillations of the string.

Let a string be stretched between two clamps separated by a fixed distance $L$. To find expressions for the resonant frequencies of the string, we note that a node must exist at each of its ends, because each end is fixed and cannot oscillate. The simplest pattern that meets this key requirement is that in Fig. 16-21a, which shows the string at both its extreme displacements (one solid and one dashed, together forming a single "loop"). There is only one antinode, which is at the center of the string. Note that half a wavelength spans the length $L$, which we take to be the string's length. Thus, for this pattern, $\lambda/2 = L$. This condition tells us that if the left-going and right-going traveling waves are to set up this pattern by their interference, they must have the wavelength $\lambda = 2L$.

A second simple pattern meeting the requirement of nodes at the fixed ends is shown in Fig. 16-21b. This pattern has three nodes and two antinodes and is said to be a two-loop pattern. For the left-going and right-going waves to set it up, they must have a wavelength $\lambda = L$. A third pattern is shown in Fig. 16-21c. It has four nodes, three antinodes, and three loops, and the wavelength is $\lambda = \frac{2}{3}L$. We could continue this progression by drawing increasingly more complicated patterns. In each step of the progression, the pattern would have one more node and one more antinode than the preceding step, and an additional $\lambda/2$ would be fitted into the distance $L$.

Thus, a standing wave can be set up on a string of length $L$ by a wave with a wavelength equal to one of the values

$$\lambda = \frac{2L}{n}, \qquad \text{for } n = 1, 2, 3, \ldots. \qquad (16\text{-}65)$$

The resonant frequencies that correspond to these wavelengths follow from Eq. 16-13:

$$f = \frac{v}{\lambda} = n\frac{v}{2L}, \qquad \text{for } n = 1, 2, 3, \ldots. \qquad (16\text{-}66)$$

Here $v$ is the speed of traveling waves on the string.

Equation 16-66 tells us that the resonant frequencies are integer multiples of the lowest resonant frequency, $f = v/2L$, which corresponds to $n = 1$. The oscillation mode with that lowest frequency is called the *fundamental mode* or the *first harmonic*. The *second harmonic* is the oscillation mode with $n = 2$, the *third harmonic* is that with $n = 3$, and so on. The frequencies associated with these modes are often labeled $f_1$, $f_2$, $f_3$, and so on. The collection of all possible oscillation modes is called the **harmonic series,** and $n$ is called the **harmonic number** of the $n$th harmonic.

For a given string under a given tension, each resonant frequency corresponds to a particular oscillation pattern. Thus, if the frequency is in the audible range, you can hear the shape of the string. Resonance can also occur in two dimensions (such as on the surface of the kettledrum in Fig. 16-22) and in three dimensions (such as in the wind-induced swaying and twisting of a tall building).

✓ **Checkpoint 6**

In the following series of resonant frequencies, one frequency (lower than 400 Hz) is missing: 150, 225, 300, 375 Hz. (a) What is the missing frequency? (b) What is the frequency of the seventh harmonic?

## Sample Problem 16.06   Resonance of transverse waves, standing waves, harmonics

Figure 16-23 shows resonant oscillation of a string of mass $m = 2.500$ g and length $L = 0.800$ m and that is under tension $\tau = 325.0$ N. What is the wavelength $\lambda$ of the transverse waves producing the standing wave pattern, and what is the harmonic number $n$? What is the frequency $f$ of the transverse waves and of the oscillations of the moving string elements? What is the maximum magnitude of the transverse velocity $u_m$ of the element oscillating at coordinate $x = 0.180$ m? At what point during the element's oscillation is the transverse velocity maximum?

### KEY IDEAS

(1) The traverse waves that produce a standing wave pattern must have a wavelength such that an integer number $n$ of half-wavelengths fit into the length $L$ of the string. (2) The frequency of those waves and of the oscillations of the string elements is given by Eq. 16-66 ($f = nv/2L$). (3) The displacement of a string element as a function of position $x$ and time $t$ is given by Eq. 16-60:

$$y'(x,t) = [2y_m \sin kx] \cos \omega t. \qquad (16\text{-}67)$$

*Wavelength and harmonic number:* In Fig. 16-23, the solid line, which is effectively a snapshot (or freeze-frame) of the oscillations, reveals that 2 full wavelengths fit into the length $L = 0.800$ m of the string. Thus, we have

$$2\lambda = L,$$

or

$$\lambda = \frac{L}{2}. \qquad (16\text{-}68)$$

$$= \frac{0.800 \text{ m}}{2} = 0.400 \text{ m}. \qquad \text{(Answer)}$$

By counting the number of loops (or half-wavelengths) in Fig. 16-23, we see that the harmonic number is

$$n = 4. \qquad \text{(Answer)}$$

We also find $n = 4$ by comparing Eqs. 16-68 and 16-65 ($\lambda = 2L/n$). Thus, the string is oscillating in its fourth harmonic.

*Frequency:* We can get the frequency $f$ of the transverse waves from Eq. 16-13 ($v = \lambda f$) if we first find the speed $v$ of the waves. That speed is given by Eq. 16-26, but we must substitute $m/L$ for the unknown linear density $\mu$. We obtain

$$v = \sqrt{\frac{\tau}{\mu}} = \sqrt{\frac{\tau}{m/L}} = \sqrt{\frac{\tau L}{m}}$$

$$= \sqrt{\frac{(325 \text{ N})(0.800 \text{ m})}{2.50 \times 10^{-3} \text{ kg}}} = 322.49 \text{ m/s}.$$

After rearranging Eq. 16-13, we write

$$f = \frac{v}{\lambda} = \frac{322.49 \text{ m/s}}{0.400 \text{ m}}$$

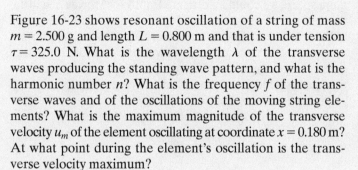
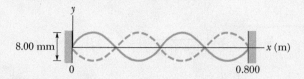

**Figure 16-23** Resonant oscillation of a string under tension.

$$= 806.2 \text{ Hz} \approx 806 \text{ Hz}. \qquad \text{(Answer)}$$

Note that we get the same answer by substituting into Eq. 16-66:

$$f = n\frac{v}{2L} = 4\frac{322.49 \text{ m/s}}{2(0.800 \text{ m})}$$

$$= 806 \text{ Hz}. \qquad \text{(Answer)}$$

Now note that this 806 Hz is not only the frequency of the waves producing the fourth harmonic but also it is said to *be* the fourth harmonic, as in the statement, "The fourth harmonic of this oscillating string is 806 Hz." It is also the frequency of the string elements as they oscillate vertically in the figure in simple harmonic motion, just as a block on a vertical spring would oscillate in simple harmonic motion. Finally, it is also the frequency of the sound you would hear as the oscillating string periodically pushes against the air.

*Transverse velocity:* The displacement $y'$ of the string element located at coordinate $x$ is given by Eq. 16-67 as a function of time $t$. The term $\cos \omega t$ contains the dependence on time and thus provides the "motion" of the standing wave. The term $2y_m \sin kx$ sets the extent of the motion — that is, the amplitude. The greatest amplitude occurs at an antinode, where $\sin kx$ is $+1$ or $-1$ and thus the greatest amplitude is $2y_m$. From Fig. 16-23, we see that $2y_m = 4.00$ mm, which tells us that $y_m = 2.00$ mm.

We want the transverse velocity — the velocity of a string element parallel to the $y$ axis. To find it, we take the time derivative of Eq. 16-67:

$$u(x,t) = \frac{\partial y'}{\partial t} = \frac{\partial}{\partial t}[(2y_m \sin kx) \cos \omega t]$$

$$= [-2y_m\omega \sin kx] \sin \omega t. \qquad (16\text{-}69)$$

Here the term $\sin \omega t$ provides the variation with time and the term $-2y_m\omega \sin kx$ provides the extent of that variation. We want the absolute magnitude of that extent:

$$u_m = |-2y_m\omega \sin kx|.$$

To evaluate this for the element at $x = 0.180$ m, we first note that $y_m = 2.00$ mm, $k = 2\pi/\lambda = 2\pi/(0.400$ m$)$, and $\omega = 2\pi f = 2\pi(806.2$ Hz$)$. Then the maximum speed of the element at $x = 0.180$ m is

$$u_m = \left| -2(2.00 \times 10^{-3}\text{ m})(2\pi)(806.2\text{ Hz}) \right.$$

$$\left. \times \sin\left(\frac{2\pi}{0.400\text{ m}}(0.180\text{ m})\right) \right|$$

$$= 6.26\text{ m/s.} \qquad \text{(Answer)}$$

To determine when the string element has this maximum speed, we could investigate Eq. 16-69. However, a little thought can save a lot of work. The element is undergoing SHM and must come to a momentary stop at its extreme upward position and extreme downward position. It has the greatest speed as it zips through the midpoint of its oscillation, just as a block does in a block–spring oscillator.

 **PLUS** Additional examples, video, and practice available at *WileyPLUS*

# Review & Summary

**Transverse and Longitudinal Waves** Mechanical waves can exist only in material media and are governed by Newton's laws. **Transverse** mechanical waves, like those on a stretched string, are waves in which the particles of the medium oscillate perpendicular to the wave's direction of travel. Waves in which the particles of the medium oscillate parallel to the wave's direction of travel are **longitudinal** waves.

**Sinusoidal Waves** A sinusoidal wave moving in the positive direction of an $x$ axis has the mathematical form

$$y(x, t) = y_m \sin(kx - \omega t), \qquad (16\text{-}2)$$

where $y_m$ is the **amplitude** of the wave, $k$ is the **angular wave number,** $\omega$ is the **angular frequency,** and $kx - \omega t$ is the **phase.** The **wavelength** $\lambda$ is related to $k$ by

$$k = \frac{2\pi}{\lambda}. \qquad (16\text{-}5)$$

The **period** $T$ and **frequency** $f$ of the wave are related to $\omega$ by

$$\frac{\omega}{2\pi} = f = \frac{1}{T}. \qquad (16\text{-}9)$$

Finally, the **wave speed** $v$ is related to these other parameters by

$$v = \frac{\omega}{k} = \frac{\lambda}{T} = \lambda f. \qquad (16\text{-}13)$$

**Equation of a Traveling Wave** Any function of the form

$$y(x, t) = h(kx \pm \omega t) \qquad (16\text{-}17)$$

can represent a **traveling wave** with a wave speed given by Eq. 16-13 and a wave shape given by the mathematical form of $h$. The plus sign denotes a wave traveling in the negative direction of the $x$ axis, and the minus sign a wave traveling in the positive direction.

**Wave Speed on Stretched String** The speed of a wave on a stretched string is set by properties of the string. The speed on a string with tension $\tau$ and linear density $\mu$ is

$$v = \sqrt{\frac{\tau}{\mu}}. \qquad (16\text{-}26)$$

**Power** The **average power** of, or average rate at which energy is transmitted by, a sinusoidal wave on a stretched string is given by

$$P_{\text{avg}} = \tfrac{1}{2}\mu v \omega^2 y_m^2. \qquad (16\text{-}33)$$

**Superposition of Waves** When two or more waves traverse the same medium, the displacement of any particle of the medium is the sum of the displacements that the individual waves would give it.

**Interference of Waves** Two sinusoidal waves on the same string exhibit **interference,** adding or canceling according to the principle of superposition. If the two are traveling in the same direction and have the same amplitude $y_m$ and frequency (hence the same wavelength) but differ in phase by a **phase constant** $\phi$, the result is a single wave with this same frequency:

$$y'(x, t) = [2y_m \cos \tfrac{1}{2}\phi] \sin(kx - \omega t + \tfrac{1}{2}\phi). \qquad (16\text{-}51)$$

If $\phi = 0$, the waves are exactly in phase and their interference is fully constructive; if $\phi = \pi$ rad, they are exactly out of phase and their interference is fully destructive.

**Phasors** A wave $y(x, t)$ can be represented with a **phasor.** This is a vector that has a magnitude equal to the amplitude $y_m$ of the wave and that rotates about an origin with an angular speed equal to the angular frequency $\omega$ of the wave. The projection of the rotating phasor on a vertical axis gives the displacement $y$ of a point along the wave's travel.

**Standing Waves** The interference of two identical sinusoidal waves moving in opposite directions produces **standing waves.** For a string with fixed ends, the standing wave is given by

$$y'(x, t) = [2y_m \sin kx] \cos \omega t. \qquad (16\text{-}60)$$

Standing waves are characterized by fixed locations of zero displacement called **nodes** and fixed locations of maximum displacement called **antinodes.**

**Resonance** Standing waves on a string can be set up by reflection of traveling waves from the ends of the string. If an end is fixed, it must be the position of a node. This limits the frequencies at which standing waves will occur on a given string. Each possible frequency is a **resonant frequency,** and the corresponding standing wave pattern is an **oscillation mode.** For a stretched string of length $L$ with fixed ends, the resonant frequencies are

$$f = \frac{v}{\lambda} = n\frac{v}{2L}, \qquad \text{for } n = 1, 2, 3, \ldots. \qquad (16\text{-}66)$$

The oscillation mode corresponding to $n = 1$ is called the *fundamental mode* or the *first harmonic*; the mode corresponding to $n = 2$ is the *second harmonic*; and so on.

## Questions

**1** The following four waves are sent along strings with the same linear densities ($x$ is in meters and $t$ is in seconds). Rank the waves according to (a) their wave speed and (b) the tension in the strings along which they travel, greatest first:

(1) $y_1 = (3\text{ mm})\sin(x - 3t)$,      (3) $y_3 = (1\text{ mm})\sin(4x - t)$,

(2) $y_2 = (6\text{ mm})\sin(2x - t)$,      (4) $y_4 = (2\text{ mm})\sin(x - 2t)$.

**2** In Fig. 16-24, wave 1 consists of a rectangular peak of height 4 units and width $d$, and a rectangular valley of depth 2 units and width $d$. The wave travels rightward along an $x$ axis. Choices 2, 3, and 4 are similar waves, with the same heights, depths, and widths, that will travel leftward along that axis and through wave 1. Right-going wave 1 and one of the left-going waves will interfere as they pass through each other. With which left-going wave will the interference give, for an instant, (a) the deepest valley, (b) a flat line, and (c) a flat peak $2d$ wide?

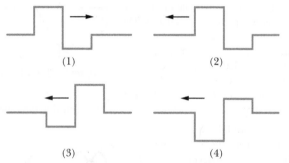

**Figure 16-24** Question 2.

**3** Figure 16-25$a$ gives a snapshot of a wave traveling in the direction of positive $x$ along a string under tension. Four string elements are indicated by the lettered points. For each of those elements, determine whether, at the instant of the snapshot, the element is moving upward or downward or is momentarily at rest. (*Hint:* Imagine the wave as it moves through the four string elements, as if you were watching a video of the wave as it traveled rightward.)

Figure 16-25$b$ gives the displacement of a string element located at, say, $x = 0$ as a function of time. At the lettered times, is the element moving upward or downward or is it momentarily at rest?

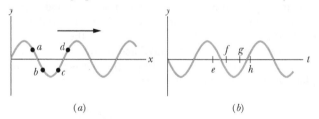

**Figure 16-25** Question 3.

**4** Figure 16-26 shows three waves that are *separately* sent along a string that is stretched under a certain tension along an $x$ axis. Rank the waves according to their (a) wavelengths, (b) speeds, and (c) angular frequencies, greatest first.

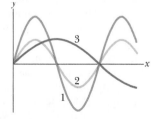

**Figure 16-26** Question 4.

**5** If you start with two sinusoidal waves of the same amplitude traveling in phase on a string and then somehow phase-shift one of them by 5.4 wavelengths, what type of interference will occur on the string?

**6** The amplitudes and phase differences for four pairs of waves of equal wavelengths are (a) 2 mm, 6 mm, and $\pi$ rad; (b) 3 mm, 5 mm, and $\pi$ rad; (c) 7 mm, 9 mm, and $\pi$ rad; (d) 2 mm, 2 mm, and 0 rad. Each pair travels in the same direction along the same string. Without written calculation, rank the four pairs according to the amplitude of their resultant wave, greatest first. (*Hint:* Construct phasor diagrams.)

**7** A sinusoidal wave is sent along a cord under tension, transporting energy at the average rate of $P_{\text{avg},1}$. Two waves, identical to that first one, are then to be sent along the cord with a phase difference $\phi$ of either 0, 0.2 wavelength, or 0.5 wavelength. (a) With only mental calculation, rank those choices of $\phi$ according to the average rate at which the waves will transport energy, greatest first. (b) For the first choice of $\phi$, what is the average rate in terms of $P_{\text{avg},1}$?

**8** (a) If a standing wave on a string is given by

$$y'(t) = (3\text{ mm})\sin(5x)\cos(4t),$$

is there a node or an antinode of the oscillations of the string at $x = 0$? (b) If the standing wave is given by

$$y'(t) = (3\text{ mm})\sin(5x + \pi/2)\cos(4t),$$

is there a node or an antinode at $x = 0$?

**9** Strings $A$ and $B$ have identical lengths and linear densities, but string $B$ is under greater tension than string $A$. Figure 16-27 shows four situations, ($a$) through ($d$), in which standing wave patterns exist on the two strings. In which situations is there the possibility that strings $A$ and $B$ are oscillating at the same resonant frequency?

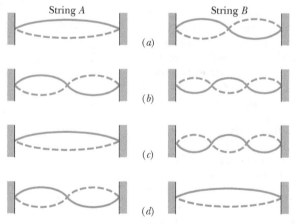

**Figure 16-27** Question 9.

**10** If you set up the seventh harmonic on a string, (a) how many nodes are present, and (b) is there a node, antinode, or some intermediate state at the midpoint? If you next set up the sixth harmonic, (c) is its resonant wavelength longer or shorter than that for the seventh harmonic, and (d) is the resonant frequency higher or lower?

**11** Figure 16-28 shows phasor diagrams for three situations in which two waves travel along the same string. All six waves have the same amplitude. Rank the situations according to the amplitude of the net wave on the string, greatest first.

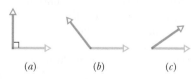

**Figure 16-28** Question 11.

# Problems

**GO** Tutoring problem available (at instructor's discretion) in *WileyPLUS* and WebAssign

**SSM** Worked-out solution available in Student Solutions Manual

**• – •••** Number of dots indicates level of problem difficulty

**WWW** Worked-out solution is at

**ILW** Interactive solution is at  http://www.wiley.com/college/halliday

Additional information available in *The Flying Circus of Physics* and at flyingcircusofphysics.com

## Module 16-1  Transverse Waves

**•1** If a wave $y(x, t) = (6.0 \text{ mm}) \sin(kx + (600 \text{ rad/s})t + \phi)$ travels along a string, how much time does any given point on the string take to move between displacements $y = +2.0$ mm and $y = -2.0$ mm?

**•2** *A human wave.* During sporting events within large, densely packed stadiums, spectators will send a wave (or pulse) around the stadium (Fig. 16-29). As the wave reaches a group of spectators, they stand with a cheer and then sit. At any instant, the width $w$ of the wave is the distance from the leading edge (people are just about to stand) to the trailing edge (people have just sat down). Suppose a human wave travels a distance of 853 seats around a stadium in 39 s, with spectators requiring about 1.8 s to respond to the wave's passage by standing and then sitting. What are (a) the wave speed $v$ (in seats per second) and (b) width $w$ (in number of seats)?

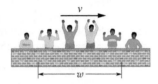

**Figure 16-29** Problem 2.

**•3** A wave has an angular frequency of 110 rad/s and a wavelength of 1.80 m. Calculate (a) the angular wave number and (b) the speed of the wave.

**•4** A sand scorpion can detect the motion of a nearby beetle (its prey) by the waves the motion sends along the sand surface (Fig. 16-30). The waves are of two types: transverse waves traveling at $v_t = 50$ m/s and longitudinal waves traveling at $v_l = 150$ m/s. If a sudden motion sends out such waves, a scorpion can tell the distance of the beetle from the difference $\Delta t$ in the arrival times of the waves at its leg nearest the beetle. If $\Delta t = 4.0$ ms, what is the beetle's distance?

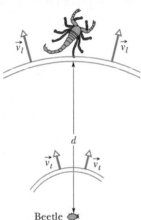

**Figure 16-30** Problem 4.

**•5** A sinusoidal wave travels along a string. The time for a particular point to move from maximum displacement to zero is 0.170 s. What are the (a) period and (b) frequency? (c) The wavelength is 1.40 m; what is the wave speed?

**••6** **GO** A sinusoidal wave travels along a string under tension. Figure 16-31 gives the slopes along the string at time $t = 0$. The scale of the $x$ axis is set by $x_s = 0.80$ m. What is the amplitude of the wave?

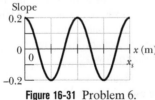

**Figure 16-31** Problem 6.

**••7** A transverse sinusoidal wave is moving along a string in the positive direction of an $x$ axis with a speed of 80 m/s. At $t = 0$, the string particle at $x = 0$ has a transverse displacement of 4.0 cm from its equilibrium position and is not moving. The maximum transverse speed of the string particle at $x = 0$ is 16 m/s. (a) What is the frequency of the wave? (b) What is the wavelength of the wave? If $y(x, t) = y_m \sin(kx \pm \omega t + \phi)$ is the form of the wave equation, what are (c) $y_m$, (d) $k$, (e) $\omega$, (f) $\phi$, and (g) the correct choice of sign in front of $\omega$?

**••8** **GO** Figure 16-32 shows the transverse velocity $u$ versus time $t$ of the point on a string at $x = 0$, as a wave passes through it. The scale on the vertical axis is set by $u_s = 4.0$ m/s. The wave has the generic form $y(x, t) = y_m \sin(kx - \omega t + \phi)$. What then is $\phi$? (*Caution:* A calculator does not always give the proper inverse trig function, so check your answer by substituting it and an assumed value of $\omega$ into $y(x, t)$ and then plotting the function.)

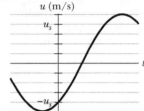

**Figure 16-32** Problem 8.

**••9** A sinusoidal wave moving along a string is shown twice in Fig. 16-33, as crest $A$ travels in the positive direction of an $x$ axis by distance $d = 6.0$ cm in 4.0 ms. The tick marks along the axis are separated by 10 cm; height $H = 6.00$ mm. The equation for the wave is in the form $y(x, t) = y_m \sin(kx \pm \omega t)$, so what are (a) $y_m$, (b) $k$, (c) $\omega$, and (d) the correct choice of sign in front of $\omega$?

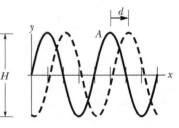

**Figure 16-33** Problem 9.

**••10** The equation of a transverse wave traveling along a very long string is $y = 6.0 \sin(0.020\pi x + 4.0\pi t)$, where $x$ and $y$ are expressed in centimeters and $t$ is in seconds. Determine (a) the amplitude, (b) the wavelength, (c) the frequency, (d) the speed, (e) the direction of propagation of the wave, and (f) the maximum transverse speed of a particle in the string. (g) What is the transverse displacement at $x = 3.5$ cm when $t = 0.26$ s?

**••11** **GO** A sinusoidal transverse wave of wavelength 20 cm travels along a string in the positive direction of an $x$ axis. The displacement $y$ of the string particle at $x = 0$ is given in Fig. 16-34 as a function of time $t$. The scale of the vertical axis is set by $y_s = 4.0$ cm. The wave equation is to be in the form $y(x, t) = y_m \sin(kx \pm \omega t + \phi)$. (a) At $t = 0$, is a plot of $y$ versus $x$ in the shape of a positive sine function or a negative sine function? What are (b) $y_m$, (c) $k$, (d) $\omega$, (e) $\phi$, (f) the sign in front of $\omega$, and (g) the speed of the wave? (h) What is the transverse velocity of the particle at $x = 0$ when $t = 5.0$ s?

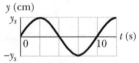

**Figure 16-34** Problem 11.

**••12** **GO** The function $y(x, t) = (15.0 \text{ cm}) \cos(\pi x - 15\pi t)$, with $x$ in meters and $t$ in seconds, describes a wave on a taut string. What is the transverse speed for a point on the string at an instant when that point has the displacement $y = +12.0$ cm?

**••13  ILW**  A sinusoidal wave of frequency 500 Hz has a speed of 350 m/s. (a) How far apart are two points that differ in phase by $\pi/3$ rad? (b) What is the phase difference between two displacements at a certain point at times 1.00 ms apart?

### Module 16-2  Wave Speed on a Stretched String
**•14**  The equation of a transverse wave on a string is

$$y = (2.0 \text{ mm}) \sin[(20 \text{ m}^{-1})x - (600 \text{ s}^{-1})t].$$

The tension in the string is 15 N. (a) What is the wave speed? (b) Find the linear density of this string in grams per meter.

**•15  SSM  WWW**  A stretched string has a mass per unit length of 5.00 g/cm and a tension of 10.0 N. A sinusoidal wave on this string has an amplitude of 0.12 mm and a frequency of 100 Hz and is traveling in the negative direction of an $x$ axis. If the wave equation is of the form $y(x, t) = y_m \sin(kx \pm \omega t)$, what are (a) $y_m$, (b) $k$, (c) $\omega$, and (d) the correct choice of sign in front of $\omega$?

**•16**  The speed of a transverse wave on a string is 170 m/s when the string tension is 120 N. To what value must the tension be changed to raise the wave speed to 180 m/s?

**•17**  The linear density of a string is $1.6 \times 10^{-4}$ kg/m. A transverse wave on the string is described by the equation

$$y = (0.021 \text{ m}) \sin[(2.0 \text{ m}^{-1})x + (30 \text{ s}^{-1})t].$$

What are (a) the wave speed and (b) the tension in the string?

**•18**  The heaviest and lightest strings on a certain violin have linear densities of 3.0 and 0.29 g/m. What is the ratio of the diameter of the heaviest string to that of the lightest string, assuming that the strings are of the same material?

**•19  SSM**  What is the speed of a transverse wave in a rope of length 2.00 m and mass 60.0 g under a tension of 500 N?

**•20**  The tension in a wire clamped at both ends is doubled without appreciably changing the wire's length between the clamps. What is the ratio of the new to the old wave speed for transverse waves traveling along this wire?

**••21  ILW**  A 100 g wire is held under a tension of 250 N with one end at $x = 0$ and the other at $x = 10.0$ m. At time $t = 0$, pulse 1 is sent along the wire from the end at $x = 10.0$ m. At time $t = 30.0$ ms, pulse 2 is sent along the wire from the end at $x = 0$. At what position $x$ do the pulses begin to meet?

**••22**  A sinusoidal wave is traveling on a string with speed 40 cm/s. The displacement of the particles of the string at $x = 10$ cm varies with time according to $y = (5.0 \text{ cm}) \sin[1.0 - (4.0 \text{ s}^{-1})t]$. The linear density of the string is 4.0 g/cm. What are (a) the frequency and (b) the wavelength of the wave? If the wave equation is of the form $y(x, t) = y_m \sin(kx \pm \omega t)$, what are (c) $y_m$, (d) $k$, (e) $\omega$, and (f) the correct choice of sign in front of $\omega$? (g) What is the tension in the string?

**••23  SSM  ILW**  A sinusoidal transverse wave is traveling along a string in the negative direction of an $x$ axis. Figure 16-35 shows a plot of the dis-

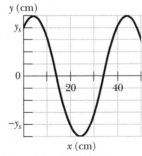

**Figure 16-35**  Problem 23.

placement as a function of position at time $t = 0$; the scale of the $y$ axis is set by $y_s = 4.0$ cm. The string tension is 3.6 N, and its linear density is 25 g/m. Find the (a) amplitude, (b) wavelength, (c) wave speed, and (d) period of the wave. (e) Find the maximum transverse speed of a particle in the string. If the wave is of the form $y(x, t) = y_m \sin(kx \pm \omega t + \phi)$, what are (f) $k$, (g) $\omega$, (h) $\phi$, and (i) the correct choice of sign in front of $\omega$?

**•••24**  In Fig. 16-36a, string 1 has a linear density of 3.00 g/m, and string 2 has a linear density of 5.00 g/m. They are under tension due to the hanging block of mass $M = 500$ g. Calculate the wave speed on (a) string 1 and (b) string 2. (*Hint:* When a string loops halfway around a pulley, it pulls on the pulley with a net force that is twice the tension in the string.) Next the block is divided into two blocks (with $M_1 + M_2 = M$) and the apparatus is rearranged as shown in Fig. 16-36b. Find (c) $M_1$ and (d) $M_2$ such that the wave speeds in the two strings are equal.

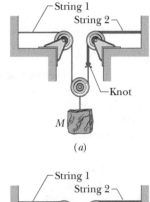

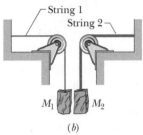

**Figure 16-36**  Problem 24.

**•••25**  A uniform rope of mass $m$ and length $L$ hangs from a ceiling. (a) Show that the speed of a transverse wave on the rope is a function of $y$, the distance from the lower end, and is given by $v = \sqrt{gy}$. (b) Show that the time a transverse wave takes to travel the length of the rope is given by $t = 2\sqrt{L/g}$.

### Module 16-3  Energy and Power of a Wave Traveling Along a String
**•26**  A string along which waves can travel is 2.70 m long and has a mass of 260 g. The tension in the string is 36.0 N. What must be the frequency of traveling waves of amplitude 7.70 mm for the average power to be 85.0 W?

**••27  GO**  A sinusoidal wave is sent along a string with a linear density of 2.0 g/m. As it travels, the kinetic energies of the mass elements along the string vary. Figure 16-37a gives the rate $dK/dt$ at which kinetic energy passes through the string elements at a particular instant, plotted as a function of distance $x$ along the string. Figure 16-37b is similar except that it gives the rate at which kinetic energy passes through a particular mass element (at a particular location), plotted as a function of time $t$. For both figures, the scale on the vertical (rate) axis is set by $R_s = 10$ W. What is the amplitude of the wave?

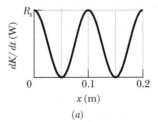

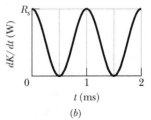

**Figure 16-37**  Problem 27.

## Module 16-4   The Wave Equation

**•28** Use the wave equation to find the speed of a wave given by

$$y(x, t) = (3.00 \text{ mm}) \sin[(4.00 \text{ m}^{-1})x - (7.00 \text{ s}^{-1})t].$$

**••29** Use the wave equation to find the speed of a wave given by

$$y(x, t) = (2.00 \text{ mm})[(20 \text{ m}^{-1})x - (4.0 \text{ s}^{-1})t]^{0.5}.$$

**•••30** Use the wave equation to find the speed of a wave given in terms of the general function $h(x, t)$:

$$y(x, t) = (4.00 \text{ mm}) \, h[(30 \text{ m}^{-1})x + (6.0 \text{ s}^{-1})t].$$

## Module 16-5   Interference of Waves

**•31  SSM** Two identical traveling waves, moving in the same direction, are out of phase by $\pi/2$ rad. What is the amplitude of the resultant wave in terms of the common amplitude $y_m$ of the two combining waves?

**•32** What phase difference between two identical traveling waves, moving in the same direction along a stretched string, results in the combined wave having an amplitude 1.50 times that of the common amplitude of the two combining waves? Express your answer in (a) degrees, (b) radians, and (c) wavelengths.

**••33  GO** Two sinusoidal waves with the same amplitude of 9.00 mm and the same wavelength travel together along a string that is stretched along an $x$ axis. Their resultant wave is shown twice in Fig. 16-38, as valley $A$ travels in the negative direction of the $x$ axis by distance $d = 56.0$ cm in 8.0 ms. The tick marks along the axis are separated by 10 cm, and height $H$ is 8.0 mm. Let the equation for one wave be of the form $y(x, t) = y_m \sin(kx \pm \omega t + \phi_1)$, where $\phi_1 = 0$ and you must choose the correct sign in front of $\omega$. For the equation for the other wave, what are (a) $y_m$, (b) $k$, (c) $\omega$, (d) $\phi_2$, and (e) the sign in front of $\omega$?

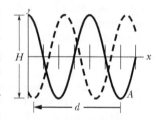

**Figure 16-38** Problem 33.

**•••34  GO** A sinusoidal wave of angular frequency 1200 rad/s and amplitude 3.00 mm is sent along a cord with linear density 2.00 g/m and tension 1200 N. (a) What is the average rate at which energy is transported by the wave to the opposite end of the cord? (b) If, simultaneously, an identical wave travels along an adjacent, identical cord, what is the total average rate at which energy is transported to the opposite ends of the two cords by the waves? If, instead, those two waves are sent along the *same* cord simultaneously, what is the total average rate at which they transport energy when their phase difference is (c) 0, (d) $0.4\pi$ rad, and (e) $\pi$ rad?

## Module 16-6   Phasors

**•35  SSM** Two sinusoidal waves of the same frequency travel in the same direction along a string. If $y_{m1} = 3.0$ cm, $y_{m2} = 4.0$ cm, $\phi_1 = 0$, and $\phi_2 = \pi/2$ rad, what is the amplitude of the resultant wave?

**••36** Four waves are to be sent along the same string, in the same direction:

$$y_1(x, t) = (4.00 \text{ mm}) \sin(2\pi x - 400\pi t)$$
$$y_2(x, t) = (4.00 \text{ mm}) \sin(2\pi x - 400\pi t + 0.7\pi)$$
$$y_3(x, t) = (4.00 \text{ mm}) \sin(2\pi x - 400\pi t + \pi)$$
$$y_4(x, t) = (4.00 \text{ mm}) \sin(2\pi x - 400\pi t + 1.7\pi).$$

What is the amplitude of the resultant wave?

**••37  GO** These two waves travel along the same string:

$$y_1(x, t) = (4.60 \text{ mm}) \sin(2\pi x - 400\pi t)$$
$$y_2(x, t) = (5.60 \text{ mm}) \sin(2\pi x - 400\pi t + 0.80\pi \text{ rad}).$$

What are (a) the amplitude and (b) the phase angle (relative to wave 1) of the resultant wave? (c) If a third wave of amplitude 5.00 mm is also to be sent along the string in the same direction as the first two waves, what should be its phase angle in order to maximize the amplitude of the new resultant wave?

**••38** Two sinusoidal waves of the same frequency are to be sent in the same direction along a taut string. One wave has an amplitude of 5.0 mm, the other 8.0 mm. (a) What phase difference $\phi_1$ between the two waves results in the smallest amplitude of the resultant wave? (b) What is that smallest amplitude? (c) What phase difference $\phi_2$ results in the largest amplitude of the resultant wave? (d) What is that largest amplitude? (e) What is the resultant amplitude if the phase angle is $(\phi_1 - \phi_2)/2$?

**••39** Two sinusoidal waves of the same period, with amplitudes of 5.0 and 7.0 mm, travel in the same direction along a stretched string; they produce a resultant wave with an amplitude of 9.0 mm. The phase constant of the 5.0 mm wave is 0. What is the phase constant of the 7.0 mm wave?

## Module 16-7   Standing Waves and Resonance

**•40** Two sinusoidal waves with identical wavelengths and amplitudes travel in opposite directions along a string with a speed of 10 cm/s. If the time interval between instants when the string is flat is 0.50 s, what is the wavelength of the waves?

**•41  SSM** A string fixed at both ends is 8.40 m long and has a mass of 0.120 kg. It is subjected to a tension of 96.0 N and set oscillating. (a) What is the speed of the waves on the string? (b) What is the longest possible wavelength for a standing wave? (c) Give the frequency of that wave.

**•42** A string under tension $\tau_i$ oscillates in the third harmonic at frequency $f_3$, and the waves on the string have wavelength $\lambda_3$. If the tension is increased to $\tau_f = 4\tau_i$ and the string is again made to oscillate in the third harmonic, what then are (a) the frequency of oscillation in terms of $f_3$ and (b) the wavelength of the waves in terms of $\lambda_3$?

**•43  SSM  WWW** What are (a) the lowest frequency, (b) the second lowest frequency, and (c) the third lowest frequency for standing waves on a wire that is 10.0 m long, has a mass of 100 g, and is stretched under a tension of 250 N?

**•44** A 125 cm length of string has mass 2.00 g and tension 7.00 N. (a) What is the wave speed for this string? (b) What is the lowest resonant frequency of this string?

**•45  SSM  ILW** A string that is stretched between fixed supports separated by 75.0 cm has resonant frequencies of 420 and 315 Hz, with no intermediate resonant frequencies. What are (a) the lowest resonant frequency and (b) the wave speed?

**•46** String $A$ is stretched between two clamps separated by distance $L$. String $B$, with the same linear density and under the same tension as string $A$, is stretched between two clamps separated by distance $4L$. Consider the first eight harmonics of string $B$. For which of these eight harmonics of $B$ (if any) does the frequency match the frequency of (a) $A$'s first harmonic, (b) $A$'s second harmonic, and (c) $A$'s third harmonic?

**•47** One of the harmonic frequencies for a particular string under tension is 325 Hz. The next higher harmonic frequency is 390 Hz.

What harmonic frequency is next higher after the harmonic frequency 195 Hz?

•48 ~~🛩️~~ If a transmission line in a cold climate collects ice, the increased diameter tends to cause vortex formation in a passing wind. The air pressure variations in the vortexes tend to cause the line to oscillate (*gallop*), especially if the frequency of the variations matches a resonant frequency of the line. In long lines, the resonant frequencies are so close that almost any wind speed can set up a resonant mode vigorous enough to pull down support towers or cause the line to *short out* with an adjacent line. If a transmission line has a length of 347 m, a linear density of 3.35 kg/m, and a tension of 65.2 MN, what are (a) the frequency of the fundamental mode and (b) the frequency difference between successive modes?

•49 ILW A nylon guitar string has a linear density of 7.20 g/m and is under a tension of 150 N. The fixed supports are distance $D = 90.0$ cm apart. The string is oscillating in the standing wave pattern shown in Fig. 16-39. Calculate the (a) speed, (b) wavelength, and (c) frequency of the traveling waves whose superposition gives this standing wave.

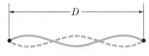
**Figure 16-39** Problem 49.

••50 For a particular transverse standing wave on a long string, one of the antinodes is at $x = 0$ and an adjacent node is at $x = 0.10$ m. The displacement $y(t)$ of the string particle at $x = 0$ is shown in Fig. 16-40, where the scale of the $y$ axis is set by $y_s = 4.0$ cm. When $t = 0.50$ s, what is the displacement of the string particle at (a) $x = 0.20$ m and (b) $x = 0.30$ m?

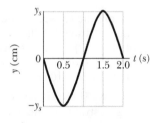
**Figure 16-40** Problem 50.

What is the transverse velocity of the string particle at $x = 0.20$ m at (c) $t = 0.50$ s and (d) $t = 1.0$ s? (e) Sketch the standing wave at $t = 0.50$ s for the range $x = 0$ to $x = 0.40$ m.

••51 SSM WWW Two waves are generated on a string of length 3.0 m to produce a three-loop standing wave with an amplitude of 1.0 cm. The wave speed is 100 m/s. Let the equation for one of the waves be of the form $y(x, t) = y_m \sin(kx + \omega t)$. In the equation for the other wave, what are (a) $y_m$, (b) $k$, (c) $\omega$, and (d) the sign in front of $\omega$?

••52 A rope, under a tension of 200 N and fixed at both ends, oscillates in a second-harmonic standing wave pattern. The displacement of the rope is given by

$$y = (0.10 \text{ m})(\sin \pi x/2) \sin 12\pi t,$$

where $x = 0$ at one end of the rope, $x$ is in meters, and $t$ is in seconds. What are (a) the length of the rope, (b) the speed of the waves on the rope, and (c) the mass of the rope? (d) If the rope oscillates in a third-harmonic standing wave pattern, what will be the period of oscillation?

••53 A string oscillates according to the equation

$$y' = (0.50 \text{ cm}) \sin\left[\left(\frac{\pi}{3} \text{ cm}^{-1}\right)x\right] \cos[(40\pi \text{ s}^{-1})t].$$

What are the (a) amplitude and (b) speed of the two waves (identical except for direction of travel) whose superposition gives this oscillation? (c) What is the distance between nodes? (d) What is the transverse speed of a particle of the string at the position $x = 1.5$ cm when $t = \frac{9}{8}$ s?

••54 🔵 Two sinusoidal waves with the same amplitude and wavelength travel through each other along a string that is stretched along an $x$ axis. Their resultant wave is shown twice in Fig. 16-41, as the antinode $A$ travels from an extreme upward displacement to an extreme downward displacement in 6.0 ms. The tick marks along the axis are separated by 10 cm; height $H$ is 1.80 cm. Let the equation for one of the two waves be of the form $y(x, t) = y_m \sin(kx + \omega t)$. In the equation for the other wave, what are (a) $y_m$, (b) $k$, (c) $\omega$, and (d) the sign in front of $\omega$?

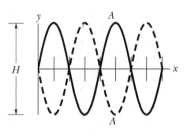
**Figure 16-41** Problem 54.

••55 🔵 The following two waves are sent in opposite directions on a horizontal string so as to create a standing wave in a vertical plane:

$$y_1(x, t) = (6.00 \text{ mm}) \sin(4.00\pi x - 400\pi t)$$
$$y_2(x, t) = (6.00 \text{ mm}) \sin(4.00\pi x + 400\pi t),$$

with $x$ in meters and $t$ in seconds. An antinode is located at point $A$. In the time interval that point takes to move from maximum upward displacement to maximum downward displacement, how far does each wave move along the string?

••56 A standing wave pattern on a string is described by

$$y(x, t) = 0.040 (\sin 5\pi x)(\cos 40\pi t),$$

where $x$ and $y$ are in meters and $t$ is in seconds. For $x \geq 0$, what is the location of the node with the (a) smallest, (b) second smallest, and (c) third smallest value of $x$? (d) What is the period of the oscillatory motion of any (nonnode) point? What are the (e) speed and (f) amplitude of the two traveling waves that interfere to produce this wave? For $t \geq 0$, what are the (g) first, (h) second, and (i) third time that all points on the string have zero transverse velocity?

••57 A generator at one end of a very long string creates a wave given by

$$y = (6.0 \text{ cm}) \cos \frac{\pi}{2} [(2.00 \text{ m}^{-1})x + (8.00 \text{ s}^{-1})t],$$

and a generator at the other end creates the wave

$$y = (6.0 \text{ cm}) \cos \frac{\pi}{2} [(2.00 \text{ m}^{-1})x - (8.00 \text{ s}^{-1})t].$$

Calculate the (a) frequency, (b) wavelength, and (c) speed of each wave. For $x \geq 0$, what is the location of the node having the (d) smallest, (e) second smallest, and (f) third smallest value of $x$? For $x \geq 0$, what is the location of the antinode having the (g) smallest, (h) second smallest, and (i) third smallest value of $x$?

••58 🔵 In Fig. 16-42, a string, tied to a sinusoidal oscillator at $P$ and running over a support at $Q$, is stretched by a block of mass $m$. Separation $L = 1.20$ m, linear density $\mu = 1.6$ g/m, and the oscillator

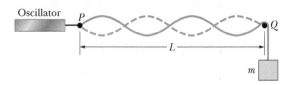
**Figure 16-42** Problems 58 and 60.

frequency $f = 120$ Hz. The amplitude of the motion at $P$ is small enough for that point to be considered a node. A node also exists at $Q$. (a) What mass $m$ allows the oscillator to set up the fourth harmonic on the string? (b) What standing wave mode, if any, can be set up if $m = 1.00$ kg?

•••59 GO In Fig. 16-43, an aluminum wire, of length $L_1 = 60.0$ cm, cross-sectional area $1.00 \times 10^{-2}$ cm², and density 2.60 g/cm³, is joined to a steel wire, of density 7.80 g/cm³ and the same cross-sectional area. The

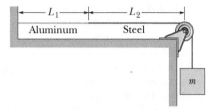

**Figure 16-43** Problem 59.

compound wire, loaded with a block of mass $m = 10.0$ kg, is arranged so that the distance $L_2$ from the joint to the supporting pulley is 86.6 cm. Transverse waves are set up on the wire by an external source of variable frequency; a node is located at the pulley. (a) Find the lowest frequency that generates a standing wave having the joint as one of the nodes. (b) How many nodes are observed at this frequency?

•••60 GO In Fig. 16-42, a string, tied to a sinusoidal oscillator at $P$ and running over a support at $Q$, is stretched by a block of mass $m$. The separation $L$ between $P$ and $Q$ is 1.20 m, and the frequency $f$ of the oscillator is fixed at 120 Hz. The amplitude of the motion at $P$ is small enough for that point to be considered a node. A node also exists at $Q$. A standing wave appears when the mass of the hanging block is 286.1 g or 447.0 g, but not for any intermediate mass. What is the linear density of the string?

## Additional Problems

61 GO In an experiment on standing waves, a string 90 cm long is attached to the prong of an electrically driven tuning fork that oscillates perpendicular to the length of the string at a frequency of 60 Hz. The mass of the string is 0.044 kg. What tension must the string be under (weights are attached to the other end) if it is to oscillate in four loops?

62 A sinusoidal transverse wave traveling in the positive direction of an $x$ axis has an amplitude of 2.0 cm, a wavelength of 10 cm, and a frequency of 400 Hz. If the wave equation is of the form $y(x, t) = y_m \sin(kx \pm \omega t)$, what are (a) $y_m$, (b) $k$, (c) $\omega$, and (d) the correct choice of sign in front of $\omega$? What are (e) the maximum transverse speed of a point on the cord and (f) the speed of the wave?

63 A wave has a speed of 240 m/s and a wavelength of 3.2 m. What are the (a) frequency and (b) period of the wave?

64 The equation of a transverse wave traveling along a string is

$$y = 0.15 \sin(0.79x - 13t),$$

in which $x$ and $y$ are in meters and $t$ is in seconds. (a) What is the displacement $y$ at $x = 2.3$ m, $t = 0.16$ s? A second wave is to be added to the first wave to produce standing waves on the string. If the second wave is of the form $y(x, t) = y_m \sin(kx \pm \omega t)$, what are (b) $y_m$, (c) $k$, (d) $\omega$, and (e) the correct choice of sign in front of $\omega$ for this second wave? (f) What is the displacement of the resultant standing wave at $x = 2.3$ m, $t = 0.16$ s?

65 The equation of a transverse wave traveling along a string is

$$y = (2.0 \text{ mm}) \sin[(20 \text{ m}^{-1})x - (600 \text{ s}^{-1})t].$$

Find the (a) amplitude, (b) frequency, (c) velocity (including

sign), and (d) wavelength of the wave. (e) Find the maximum transverse speed of a particle in the string.

66 Figure 16-44 shows the displacement $y$ versus time $t$ of the point on a string at $x = 0$, as a wave passes through that point. The scale of the $y$ axis is set by $y_s = 6.0$ mm. The wave is given by $y(x, t) = y_m \sin(kx - \omega t + \phi)$. What is $\phi$? (*Caution:* A calculator does not always give the proper inverse trig function, so check your answer by substituting it and an assumed value of $\omega$ into $y(x, t)$ and then plotting the function.)

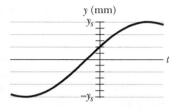

**Figure 16-44** Problem 66.

67 Two sinusoidal waves, identical except for phase, travel in the same direction along a string, producing the net wave $y'(x, t) = (3.0 \text{ mm}) \sin(20x - 4.0t + 0.820 \text{ rad})$, with $x$ in meters and $t$ in seconds. What are (a) the wavelength $\lambda$ of the two waves, (b) the phase difference between them, and (c) their amplitude $y_m$?

68 A single pulse, given by $h(x - 5.0t)$, is shown in Fig. 16-45 for $t = 0$. The scale of the vertical axis is set by $h_s = 2$. Here $x$ is in centimeters and $t$ is in seconds. What are the (a) speed and (b) direction of travel of the pulse? (c) Plot $h(x - 5t)$ as a function of $x$ for $t = 2$ s. (d) Plot $h(x - 5t)$ as a function of $t$ for $x = 10$ cm.

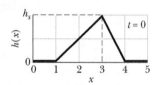

**Figure 16-45** Problem 68.

69 SSM Three sinusoidal waves of the same frequency travel along a string in the positive direction of an $x$ axis. Their amplitudes are $y_1$, $y_1/2$, and $y_1/3$, and their phase constants are $0$, $\pi/2$, and $\pi$, respectively. What are the (a) amplitude and (b) phase constant of the resultant wave? (c) Plot the wave form of the resultant wave at $t = 0$, and discuss its behavior as $t$ increases.

70 GO Figure 16-46 shows transverse acceleration $a_y$ versus time $t$ of the point on a string at $x = 0$, as a wave in the form of $y(x, t) = y_m \sin(kx - \omega t + \phi)$ passes through that point. The scale of the vertical axis is set by $a_s = 400$ m/s². What is $\phi$? (*Caution:* A calculator does not always give the proper inverse trig function, so check your answer by substituting it and an assumed value of $\omega$ into $y(x, t)$ and then plotting the function.)

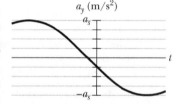

**Figure 16-46** Problem 70.

71 A transverse sinusoidal wave is generated at one end of a long, horizontal string by a bar that moves up and down through a distance of 1.00 cm. The motion is continuous and is repeated regularly 120 times per second. The string has linear density 120 g/m and is kept under a tension of 90.0 N. Find the maximum value of (a) the transverse speed $u$ and (b) the transverse component of the tension $\tau$.

(c) Show that the two maximum values calculated above occur at the same phase values for the wave. What is the transverse displacement $y$ of the string at these phases? (d) What is the maximum rate of energy transfer along the string? (e) What is the transverse displacement $y$ when this maximum transfer occurs? (f) What is the minimum rate of energy transfer along the string? (g) What is the transverse displacement $y$ when this minimum transfer occurs?

**72** Two sinusoidal 120 Hz waves, of the same frequency and amplitude, are to be sent in the positive direction of an $x$ axis that is directed along a cord under tension. The waves can be sent in phase, or they can be phase-shifted. Figure 16-47 shows the amplitude $y'$ of the resulting wave versus the distance of the shift (how far one wave is shifted from the other wave). The scale of the vertical axis is set by $y'_s = 6.0$ mm. If the equations for the two waves are of the form $y(x, t) = y_m \sin(kx \pm \omega t)$, what are (a) $y_m$, (b) $k$, (c) $\omega$, and (d) the correct choice of sign in front of $\omega$?

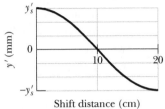

**Figure 16-47** Problem 72.

**73** At time $t = 0$ and at position $x = 0$ m along a string, a traveling sinusoidal wave with an angular frequency of 440 rad/s has displacement $y = +4.5$ mm and transverse velocity $u = -0.75$ m/s. If the wave has the general form $y(x, t) = y_m \sin(kx - \omega t + \phi)$, what is phase constant $\phi$?

**74** Energy is transmitted at rate $P_1$ by a wave of frequency $f_1$ on a string under tension $\tau_1$. What is the new energy transmission rate $P_2$ in terms of $P_1$ (a) if the tension is increased to $\tau_2 = 4\tau_1$ and (b) if, instead, the frequency is decreased to $f_2 = f_1/2$?

**75** (a) What is the fastest transverse wave that can be sent along a steel wire? For safety reasons, the maximum tensile stress to which steel wires should be subjected is $7.00 \times 10^8$ N/m$^2$. The density of steel is 7800 kg/m$^3$. (b) Does your answer depend on the diameter of the wire?

**76** A standing wave results from the sum of two transverse traveling waves given by

$$y_1 = 0.050 \cos(\pi x - 4\pi t)$$

and $$y_2 = 0.050 \cos(\pi x + 4\pi t),$$

where $x$, $y_1$, and $y_2$ are in meters and $t$ is in seconds. (a) What is the smallest positive value of $x$ that corresponds to a node? Beginning at $t = 0$, what is the value of the (b) first, (c) second, and (d) third time the particle at $x = 0$ has zero velocity?

**77** SSM The type of rubber band used inside some baseballs and golf balls obeys Hooke's law over a wide range of elongation of the band. A segment of this material has an unstretched length $\ell$ and a mass $m$. When a force $F$ is applied, the band stretches an additional length $\Delta\ell$. (a) What is the speed (in terms of $m$, $\Delta\ell$, and the spring constant $k$) of transverse waves on this stretched rubber band? (b) Using your answer to (a), show that the time required for a transverse pulse to travel the length of the rubber band is proportional to $1/\sqrt{\Delta\ell}$ if $\Delta\ell \ll \ell$ and is constant if $\Delta\ell \gg \ell$.

**78** The speed of electromagnetic waves (which include visible light, radio, and x rays) in vacuum is $3.0 \times 10^8$ m/s. (a) Wavelengths of visible light waves range from about 400 nm in the violet to about 700 nm in the red. What is the range of frequencies of these waves? (b) The range of frequencies for shortwave radio (for example, FM radio and VHF television) is 1.5 to 300 MHz. What is the corresponding wavelength range? (c) X-ray wavelengths range from about 5.0 nm to about $1.0 \times 10^{-2}$ nm. What is the frequency range for x rays?

**79** SSM A 1.50 m wire has a mass of 8.70 g and is under a tension of 120 N. The wire is held rigidly at both ends and set into oscillation. (a) What is the speed of waves on the wire? What is the wavelength of the waves that produce (b) one-loop and (c) two-loop standing waves? What is the frequency of the waves that produce (d) one-loop and (e) two-loop standing waves?

**80** When played in a certain manner, the lowest resonant frequency of a certain violin string is concert A (440 Hz). What is the frequency of the (a) second and (b) third harmonic of the string?

**81** A sinusoidal transverse wave traveling in the negative direction of an $x$ axis has an amplitude of 1.00 cm, a frequency of 550 Hz, and a speed of 330 m/s. If the wave equation is of the form $y(x, t) = y_m \sin(kx \pm \omega t)$, what are (a) $y_m$, (b) $\omega$, (c) $k$, and (d) the correct choice of sign in front of $\omega$?

**82** Two sinusoidal waves of the same wavelength travel in the same direction along a stretched string. For wave 1, $y_m = 3.0$ mm and $\phi = 0$; for wave 2, $y_m = 5.0$ mm and $\phi = 70°$. What are the (a) amplitude and (b) phase constant of the resultant wave?

**83** SSM A sinusoidal transverse wave of amplitude $y_m$ and wavelength $\lambda$ travels on a stretched cord. (a) Find the ratio of the maximum particle speed (the speed with which a single particle in the cord moves transverse to the wave) to the wave speed. (b) Does this ratio depend on the material of which the cord is made?

**84** Oscillation of a 600 Hz tuning fork sets up standing waves in a string clamped at both ends. The wave speed for the string is 400 m/s. The standing wave has four loops and an amplitude of 2.0 mm. (a) What is the length of the string? (b) Write an equation for the displacement of the string as a function of position and time.

**85** A 120 cm length of string is stretched between fixed supports. What are the (a) longest, (b) second longest, and (c) third longest wavelength for waves traveling on the string if standing waves are to be set up? (d) Sketch those standing waves.

**86** (a) Write an equation describing a sinusoidal transverse wave traveling on a cord in the positive direction of a $y$ axis with an angular wave number of 60 cm$^{-1}$, a period of 0.20 s, and an amplitude of 3.0 mm. Take the transverse direction to be the $z$ direction. (b) What is the maximum transverse speed of a point on the cord?

**87** A wave on a string is described by

$$y(x, t) = 15.0 \sin(\pi x/8 - 4\pi t),$$

where $x$ and $y$ are in centimeters and $t$ is in seconds. (a) What is the transverse speed for a point on the string at $x = 6.00$ cm when $t = 0.250$ s? (b) What is the maximum transverse speed of any point on the string? (c) What is the magnitude of the transverse acceleration for a point on the string at $x = 6.00$ cm when $t = 0.250$ s? (d) What is the magnitude of the maximum transverse acceleration for any point on the string?

**88** ✈ *Body armor.* When a high-speed projectile such as a bullet or bomb fragment strikes modern body armor, the fabric of the armor stops the projectile and prevents penetration by quickly spreading the projectile's energy over a large area. This spreading is done by longitudinal and transverse pulses that move *radially* from the impact point, where the projectile pushes a cone-shaped dent into the fabric. The longitudinal pulse, racing along the fibers of the fabric at speed $v_l$ ahead of the denting, causes the fibers to thin and stretch, with material flowing radially inward into the dent. One such radial fiber is shown in Fig. 16-48$a$. Part of the projectile's energy goes into this motion and stretching. The transverse pulse, moving at a slower speed $v_t$, is due to the denting. As the projectile increases the dent's depth, the dent increases in radius, causing the material in the fibers to move in the same direction as the projectile (perpendicular to the transverse pulse's direction of travel). The rest of the projectile's energy goes into this motion. All the energy that does not eventually go into permanently deforming the fibers ends up as thermal energy.

Figure 16-48b is a graph of speed $v$ versus time $t$ for a bullet of mass 10.2 g fired from a .38 Special revolver directly into body armor. The scales of the vertical and horizontal axes are set by $v_s = 300$ m/s and $t_s = 40.0$ μs. Take $v_l = 2000$ m/s, and assume that the half-angle $\theta$ of the conical dent is 60°. At the end of the collision, what are the radii of (a) the thinned region and (b) the dent (assuming that the person wearing the armor remains stationary)?

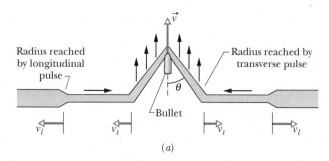

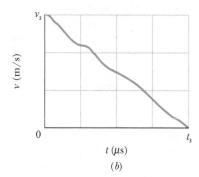

**Figure 16-48** Problem 88.

**89** Two waves are described by

$$y_1 = 0.30 \sin[\pi(5x - 200t)]$$

and $\qquad y_2 = 0.30 \sin[\pi(5x - 200t) + \pi/3],$

where $y_1$, $y_2$, and $x$ are in meters and $t$ is in seconds. When these two waves are combined, a traveling wave is produced. What are the (a) amplitude, (b) wave speed, and (c) wavelength of that traveling wave?

**90** A certain transverse sinusoidal wave of wavelength 20 cm is moving in the positive direction of an $x$ axis. The transverse velocity of the particle at $x = 0$ as a function of time is shown in Fig. 16-49, where the scale of the vertical axis is set by $u_s = 5.0$ cm/s. What are the (a) wave speed, (b) amplitude, and (c) frequency? (d) Sketch the wave between $x = 0$ and $x = 20$ cm at $t = 2.0$ s.

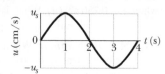

**Figure 16-49** Problem 90.

**91** **SSM** In a demonstration, a 1.2 kg horizontal rope is fixed in place at its two ends ($x = 0$ and $x = 2.0$ m) and made to oscillate up and down in the fundamental mode, at frequency 5.0 Hz. At $t = 0$, the point at $x = 1.0$ m has zero displacement and is moving upward in the positive direction of a $y$ axis with a transverse velocity of 5.0 m/s. What are (a) the amplitude of the motion of that point and (b) the tension in the rope? (c) Write the standing wave equation for the fundamental mode.

**92** Two waves,

$$y_1 = (2.50 \text{ mm}) \sin[(25.1 \text{ rad/m})x - (440 \text{ rad/s})t]$$

and $\quad y_2 = (1.50 \text{ mm}) \sin[(25.1 \text{ rad/m})x + (440 \text{ rad/s})t],$

travel along a stretched string. (a) Plot the resultant wave as a function of $t$ for $x = 0$, $\lambda/8$, $\lambda/4$, $3\lambda/8$, and $\lambda/2$, where $\lambda$ is the wavelength. The graphs should extend from $t = 0$ to a little over one period. (b) The resultant wave is the superposition of a standing wave and a traveling wave. In which direction does the traveling wave move? (c) How can you change the original waves so the resultant wave is the superposition of standing and traveling waves with the same amplitudes as before but with the traveling wave moving in the opposite direction? Next, use your graphs to find the place at which the oscillation amplitude is (d) maximum and (e) minimum. (f) How is the maximum amplitude related to the amplitudes of the original two waves? (g) How is the minimum amplitude related to the amplitudes of the original two waves?

**93** A traveling wave on a string is described by

$$y = 2.0 \sin\left[2\pi\left(\frac{t}{0.40} + \frac{x}{80}\right)\right],$$

where $x$ and $y$ are in centimeters and $t$ is in seconds. (a) For $t = 0$, plot $y$ as a function of $x$ for $0 \le x \le 160$ cm. (b) Repeat (a) for $t = 0.05$ s and $t = 0.10$ s. From your graphs, determine (c) the wave speed and (d) the direction in which the wave is traveling.

**94** In Fig. 16-50, a circular loop of string is set spinning about the center point in a place with negligible gravity. The radius is 4.00 cm and the tangential speed of a string segment is 5.00 cm/s. The string is plucked. At what speed do transverse waves move along the string? (*Hint:* Apply Newton's second law to a small, but finite, section of the string.)

**Figure 16-50** Problem 94.

**95** A continuous traveling wave with amplitude $A$ is incident on a boundary. The continuous reflection, with a smaller amplitude $B$, travels back through the incoming wave. The resulting interference pattern is displayed in Fig. 16-51. The standing wave ratio is defined to be

$$\text{SWR} = \frac{A + B}{A - B}.$$

The reflection coefficient $R$ is the ratio of the power of the reflected wave to the power of the incoming wave and is thus proportional to the ratio $(B/A)^2$. What is the SWR for (a) total reflection

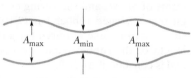

**Figure 16-51** Problem 95.

and (b) no reflection? (c) For SWR = 1.50, what is $R$ expressed as a percentage?

**96** Consider a loop in the standing wave created by two waves (amplitude 5.00 mm and frequency 120 Hz) traveling in opposite directions along a string with length 2.25 m and mass 125 g and under tension 40 N. At what rate does energy enter the loop from (a) each side and (b) both sides? (c) What is the maximum kinetic energy of the string in the loop during its oscillation?

# Waves—II

## 17-1 SPEED OF SOUND

### Learning Objectives

*After reading this module, you should be able to . . .*

**17.01** Distinguish between a longitudinal wave and a transverse wave.

**17.02** Explain wavefronts and rays.

**17.03** Apply the relationship between the speed of sound through a material, the material's bulk modulus, and the material's density.

**17.04** Apply the relationship between the speed of sound, the distance traveled by a sound wave, and the time required to travel that distance.

### Key Idea

● Sound waves are longitudinal mechanical waves that can travel through solids, liquids, or gases. The speed $v$ of a sound wave in a medium having bulk modulus $B$ and density $\rho$ is

$$v = \sqrt{\frac{B}{\rho}} \quad \text{(speed of sound).}$$

In air at 20°C, the speed of sound is 343 m/s.

## What Is Physics?

The physics of sound waves is the basis of countless studies in the research journals of many fields. Here are just a few examples. Some physiologists are concerned with how speech is produced, how speech impairment might be corrected, how hearing loss can be alleviated, and even how snoring is produced. Some acoustic engineers are concerned with improving the acoustics of cathedrals and concert halls, with reducing noise near freeways and road construction, and with reproducing music by speaker systems. Some aviation engineers are concerned with the shock waves produced by supersonic aircraft and the aircraft noise produced in communities near an airport. Some medical researchers are concerned with how noises produced by the heart and lungs can signal a medical problem in a patient. Some paleontologists are concerned with how a dinosaur's fossil might reveal the dinosaur's vocalizations. Some military engineers are concerned with how the sounds of sniper fire might allow a soldier to pinpoint the sniper's location, and, on the gentler side, some biologists are concerned with how a cat purrs.

To begin our discussion of the physics of sound, we must first answer the question "What *are* sound waves?"

## Sound Waves

As we saw in Chapter 16, mechanical waves are waves that require a material medium to exist. There are two types of mechanical waves: *Transverse waves* involve oscillations perpendicular to the direction in which the wave travels; *longitudinal waves* involve oscillations parallel to the direction of wave travel.

In this book, a **sound wave** is defined roughly as any longitudinal wave. Seismic prospecting teams use such waves to probe Earth's crust for oil. Ships

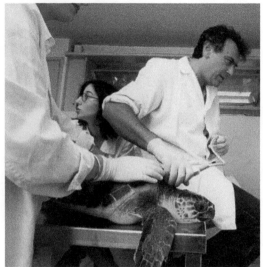

Mauro Fermariello/SPL/Photo Researchers, Inc.

**Figure 17-1** A loggerhead turtle is being checked with ultrasound (which has a frequency above your hearing range); an image of its interior is being produced on a monitor off to the right.

carry sound-ranging gear (sonar) to detect underwater obstacles. Submarines use sound waves to stalk other submarines, largely by listening for the characteristic noises produced by the propulsion system. Figure 17-1 suggests how sound waves can be used to explore the soft tissues of an animal or human body. In this chapter we shall focus on sound waves that travel through the air and that are audible to people.

Figure 17-2 illustrates several ideas that we shall use in our discussions. Point $S$ represents a tiny sound source, called a *point source,* that emits sound waves in all directions. The *wavefronts* and *rays* indicate the direction of travel and the spread of the sound waves. **Wavefronts** are surfaces over which the oscillations due to the sound wave have the same value; such surfaces are represented by whole or partial circles in a two-dimensional drawing for a point source. **Rays** are directed lines perpendicular to the wavefronts that indicate the direction of travel of the wavefronts. The short double arrows superimposed on the rays of Fig. 17-2 indicate that the longitudinal oscillations of the air are parallel to the rays.

Near a point source like that of Fig. 17-2, the wavefronts are spherical and spread out in three dimensions, and there the waves are said to be *spherical.* As the wavefronts move outward and their radii become larger, their curvature decreases. Far from the source, we approximate the wavefronts as planes (or lines on two-dimensional drawings), and the waves are said to be *planar.*

## The Speed of Sound

The speed of any mechanical wave, transverse or longitudinal, depends on both an inertial property of the medium (to store kinetic energy) and an elastic property of the medium (to store potential energy). Thus, we can generalize Eq. 16-26, which gives the speed of a transverse wave along a stretched string, by writing

$$v = \sqrt{\frac{\tau}{\mu}} = \sqrt{\frac{\text{elastic property}}{\text{inertial property}}}, \qquad (17\text{-}1)$$

where (for transverse waves) $\tau$ is the tension in the string and $\mu$ is the string's linear density. If the medium is air and the wave is longitudinal, we can guess that the inertial property, corresponding to $\mu$, is the volume density $\rho$ of air. What shall we put for the elastic property?

In a stretched string, potential energy is associated with the periodic stretching of the string elements as the wave passes through them. As a sound wave passes through air, potential energy is associated with periodic compressions and expansions of small volume elements of the air. The property that determines the extent to which an element of a medium changes in volume when the pressure (force per unit area) on it changes is the **bulk modulus** $B$, defined (from Eq. 12-25) as

$$B = -\frac{\Delta p}{\Delta V/V} \qquad \text{(definition of bulk modulus)}. \qquad (17\text{-}2)$$

Here $\Delta V/V$ is the fractional change in volume produced by a change in pressure $\Delta p$. As explained in Module 14-1, the SI unit for pressure is the newton per square meter, which is given a special name, the *pascal* (Pa). From Eq. 17-2 we see that the unit for $B$ is also the pascal. The signs of $\Delta p$ and $\Delta V$ are always opposite: When we increase the pressure on an element ($\Delta p$ is positive), its volume decreases ($\Delta V$ is negative). We include a minus sign in Eq. 17-2 so that $B$ is always a positive quantity. Now substituting $B$ for $\tau$ and $\rho$ for $\mu$ in Eq. 17-1 yields

$$v = \sqrt{\frac{B}{\rho}} \qquad \text{(speed of sound)} \qquad (17\text{-}3)$$

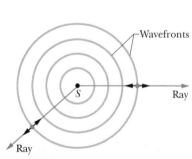

**Figure 17-2** A sound wave travels from a point source $S$ through a three-dimensional medium. The wavefronts form spheres centered on $S$; the rays are radial to $S$. The short, double-headed arrows indicate that elements of the medium oscillate parallel to the rays.

as the speed of sound in a medium with bulk modulus $B$ and density $\rho$. Table 17-1 lists the speed of sound in various media.

The density of water is almost 1000 times greater than the density of air. If this were the only relevant factor, we would expect from Eq. 17-3 that the speed of sound in water would be considerably less than the speed of sound in air. However, Table 17-1 shows us that the reverse is true. We conclude (again from Eq. 17-3) that the bulk modulus of water must be more than 1000 times greater than that of air. This is indeed the case. Water is much more incompressible than air, which (see Eq. 17-2) is another way of saying that its bulk modulus is much greater.

### Formal Derivation of Eq. 17-3

We now derive Eq. 17-3 by direct application of Newton's laws. Let a single pulse in which air is compressed travel (from right to left) with speed $v$ through the air in a long tube, like that in Fig. 16-2. Let us run along with the pulse at that speed, so that the pulse appears to stand still in our reference frame. Figure 17-3a shows the situation as it is viewed from that frame. The pulse is standing still, and air is moving at speed $v$ through it from left to right.

Let the pressure of the undisturbed air be $p$ and the pressure inside the pulse be $p + \Delta p$, where $\Delta p$ is positive due to the compression. Consider an element of air of thickness $\Delta x$ and face area $A$, moving toward the pulse at speed $v$. As this element enters the pulse, the leading face of the element encounters a region of higher pressure, which slows the element to speed $v + \Delta v$, in which $\Delta v$ is negative. This slowing is complete when the rear face of the element reaches the pulse, which requires time interval

$$\Delta t = \frac{\Delta x}{v}. \tag{17-4}$$

Let us apply Newton's second law to the element. During $\Delta t$, the average force on the element's trailing face is $pA$ toward the right, and the average force on the leading face is $(p + \Delta p)A$ toward the left (Fig. 17-3b). Therefore, the average net force on the element during $\Delta t$ is

$$F = pA - (p + \Delta p)A$$

$$= -\Delta p\, A \quad \text{(net force)}. \tag{17-5}$$

The minus sign indicates that the net force on the air element is directed to the left in Fig. 17-3b. The volume of the element is $A\, \Delta x$, so with the aid of Eq. 17-4, we can write its mass as

$$\Delta m = \rho\, \Delta V = \rho A\, \Delta x = \rho A v\, \Delta t \quad \text{(mass)}. \tag{17-6}$$

The average acceleration of the element during $\Delta t$ is

$$a = \frac{\Delta v}{\Delta t} \quad \text{(acceleration)}. \tag{17-7}$$

| Medium | Speed (m/s) |
|---|---|
| *Gases* | |
| Air (0°C) | 331 |
| Air (20°C) | 343 |
| Helium | 965 |
| Hydrogen | 1284 |
| *Liquids* | |
| Water (0°C) | 1402 |
| Water (20°C) | 1482 |
| Seawater[b] | 1522 |
| *Solids* | |
| Aluminum | 6420 |
| Steel | 5941 |
| Granite | 6000 |

**Table 17-1 The Speed of Sound[a]**

[a]At 0°C and 1 atm pressure, except where noted.
[b]At 20°C and 3.5% salinity.

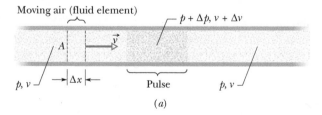

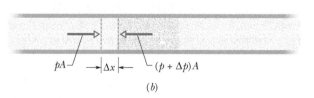

(a)  (b)

**Figure 17-3** A compression pulse is sent from right to left down a long air-filled tube. The reference frame of the figure is chosen so that the pulse is at rest and the air moves from left to right. (a) An element of air of width $\Delta x$ moves toward the pulse with speed $v$. (b) The leading face of the element enters the pulse. The forces acting on the leading and trailing faces (due to air pressure) are shown.

Thus, from Newton's second law ($F = ma$), we have, from Eqs. 17-5, 17-6, and 17-7,

$$-\Delta p\, A = (\rho A v\, \Delta t)\,\frac{\Delta v}{\Delta t}, \tag{17-8}$$

which we can write as

$$\rho v^2 = -\frac{\Delta p}{\Delta v/v}. \tag{17-9}$$

The air that occupies a volume $V\,(= Av\,\Delta t)$ outside the pulse is compressed by an amount $\Delta V\,(= A\,\Delta v\,\Delta t)$ as it enters the pulse. Thus,

$$\frac{\Delta V}{V} = \frac{A\,\Delta v\,\Delta t}{Av\,\Delta t} = \frac{\Delta v}{v}. \tag{17-10}$$

Substituting Eq. 17-10 and then Eq. 17-2 into Eq. 17-9 leads to

$$\rho v^2 = -\frac{\Delta p}{\Delta v/v} = -\frac{\Delta p}{\Delta V/V} = B. \tag{17-11}$$

Solving for $v$ yields Eq. 17-3 for the speed of the air toward the right in Fig. 17-3, and thus for the actual speed of the pulse toward the left.

# 17-2 TRAVELING SOUND WAVES

## Learning Objectives

*After reading this module, you should be able to . . .*

**17.05** For any particular time and position, calculate the displacement $s(x, t)$ of an element of air as a sound wave travels through its location.

**17.06** Given a displacement function $s(x, t)$ for a sound wave, calculate the time between two given displacements.

**17.07** Apply the relationships between wave speed $v$, angular frequency $\omega$, angular wave number $k$, wavelength $\lambda$, period $T$, and frequency $f$.

**17.08** Sketch a graph of the displacement $s(x)$ of an element of air as a function of position, and identify the amplitude $s_m$ and wavelength $\lambda$.

**17.09** For any particular time and position, calculate the pressure variation $\Delta p$ (variation from atmospheric pressure) of an element of air as a sound wave travels through its location.

**17.10** Sketch a graph of the pressure variation $\Delta p(x)$ of an element as a function of position, and identify the amplitude $\Delta p_m$ and wavelength $\lambda$.

**17.11** Apply the relationship between pressure-variation amplitude $\Delta p_m$ and displacement amplitude $s_m$.

**17.12** Given a graph of position $s$ versus time for a sound wave, determine the amplitude $s_m$ and the period $T$.

**17.13** Given a graph of pressure variation $\Delta p$ versus time for a sound wave, determine the amplitude $\Delta p_m$ and the period $T$.

## Key Ideas

● A sound wave causes a longitudinal displacement $s$ of a mass element in a medium as given by

$$s = s_m \cos(kx - \omega t),$$

where $s_m$ is the displacement amplitude (maximum displacement) from equilibrium, $k = 2\pi/\lambda$, and $\omega = 2\pi f$, $\lambda$ and $f$ being the wavelength and frequency, respectively, of the sound wave.

● The sound wave also causes a pressure change $\Delta p$ of the medium from the equilibrium pressure:

$$\Delta p = \Delta p_m \sin(kx - \omega t),$$

where the pressure amplitude is

$$\Delta p_m = (v\rho\omega)s_m.$$

## Traveling Sound Waves

Here we examine the displacements and pressure variations associated with a sinusoidal sound wave traveling through air. Figure 17-4$a$ displays such a wave traveling rightward through a long air-filled tube. Recall from Chapter 16 that we can produce such a wave by sinusoidally moving a piston at the left end of

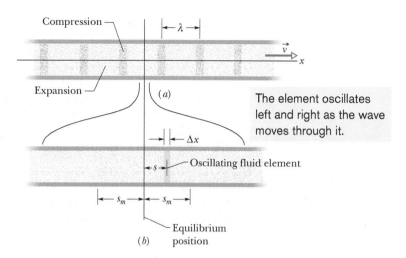

**Figure 17-4** (*a*) A sound wave, traveling through a long air-filled tube with speed *v*, consists of a moving, periodic pattern of expansions and compressions of the air. The wave is shown at an arbitrary instant. (*b*) A horizontally expanded view of a short piece of the tube. As the wave passes, an air element of thickness Δ*x* oscillates left and right in simple harmonic motion about its equilibrium position. At the instant shown in (*b*), the element happens to be displaced a distance *s* to the right of its equilibrium position. Its maximum displacement, either right or left, is $s_m$.

the tube (as in Fig. 16-2). The piston's rightward motion moves the element of air next to the piston face and compresses that air; the piston's leftward motion allows the element of air to move back to the left and the pressure to decrease. As each element of air pushes on the next element in turn, the right–left motion of the air and the change in its pressure travel along the tube as a sound wave.

Consider the thin element of air of thickness Δ*x* shown in Fig. 17-4*b*. As the wave travels through this portion of the tube, the element of air oscillates left and right in simple harmonic motion about its equilibrium position. Thus, the oscillations of each air element due to the traveling sound wave are like those of a string element due to a transverse wave, except that the air element oscillates *longitudinally* rather than *transversely*. Because string elements oscillate parallel to the *y* axis, we write their displacements in the form *y*(*x*, *t*). Similarly, because air elements oscillate parallel to the *x* axis, we could write their displacements in the confusing form *x*(*x*, *t*), but we shall use *s*(*x*, *t*) instead.

***Displacement.*** To show that the displacements *s*(*x*, *t*) are sinusoidal functions of *x* and *t*, we can use either a sine function or a cosine function. In this chapter we use a cosine function, writing

$$s(x, t) = s_m \cos(kx - \omega t). \tag{17-12}$$

Figure 17-5*a* labels the various parts of this equation. In it, $s_m$ is the **displacement amplitude**—that is, the maximum displacement of the air element to either side of its equilibrium position (see Fig. 17-4*b*). The angular wave number *k*, angular frequency *ω*, frequency *f*, wavelength *λ*, speed *v*, and period *T* for a sound (longitudinal) wave are defined and interrelated exactly as for a transverse wave, except that *λ* is now the distance (again along the direction of travel) in which the pattern of compression and expansion due to the wave begins to repeat itself (see Fig. 17-4*a*). (We assume $s_m$ is much less than *λ*.)

***Pressure.*** As the wave moves, the air pressure at any position *x* in Fig. 17-4*a* varies sinusoidally, as we prove next. To describe this variation we write

$$\Delta p(x, t) = \Delta p_m \sin(kx - \omega t). \tag{17-13}$$

Figure 17-5*b* labels the various parts of this equation. A negative value of Δ*p* in Eq. 17-13 corresponds to an expansion of the air, and a positive value to a compression. Here $\Delta p_m$ is the **pressure amplitude,** which is the maximum increase or decrease in pressure due to the wave; $\Delta p_m$ is normally very much less than the pressure *p* present when there is no wave. As we shall prove, the pressure ampli-

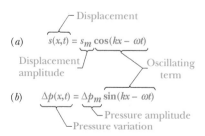

**Figure 17-5** (*a*) The displacement function and (*b*) the pressure-variation function of a traveling sound wave consist of an amplitude and an oscillating term.

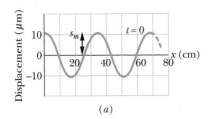

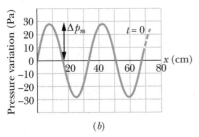

**Figure 17-6** (*a*) A plot of the displacement function (Eq. 17-12) for *t* = 0. (*b*) A similar plot of the pressure-variation function (Eq. 17-13). Both plots are for a 1000 Hz sound wave whose pressure amplitude is at the threshold of pain.

tude $\Delta p_m$ is related to the displacement amplitude $s_m$ in Eq. 17-12 by

$$\Delta p_m = (v\rho\omega)s_m. \tag{17-14}$$

Figure 17-6 shows plots of Eqs. 17-12 and 17-13 at $t = 0$; with time, the two curves would move rightward along the horizontal axes. Note that the displacement and pressure variation are $\pi/2$ rad (or 90°) out of phase. Thus, for example, the pressure variation $\Delta p$ at any point along the wave is zero when the displacement there is a maximum.

### ☑ Checkpoint 1

When the oscillating air element in Fig. 17-4b is moving rightward through the point of zero displacement, is the pressure in the element at its equilibrium value, just beginning to increase, or just beginning to decrease?

### Derivation of Eqs. 17-13 and 17-14

Figure 17-4b shows an oscillating element of air of cross-sectional area $A$ and thickness $\Delta x$, with its center displaced from its equilibrium position by distance $s$. From Eq. 17-2 we can write, for the pressure variation in the displaced element,

$$\Delta p = -B\frac{\Delta V}{V}. \tag{17-15}$$

The quantity $V$ in Eq. 17-15 is the volume of the element, given by

$$V = A \Delta x. \tag{17-16}$$

The quantity $\Delta V$ in Eq. 17-15 is the change in volume that occurs when the element is displaced. This volume change comes about because the displacements of the two faces of the element are not quite the same, differing by some amount $\Delta s$. Thus, we can write the change in volume as

$$\Delta V = A \Delta s. \tag{17-17}$$

Substituting Eqs. 17-16 and 17-17 into Eq. 17-15 and passing to the differential limit yield

$$\Delta p = -B\frac{\Delta s}{\Delta x} = -B\frac{\partial s}{\partial x}. \tag{17-18}$$

The symbols $\partial$ indicate that the derivative in Eq. 17-18 is a *partial derivative*, which tells us how $s$ changes with $x$ when the time $t$ is fixed. From Eq. 17-12 we then have, treating $t$ as a constant,

$$\frac{\partial s}{\partial x} = \frac{\partial}{\partial x}[s_m\cos(kx - \omega t)] = -ks_m\sin(kx - \omega t).$$

Substituting this quantity for the partial derivative in Eq. 17-18 yields

$$\Delta p = Bks_m\sin(kx - \omega t).$$

This tells us that the pressure varies as a sinusoidal function of time and that the amplitude of the variation is equal to the terms in front of the sine function. Setting $\Delta p_m = Bks_m$, this yields Eq. 17-13, which we set out to prove.

Using Eq. 17-3, we can now write

$$\Delta p_m = (Bk)s_m = (v^2\rho k)s_m.$$

Equation 17-14, which we also wanted to prove, follows at once if we substitute $\omega/v$ for $k$ from Eq. 16-12.

## Sample Problem 17.01    Pressure amplitude, displacement amplitude

The maximum pressure amplitude $\Delta p_m$ that the human ear can tolerate in loud sounds is about 28 Pa (which is very much less than the normal air pressure of about $10^5$ Pa). What is the displacement amplitude $s_m$ for such a sound in air of density $\rho = 1.21$ kg/m³, at a frequency of 1000 Hz and a speed of 343 m/s?

### KEY IDEA

The displacement amplitude $s_m$ of a sound wave is related to the pressure amplitude $\Delta p_m$ of the wave according to Eq. 17-14.

*Calculations:*  Solving that equation for $s_m$ yields

$$s_m = \frac{\Delta p_m}{v\rho\omega} = \frac{\Delta p_m}{v\rho(2\pi f)}.$$

Substituting known data then gives us

$$s_m = \frac{28 \text{ Pa}}{(343 \text{ m/s})(1.21 \text{ kg/m}^3)(2\pi)(1000 \text{ Hz})}$$
$$= 1.1 \times 10^{-5} \text{ m} = 11 \; \mu\text{m}. \qquad \text{(Answer)}$$

That is only about one-seventh the thickness of a book page. Obviously, the displacement amplitude of even the loudest sound that the ear can tolerate is very small. Temporary exposure to such loud sound produces temporary hearing loss, probably due to a decrease in blood supply to the inner ear. Prolonged exposure produces permanent damage.

The pressure amplitude $\Delta p_m$ for the *faintest* detectable sound at 1000 Hz is $2.8 \times 10^{-5}$ Pa. Proceeding as above leads to $s_m = 1.1 \times 10^{-11}$ m or 11 pm, which is about one-tenth the radius of a typical atom. The ear is indeed a sensitive detector of sound waves.

 Additional examples, video, and practice available at *WileyPLUS*

# 17-3 INTERFERENCE

## Learning Objectives

*After reading this module, you should be able to . . .*

**17.14** If two waves with the same wavelength begin in phase but reach a common point by traveling along different paths, calculate their phase difference $\phi$ at that point by relating the path length difference $\Delta L$ to the wavelength $\lambda$.

**17.15** Given the phase difference between two sound waves with the same amplitude, wavelength, and travel direction, determine the type of interference between the waves (fully destructive interference, fully constructive interference, or indeterminate interference).

**17.16** Convert a phase difference between radians, degrees, and number of wavelengths.

## Key Ideas

● The interference of two sound waves with identical wavelengths passing through a common point depends on their phase difference $\phi$ there. If the sound waves were emitted in phase and are traveling in approximately the same direction, $\phi$ is given by

$$\phi = \frac{\Delta L}{\lambda} 2\pi,$$

where $\Delta L$ is their path length difference.

● Fully constructive interference occurs when $\phi$ is an integer multiple of $2\pi$,

$$\phi = m(2\pi), \qquad \text{for } m = 0, 1, 2, \ldots ,$$

and, equivalently, when $\Delta L$ is related to wavelength $\lambda$ by

$$\frac{\Delta L}{\lambda} = 0, 1, 2, \ldots .$$

● Fully destructive interference occurs when $\phi$ is an odd multiple of $\pi$,

$$\phi = (2m + 1)\pi, \qquad \text{for } m = 0, 1, 2, \ldots ,$$

and

$$\frac{\Delta L}{\lambda} = 0.5, 1.5, 2.5, \ldots .$$

## Interference

Like transverse waves, sound waves can undergo interference. In fact, we can write equations for the interference as we did in Module 16-5 for transverse waves. Suppose two sound waves with the same amplitude and wavelength are traveling in the positive direction of an $x$ axis with a phase difference of $\phi$. We can express the waves in the form of Eqs. 16-47 and 16-48 but, to be consistent with Eq. 17-12, we use cosine functions instead of sine functions:

$$s_1(x, t) = s_m \cos(kx - \omega t)$$

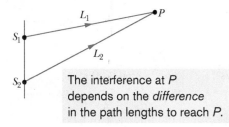

The interference at $P$ depends on the *difference* in the path lengths to reach $P$.

(a)

If the difference is equal to, say, $2.0\lambda$, then the waves arrive exactly in phase. This is how transverse waves would look.

(b)

If the difference is equal to, say, $2.5\lambda$, then the waves arrive exactly out of phase. This is how transverse waves would look.

(c)

**Figure 17-7** (a) Two point sources $S_1$ and $S_2$ emit spherical sound waves in phase. The rays indicate that the waves pass through a common point $P$. The waves (represented with *transverse* waves) arrive at $P$ (b) exactly in phase and (c) exactly out of phase.

and

$$s_2(x, t) = s_m \cos(kx - \omega t + \phi).$$

These waves overlap and interfere. From Eq. 16-51, we can write the resultant wave as

$$s' = [2s_m \cos\tfrac{1}{2}\phi] \cos(kx - \omega t + \tfrac{1}{2}\phi).$$

As we saw with transverse waves, the resultant wave is itself a traveling wave. Its amplitude is the magnitude

$$s'_m = |2s_m \cos\tfrac{1}{2}\phi|. \tag{17-19}$$

As with transverse waves, the value of $\phi$ determines what type of interference the individual waves undergo.

One way to control $\phi$ is to send the waves along paths with different lengths. Figure 17-7a shows how we can set up such a situation: Two point sources $S_1$ and $S_2$ emit sound waves that are in phase and of identical wavelength $\lambda$. Thus, the *sources* themselves are said to be in phase; that is, as the waves emerge from the sources, their displacements are always identical. We are interested in the waves that then travel through point $P$ in Fig. 17-7a. We assume that the distance to $P$ is much greater than the distance between the sources so that we can approximate the waves as traveling in the same direction at $P$.

If the waves traveled along paths with identical lengths to reach point $P$, they would be in phase there. As with transverse waves, this means that they would undergo fully constructive interference there. However, in Fig. 17-7a, path $L_2$ traveled by the wave from $S_2$ is longer than path $L_1$ traveled by the wave from $S_1$. The difference in path lengths means that the waves may not be in phase at point $P$. In other words, their phase difference $\phi$ at $P$ depends on their **path length difference** $\Delta L = |L_2 - L_1|$.

To relate phase difference $\phi$ to path length difference $\Delta L$, we recall (from Module 16-1) that a phase difference of $2\pi$ rad corresponds to one wavelength. Thus, we can write the proportion

$$\frac{\phi}{2\pi} = \frac{\Delta L}{\lambda}, \tag{17-20}$$

from which

$$\phi = \frac{\Delta L}{\lambda} 2\pi. \tag{17-21}$$

Fully constructive interference occurs when $\phi$ is zero, $2\pi$, or any integer multiple of $2\pi$. We can write this condition as

$$\phi = m(2\pi), \quad \text{for } m = 0, 1, 2, \ldots \quad \text{(fully constructive interference)}. \tag{17-22}$$

From Eq. 17-21, this occurs when the ratio $\Delta L/\lambda$ is

$$\frac{\Delta L}{\lambda} = 0, 1, 2, \ldots \quad \text{(fully constructive interference)}. \tag{17-23}$$

For example, if the path length difference $\Delta L = |L_2 - L_1|$ in Fig. 17-7a is equal to $2\lambda$, then $\Delta L/\lambda = 2$ and the waves undergo fully constructive interference at point $P$ (Fig. 17-7b). The interference is fully constructive because the wave from $S_2$ is phase-shifted relative to the wave from $S_1$ by $2\lambda$, putting the two waves *exactly in phase* at $P$.

Fully destructive interference occurs when $\phi$ is an odd multiple of $\pi$:

$$\phi = (2m + 1)\pi, \quad \text{for } m = 0, 1, 2, \ldots \quad \text{(fully destructive interference)}. \tag{17-24}$$

From Eq. 17-21, this occurs when the ratio $\Delta L/\lambda$ is

$$\frac{\Delta L}{\lambda} = 0.5, 1.5, 2.5, \ldots \quad \text{(fully destructive interference).} \quad (17\text{-}25)$$

For example, if the path length difference $\Delta L = |L_2 - L_1|$ in Fig. 17-7a is equal to 2.5$\lambda$, then $\Delta L/\lambda = 2.5$ and the waves undergo fully destructive interference at point $P$ (Fig. 17-7c). The interference is fully destructive because the wave from $S_2$ is phase-shifted relative to the wave from $S_1$ by 2.5 wavelengths, which puts the two waves *exactly out of phase* at P.

Of course, two waves could produce intermediate interference as, say, when $\Delta L/\lambda = 1.2$. This would be closer to fully constructive interference ($\Delta L/\lambda = 1.0$) than to fully destructive interference ($\Delta L/\lambda = 1.5$).

## Sample Problem 17.02    Interference points along a big circle

In Fig. 17-8a, two point sources $S_1$ and $S_2$, which are in phase and separated by distance $D = 1.5\lambda$, emit identical sound waves of wavelength $\lambda$.

(a) What is the path length difference of the waves from $S_1$ and $S_2$ at point $P_1$, which lies on the perpendicular bisector of distance $D$, at a distance greater than $D$ from the sources (Fig. 17-8b)? (That is, what is the difference in the distance from source $S_1$ to point $P_1$ and the distance from source $S_2$ to $P_1$?) What type of interference occurs at $P_1$?

**Reasoning:** Because the waves travel identical distances to reach $P_1$, their path length difference is

$$\Delta L = 0. \quad \text{(Answer)}$$

From Eq. 17-23, this means that the waves undergo fully constructive interference at $P_1$ because they start in phase at the sources and reach $P_1$ in phase.

(b) What are the path length difference and type of interference at point $P_2$ in Fig. 17-8c?

**Reasoning:** The wave from $S_1$ travels the extra distance $D$ (= 1.5$\lambda$) to reach $P_2$. Thus, the path length difference is

$$\Delta L = 1.5\lambda. \quad \text{(Answer)}$$

From Eq. 17-25, this means that the waves are exactly out of phase at $P_2$ and undergo fully destructive interference there.

(c) Figure 17-8d shows a circle with a radius much greater than $D$, centered on the midpoint between sources $S_1$ and $S_2$. What is the number of points $N$ around this circle at which the interference is fully constructive? (That is, at how many points do the waves arrive exactly in phase?)

**Reasoning:** Starting at point $a$, let's move clockwise along the circle to point $d$. As we move, path length difference $\Delta L$ increases and so the type of interference changes. From (a), we know that is $\Delta L = 0\lambda$ at point $a$. From (b), we know that $\Delta L = 1.5\lambda$ at point $d$. Thus, there must be

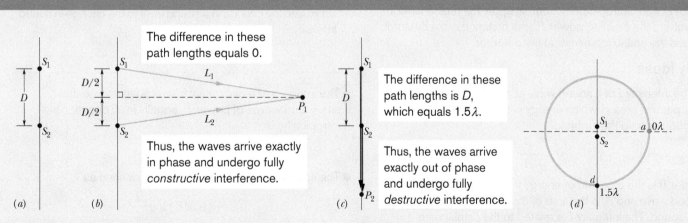

(a)  (b)  The difference in these path lengths equals 0.  Thus, the waves arrive exactly in phase and undergo fully *constructive* interference.  (c)  The difference in these path lengths is $D$, which equals 1.5$\lambda$.  Thus, the waves arrive exactly out of phase and undergo fully *destructive* interference.  (d)

**Figure 17-8** (a) Two point sources $S_1$ and $S_2$, separated by distance $D$, emit spherical sound waves in phase. (b) The waves travel equal distances to reach point $P_1$. (c) Point $P_2$ is on the line extending through $S_1$ and $S_2$. (d) We move around a large circle. (*Figure continues*)

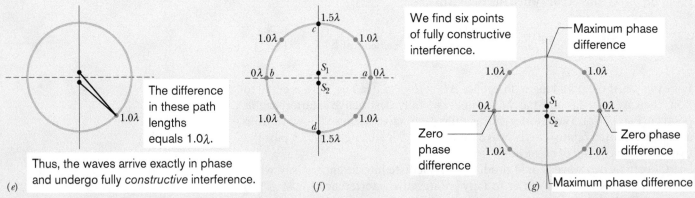

**Figure 17-8** (*continued*) (*e*) Another point of fully constructive interference. (*f*) Using symmetry to determine other points. (*g*) The six points of fully constructive interference.

one point between *a* and *d* at which $\Delta L = \lambda$ (Fig. 17-8*e*). From Eq. 17-23, fully constructive interference occurs at that point. Also, there can be no other point along the way from point *a* to point *d* at which fully constructive interference occurs, because there is no other integer than 1 between 0 at point *a* and 1.5 at point *d*.

We can now use symmetry to locate other points of fully constructive or destructive interference (Fig. 17-8*f*). Symmetry about line *cd* gives us point *b*, at which $\Delta L = 0\lambda$. Also, there are three more points at which $\Delta L = \lambda$. In all (Fig. 17-8*g*) we have

$$N = 6. \qquad \text{(Answer)}$$

 **PLUS** Additional examples, video, and practice available at *WileyPLUS*

# 17-4 INTENSITY AND SOUND LEVEL

## Learning Objectives

*After reading this module, you should be able to . . .*

**17.17** Calculate the sound intensity *I* at a surface as the ratio of the power *P* to the surface area *A*.

**17.18** Apply the relationship between the sound intensity *I* and the displacement amplitude $s_m$ of the sound wave.

**17.19** Identify an isotropic point source of sound.

**17.20** For an isotropic point source, apply the relationship involving the emitting power $P_s$, the distance *r* to a detector, and the sound intensity *I* at the detector.

**17.21** Apply the relationship between the sound level $\beta$, the sound intensity *I*, and the standard reference intensity $I_0$.

**17.22** Evaluate a logarithm function (log) and an antilogarithm function ($\log^{-1}$).

**17.23** Relate the change in a sound level to the change in sound intensity.

## Key Ideas

● The intensity *I* of a sound wave at a surface is the average rate per unit area at which energy is transferred by the wave through or onto the surface:

$$I = \frac{P}{A},$$

where *P* is the time rate of energy transfer (power) of the sound wave and *A* is the area of the surface intercepting the sound. The intensity *I* is related to the displacement amplitude $s_m$ of the sound wave by

$$I = \tfrac{1}{2}\rho v \omega^2 s_m^2.$$

● The intensity at a distance *r* from a point source that emits sound waves of power $P_s$ equally in all directions (isotropically) is

$$I = \frac{P_s}{4\pi r^2}.$$

● The sound level $\beta$ in decibels (dB) is defined as

$$\beta = (10 \text{ dB}) \log \frac{I}{I_0},$$

where $I_0 \, (= 10^{-12} \text{ W/m}^2)$ is a reference intensity level to which all intensities are compared. For every factor-of-10 increase in intensity, 10 dB is added to the sound level.

## Intensity and Sound Level

If you have ever tried to sleep while someone played loud music nearby, you are well aware that there is more to sound than frequency, wavelength, and speed. There is also intensity. The **intensity** $I$ of a sound wave at a surface is the average rate per unit area at which energy is transferred by the wave through or onto the surface. We can write this as

$$I = \frac{P}{A},\qquad(17\text{-}26)$$

where $P$ is the time rate of energy transfer (the power) of the sound wave and $A$ is the area of the surface intercepting the sound. As we shall derive shortly, the intensity $I$ is related to the displacement amplitude $s_m$ of the sound wave by

$$I = \tfrac{1}{2}\rho v \omega^2 s_m^2.\qquad(17\text{-}27)$$

Intensity can be measured on a detector. *Loudness* is a perception, something that you sense. The two can differ because your perception depends on factors such as the sensitivity of your hearing mechanism to various frequencies.

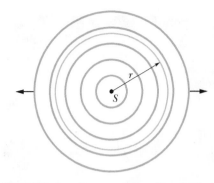

**Figure 17-9** A point source $S$ emits sound waves uniformly in all directions. The waves pass through an imaginary sphere of radius $r$ that is centered on $S$.

### Variation of Intensity with Distance

How intensity varies with distance from a real sound source is often complex. Some real sources (like loudspeakers) may transmit sound only in particular directions, and the environment usually produces echoes (reflected sound waves) that overlap the direct sound waves. In some situations, however, we can ignore echoes and assume that the sound source is a point source that emits the sound *isotropically*—that is, with equal intensity in all directions. The wavefronts spreading from such an isotropic point source $S$ at a particular instant are shown in Fig. 17-9.

Let us assume that the mechanical energy of the sound waves is conserved as they spread from this source. Let us also center an imaginary sphere of radius $r$ on the source, as shown in Fig. 17-9. All the energy emitted by the source must pass through the surface of the sphere. Thus, the time rate at which energy is transferred through the surface by the sound waves must equal the time rate at which energy is emitted by the source (that is, the power $P_s$ of the source). From Eq. 17-26, the intensity $I$ at the sphere must then be

$$I = \frac{P_s}{4\pi r^2},\qquad(17\text{-}28)$$

where $4\pi r^2$ is the area of the sphere. Equation 17-28 tells us that the intensity of sound from an isotropic point source decreases with the square of the distance $r$ from the source.

 **Checkpoint 2**

The figure indicates three small patches 1, 2, and 3 that lie on the surfaces of two imaginary spheres; the spheres are centered on an isotropic point source $S$ of sound. The rates at which energy is transmitted through the three patches by the sound waves are equal. Rank the patches according to (a) the intensity of the sound on them and (b) their area, greatest first.

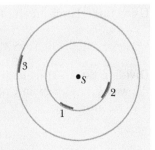

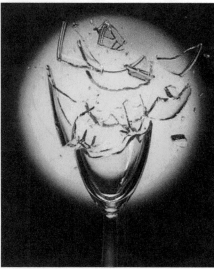

© Ben Rose

Sound can cause the wall of a drinking glass to oscillate. If the sound produces a standing wave of oscillations and if the intensity of the sound is large enough, the glass will shatter.

## The Decibel Scale

The displacement amplitude at the human ear ranges from about $10^{-5}$ m for the loudest tolerable sound to about $10^{-11}$ m for the faintest detectable sound, a ratio of $10^6$. From Eq. 17-27 we see that the intensity of a sound varies as the *square* of its amplitude, so the ratio of intensities at these two limits of the human auditory system is $10^{12}$. Humans can hear over an enormous range of intensities.

We deal with such an enormous range of values by using logarithms. Consider the relation

$$y = \log x,$$

in which $x$ and $y$ are variables. It is a property of this equation that if we *multiply* $x$ by 10, then $y$ increases by 1. To see this, we write

$$y' = \log(10x) = \log 10 + \log x = 1 + y.$$

Similarly, if we multiply $x$ by $10^{12}$, $y$ increases by only 12.

Thus, instead of speaking of the intensity $I$ of a sound wave, it is much more convenient to speak of its **sound level** $\beta$, defined as

$$\beta = (10\ \text{dB}) \log \frac{I}{I_0}. \qquad (17\text{-}29)$$

Here dB is the abbreviation for **decibel,** the unit of sound level, a name that was chosen to recognize the work of Alexander Graham Bell. $I_0$ in Eq. 17-29 is a standard reference intensity ($= 10^{-12}$ W/m$^2$), chosen because it is near the lower limit of the human range of hearing. For $I = I_0$, Eq. 17-29 gives $\beta = 10 \log 1 = 0$, so our standard reference level corresponds to zero decibels. Then $\beta$ increases by 10 dB every time the sound intensity increases by an order of magnitude (a factor of 10). Thus, $\beta = 40$ corresponds to an intensity that is $10^4$ times the standard reference level. Table 17-2 lists the sound levels for a variety of environments.

### Derivation of Eq. 17-27

Consider, in Fig. 17-4a, a thin slice of air of thickness $dx$, area $A$, and mass $dm$, oscillating back and forth as the sound wave of Eq. 17-12 passes through it. The kinetic energy $dK$ of the slice of air is

$$dK = \tfrac{1}{2} dm\, v_s^2. \qquad (17\text{-}30)$$

Here $v_s$ is not the speed of the wave but the speed of the oscillating element of air, obtained from Eq. 17-12 as

$$v_s = \frac{\partial s}{\partial t} = -\omega s_m \sin(kx - \omega t).$$

Using this relation and putting $dm = \rho A\, dx$ allow us to rewrite Eq. 17-30 as

$$dK = \tfrac{1}{2}(\rho A\, dx)(-\omega s_m)^2 \sin^2(kx - \omega t). \qquad (17\text{-}31)$$

Dividing Eq. 17-31 by $dt$ gives the rate at which kinetic energy moves along with the wave. As we saw in Chapter 16 for transverse waves, $dx/dt$ is the wave speed $v$, so we have

$$\frac{dK}{dt} = \tfrac{1}{2}\rho A v \omega^2 s_m^2 \sin^2(kx - \omega t). \qquad (17\text{-}32)$$

**Table 17-2 Some Sound Levels (dB)**

| | |
|---|---|
| Hearing threshold | 0 |
| Rustle of leaves | 10 |
| Conversation | 60 |
| Rock concert | 110 |
| Pain threshold | 120 |
| Jet engine | 130 |

The *average* rate at which kinetic energy is transported is

$$\left(\frac{dK}{dt}\right)_{avg} = \tfrac{1}{2}\rho A v\omega^2 s_m^2[\sin^2(kx - \omega t)]_{avg}$$

$$= \tfrac{1}{4}\rho A v\omega^2 s_m^2. \tag{17-33}$$

To obtain this equation, we have used the fact that the average value of the square of a sine (or a cosine) function over one full oscillation is $\tfrac{1}{2}$.

We assume that *potential* energy is carried along with the wave at this same average rate. The wave intensity $I$, which is the average rate per unit area at which energy of both kinds is transmitted by the wave, is then, from Eq. 17-33,

$$I = \frac{2(dK/dt)_{avg}}{A} = \tfrac{1}{2}\rho v\omega^2 s_m^2,$$

which is Eq. 17-27, the equation we set out to derive.

---

## Sample Problem 17.03   Intensity change with distance, cylindrical sound wave

An electric spark jumps along a straight line of length $L = 10$ m, emitting a pulse of sound that travels radially outward from the spark. (The spark is said to be a *line source* of sound.) The power of this acoustic emission is $P_s = 1.6 \times 10^4$ W.

(a) What is the intensity $I$ of the sound when it reaches a distance $r = 12$ m from the spark?

### KEY IDEAS

(1) Let us center an imaginary cylinder of radius $r = 12$ m and length $L = 10$ m (open at both ends) on the spark, as shown in Fig. 17-10. Then the intensity $I$ at the cylindrical surface is the ratio $P/A$, where $P$ is the time rate at which sound energy passes through the surface and $A$ is the surface area. (2) We assume that the principle of conservation of energy applies to the sound energy. This means that the rate $P$ at which energy is transferred through the cylinder must equal the rate $P_s$ at which energy is emitted by the source.

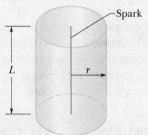

**Figure 17-10** A spark along a straight line of length $L$ emits sound waves radially outward. The waves pass through an imaginary cylinder of radius $r$ and length $L$ that is centered on the spark.

*Calculations:* Putting these ideas together and noting that the area of the cylindrical surface is $A = 2\pi rL$, we have

$$I = \frac{P}{A} = \frac{P_s}{2\pi rL}. \tag{17-34}$$

This tells us that the intensity of the sound from a line source decreases with distance $r$ (and not with the square of distance $r$ as for a point source). Substituting the given data, we find

$$I = \frac{1.6 \times 10^4 \text{ W}}{2\pi(12 \text{ m})(10 \text{ m})}$$

$$= 21.2 \text{ W/m}^2 \approx 21 \text{ W/m}^2. \qquad \text{(Answer)}$$

(b) At what time rate $P_d$ is sound energy intercepted by an acoustic detector of area $A_d = 2.0$ cm$^2$, aimed at the spark and located a distance $r = 12$ m from the spark?

*Calculations:* We know that the intensity of sound at the detector is the ratio of the energy transfer rate $P_d$ there to the detector's area $A_d$:

$$I = \frac{P_d}{A_d}. \tag{17-35}$$

We can imagine that the detector lies on the cylindrical surface of (a). Then the sound intensity at the detector is the intensity $I\ (= 21.2$ W/m$^2)$ at the cylindrical surface. Solving Eq. 17-35 for $P_d$ gives us

$$P_d = (21.2 \text{ W/m}^2)(2.0 \times 10^{-4} \text{ m}^2) = 4.2 \text{ mW}. \qquad \text{(Answer)}$$

**Sample Problem 17.04** Decibels, sound level, change in intensity

Many veteran rockers suffer from acute hearing damage because of the high sound levels they endured for years. Many rockers now wear special earplugs to protect their hearing during performances (Fig. 17-11). If an earplug decreases the sound level of the sound waves by 20 dB, what is the ratio of the final intensity $I_f$ of the waves to their initial intensity $I_i$?

### KEY IDEA

For both the final and initial waves, the sound level $\beta$ is related to the intensity by the definition of sound level in Eq. 17-29.

*Calculations:* For the final waves we have

$$\beta_f = (10 \text{ dB}) \log \frac{I_f}{I_0},$$

and for the initial waves we have

$$\beta_i = (10 \text{ dB}) \log \frac{I_i}{I_0}.$$

The difference in the sound levels is

$$\beta_f - \beta_i = (10 \text{ dB}) \left( \log \frac{I_f}{I_0} - \log \frac{I_i}{I_0} \right). \quad (17\text{-}36)$$

Using the identity

$$\log \frac{a}{b} - \log \frac{c}{d} = \log \frac{ad}{bc},$$

we can rewrite Eq. 17-36 as

$$\beta_f - \beta_i = (10 \text{ dB}) \log \frac{I_f}{I_i}. \quad (17\text{-}37)$$

Rearranging and then substituting the given decrease in

**Figure 17-11** Lars Ulrich of Metallica is an advocate for the organization HEAR (Hearing Education and Awareness for Rockers), which warns about the damage high sound levels can have on hearing.

Tim Mosenfelder/Getty Images, Inc.

sound level as $\beta_f - \beta_i = -20$ dB, we find

$$\log \frac{I_f}{I_i} = \frac{\beta_f - \beta_i}{10 \text{ dB}} = \frac{-20 \text{ dB}}{10 \text{ dB}} = -2.0.$$

We next take the antilog of the far left and far right sides of this equation. (Although the antilog $10^{-2.0}$ can be evaluated mentally, you could use a calculator by keying in 10^-2.0 or using the $10^x$ key.) We find

$$\frac{I_f}{I_i} = \log^{-1}(-2.0) = 0.010. \quad \text{(Answer)}$$

Thus, the earplug reduces the intensity of the sound waves to 0.010 of their initial intensity (two orders of magnitude).

PLUS Additional examples, video, and practice available at *WileyPLUS*

# 17-5 SOURCES OF MUSICAL SOUND

## Learning Objectives

*After reading this module, you should be able to . . .*

**17.24** Using standing wave patterns for string waves, sketch the standing wave patterns for the first several acoustical harmonics of a pipe with only one open end and with two open ends.

**17.25** For a standing wave of sound, relate the distance between nodes and the wavelength.

**17.26** Identify which type of pipe has even harmonics.

**17.27** For any given harmonic and for a pipe with only one open end or with two open ends, apply the relationships between the pipe length $L$, the speed of sound $v$, the wavelength $\lambda$, the harmonic frequency $f$, and the harmonic number $n$.

## Key Ideas

● Standing sound wave patterns can be set up in pipes (that is, resonance can be set up) if sound of the proper wavelength is introduced in the pipe.

● A pipe open at both ends will resonate at frequencies

$$f = \frac{v}{\lambda} = \frac{nv}{2L}, \quad n = 1, 2, 3, \ldots,$$

where $v$ is the speed of sound in the air in the pipe.

● For a pipe closed at one end and open at the other, the resonant frequencies are

$$f = \frac{v}{\lambda} = \frac{nv}{4L}, \quad n = 1, 3, 5, \ldots.$$

## Sources of Musical Sound

Musical sounds can be set up by oscillating strings (guitar, piano, violin), membranes (kettledrum, snare drum), air columns (flute, oboe, pipe organ, and the didgeridoo of Fig. 17-12), wooden blocks or steel bars (marimba, xylophone), and many other oscillating bodies. Most common instruments involve more than a single oscillating part.

Recall from Chapter 16 that standing waves can be set up on a stretched string that is fixed at both ends. They arise because waves traveling along the string are reflected back onto the string at each end. If the wavelength of the waves is suitably matched to the length of the string, the superposition of waves traveling in opposite directions produces a standing wave pattern (or oscillation mode). The wavelength required of the waves for such a match is one that corresponds to a *resonant frequency* of the string. The advantage of setting up standing waves is that the string then oscillates with a large, sustained amplitude, pushing back and forth against the surrounding air and thus generating a noticeable sound wave with the same frequency as the oscillations of the string. This production of sound is of obvious importance to, say, a guitarist.

***Sound Waves.*** We can set up standing waves of sound in an air-filled pipe in a similar way. As sound waves travel through the air in the pipe, they are reflected at each end and travel back through the pipe. (The reflection occurs even if an end is open, but the reflection is not as complete as when the end is closed.) If the wavelength of the sound waves is suitably matched to the length of the pipe, the superposition of waves traveling in opposite directions through the pipe sets up a standing wave pattern. The wavelength required of the sound waves for such a match is one that corresponds to a resonant frequency of the pipe. The advantage of such a standing wave is that the air in the pipe oscillates with a large, sustained amplitude, emitting at any open end a sound wave that has the same frequency as the oscillations in the pipe. This emission of sound is of obvious importance to, say, an organist.

Many other aspects of standing sound wave patterns are similar to those of string waves: The closed end of a pipe is like the fixed end of a string in that there must be a node (zero displacement) there, and the open end of a pipe is like the end of a string attached to a freely moving ring, as in Fig. 16-19b, in that there must be an antinode there. (Actually, the antinode for the open end of a pipe is located slightly beyond the end, but we shall not dwell on that detail.)

***Two Open Ends.*** The simplest standing wave pattern that can be set up in a pipe with two open ends is shown in Fig. 17-13a. There is an antinode across each

**Figure 17-12** The air column within a didgeridoo ("a pipe") oscillates when the instrument is played.

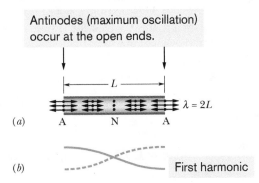

Antinodes (maximum oscillation) occur at the open ends.

$L$

$\lambda = 2L$

(a)  A  N  A

(b)  First harmonic

**Figure 17-13** (a) The simplest standing wave pattern of displacement for (longitudinal) sound waves in a pipe with both ends open has an antinode (A) across each end and a node (N) across the middle. (The longitudinal displacements represented by the double arrows are greatly exaggerated.) (b) The corresponding standing wave pattern for (transverse) string waves.

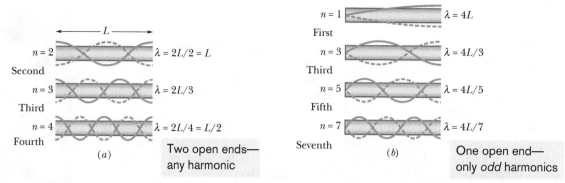

**Figure 17-14** Standing wave patterns for string waves superimposed on pipes to represent standing sound wave patterns in the pipes. (*a*) With *both* ends of the pipe open, any harmonic can be set up in the pipe. (*b*) With only *one* end open, only odd harmonics can be set up.

open end, as required. There is also a node across the middle of the pipe. An easier way of representing this standing longitudinal sound wave is shown in Fig. 17-13*b* — by drawing it as a standing transverse string wave.

The standing wave pattern of Fig. 17-13*a* is called the *fundamental mode* or *first harmonic*. For it to be set up, the sound waves in a pipe of length $L$ must have a wavelength given by $L = \lambda/2$, so that $\lambda = 2L$. Several more standing sound wave patterns for a pipe with two open ends are shown in Fig. 17-14*a* using string wave representations. The *second harmonic* requires sound waves of wavelength $\lambda = L$, the *third harmonic* requires wavelength $\lambda = 2L/3$, and so on.

More generally, the resonant frequencies for a pipe of length $L$ with two open ends correspond to the wavelengths

$$\lambda = \frac{2L}{n}, \quad \text{for } n = 1, 2, 3, \ldots, \tag{17-38}$$

where $n$ is called the *harmonic number*. Letting $v$ be the speed of sound, we write the resonant frequencies for a pipe with two open ends as

$$f = \frac{v}{\lambda} = \frac{nv}{2L}, \quad \text{for } n = 1, 2, 3, \ldots \quad \text{(pipe, two open ends).} \tag{17-39}$$

**One Open End.** Figure 17-14*b* shows (using string wave representations) some of the standing sound wave patterns that can be set up in a pipe with only one open end. As required, across the open end there is an antinode and across the closed end there is a node. The simplest pattern requires sound waves having a wavelength given by $L = \lambda/4$, so that $\lambda = 4L$. The next simplest pattern requires a wavelength given by $L = 3\lambda/4$, so that $\lambda = 4L/3$, and so on.

More generally, the resonant frequencies for a pipe of length $L$ with only one open end correspond to the wavelengths

$$\lambda = \frac{4L}{n}, \quad \text{for } n = 1, 3, 5, \ldots, \tag{17-40}$$

in which the harmonic number $n$ *must be an odd number*. The resonant frequencies are then given by

$$f = \frac{v}{\lambda} = \frac{nv}{4L}, \quad \text{for } n = 1, 3, 5, \ldots \quad \text{(pipe, one open end).} \tag{17-41}$$

Note again that only odd harmonics can exist in a pipe with one open end. For example, the second harmonic, with $n = 2$, cannot be set up in such a pipe. Note also that for such a pipe the adjective in a phrase such as "the third harmonic" still refers to the harmonic number $n$ (and not to, say, the third possible harmonic). Finally note that Eqs. 17-38 and 17-39 for two open ends contain the

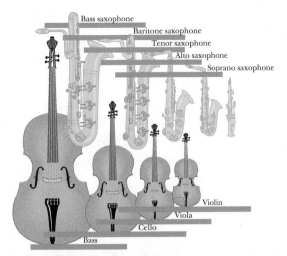

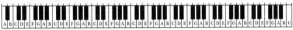

**Figure 17-15** The saxophone and violin families, showing the relations between instrument length and frequency range. The frequency range of each instrument is indicated by a horizontal bar along a frequency scale suggested by the keyboard at the bottom; the frequency increases toward the right.

number 2 and any integer value of $n$, but Eqs. 17-40 and 17-41 for one open end contain the number 4 and only odd values of $n$.

*Length.* The length of a musical instrument reflects the range of frequencies over which the instrument is designed to function, and smaller length implies higher frequencies, as we can tell from Eq. 16-66 for string instruments and Eqs. 17-39 and 17-41 for instruments with air columns. Figure 17-15, for example, shows the saxophone and violin families, with their frequency ranges suggested by the piano keyboard. Note that, for every instrument, there is overlap with its higher- and lower-frequency neighbors.

*Net Wave.* In any oscillating system that gives rise to a musical sound, whether it is a violin string or the air in an organ pipe, the fundamental and one or more of the higher harmonics are usually generated simultaneously. Thus, you hear them together—that is, superimposed as a net wave. When different instruments are played at the same note, they produce the same fundamental frequency but different intensities for the higher harmonics. For example, the fourth harmonic of middle C might be relatively loud on one instrument and relatively quiet or even missing on another. Thus, because different instruments produce different net waves, they sound different to you even when they are played at the same note. That would be the case for the two net waves shown in Fig. 17-16, which were produced at the same note by different instruments. If you heard only the fundamentals, the music would not be musical.

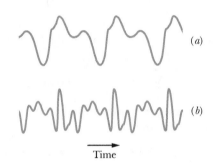

**Figure 17-16** The wave forms produced by (*a*) a flute and (*b*) an oboe when played at the same note, with the same first harmonic frequency.

 **Checkpoint 3**

Pipe $A$, with length $L$, and pipe $B$, with length $2L$, both have two open ends. Which harmonic of pipe $B$ has the same frequency as the fundamental of pipe $A$?

**Sample Problem 17.05  Resonance between pipes of different lengths**

Pipe $A$ is open at both ends and has length $L_A = 0.343$ m. We want to place it near three other pipes in which standing waves have been set up, so that the sound can set up a standing wave in pipe $A$. Those other three pipes are each closed at one end and have lengths $L_B = 0.500L_A$, $L_C = 0.250L_A$, and $L_D = 2.00L_A$. For each of these three pipes, which of their harmonics can excite a harmonic in pipe $A$?

**KEY IDEAS**

(1) The sound from one pipe can set up a standing wave in another pipe only if the harmonic frequencies match. (2) Equation 17-39 gives the harmonic frequencies in a pipe with two open ends (a symmetric pipe) as $f = nv/2L$, for $n = 1, 2, 3, \ldots$, that is, for any positive integer. (3) Equation

17-41 gives the harmonic frequencies in a pipe with only one open end (an asymmetric pipe) as $f = nv/4L$, for $n = 1, 3, 5, \ldots$, that is, for only odd positive integers.

**Pipe A:** Let's first find the resonant frequencies of symmetric pipe $A$ (with two open ends) by evaluating Eq. 17-39:

$$f_A = \frac{n_A v}{2L_A} = \frac{n_A (343 \text{ m/s})}{2(0.343 \text{ m})}$$

$$= n_A (500 \text{ Hz}) = n_A (0.50 \text{ kHz}), \quad \text{for } n_A = 1, 2, 3, \ldots.$$

The first six harmonic frequencies are shown in the top plot in Fig. 17-17.

**Pipe B:** Next let's find the resonant frequencies of asymmetric pipe $B$ (with only one open end) by evaluating Eq. 17-41, being careful to use only odd integers for the harmonic numbers:

$$f_B = \frac{n_B v}{4L_B} = \frac{n_B v}{4(0.500 L_A)} = \frac{n_B (343 \text{ m/s})}{2(0.343 \text{ m})}$$

$$= n_B (500 \text{ Hz}) = n_B (0.500 \text{ kHz}), \quad \text{for } n_B = 1, 3, 5, \ldots.$$

Comparing our two results, we see that we get a match for each choice of $n_B$:

$$f_A = f_B \quad \text{for } n_A = n_B \quad \text{with } n_B = 1, 3, 5, \ldots. \quad \text{(Answer)}$$

For example, as shown in Fig. 17-17, if we set up the fifth harmonic in pipe $B$ and bring the pipe close to pipe $A$, the fifth harmonic will then be set up in pipe $A$. However, no harmonic in $B$ can set up an even harmonic in $A$.

**Pipe C:** Let's continue with pipe $C$ (with only one end) by writing Eq. 17-41 as

$$f_C = \frac{n_C v}{4L_C} = \frac{n_C v}{4(0.250 L_A)} = \frac{n_C (343 \text{ m/s})}{0.343 \text{ m/s}}$$

$$= n_C (1000 \text{ Hz}) = n_C (1.00 \text{ kHz}), \quad \text{for } n_C = 1, 3, 5, \ldots.$$

From this we see that $C$ can excite some of the harmonics of $A$ but only those with harmonic numbers $n_A$ that are twice an odd integer:

$$f_A = f_C \quad \text{for } n_A = 2n_C, \quad \text{with } n_C = 1, 3, 5, \ldots. \quad \text{(Answer)}$$

**Pipe D:** Finally, let's check $D$ with our same procedure:

$$f_D = \frac{n_D v}{4L_D} = \frac{n_D v}{4(2L_A)} = \frac{n_D (343 \text{ m/s})}{8(0.343 \text{ m/s})}$$

$$= n_D (125 \text{ Hz}) = n_D (0.125 \text{ kHz}), \quad \text{for } n_D = 1, 3, 5, \ldots.$$

As shown in Fig. 17-17, none of these frequencies match a harmonic frequency of $A$. (Can you see that we would get a match if $n_D = 4n_A$? But that is impossible because $4n_A$ cannot yield an odd integer, as required of $n_D$.) Thus $D$ cannot set up a standing wave in $A$.

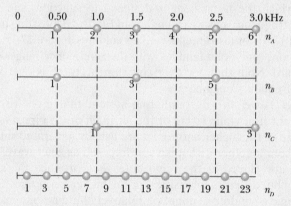

**Figure 17-17** Harmonic frequencies of four pipes.

# 17-6 BEATS

## Learning Objectives

*After reading this module, you should be able to . . .*

**17.28** Explain how beats are produced.

**17.29** Add the displacement equations for two sound waves of the same amplitude and slightly different angular frequencies to find the displacement equation of the resultant wave and identify the time-varying amplitude.

**17.30** Apply the relationship between the beat frequency and the frequencies of two sound waves that have the same amplitude when the frequencies (or, equivalently, the angular frequencies) differ by a small amount.

## Key Idea

● Beats arise when two waves having slightly different frequencies, $f_1$ and $f_2$, are detected together. The beat frequency is

$$f_{\text{beat}} = f_1 - f_2.$$

## Beats

If we listen, a few minutes apart, to two sounds whose frequencies are, say, 552 and 564 Hz, most of us cannot tell one from the other because the frequencies are so close to each other. However, if the sounds reach our ears simultaneously, what we hear is a sound whose frequency is 558 Hz, the *average* of the two combining frequencies. We also hear a striking variation in the intensity of this sound—it increases and decreases in slow, wavering **beats** that repeat at a frequency of 12 Hz, the *difference* between the two combining frequencies. Figure 17-18 shows this beat phenomenon.

Let the time-dependent variations of the displacements due to two sound waves of equal amplitude $s_m$ be

$$s_1 = s_m \cos \omega_1 t \qquad \text{and} \qquad s_2 = s_m \cos \omega_2 t, \tag{17-42}$$

where $\omega_1 > \omega_2$. From the superposition principle, the resultant displacement is the sum of the individual displacements:

$$s = s_1 + s_2 = s_m(\cos \omega_1 t + \cos \omega_2 t).$$

Using the trigonometric identity (see Appendix E)

$$\cos \alpha + \cos \beta = 2 \cos[\tfrac{1}{2}(\alpha - \beta)] \cos[\tfrac{1}{2}(\alpha + \beta)]$$

allows us to write the resultant displacement as

$$s = 2 s_m \cos[\tfrac{1}{2}(\omega_1 - \omega_2)t] \cos[\tfrac{1}{2}(\omega_1 + \omega_2)t]. \tag{17-43}$$

If we write

$$\omega' = \tfrac{1}{2}(\omega_1 - \omega_2) \qquad \text{and} \qquad \omega = \tfrac{1}{2}(\omega_1 + \omega_2), \tag{17-44}$$

we can then write Eq. 17-43 as

$$s(t) = [2 s_m \cos \omega' t] \cos \omega t. \tag{17-45}$$

We now assume that the angular frequencies $\omega_1$ and $\omega_2$ of the combining waves are almost equal, which means that $\omega \gg \omega'$ in Eq. 17-44. We can then regard Eq. 17-45 as a cosine function whose angular frequency is $\omega$ and whose amplitude (which is not constant but varies with angular frequency $\omega'$) is the absolute value of the quantity in the brackets.

A maximum amplitude will occur whenever $\cos \omega' t$ in Eq. 17-45 has the value $+1$ or $-1$, which happens twice in each repetition of the cosine function. Because $\cos \omega' t$ has angular frequency $\omega'$, the angular frequency $\omega_{\text{beat}}$ at which beats occur is $\omega_{\text{beat}} = 2\omega'$. Then, with the aid of Eq. 17-44, we can write the beat angular frequency as

$$\omega_{\text{beat}} = 2\omega' = (2)(\tfrac{1}{2})(\omega_1 - \omega_2) = \omega_1 - \omega_2.$$

Because $\omega = 2\pi f$, we can recast this as

$$f_{\text{beat}} = f_1 - f_2 \quad \text{(beat frequency)}. \tag{17-46}$$

Musicians use the beat phenomenon in tuning instruments. If an instrument is sounded against a standard frequency (for example, the note called "concert A" played on an orchestra's first oboe) and tuned until the beat disappears, the instrument is in tune with that standard. In musical Vienna, concert A (440 Hz) is available as a convenient telephone service for the city's many musicians.

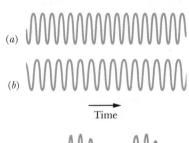

(a)

(b)

Time

(c)

---

**Figure 17-18** $(a, b)$ The pressure variations $\Delta p$ of two sound waves as they would be detected separately. The frequencies of the waves are nearly equal. $(c)$ The resultant pressure variation if the two waves are detected simultaneously.

## Sample Problem 17.06 Beat frequencies and penguins finding one another

When an emperor penguin returns from a search for food, how can it find its mate among the thousands of penguins huddled together for warmth in the harsh Antarctic weather? It is not by sight, because penguins all look alike, even to a penguin.

The answer lies in the way penguins vocalize. Most birds vocalize by using only one side of their two-sided vocal organ, called the *syrinx*. Emperor penguins, however, vocalize by using both sides simultaneously. Each side sets up acoustic standing waves in the bird's throat and mouth, much like in a pipe with two open ends. Suppose that the frequency of the first harmonic produced by side A is $f_{A1} =$ 432 Hz and the frequency of the first harmonic produced by side B is $f_{B1}$ = 371 Hz. What is the beat frequency between those two first-harmonic frequencies and between the two second-harmonic frequencies?

### KEY IDEA

The beat frequency between two frequencies is their difference, as given by Eq. 17-46 ($f_{beat} = f_1 - f_2$).

**Calculations:** For the two first-harmonic frequencies $f_{A1}$ and $f_{B1}$, the beat frequency is

$$f_{beat,1} = f_{A1} - f_{B1} = 432 \text{ Hz} - 371 \text{ Hz}$$
$$= 61 \text{ Hz}. \qquad \text{(Answer)}$$

Because the standing waves in the penguin are effectively in a pipe with two open ends, the resonant frequencies are given by Eq. 17-39 ($f = nv/2L$), in which $L$ is the (unknown) length of the effective pipe. The first-harmonic frequency is $f_1 = v/2L$, and the second-harmonic frequency is $f_2 = 2v/2L$. Comparing these two frequencies, we see that, in general,

$$f_2 = 2f_1.$$

For the penguin, the second harmonic of side A has frequency $f_{A2} = 2f_{A1}$ and the second harmonic of side B has frequency $f_{B2} = 2f_{B1}$. Using Eq. 17-46 with frequencies $f_{A2}$ and $f_{B2}$, we find that the corresponding beat frequency associated with the second harmonics is

$$f_{beat,2} = f_{A2} - f_{B2} = 2f_{A1} - 2f_{B1}$$
$$= 2(432 \text{ Hz}) - 2(371 \text{ Hz})$$
$$= 122 \text{ Hz}. \qquad \text{(Answer)}$$

Experiments indicate that penguins can perceive such large beat frequencies. (Humans cannot hear a beat frequency any higher than about 12 Hz — we perceive the two separate frequencies.) Thus, a penguin's cry can be rich with different harmonics and different beat frequencies, allowing the voice to be recognized even among the voices of thousands of other, closely huddled penguins.

  Additional examples, video, and practice available at *WileyPLUS*

# 17-7 THE DOPPLER EFFECT

## Learning Objectives

*After reading this module, you should be able to . . .*

**17.31** Identify that the Doppler effect is the shift in the detected frequency from the frequency emitted by a sound source due to the relative motion between the source and the detector.

**17.32** Identify that in calculating the Doppler shift in sound, the speeds are measured relative to the medium (such as air or water), which may be moving.

**17.33** Calculate the shift in sound frequency for (a) a source

moving either directly toward or away from a stationary detector, (b) a detector moving either directly toward or away from a stationary source, and (c) both source and detector moving either directly toward each other or directly away from each other.

**17.34** Identify that for relative motion between a sound source and a sound detector, motion *toward* tends to shift the frequency up and motion *away* tends to shift it down.

## Key Ideas

● The Doppler effect is a change in the observed frequency of a wave when the source or the detector moves relative to the transmitting medium (such as air). For sound the observed frequency $f'$ is given in terms of the source frequency $f$ by

$$f' = f\frac{v \pm v_D}{v \pm v_S} \quad \text{(general Doppler effect)},$$

where $v_D$ is the speed of the detector relative to the medium, $v_S$ is that of the source, and $v$ is the speed of sound in the medium.

● The signs are chosen such that $f'$ tends to be *greater* for relative motion toward (one of the objects moves toward the other) and *less* for motion away.

## The Doppler Effect

A police car is parked by the side of the highway, sounding its 1000 Hz siren. If you are also parked by the highway, you will hear that same frequency. However, if there is relative motion between you and the police car, either toward or away from each other, you will hear a different frequency. For example, if you are driving *toward* the police car at 120 km/h (about 75 mi/h), you will hear a *higher* frequency (1096 Hz, an *increase* of 96 Hz). If you are driving *away from* the police car at that same speed, you will hear a *lower* frequency (904 Hz, a *decrease* of 96 Hz).

These motion-related frequency changes are examples of the **Doppler effect.** The effect was proposed (although not fully worked out) in 1842 by Austrian physicist Johann Christian Doppler. It was tested experimentally in 1845 by Buys Ballot in Holland, "using a locomotive drawing an open car with several trumpeters."

The Doppler effect holds not only for sound waves but also for electromagnetic waves, including microwaves, radio waves, and visible light. Here, however, we shall consider only sound waves, and we shall take as a reference frame the body of air through which these waves travel. This means that we shall measure the speeds of a source $S$ of sound waves and a detector $D$ of those waves *relative to that body of air.* (Unless otherwise stated, the body of air is stationary relative to the ground, so the speeds can also be measured relative to the ground.) We shall assume that $S$ and $D$ move either directly toward or directly away from each other, at speeds less than the speed of sound.

***General Equation.*** If either the detector or the source is moving, or both are moving, the emitted frequency $f$ and the detected frequency $f'$ are related by

$$f' = f\frac{v \pm v_D}{v \pm v_S} \quad \text{(general Doppler effect)}, \quad (17\text{-}47)$$

where $v$ is the speed of sound through the air, $v_D$ is the detector's speed relative to the air, and $v_S$ is the source's speed relative to the air. The choice of plus or minus signs is set by this rule:

When the motion of detector or source is toward the other, the sign on its speed must give an upward shift in frequency. When the motion of detector or source is away from the other, the sign on its speed must give a downward shift in frequency.

In short, *toward* means *shift up,* and *away* means *shift down.*

Here are some examples of the rule. If the detector moves toward the source, use the plus sign in the numerator of Eq. 17-47 to get a shift up in the frequency. If it moves away, use the minus sign in the numerator to get a shift down. If it is stationary, substitute 0 for $v_D$. If the source moves toward the detector, use the minus sign in the denominator of Eq. 17-47 to get a shift up in the frequency. If it moves away, use the plus sign in the denominator to get a shift down. If the source is stationary, substitute 0 for $v_S$.

Next, we derive equations for the Doppler effect for the following two specific situations and then derive Eq. 17-47 for the general situation.

1. When the detector moves relative to the air and the source is stationary relative to the air, the motion changes the frequency at which the detector intercepts wavefronts and thus changes the detected frequency of the sound wave.

2. When the source moves relative to the air and the detector is stationary relative to the air, the motion changes the wavelength of the sound wave and thus changes the detected frequency (recall that frequency is related to wavelength).

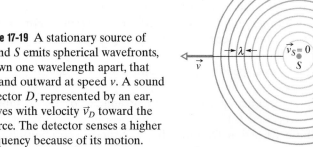

**Figure 17-19** A stationary source of sound $S$ emits spherical wavefronts, shown one wavelength apart, that expand outward at speed $v$. A sound detector $D$, represented by an ear, moves with velocity $\vec{v}_D$ toward the source. The detector senses a higher frequency because of its motion.

Shift up: The detector moves *toward* the source.

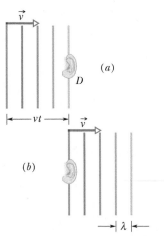

**Figure 17-20** The wavefronts of Fig. 17-19, assumed planar, (a) reach and (b) pass a stationary detector $D$; they move a distance $vt$ to the right in time $t$.

### Detector Moving, Source Stationary

In Fig. 17-19, a detector $D$ (represented by an ear) is moving at speed $v_D$ toward a stationary source $S$ that emits spherical wavefronts, of wavelength $\lambda$ and frequency $f$, moving at the speed $v$ of sound in air. The wavefronts are drawn one wavelength apart. The frequency detected by detector $D$ is the rate at which $D$ intercepts wavefronts (or individual wavelengths). If $D$ were stationary, that rate would be $f$, but since $D$ is moving into the wavefronts, the rate of interception is greater, and thus the detected frequency $f'$ is greater than $f$.

Let us for the moment consider the situation in which $D$ is stationary (Fig. 17-20). In time $t$, the wavefronts move to the right a distance $vt$. The number of wavelengths in that distance $vt$ is the number of wavelengths intercepted by $D$ in time $t$, and that number is $vt/\lambda$. The rate at which $D$ intercepts wavelengths, which is the frequency $f$ detected by $D$, is

$$f = \frac{vt/\lambda}{t} = \frac{v}{\lambda}. \qquad (17\text{-}48)$$

In this situation, with $D$ stationary, there is no Doppler effect—the frequency detected by $D$ is the frequency emitted by $S$.

Now let us again consider the situation in which $D$ moves in the direction opposite the wavefront velocity (Fig. 17-21). In time $t$, the wavefronts move to the right a distance $vt$ as previously, but now $D$ moves to the left a distance $v_D t$. Thus, in this time $t$, the distance moved by the wavefronts relative to $D$ is $vt + v_D t$. The number of wavelengths in this relative distance $vt + v_D t$ is the number of wavelengths intercepted by $D$ in time $t$ and is $(vt + v_D t)/\lambda$. The *rate* at which $D$ intercepts wavelengths in this situation is the frequency $f'$, given by

$$f' = \frac{(vt + v_D t)/\lambda}{t} = \frac{v + v_D}{\lambda}. \qquad (17\text{-}49)$$

From Eq. 17-48, we have $\lambda = v/f$. Then Eq. 17-49 becomes

$$f' = \frac{v + v_D}{v/f} = f\frac{v + v_D}{v}. \qquad (17\text{-}50)$$

Note that in Eq. 17-50, $f' > f$ unless $v_D = 0$ (the detector is stationary).

Similarly, we can find the frequency detected by $D$ if $D$ moves away from the source. In this situation, the wavefronts move a distance $vt - v_D t$ relative to $D$ in time $t$, and $f'$ is given by

$$f' = f\frac{v - v_D}{v}. \qquad (17\text{-}51)$$

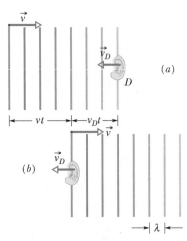

**Figure 17-21** Wavefronts traveling to the right (a) reach and (b) pass detector $D$, which moves in the opposite direction. In time $t$, the wavefronts move a distance $vt$ to the right and $D$ moves a distance $v_D t$ to the left.

In Eq. 17-51, $f' < f$ unless $v_D = 0$. We can summarize Eqs. 17-50 and 17-51 with

$$f' = f\frac{v \pm v_D}{v} \qquad \text{(detector moving, source stationary)}. \qquad (17\text{-}52)$$

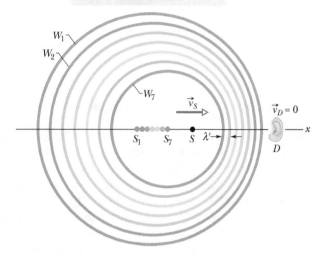

**Figure 17-22** A detector $D$ is stationary, and a source $S$ is moving toward it at speed $v_S$. Wavefront $W_1$ was emitted when the source was at $S_1$, wavefront $W_7$ when it was at $S_7$. At the moment depicted, the source is at $S$. The detector senses a higher frequency because the moving source, chasing its own wavefronts, emits a reduced wavelength $\lambda'$ in the direction of its motion.

### Source Moving, Detector Stationary

Let detector $D$ be stationary with respect to the body of air, and let source $S$ move toward $D$ at speed $v_S$ (Fig. 17-22). The motion of $S$ changes the wavelength of the sound waves it emits and thus the frequency detected by $D$.

To see this change, let $T\ (= 1/f)$ be the time between the emission of any pair of successive wavefronts $W_1$ and $W_2$. During $T$, wavefront $W_1$ moves a distance $vT$ and the source moves a distance $v_S T$. At the end of $T$, wavefront $W_2$ is emitted. In the direction in which $S$ moves, the distance between $W_1$ and $W_2$, which is the wavelength $\lambda'$ of the waves moving in that direction, is $vT - v_S T$. If $D$ detects those waves, it detects frequency $f'$ given by

$$f' = \frac{v}{\lambda'} = \frac{v}{vT - v_S T} = \frac{v}{v/f - v_S/f}$$

$$= f\frac{v}{v - v_S}. \tag{17-53}$$

Note that $f'$ must be greater than $f$ unless $v_S = 0$.

In the direction opposite that taken by $S$, the wavelength $\lambda'$ of the waves is again the distance between successive waves but now that distance is $vT + v_S T$. If $D$ detects those waves, it detects frequency $f'$ given by

$$f' = f\frac{v}{v + v_S}. \tag{17-54}$$

Now $f'$ must be less than $f$ unless $v_S = 0$.

We can summarize Eqs. 17-53 and 17-54 with

$$f' = f\frac{v}{v \pm v_S} \qquad \text{(source moving, detector stationary).} \tag{17-55}$$

### General Doppler Effect Equation

We can now derive the general Doppler effect equation by replacing $f$ in Eq. 17-55 (the source frequency) with $f'$ of Eq. 17-52 (the frequency associated with motion of the detector). That simple replacement gives us Eq. 17-47 for the general Doppler effect. That general equation holds not only when both detector and source are moving but also in the two specific situations we just discussed. For the situation in which the detector is moving and the source is stationary, substitution of $v_S = 0$ into Eq. 17-47 gives us Eq. 17-52, which we previously found. For the situation in which the source is moving and the detector is stationary, substitution of $v_D = 0$ into Eq. 17-47 gives us Eq. 17-55, which we previously found. Thus, Eq. 17-47 is the equation to remember.

## Checkpoint 4

The figure indicates the directions of motion of a sound source and a detector for six situations in stationary air. For each situation, is the detected frequency greater than or less than the emitted frequency, or can't we tell without more information about the actual speeds?

| | Source | Detector | | Source | Detector |
|---|---|---|---|---|---|
| (a) | → | • 0 speed | (d) | ← | ← |
| (b) | ← | • 0 speed | (e) | → | ← |
| (c) | → | → | (f) | ← | → |

### Sample Problem 17.07   Double Doppler shift in the echoes used by bats

Bats navigate and search out prey by emitting, and then detecting reflections of, ultrasonic waves, which are sound waves with frequencies greater than can be heard by a human. Suppose a bat emits ultrasound at frequency $f_{be}$ = 82.52 kHz while flying with velocity $\vec{v}_b$ = (9.00 m/s)$\hat{i}$ as it chases a moth that flies with velocity $\vec{v}_m$ = (8.00 m/s)$\hat{i}$. What frequency $f_{md}$ does the moth detect? What frequency $f_{bd}$ does the bat detect in the returning echo from the moth?

#### KEY IDEAS

The frequency is shifted by the relative motion of the bat and moth. Because they move along a single axis, the shifted frequency is given by Eq. 17-47. Motion *toward* tends to shift the frequency *up*, and motion *away* tends to shift it *down*.

**Detection by moth:** The general Doppler equation is

$$f' = f\frac{v \pm v_D}{v \pm v_S}. \qquad (17\text{-}56)$$

Here, the detected frequency $f'$ that we want to find is the frequency $f_{md}$ detected by the moth. On the right side, the emitted frequency $f$ is the bat's emission frequency $f_{be}$ = 82.52 kHz, the speed of sound is $v$ = 343 m/s, the speed $v_D$ of the detector is the moth's speed $v_m$ = 8.00 m/s, and the speed $v_S$ of the source is the bat's speed $v_b$ = 9.00 m/s.

The decisions about the plus and minus signs can be tricky. Think in terms of *toward* and *away*. We have the speed of the moth (the detector) in the numerator of Eq. 17-56. The moth moves *away* from the bat, which tends to lower the detected frequency. Because the speed is in the numerator, we choose the minus sign to meet that tendency (the numerator becomes smaller). These reasoning steps are shown in Table 17-3.

We have the speed of the bat in the denominator of Eq. 17-56. The bat moves *toward* the moth, which tends to increase the detected frequency. Because the speed is in the denominator, we choose the minus sign to meet that tendency (the denominator becomes smaller).

With these substitutions and decisions, we have

$$f_{md} = f_{be}\frac{v - v_m}{v - v_b}$$
$$= (82.52 \text{ kHz})\frac{343 \text{ m/s} - 8.00 \text{ m/s}}{343 \text{ m/s} - 9.00 \text{ m/s}}$$
$$= 82.767 \text{ kHz} \approx 82.8 \text{ kHz}. \qquad \text{(Answer)}$$

**Detection of echo by bat:** In the echo back to the bat, the moth acts as a source of sound, emitting at the frequency $f_{md}$ we just calculated. So now the moth is the source (moving *away*) and the bat is the detector (moving *toward*). The reasoning steps are shown in Table 17-3. To find the frequency $f_{bd}$ detected by the bat, we write Eq. 17-56 as

$$f_{bd} = f_{md}\frac{v + v_b}{v + v_m}$$
$$= (82.767 \text{ kHz})\frac{343 \text{ m/s} + 9.00 \text{ m/s}}{343 \text{ m/s} + 8.00 \text{ m/s}}$$
$$= 83.00 \text{ kHz} \approx 83.0 \text{ kHz}. \qquad \text{(Answer)}$$

Some moths evade bats by "jamming" the detection system with ultrasonic clicks.

**Table 17-3**

| Bat to Moth | | Echo Back to Bat | |
|---|---|---|---|
| Detector | Source | Detector | Source |
| moth | bat | bat | moth |
| speed $v_D = v_m$ | speed $v_S = v_b$ | speed $v_D = v_b$ | speed $v_S = v_m$ |
| away | toward | toward | away |
| shift down | shift up | shift up | shift down |
| numerator | denominator | numerator | denominator |
| minus | minus | plus | plus |

**WILEY PLUS** Additional examples, video, and practice available at *WileyPLUS*

# 17-8 SUPERSONIC SPEEDS, SHOCK WAVES

## Learning Objectives

*After reading this module, you should be able to . . .*

**17.35** Sketch the bunching of wavefronts for a sound source traveling at the speed of sound or faster.

**17.36** Calculate the Mach number for a sound source exceeding the speed of sound.

**17.37** For a sound source exceeding the speed of sound, apply the relationship between the Mach cone angle, the speed of sound, and the speed of the source.

## Key Idea

● If the speed of a source relative to the medium exceeds the speed of sound in the medium, the Doppler equation no longer applies. In such a case, shock waves result. The half-angle $\theta$ of the Mach cone is given by

$$\sin \theta = \frac{v}{v_S} \quad \text{(Mach cone angle)}.$$

## Supersonic Speeds, Shock Waves

If a source is moving toward a stationary detector at a speed $v_S$ equal to the speed of sound $v$, Eqs. 17-47 and 17-55 predict that the detected frequency $f'$ will be infinitely great. This means that the source is moving so fast that it keeps pace with its own spherical wavefronts (Fig. 17-23a). What happens when $v_S > v$? For such *supersonic* speeds, Eqs. 17-47 and 17-55 no longer apply. Figure 17-23b depicts the spherical wavefronts that originated at various positions of the source. The radius of any wavefront is $vt$, where $t$ is the time that has elapsed since the source emitted that wavefront. Note that all the wavefronts bunch along a V-shaped envelope in this two-dimensional drawing. The wavefronts actually extend in three dimensions, and the bunching actually forms a cone called the *Mach cone*. A *shock wave* exists along the surface of this cone, because the bunching of wavefronts causes an abrupt rise and fall of air pressure as the surface passes through any point. From Fig. 17-23b, we see that the half-angle $\theta$ of the cone (the *Mach cone angle*) is given by

$$\sin \theta = \frac{vt}{v_S t} = \frac{v}{v_S} \quad \text{(Mach cone angle)}. \quad (17\text{-}57)$$

The ratio $v_S/v$ is the *Mach number*. If a plane flies at Mach 2.3, its speed is 2.3 times the speed of sound in the air through which the plane is flying. The shock wave generated by a supersonic aircraft (Fig. 17-24)

U.S. Navy photo by Ensign John Gay

**Figure 17-24** Shock waves produced by the wings of a Navy FA 18 jet. The shock waves are visible because the sudden decrease in air pressure in them caused water molecules in the air to condense, forming a fog.

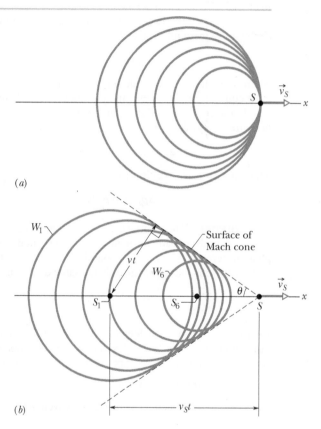

(a)

(b)

**Figure 17-23** (a) A source of sound $S$ moves at speed $v_S$ equal to the speed of sound and thus as fast as the wavefronts it generates. (b) A source $S$ moves at speed $v_S$ faster than the speed of sound and thus faster than the wavefronts. When the source was at position $S_1$ it generated wavefront $W_1$, and at position $S_6$ it generated $W_6$. All the spherical wavefronts expand at the speed of sound $v$ and bunch along the surface of a cone called the Mach cone, forming a shock wave. The surface of the cone has half-angle $\theta$ and is tangent to all the wavefronts.

or projectile produces a burst of sound, called a *sonic boom,* in which the air pressure first suddenly increases and then suddenly decreases below normal before returning to normal. Part of the sound that is heard when a rifle is fired is the sonic boom produced by the bullet. When a long bull whip is snapped, its tip is moving faster than sound and produces a small sonic boom—the *crack* of the whip.

# Review & Summary

**Sound Waves** Sound waves are longitudinal mechanical waves that can travel through solids, liquids, or gases. The speed $v$ of a sound wave in a medium having **bulk modulus** $B$ and density $\rho$ is

$$v = \sqrt{\frac{B}{\rho}} \quad \text{(speed of sound).} \tag{17-3}$$

In air at 20°C, the speed of sound is 343 m/s.

A sound wave causes a longitudinal displacement $s$ of a mass element in a medium as given by

$$s = s_m \cos(kx - \omega t), \tag{17-12}$$

where $s_m$ is the **displacement amplitude** (maximum displacement) from equilibrium, $k = 2\pi/\lambda$, and $\omega = 2\pi f$, $\lambda$ and $f$ being the wavelength and frequency of the sound wave. The wave also causes a pressure change $\Delta p$ from the equilibrium pressure:

$$\Delta p = \Delta p_m \sin(kx - \omega t), \tag{17-13}$$

where the **pressure amplitude** is

$$\Delta p_m = (v\rho\omega)s_m. \tag{17-14}$$

**Interference** The interference of two sound waves with identical wavelengths passing through a common point depends on their phase difference $\phi$ there. If the sound waves were emitted in phase and are traveling in approximately the same direction, $\phi$ is given by

$$\phi = \frac{\Delta L}{\lambda} 2\pi, \tag{17-21}$$

where $\Delta L$ is their **path length difference** (the difference in the distances traveled by the waves to reach the common point). Fully constructive interference occurs when $\phi$ is an integer multiple of $2\pi$,

$$\phi = m(2\pi), \quad \text{for } m = 0, 1, 2, \dots, \tag{17-22}$$

and, equivalently, when $\Delta L$ is related to wavelength $\lambda$ by

$$\frac{\Delta L}{\lambda} = 0, 1, 2, \dots. \tag{17-23}$$

Fully destructive interference occurs when $\phi$ is an odd multiple of $\pi$,

$$\phi = (2m + 1)\pi, \quad \text{for } m = 0, 1, 2, \dots, \tag{17-24}$$

and, equivalently, when $\Delta L$ is related to $\lambda$ by

$$\frac{\Delta L}{\lambda} = 0.5, 1.5, 2.5, \dots. \tag{17-25}$$

**Sound Intensity** The **intensity** $I$ of a sound wave at a surface is the average rate per unit area at which energy is transferred by the wave through or onto the surface:

$$I = \frac{P}{A}, \tag{17-26}$$

where $P$ is the time rate of energy transfer (power) of the sound wave

and $A$ is the area of the surface intercepting the sound. The intensity $I$ is related to the displacement amplitude $s_m$ of the sound wave by

$$I = \tfrac{1}{2}\rho v\omega^2 s_m^2. \tag{17-27}$$

The intensity at a distance $r$ from a point source that emits sound waves of power $P_s$ is

$$I = \frac{P_s}{4\pi r^2}. \tag{17-28}$$

**Sound Level in Decibels** The *sound level* $\beta$ in *decibels* (dB) is defined as

$$\beta = (10 \text{ dB}) \log \frac{I}{I_0}, \tag{17-29}$$

where $I_0$ ($= 10^{-12}$ W/m²) is a reference intensity level to which all intensities are compared. For every factor-of-10 increase in intensity, 10 dB is added to the sound level.

**Standing Wave Patterns in Pipes** Standing sound wave patterns can be set up in pipes. A pipe open at both ends will resonate at frequencies

$$f = \frac{v}{\lambda} = \frac{nv}{2L}, \quad n = 1, 2, 3, \dots, \tag{17-39}$$

where $v$ is the speed of sound in the air in the pipe. For a pipe closed at one end and open at the other, the resonant frequencies are

$$f = \frac{v}{\lambda} = \frac{nv}{4L}, \quad n = 1, 3, 5, \dots. \tag{17-41}$$

**Beats** *Beats* arise when two waves having slightly different frequencies, $f_1$ and $f_2$, are detected together. The beat frequency is

$$f_{\text{beat}} = f_1 - f_2. \tag{17-46}$$

**The Doppler Effect** The *Doppler effect* is a change in the observed frequency of a wave when the source or the detector moves relative to the transmitting medium (such as air). For sound the observed frequency $f'$ is given in terms of the source frequency $f$ by

$$f' = f\frac{v \pm v_D}{v \pm v_S} \quad \text{(general Doppler effect),} \tag{17-47}$$

where $v_D$ is the speed of the detector relative to the medium, $v_S$ is that of the source, and $v$ is the speed of sound in the medium. The signs are chosen such that $f'$ tends to be *greater* for motion toward and *less* for motion away.

**Shock Wave** If the speed of a source relative to the medium exceeds the speed of sound in the medium, the Doppler equation no longer applies. In such a case, shock waves result. The half-angle $\theta$ of the Mach cone is given by

$$\sin \theta = \frac{v}{v_S} \quad \text{(Mach cone angle).} \tag{17-57}$$

# Questions

**1** In a first experiment, a sinusoidal sound wave is sent through a long tube of air, transporting energy at the average rate of $P_{avg,1}$. In a second experiment, two other sound waves, identical to the first one, are to be sent simultaneously through the tube with a phase difference $\phi$ of either 0, 0.2 wavelength, or 0.5 wavelength between the waves. (a) With only mental calculation, rank those choices of $\phi$ according to the average rate at which the waves will transport energy, greatest first. (b) For the first choice of $\phi$, what is the average rate in terms of $P_{avg,1}$?

**2** In Fig. 17-25, two point sources $S_1$ and $S_2$, which are in phase, emit identical sound waves of wavelength 2.0 m. In terms of wavelengths, what is the phase difference between the waves arriving at point $P$ if (a) $L_1 = 38$ m and $L_2 = 34$ m, and (b) $L_1 = 39$ m and $L_2 = 36$ m? (c) Assuming that the source separation is much smaller than $L_1$ and $L_2$, what type of interference occurs at $P$ in situations (a) and (b)?

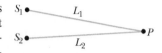

**Figure 17-25** Question 2.

**3** In Fig. 17-26, three long tubes ($A$, $B$, and $C$) are filled with different gases under different pressures. The ratio of the bulk modulus to the density is indicated for each gas in terms of a basic value $B_0/\rho_0$. Each tube has a piston at its left end that can send a sound pulse through the tube (as in Fig. 16-2). The three pulses are sent simultaneously. Rank the tubes according to the time of arrival of the pulses at the open right ends of the tubes, earliest first.

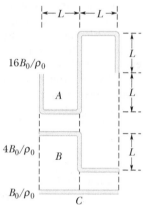

**Figure 17-26** Question 3.

**4** The sixth harmonic is set up in a pipe. (a) How many open ends does the pipe have (it has at least one)? (b) Is there a node, antinode, or some intermediate state at the midpoint?

**5** In Fig. 17-27, pipe $A$ is made to oscillate in its third harmonic by a small internal sound source. Sound emitted at the right end happens to resonate four nearby pipes, each with only one open end (they are *not* drawn to scale). Pipe $B$ oscillates in its lowest harmonic, pipe $C$ in its second lowest harmonic, pipe $D$ in its third lowest harmonic, and pipe $E$ in its fourth lowest harmonic. Without computation, rank all five pipes according to their length, greatest first. (*Hint:* Draw the standing waves to scale and then draw the pipes to scale.)

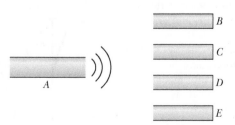

**Figure 17-27** Question 5.

**6** Pipe $A$ has length $L$ and one open end. Pipe $B$ has length $2L$ and two open ends. Which harmonics of pipe $B$ have a frequency that matches a resonant frequency of pipe $A$?

**7** Figure 17-28 shows a moving sound source $S$ that emits at a certain frequency, and four stationary sound detectors. Rank the detectors according to the frequency of the sound they detect from the source, greatest first.

**Figure 17-28** Question 7.

**8** A friend rides, in turn, the rims of three fast merry-go-rounds while holding a sound source that emits isotropically at a certain frequency. You stand far from each merry-go-round. The frequency you hear for each of your friend's three rides varies as the merry-go-round rotates. The variations in frequency for the three rides are given by the three curves in Fig. 17-29. Rank the curves according to (a) the linear speed $v$ of the sound source, (b) the angular speeds $\omega$ of the merry-go-rounds, and (c) the radii $r$ of the merry-go-rounds, greatest first.

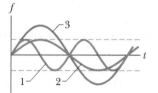

**Figure 17-29** Question 8.

**9** For a particular tube, here are four of the six harmonic frequencies below 1000 Hz: 300, 600, 750, and 900 Hz. What two frequencies are missing from the list?

**10** Figure 17-30 shows a stretched string of length $L$ and pipes $a$, $b$, $c$, and $d$ of lengths $L$, $2L$, $L/2$, and $L/2$, respectively. The string's tension is adjusted until the speed of waves on the string equals the speed of sound waves in the air. The fundamental mode of oscillation is then set up on the string. In which pipe will the sound produced by the string cause resonance, and what oscillation mode will that sound set up?

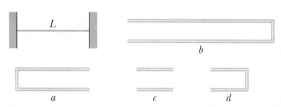

**Figure 17-30** Question 10.

**11** You are given four tuning forks. The fork with the lowest frequency oscillates at 500 Hz. By striking two tuning forks at a time, you can produce the following beat frequencies, 1, 2, 3, 5, 7, and 8 Hz. What are the possible frequencies of the other three forks? (There are two sets of answers.)

# Problems

*Where needed in the problems, use*

speed of sound in air = 343 m/s

*and* density of air = 1.21 kg/m³

*unless otherwise specified.*

## Module 17-1 Speed of Sound

•1 Two spectators at a soccer game see, and a moment later hear, the ball being kicked on the playing field. The time delay for spectator *A* is 0.23 s, and for spectator *B* it is 0.12 s. Sight lines from the two spectators to the player kicking the ball meet at an angle of 90°. How far are (a) spectator *A* and (b) spectator *B* from the player? (c) How far are the spectators from each other?

•2 What is the bulk modulus of oxygen if 32.0 g of oxygen occupies 22.4 L and the speed of sound in the oxygen is 317 m/s?

•3 When the door of the Chapel of the Mausoleum in Hamilton, Scotland, is slammed shut, the last echo heard by someone standing just inside the door reportedly comes 15 s later. (a) If that echo were due to a single reflection off a wall opposite the door, how far from the door is the wall? (b) If, instead, the wall is 25.7 m away, how many reflections (back and forth) occur?

•4 A column of soldiers, marching at 120 paces per minute, keep in step with the beat of a drummer at the head of the column. The soldiers in the rear end of the column are striding forward with the left foot when the drummer is advancing with the right foot. What is the approximate length of the column?

••5 **SSM** **ILW** Earthquakes generate sound waves inside Earth. Unlike a gas, Earth can experience both transverse (S) and longitudinal (P) sound waves. Typically, the speed of S waves is about 4.5 km/s, and that of P waves 8.0 km/s. A seismograph records P and S waves from an earthquake. The first P waves arrive 3.0 min before the first S waves. If the waves travel in a straight line, how far away did the earthquake occur?

••6 A man strikes one end of a thin rod with a hammer. The speed of sound in the rod is 15 times the speed of sound in air. A woman, at the other end with her ear close to the rod, hears the sound of the blow twice with a 0.12 s interval between; one sound comes through the rod and the other comes through the air alongside the rod. If the speed of sound in air is 343 m/s, what is the length of the rod?

••7 **SSM** **WWW** A stone is dropped into a well. The splash is heard 3.00 s later. What is the depth of the well?

••8 **GO** *Hot chocolate effect.* Tap a metal spoon inside a mug of water and note the frequency $f_i$ you hear. Then add a spoonful of powder (say, chocolate mix or instant coffee) and tap again as you stir the powder. The frequency you hear has a lower value $f_s$ because the tiny air bubbles released by the powder change the water's bulk modulus. As the bubbles reach the water surface and disappear, the frequency gradually shifts back to its initial value. During the effect, the bubbles don't appreciably change the water's density or volume or the sound's wavelength.

Rather, they change the value of $dV/dp$ —that is, the differential change in volume due to the differential change in the pressure caused by the sound wave in the water. If $f_s/f_i = 0.333$, what is the ratio $(dV/dp)_s/(dV/dp)_i$?

## Module 17-2 Traveling Sound Waves

•9 If the form of a sound wave traveling through air is

$$s(x, t) = (6.0\ \text{nm}) \cos(kx + (3000\ \text{rad/s})t + \phi),$$

how much time does any given air molecule along the path take to move between displacements $s = +2.0$ nm and $s = -2.0$ nm?

•10 *Underwater illusion.* One clue used by your brain to determine the direction of a source of sound is the time delay $\Delta t$ between the arrival of the sound at the ear closer to the source and the arrival at the farther ear. Assume that the source is distant so that a wavefront from it is approximately planar when it reaches you, and let $D$ represent the separation between your ears. (a) If the source is located at angle $\theta$ in front of you (Fig. 17-31), what is $\Delta t$ in terms of $D$ and the speed of sound $v$ in air? (b) If you are submerged in water and the sound source is directly to your right, what is $\Delta t$ in terms of $D$ and the speed of sound $v_w$ in water? (c) Based on the time-delay clue, your brain interprets the submerged sound to arrive at an angle $\theta$ from the forward direction. Evaluate $\theta$ for fresh water at 20°C.

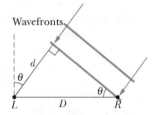

**Figure 17-31** Problem 10.

•11 **SSM** Diagnostic ultrasound of frequency 4.50 MHz is used to examine tumors in soft tissue. (a) What is the wavelength in air of such a sound wave? (b) If the speed of sound in tissue is 1500 m/s, what is the wavelength of this wave in tissue?

•12 The pressure in a traveling sound wave is given by the equation

$$\Delta p = (1.50\ \text{Pa}) \sin \pi[(0.900\ \text{m}^{-1})x - (315\ \text{s}^{-1})t].$$

Find the (a) pressure amplitude, (b) frequency, (c) wavelength, and (d) speed of the wave.

••13 A sound wave of the form $s = s_m \cos(kx - \omega t + \phi)$ travels at 343 m/s through air in a long horizontal tube. At one instant, air molecule *A* at $x = 2.000$ m is at its maximum positive displacement of 6.00 nm and air molecule *B* at $x = 2.070$ m is at a positive displacement of 2.00 nm. All the molecules between *A* and *B* are at intermediate displacements. What is the frequency of the wave?

••14 Figure 17-32 shows the output from a pressure monitor mounted at a point along the

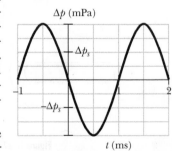

**Figure 17-32** Problem 14.

path taken by a sound wave of a single frequency traveling at 343 m/s through air with a uniform density of 1.21 kg/m³. The vertical axis scale is set by $\Delta p_s = 4.0$ mPa. If the *displacement* function of the wave is $s(x, t) = s_m \cos(kx - \omega t)$, what are (a) $s_m$, (b) $k$, and (c) $\omega$? The air is then cooled so that its density is 1.35 kg/m³ and the speed of a sound wave through it is 320 m/s. The sound source again emits the sound wave at the same frequency and same pressure amplitude. What now are (d) $s_m$, (e) $k$, and (f) $\omega$?

**••15** GO A handclap on stage in an amphitheater sends out sound waves that scatter from terraces of width $w = 0.75$ m (Fig. 17-33). The sound returns to the stage as a periodic series of pulses, one from each terrace; the parade of pulses sounds like a played note. (a) Assuming that all the rays in Fig. 17-33 are horizontal, find the frequency at which the pulses return (that is, the frequency of the perceived note). (b) If the width $w$ of the terraces were smaller, would the frequency be higher or lower?

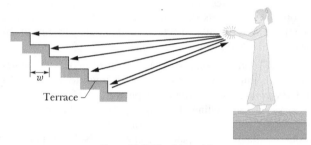

**Figure 17-33** Problem 15.

## Module 17-3   **Interference**

**•16** Two sound waves, from two different sources with the same frequency, 540 Hz, travel in the same direction at 330 m/s. The sources are in phase. What is the phase difference of the waves at a point that is 4.40 m from one source and 4.00 m from the other?

**••17** ILW Two loud speakers are located 3.35 m apart on an outdoor stage. A listener is 18.3 m from one and 19.5 m from the other. During the sound check, a signal generator drives the two speakers in phase with the same amplitude and frequency. The transmitted frequency is swept through the audible range (20 Hz to 20 kHz). (a) What is the lowest frequency $f_{min,1}$ that gives minimum signal (destructive interference) at the listener's location? By what number must $f_{min,1}$ be multiplied to get (b) the second lowest frequency $f_{min,2}$ that gives minimum signal and (c) the third lowest frequency $f_{min,3}$ that gives minimum signal? (d) What is the lowest frequency $f_{max,1}$ that gives maximum signal (constructive interference) at the listener's location? By what number must $f_{max,1}$ be multiplied to get (e) the second lowest frequency $f_{max,2}$ that gives maximum signal and (f) the third lowest frequency $f_{max,3}$ that gives maximum signal?

**••18** GO In Fig. 17-34, sound waves $A$ and $B$, both of wavelength $\lambda$, are initially in phase and traveling rightward, as indicated by the two rays. Wave $A$ is reflected from four surfaces but ends up traveling in its original direction. Wave $B$ ends in that direction after reflecting from two surfaces. Let distance $L$ in the figure be expressed as a multiple $q$ of $\lambda$: $L =$

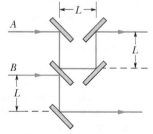

**Figure 17-34** Problem 18.

$q\lambda$. What are the (a) smallest and (b) second smallest values of $q$ that put $A$ and $B$ exactly out of phase with each other after the reflections?

**••19** GO Figure 17-35 shows two isotropic point sources of sound, $S_1$ and $S_2$. The sources emit waves in phase at wavelength 0.50 m; they are separated by $D = 1.75$ m. If we move a sound detector along a large circle centered at the midpoint between the sources, at how many points do waves arrive at the detector (a) exactly in phase and (b) exactly out of phase?

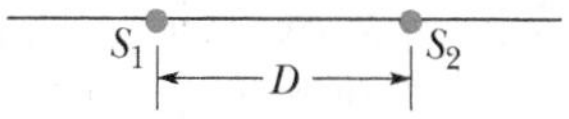

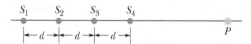

**Figure 17-35**
Problems 19 and 105.

**••20** Figure 17-36 shows four isotropic point sources of sound that are uniformly spaced on an $x$ axis. The sources emit sound at the same wavelength $\lambda$ and same amplitude $s_m$, and they emit in phase. A point $P$ is shown on the $x$ axis. Assume that as the sound waves travel to $P$, the decrease in their amplitude is negligible. What multiple of $s_m$ is the amplitude of the net wave at $P$ if distance $d$ in the figure is (a) $\lambda/4$, (b) $\lambda/2$, and (c) $\lambda$?

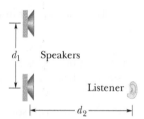

**Figure 17-36** Problem 20.

**••21** SSM In Fig. 17-37, two speakers separated by distance $d_1 = 2.00$ m are in phase. Assume the amplitudes of the sound waves from the speakers are approximately the same at the listener's ear at distance $d_2 = 3.75$ m directly in front of one speaker. Consider the full audible range for normal hearing, 20 Hz to 20 kHz. (a) What is the lowest frequency $f_{min,1}$ that gives minimum signal (destructive interference) at the listener's ear? By what number must $f_{min,1}$ be multiplied to get (b) the second lowest frequency $f_{min,2}$ that gives minimum signal and (c) the third lowest frequency $f_{min,3}$ that gives minimum signal? (d) What is the lowest frequency $f_{max,1}$ that gives maximum signal (constructive interference) at the listener's ear? By what number must $f_{max,1}$ be multiplied to get (e) the second lowest frequency $f_{max,2}$ that gives maximum signal and (f) the third lowest frequency $f_{max,3}$ that gives maximum signal?

**Figure 17-37** Problem 21.

**••22** In Fig. 17-38, sound with a 40.0 cm wavelength travels rightward from a source and through a tube that consists of a straight portion and a half-circle. Part of the sound wave travels through the half-circle and then rejoins the rest of the wave, which goes directly through the straight portion. This rejoining results in interference. What is the smallest radius $r$ that results in an intensity minimum at the detector?

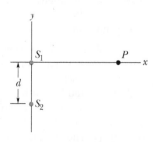

**Figure 17-38** Problem 22.

**•••23** GO Figure 17-39 shows two point sources $S_1$ and $S_2$ that emit sound of wavelength $\lambda = 2.00$ m. The emissions are isotropic and in phase, and the separation between

**Figure 17-39** Problem 23.

the sources is $d = 16.0$ m. At any point $P$ on the $x$ axis, the wave from $S_1$ and the wave from $S_2$ interfere. When $P$ is very far away ($x \approx \infty$), what are (a) the phase difference between the arriving waves from $S_1$ and $S_2$ and (b) the type of interference they produce? Now move point $P$ along the $x$ axis toward $S_1$. (c) Does the phase difference between the waves increase or decrease? At what distance $x$ do the waves have a phase difference of (d) $0.50\lambda$, (e) $1.00\lambda$, and (f) $1.50\lambda$?

### Module 17-4  Intensity and Sound Level

•24  Suppose that the sound level of a conversation is initially at an angry 70 dB and then drops to a soothing 50 dB. Assuming that the frequency of the sound is 500 Hz, determine the (a) initial and (b) final sound intensities and the (c) initial and (d) final sound wave amplitudes.

•25  A sound wave of frequency 300 Hz has an intensity of $1.00~\mu$W/m². What is the amplitude of the air oscillations caused by this wave?

•26  A 1.0 W point source emits sound waves isotropically. Assuming that the energy of the waves is conserved, find the intensity (a) 1.0 m from the source and (b) 2.5 m from the source.

•27 SSM WWW  A certain sound source is increased in sound level by 30.0 dB. By what multiple is (a) its intensity increased and (b) its pressure amplitude increased?

•28  Two sounds differ in sound level by 1.00 dB. What is the ratio of the greater intensity to the smaller intensity?

•29 SSM  A point source emits sound waves isotropically. The intensity of the waves 2.50 m from the source is $1.91 \times 10^{-4}$ W/m². Assuming that the energy of the waves is conserved, find the power of the source.

•30  The source of a sound wave has a power of 1.00 $\mu$W. If it is a point source, (a) what is the intensity 3.00 m away and (b) what is the sound level in decibels at that distance?

•31 GO  When you "crack" a knuckle, you suddenly widen the knuckle cavity, allowing more volume for the synovial fluid inside it and causing a gas bubble suddenly to appear in the fluid. The sudden production of the bubble, called "cavitation," produces a sound pulse—the cracking sound. Assume that the sound is transmitted uniformly in all directions and that it fully passes from the knuckle interior to the outside. If the pulse has a sound level of 62 dB at your ear, estimate the rate at which energy is produced by the cavitation.

•32  Approximately a third of people with normal hearing have ears that continuously emit a low-intensity sound outward through the ear canal. A person with such *spontaneous otoacoustic emission* is rarely aware of the sound, except perhaps in a noise-free environment, but occasionally the emission is loud enough to be heard by someone else nearby. In one observation, the sound wave had a frequency of 1665 Hz and a pressure amplitude of $1.13 \times 10^{-3}$ Pa. What were (a) the displacement amplitude and (b) the intensity of the wave emitted by the ear?

•33  Male *Rana catesbeiana* bullfrogs are known for their loud mating call. The call is emitted not by the frog's mouth but by its eardrums, which lie on the surface of the head. And, surprisingly, the sound has nothing to do with the frog's inflated throat. If the emitted sound has a frequency of 260 Hz and a sound level of 85 dB (near the eardrum), what is the amplitude of the eardrum's oscillation? The air density is 1.21 kg/m³.

••34 GO  Two atmospheric sound sources $A$ and $B$ emit isotropically at constant power. The sound levels $\beta$ of their emissions are plotted in Fig. 17-40 versus the radial distance $r$ from the sources. The vertical axis scale is set by $\beta_1 = 85.0$ dB and $\beta_2 = 65.0$ dB. What are (a) the ratio of the larger power to the smaller power and (b) the sound level difference at $r = 10$ m?

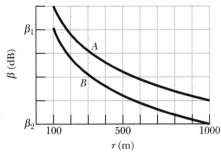

**Figure 17-40** Problem 34.

••35  A point source emits 30.0 W of sound isotropically. A small microphone intercepts the sound in an area of 0.750 cm², 200 m from the source. Calculate (a) the sound intensity there and (b) the power intercepted by the microphone.

••36  *Party hearing.* As the number of people at a party increases, you must raise your voice for a listener to hear you against the *background noise* of the other partygoers. However, once you reach the level of yelling, the only way you can be heard is if you move closer to your listener, into the listener's "personal space." Model the situation by replacing you with an isotropic point source of fixed power $P$ and replacing your listener with a point that absorbs part of your sound waves. These points are initially separated by $r_i = 1.20$ m. If the background noise increases by $\Delta\beta = 5$ dB, the sound level at your listener must also increase. What separation $r_f$ is then required?

•••37 GO  A sound source sends a sinusoidal sound wave of angular frequency 3000 rad/s and amplitude 12.0 nm through a tube of air. The internal radius of the tube is 2.00 cm. (a) What is the average rate at which energy (the sum of the kinetic and potential energies) is transported to the opposite end of the tube? (b) If, simultaneously, an identical wave travels along an adjacent, identical tube, what is the total average rate at which energy is transported to the opposite ends of the two tubes by the waves? If, instead, those two waves are sent along the *same* tube simultaneously, what is the total average rate at which they transport energy when their phase difference is (c) 0, (d) $0.40\pi$ rad, and (e) $\pi$ rad?

### Module 17-5  Sources of Musical Sound

•38  The water level in a vertical glass tube 1.00 m long can be adjusted to any position in the tube. A tuning fork vibrating at 686 Hz is held just over the open top end of the tube, to set up a standing wave of sound in the air-filled top portion of the tube. (That air-filled top portion acts as a tube with one end closed and the other end open.) (a) For how many different positions of the water level will sound from the fork set up resonance in the tube's air-filled portion? What are the (b) least and (c) second least water heights in the tube for resonance to occur?

•39 SSM ILW  (a) Find the speed of waves on a violin string of mass 800 mg and length 22.0 cm if the fundamental frequency is 920 Hz. (b) What is the tension in the string? For the fundamental, what is the wavelength of (c) the waves on the string and (d) the sound waves emitted by the string?

•**40** Organ pipe *A*, with both ends open, has a fundamental frequency of 300 Hz. The third harmonic of organ pipe *B*, with one end open, has the same frequency as the second harmonic of pipe *A*. How long are (a) pipe *A* and (b) pipe *B*?

•**41** A violin string 15.0 cm long and fixed at both ends oscillates in its *n* = 1 mode. The speed of waves on the string is 250 m/s, and the speed of sound in air is 348 m/s. What are the (a) frequency and (b) wavelength of the emitted sound wave?

•**42** A sound wave in a fluid medium is reflected at a barrier so that a standing wave is formed. The distance between nodes is 3.8 cm, and the speed of propagation is 1500 m/s. Find the frequency of the sound wave.

•**43** SSM In Fig. 17-41, *S* is a small loudspeaker driven by an audio oscillator with a frequency that is varied from 1000 Hz to 2000 Hz, and *D* is a cylindrical pipe with two open ends and a length of 45.7 cm. The speed of sound in the air-filled pipe is 344 m/s. (a) At how many frequencies does the sound from the loudspeaker set up resonance in the pipe? What are the (b) lowest and (c) second lowest frequencies at which resonance occurs?

**Figure 17-41** Problem 43.

•**44** The crest of a *Parasaurolophus* dinosaur skull is shaped somewhat like a trombone and contains a nasal passage in the form of a long, bent tube open at both ends. The dinosaur may have used the passage to produce sound by setting up the fundamental mode in it. (a) If the nasal passage in a certain *Parasaurolophus* fossil is 2.0 m long, what frequency would have been produced? (b) If that dinosaur could be recreated (as in *Jurassic Park*), would a person with a hearing range of 60 Hz to 20 kHz be able to hear that fundamental mode and, if so, would the sound be high or low frequency? Fossil skulls that contain shorter nasal passages are thought to be those of the female *Parasaurolophus*. (c) Would that make the female's fundamental frequency higher or lower than the male's?

•**45** In pipe *A*, the ratio of a particular harmonic frequency to the next lower harmonic frequency is 1.2. In pipe *B*, the ratio of a particular harmonic frequency to the next lower harmonic frequency is 1.4. How many open ends are in (a) pipe *A* and (b) pipe *B*?

••**46** Pipe *A*, which is 1.20 m long and open at both ends, oscillates at its third lowest harmonic frequency. It is filled with air for which the speed of sound is 343 m/s. Pipe *B*, which is closed at one end, oscillates at its second lowest harmonic frequency. This frequency of *B* happens to match the frequency of *A*. An *x* axis extends along the interior of *B*, with *x* = 0 at the closed end. (a) How many nodes are along that axis? What are the (b) smallest and (c) second smallest value of *x* locating those nodes? (d) What is the fundamental frequency of *B*?

••**47** A well with vertical sides and water at the bottom resonates at 7.00 Hz and at no lower frequency. The air-filled portion of the well acts as a tube with one closed end (at the bottom) and one open end (at the top). The air in the well has a density of 1.10 kg/m³ and a bulk modulus of 1.33 × 10⁵ Pa. How far down in the well is the water surface?

••**48** One of the harmonic frequencies of tube *A* with two open ends is 325 Hz. The next-highest harmonic frequency is 390 Hz. (a) What harmonic frequency is next highest after the harmonic frequency 195 Hz? (b) What is the number of this next-highest harmonic? One of the harmonic frequencies of tube *B* with only

one open end is 1080 Hz. The next-highest harmonic frequency is 1320 Hz. (c) What harmonic frequency is next highest after the harmonic frequency 600 Hz? (d) What is the number of this next-highest harmonic?

••**49** SSM A violin string 30.0 cm long with linear density 0.650 g/m is placed near a loudspeaker that is fed by an audio oscillator of variable frequency. It is found that the string is set into oscillation only at the frequencies 880 and 1320 Hz as the frequency of the oscillator is varied over the range 500–1500 Hz. What is the tension in the string?

••**50** A tube 1.20 m long is closed at one end. A stretched wire is placed near the open end. The wire is 0.330 m long and has a mass of 9.60 g. It is fixed at both ends and oscillates in its fundamental mode. By resonance, it sets the air column in the tube into oscillation at that column's fundamental frequency. Find (a) that frequency and (b) the tension in the wire.

### Module 17-6 Beats

•**51** The A string of a violin is a little too tightly stretched. Beats at 4.00 per second are heard when the string is sounded together with a tuning fork that is oscillating accurately at concert A (440 Hz). What is the period of the violin string oscillation?

•**52** A tuning fork of unknown frequency makes 3.00 beats per second with a standard fork of frequency 384 Hz. The beat frequency decreases when a small piece of wax is put on a prong of the first fork. What is the frequency of this fork?

••**53** SSM Two identical piano wires have a fundamental frequency of 600 Hz when kept under the same tension. What fractional increase in the tension of one wire will lead to the occurrence of 6.0 beats/s when both wires oscillate simultaneously?

••**54** You have five tuning forks that oscillate at close but different resonant frequencies. What are the (a) maximum and (b) minimum number of different beat frequencies you can produce by sounding the forks two at a time, depending on how the resonant frequencies differ?

### Module 17-7 The Doppler Effect

•**55** ILW A whistle of frequency 540 Hz moves in a circle of radius 60.0 cm at an angular speed of 15.0 rad/s. What are the (a) lowest and (b) highest frequencies heard by a listener a long distance away, at rest with respect to the center of the circle?

•**56** An ambulance with a siren emitting a whine at 1600 Hz overtakes and passes a cyclist pedaling a bike at 2.44 m/s. After being passed, the cyclist hears a frequency of 1590 Hz. How fast is the ambulance moving?

•**57** A state trooper chases a speeder along a straight road; both vehicles move at 160 km/h. The siren on the trooper's vehicle produces sound at a frequency of 500 Hz. What is the Doppler shift in the frequency heard by the speeder?

••**58** A sound source *A* and a reflecting surface *B* move directly toward each other. Relative to the air, the speed of source *A* is 29.9 m/s, the speed of surface *B* is 65.8 m/s, and the speed of sound is 329 m/s. The source emits waves at frequency 1200 Hz as measured in the source frame. In the reflector frame, what are the (a) frequency and (b) wavelength of the arriving sound waves? In the source frame, what are the (c) frequency and (d) wavelength of the sound waves reflected back to the source?

**••59** 🔵 In Fig. 17-42, a French submarine and a U.S. submarine move toward each other during maneuvers in motionless water in the North Atlantic. The French sub moves at speed $v_F = 50.00$ km/h, and the U.S. sub at $v_{US} = 70.00$ km/h. The French sub sends out a sonar signal (sound wave in water) at $1.000 \times 10^3$ Hz. Sonar waves travel at 5470 km/h. (a) What is the signal's frequency as detected by the U.S. sub? (b) What frequency is detected by the French sub in the signal reflected back to it by the U.S. sub?

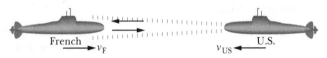

**Figure 17-42** Problem 59.

**••60** A stationary motion detector sends sound waves of frequency 0.150 MHz toward a truck approaching at a speed of 45.0 m/s. What is the frequency of the waves reflected back to the detector?

**••61** 🔵 ✈️ A bat is flitting about in a cave, navigating via ultrasonic bleeps. Assume that the sound emission frequency of the bat is 39 000 Hz. During one fast swoop directly toward a flat wall surface, the bat is moving at 0.025 times the speed of sound in air. What frequency does the bat hear reflected off the wall?

**••62** Figure 17-43 shows four tubes with lengths 1.0 m or 2.0 m, with one or two open ends as drawn. The third harmonic is set up in each tube, and some of the sound that escapes from them is detected by detector $D$, which moves directly away from the tubes. In terms of the speed of sound $v$, what speed must the detector have such that the detected frequency of the sound from (a) tube 1, (b) tube 2, (c) tube 3, and (d) tube 4 is equal to the tube's fundamental frequency?

**Figure 17-43** Problem 62.

**••63** ILW An acoustic burglar alarm consists of a source emitting waves of frequency 28.0 kHz. What is the beat frequency between the source waves and the waves reflected from an intruder walking at an average speed of 0.950 m/s directly away from the alarm?

**••64** A stationary detector measures the frequency of a sound source that first moves at constant velocity directly toward the detector and then (after passing the detector) directly away from it. The emitted frequency is $f$. During the approach the detected frequency is $f'_{app}$ and during the recession it is $f'_{rec}$. If $(f'_{app} - f'_{rec})/f = 0.500$, what is the ratio $v_s/v$ of the speed of the source to the speed of sound?

**•••65** 🔵 A 2000 Hz siren and a civil defense official are both at rest with respect to the ground. What frequency does the official hear if the wind is blowing at 12 m/s (a) from source to official and (b) from official to source?

**•••66** 🔵 Two trains are traveling toward each other at 30.5 m/s relative to the ground. One train is blowing a whistle at 500 Hz. (a) What frequency is heard on the other train in still air? (b) What frequency is heard on the other train if the wind is blowing at 30.5 m/s toward the whistle and away from the listener? (c) What frequency is heard if the wind direction is reversed?

**•••67** SSM WWW A girl is sitting near the open window of a train that is moving at a velocity of 10.00 m/s to the east. The girl's uncle stands near the tracks and watches the train move away. The

locomotive whistle emits sound at frequency 500.0 Hz. The air is still. (a) What frequency does the uncle hear? (b) What frequency does the girl hear? A wind begins to blow from the east at 10.00 m/s. (c) What frequency does the uncle now hear? (d) What frequency does the girl now hear?

## Module 17-8 Supersonic Speeds, Shock Waves

**•68** The shock wave off the cockpit of the FA 18 in Fig. 17-24 has an angle of about 60°. The airplane was traveling at about 1350 km/h when the photograph was taken. Approximately what was the speed of sound at the airplane's altitude?

**••69** SSM ✈️ A jet plane passes over you at a height of 5000 m and a speed of Mach 1.5. (a) Find the Mach cone angle (the sound speed is 331 m/s). (b) How long after the jet passes directly overhead does the shock wave reach you?

**••70** A plane flies at 1.25 times the speed of sound. Its sonic boom reaches a man on the ground 1.00 min after the plane passes directly overhead. What is the altitude of the plane? Assume the speed of sound to be 330 m/s.

## Additional Problems

**71** At a distance of 10 km, a 100 Hz horn, assumed to be an isotropic point source, is barely audible. At what distance would it begin to cause pain?

**72** A bullet is fired with a speed of 685 m/s. Find the angle made by the shock cone with the line of motion of the bullet.

**73** ✈️ A sperm whale (Fig. 17-44a) vocalizes by producing a series of clicks. Actually, the whale makes only a single sound near the front of its head to start the series. Part of that sound then emerges from the head into the water to become the first click of the series. The rest of the sound travels backward through the spermaceti sac (a body of fat), reflects from the frontal sac (an air layer), and then travels forward through the spermaceti sac. When it reaches the distal sac (another air layer) at the front of the head, some of the sound escapes into the water to form the second click, and the rest is sent back through the spermaceti sac (and ends up forming later clicks).

Figure 17-44b shows a strip-chart recording of a series of clicks. A unit time interval of 1.0 ms is indicated on the chart. Assuming that the speed of sound in the spermaceti sac is 1372 m/s, find the length of the spermaceti sac. From such a calculation, marine scientists estimate the length of a whale from its click series.

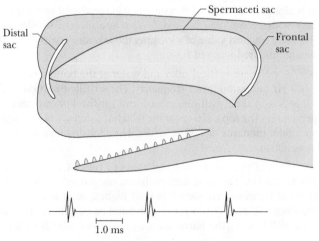

**Figure 17-44** Problem 73.

**74** The average density of Earth's crust 10 km beneath the continents is 2.7 g/cm³. The speed of longitudinal seismic waves at that depth, found by timing their arrival from distant earthquakes, is 5.4 km/s. Find the bulk modulus of Earth's crust at that depth. For comparison, the bulk modulus of steel is about $16 \times 10^{10}$ Pa.

**75** A certain loudspeaker system emits sound isotropically with a frequency of 2000 Hz and an intensity of 0.960 mW/m² at a distance of 6.10 m. Assume that there are no reflections. (a) What is the intensity at 30.0 m? At 6.10 m, what are (b) the displacement amplitude and (c) the pressure amplitude?

**76** Find the ratios (greater to smaller) of the (a) intensities, (b) pressure amplitudes, and (c) particle displacement amplitudes for two sounds whose sound levels differ by 37 dB.

**77** In Fig. 17-45, sound waves $A$ and $B$, both of wavelength $\lambda$, are initially in phase and traveling rightward, as indicated by the two rays. Wave $A$ is reflected from four surfaces but ends up traveling in its original direction. What multiple of wavelength $\lambda$ is the smallest value of distance $L$ in the figure that puts $A$ and $B$ exactly out of phase with each other after the reflections?

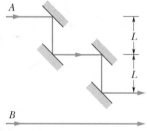

**Figure 17-45** Problem 77.

**78** A trumpet player on a moving railroad flatcar moves toward a second trumpet player standing alongside the track while both play a 440 Hz note. The sound waves heard by a stationary observer between the two players have a beat frequency of 4.0 beats/s. What is the flatcar's speed?

**79** GO In Fig. 17-46, sound of wavelength 0.850 m is emitted isotropically by point source $S$. Sound ray 1 extends directly to detector $D$, at distance $L = 10.0$ m. Sound ray 2 extends to $D$ via a reflection (effectively, a "bouncing") of the sound at a flat surface. That reflection occurs on a perpendicular bisector to the $SD$ line, at distance $d$ from the line. Assume that the reflection shifts the sound wave by $0.500\lambda$. For what least value of $d$ (other than zero) do the direct sound and the reflected sound arrive at $D$ (a) exactly out of phase and (b) exactly in phase?

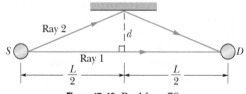

**Figure 17-46** Problem 79.

**80** GO A detector initially moves at constant velocity directly toward a stationary sound source and then (after passing it) directly from it. The emitted frequency is $f$. During the approach the detected frequency is $f'_{app}$ and during the recession it is $f'_{rec}$. If the frequencies are related by $(f'_{app} - f'_{rec})/f = 0.500$, what is the ratio $v_D/v$ of the speed of the detector to the speed of sound?

**81** SSM (a) If two sound waves, one in air and one in (fresh) water, are equal in intensity and angular frequency, what is the ratio of the pressure amplitude of the wave in water to that of the wave in air? Assume the water and the air are at 20°C. (See Table 14-1.) (b) If the pressure amplitudes are equal instead, what is the ratio of the intensities of the waves?

**82** A continuous sinusoidal longitudinal wave is sent along a very long coiled spring from an attached oscillating source. The wave travels in the negative direction of an $x$ axis; the source frequency is 25 Hz; at any instant the distance between successive points of maximum expansion in the spring is 24 cm; the maximum longitudinal displacement of a spring particle is 0.30 cm; and the particle at $x = 0$ has zero displacement at time $t = 0$. If the wave is written in the form $s(x, t) = s_m \cos(kx \pm \omega t)$, what are (a) $s_m$, (b) $k$, (c) $\omega$, (d) the wave speed, and (e) the correct choice of sign in front of $\omega$?

**83** SSM Ultrasound, which consists of sound waves with frequencies above the human audible range, can be used to produce an image of the interior of a human body. Moreover, ultrasound can be used to measure the speed of the blood in the body; it does so by comparing the frequency of the ultrasound sent into the body with the frequency of the ultrasound reflected back to the body's surface by the blood. As the blood pulses, this detected frequency varies.

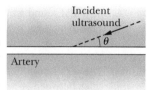

**Figure 17-47** Problem 83.

Suppose that an ultrasound image of the arm of a patient shows an artery that is angled at $\theta = 20°$ to the ultrasound's line of travel (Fig. 17-47). Suppose also that the frequency of the ultrasound reflected by the blood in the artery is increased by a maximum of 5495 Hz from the original ultrasound frequency of 5.000 000 MHz. (a) In Fig. 17-47, is the direction of the blood flow rightward or leftward? (b) The speed of sound in the human arm is 1540 m/s. What is the maximum speed of the blood? (*Hint:* The Doppler effect is caused by the component of the blood's velocity along the ultrasound's direction of travel.) (c) If angle $\theta$ were greater, would the reflected frequency be greater or less?

**84** The speed of sound in a certain metal is $v_m$. One end of a long pipe of that metal of length $L$ is struck a hard blow. A listener at the other end hears two sounds, one from the wave that travels along the pipe's metal wall and the other from the wave that travels through the air inside the pipe. (a) If $v$ is the speed of sound in air, what is the time interval $\Delta t$ between the arrivals of the two sounds at the listener's ear? (b) If $\Delta t = 1.00$ s and the metal is steel, what is the length $L$?

**85** An avalanche of sand along some rare desert sand dunes can produce a booming that is loud enough to be heard 10 km away. The booming apparently results from a periodic oscillation of the sliding layer of sand—the layer's thickness expands and contracts. If the emitted frequency is 90 Hz, what are (a) the period of the thickness oscillation and (b) the wavelength of the sound?

**86** A sound source moves along an $x$ axis, between detectors $A$ and $B$. The wavelength of the sound detected at $A$ is 0.500 that of the sound detected at $B$. What is the ratio $v_s/v$ of the speed of the source to the speed of sound?

**87** SSM A siren emitting a sound of frequency 1000 Hz moves away from you toward the face of a cliff at a speed of 10 m/s. Take the speed of sound in air as 330 m/s. (a) What is the frequency of the sound you hear coming directly from the siren? (b) What is the frequency of the sound you hear reflected off the cliff? (c) What is the beat frequency between the two sounds? Is it perceptible (less than 20 Hz)?

**88** At a certain point, two waves produce pressure variations given by $\Delta p_1 = \Delta p_m \sin \omega t$ and $\Delta p_2 = \Delta p_m \sin(\omega t - \phi)$. At this point, what is the ratio $\Delta p_r / \Delta p_m$, where $\Delta p_r$ is the pressure amplitude of the resultant wave, if $\phi$ is (a) 0, (b) $\pi/2$, (c) $\pi/3$, and (d) $\pi/4$?

**89** Two sound waves with an amplitude of 12 nm and a wavelength of 35 cm travel in the same direction through a long tube, with a phase difference of $\pi/3$ rad. What are the (a) amplitude and (b) wavelength of the net sound wave produced by their interference? If, instead, the sound waves travel through the tube in opposite directions, what are the (c) amplitude and (d) wavelength of the net wave?

**90** A sinusoidal sound wave moves at 343 m/s through air in the positive direction of an $x$ axis. At one instant during the oscillations, air molecule $A$ is at its maximum displacement in the negative direction of the axis while air molecule $B$ is at its equilibrium position. The separation between those molecules is 15.0 cm, and the molecules between $A$ and $B$ have intermediate displacements in the negative direction of the axis. (a) What is the frequency of the sound wave?

In a similar arrangement but for a different sinusoidal sound wave, at one instant air molecule $C$ is at its maximum displacement in the positive direction while molecule $D$ is at its maximum displacement in the negative direction. The separation between the molecules is again 15.0 cm, and the molecules between $C$ and $D$ have intermediate displacements. (b) What is the frequency of the sound wave?

**91** Two identical tuning forks can oscillate at 440 Hz. A person is located somewhere on the line between them. Calculate the beat frequency as measured by this individual if (a) she is standing still and the tuning forks move in the same direction along the line at 3.00 m/s, and (b) the tuning forks are stationary and the listener moves along the line at 3.00 m/s.

**92** You can estimate your distance from a lightning stroke by counting the seconds between the flash you see and the thunder you later hear. By what integer should you divide the number of seconds to get the distance in kilometers?

**93 SSM** Figure 17-48 shows an air-filled, acoustic interferometer, used to demonstrate the interference of sound waves. Sound source $S$ is an oscillating diaphragm; $D$ is a sound detector, such as the ear or a microphone. Path $SBD$ can be varied in length, but path $SAD$ is fixed. At $D$, the sound wave coming along path

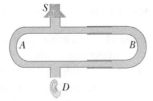

**Figure 17-48** Problem 93.

$SBD$ interferes with that coming along path $SAD$. In one demonstration, the sound intensity at $D$ has a minimum value of 100 units at one position of the movable arm and continuously climbs to a maximum value of 900 units when that arm is shifted by 1.65 cm. Find (a) the frequency of the sound emitted by the source and (b) the ratio of the amplitude at $D$ of the $SAD$ wave to that of the $SBD$ wave. (c) How can it happen that these waves have different amplitudes, considering that they originate at the same source?

**94** On July 10, 1996, a granite block broke away from a wall in Yosemite Valley and, as it began to slide down the wall, was launched into projectile motion. Seismic waves produced by its impact with the ground triggered seismographs as far away as 200 km. Later measurements indicated that the block had a mass between $7.3 \times 10^7$ kg and $1.7 \times 10^8$ kg and that it landed 500 m vertically below the launch point and 30 m horizontally from it.

(The launch angle is not known.) (a) Estimate the block's kinetic energy just before it landed.

Consider two types of seismic waves that spread from the impact point—a hemispherical *body wave* traveled through the ground in an expanding hemisphere and a cylindrical *surface wave* traveled along the ground in an expanding shallow vertical cylinder (Fig. 17-49). Assume that the impact lasted 0.50 s, the vertical cylinder had a depth $d$ of 5.0 m, and each wave type received 20% of the energy the block had just before impact. Neglecting any mechanical energy loss the waves experienced as they traveled, determine the intensities of (b) the body wave and (c) the surface wave when they reached a seismograph 200 km away. (d) On the basis of these results, which wave is more easily detected on a distant seismograph?

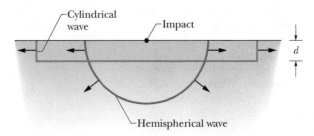

**Figure 17-49** Problem 94.

**95 SSM** The sound intensity is 0.0080 W/m² at a distance of 10 m from an isotropic point source of sound. (a) What is the power of the source? (b) What is the sound intensity 5.0 m from the source? (c) What is the sound level 10 m from the source?

**96** Four sound waves are to be sent through the same tube of air, in the same direction:

$$s_1(x, t) = (9.00 \text{ nm}) \cos(2\pi x - 700\pi t)$$
$$s_2(x, t) = (9.00 \text{ nm}) \cos(2\pi x - 700\pi t + 0.7\pi)$$
$$s_3(x, t) = (9.00 \text{ nm}) \cos(2\pi x - 700\pi t + \pi)$$
$$s_4(x, t) = (9.00 \text{ nm}) \cos(2\pi x - 700\pi t + 1.7\pi).$$

What is the amplitude of the resultant wave? (*Hint:* Use a phasor diagram to simplify the problem.)

**97** Straight line $AB$ connects two point sources that are 5.00 m apart, emit 300 Hz sound waves of the same amplitude, and emit exactly out of phase. (a) What is the shortest distance between the midpoint of $AB$ and a point on $AB$ where the interfering waves cause maximum oscillation of the air molecules? What are the (b) second and (c) third shortest distances?

**98** A point source that is stationary on an $x$ axis emits a sinusoidal sound wave at a frequency of 686 Hz and speed 343 m/s. The wave travels radially outward from the source, causing air molecules to oscillate radially inward and outward. Let us define a wavefront as a line that connects points where the air molecules have the maximum, radially outward displacement. At any given instant, the wavefronts are concentric circles that are centered on the source. (a) Along $x$, what is the adjacent wavefront separation? Next, the source moves along $x$ at a speed of 110 m/s. Along $x$, what are the wavefront separations (b) in front of and (c) behind the source?

**99** You are standing at a distance $D$ from an isotropic point source of sound. You walk 50.0 m toward the source and observe that the intensity of the sound has doubled. Calculate the distance $D$.

**100**   Pipe $A$ has only one open end; pipe $B$ is four times as long and has two open ends. Of the lowest 10 harmonic numbers $n_B$ of pipe $B$, what are the (a) smallest, (b) second smallest, and (c) third smallest values at which a harmonic frequency of $B$ matches one of the harmonic frequencies of $A$?

**101**   A pipe 0.60 m long and closed at one end is filled with an unknown gas. The third lowest harmonic frequency for the pipe is 750 Hz. (a) What is the speed of sound in the unknown gas? (b) What is the fundamental frequency for this pipe when it is filled with the unknown gas?

**102**   A sound wave travels out uniformly in all directions from a point source. (a) Justify the following expression for the displacement $s$ of the transmitting medium at any distance $r$ from the source:

$$s = \frac{b}{r} \sin k(r - vt),$$

where $b$ is a constant. Consider the speed, direction of propagation, periodicity, and intensity of the wave. (b) What is the dimension of the constant $b$?

**103**   A police car is chasing a speeding Porsche 911. Assume that the Porsche's maximum speed is 80.0 m/s and the police car's is 54.0 m/s. At the moment both cars reach their maximum speed, what frequency will the Porsche driver hear if the frequency of the police car's siren is 440 Hz? Take the speed of sound in air to be 340 m/s.

**104**   Suppose a spherical loudspeaker emits sound isotropically at 10 W into a room with completely absorbent walls, floor, and ceiling (an *anechoic chamber*). (a) What is the intensity of the sound at distance $d = 3.0$ m from the center of the source? (b) What is the ratio of the wave amplitude at $d = 4.0$ m to that at $d = 3.0$ m?

**105**   In Fig. 17-35, $S_1$ and $S_2$ are two isotropic point sources of sound. They emit waves in phase at wavelength 0.50 m; they are separated by $D = 1.60$ m. If we move a sound detector along a large circle centered at the midpoint between the sources, at how many points do waves arrive at the detector (a) exactly in phase and (b) exactly out of phase?

**106**   Figure 17-50 shows a transmitter and receiver of waves contained in a single instrument. It is used to measure the speed $u$ of a target object (idealized as a flat plate) that is moving directly toward the unit, by analyzing the waves reflected from the target. What is $u$ if the emitted frequency is 18.0 kHz and the detected frequency (of the returning waves) is 22.2 kHz?

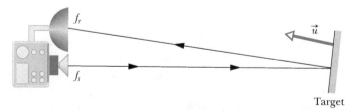

Target

**Figure 17-50** Problem 106.

**107**   *Kundt's method for measuring the speed of sound.* In Fig. 17-51, a rod $R$ is clamped at its center; a disk $D$ at its end projects into a glass tube that has cork filings spread over its interior. A

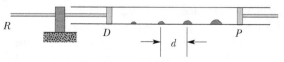

**Figure 17-51** Problem 107.

plunger $P$ is provided at the other end of the tube, and the tube is filled with a gas. The rod is made to oscillate longitudinally at frequency $f$ to produce sound waves inside the gas, and the location of the plunger is adjusted until a standing sound wave pattern is set up inside the tube. Once the standing wave is set up, the motion of the gas molecules causes the cork filings to collect in a pattern of ridges at the displacement nodes. If $f = 4.46 \times 10^3$ Hz and the separation between ridges is 9.20 cm, what is the speed of sound in the gas?

**108**   A source S and a detector D of radio waves are a distance $d$ apart on level ground (Fig. 17-52). Radio waves of wavelength $\lambda$ reach D either along a straight path or by reflecting (bouncing) from a certain layer in the atmosphere. When the layer is at height $H$, the two waves reaching D are exactly in phase. If the layer gradually rises, the phase difference between the two waves gradually shifts, until they are exactly out of phase when the layer is at height $H + h$. Express $\lambda$ in terms of $d, h$, and $H$.

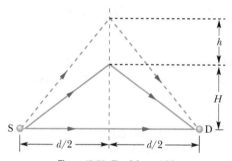

**Figure 17-52** Problem 108.

**109**   In Fig. 17-53, a point source $S$ of sound waves lies near a reflecting wall $AB$. A sound detector $D$ intercepts sound ray $R_1$ traveling directly from $S$. It also intercepts sound ray $R_2$ that reflects from the wall such that the angle of incidence $\theta_i$ is equal to the angle of reflection $\theta_r$. Assume that the reflection of sound by the wall causes a phase shift of $0.500\lambda$. If the distances are $d_1 = 2.50$ m, $d_2 = 20.0$ m, and $d_3 = 12.5$ m, what are the (a) lowest and (b) second lowest frequency at which $R_1$ and $R_2$ are in phase at $D$?

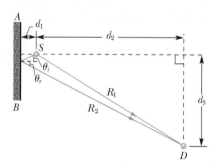

**Figure 17-53** Problem 109.

**110**   A person on a railroad car blows a trumpet note at 440 Hz. The car is moving toward a wall at 20.0 m/s. Find the sound frequency (a) at the wall and (b) reflected back to the trumpeter.

**111**   A listener at rest (with respect to the air and the ground) hears a signal of frequency $f_1$ from a source moving toward him with a velocity of 15 m/s, due east. If the listener then moves toward the approaching source with a velocity of 25 m/s, due west, he hears a frequency $f_2$ that differs from $f_1$ by 37 Hz. What is the frequency of the source? (Take the speed of sound in air to be 340 m/s.)

## CHAPTER 18

# Temperature, Heat, and the First Law of Thermodynamics

## 18-1 TEMPERATURE

### Learning Objectives

*After reading this module, you should be able to . . .*

**18.01** Identify the lowest temperature as 0 on the Kelvin scale (absolute zero).

**18.02** Explain the zeroth law of thermodynamics.

**18.03** Explain the conditions for the triple-point temperature.

**18.04** Explain the conditions for measuring a temperature with a constant-volume gas thermometer.

**18.05** For a constant-volume gas thermometer, relate the pressure and temperature of the gas in some given state to the pressure and temperature at the triple point.

### Key Ideas

● Temperature is an SI base quantity related to our sense of hot and cold. It is measured with a thermometer, which contains a working substance with a measurable property, such as length or pressure, that changes in a regular way as the substance becomes hotter or colder.

● When a thermometer and some other object are placed in contact with each other, they eventually reach thermal equilibrium. The reading of the thermometer is then taken to be the temperature of the other object. The process provides consistent and useful temperature measurements because of the zeroth law of thermodynamics: If bodies $A$ and $B$ are each in thermal equilibrium with a third body $C$ (the thermometer), then $A$ and $B$ are in thermal equilibrium with each other.

● In the SI system, temperature is measured on the Kelvin scale, which is based on the triple point of water (273.16 K). Other temperatures are then defined by use of a constant-volume gas thermometer, in which a sample of gas is maintained at constant volume so its pressure is proportional to its temperature. We define the temperature $T$ as measured with a gas thermometer to be

$$T = (273.16 \text{ K}) \left( \lim_{\text{gas} \to 0} \frac{p}{p_3} \right).$$

Here $T$ is in kelvins, and $p_3$ and $p$ are the pressures of the gas at 273.16 K and the measured temperature, respectively.

## What Is Physics?

One of the principal branches of physics and engineering is **thermodynamics,** which is the study and application of the *thermal energy* (often called the *internal energy*) of systems. One of the central concepts of thermodynamics is temperature. Since childhood, you have been developing a working knowledge of thermal energy and temperature. For example, you know to be cautious with hot foods and hot stoves and to store perishable foods in cool or cold compartments. You also know how to control the temperature inside home and car, and how to protect yourself from wind chill and heat stroke.

Examples of how thermodynamics figures into everyday engineering and science are countless. Automobile engineers are concerned with the heating of a car engine, such as during a NASCAR race. Food engineers are concerned both with the proper heating of foods, such as pizzas being microwaved, and with the proper cooling of foods, such as TV dinners being quickly frozen at a processing plant. Geologists are concerned with the transfer of thermal energy in an El Niño event and in the gradual warming of ice expanses in the Arctic and Antarctic.

Agricultural engineers are concerned with the weather conditions that determine whether the agriculture of a country thrives or vanishes. Medical engineers are concerned with how a patient's temperature might distinguish between a benign viral infection and a cancerous growth.

The starting point in our discussion of thermodynamics is the concept of temperature and how it is measured.

## Temperature

Temperature is one of the seven SI base quantities. Physicists measure temperature on the **Kelvin scale,** which is marked in units called *kelvins.* Although the temperature of a body apparently has no upper limit, it does have a lower limit; this limiting low temperature is taken as the zero of the Kelvin temperature scale. Room temperature is about 290 kelvins, or 290 K as we write it, above this *absolute zero.* Figure 18-1 shows a wide range of temperatures.

When the universe began 13.7 billion years ago, its temperature was about $10^{39}$ K. As the universe expanded it cooled, and it has now reached an average temperature of about 3 K. We on Earth are a little warmer than that because we happen to live near a star. Without our Sun, we too would be at 3 K (or, rather, we could not exist).

## The Zeroth Law of Thermodynamics

The properties of many bodies change as we alter their temperature, perhaps by moving them from a refrigerator to a warm oven. To give a few examples: As their temperature increases, the volume of a liquid increases, a metal rod grows a little longer, and the electrical resistance of a wire increases, as does the pressure exerted by a confined gas. We can use any one of these properties as the basis of an instrument that will help us pin down the concept of temperature.

Figure 18-2 shows such an instrument. Any resourceful engineer could design and construct it, using any one of the properties listed above. The instrument is fitted with a digital readout display and has the following properties: If you heat it (say, with a Bunsen burner), the displayed number starts to increase; if you then put it into a refrigerator, the displayed number starts to decrease. The instrument is not calibrated in any way, and the numbers have (as yet) no physical meaning. The device is a *thermoscope* but not (as yet) a *thermometer.*

Suppose that, as in Fig. 18-3a, we put the thermoscope (which we shall call body $T$) into intimate contact with another body (body $A$). The entire system is confined within a thick-walled insulating box. The numbers displayed by the thermoscope roll by until, eventually, they come to rest (let us say the reading is "137.04") and no further change takes place. In fact, we suppose that every measurable property of body $T$ and of body $A$ has assumed a stable, unchanging value. Then we say that the two bodies are in *thermal equilibrium* with each other. Even though the displayed readings for body $T$ have not been calibrated, we conclude that bodies $T$ and $A$ must be at the same (unknown) temperature.

Suppose that we next put body $T$ into intimate contact with body $B$ (Fig. 18-3b) and find that the two bodies come to thermal equilibrium *at the same reading of the thermoscope.* Then bodies $T$ and $B$ must be at the same (still unknown) temperature. If we now put bodies $A$ and $B$ into intimate contact (Fig. 18-3c), are they immediately in thermal equilibrium with each other? Experimentally, we find that they are.

The experimental fact shown in Fig. 18-3 is summed up in the **zeroth law of thermodynamics:**

> If bodies $A$ and $B$ are each in thermal equilibrium with a third body $T$, then $A$ and $B$ are in thermal equilibrium with each other.

In less formal language, the message of the zeroth law is: "Every body has a property called **temperature.** When two bodies are in thermal equilibrium, their temperatures are equal. And vice versa." We can now make our thermoscope

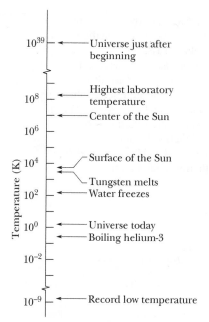

**Figure 18-1** Some temperatures on the Kelvin scale. Temperature $T = 0$ corresponds to $10^{-\infty}$ and cannot be plotted on this logarithmic scale.

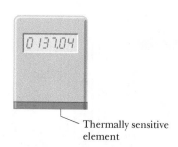

**Figure 18-2** A thermoscope. The numbers increase when the device is heated and decrease when it is cooled. The thermally sensitive element could be — among many possibilities — a coil of wire whose electrical resistance is measured and displayed.

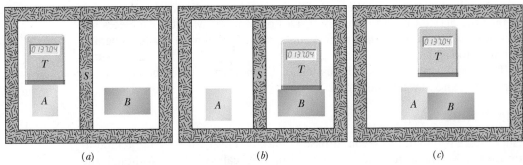

**Figure 18-3** (*a*) Body *T* (a thermoscope) and body *A* are in thermal equilibrium. (Body *S* is a thermally insulating screen.) (*b*) Body *T* and body *B* are also in thermal equilibrium, at the same reading of the thermoscope. (*c*) If (*a*) and (*b*) are true, the zeroth law of thermodynamics states that body *A* and body *B* are also in thermal equilibrium.

(the third body *T*) into a thermometer, confident that its readings will have physical meaning. All we have to do is calibrate it.

We use the zeroth law constantly in the laboratory. If we want to know whether the liquids in two beakers are at the same temperature, we measure the temperature of each with a thermometer. We do not need to bring the two liquids into intimate contact and observe whether they are or are not in thermal equilibrium.

The zeroth law, which has been called a logical afterthought, came to light only in the 1930s, long after the first and second laws of thermodynamics had been discovered and numbered. Because the concept of temperature is fundamental to those two laws, the law that establishes temperature as a valid concept should have the lowest number—hence the zero.

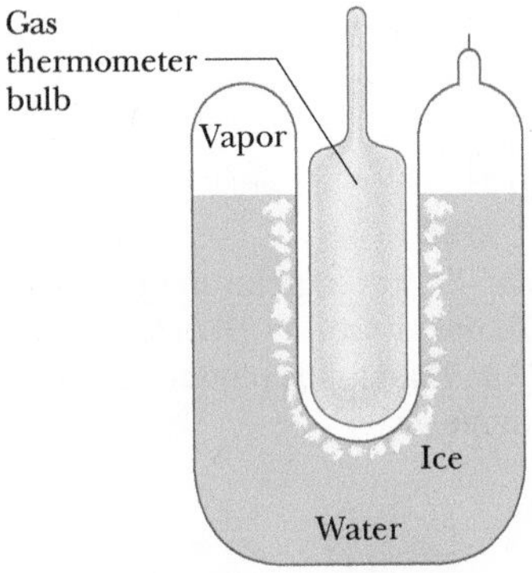

## Measuring Temperature

Here we first define and measure temperatures on the Kelvin scale. Then we calibrate a thermoscope so as to make it a thermometer.

### The Triple Point of Water

To set up a temperature scale, we pick some reproducible thermal phenomenon and, quite arbitrarily, assign a certain Kelvin temperature to its environment; that is, we select a *standard fixed point* and give it a standard fixed-point *temperature*. We could, for example, select the freezing point or the boiling point of water but, for technical reasons, we select instead the **triple point of water.**

Liquid water, solid ice, and water vapor (gaseous water) can coexist, in thermal equilibrium, at only one set of values of pressure and temperature. Figure 18-4 shows a triple-point cell, in which this so-called triple point of water can be achieved in the laboratory. By international agreement, the triple point of water has been assigned a value of 273.16 K as the standard fixed-point temperature for the calibration of thermometers; that is,

$$T_3 = 273.16 \text{ K} \quad \text{(triple-point temperature),} \quad (18\text{-}1)$$

**Figure 18-4** A triple-point cell, in which solid ice, liquid water, and water vapor coexist in thermal equilibrium. By international agreement, the temperature of this mixture has been defined to be 273.16 K. The bulb of a constant-volume gas thermometer is shown inserted into the well of the cell.

in which the subscript 3 means "triple point." This agreement also sets the size of the kelvin as 1/273.16 of the difference between the triple-point temperature of water and absolute zero.

Note that we do not use a degree mark in reporting Kelvin temperatures. It is 300 K (not 300°K), and it is read "300 kelvins" (not "300 degrees Kelvin"). The usual SI prefixes apply. Thus, 0.0035 K is 3.5 mK. No distinction in nomenclature is made between Kelvin temperatures and temperature differences, so we can write, "the boiling point of sulfur is 717.8 K" and "the temperature of this water bath was raised by 8.5 K."

### The Constant-Volume Gas Thermometer

The standard thermometer, against which all other thermometers are calibrated, is based on the pressure of a gas in a fixed volume. Figure 18-5 shows such a **constant-volume gas thermometer;** it consists of a gas-filled bulb connected by a tube to a mercury manometer. By raising and lowering reservoir *R*, the mercury

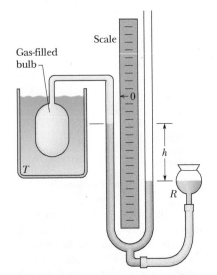

**Figure 18-5** A constant-volume gas thermometer, its bulb immersed in a liquid whose temperature *T* is to be measured.

**Figure 18-6** Temperatures measured by a constant-volume gas thermometer, with its bulb immersed in boiling water. For temperature calculations using Eq. 18-5, pressure $p_3$ was measured at the triple point of water. Three different gases in the thermometer bulb gave generally different results at different gas pressures, but as the amount of gas was decreased (decreasing $p_3$), all three curves converged to 373.125 K.

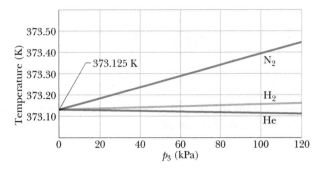

level in the left arm of the U-tube can always be brought to the zero of the scale to keep the gas volume constant (variations in the gas volume can affect temperature measurements).

The temperature of any body in thermal contact with the bulb (such as the liquid surrounding the bulb in Fig. 18-5) is then defined to be

$$T = Cp, \tag{18-2}$$

in which $p$ is the pressure exerted by the gas and $C$ is a constant. From Eq. 14-10, the pressure $p$ is

$$p = p_0 - \rho g h, \tag{18-3}$$

in which $p_0$ is the atmospheric pressure, $\rho$ is the density of the mercury in the manometer, and $h$ is the measured difference between the mercury levels in the two arms of the tube.* (The minus sign is used in Eq. 18-3 because pressure $p$ is measured *above* the level at which the pressure is $p_0$.)

If we next put the bulb in a triple-point cell (Fig. 18-4), the temperature now being measured is

$$T_3 = Cp_3, \tag{18-4}$$

in which $p_3$ is the gas pressure now. Eliminating $C$ between Eqs. 18-2 and 18-4 gives us the temperature as

$$T = T_3 \left( \frac{p}{p_3} \right) = (273.16 \text{ K}) \left( \frac{p}{p_3} \right) \qquad \text{(provisional)}. \tag{18-5}$$

We still have a problem with this thermometer. If we use it to measure, say, the boiling point of water, we find that different gases in the bulb give slightly different results. However, as we use smaller and smaller amounts of gas to fill the bulb, the readings converge nicely to a single temperature, no matter what gas we use. Figure 18-6 shows this convergence for three gases.

Thus the recipe for measuring a temperature with a gas thermometer is

$$T = (273.16 \text{ K}) \left( \lim_{\text{gas} \to 0} \frac{p}{p_3} \right). \tag{18-6}$$

The recipe instructs us to measure an unknown temperature $T$ as follows: Fill the thermometer bulb with an arbitrary amount of *any* gas (for example, nitrogen) and measure $p_3$ (using a triple-point cell) and $p$, the gas pressure at the temperature being measured. (Keep the gas volume the same.) Calculate the ratio $p/p_3$. Then repeat both measurements with a smaller amount of gas in the bulb, and again calculate this ratio. Continue this way, using smaller and smaller amounts of gas, until you can extrapolate to the ratio $p/p_3$ that you would find if there were approximately no gas in the bulb. Calculate the temperature $T$ by substituting that extrapolated ratio into Eq. 18-6. (The temperature is called the *ideal gas temperature*.)

---

*For pressure units, we shall use units introduced in Module 14-1. The SI unit for pressure is the newton per square meter, which is called the pascal (Pa). The pascal is related to other common pressure units by

$$1 \text{ atm} = 1.01 \times 10^5 \text{ Pa} = 760 \text{ torr} = 14.7 \text{ lb/in.}^2.$$

# 18-2 THE CELSIUS AND FAHRENHEIT SCALES

## Learning Objectives

*After reading this module, you should be able to . . .*

**18.06** Convert a temperature between any two (linear) temperature scales, including the Celsius, Fahrenheit, and Kelvin scales.

**18.07** Identify that a change of one degree is the same on the Celsius and Kelvin scales.

## Key Idea

● The Celsius temperature scale is defined by

$$T_C = T - 273.15°,$$

with $T$ in kelvins. The Fahrenheit temperature scale is defined by

$$T_F = \tfrac{9}{5}T_C + 32°.$$

## The Celsius and Fahrenheit Scales

So far, we have discussed only the Kelvin scale, used in basic scientific work. In nearly all countries of the world, the Celsius scale (formerly called the centigrade scale) is the scale of choice for popular and commercial use and much scientific use. Celsius temperatures are measured in degrees, and the Celsius degree has the same size as the kelvin. However, the zero of the Celsius scale is shifted to a more convenient value than absolute zero. If $T_C$ represents a Celsius temperature and $T$ a Kelvin temperature, then

$$T_C = T - 273.15°. \tag{18-7}$$

In expressing temperatures on the Celsius scale, the degree symbol is commonly used. Thus, we write 20.00°C for a Celsius reading but 293.15 K for a Kelvin reading.

The Fahrenheit scale, used in the United States, employs a smaller degree than the Celsius scale and a different zero of temperature. You can easily verify both these differences by examining an ordinary room thermometer on which both scales are marked. The relation between the Celsius and Fahrenheit scales is

$$T_F = \tfrac{9}{5}T_C + 32°, \tag{18-8}$$

where $T_F$ is Fahrenheit temperature. Converting between these two scales can be done easily by remembering a few corresponding points, such as the freezing and boiling points of water (Table 18-1). Figure 18-7 compares the Kelvin, Celsius, and Fahrenheit scales.

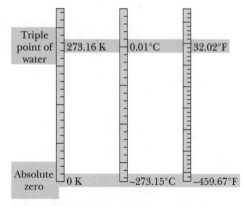

**Figure 18-7** The Kelvin, Celsius, and Fahrenheit temperature scales compared.

**Table 18-1** Some Corresponding Temperatures

| Temperature | °C | °F |
|---|---|---|
| Boiling point of water[a] | 100 | 212 |
| Normal body temperature | 37.0 | 98.6 |
| Accepted comfort level | 20 | 68 |
| Freezing point of water[a] | 0 | 32 |
| Zero of Fahrenheit scale | ≈ −18 | 0 |
| Scales coincide | −40 | −40 |

[a]Strictly, the boiling point of water on the Celsius scale is 99.975°C, and the freezing point is 0.00°C. Thus, there is slightly less than 100 C° between those two points.

We use the letters C and F to distinguish measurements and degrees on the two scales. Thus,

$$0°C = 32°F$$

means that 0° on the Celsius scale measures the same temperature as 32° on the Fahrenheit scale, whereas

$$5\ C° = 9\ F°$$

means that a temperature difference of 5 Celsius degrees (note the degree symbol appears *after* C) is equivalent to a temperature difference of 9 Fahrenheit degrees.

## ☑ Checkpoint 1

The figure here shows three linear temperature scales with the freezing and boiling points of water indicated. (a) Rank the degrees on these scales by size, greatest first. (b) Rank the following temperatures, highest first: 50°X, 50°W, and 50°Y.

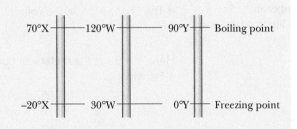

## Sample Problem 18.01   Conversion between two temperature scales

Suppose you come across old scientific notes that describe a temperature scale called Z on which the boiling point of water is 65.0°Z and the freezing point is −14.0°Z. To what temperature on the Fahrenheit scale would a temperature of $T = -98.0°Z$ correspond? Assume that the Z scale is linear; that is, the size of a Z degree is the same everywhere on the Z scale.

### KEY IDEA

A conversion factor between two (linear) temperature scales can be calculated by using two known (benchmark) temperatures, such as the boiling and freezing points of water. The number of degrees between the known temperatures on one scale is equivalent to the number of degrees between them on the other scale.

*Calculations:* We begin by relating the given temperature $T$ to *either* known temperature on the Z scale. Since $T = -98.0°Z$ is closer to the freezing point (−14.0°Z) than to the boiling point (65.0°Z), we use the freezing point. Then we note that the $T$ we seek is *below this point* by −14.0°Z − (−98.0°Z) = 84.0 Z° (Fig. 18-8). (Read this difference as "84.0 Z degrees.")

Next, we set up a conversion factor between the Z and Fahrenheit scales to convert this difference. To do so, we use *both* known temperatures on the Z scale and the

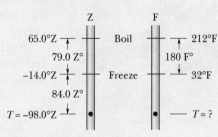

**Figure 18-8** An unknown temperature scale compared with the Fahrenheit temperature scale.

corresponding temperatures on the Fahrenheit scale. On the Z scale, the difference between the boiling and freezing points is 65.0°Z − (−14.0°Z) = 79.0 Z°. On the Fahrenheit scale, it is 212°F − 32.0°F = 180 F°. Thus, a temperature difference of 79.0 Z° is equivalent to a temperature difference of 180 F° (Fig. 18-8), and we can use the ratio (180 F°)/(79.0 Z°) as our conversion factor.

Now, since $T$ is below the freezing point by 84.0 Z°, it must also be below the freezing point by

$$(84.0\ Z°)\,\frac{180\ F°}{79.0\ Z°} = 191\ F°.$$

Because the freezing point is at 32.0°F, this means that

$$T = 32.0°F - 191\ F° = -159°F. \qquad \text{(Answer)}$$

# 18-3 THERMAL EXPANSION

## Learning Objectives

*After reading this module, you should be able to . . .*

**18.08** For one-dimensional thermal expansion, apply the relationship between the temperature change $\Delta T$, the length change $\Delta L$, the initial length $L$, and the coefficient of linear expansion $\alpha$.

**18.09** For two-dimensional thermal expansion, use one-dimensional thermal expansion to find the change in area.

**18.10** For three-dimensional thermal expansion, apply the relationship between the temperature change $\Delta T$, the volume change $\Delta V$, the initial volume $V$, and the coefficient of volume expansion $\beta$.

## Key Ideas

● All objects change size with changes in temperature. For a temperature change $\Delta T$, a change $\Delta L$ in any linear dimension $L$ is given by

$$\Delta L = L\alpha\,\Delta T,$$

in which $\alpha$ is the coefficient of linear expansion.

● The change $\Delta V$ in the volume $V$ of a solid or liquid is

$$\Delta V = V\beta\,\Delta T.$$

Here $\beta = 3\alpha$ is the material's coefficient of volume expansion.

## Thermal Expansion

Hugh Thomas/BWP Media/Getty Images, Inc.

**Figure 18-9** When a Concorde flew faster than the speed of sound, thermal expansion due to the rubbing by passing air increased the aircraft's length by about 12.5 cm. (The temperature increased to about 128°C at the aircraft nose and about 90°C at the tail, and cabin windows were noticeably warm to the touch.)

You can often loosen a tight metal jar lid by holding it under a stream of hot water. Both the metal of the lid and the glass of the jar expand as the hot water adds energy to their atoms. (With the added energy, the atoms can move a bit farther from one another than usual, against the spring-like interatomic forces that hold every solid together.) However, because the atoms in the metal move farther apart than those in the glass, the lid expands more than the jar and thus is loosened.

Such **thermal expansion** of materials with an increase in temperature must be anticipated in many common situations. When a bridge is subject to large seasonal changes in temperature, for example, sections of the bridge are separated by *expansion slots* so that the sections have room to expand on hot days without the bridge buckling. When a dental cavity is filled, the filling material must have the same thermal expansion properties as the surrounding tooth; otherwise, consuming cold ice cream and then hot coffee would be very painful. When the Concorde aircraft (Fig. 18-9) was built, the design had to allow for the thermal expansion of the fuselage during supersonic flight because of frictional heating by the passing air.

The thermal expansion properties of some materials can be put to common use. Thermometers and thermostats may be based on the differences in expansion between the components of a *bimetal strip* (Fig. 18-10). Also, the familiar liquid-in-glass thermometers are based on the fact that liquids such as mercury and alcohol expand to a different (greater) extent than their glass containers.

### Linear Expansion

If the temperature of a metal rod of length $L$ is raised by an amount $\Delta T$, its length is found to increase by an amount

$$\Delta L = L\alpha\,\Delta T, \tag{18-9}$$

**Figure 18-10** (*a*) A bimetal strip, consisting of a strip of brass and a strip of steel welded together, at temperature $T_0$. (*b*) The strip bends as shown at temperatures above this reference temperature. Below the reference temperature the strip bends the other way. Many thermostats operate on this principle, making and breaking an electrical contact as the temperature rises and falls.

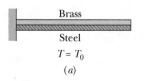

Brass
Steel
$T = T_0$
(*a*)

Different amounts of expansion or contraction can produce bending.

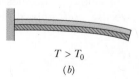

$T > T_0$
(*b*)

in which $\alpha$ is a constant called the **coefficient of linear expansion.** The coefficient $\alpha$ has the unit "per degree" or "per kelvin" and depends on the material. Although $\alpha$ varies somewhat with temperature, for most practical purposes it can be taken as constant for a particular material. Table 18-2 shows some coefficients of linear expansion. Note that the unit C° there could be replaced with the unit K.

The thermal expansion of a solid is like photographic enlargement except it is in three dimensions. Figure 18-11b shows the (exaggerated) thermal expansion of a steel ruler. Equation 18-9 applies to every linear dimension of the ruler, including its edge, thickness, diagonals, and the diameters of the circle etched on it and the circular hole cut in it. If the disk cut from that hole originally fits snugly in the hole, it will continue to fit snugly if it undergoes the same temperature increase as the ruler.

## Volume Expansion

If all dimensions of a solid expand with temperature, the volume of that solid must also expand. For liquids, volume expansion is the only meaningful expansion parameter. If the temperature of a solid or liquid whose volume is $V$ is increased by an amount $\Delta T$, the increase in volume is found to be

$$\Delta V = V\beta\,\Delta T, \tag{18-10}$$

where $\beta$ is the **coefficient of volume expansion** of the solid or liquid. The coefficients of volume expansion and linear expansion for a solid are related by

$$\beta = 3\alpha. \tag{18-11}$$

The most common liquid, water, does not behave like other liquids. Above about 4°C, water expands as the temperature rises, as we would expect. Between 0 and about 4°C, however, water *contracts* with increasing temperature. Thus, at about 4°C, the density of water passes through a maximum. At all other temperatures, the density of water is less than this maximum value.

This behavior of water is the reason lakes freeze from the top down rather than from the bottom up. As water on the surface is cooled from, say, 10°C toward the freezing point, it becomes denser ("heavier") than lower water and sinks to the bottom. Below 4°C, however, further cooling makes the water then on the surface *less* dense ("lighter") than the lower water, so it stays on the surface until it freezes. Thus the surface freezes while the lower water is still liquid. If lakes froze from the bottom up, the ice so formed would tend not to melt completely during the summer, because it would be insulated by the water above. After a few years, many bodies of open water in the temperate zones of Earth would be frozen solid all year round—and aquatic life could not exist.

**Table 18-2** Some Coefficients of Linear Expansion[a]

| Substance | $\alpha\,(10^{-6}/\text{C}°)$ |
|---|---|
| Ice (at 0°C) | 51 |
| Lead | 29 |
| Aluminum | 23 |
| Brass | 19 |
| Copper | 17 |
| Concrete | 12 |
| Steel | 11 |
| Glass (ordinary) | 9 |
| Glass (Pyrex) | 3.2 |
| Diamond | 1.2 |
| Invar[b] | 0.7 |
| Fused quartz | 0.5 |

[a]Room temperature values except for the listing for ice.

[b]This alloy was designed to have a low coefficient of expansion. The word is a shortened form of "invariable."

**Figure 18-11** The same steel ruler at two different temperatures. When it expands, the scale, the numbers, the thickness, and the diameters of the circle and circular hole are all increased by the same factor. (The expansion has been exaggerated for clarity.)

## ☑ Checkpoint 2

The figure here shows four rectangular metal plates, with sides of $L$, $2L$, or $3L$. They are all made of the same material, and their temperature is to be increased by the same amount. Rank the plates according to the expected increase in (a) their vertical heights and (b) their areas, greatest first.

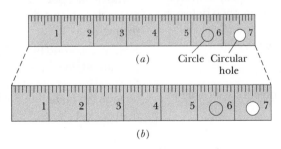

## Sample Problem 18.02 Thermal expansion of a volume

On a hot day in Las Vegas, an oil trucker loaded 37 000 L of diesel fuel. He encountered cold weather on the way to Payson, Utah, where the temperature was 23.0 K lower than in Las Vegas, and where he delivered his entire load. How many liters did he deliver? The coefficient of volume expansion for diesel fuel is $9.50 \times 10^{-4}/\text{C}°$, and the coefficient of linear expansion for his steel truck tank is $11 \times 10^{-6}/\text{C}°$.

### KEY IDEA

The volume of the diesel fuel depends directly on the temperature. Thus, because the temperature decreased, the

volume of the fuel did also, as given by Eq. 18-10 ($\Delta V = V\beta\,\Delta T$).

*Calculations:* We find
$$\Delta V = (37\ 000\ \text{L})(9.50 \times 10^{-4}/\text{C}°)(-23.0\ \text{K}) = -808\ \text{L}.$$
Thus, the amount delivered was
$$V_{\text{del}} = V + \Delta V = 37\ 000\ \text{L} - 808\ \text{L}$$
$$= 36\ 190\ \text{L}. \tag{Answer}$$
Note that the thermal expansion of the steel tank has nothing to do with the problem. Question: Who paid for the "missing" diesel fuel?

**PLUS** Additional examples, video, and practice available at *WileyPLUS*

# 18-4 ABSORPTION OF HEAT

## Learning Objectives

*After reading this module, you should be able to . . .*

**18.11** Identify that *thermal energy* is associated with the random motions of the microscopic bodies in an object.

**18.12** Identify that *heat Q* is the amount of transferred energy (either to or from an object's thermal energy) due to a temperature difference between the object and its environment.

**18.13** Convert energy units between various measurement systems.

**18.14** Convert between mechanical or electrical energy and thermal energy.

**18.15** For a temperature change $\Delta T$ of a substance, relate the change to the heat transfer $Q$ and the substance's heat capacity $C$.

**18.16** For a temperature change $\Delta T$ of a substance, relate the

change to the heat transfer $Q$ and the substance's specific heat $c$ and mass $m$.

**18.17** Identify the three phases of matter.

**18.18** For a phase change of a substance, relate the heat transfer $Q$, the heat of transformation $L$, and the amount of mass $m$ transformed.

**18.19** Identify that if a heat transfer $Q$ takes a substance across a phase-change temperature, the transfer must be calculated in steps: (a) a temperature change to reach the phase-change temperature, (b) the phase change, and then (c) any temperature change that moves the substance away from the phase-change temperature.

## Key Ideas

● Heat $Q$ is energy that is transferred between a system and its environment because of a temperature difference between them. It can be measured in joules (J), calories (cal), kilocalories (Cal or kcal), or British thermal units (Btu), with

$$1\ \text{cal} = 3.968 \times 10^{-3}\ \text{Btu} = 4.1868\ \text{J}.$$

● If heat $Q$ is absorbed by an object, the object's temperature change $T_f - T_i$ is related to $Q$ by

$$Q = C(T_f - T_i),$$

in which $C$ is the heat capacity of the object. If the object has mass $m$, then

$$Q = cm(T_f - T_i),$$

where $c$ is the specific heat of the material making up the object.

● The molar specific heat of a material is the heat capacity per mole, which means per $6.02 \times 10^{23}$ elementary units of the material.

● Heat absorbed by a material may change the material's physical state —for example, from solid to liquid or from liquid to gas. The amount of energy required per unit mass to change the state (but not the temperature) of a particular material is its heat of transformation $L$. Thus,

$$Q = Lm.$$

● The heat of vaporization $L_V$ is the amount of energy per unit mass that must be added to vaporize a liquid or that must be removed to condense a gas.

● The heat of fusion $L_F$ is the amount of energy per unit mass that must be added to melt a solid or that must be removed to freeze a liquid.

## Temperature and Heat

If you take a can of cola from the refrigerator and leave it on the kitchen table, its temperature will rise—rapidly at first but then more slowly—until the temperature of the cola equals that of the room (the two are then in thermal equilibrium). In the same way, the temperature of a cup of hot coffee, left sitting on the table, will fall until it also reaches room temperature.

In generalizing this situation, we describe the cola or the coffee as a *system* (with temperature $T_S$) and the relevant parts of the kitchen as the *environment* (with temperature $T_E$) of that system. Our observation is that if $T_S$ is not equal to $T_E$, then $T_S$ will change ($T_E$ can also change some) until the two temperatures are equal and thus thermal equilibrium is reached.

Such a change in temperature is due to a change in the thermal energy of the system because of a transfer of energy between the system and the system's environment. (Recall that *thermal energy* is an internal energy that consists of the kinetic and potential energies associated with the random motions of the atoms, molecules, and other microscopic bodies within an object.) The transferred energy is called **heat** and is symbolized $Q$. Heat is *positive* when energy is transferred to a system's thermal energy from its environment (we say that heat is absorbed by the system). Heat is *negative* when energy is transferred from a system's thermal energy to its environment (we say that heat is released or lost by the system).

This transfer of energy is shown in Fig. 18-12. In the situation of Fig. 18-12a, in which $T_S > T_E$, energy is transferred from the system to the environment, so $Q$ is negative. In Fig. 18-12b, in which $T_S = T_E$, there is no such transfer, $Q$ is zero, and heat is neither released nor absorbed. In Fig. 18-12c, in which $T_S < T_E$, the transfer is to the system from the environment; so $Q$ is positive.

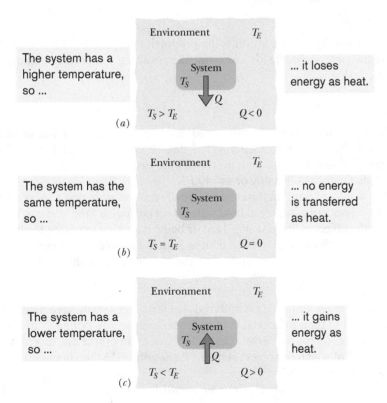

**Figure 18-12** If the temperature of a system exceeds that of its environment as in (a), heat $Q$ is lost by the system to the environment until thermal equilibrium (b) is established. (c) If the temperature of the system is below that of the environment, heat is absorbed by the system until thermal equilibrium is established.

We are led then to this definition of heat:

> Heat is the energy transferred between a system and its environment because of a temperature difference that exists between them.

***Language.*** Recall that energy can also be transferred between a system and its environment as *work W* via a force acting on a system. Heat and work, unlike temperature, pressure, and volume, are not intrinsic properties of a system. They have meaning only as they describe the transfer of energy into or out of a system. Similarly, the phrase "a $600 transfer" has meaning if it describes the transfer to or from an account, not what is in the account, because the account holds money, not a transfer.

***Units.*** Before scientists realized that heat is transferred energy, heat was measured in terms of its ability to raise the temperature of water. Thus, the **calorie** (cal) was defined as the amount of heat that would raise the temperature of 1 g of water from 14.5°C to 15.5°C. In the British system, the corresponding unit of heat was the **British thermal unit** (Btu), defined as the amount of heat that would raise the temperature of 1 lb of water from 63°F to 64°F.

In 1948, the scientific community decided that since heat (like work) is transferred energy, the SI unit for heat should be the one we use for energy—namely, the **joule.** The calorie is now defined to be 4.1868 J (exactly), with no reference to the heating of water. (The "calorie" used in nutrition, sometimes called the Calorie (Cal), is really a kilocalorie.) The relations among the various heat units are

$$1 \text{ cal} = 3.968 \times 10^{-3} \text{ Btu} = 4.1868 \text{ J}. \qquad (18\text{-}12)$$

## The Absorption of Heat by Solids and Liquids

### Heat Capacity

The **heat capacity** $C$ of an object is the proportionality constant between the heat $Q$ that the object absorbs or loses and the resulting temperature change $\Delta T$ of the object; that is,

$$Q = C \Delta T = C(T_f - T_i), \qquad (18\text{-}13)$$

in which $T_i$ and $T_f$ are the initial and final temperatures of the object. Heat capacity $C$ has the unit of energy per degree or energy per kelvin. The heat capacity $C$ of, say, a marble slab used in a bun warmer might be 179 cal/C°, which we can also write as 179 cal/K or as 749 J/K.

The word "capacity" in this context is really misleading in that it suggests analogy with the capacity of a bucket to hold water. *That analogy is false,* and you should not think of the object as "containing" heat or being limited in its ability to absorb heat. Heat transfer can proceed without limit as long as the necessary temperature difference is maintained. The object may, of course, melt or vaporize during the process.

### Specific Heat

Two objects made of the same material—say, marble—will have heat capacities proportional to their masses. It is therefore convenient to define a "heat capacity per unit mass" or **specific heat** $c$ that refers not to an object but to a unit mass of the material of which the object is made. Equation 18-13 then becomes

$$Q = cm \Delta T = cm(T_f - T_i). \qquad (18\text{-}14)$$

Through experiment we would find that although the heat capacity of a particular marble slab might be 179 cal/C° (or 749 J/K), the specific heat of marble itself (in that slab or in any other marble object) is 0.21 cal/g·C° (or 880 J/kg·K).

From the way the calorie and the British thermal unit were initially defined, the specific heat of water is

$$c = 1 \text{ cal/g} \cdot C° = 1 \text{ Btu/lb} \cdot F° = 4186.8 \text{ J/kg} \cdot \text{K}. \qquad (18\text{-}15)$$

Table 18-3 shows the specific heats of some substances at room temperature. Note that the value for water is relatively high. The specific heat of any substance actually depends somewhat on temperature, but the values in Table 18-3 apply reasonably well in a range of temperatures near room temperature.

 **Checkpoint 3**

A certain amount of heat $Q$ will warm 1 g of material $A$ by 3 C° and 1 g of material $B$ by 4 C°. Which material has the greater specific heat?

## Molar Specific Heat

In many instances the most convenient unit for specifying the amount of a substance is the mole (mol), where

$$1 \text{ mol} = 6.02 \times 10^{23} \text{ elementary units}$$

of *any* substance. Thus 1 mol of aluminum means $6.02 \times 10^{23}$ atoms (the atom is the elementary unit), and 1 mol of aluminum oxide means $6.02 \times 10^{23}$ molecules (the molecule is the elementary unit of the compound).

When quantities are expressed in moles, specific heats must also involve moles (rather than a mass unit); they are then called **molar specific heats.** Table 18-3 shows the values for some elemental solids (each consisting of a single element) at room temperature.

## An Important Point

In determining and then using the specific heat of any substance, we need to know the conditions under which energy is transferred as heat. For solids and liquids, we usually assume that the sample is under constant pressure (usually atmospheric) during the transfer. It is also conceivable that the sample is held at constant volume while the heat is absorbed. This means that thermal expansion of the sample is prevented by applying external pressure. For solids and liquids, this is very hard to arrange experimentally, but the effect can be calculated, and it turns out that the specific heats under constant pressure and constant volume for any solid or liquid differ usually by no more than a few percent. Gases, as you will see, have quite different values for their specific heats under constant-pressure conditions and under constant-volume conditions.

## Heats of Transformation

When energy is absorbed as heat by a solid or liquid, the temperature of the sample does not necessarily rise. Instead, the sample may change from one *phase,* or *state,* to another. Matter can exist in three common states: In the *solid state,* the molecules of a sample are locked into a fairly rigid structure by their mutual attraction. In the *liquid state,* the molecules have more energy and move about more. They may form brief clusters, but the sample does not have a rigid structure and can flow or settle into a container. In the *gas,* or *vapor, state,* the molecules have even more energy, are free of one another, and can fill up the full volume of a container.

*Melting.* To *melt* a solid means to change it from the solid state to the liquid state. The process requires energy because the molecules of the solid must be freed from their rigid structure. Melting an ice cube to form liquid water is a common example. To *freeze* a liquid to form a solid is the reverse of melting and requires that energy be removed from the liquid, so that the molecules can settle into a rigid structure.

**Table 18-3** Some Specific Heats and Molar Specific Heats at Room Temperature

| Substance | Specific Heat cal / g·K | Specific Heat J / kg·K | Molar Specific Heat J / mol·K |
|---|---|---|---|
| *Elemental Solids* | | | |
| Lead | 0.0305 | 128 | 26.5 |
| Tungsten | 0.0321 | 134 | 24.8 |
| Silver | 0.0564 | 236 | 25.5 |
| Copper | 0.0923 | 386 | 24.5 |
| Aluminum | 0.215 | 900 | 24.4 |
| *Other Solids* | | | |
| Brass | 0.092 | 380 | |
| Granite | 0.19 | 790 | |
| Glass | 0.20 | 840 | |
| Ice (−10°C) | 0.530 | 2220 | |
| *Liquids* | | | |
| Mercury | 0.033 | 140 | |
| Ethyl alcohol | 0.58 | 2430 | |
| Seawater | 0.93 | 3900 | |
| Water | 1.00 | 4187 | |

**Table 18-4** Some Heats of Transformation

| Substance | Melting | | Boiling | |
|---|---|---|---|---|
| | Melting Point (K) | Heat of Fusion $L_F$ (kJ/kg) | Boiling Point (K) | Heat of Vaporization $L_V$ (kJ/kg) |
| Hydrogen | 14.0 | 58.0 | 20.3 | 455 |
| Oxygen | 54.8 | 13.9 | 90.2 | 213 |
| Mercury | 234 | 11.4 | 630 | 296 |
| Water | 273 | 333 | 373 | 2256 |
| Lead | 601 | 23.2 | 2017 | 858 |
| Silver | 1235 | 105 | 2323 | 2336 |
| Copper | 1356 | 207 | 2868 | 4730 |

**Vaporizing.** To *vaporize* a liquid means to change it from the liquid state to the vapor (gas) state. This process, like melting, requires energy because the molecules must be freed from their clusters. Boiling liquid water to transfer it to water vapor (or steam—a gas of individual water molecules) is a common example. *Condensing* a gas to form a liquid is the reverse of vaporizing; it requires that energy be removed from the gas, so that the molecules can cluster instead of flying away from one another.

The amount of energy per unit mass that must be transferred as heat when a sample completely undergoes a phase change is called the **heat of transformation** $L$. Thus, when a sample of mass $m$ completely undergoes a phase change, the total energy transferred is

$$Q = Lm. \qquad (18\text{-}16)$$

When the phase change is from liquid to gas (then the sample must absorb heat) or from gas to liquid (then the sample must release heat), the heat of transformation is called the **heat of vaporization** $L_V$. For water at its normal boiling or condensation temperature,

$$L_V = 539 \text{ cal/g} = 40.7 \text{ kJ/mol} = 2256 \text{ kJ/kg}. \qquad (18\text{-}17)$$

When the phase change is from solid to liquid (then the sample must absorb heat) or from liquid to solid (then the sample must release heat), the heat of transformation is called the **heat of fusion** $L_F$. For water at its normal freezing or melting temperature,

$$L_F = 79.5 \text{ cal/g} = 6.01 \text{ kJ/mol} = 333 \text{ kJ/kg}. \qquad (18\text{-}18)$$

Table 18-4 shows the heats of transformation for some substances.

## Sample Problem 18.03  Hot slug in water, coming to equilibrium

A copper slug whose mass $m_c$ is 75 g is heated in a laboratory oven to a temperature $T$ of 312°C. The slug is then dropped into a glass beaker containing a mass $m_w = 220$ g of water. The heat capacity $C_b$ of the beaker is 45 cal/K. The initial temperature $T_i$ of the water and the beaker is 12°C. Assuming that the slug, beaker, and water are an isolated system and the water does not vaporize, find the final temperature $T_f$ of the system at thermal equilibrium.

### KEY IDEAS

(1) Because the system is isolated, the system's total energy cannot change and only internal transfers of thermal energy

can occur. (2) Because nothing in the system undergoes a phase change, the thermal energy transfers can only change the temperatures.

*Calculations:* To relate the transfers to the temperature changes, we can use Eqs. 18-13 and 18-14 to write

$$\text{for the water: } Q_w = c_w m_w (T_f - T_i); \qquad (18\text{-}19)$$
$$\text{for the beaker: } Q_b = C_b (T_f - T_i); \qquad (18\text{-}20)$$
$$\text{for the copper: } Q_c = c_c m_c (T_f - T). \qquad (18\text{-}21)$$

Because the total energy of the system cannot change, the sum of these three energy transfers is zero:

$$Q_w + Q_b + Q_c = 0. \qquad (18\text{-}22)$$

Substituting Eqs. 18-19 through 18-21 into Eq. 18-22 yields

$$c_w m_w(T_f - T_i) + C_b(T_f - T_i) + c_c m_c(T_f - T) = 0. \qquad (18\text{-}23)$$

Temperatures are contained in Eq. 18-23 only as differences. Thus, because the differences on the Celsius and Kelvin scales are identical, we can use either of those scales in this equation. Solving it for $T_f$, we obtain

$$T_f = \frac{c_c m_c T + C_b T_i + c_w m_w T_i}{c_w m_w + C_b + c_c m_c}.$$

Using Celsius temperatures and taking values for $c_c$ and $c_w$ from Table 18-3, we find the numerator to be

$(0.0923 \text{ cal/g} \cdot \text{K})(75 \text{ g})(312°\text{C}) + (45 \text{ cal/K})(12°\text{C})$

$+ (1.00 \text{ cal/g} \cdot \text{K})(220 \text{ g})(12°\text{C}) = 5339.8 \text{ cal,}$

and the denominator to be

$(1.00 \text{ cal/g} \cdot \text{K})(220 \text{ g}) + 45 \text{ cal/K}$

$+ (0.0923 \text{ cal/g} \cdot \text{K})(75 \text{ g}) = 271.9 \text{ cal/C°.}$

We then have

$$T_f = \frac{5339.8 \text{ cal}}{271.9 \text{ cal/C°}} = 19.6°\text{C} \approx 20°\text{C.} \qquad \text{(Answer)}$$

From the given data you can show that

$$Q_w \approx 1670 \text{ cal}, \qquad Q_b \approx 342 \text{ cal}, \qquad Q_c \approx -2020 \text{ cal.}$$

Apart from rounding errors, the algebraic sum of these three heat transfers is indeed zero, as required by the conservation of energy (Eq. 18-22).

---

## Sample Problem 18.04  Heat to change temperature and state

(a) How much heat must be absorbed by ice of mass $m = 720$ g at $-10°$C to take it to the liquid state at $15°$C?

### KEY IDEAS

The heating process is accomplished in three steps: (1) The ice cannot melt at a temperature below the freezing point—so initially, any energy transferred to the ice as heat can only increase the temperature of the ice, until $0°$C is reached. (2) The temperature then cannot increase until all the ice melts—so any energy transferred to the ice as heat now can only change ice to liquid water, until all the ice melts. (3) Now the energy transferred to the liquid water as heat can only increase the temperature of the liquid water.

*Warming the ice:* The heat $Q_1$ needed to take the ice from the initial $T_i = -10°$C to the final $T_f = 0°$C (so that the ice can then melt) is given by Eq. 18-14 ($Q = cm \, \Delta T$). Using the specific heat of ice $c_{ice}$ in Table 18-3 gives us

$$Q_1 = c_{ice} m(T_f - T_i)$$
$$= (2220 \text{ J/kg} \cdot \text{K})(0.720 \text{ kg})[0°\text{C} - (-10°\text{C})]$$
$$= 15\,984 \text{ J} \approx 15.98 \text{ kJ.}$$

*Melting the ice:* The heat $Q_2$ needed to melt all the ice is given by Eq. 18-16 ($Q = Lm$). Here $L$ is the heat of fusion $L_F$, with the value given in Eq. 18-18 and Table 18-4. We find

$$Q_2 = L_F m = (333 \text{ kJ/kg})(0.720 \text{ kg}) \approx 239.8 \text{ kJ.}$$

*Warming the liquid:* The heat $Q_3$ needed to increase the temperature of the water from the initial value $T_i = 0°$C to the final value $T_f = 15°$C is given by Eq. 18-14 (with the specific heat of liquid water $c_{liq}$):

$$Q_3 = c_{liq} m(T_f - T_i)$$
$$= (4186.8 \text{ J/kg} \cdot \text{K})(0.720 \text{ kg})(15°\text{C} - 0°\text{C})$$
$$= 45\,217 \text{ J} \approx 45.22 \text{ kJ.}$$

*Total:* The total required heat $Q_{tot}$ is the sum of the amounts required in the three steps:

$$Q_{tot} = Q_1 + Q_2 + Q_3$$
$$= 15.98 \text{ kJ} + 239.8 \text{ kJ} + 45.22 \text{ kJ}$$
$$\approx 300 \text{ kJ.} \qquad \text{(Answer)}$$

Note that most of the energy goes into melting the ice rather than raising the temperature.

(b) If we supply the ice with a total energy of only 210 kJ (as heat), what are the final state and temperature of the water?

### KEY IDEA

From step 1, we know that 15.98 kJ is needed to raise the temperature of the ice to the melting point. The remaining heat $Q_{rem}$ is then 210 kJ − 15.98 kJ, or about 194 kJ. From step 2, we can see that this amount of heat is insufficient to melt all the ice. Because the melting of the ice is incomplete, we must end up with a mixture of ice and liquid; the temperature of the mixture must be the freezing point, $0°$C.

*Calculations:* We can find the mass $m$ of ice that is melted by the available energy $Q_{rem}$ by using Eq. 18-16 with $L_F$:

$$m = \frac{Q_{rem}}{L_F} = \frac{194 \text{ kJ}}{333 \text{ kJ/kg}} = 0.583 \text{ kg} \approx 580 \text{ g.}$$

Thus, the mass of the ice that remains is 720 g − 580 g, or 140 g, and we have

$$580 \text{ g water} \quad \text{and} \quad 140 \text{ g ice,} \quad \text{at } 0°\text{C.} \qquad \text{(Answer)}$$

**PLUS** Additional examples, video, and practice available at *WileyPLUS*

# 18-5 THE FIRST LAW OF THERMODYNAMICS

## Learning Objectives

*After reading this module, you should be able to . . .*

**18.20** If an enclosed gas expands or contracts, calculate the work $W$ done by the gas by integrating the gas pressure with respect to the volume of the enclosure.

**18.21** Identify the algebraic sign of work $W$ associated with expansion and contraction of a gas.

**18.22** Given a $p$-$V$ graph of pressure versus volume for a process, identify the starting point (the initial state) and the final point (the final state) and calculate the work by using graphical integration.

**18.23** On a $p$-$V$ graph of pressure versus volume for a gas, identify the algebraic sign of the work associated with a right-going process and a left-going process.

**18.24** Apply the first law of thermodynamics to relate the change in the internal energy $\Delta E_{int}$ of a gas, the energy $Q$ transferred as heat to or from the gas, and the work $W$ done on or by the gas.

**18.25** Identify the algebraic sign of a heat transfer $Q$ that is associated with a transfer to a gas and a transfer from the gas.

**18.26** Identify that the internal energy $\Delta E_{int}$ of a gas tends to increase if the heat transfer is *to* the gas, and it tends to decrease if the gas does work on its environment.

**18.27** Identify that in an adiabatic process with a gas, there is no heat transfer $Q$ with the environment.

**18.28** Identify that in a constant-volume process with a gas, there is no work $W$ done by the gas.

**18.29** Identify that in a cyclical process with a gas, there is no net change in the internal energy $\Delta E_{int}$.

**18.30** Identify that in a free expansion with a gas, the heat transfer $Q$, work done $W$, and change in internal energy $\Delta E_{int}$ are each zero.

## Key Ideas

● A gas may exchange energy with its surroundings through work. The amount of work $W$ done *by* a gas as it expands or contracts from an initial volume $V_i$ to a final volume $V_f$ is given by

$$W = \int dW = \int_{V_i}^{V_f} p\, dV.$$

The integration is necessary because the pressure $p$ may vary during the volume change.

● The principle of conservation of energy for a thermodynamic process is expressed in the first law of thermodynamics, which may assume either of the forms

$$\Delta E_{int} = E_{int,f} - E_{int,i} = Q - W \quad \text{(first law)}$$

or $\qquad dE_{int} = dQ - dW \quad \text{(first law)}.$

$E_{int}$ represents the internal energy of the material, which depends only on the material's state (temperature,

pressure, and volume). $Q$ represents the energy exchanged as heat between the system and its surroundings; $Q$ is positive if the system absorbs heat and negative if the system loses heat. $W$ is the work done *by* the system; $W$ is positive if the system expands against an external force from the surroundings and negative if the system contracts because of an external force.

● $Q$ and $W$ are path dependent; $\Delta E_{int}$ is path independent.

● The first law of thermodynamics finds application in several special cases:

| | | |
|---|---|---|
| adiabatic processes: | $Q = 0,$ | $\Delta E_{int} = -W$ |
| constant-volume processes: | $W = 0,$ | $\Delta E_{int} = Q$ |
| cyclical processes: | $\Delta E_{int} = 0,$ | $Q = W$ |
| free expansions: | $Q = W = \Delta E_{int} = 0$ | |

## A Closer Look at Heat and Work

Here we look in some detail at how energy can be transferred as heat and work between a system and its environment. Let us take as our system a gas confined to a cylinder with a movable piston, as in Fig. 18-13. The upward force on the piston due to the pressure of the confined gas is equal to the weight of lead shot loaded onto the top of the piston. The walls of the cylinder are made of insulating material that does not allow any transfer of energy as heat. The bottom of the cylinder rests on a reservoir for thermal energy, a *thermal reservoir* (perhaps a hot plate) whose temperature $T$ you can control by turning a knob.

The system (the gas) starts from an *initial state i*, described by a pressure $p_i$, a volume $V_i$, and a temperature $T_i$. You want to change the system to a *final state f*, described by a pressure $p_f$, a volume $V_f$, and a temperature $T_f$. The procedure by which you change the system from its initial state to its final state is called a *thermodynamic process*. During such a process, energy may be transferred into the system from the thermal reservoir (positive heat) or vice versa (negative heat). Also, work can be done by the system to raise the loaded piston (positive work) or lower it (negative work). We assume that all such changes occur slowly, with the result that the system is always in (approximate) thermal equilibrium (every part is always in thermal equilibrium).

Suppose that you remove a few lead shot from the piston of Fig. 18-13, allowing the gas to push the piston and remaining shot upward through a differential displacement $d\vec{s}$ with an upward force $\vec{F}$. Since the displacement is tiny, we can assume that $\vec{F}$ is constant during the displacement. Then $\vec{F}$ has a magnitude that is equal to $pA$, where $p$ is the pressure of the gas and $A$ is the face area of the piston. The differential work $dW$ done by the gas during the displacement is

$$dW = \vec{F} \cdot d\vec{s} = (pA)(ds) = p(A\ ds)$$
$$= p\ dV,  \qquad (18\text{-}24)$$

in which $dV$ is the differential change in the volume of the gas due to the movement of the piston. When you have removed enough shot to allow the gas to change its volume from $V_i$ to $V_f$, the total work done by the gas is

$$W = \int dW = \int_{V_i}^{V_f} p\ dV.  \qquad (18\text{-}25)$$

During the volume change, the pressure and temperature may also change. To evaluate Eq. 18-25 directly, we would need to know how pressure varies with volume for the actual process by which the system changes from state $i$ to state $f$.

**One Path.**   There are actually many ways to take the gas from state $i$ to state $f$. One way is shown in Fig. 18-14$a$, which is a plot of the pressure of the gas versus its volume and which is called a $p$-$V$ diagram. In Fig. 18-14$a$, the curve indicates that the

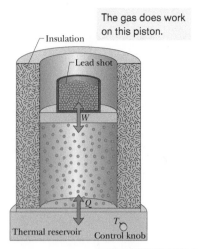

The gas does work on this piston.

**Figure 18-13**   A gas is confined to a cylinder with a movable piston. Heat $Q$ can be added to or withdrawn from the gas by regulating the temperature $T$ of the adjustable thermal reservoir. Work $W$ can be done by the gas by raising or lowering the piston.

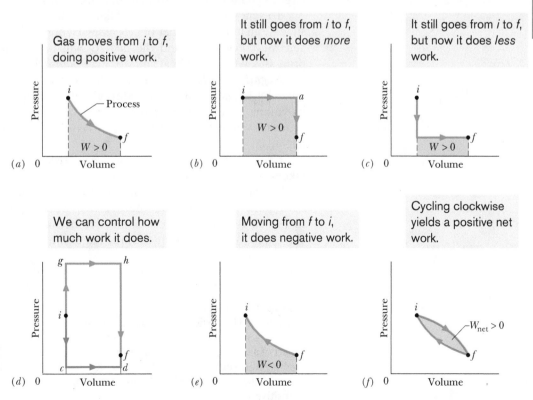

**Figure 18-14**   ($a$) The shaded area represents the work $W$ done by a system as it goes from an initial state $i$ to a final state $f$. Work $W$ is positive because the system's volume increases. ($b$) $W$ is still positive, but now greater. ($c$) $W$ is still positive, but now smaller. ($d$) $W$ can be even smaller (path $icdf$) or larger (path $ighf$). ($e$) Here the system goes from state $f$ to state $i$ as the gas is compressed to less volume by an external force. The work $W$ done by the system is now negative. ($f$) The net work $W_{net}$ done by the system during a complete cycle is represented by the shaded area.

pressure decreases as the volume increases. The integral in Eq. 18-25 (and thus the work $W$ done by the gas) is represented by the shaded area under the curve between points $i$ and $f$. Regardless of what exactly we do to take the gas along the curve, that work is positive, due to the fact that the gas increases its volume by forcing the piston upward.

***Another Path.*** Another way to get from state $i$ to state $f$ is shown in Fig. 18-14$b$. There the change takes place in two steps—the first from state $i$ to state $a$, and the second from state $a$ to state $f$.

Step $ia$ of this process is carried out at constant pressure, which means that you leave undisturbed the lead shot that ride on top of the piston in Fig. 18-13. You cause the volume to increase (from $V_i$ to $V_f$) by slowly turning up the temperature control knob, raising the temperature of the gas to some higher value $T_a$. (Increasing the temperature increases the force from the gas on the piston, moving it upward.) During this step, positive work is done by the expanding gas (to lift the loaded piston) and heat is absorbed by the system from the thermal reservoir (in response to the arbitrarily small temperature differences that you create as you turn up the temperature). This heat is positive because it is added to the system.

Step $af$ of the process of Fig. 18-14$b$ is carried out at constant volume, so you must wedge the piston, preventing it from moving. Then as you use the control knob to decrease the temperature, you find that the pressure drops from $p_a$ to its final value $p_f$. During this step, heat is lost by the system to the thermal reservoir.

For the overall process $iaf$, the work $W$, which is positive and is carried out only during step $ia$, is represented by the shaded area under the curve. Energy is transferred as heat during both steps $ia$ and $af$, with a net energy transfer $Q$.

***Reversed Steps.*** Figure 18-14$c$ shows a process in which the previous two steps are carried out in reverse order. The work $W$ in this case is smaller than for Fig. 18-14$b$, as is the net heat absorbed. Figure 18-14$d$ suggests that you can make the work done by the gas as small as you want (by following a path like $icdf$) or as large as you want (by following a path like $ighf$).

To sum up: A system can be taken from a given initial state to a given final state by an infinite number of processes. Heat may or may not be involved, and in general, the work $W$ and the heat $Q$ will have different values for different processes. We say that heat and work are *path-dependent* quantities.

***Negative Work.*** Figure 18-14$e$ shows an example in which negative work is done by a system as some external force compresses the system, reducing its volume. The absolute value of the work done is still equal to the area beneath the curve, but because the gas is *compressed*, the work done by the gas is negative.

***Cycle.*** Figure 18-14$f$ shows a *thermodynamic cycle* in which the system is taken from some initial state $i$ to some other state $f$ and then back to $i$. The net work done by the system during the cycle is the sum of the *positive* work done during the expansion and the *negative* work done during the compression. In Fig. 18-14$f$, the net work is positive because the area under the expansion curve ($i$ to $f$) is greater than the area under the compression curve ($f$ to $i$).

**✓ Checkpoint 4**

The $p$-$V$ diagram here shows six curved paths (connected by vertical paths) that can be followed by a gas. Which two of the curved paths should be part of a closed cycle (those curved paths plus connecting vertical paths) if the net work done by the gas during the cycle is to be at its maximum positive value?

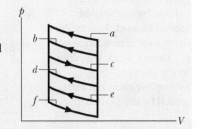

## The First Law of Thermodynamics

You have just seen that when a system changes from a given initial state to a given final state, both the work $W$ and the heat $Q$ depend on the nature of the process. Experimentally, however, we find a surprising thing. *The quantity $Q - W$ is the same for all processes.* It depends only on the initial and final states and does not depend at all on how the system gets from one to the other. All other combinations of $Q$ and $W$, including $Q$ alone, $W$ alone, $Q + W$, and $Q - 2W$, are *path dependent;* only the quantity $Q - W$ is not.

The quantity $Q - W$ must represent a change in some intrinsic property of the system. We call this property the *internal energy $E_{int}$* and we write

$$\Delta E_{int} = E_{int,f} - E_{int,i} = Q - W \quad \text{(first law).} \tag{18-26}$$

Equation 18-26 is the **first law of thermodynamics.** If the thermodynamic system undergoes only a differential change, we can write the first law as*

$$dE_{int} = dQ - dW \quad \text{(first law).} \tag{18-27}$$

The internal energy $E_{int}$ of a system tends to increase if energy is added as heat $Q$ and tends to decrease if energy is lost as work $W$ done by the system.

In Chapter 8, we discussed the principle of energy conservation as it applies to isolated systems—that is, to systems in which no energy enters or leaves the system. The first law of thermodynamics is an extension of that principle to systems that are *not* isolated. In such cases, energy may be transferred into or out of the system as either work $W$ or heat $Q$. In our statement of the first law of thermodynamics above, we assume that there are no changes in the kinetic energy or the potential energy of the system as a whole; that is, $\Delta K = \Delta U = 0$.

***Rules.*** Before this chapter, the term *work* and the symbol $W$ always meant the work done *on* a system. However, starting with Eq. 18-24 and continuing through the next two chapters about thermodynamics, we focus on the work done *by* a system, such as the gas in Fig. 18-13.

The work done *on* a system is always the negative of the work done *by* the system, so if we rewrite Eq. 18-26 in terms of the work $W_{on}$ done *on* the system, we have $\Delta E_{int} = Q + W_{on}$. This tells us the following: The internal energy of a system tends to increase if heat is absorbed by the system or if positive work is done *on* the system. Conversely, the internal energy tends to decrease if heat is lost by the system or if negative work is done *on* the system.

### Checkpoint 5

The figure here shows four paths on a $p$-$V$ diagram along which a gas can be taken from state $i$ to state $f$. Rank the paths according to (a) the change $\Delta E_{int}$ in the internal energy of the gas, (b) the work $W$ done by the gas, and (c) the magnitude of the energy transferred as heat $Q$ between the gas and its environment, greatest first.

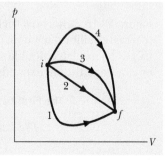

---

*Here $dQ$ and $dW$, unlike $dE_{int}$, are not true differentials; that is, there are no such functions as $Q(p, V)$ and $W(p, V)$ that depend only on the state of the system. The quantities $dQ$ and $dW$ are called *inexact differentials* and are usually represented by the symbols $đQ$ and $đW$. For our purposes, we can treat them simply as infinitesimally small energy transfers.

We slowly remove lead shot, allowing an expansion without any heat transfer.

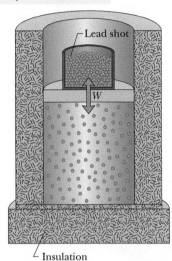

**Figure 18-15** An adiabatic expansion can be carried out by slowly removing lead shot from the top of the piston. Adding lead shot reverses the process at any stage.

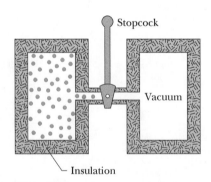

**Figure 18-16** The initial stage of a free-expansion process. After the stopcock is opened, the gas fills both chambers and eventually reaches an equilibrium state.

## Some Special Cases of the First Law of Thermodynamics

Here are four thermodynamic processes as summarized in Table 18-5.

1. ***Adiabatic processes.*** An adiabatic process is one that occurs so rapidly or occurs in a system that is so well insulated that *no transfer of energy as heat* occurs between the system and its environment. Putting $Q = 0$ in the first law (Eq. 18-26) yields

$$\Delta E_{int} = -W \qquad \text{(adiabatic process)}. \qquad (18\text{-}28)$$

This tells us that if work is done *by* the system (that is, if $W$ is positive), the internal energy of the system decreases by the amount of work. Conversely, if work is done *on* the system (that is, if $W$ is negative), the internal energy of the system increases by that amount.

Figure 18-15 shows an idealized adiabatic process. Heat cannot enter or leave the system because of the insulation. Thus, the only way energy can be transferred between the system and its environment is by work. If we remove shot from the piston and allow the gas to expand, the work done by the system (the gas) is positive and the internal energy of the gas decreases. If, instead, we add shot and compress the gas, the work done by the system is negative and the internal energy of the gas increases.

2. ***Constant-volume processes.*** If the volume of a system (such as a gas) is held constant, that system can do no work. Putting $W = 0$ in the first law (Eq. 18-26) yields

$$\Delta E_{int} = Q \qquad \text{(constant-volume process)}. \qquad (18\text{-}29)$$

Thus, if heat is absorbed by a system (that is, if $Q$ is positive), the internal energy of the system increases. Conversely, if heat is lost during the process (that is, if $Q$ is negative), the internal energy of the system must decrease.

3. ***Cyclical processes.*** There are processes in which, after certain interchanges of heat and work, the system is restored to its initial state. In that case, no intrinsic property of the system—including its internal energy—can possibly change. Putting $\Delta E_{int} = 0$ in the first law (Eq. 18-26) yields

$$Q = W \qquad \text{(cyclical process)}. \qquad (18\text{-}30)$$

Thus, the net work done during the process must exactly equal the net amount of energy transferred as heat; the store of internal energy of the system remains unchanged. Cyclical processes form a closed loop on a *p-V* plot, as shown in Fig. 18-14*f*. We discuss such processes in detail in Chapter 20.

4. ***Free expansions.*** These are adiabatic processes in which no transfer of heat occurs between the system and its environment and no work is done on or by the system. Thus, $Q = W = 0$, and the first law requires that

$$\Delta E_{int} = 0 \qquad \text{(free expansion)}. \qquad (18\text{-}31)$$

Figure 18-16 shows how such an expansion can be carried out. A gas, which is in thermal equilibrium within itself, is initially confined by a closed stopcock to one half of an insulated double chamber; the other half is evacuated. The stopcock is opened, and the gas expands freely to fill both halves of the chamber. No heat is

**Table 18-5** The First Law of Thermodynamics: Four Special Cases

| *The Law:* $\Delta E_{int} = Q - W$ (Eq. 18-26) | | |
|---|---|---|
| Process | Restriction | Consequence |
| Adiabatic | $Q = 0$ | $\Delta E_{int} = -W$ |
| Constant volume | $W = 0$ | $\Delta E_{int} = Q$ |
| Closed cycle | $\Delta E_{int} = 0$ | $Q = W$ |
| Free expansion | $Q = W = 0$ | $\Delta E_{int} = 0$ |

transferred to or from the gas because of the insulation. No work is done by the gas because it rushes into a vacuum and thus does not meet any pressure.

A free expansion differs from all other processes we have considered because it cannot be done slowly and in a controlled way. As a result, at any given instant during the sudden expansion, the gas is not in thermal equilibrium and its pressure is not uniform. Thus, although we can plot the initial and final states on a *p-V* diagram, we cannot plot the expansion itself.

### Checkpoint 6

For one complete cycle as shown in the *p-V* diagram here, are (a) $\Delta E_{int}$ for the gas and (b) the net energy transferred as heat $Q$ positive, negative, or zero?

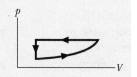

---

### Sample Problem 18.05  First law of thermodynamics: work, heat, internal energy change

Let 1.00 kg of liquid water at 100°C be converted to steam at 100°C by boiling at standard atmospheric pressure (which is 1.00 atm or $1.01 \times 10^5$ Pa) in the arrangement of Fig. 18-17. The volume of that water changes from an initial value of $1.00 \times 10^{-3}$ m³ as a liquid to 1.671 m³ as steam.

(a) How much work is done by the system during this process?

**KEY IDEAS**

(1) The system must do positive work because the volume increases. (2) We calculate the work $W$ done by integrating the pressure with respect to the volume (Eq. 18-25).

*Calculation:* Because here the pressure is constant at $1.01 \times 10^5$ Pa, we can take $p$ outside the integral. Thus,

$$W = \int_{V_i}^{V_f} p\,dV = p\int_{V_i}^{V_f} dV = p(V_f - V_i)$$

$$= (1.01 \times 10^5 \text{ Pa})(1.671 \text{ m}^3 - 1.00 \times 10^{-3} \text{ m}^3)$$

$$= 1.69 \times 10^5 \text{ J} = 169 \text{ kJ.} \qquad \text{(Answer)}$$

(b) How much energy is transferred as heat during the process?

**KEY IDEA**

Because the heat causes only a phase change and not a change in temperature, it is given fully by Eq. 18-16 ($Q = Lm$).

*Calculation:* Because the change is from liquid to gaseous phase, $L$ is the heat of vaporization $L_V$, with the value given in Eq. 18-17 and Table 18-4. We find

$$Q = L_V m = (2256 \text{ kJ/kg})(1.00 \text{ kg})$$

$$= 2256 \text{ kJ} \approx 2260 \text{ kJ.} \qquad \text{(Answer)}$$

(c) What is the change in the system's internal energy during the process?

**KEY IDEA**

The change in the system's internal energy is related to the heat (here, this is energy transferred into the system) and the work (here, this is energy transferred out of the system) by the first law of thermodynamics (Eq. 18-26).

*Calculation:* We write the first law as

$$\Delta E_{int} = Q - W = 2256 \text{ kJ} - 169 \text{ kJ}$$

$$\approx 2090 \text{ kJ} = 2.09 \text{ MJ.} \qquad \text{(Answer)}$$

This quantity is positive, indicating that the internal energy of the system has increased during the boiling process. The added energy goes into separating the $H_2O$ molecules, which strongly attract one another in the liquid state. We see that, when water is boiled, about 7.5% (= 169 kJ/2260 kJ) of the heat goes into the work of pushing back the atmosphere. The rest of the heat goes into the internal energy of the system.

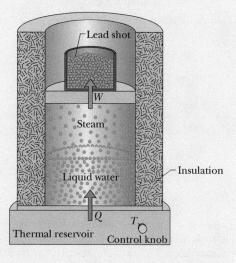

**Figure 18-17** Water boiling at constant pressure. Energy is transferred from the thermal reservoir as heat until the liquid water has changed completely into steam. Work is done by the expanding gas as it lifts the loaded piston.

 Additional examples, video, and practice available at *WileyPLUS*

# 18-6 HEAT TRANSFER MECHANISMS

## Learning Objectives

*After reading this module, you should be able to . . .*

**18.31** For thermal conduction through a layer, apply the relationship between the energy-transfer rate $P_{cond}$ and the layer's area $A$, thermal conductivity $k$, thickness $L$, and temperature difference $\Delta T$ (between its two sides).

**18.32** For a composite slab (two or more layers) that has reached the steady state in which temperatures are no longer changing, identify that (by the conservation of energy) the rates of thermal conduction $P_{cond}$ through the layers must be equal.

**18.33** For thermal conduction through a layer, apply the relationship between thermal resistance $R$, thickness $L$, and thermal conductivity $k$.

**18.34** Identify that thermal energy can be transferred by

convection, in which a warmer fluid (gas or liquid) tends to rise in a cooler fluid.

**18.35** In the *emission* of thermal radiation by an object, apply the relationship between the energy-transfer rate $P_{rad}$ and the object's surface area $A$, emissivity $\varepsilon$, and *surface* temperature $T$ (in kelvins).

**18.36** In the *absorption* of thermal radiation by an object, apply the relationship between the energy-transfer rate $P_{abs}$ and the object's surface area $A$ and emissivity $\varepsilon$, and the *environmental* temperature $T$ (in kelvins).

**18.37** Calculate the net energy-transfer rate $P_{net}$ of an object emitting radiation to its environment and absorbing radiation from that environment.

## Key Ideas

● The rate $P_{cond}$ at which energy is conducted through a slab for which one face is maintained at the higher temperature $T_H$ and the other face is maintained at the lower temperature $T_C$ is

$$P_{cond} = \frac{Q}{t} = kA\,\frac{T_H - T_C}{L}.$$

Here each face of the slab has area $A$, the length of the slab (the distance between the faces) is $L$, and $k$ is the thermal conductivity of the material.

● Convection occurs when temperature differences cause an energy transfer by motion within a fluid.

● Radiation is an energy transfer via the emission of electromagnetic energy. The rate $P_{rad}$ at which an object emits energy via thermal radiation is

$$P_{rad} = \sigma\varepsilon AT^4,$$

where $\sigma\ (= 5.6704 \times 10^{-8}\ \mathrm{W/m^2 \cdot K^4})$ is the Stefan–Boltzmann constant, $\varepsilon$ is the emissivity of the object's surface, $A$ is its surface area, and $T$ is its surface temperature (in kelvins). The rate $P_{abs}$ at which an object absorbs energy via thermal radiation from its environment, which is at the uniform temperature $T_{env}$ (in kelvins), is

$$P_{abs} = \sigma\varepsilon AT^4_{env}.$$

## Heat Transfer Mechanisms

We have discussed the transfer of energy as heat between a system and its environment, but we have not yet described how that transfer takes place. There are three transfer mechanisms: conduction, convection, and radiation. Let's next examine these mechanisms in turn.

### Conduction

If you leave the end of a metal poker in a fire for enough time, its handle will get hot. Energy is transferred from the fire to the handle by (thermal) **conduction** along the length of the poker. The vibration amplitudes of the atoms and electrons of the metal at the fire end of the poker become relatively large because of the high temperature of their environment. These increased vibrational amplitudes, and thus the associated energy, are passed along the poker, from atom to atom, during collisions between adjacent atoms. In this way, a region of rising temperature extends itself along the poker to the handle.

Consider a slab of face area $A$ and thickness $L$, whose faces are maintained at temperatures $T_H$ and $T_C$ by a hot reservoir and a cold reservoir, as in Fig. 18-18. Let $Q$ be the energy that is transferred as heat through the slab, from its hot face to its cold face, in time $t$. Experiment shows that the *conduction rate* $P_{cond}$ (the

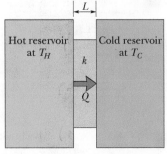

We assume a steady transfer of energy as heat.

$T_H > T_C$

**Figure 18-18** Thermal conduction. Energy is transferred as heat from a reservoir at temperature $T_H$ to a cooler reservoir at temperature $T_C$ through a conducting slab of thickness $L$ and thermal conductivity $k$.

amount of energy transferred per unit time) is

$$P_{cond} = \frac{Q}{t} = kA\frac{T_H - T_C}{L},\qquad(18\text{-}32)$$

in which $k$, called the *thermal conductivity*, is a constant that depends on the material of which the slab is made. A material that readily transfers energy by conduction is a *good thermal conductor* and has a high value of $k$. Table 18-6 gives the thermal conductivities of some common metals, gases, and building materials.

### Thermal Resistance to Conduction (*R*-Value)

If you are interested in insulating your house or in keeping cola cans cold on a picnic, you are more concerned with poor heat conductors than with good ones. For this reason, the concept of *thermal resistance R* has been introduced into engineering practice. The $R$-value of a slab of thickness $L$ is defined as

$$R = \frac{L}{k}.\qquad(18\text{-}33)$$

The lower the thermal conductivity of the material of which a slab is made, the higher the $R$-value of the slab; so something that has a high $R$-value is a *poor thermal conductor* and thus a *good thermal insulator.*

Note that $R$ is a property attributed to a slab of a specified thickness, not to a material. The commonly used unit for $R$ (which, in the United States at least, is almost never stated) is the square foot–Fahrenheit degree–hour per British thermal unit ($ft^2 \cdot F° \cdot h/Btu$). (Now you know why the unit is rarely stated.)

### Conduction Through a Composite Slab

Figure 18-19 shows a composite slab, consisting of two materials having different thicknesses $L_1$ and $L_2$ and different thermal conductivities $k_1$ and $k_2$. The temperatures of the outer surfaces of the slab are $T_H$ and $T_C$. Each face of the slab has area $A$. Let us derive an expression for the conduction rate through the slab under the assumption that the transfer is a *steady-state* process; that is, the temperatures everywhere in the slab and the rate of energy transfer do not change with time.

In the steady state, the conduction rates through the two materials must be equal. This is the same as saying that the energy transferred through one material in a certain time must be equal to that transferred through the other material in the same time. If this were not true, temperatures in the slab would be changing and we would not have a steady-state situation. Letting $T_X$ be the temperature of the interface between the two materials, we can now use Eq. 18-32 to write

$$P_{cond} = \frac{k_2A(T_H - T_X)}{L_2} = \frac{k_1A(T_X - T_C)}{L_1}.\qquad(18\text{-}34)$$

Solving Eq. 18-34 for $T_X$ yields, after a little algebra,

$$T_X = \frac{k_1L_2T_C + k_2L_1T_H}{k_1L_2 + k_2L_1}.\qquad(18\text{-}35)$$

Substituting this expression for $T_X$ into either equality of Eq. 18-34 yields

$$P_{cond} = \frac{A(T_H - T_C)}{L_1/k_1 + L_2/k_2}.\qquad(18\text{-}36)$$

We can extend Eq. 18-36 to apply to any number $n$ of materials making up a slab:

$$P_{cond} = \frac{A(T_H - T_C)}{\Sigma\,(L/k)}.\qquad(18\text{-}37)$$

The summation sign in the denominator tells us to add the values of $L/k$ for all the materials.

**Table 18-6** **Some Thermal Conductivities**

| Substance | $k$ (W/m·K) |
|---|---|
| *Metals* | |
| Stainless steel | 14 |
| Lead | 35 |
| Iron | 67 |
| Brass | 109 |
| Aluminum | 235 |
| Copper | 401 |
| Silver | 428 |
| *Gases* | |
| Air (dry) | 0.026 |
| Helium | 0.15 |
| Hydrogen | 0.18 |
| *Building Materials* | |
| Polyurethane foam | 0.024 |
| Rock wool | 0.043 |
| Fiberglass | 0.048 |
| White pine | 0.11 |
| Window glass | 1.0 |

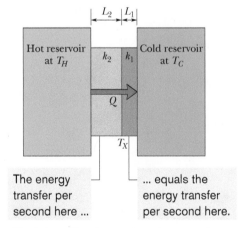

**Figure 18-19** Heat is transferred at a steady rate through a composite slab made up of two different materials with different thicknesses and different thermal conductivities. The steady-state temperature at the interface of the two materials is $T_X$.

## ✓ Checkpoint 7

The figure shows the face and interface temperatures of a composite slab consisting of four materials, of identical thicknesses, through which the heat transfer is steady. Rank the materials according to their thermal conductivities, greatest first.

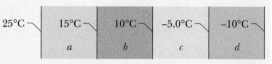

Edward Kinsman/Photo Researchers, Inc.

**Figure 18-20** A false-color thermogram reveals the rate at which energy is radiated by a cat. The rate is color-coded, with white and red indicating the greatest radiation rate. The nose is cool.

## Convection

When you look at the flame of a candle or a match, you are watching thermal energy being transported upward by **convection.** Such energy transfer occurs when a fluid, such as air or water, comes in contact with an object whose temperature is higher than that of the fluid. The temperature of the part of the fluid that is in contact with the hot object increases, and (in most cases) that fluid expands and thus becomes less dense. Because this expanded fluid is now lighter than the surrounding cooler fluid, buoyant forces cause it to rise. Some of the surrounding cooler fluid then flows so as to take the place of the rising warmer fluid, and the process can then continue.

Convection is part of many natural processes. Atmospheric convection plays a fundamental role in determining global climate patterns and daily weather variations. Glider pilots and birds alike seek rising thermals (convection currents of warm air) that keep them aloft. Huge energy transfers take place within the oceans by the same process. Finally, energy is transported to the surface of the Sun from the nuclear furnace at its core by enormous cells of convection, in which hot gas rises to the surface along the cell core and cooler gas around the core descends below the surface.

## Radiation

The third method by which an object and its environment can exchange energy as heat is via electromagnetic waves (visible light is one kind of electromagnetic wave). Energy transferred in this way is often called **thermal radiation** to distinguish it from electromagnetic *signals* (as in, say, television broadcasts) and from nuclear radiation (energy and particles emitted by nuclei). (To "radiate" generally means to emit.) When you stand in front of a big fire, you are warmed by absorbing thermal radiation from the fire; that is, your thermal energy increases as the fire's thermal energy decreases. No medium is required for heat transfer via radiation—the radiation can travel through vacuum from, say, the Sun to you.

The rate $P_{rad}$ at which an object emits energy via electromagnetic radiation depends on the object's surface area $A$ and the temperature $T$ of that area in kelvins and is given by

$$P_{rad} = \sigma \varepsilon A T^4. \tag{18-38}$$

Here $\sigma = 5.6704 \times 10^{-8}$ W/m²·K⁴ is called the *Stefan–Boltzmann constant* after Josef Stefan (who discovered Eq. 18-38 experimentally in 1879) and Ludwig Boltzmann (who derived it theoretically soon after). The symbol $\varepsilon$ represents the *emissivity* of the object's surface, which has a value between 0 and 1, depending on the composition of the surface. A surface with the maximum emissivity of 1.0 is said to be a *blackbody radiator,* but such a surface is an ideal limit and does not occur in nature. Note again that the temperature in Eq. 18-38 must be in kelvins so that a temperature of absolute zero corresponds to no radiation. Note also that every object whose temperature is above 0 K—including you—emits thermal radiation. (See Fig. 18-20.)

The rate $P_{abs}$ at which an object absorbs energy via thermal radiation from its environment, which we take to be at uniform temperature $T_{env}$ (in kelvins), is

$$P_{abs} = \sigma \varepsilon A T_{env}^4. \qquad (18\text{-}39)$$

The emissivity $\varepsilon$ in Eq. 18-39 is the same as that in Eq. 18-38. An idealized blackbody radiator, with $\varepsilon = 1$, will absorb all the radiated energy it intercepts (rather than sending a portion back away from itself through reflection or scattering).

Because an object both emits and absorbs thermal radiation, its net rate $P_{net}$ of energy exchange due to thermal radiation is

$$P_{net} = P_{abs} - P_{rad} = \sigma \varepsilon A (T_{env}^4 - T^4). \qquad (18\text{-}40)$$

$P_{net}$ is positive if net energy is being absorbed via radiation and negative if it is being lost via radiation.

Thermal radiation is involved in the numerous medical cases of a *dead* rattlesnake striking a hand reaching toward it. Pits between each eye and nostril of a rattlesnake (Fig. 18-21) serve as sensors of thermal radiation. When, say, a mouse moves close to a rattlesnake's head, the thermal radiation from the mouse triggers these sensors, causing a reflex action in which the snake strikes the mouse with its fangs and injects its venom. The thermal radiation from a reaching hand can cause the same reflex action even if the snake has been dead for as long as 30 min because the snake's nervous system continues to function. As one snake expert advised, if you must remove a recently killed rattlesnake, use a long stick rather than your hand.

© David A. Northcott/Corbis Images

**Figure 18-21** A rattlesnake's face has thermal radiation detectors, allowing the snake to strike at an animal even in complete darkness.

## Sample Problem 18.06  Thermal conduction through a layered wall

Figure 18-22 shows the cross section of a wall made of white pine of thickness $L_a$ and brick of thickness $L_d$ ($= 2.0L_a$), sandwiching two layers of unknown material with identical thicknesses and thermal conductivities. The thermal conductivity of the pine is $k_a$ and that of the brick is $k_d$ ($= 5.0k_a$). The face area $A$ of the wall is unknown. Thermal conduction through the wall has reached the steady state; the only known interface temperatures are $T_1 = 25°C$, $T_2 = 20°C$, and $T_5 = -10°C$. What is interface temperature $T_4$?

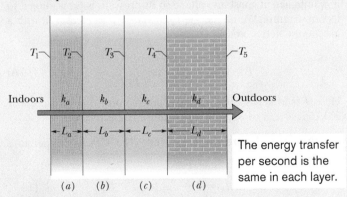

The energy transfer per second is the same in each layer.

**Figure 18-22** Steady-state heat transfer through a wall.

### KEY IDEAS

(1) Temperature $T_4$ helps determine the rate $P_d$ at which energy is conducted through the brick, as given by Eq. 18-32. However, we lack enough data to solve Eq. 18-32 for $T_4$. (2) Because the conduction is steady, the conduction rate $P_d$ through the brick must equal the conduction rate $P_a$ through the pine. That gets us going.

*Calculations:* From Eq. 18-32 and Fig. 18-22, we can write

$$P_a = k_a A \frac{T_1 - T_2}{L_a} \quad \text{and} \quad P_d = k_d A \frac{T_4 - T_5}{L_d}.$$

Setting $P_a = P_d$ and solving for $T_4$ yield

$$T_4 = \frac{k_a L_d}{k_d L_a} (T_1 - T_2) + T_5.$$

Letting $L_d = 2.0L_a$ and $k_d = 5.0k_a$, and inserting the known temperatures, we find

$$T_4 = \frac{k_a(2.0L_a)}{(5.0k_a)L_a} (25°C - 20°C) + (-10°C)$$

$$= -8.0°C. \qquad \text{(Answer)}$$

## Sample Problem 18.07 Thermal radiation by a skunk cabbage can melt surrounding snow

Unlike most other plants, a skunk cabbage can regulate its internal temperature (set at $T = 22°C$) by altering the rate at which it produces energy. If it becomes covered with snow, it can increase that production so that its thermal radiation melts the snow in order to re-expose the plant to sunlight. Let's model a skunk cabbage with a cylinder of height $h = 5.0$ cm and radius $R = 1.5$ cm and assume it is surrounded by a snow wall at temperature $T_{env} = -3.0°C$ (Fig. 18-23). If the emissivity $\varepsilon$ is 0.80, what is the net rate of energy exchange via thermal radiation between the plant's curved side and the snow?

**Figure 18-23** Model of skunk cabbage that has melted snow to uncover itself.

### KEY IDEAS

(1) In a steady-state situation, a surface with area $A$, emissivity $\varepsilon$, and temperature $T$ loses energy to thermal radiation at the rate given by Eq. 18-38 ($P_{rad} = \sigma\varepsilon AT^4$). (2) Simultaneously, it gains energy by thermal radiation from its environment at temperature $T_{env}$ at the rate given by Eq. 18-39 ($P_{abs} = \sigma\varepsilon AT_{env}^4$).

*Calculations:* To find the net rate of energy exchange, we subtract Eq. 18-38 from Eq. 18-39 to write

$$P_{net} = P_{abs} - P_{rad}$$
$$= \sigma\varepsilon A(T_{env}^4 - T^4). \quad (18\text{-}41)$$

We need the area of the curved surface of the cylinder, which is $A = h(2\pi R)$. We also need the temperatures in kelvins: $T_{env} = 273$ K $- 3$ K $= 270$ K and $T = 273$ K $+ 22$ K $= 295$ K. Substituting in Eq. 18-41 for $A$ and then substituting known values in SI units (which are not displayed here), we find

$$P_{net} = (5.67 \times 10^{-8})(0.80)(0.050)(2\pi)(0.015)(270^4 - 295^4)$$
$$= -0.48\,\text{W}. \quad \text{(Answer)}$$

Thus, the plant has a net loss of energy via thermal radiation of 0.48 W. The plant's energy production rate is comparable to that of a hummingbird in flight.

WILEY**PLUS** Additional examples, video, and practice available at *WileyPLUS*

# Review & Summary

**Temperature; Thermometers** Temperature is an SI base quantity related to our sense of hot and cold. It is measured with a thermometer, which contains a working substance with a measurable property, such as length or pressure, that changes in a regular way as the substance becomes hotter or colder.

**Zeroth Law of Thermodynamics** When a thermometer and some other object are placed in contact with each other, they eventually reach thermal equilibrium. The reading of the thermometer is then taken to be the temperature of the other object. The process provides consistent and useful temperature measurements because of the **zeroth law of thermodynamics:** If bodies $A$ and $B$ are each in thermal equilibrium with a third body $C$ (the thermometer), then $A$ and $B$ are in thermal equilibrium with each other.

**The Kelvin Temperature Scale** In the SI system, temperature is measured on the **Kelvin scale,** which is based on the *triple point* of water (273.16 K). Other temperatures are then defined by use of a *constant-volume gas thermometer*, in which a sample of gas is maintained at constant volume so its pressure is proportional to its temperature. We define the *temperature T* as measured with a gas thermometer to be

$$T = (273.16\,\text{K})\left(\lim_{gas \to 0} \frac{p}{p_3}\right). \quad (18\text{-}6)$$

Here $T$ is in kelvins, and $p_3$ and $p$ are the pressures of the gas at 273.16 K and the measured temperature, respectively.

**Celsius and Fahrenheit Scales** The Celsius temperature scale is defined by

$$T_C = T - 273.15°, \quad (18\text{-}7)$$

with $T$ in kelvins. The Fahrenheit temperature scale is defined by

$$T_F = \tfrac{9}{5}T_C + 32°. \quad (18\text{-}8)$$

**Thermal Expansion**    All objects change size with changes in temperature. For a temperature change $\Delta T$, a change $\Delta L$ in any linear dimension $L$ is given by

$$\Delta L = L\alpha\,\Delta T, \tag{18-9}$$

in which $\alpha$ is the **coefficient of linear expansion.** The change $\Delta V$ in the volume $V$ of a solid or liquid is

$$\Delta V = V\beta\,\Delta T. \tag{18-10}$$

Here $\beta = 3\alpha$ is the material's **coefficient of volume expansion.**

**Heat**    Heat $Q$ is energy that is transferred between a system and its environment because of a temperature difference between them. It can be measured in **joules** (J), **calories** (cal), **kilocalories** (Cal or kcal), or **British thermal units** (Btu), with

$$1\ \text{cal} = 3.968 \times 10^{-3}\ \text{Btu} = 4.1868\ \text{J}. \tag{18-12}$$

**Heat Capacity and Specific Heat**    If heat $Q$ is absorbed by an object, the object's temperature change $T_f - T_i$ is related to $Q$ by

$$Q = C(T_f - T_i), \tag{18-13}$$

in which $C$ is the **heat capacity** of the object. If the object has mass $m$, then

$$Q = cm(T_f - T_i), \tag{18-14}$$

where $c$ is the **specific heat** of the material making up the object. The **molar specific heat** of a material is the heat capacity per mole, which means per $6.02 \times 10^{23}$ elementary units of the material.

**Heat of Transformation**    Matter can exist in three common states: solid, liquid, and vapor. Heat absorbed by a material may change the material's physical state—for example, from solid to liquid or from liquid to gas. The amount of energy required per unit mass to change the state (but not the temperature) of a particular material is its **heat of transformation** $L$. Thus,

$$Q = Lm. \tag{18-16}$$

The **heat of vaporization** $L_V$ is the amount of energy per unit mass that must be added to vaporize a liquid or that must be removed to condense a gas. The **heat of fusion** $L_F$ is the amount of energy per unit mass that must be added to melt a solid or that must be removed to freeze a liquid.

**Work Associated with Volume Change**    A gas may exchange energy with its surroundings through work. The amount of work $W$ done by a gas as it expands or contracts from an initial volume $V_i$ to a final volume $V_f$ is given by

$$W = \int dW = \int_{V_i}^{V_f} p\,dV. \tag{18-25}$$

The integration is necessary because the pressure $p$ may vary during the volume change.

**First Law of Thermodynamics**    The principle of conservation of energy for a thermodynamic process is expressed in the **first law of thermodynamics,** which may assume either of the forms

$$\Delta E_{\text{int}} = E_{\text{int},f} - E_{\text{int},i} = Q - W \quad \text{(first law)} \tag{18-26}$$

or

$$dE_{\text{int}} = dQ - dW \quad \text{(first law).} \tag{18-27}$$

$E_{\text{int}}$ represents the internal energy of the material, which depends only on the material's state (temperature, pressure, and volume). $Q$ represents the energy exchanged as heat between the system and its surroundings; $Q$ is positive if the system absorbs heat and negative if the system loses heat. $W$ is the work done by the system; $W$ is positive if the system expands against an external force from the surroundings and negative if the system contracts because of an external force. $Q$ and $W$ are path dependent; $\Delta E_{\text{int}}$ is path independent.

**Applications of the First Law**    The first law of thermodynamics finds application in several special cases:

$$\text{adiabatic processes:} \quad Q = 0, \quad \Delta E_{\text{int}} = -W$$

$$\text{constant-volume processes:} \quad W = 0, \quad \Delta E_{\text{int}} = Q$$

$$\text{cyclical processes:} \quad \Delta E_{\text{int}} = 0, \quad Q = W$$

$$\text{free expansions:} \quad Q = W = \Delta E_{\text{int}} = 0$$

**Conduction, Convection, and Radiation**    The rate $P_{\text{cond}}$ at which energy is *conducted* through a slab for which one face is maintained at the higher temperature $T_H$ and the other face is maintained at the lower temperature $T_C$ is

$$P_{\text{cond}} = \frac{Q}{t} = kA\,\frac{T_H - T_C}{L} \tag{18-32}$$

Here each face of the slab has area $A$, the length of the slab (the distance between the faces) is $L$, and $k$ is the thermal conductivity of the material.

*Convection* occurs when temperature differences cause an energy transfer by motion within a fluid.

*Radiation* is an energy transfer via the emission of electromagnetic energy. The rate $P_{\text{rad}}$ at which an object emits energy via thermal radiation is

$$P_{\text{rad}} = \sigma\varepsilon A T^4, \tag{18-38}$$

where $\sigma$ $(= 5.6704 \times 10^{-8}\ \text{W/m}^2\cdot\text{K}^4)$ is the Stefan–Boltzmann constant, $\varepsilon$ is the emissivity of the object's surface, $A$ is its surface area, and $T$ is its surface temperature (in kelvins). The rate $P_{\text{abs}}$ at which an object absorbs energy via thermal radiation from its environment, which is at the uniform temperature $T_{\text{env}}$ (in kelvins), is

$$P_{\text{abs}} = \sigma\varepsilon A T_{\text{env}}^4. \tag{18-39}$$

# Questions

**1** The initial length $L$, change in temperature $\Delta T$, and change in length $\Delta L$ of four rods are given in the following table. Rank the rods according to their coefficients of thermal expansion, greatest first.

| Rod | $L$ (m) | $\Delta T$ (C°) | $\Delta L$ (m) |
|-----|---------|-----------------|----------------|
| a | 2 | 10 | $4 \times 10^{-4}$ |
| b | 1 | 20 | $4 \times 10^{-4}$ |
| c | 2 | 10 | $8 \times 10^{-4}$ |
| d | 4 | 5 | $4 \times 10^{-4}$ |

**2** Figure 18-24 shows three linear temperature scales, with the freezing and boiling points of water indicated. Rank the three scales according to the size of one degree on them, greatest first.

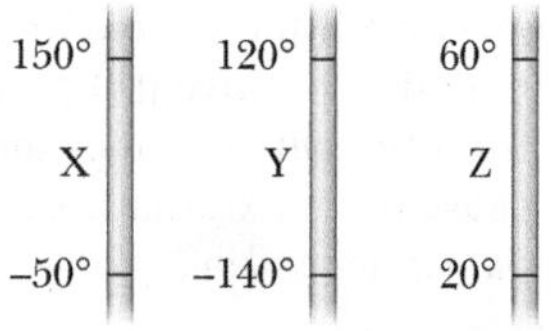

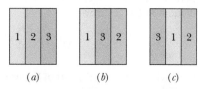

Figure 18-24 Question 2.

**3** Materials $A$, $B$, and $C$ are solids that are at their melting temperatures. Material $A$ requires 200 J to melt 4 kg, material $B$ requires 300 J to melt 5 kg, and material $C$ requires 300 J to melt 6 kg. Rank the materials according to their heats of fusion, greatest first.

**4** A sample $A$ of liquid water and a sample $B$ of ice, of identical mass, are placed in a thermally insulated container and allowed to come to thermal equilibrium. Figure 18-25a is a sketch of the temperature $T$ of the samples versus time $t$. (a) Is the equilibrium temperature above, below, or at the freezing point of water? (b) In reaching equilibrium, does the liquid partly freeze, fully freeze, or undergo no freezing? (c) Does the ice partly melt, fully melt, or undergo no melting?

**5** Question 4 continued: Graphs $b$ through $f$ of Fig. 18-25 are additional sketches of $T$ versus $t$, of which one or more are impossible to produce. (a) Which is impossible and why? (b) In the possible ones, is the equilibrium temperature above, below, or at the freezing point of water? (c) As the possible situations reach equilibrium, does the liquid partly freeze, fully freeze, or undergo no freezing? Does the ice partly melt, fully melt, or undergo no melting?

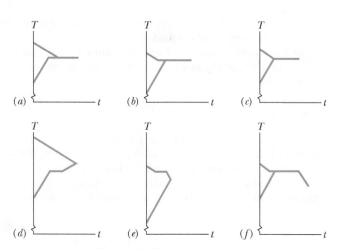

Figure 18-25 Questions 4 and 5.

**6** Figure 18-26 shows three different arrangements of materials 1, 2, and 3 to form a wall. The thermal conductivities are $k_1 > k_2 > k_3$. The left side of the wall is 20 C° higher than the right side. Rank the arrangements according to (a) the (steady state) rate of energy conduction through the wall and (b) the temperature difference across material 1, greatest first.

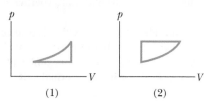

Figure 18-26 Question 6.

**7** Figure 18-27 shows two closed cycles on $p$-$V$ diagrams for a gas. The three parts of cycle 1 are of the same length and shape as those of cycle 2. For each cycle, should the cycle be traversed clockwise or counterclockwise if (a) the net work $W$ done by the gas is to be positive and (b) the net energy transferred by the gas as heat $Q$ is to be positive?

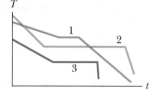

Figure 18-27 Questions 7 and 8.

**8** For which cycle in Fig. 18-27, traversed clockwise, is (a) $W$ greater and (b) $Q$ greater?

**9** Three different materials of identical mass are placed one at a time in a special freezer that can extract energy from a material at a certain constant rate. During the cooling process, each material begins in the liquid state and ends in the solid state; Fig. 18-28 shows the temperature $T$ versus time $t$. (a) For material 1, is the specific heat for the liquid state greater than or less than that for the solid state? Rank the materials according to (b) freezing-point temperature, (c) specific heat in the liquid state, (d) specific heat in the solid state, and (e) heat of fusion, all greatest first.

Figure 18-28 Question 9.

**10** A solid cube of edge length $r$, a solid sphere of radius $r$, and a solid hemisphere of radius $r$, all made of the same material, are maintained at temperature 300 K in an environment at temperature 350 K. Rank the objects according to the net rate at which thermal radiation is exchanged with the environment, greatest first.

**11** A hot object is dropped into a thermally insulated container of water, and the object and water are then allowed to come to thermal equilibrium. The experiment is repeated twice, with different hot objects. All three objects have the same mass and initial temperature, and the mass and initial temperature of the water are the same in the three experiments. For each of the experiments, Fig. 18-29 gives graphs of the temperatures $T$ of the object and the water versus time $t$. Rank the graphs according to the specific heats of the objects, greatest first.

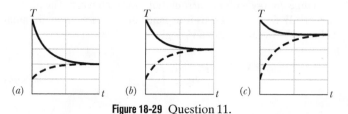

Figure 18-29 Question 11.

# Problems

## Module 18-1 Temperature

•1 Suppose the temperature of a gas is 373.15 K when it is at the boiling point of water. What then is the limiting value of the ratio of the pressure of the gas at that boiling point to its pressure at the triple point of water? (Assume the volume of the gas is the same at both temperatures.)

•2 Two constant-volume gas thermometers are assembled, one with nitrogen and the other with hydrogen. Both contain enough gas so that $p_3 = 80$ kPa. (a) What is the difference between the pressures in the two thermometers if both bulbs are in boiling water? (*Hint:* See Fig. 18-6.) (b) Which gas is at higher pressure?

•3 A gas thermometer is constructed of two gas-containing bulbs, each in a water bath, as shown in Fig. 18-30. The pressure difference between the two bulbs is measured by a mercury manometer as shown. Appropriate reservoirs, not shown in the diagram, maintain constant gas volume in the two bulbs. There is no difference in pressure when both baths are at the triple point of water. The pressure difference is 120 torr when one bath is at the triple point and the other is at the boiling point of water. It is 90.0 torr when one bath is at the triple point and the other is at an unknown temperature to be measured. What is the unknown temperature?

**Figure 18-30** Problem 3.

## Module 18-2 The Celsius and Fahrenheit Scales

•4 (a) In 1964, the temperature in the Siberian village of Oymyakon reached −71°C. What temperature is this on the Fahrenheit scale? (b) The highest officially recorded temperature in the continental United States was 134°F in Death Valley, California. What is this temperature on the Celsius scale?

•5 At what temperature is the Fahrenheit scale reading equal to (a) twice that of the Celsius scale and (b) half that of the Celsius scale?

••6 On a linear X temperature scale, water freezes at −125.0°X and boils at 375.0°X. On a linear Y temperature scale, water freezes at −70.00°Y and boils at −30.00°Y. A temperature of 50.00°Y corresponds to what temperature on the X scale?

••7 **ILW** Suppose that on a linear temperature scale X, water boils at −53.5°X and freezes at −170°X. What is a temperature of 340 K on the X scale? (Approximate water's boiling point as 373 K.)

## Module 18-3 Thermal Expansion

•8 At 20°C, a brass cube has edge length 30 cm. What is the increase in the surface area when it is heated from 20°C to 75°C?

•9 **ILW** A circular hole in an aluminum plate is 2.725 cm in diameter at 0.000°C. What is its diameter when the temperature of the plate is raised to 100.0°C?

•10 An aluminum flagpole is 33 m high. By how much does its length increase as the temperature increases by 15 C°?

•11 What is the volume of a lead ball at 30.00°C if the ball's volume at 60.00°C is 50.00 cm³?

•12 An aluminum-alloy rod has a length of 10.000 cm at 20.000°C and a length of 10.015 cm at the boiling point of water. (a) What is the length of the rod at the freezing point of water? (b) What is the temperature if the length of the rod is 10.009 cm?

•13 **SSM** Find the change in volume of an aluminum sphere with an initial radius of 10 cm when the sphere is heated from 0.0°C to 100°C.

••14 When the temperature of a copper coin is raised by 100 C°, its diameter increases by 0.18%. To two significant figures, give the percent increase in (a) the area of a face, (b) the thickness, (c) the volume, and (d) the mass of the coin. (e) Calculate the coefficient of linear expansion of the coin.

••15 **ILW** A steel rod is 3.000 cm in diameter at 25.00°C. A brass ring has an interior diameter of 2.992 cm at 25.00°C. At what common temperature will the ring just slide onto the rod?

••16 When the temperature of a metal cylinder is raised from 0.0°C to 100°C, its length increases by 0.23%. (a) Find the percent change in density. (b) What is the metal? Use Table 18-2.

••17 **SSM** **WWW** An aluminum cup of 100 cm³ capacity is completely filled with glycerin at 22°C. How much glycerin, if any, will spill out of the cup if the temperature of both the cup and the glycerin is increased to 28°C? (The coefficient of volume expansion of glycerin is $5.1 \times 10^{-4}$/C°.)

••18 At 20°C, a rod is exactly 20.05 cm long on a steel ruler. Both are placed in an oven at 270°C, where the rod now measures 20.11 cm on the same ruler. What is the coefficient of linear expansion for the material of which the rod is made?

••19 **GO** A vertical glass tube of length $L = 1.280\,000$ m is half filled with a liquid at 20.000 000°C. How much will the height of the liquid column change when the tube and liquid are heated to 30.000 000°C? Use coefficients $\alpha_{\text{glass}} = 1.000\,000 \times 10^{-5}$/K and $\beta_{\text{liquid}} = 4.000\,000 \times 10^{-5}$/K.

••20 **GO** In a certain experiment, a small radioactive source must move at selected, extremely slow speeds. This motion is accomplished by fastening the source to one end of an aluminum rod and heating the central section of the rod in a controlled way. If the effective heated section of the rod in Fig. 18-31 has length $d = 2.00$ cm, at what constant rate must the temperature of the rod be changed if the source is to move at a constant speed of 100 nm/s?

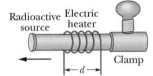

**Figure 18-31** Problem 20.

•••21 **SSM** **ILW** As a result of a temperature rise of 32 C°, a bar with a crack at its center buckles upward (Fig. 18-32). The fixed distance $L_0$ is 3.77 m and the coefficient of linear expansion of the bar is $25 \times 10^{-6}$/C°. Find the rise $x$ of the center.

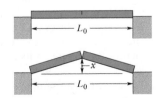

**Figure 18-32** Problem 21.

**Module 18-4   Absorption of Heat**

•22 ╾╼╾ One way to keep the contents of a garage from becoming too cold on a night when a severe subfreezing temperature is forecast is to put a tub of water in the garage. If the mass of the water is 125 kg and its initial temperature is 20°C, (a) how much energy must the water transfer to its surroundings in order to freeze completely and (b) what is the lowest possible temperature of the water and its surroundings until that happens?

•23 SSM A small electric immersion heater is used to heat 100 g of water for a cup of instant coffee. The heater is labeled "200 watts" (it converts electrical energy to thermal energy at this rate). Calculate the time required to bring all this water from 23.0°C to 100°C, ignoring any heat losses.

•24 A certain substance has a mass per mole of 50.0 g/mol. When 314 J is added as heat to a 30.0 g sample, the sample's temperature rises from 25.0°C to 45.0°C. What are the (a) specific heat and (b) molar specific heat of this substance? (c) How many moles are in the sample?

•25 A certain diet doctor encourages people to diet by drinking ice water. His theory is that the body must burn off enough fat to raise the temperature of the water from 0.00°C to the body temperature of 37.0°C. How many liters of ice water would have to be consumed to burn off 454 g (about 1 lb) of fat, assuming that burning this much fat requires 3500 Cal be transferred to the ice water? Why is it not advisable to follow this diet? (One liter = $10^3$ cm$^3$. The density of water is 1.00 g/cm$^3$.)

•26 What mass of butter, which has a usable energy content of 6.0 Cal/g (= 6000 cal/g), would be equivalent to the change in gravitational potential energy of a 73.0 kg man who ascends from sea level to the top of Mt. Everest, at elevation 8.84 km? Assume that the average g for the ascent is 9.80 m/s$^2$.

•27 SSM Calculate the minimum amount of energy, in joules, required to completely melt 130 g of silver initially at 15.0°C.

•28 How much water remains unfrozen after 50.2 kJ is transferred as heat from 260 g of liquid water initially at its freezing point?

••29 In a solar water heater, energy from the Sun is gathered by water that circulates through tubes in a rooftop collector. The solar radiation enters the collector through a transparent cover and warms the water in the tubes; this water is pumped into a holding tank. Assume that the efficiency of the overall system is 20% (that is, 80% of the incident solar energy is lost from the system). What collector area is necessary to raise the temperature of 200 L of water in the tank from 20°C to 40°C in 1.0 h when the intensity of incident sunlight is 700 W/m$^2$?

••30 A 0.400 kg sample is placed in a cooling apparatus that removes energy as heat at a constant rate. Figure 18-33 gives the temperature T of the sample versus time t; the horizontal scale is set by $t_s$ = 80.0 min. The sample freezes during the energy removal. The specific heat of the sample in its initial liquid phase is 3000 J/kg·K. What are (a) the sample's heat of fusion and (b) its specific heat in the frozen phase?

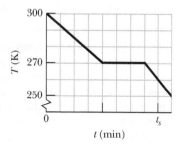

**Figure 18-33** Problem 30.

••31 ILW What mass of steam at 100°C must be mixed with 150 g of ice at its melting point, in a thermally insulated container, to produce liquid water at 50°C?

••32 The specific heat of a substance varies with temperature according to the function c = 0.20 + 0.14T + 0.023T$^2$, with T in °C and c in cal/g·K. Find the energy required to raise the temperature of 2.0 g of this substance from 5.0°C to 15°C.

••33 Nonmetric version: (a) How long does a 2.0 × 10$^5$ Btu/h water heater take to raise the temperature of 40 gal of water from 70°F to 100°F? Metric version: (b) How long does a 59 kW water heater take to raise the temperature of 150 L of water from 21°C to 38°C?

••34 GO Samples A and B are at different initial temperatures when they are placed in a thermally insulated container and allowed to come to thermal equilibrium. Figure 18-34a gives their temperatures T versus time t. Sample A has a mass of 5.0 kg; sample B has a mass of 1.5 kg. Figure 18-34b is a general plot for the material of sample B. It shows the temperature change ΔT that the material undergoes when energy is transferred to it as heat Q. The change ΔT is plotted versus the energy Q per unit mass of the material, and the scale of the vertical axis is set by $\Delta T_s$ = 4.0 C°. What is the specific heat of sample A?

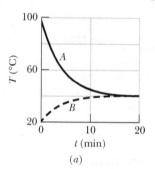
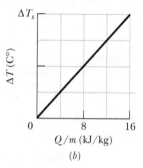

(a)
(b)

**Figure 18-34** Problem 34.

••35 An insulated Thermos contains 130 cm$^3$ of hot coffee at 80.0°C. You put in a 12.0 g ice cube at its melting point to cool the coffee. By how many degrees has your coffee cooled once the ice has melted and equilibrium is reached? Treat the coffee as though it were pure water and neglect energy exchanges with the environment.

••36 A 150 g copper bowl contains 220 g of water, both at 20.0°C. A very hot 300 g copper cylinder is dropped into the water, causing the water to boil, with 5.00 g being converted to steam. The final temperature of the system is 100°C. Neglect energy transfers with the environment. (a) How much energy (in calories) is transferred to the water as heat? (b) How much to the bowl? (c) What is the original temperature of the cylinder?

••37 A person makes a quantity of iced tea by mixing 500 g of hot tea (essentially water) with an equal mass of ice at its melting point. Assume the mixture has negligible energy exchanges with its environment. If the tea's initial temperature is $T_i$ = 90°C, when thermal equilibrium is reached what are (a) the mixture's temperature $T_f$ and (b) the remaining mass $m_f$ of ice? If $T_i$ = 70°C, when thermal equilibrium is reached what are (c) $T_f$ and (d) $m_f$?

••38 A 0.530 kg sample of liquid water and a sample of ice are placed in a thermally insulated container. The container also contains a device that transfers energy as heat from the liquid water to the ice at a constant rate P, until thermal equilibrium is reached. The temperatures $T$ of the liquid water and the ice are given in Fig. 18-35 as functions of time $t$; the horizontal scale is set by $t_s = 80.0$ min. (a) What is rate $P$? (b) What is the initial mass of the ice in the container? (c) When thermal equilibrium is reached, what is the mass of the ice produced in this process?

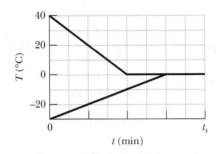

**Figure 18-35** Problem 38.

**••39** GO Ethyl alcohol has a boiling point of 78.0°C, a freezing point of −114°C, a heat of vaporization of 879 kJ/kg, a heat of fusion of 109 kJ/kg, and a specific heat of 2.43 kJ/kg·K. How much energy must be removed from 0.510 kg of ethyl alcohol that is initially a gas at 78.0°C so that it becomes a solid at −114°C?

**••40** GO Calculate the specific heat of a metal from the following data. A container made of the metal has a mass of 3.6 kg and contains 14 kg of water. A 1.8 kg piece of the metal initially at a temperature of 180°C is dropped into the water. The container and water initially have a temperature of 16.0°C, and the final temperature of the entire (insulated) system is 18.0°C.

**•••41** SSM WWW (a) Two 50 g ice cubes are dropped into 200 g of water in a thermally insulated container. If the water is initially at 25°C, and the ice comes directly from a freezer at −15°C, what is the final temperature at thermal equilibrium? (b) What is the final temperature if only one ice cube is used?

**•••42** GO A 20.0 g copper ring at 0.000°C has an inner diameter of $D = 2.54000$ cm. An aluminum sphere at 100.0°C has a diameter of $d = 2.545\,08$ cm. The sphere is put on top of the ring (Fig. 18-36), and the two are allowed to come to thermal equilibrium, with no heat lost to the surroundings. The sphere just passes through the ring at the equilibrium temperature. What is the mass of the sphere?

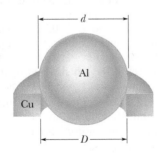

**Figure 18-36** Problem 42.

## Module 18-5   The First Law of Thermodynamics

**•43** In Fig. 18-37, a gas sample expands from $V_0$ to $4.0V_0$ while its pressure decreases from $p_0$ to $p_0/4.0$. If $V_0 = 1.0$ m³ and $p_0 = 40$ Pa, how much work is done by the gas if its pressure changes with volume via (a) path $A$, (b) path $B$, and (c) path $C$?

**•44** GO A thermodynamic system is taken from state $A$ to state $B$ to

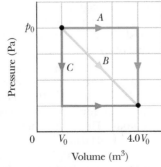

**Figure 18-37** Problem 43.

state $C$, and then back to $A$, as shown in the $p$-$V$ diagram of Fig. 18-38a. The vertical scale is set by $p_s = 40$ Pa, and the horizontal scale is set by $V_s = 4.0$ m³. (a)–(g) Complete the table in Fig. 18-38b by inserting a plus sign, a minus sign, or a zero in each indicated cell. (h) What is the net work done by the system as it moves once through the cycle $ABCA$?

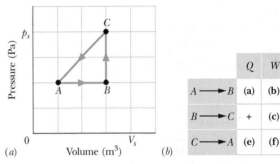

|  | $Q$ | $W$ | $\Delta E_{int}$ |
|---|---|---|---|
| $A \longrightarrow B$ | (a) | (b) | + |
| $B \longrightarrow C$ | + | (c) | (d) |
| $C \longrightarrow A$ | (e) | (f) | (g) |

(a) Volume (m³) (b)

**Figure 18-38** Problem 44.

**•45** SSM ILW A gas within a closed chamber undergoes the cycle shown in the $p$-$V$ diagram of Fig. 18-39. The horizontal scale is set by $V_s = 4.0$ m³. Calculate the net energy added to the system as heat during one complete cycle.

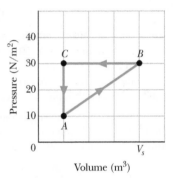

**Figure 18-39** Problem 45.

**•46** Suppose 200 J of work is done on a system and 70.0 cal is extracted from the system as heat. In the sense of the first law of thermodynamics, what are the values (including algebraic signs) of (a) $W$, (b) $Q$, and (c) $\Delta E_{int}$?

**••47** SSM WWW When a system is taken from state $i$ to state $f$ along path $iaf$ in Fig. 18-40, $Q = 50$ cal and $W = 20$ cal. Along path $ibf$, $Q = 36$ cal. (a) What is $W$ along path $ibf$? (b) If $W = -13$ cal for the return path $fi$, what is $Q$ for this path? (c) If $E_{int,i} = 10$ cal, what is $E_{int,f}$? If $E_{int,b} = 22$ cal, what is $Q$ for (d) path $ib$ and (e) path $bf$?

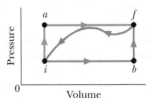

**Figure 18-40** Problem 47.

**••48** GO As a gas is held within a closed chamber, it passes through the cycle shown in Fig. 18-41. Determine the energy transferred by the system as heat during constant-pressure process $CA$ if the energy added as heat $Q_{AB}$ during constant-volume process $AB$ is 20.0 J, no energy is transferred as heat during adiabatic process $BC$, and the net work done during the cycle is 15.0 J.

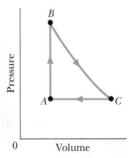

**Figure 18-41** Problem 48.

**••49** GO Figure 18-42 represents a closed cycle for a gas (the figure is not drawn to scale). The change in the internal energy of the gas as it moves from $a$ to $c$ along the path $abc$ is $-200$ J. As it moves from $c$ to $d$, 180 J must be transferred to it as heat. An additional transfer of 80 J to it as heat is needed as it moves from $d$ to $a$. How much work is done on the gas as it moves from $c$ to $d$?

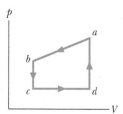

**Figure 18-42** Problem 49.

**••50** GO A lab sample of gas is taken through cycle $abca$ shown in the $p$-$V$ diagram of Fig. 18-43. The net work done is $+1.2$ J. Along path $ab$, the change in the internal energy is $+3.0$ J and the magnitude of the work done is 5.0 J. Along path $ca$, the energy transferred to the gas as heat is $+2.5$ J. How much energy is transferred as heat along (a) path $ab$ and (b) path $bc$?

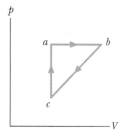

**Figure 18-43** Problem 50.

### Module 18-6 Heat Transfer Mechanisms

**•51** A sphere of radius 0.500 m, temperature 27.0°C, and emissivity 0.850 is located in an environment of temperature 77.0°C. At what rate does the sphere (a) emit and (b) absorb thermal radiation? (c) What is the sphere's net rate of energy exchange?

**•52** The ceiling of a single-family dwelling in a cold climate should have an $R$-value of 30. To give such insulation, how thick would a layer of (a) polyurethane foam and (b) silver have to be?

**•53** SSM Consider the slab shown in Fig. 18-18. Suppose that $L = 25.0$ cm, $A = 90.0$ cm², and the material is copper. If $T_H = 125$°C, $T_C = 10.0$°C, and a steady state is reached, find the conduction rate through the slab.

**•54** If you were to walk briefly in space without a spacesuit while far from the Sun (as an astronaut does in the movie *2001, A Space Odyssey*), you would feel the cold of space—while you radiated energy, you would absorb almost none from your environment. (a) At what rate would you lose energy? (b) How much energy would you lose in 30 s? Assume that your emissivity is 0.90, and estimate other data needed in the calculations.

**•55** ILW A cylindrical copper rod of length 1.2 m and cross-sectional area 4.8 cm² is insulated along its side. The ends are held at a temperature difference of 100 C° by having one end in a water–ice mixture and the other in a mixture of boiling water and steam. At what rate (a) is energy conducted by the rod and (b) does the ice melt?

**••56** The giant hornet *Vespa mandarinia japonica* preys on Japanese bees. However, if one of the hornets attempts to invade

**Figure 18-44**
Problem 56.
© Dr. Masato Ono, Tamagawa University

a beehive, several hundred of the bees quickly form a compact ball around the hornet to stop it. They don't sting, bite, crush, or suffocate it. Rather they overheat it by quickly raising their body temperatures from the normal 35°C to 47°C or 48°C, which is lethal to the hornet but not to the bees (Fig. 18-44). Assume the following: 500 bees form a ball of radius $R = 2.0$ cm for a time $t = 20$ min, the primary loss of energy by the ball is by thermal radiation, the ball's surface has emissivity $\varepsilon = 0.80$, and the ball has a uniform temperature. On average, how much additional energy must each bee produce during the 20 min to maintain 47°C?

**••57** (a) What is the rate of energy loss in watts per square meter through a glass window 3.0 mm thick if the outside temperature is $-20$°F and the inside temperature is $+72$°F? (b) A storm window having the same thickness of glass is installed parallel to the first window, with an air gap of 7.5 cm between the two windows. What now is the rate of energy loss if conduction is the only important energy-loss mechanism?

**••58** A solid cylinder of radius $r_1 = 2.5$ cm, length $h_1 = 5.0$ cm, emissivity 0.85, and temperature 30°C is suspended in an environment of temperature 50°C. (a) What is the cylinder's net thermal radiation transfer rate $P_1$? (b) If the cylinder is stretched until its radius is $r_2 = 0.50$ cm, its net thermal radiation transfer rate becomes $P_2$. What is the ratio $P_2/P_1$?

**••59** In Fig. 18-45a, two identical rectangular rods of metal are welded end to end, with a temperature of $T_1 = 0$°C on the left side and a temperature of $T_2 = 100$°C on the right side. In 2.0 min, 10 J is conducted at a constant rate from the right side to the left side. How much time would be required to conduct 10 J if the rods were welded side to side as in Fig. 18-45b?

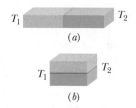

**Figure 18-45** Problem 59.

**••60** GO Figure 18-46 shows the cross section of a wall made of three layers. The layer thicknesses are $L_1$, $L_2 = 0.700L_1$, and $L_3 = 0.350L_1$. The thermal conductivities are $k_1$, $k_2 = 0.900k_1$, and $k_3 = 0.800k_1$. The temperatures at the left side and right side of the wall are $T_H = 30.0$°C and $T_C = -15.0$°C, respectively. Thermal conduction is steady. (a) What is the temperature difference $\Delta T_2$ across layer 2 (between the left and right sides of the layer)? If $k_2$ were, instead, equal to $1.1k_1$, (b) would the rate at which energy is conducted through the wall be greater than, less than, or the same as previously, and (c) what would be the value of $\Delta T_2$?

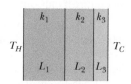

**Figure 18-46** Problem 60.

**••61** SSM A 5.0 cm slab has formed on an outdoor tank of water (Fig. 18-47). The air is at $-10$°C. Find the rate of ice formation (centimeters per hour). The ice has thermal conductivity 0.0040 cal/s·cm·C° and density 0.92 g/cm³. Assume there is no energy transfer through the walls or bottom.

**Figure 18-47** Problem 61.

**••62**  *Leidenfrost effect.* A water drop will last about 1 s on a hot skillet with a temperature between 100°C and about 200°C. However, if the skillet is much hotter, the drop can last several min-

**Figure 18-48** Problem 62.

utes, an effect named after an early investigator. The longer lifetime is due to the support of a thin layer of air and water vapor that separates the drop from the metal (by distance $L$ in Fig. 18-48). Let $L = 0.100$ mm, and assume that the drop is flat with height $h = 1.50$ mm and bottom face area $A = 4.00 \times 10^{-6}$ m². Also assume that the skillet has a constant temperature $T_s = 300$°C and the drop has a temperature of 100°C. Water has density $\rho = 1000$ kg/m³, and the supporting layer has thermal conductivity $k = 0.026$ W/m·K. (a) At what rate is energy conducted from the skillet to the drop through the drop's bottom surface? (b) If conduction is the primary way energy moves from the skillet to the drop, how long will the drop last?

**••63** Figure 18-49 shows (in cross section) a wall consisting of four layers, with thermal conductivities $k_1 = 0.060$ W/m·K, $k_3 = 0.040$ W/m·K, and $k_4 = 0.12$ W/m·K ($k_2$ is not known). The layer thicknesses are $L_1 = 1.5$ cm, $L_3 = 2.8$ cm, and $L_4 = 3.5$ cm ($L_2$ is not known). The known temperatures are $T_1 = 30$°C, $T_{12} = 25$°C, and $T_4 = -10$°C. Energy transfer through the wall is steady. What is interface temperature $T_{34}$?

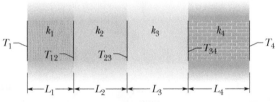

**Figure 18-49** Problem 63.

**••64**  *Penguin huddling.* To withstand the harsh weather of the Antarctic, emperor penguins huddle in groups (Fig. 18-50). Assume that a penguin is a circular cylinder with a top surface area $a = 0.34$ m² and height $h = 1.1$ m. Let $P_r$ be the rate at which an individual penguin radiates energy to the environment (through the top and the sides); thus $NP_r$ is the rate at which $N$ identical, well-separated penguins radiate. If the penguins huddle closely to form

Alain Torterotot/Peter Arnold/Photolibrary

**Figure 18-50** Problem 64.

a *huddled cylinder* with top surface area $Na$ and height $h$, the cylinder radiates at the rate $P_h$. If $N = 1000$, (a) what is the value of the fraction $P_h/NP_r$ and (b) by what percentage does huddling reduce the total radiation loss?

**••65** Ice has formed on a shallow pond, and a steady state has been reached, with the air above the ice at $-5.0$°C and the bottom of the pond at 4.0°C. If the total depth of *ice + water* is 1.4 m, how thick is the ice? (Assume that the thermal conductivities of ice and water are 0.40 and 0.12 cal/m·C°·s, respectively.)

**•••66** GO  *Evaporative cooling of beverages.* A cold beverage can be kept cold even on a warm day if it is slipped into a porous ceramic container that has been soaked in water. Assume that energy lost to evaporation matches the net energy gained via the radiation exchange through the top and side surfaces. The container and beverage have temperature $T = 15$°C, the environment has temperature $T_{env} = 32$°C, and the container is a cylinder with radius $r = 2.2$ cm and height 10 cm. Approximate the emissivity as $\varepsilon = 1$, and neglect other energy exchanges. At what rate $dm/dt$ is the container losing water mass?

**Additional Problems**

**67** In the extrusion of cold chocolate from a tube, work is done on the chocolate by the pressure applied by a ram forcing the chocolate through the tube. The work per unit mass of extruded chocolate is equal to $p/\rho$, where $p$ is the difference between the applied pressure and the pressure where the chocolate emerges from the tube, and $\rho$ is the density of the chocolate. Rather than increasing the temperature of the chocolate, this work melts cocoa fats in the chocolate. These fats have a heat of fusion of 150 kJ/kg. Assume that all of the work goes into that melting and that these fats make up 30% of the chocolate's mass. What percentage of the fats melt during the extrusion if $p = 5.5$ MPa and $\rho = 1200$ kg/m³?

**68** Icebergs in the North Atlantic present hazards to shipping, causing the lengths of shipping routes to be increased by about 30% during the iceberg season. Attempts to destroy icebergs include planting explosives, bombing, torpedoing, shelling, ramming, and coating with black soot. Suppose that direct melting of the iceberg, by placing heat sources in the ice, is tried. How much energy as heat is required to melt 10% of an iceberg that has a mass of 200 000 metric tons? (Use 1 metric ton = 1000 kg.)

**69** Figure 18-51 displays a closed cycle for a gas. The change in internal energy along path $ca$ is $-160$ J. The energy transferred to the gas as heat is 200 J along path $ab$, and 40 J along path $bc$. How much work is done by the gas along (a) path $abc$ and (b) path $ab$?

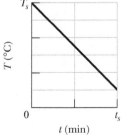

**Figure 18-51**
Problem 69.

**70** In a certain solar house, energy from the Sun is stored in barrels filled with water. In a particular winter stretch of five cloudy days, $1.00 \times 10^6$ kcal is needed to maintain the inside of the house at 22.0°C. Assuming that the water in the barrels is at 50.0°C and that the water has a density of $1.00 \times 10^3$ kg/m³, what volume of water is required?

**71** A 0.300 kg sample is placed in a cooling apparatus that removes energy as heat at a constant rate of 2.81 W. Figure 18-52 gives the temperature $T$ of the sam-

**Figure 18-52** Problem 71.

ple versus time $t$. The temperature scale is set by $T_s = 30°C$ and the time scale is set by $t_s = 20$ min. What is the specific heat of the sample?

**72** The average rate at which energy is conducted outward through the ground surface in North America is 54.0 mW/m², and the average thermal conductivity of the near-surface rocks is 2.50 W/m·K. Assuming a surface temperature of 10.0°C, find the temperature at a depth of 35.0 km (near the base of the crust). Ignore the heat generated by the presence of radioactive elements.

**73** What is the volume increase of an aluminum cube 5.00 cm on an edge when heated from 10.0°C to 60.0°C?

**74** In a series of experiments, block $B$ is to be placed in a thermally insulated container with block $A$, which has the same mass as block $B$. In each experiment, block $B$ is initially at a certain temperature $T_B$, but temperature $T_A$ of block $A$ is changed from experiment to experiment. Let $T_f$ represent the final temperature of the two blocks when they reach thermal equilibrium in any of the experiments. Figure 18-53 gives temperature $T_f$ versus the initial temperature $T_A$ for a range of possible values of $T_A$, from $T_{A1} = 0$ K to $T_{A2} = 500$ K. The vertical axis scale is set by $T_{fs} = 400$ K. What are (a) temperature $T_B$ and (b) the ratio $c_B/c_A$ of the specific heats of the blocks?

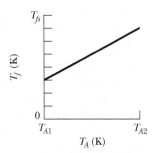

**Figure 18-53** Problem 74.

**75** Figure 18-54 displays a closed cycle for a gas. From $c$ to $b$, 40 J is transferred from the gas as heat. From $b$ to $a$, 130 J is transferred from the gas as heat, and the magnitude of the work done by the gas is 80 J. From $a$ to $c$, 400 J is transferred to the gas as heat. What is the work done by the gas from $a$ to $c$? (*Hint:* You need to supply the plus and minus signs for the given data.)

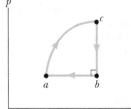

**Figure 18-54** Problem 75.

**76** Three equal-length straight rods, of aluminum, Invar, and steel, all at 20.0°C, form an equilateral triangle with hinge pins at the vertices. At what temperature will the angle opposite the Invar rod be 59.95°? See Appendix E for needed trigonometric formulas and Table 18-2 for needed data.

**77** SSM The temperature of a 0.700 kg cube of ice is decreased to −150°C. Then energy is gradually transferred to the cube as heat while it is otherwise thermally isolated from its environment. The total transfer is 0.6993 MJ. Assume the value of $c_{ice}$ given in Table 18-3 is valid for temperatures from −150°C to 0°C. What is the final temperature of the water?

**78** GO ✈ *Icicles.* Liquid water coats an active (growing) icicle and extends up a short, narrow tube along the central axis (Fig. 18-55). Because the water–ice interface must have a temperature of 0°C, the water in the tube cannot lose energy through the

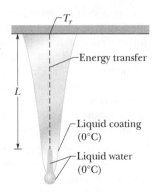

**Figure 18-55** Problem 78.

sides of the icicle or down through the tip because there is no temperature change in those directions. It can lose energy and freeze only by sending energy up (through distance $L$) to the top of the icicle, where the temperature $T_r$ can be below 0°C. Take $L = 0.12$ m and $T_r = −5°C$. Assume that the central tube and the upward conduction path both have cross-sectional area $A$. In terms of $A$, what rate is (a) energy conducted upward and (b) mass converted from liquid to ice at the top of the central tube? (c) At what rate does the top of the tube move downward because of water freezing there? The thermal conductivity of ice is 0.400 W/m·K, and the density of liquid water is 1000 kg/m³.

**79** SSM A sample of gas expands from an initial pressure and volume of 10 Pa and 1.0 m³ to a final volume of 2.0 m³. During the expansion, the pressure and volume are related by the equation $p = aV^2$, where $a = 10$ N/m⁸. Determine the work done by the gas during this expansion.

**80** Figure 18-56a shows a cylinder containing gas and closed by a movable piston. The cylinder is kept submerged in an ice–water mixture. The piston is *quickly* pushed down from position 1 to position 2 and then held at position 2 until the gas is again at the temperature of the ice–water mixture; it then is *slowly* raised back to position 1. Figure 18-56b is a p-V diagram for the process. If 100 g of ice is melted during the cycle, how much work has been done *on* the gas?

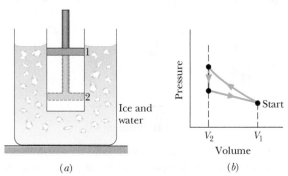

**Figure 18-56** Problem 80.

**81** SSM A sample of gas undergoes a transition from an initial state $a$ to a final state $b$ by three different paths (processes), as shown in the p-V diagram in Fig. 18-57, where $V_b = 5.00V_i$. The energy transferred to the gas as heat in process 1 is $10p_iV_i$. In terms of $p_iV_i$, what are (a) the energy transferred to the gas as heat in process 2 and (b) the change in internal energy that the gas undergoes in process 3?

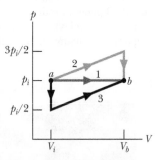

**Figure 18-57** Problem 81.

**82** A copper rod, an aluminum rod, and a brass rod, each of 6.00 m length and 1.00 cm diameter, are placed end to end with the aluminum rod between the other two. The free end of the copper rod is maintained at water's boiling point, and the free end of the brass rod is maintained at water's freezing point. What is the steady-state temperature of (a) the copper–aluminum junction and (b) the aluminum–brass junction?

**83** SSM The temperature of a Pyrex disk is changed from 10.0°C to 60.0°C. Its initial radius is 8.00 cm; its initial thickness is 0.500 cm. Take these data as being exact. What is the change in the volume of the disk? (See Table 18-2.)

**84** (a) Calculate the rate at which body heat is conducted through the clothing of a skier in a steady-state process, given the following data: the body surface area is 1.8 m², and the clothing is 1.0 cm thick; the skin surface temperature is 33°C and the outer surface of the clothing is at 1.0°C; the thermal conductivity of the clothing is 0.040 W/m·K. (b) If, after a fall, the skier's clothes became soaked with water of thermal conductivity 0.60 W/m·K, by how much is the rate of conduction multiplied?

**85** SSM A 2.50 kg lump of aluminum is heated to 92.0°C and then dropped into 8.00 kg of water at 5.00°C. Assuming that the lump–water system is thermally isolated, what is the system's equilibrium temperature?

**86** A glass window pane is exactly 20 cm by 30 cm at 10°C. By how much has its area increased when its temperature is 40°C, assuming that it can expand freely?

**87** A recruit can join the semi-secret "300 F" club at the Amundsen–Scott South Pole Station only when the outside temperature is below −70°C. On such a day, the recruit first basks in a hot sauna and then runs outside wearing only shoes. (This is, of course, extremely dangerous, but the rite is effectively a protest against the constant danger of the cold.)

Assume that upon stepping out of the sauna, the recruit's skin temperature is 102°F and the walls, ceiling, and floor of the sauna room have a temperature of 30°C. Estimate the recruit's surface area, and take the skin emissivity to be 0.80. (a) What is the approximate net rate $P_{net}$ at which the recruit loses energy via thermal radiation exchanges with the room? Next, assume that when outdoors, half the recruit's surface area exchanges thermal radiation with the sky at a temperature of −25°C and the other half exchanges thermal radiation with the snow and ground at a temperature of −80°C. What is the approximate net rate at which the recruit loses energy via thermal radiation exchanges with (b) the sky and (c) the snow and ground?

**88** A steel rod at 25.0°C is bolted at both ends and then cooled. At what temperature will it rupture? Use Table 12-1.

**89** An athlete needs to lose weight and decides to do it by "pumping iron." (a) How many times must an 80.0 kg weight be lifted a distance of 1.00 m in order to burn off 1.00 lb of fat, assuming that that much fat is equivalent to 3500 Cal? (b) If the weight is lifted once every 2.00 s, how long does the task take?

**90** Soon after Earth was formed, heat released by the decay of radioactive elements raised the average internal temperature from 300 to 3000 K, at about which value it remains today. Assuming an average coefficient of volume expansion of $3.0 \times 10^{-5}$ K⁻¹, by how much has the radius of Earth increased since the planet was formed?

**91** It is possible to melt ice by rubbing one block of it against another. How much work, in joules, would you have to do to get 1.00 g of ice to melt?

**92** A rectangular plate of glass initially has the dimensions 0.200 m by 0.300 m. The coefficient of linear expansion for the glass is $9.00 \times 10^{-6}$/K. What is the change in the plate's area if its temperature is increased by 20.0 K?

**93** Suppose that you intercept $5.0 \times 10^{-3}$ of the energy radiated by a hot sphere that has a radius of 0.020 m, an emissivity of 0.80, and a surface temperature of 500 K. How much energy do you intercept in 2.0 min?

**94** A thermometer of mass 0.0550 kg and of specific heat 0.837 kJ/kg·K reads 15.0°C. It is then completely immersed in

0.300 kg of water, and it comes to the same final temperature as the water. If the thermometer then reads 44.4°C, what was the temperature of the water before insertion of the thermometer?

**95** A sample of gas expands from $V_1 = 1.0$ m³ and $p_1 = 40$ Pa to $V_2 = 4.0$ m³ and $p_2 = 10$ Pa along path $B$ in the p-V diagram in Fig. 18-58. It is then compressed back to $V_1$ along either path $A$ or path $C$. Compute the net work done by the gas for the complete cycle along (a) path $BA$ and (b) path $BC$.

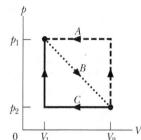

**Figure 18-58** Problem 95.

**96** Figure 18-59 shows a composite bar of length $L = L_1 + L_2$ and consisting of two materials. One material has length $L_1$ and coefficient of linear expansion $\alpha_1$; the other has length $L_2$ and coefficient of linear expansion $\alpha_2$. (a) What is the coefficient of linear expansion $\alpha$ for the composite bar? For a particular composite bar, $L$ is 52.4 cm, material 1 is steel, and material 2 is brass. If $\alpha = 1.3 \times 10^{-5}$/C°, what are the lengths (b) $L_1$ and (c) $L_2$?

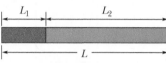

**Figure 18-59** Problem 96.

**97** On finding your stove out of order, you decide to boil the water for a cup of tea by shaking it in a thermos flask. Suppose that you use tap water at 19°C, the water falls 32 cm each shake, and you make 27 shakes each minute. Neglecting any loss of thermal energy by the flask, how long (in minutes) must you shake the flask until the water reaches 100°C?

**98** The p-V diagram in Fig. 18-60 shows two paths along which a sample of gas can be taken from state $a$ to state $b$, where $V_b = 3.0V_1$. Path 1 requires that energy equal to $5.0p_1V_1$ be transferred to the gas as heat. Path 2 requires that energy equal to $5.5p_1V_1$ be transferred to the gas as heat. What is the ratio $p_2/p_1$?

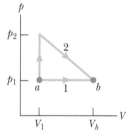

**Figure 18-60** Problem 98.

**99** A cube of edge length $6.0 \times 10^{-6}$ m, emissivity 0.75, and temperature −100°C floats in an environment at −150°C. What is the cube's net thermal radiation transfer rate?

**100** A *flow calorimeter* is a device used to measure the specific heat of a liquid. Energy is added as heat at a known rate to a stream of the liquid as it passes through the calorimeter at a known rate. Measurement of the resulting temperature difference between the inflow and the outflow points of the liquid stream enables us to compute the specific heat of the liquid. Suppose a liquid of density 0.85 g/cm³ flows through a calorimeter at the rate of 8.0 cm³/s. When energy is added at the rate of 250 W by means of an electric heating coil, a temperature difference of 15 C° is established in steady-state conditions between the inflow and the outflow points. What is the specific heat of the liquid?

**101** An object of mass 6.00 kg falls through a height of 50.0 m and, by means of a mechanical linkage, rotates a paddle wheel that stirs 0.600 kg of water. Assume that the initial gravitational potential energy of the object is fully transferred to thermal energy of the water, which is initially at 15.0°C. What is the temperature rise of the water?

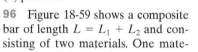

**102**   The Pyrex glass mirror in a telescope has a diameter of 170 in. The temperature ranges from $-16°C$ to $32°C$ on the location of the telescope. What is the maximum change in the diameter of the mirror, assuming that the glass can freely expand and contract?

**103**   The area $A$ of a rectangular plate is $ab = 1.4 \text{ m}^2$. Its coefficient of linear expansion is $\alpha = 32 \times 10^{-6}/C°$. After a temperature rise $\Delta T = 89°C$, side $a$ is longer by $\Delta a$ and side $b$ is longer by $\Delta b$ (Fig. 18-61). Neglecting the small quantity $(\Delta a \, \Delta b)/ab$, find $\Delta A$.

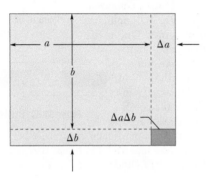

**Figure 18-61**   Problem 103.

**104**   Consider the liquid in a barometer whose coefficient of volume expansion is $6.6 \times 10^{-4}/C°$. Find the relative change in the liquid's height if the temperature changes by $12 \text{ C}°$ while the pressure remains constant. Neglect the expansion of the glass tube.

**105**   A pendulum clock with a pendulum made of brass is designed to keep accurate time at $23°C$. Assume it is a simple pendulum consisting of a bob at one end of a brass rod of negligible mass that is pivoted about the other end. If the clock operates at $0.0°C$, (a) does it run too fast or too slow, and (b) what is the magnitude of its error in seconds per hour?

**106**   A room is lighted by four 100 W incandescent lightbulbs. (The power of 100 W is the rate at which a bulb converts electrical energy to heat and the energy of visible light.) Assuming that 73% of the energy is converted to heat, how much heat does the room receive in 6.9 h?

**107**   An energetic athlete can use up all the energy from a diet of 4000 Cal/day. If he were to use up this energy at a steady rate, what is the ratio of the rate of energy use compared to that of a 100 W bulb? (The power of 100 W is the rate at which the bulb converts electrical energy to heat and the energy of visible light.)

**108**   A 1700 kg Buick moving at 83 km/h brakes to a stop, at uniform deceleration and without skidding, over a distance of 93 m. At what average rate is mechanical energy transferred to thermal energy in the brake system?

# CHAPTER 19

# The Kinetic Theory of Gases

## 19-1 AVOGADRO'S NUMBER

### Learning Objectives

*After reading this module, you should be able to . . .*

**19.01** Identify Avogadro's number $N_A$.

**19.02** Apply the relationship between the number of moles $n$, the number of molecules $N$, and Avogadro's number $N_A$.

**19.03** Apply the relationships between the mass $m$ of a sample, the molar mass $M$ of the molecules in the sample, the number of moles $n$ in the sample, and Avogadro's number $N_A$.

### Key Ideas

● The kinetic theory of gases relates the macroscopic properties of gases (for example, pressure and temperature) to the microscopic properties of gas molecules (for example, speed and kinetic energy).

● One mole of a substance contains $N_A$ (Avogadro's number) elementary units (usually atoms or molecules), where $N_A$ is found experimentally to be

$$N_A = 6.02 \times 10^{23} \text{ mol}^{-1} \quad \text{(Avogadro's number)}.$$

One molar mass $M$ of any substance is the mass of one mole of the substance.

● A mole is related to the mass $m$ of the individual molecules of the substance by

$$M = mN_A.$$

● The number of moles $n$ contained in a sample of mass $M_{sam}$, consisting of $N$ molecules, is related to the molar mass $M$ of the molecules and to Avogadro's number $N_A$ as given by

$$n = \frac{N}{N_A} = \frac{M_{sam}}{M} = \frac{M_{sam}}{mN_A}.$$

## What Is Physics?

One of the main subjects in thermodynamics is the physics of gases. A gas consists of atoms (either individually or bound together as molecules) that fill their container's volume and exert pressure on the container's walls. We can usually assign a temperature to such a contained gas. These three variables associated with a gas—volume, pressure, and temperature—are all a consequence of the motion of the atoms. The volume is a result of the freedom the atoms have to spread throughout the container, the pressure is a result of the collisions of the atoms with the container's walls, and the temperature has to do with the kinetic energy of the atoms. The **kinetic theory of gases,** the focus of this chapter, relates the motion of the atoms to the volume, pressure, and temperature of the gas.

Applications of the kinetic theory of gases are countless. Automobile engineers are concerned with the combustion of vaporized fuel (a gas) in the automobile engines. Food engineers are concerned with the production rate of the fermentation gas that causes bread to rise as it bakes. Beverage engineers are concerned with how gas can produce the head in a glass of beer or shoot a cork from a champagne bottle. Medical engineers and physiologists are concerned with calculating how long a scuba diver must pause during ascent to eliminate nitrogen gas from the bloodstream (to avoid the *bends*). Environmental scientists are concerned with how heat exchanges between the oceans and the atmosphere can affect weather conditions.

The first step in our discussion of the kinetic theory of gases deals with measuring the amount of a gas present in a sample, for which we use Avogadro's number.

## Avogadro's Number

When our thinking is slanted toward atoms and molecules, it makes sense to measure the sizes of our samples in moles. If we do so, we can be certain that we are comparing samples that contain the same number of atoms or molecules. The *mole* is one of the seven SI base units and is defined as follows:

> One mole is the number of atoms in a 12 g sample of carbon-12.

The obvious question now is: "How many atoms or molecules are there in a mole?" The answer is determined experimentally and, as you saw in Chapter 18, is

$$N_A = 6.02 \times 10^{23} \text{ mol}^{-1} \quad \text{(Avogadro's number)}, \tag{19-1}$$

where $\text{mol}^{-1}$ represents the inverse mole or "per mole," and mol is the abbreviation for mole. The number $N_A$ is called **Avogadro's number** after Italian scientist Amedeo Avogadro (1776–1856), who suggested that all gases occupying the same volume under the same conditions of temperature and pressure contain the same number of atoms or molecules.

The number of moles $n$ contained in a sample of any substance is equal to the ratio of the number of molecules $N$ in the sample to the number of molecules $N_A$ in 1 mol:

$$n = \frac{N}{N_A}. \tag{19-2}$$

(*Caution:* The three symbols in this equation can easily be confused with one another, so you should sort them with their meanings now, before you end in "N-confusion.") We can find the number of moles $n$ in a sample from the mass $M_{sam}$ of the sample and either the *molar mass M* (the mass of 1 mol) or the molecular mass $m$ (the mass of one molecule):

$$n = \frac{M_{sam}}{M} = \frac{M_{sam}}{mN_A}. \tag{19-3}$$

In Eq. 19-3, we used the fact that the mass $M$ of 1 mol is the product of the mass $m$ of one molecule and the number of molecules $N_A$ in 1 mol:

$$M = mN_A. \tag{19-4}$$

# 19-2 IDEAL GASES

## Learning Objectives

*After reading this module, you should be able to . . .*

**19.04** Identify why an ideal gas is said to be ideal.

**19.05** Apply either of the two forms of the ideal gas law, written in terms of the number of moles $n$ or the number of molecules $N$.

**19.06** Relate the ideal gas constant $R$ and the Boltzmann constant $k$.

**19.07** Identify that the temperature in the ideal gas law must be in kelvins.

**19.08** Sketch $p$-$V$ diagrams for a constant-temperature expansion of a gas and a constant-temperature contraction.

**19.09** Identify the term isotherm.

**19.10** Calculate the work done by a gas, including the algebraic sign, for an expansion and a contraction along an isotherm.

**19.11** For an isothermal process, identify that the change in internal energy $\Delta E$ is zero and that the energy $Q$ transferred as heat is equal to the work $W$ done.

**19.12** On a $p$-$V$ diagram, sketch a constant-volume process and identify the amount of work done in terms of area on the diagram.

**19.13** On a $p$-$V$ diagram, sketch a constant-pressure process and determine the work done in terms of area on the diagram.

## Key Ideas

● An ideal gas is one for which the pressure $p$, volume $V$, and temperature $T$ are related by

$$pV = nRT \quad \text{(ideal gas law)}.$$

Here $n$ is the number of moles of the gas present and $R$ is a constant (8.31 J/mol·K) called the gas constant.

● The ideal gas law can also be written as

$$pV = NkT,$$

where the Boltzmann constant $k$ is

$$k = \frac{R}{N_A} = 1.38 \times 10^{-23} \text{ J/K}.$$

● The work done *by* an ideal gas during an isothermal (constant-temperature) change from volume $V_i$ to volume $V_f$ is

$$W = nRT \ln \frac{V_f}{V_i} \quad \text{(ideal gas, isothermal process)}.$$

## Ideal Gases

Our goal in this chapter is to explain the macroscopic properties of a gas—such as its pressure and its temperature—in terms of the behavior of the molecules that make it up. However, there is an immediate problem: which gas? Should it be hydrogen, oxygen, or methane, or perhaps uranium hexafluoride? They are all different. Experimenters have found, though, that if we confine 1 mol samples of various gases in boxes of identical volume and hold the gases at the same temperature, then their measured pressures are almost the same, and at lower densities the differences tend to disappear. Further experiments show that, at low enough densities, all real gases tend to obey the relation

$$pV = nRT \quad \text{(ideal gas law)}, \tag{19-5}$$

in which $p$ is the absolute (not gauge) pressure, $n$ is the number of moles of gas present, and $T$ is the temperature in kelvins. The symbol $R$ is a constant called the **gas constant** that has the same value for all gases—namely,

$$R = 8.31 \text{ J/mol·K}. \tag{19-6}$$

Equation 19-5 is called the **ideal gas law.** Provided the gas density is low, this law holds for any single gas or for any mixture of different gases. (For a mixture, $n$ is the total number of moles in the mixture.)

We can rewrite Eq. 19-5 in an alternative form, in terms of a constant called the **Boltzmann constant** $k$, which is defined as

$$k = \frac{R}{N_A} = \frac{8.31 \text{ J/mol·K}}{6.02 \times 10^{23} \text{ mol}^{-1}} = 1.38 \times 10^{-23} \text{ J/K}. \tag{19-7}$$

This allows us to write $R = kN_A$. Then, with Eq. 19-2 ($n = N/N_A$), we see that

$$nR = Nk. \tag{19-8}$$

Substituting this into Eq. 19-5 gives a second expression for the ideal gas law:

$$pV = NkT \quad \text{(ideal gas law)}. \tag{19-9}$$

(*Caution*: Note the difference between the two expressions for the ideal gas law—Eq. 19-5 involves the number of moles $n$, and Eq. 19-9 involves the number of molecules $N$.)

You may well ask, "What is an *ideal gas,* and what is so 'ideal' about it?" The answer lies in the simplicity of the law (Eqs. 19-5 and 19-9) that governs its macroscopic properties. Using this law—as you will see—we can deduce many properties of the ideal gas in a simple way. Although there is no such thing in nature as a truly ideal gas, *all real* gases approach the ideal state at low enough densities—that is, under conditions in which their molecules are far enough apart that they do not interact with one another. Thus, the ideal gas concept allows us to gain useful insights into the limiting behavior of real gases.

(a)

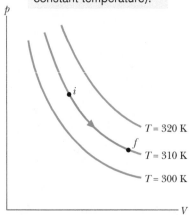

(b)

Courtesy www.doctorslime.com

**Figure 19-1** (a) Before and (b) after images of a large steel tank crushed by atmospheric pressure after internal steam cooled and condensed.

Figure 19-1 gives a dramatic example of the ideal gas law. A stainless-steel tank with a volume of 18 m³ was filled with steam at a temperature of 110°C through a valve at one end. The steam supply was then turned off and the valve closed, so that the steam was trapped inside the tank (Fig. 19-1a). Water from a fire hose was then poured onto the tank to rapidly cool it. Within less than a minute, the enormously sturdy tank was crushed (Fig. 19-1b), as if some giant invisible creature from a grade B science fiction movie had stepped on it during a rampage.

Actually, it was the atmosphere that crushed the tank. As the tank was cooled by the water stream, the steam cooled and much of it condensed, which means that the number $N$ of gas molecules and the temperature $T$ of the gas inside the tank both decreased. Thus, the right side of Eq. 19-9 decreased, and because volume $V$ was constant, the gas pressure $p$ on the left side also decreased. The gas pressure decreased so much that the external atmospheric pressure was able to crush the tank's steel wall. Figure 19-1 was staged, but this type of crushing sometimes occurs in industrial accidents (photos and videos can be found on the web).

### Work Done by an Ideal Gas at Constant Temperature

Suppose we put an ideal gas in a piston–cylinder arrangement like those in Chapter 18. Suppose also that we allow the gas to expand from an initial volume $V_i$ to a final volume $V_f$ while we keep the temperature $T$ of the gas constant. Such a process, at *constant temperature,* is called an **isothermal expansion** (and the reverse is called an **isothermal compression**).

On a *p-V* diagram, an *isotherm* is a curve that connects points that have the same temperature. Thus, it is a graph of pressure versus volume for a gas whose temperature $T$ is held constant. For $n$ moles of an ideal gas, it is a graph of the equation

$$p = nRT\frac{1}{V} = (\text{a constant})\frac{1}{V}. \tag{19-10}$$

Figure 19-2 shows three isotherms, each corresponding to a different (constant) value of $T$. (Note that the values of $T$ for the isotherms increase upward to the right.) Superimposed on the middle isotherm is the path followed by a gas during an isothermal expansion from state $i$ to state $f$ at a constant temperature of 310 K.

To find the work done by an ideal gas during an isothermal expansion, we start with Eq. 18-25,

$$W = \int_{V_i}^{V_f} p\, dV. \tag{19-11}$$

This is a general expression for the work done during any change in volume of any gas. For an ideal gas, we can use Eq. 19-5 ($pV = nRT$) to substitute for $p$, obtaining

$$W = \int_{V_i}^{V_f} \frac{nRT}{V}\, dV. \tag{19-12}$$

Because we are considering an isothermal expansion, $T$ is constant, so we can move it in front of the integral sign to write

$$W = nRT\int_{V_i}^{V_f} \frac{dV}{V} = nRT\left[\ln V\right]_{V_i}^{V_f}. \tag{19-13}$$

By evaluating the expression in brackets at the limits and then using the relationship $\ln a - \ln b = \ln(a/b)$, we find that

$$W = nRT\ln\frac{V_f}{V_i} \qquad \text{(ideal gas, isothermal process).} \tag{19-14}$$

Recall that the symbol ln specifies a *natural* logarithm, which has base $e$.

The expansion is along an isotherm (the gas has constant temperature).

$T = 320$ K

$T = 310$ K

$T = 300$ K

**Figure 19-2** Three isotherms on a *p-V* diagram. The path shown along the middle isotherm represents an isothermal expansion of a gas from an initial state $i$ to a final state $f$. The path from $f$ to $i$ along the isotherm would represent the reverse process—that is, an isothermal compression.

For an expansion, $V_f$ is greater than $V_i$, so the ratio $V_f/V_i$ in Eq. 19-14 is greater than unity. The natural logarithm of a quantity greater than unity is positive, and so the work $W$ done by an ideal gas during an isothermal expansion is positive, as we expect. For a compression, $V_f$ is less than $V_i$, so the ratio of volumes in Eq. 19-14 is less than unity. The natural logarithm in that equation—hence the work $W$—is negative, again as we expect.

### Work Done at Constant Volume and at Constant Pressure

Equation 19-14 does not give the work $W$ done by an ideal gas during *every* thermodynamic process. Instead, it gives the work only for a process in which the temperature is held constant. If the temperature varies, then the symbol $T$ in Eq. 19-12 cannot be moved in front of the integral symbol as in Eq. 19-13, and thus we do not end up with Eq. 19-14.

However, we can always go back to Eq. 19-11 to find the work $W$ done by an ideal gas (or any other gas) during any process, such as a constant-volume process and a constant-pressure process. If the volume of the gas is constant, then Eq. 19-11 yields

$$W = 0 \quad \text{(constant-volume process).} \qquad (19\text{-}15)$$

If, instead, the volume changes while the pressure $p$ of the gas is held constant, then Eq. 19-11 becomes

$$W = p(V_f - V_i) = p\,\Delta V \quad \text{(constant-pressure process).} \qquad (19\text{-}16)$$

 **Checkpoint 1**

An ideal gas has an initial pressure of 3 pressure units and an initial volume of 4 volume units. The table gives the final pressure and volume of the gas (in those same units) in five processes. Which processes start and end on the same isotherm?

|   | a | b | c | d | e |
|---|---|---|---|---|---|
| $p$ | 12 | 6 | 5 | 4 | 1 |
| $V$ | 1 | 2 | 7 | 3 | 12 |

---

## Sample Problem 19.01    Ideal gas and changes of temperature, volume, and pressure

A cylinder contains 12 L of oxygen at 20°C and 15 atm. The temperature is raised to 35°C, and the volume is reduced to 8.5 L. What is the final pressure of the gas in atmospheres? Assume that the gas is ideal.

### KEY IDEA

Because the gas is ideal, we can use the ideal gas law to relate its parameters, both in the initial state $i$ and in the final state $f$.

**Calculations:** From Eq. 19-5 we can write

$$p_iV_i = nRT_i \quad \text{and} \quad p_fV_f = nRT_f.$$

Dividing the second equation by the first equation and solving for $p_f$ yields

$$p_f = \frac{p_iT_fV_i}{T_iV_f}. \qquad (19\text{-}17)$$

Note here that if we converted the given initial and final volumes from liters to the proper units of cubic meters, the multiplying conversion factors would cancel out of Eq. 19-17. The same would be true for conversion factors that convert the pressures from atmospheres to the proper pascals. However, to convert the given temperatures to kelvins requires the addition of an amount that would not cancel and thus must be included. Hence, we must write

$$T_i = (273 + 20) \text{ K} = 293 \text{ K}$$

and

$$T_f = (273 + 35) \text{ K} = 308 \text{ K}.$$

Inserting the given data into Eq. 19-17 then yields

$$p_f = \frac{(15 \text{ atm})(308 \text{ K})(12 \text{ L})}{(293 \text{ K})(8.5 \text{ L})} = 22 \text{ atm.} \qquad \text{(Answer)}$$

## Sample Problem 19.02  Work by an ideal gas

One mole of oxygen (assume it to be an ideal gas) expands at a constant temperature $T$ of 310 K from an initial volume $V_i$ of 12 L to a final volume $V_f$ of 19 L. How much work is done by the gas during the expansion?

### KEY IDEA

Generally we find the work by integrating the gas pressure with respect to the gas volume, using Eq. 19-11. However, because the gas here is ideal and the expansion is isothermal, that integration leads to Eq. 19-14.

*Calculation:* Therefore, we can write

$$W = nRT \ln \frac{V_f}{V_i}$$

$$= (1 \text{ mol})(8.31 \text{ J/mol} \cdot \text{K})(310 \text{ K}) \ln \frac{19 \text{ L}}{12 \text{ L}}$$

$$= 1180 \text{ J.} \qquad \text{(Answer)}$$

The expansion is graphed in the $p$-$V$ diagram of Fig. 19-3. The work done by the gas during the expansion is represented by the area beneath the curve *if*.

You can show that if the expansion is now reversed, with the gas undergoing an isothermal compression from 19 L to 12 L, the work done by the gas will be −1180 J. Thus, an external force would have to do 1180 J of work on the gas to compress it.

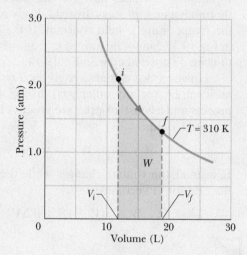

**Figure 19-3** The shaded area represents the work done by 1 mol of oxygen in expanding from $V_i$ to $V_f$ at a temperature $T$ of 310 K.

**WILEY PLUS** Additional examples, video, and practice available at *WileyPLUS*

# 19-3 PRESSURE, TEMPERATURE, AND RMS SPEED

## Learning Objectives

*After reading this module, you should be able to . . .*

**19.14** Identify that the pressure on the interior walls of a gas container is due to the molecular collisions with the walls.

**19.15** Relate the pressure on a container wall to the momentum of the gas molecules and the time intervals between their collisions with the wall.

**19.16** For the molecules of an ideal gas, relate the root-

mean-square speed $v_{rms}$ and the average speed $v_{avg}$.

**19.17** Relate the pressure of an ideal gas to the rms speed $v_{rms}$ of the molecules.

**19.18** For an ideal gas, apply the relationship between the gas temperature $T$ and the rms speed $v_{rms}$ and molar mass $M$ of the molecules.

## Key Ideas

● In terms of the speed of the gas molecules, the pressure exerted by $n$ moles of an ideal gas is

$$p = \frac{nMv_{rms}^2}{3V},$$

where $v_{rms} = \sqrt{(v^2)_{avg}}$ is the root-mean-square speed of the

molecules, $M$ is the molar mass, and $V$ is the volume.

● The rms speed can be written in terms of the temperature as

$$v_{rms} = \sqrt{\frac{3RT}{M}}.$$

## Pressure, Temperature, and RMS Speed

Here is our first kinetic theory problem. Let $n$ moles of an ideal gas be confined in a cubical box of volume $V$, as in Fig. 19-4. The walls of the box are held at temperature $T$. What is the connection between the pressure $p$ exerted by the gas on the walls and the speeds of the molecules?

The molecules of gas in the box are moving in all directions and with various speeds, bumping into one another and bouncing from the walls of the box like balls in a racquetball court. We ignore (for the time being) collisions of the molecules with one another and consider only elastic collisions with the walls.

Figure 19-4 shows a typical gas molecule, of mass $m$ and velocity $\vec{v}$, that is about to collide with the shaded wall. Because we assume that any collision of a molecule with a wall is elastic, when this molecule collides with the shaded wall, the only component of its velocity that is changed is the $x$ component, and that component is reversed. This means that the only change in the particle's momentum is along the $x$ axis, and that change is

$$\Delta p_x = (-mv_x) - (mv_x) = -2mv_x.$$

Hence, the momentum $\Delta p_x$ delivered to the wall by the molecule during the collision is $+2mv_x$. (Because in this book the symbol $p$ represents both momentum and pressure, we must be careful to note that here $p$ represents momentum and is a vector quantity.)

The molecule of Fig. 19-4 will hit the shaded wall repeatedly. The time $\Delta t$ between collisions is the time the molecule takes to travel to the opposite wall and back again (a distance $2L$) at speed $v_x$. Thus, $\Delta t$ is equal to $2L/v_x$. (Note that this result holds even if the molecule bounces off any of the other walls along the way, because those walls are parallel to $x$ and so cannot change $v_x$.) Therefore, the average rate at which momentum is delivered to the shaded wall by this single molecule is

$$\frac{\Delta p_x}{\Delta t} = \frac{2mv_x}{2L/v_x} = \frac{mv_x^2}{L}.$$

From Newton's second law ($\vec{F} = d\vec{p}/dt$), the rate at which momentum is delivered to the wall is the force acting on that wall. To find the total force, we must add up the contributions of all the molecules that strike the wall, allowing for the possibility that they all have different speeds. Dividing the magnitude of the total force $F_x$ by the area of the wall ($= L^2$) then gives the pressure $p$ on that wall, where now and in the rest of this discussion, $p$ represents pressure. Thus, using the expression for $\Delta p_x/\Delta t$, we can write this pressure as

$$p = \frac{F_x}{L^2} = \frac{mv_{x1}^2/L + mv_{x2}^2/L + \cdots + mv_{xN}^2/L}{L^2}$$

$$= \left(\frac{m}{L^3}\right)(v_{x1}^2 + v_{x2}^2 + \cdots + v_{xN}^2), \qquad (19\text{-}18)$$

where $N$ is the number of molecules in the box.

Since $N = nN_A$, there are $nN_A$ terms in the second set of parentheses of Eq. 19-18. We can replace that quantity by $nN_A(v_x^2)_{\text{avg}}$, where $(v_x^2)_{\text{avg}}$ is the average value of the square of the $x$ components of all the molecular speeds. Equation 19-18 then becomes

$$p = \frac{nmN_A}{L^3}(v_x^2)_{\text{avg}}.$$

However, $mN_A$ is the molar mass $M$ of the gas (that is, the mass of 1 mol of the gas). Also, $L^3$ is the volume of the box, so

$$p = \frac{nM(v_x^2)_{\text{avg}}}{V}. \qquad (19\text{-}19)$$

For any molecule, $v^2 = v_x^2 + v_y^2 + v_z^2$. Because there are many molecules and because they are all moving in random directions, the average values of the squares of their velocity components are equal, so that $v_x^2 = \frac{1}{3}v^2$. Thus, Eq. 19-19 becomes

$$p = \frac{nM(v^2)_{\text{avg}}}{3V}. \qquad (19\text{-}20)$$

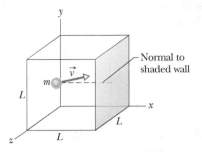

**Figure 19-4** A cubical box of edge length $L$, containing $n$ moles of an ideal gas. A molecule of mass $m$ and velocity $\vec{v}$ is about to collide with the shaded wall of area $L^2$. A normal to that wall is shown.

**Table 19-1** Some RMS Speeds at Room Temperature ($T = 300$ K)[a]

| Gas | Molar Mass ($10^{-3}$ kg/mol) | $v_{rms}$ (m/s) |
|---|---|---|
| Hydrogen ($H_2$) | 2.02 | 1920 |
| Helium (He) | 4.0 | 1370 |
| Water vapor ($H_2O$) | 18.0 | 645 |
| Nitrogen ($N_2$) | 28.0 | 517 |
| Oxygen ($O_2$) | 32.0 | 483 |
| Carbon dioxide ($CO_2$) | 44.0 | 412 |
| Sulfur dioxide ($SO_2$) | 64.1 | 342 |

[a]For convenience, we often set room temperature equal to 300 K even though (at 27°C or 81°F) that represents a fairly warm room.

The square root of $(v^2)_{avg}$ is a kind of average speed, called the **root-mean-square speed** of the molecules and symbolized by $v_{rms}$. Its name describes it rather well: You *square* each speed, you find the *mean* (that is, the average) of all these squared speeds, and then you take the square *root* of that mean. With $\sqrt{(v^2)_{avg}} = v_{rms}$, we can then write Eq. 19-20 as

$$p = \frac{nMv_{rms}^2}{3V}. \tag{19-21}$$

This tells us how the pressure of the gas (a purely macroscopic quantity) depends on the speed of the molecules (a purely microscopic quantity).

We can turn Eq. 19-21 around and use it to calculate $v_{rms}$. Combining Eq. 19-21 with the ideal gas law ($pV = nRT$) leads to

$$v_{rms} = \sqrt{\frac{3RT}{M}}. \tag{19-22}$$

Table 19-1 shows some rms speeds calculated from Eq. 19-22. The speeds are surprisingly high. For hydrogen molecules at room temperature (300 K), the rms speed is 1920 m/s, or 4300 mi/h—faster than a speeding bullet! On the surface of the Sun, where the temperature is $2 \times 10^6$ K, the rms speed of hydrogen molecules would be 82 times greater than at room temperature were it not for the fact that at such high speeds, the molecules cannot survive collisions among themselves. Remember too that the rms speed is only a kind of average speed; many molecules move much faster than this, and some much slower.

The speed of sound in a gas is closely related to the rms speed of the molecules of that gas. In a sound wave, the disturbance is passed on from molecule to molecule by means of collisions. The wave cannot move any faster than the "average" speed of the molecules. In fact, the speed of sound must be somewhat less than this "average" molecular speed because not all molecules are moving in exactly the same direction as the wave. As examples, at room temperature, the rms speeds of hydrogen and nitrogen molecules are 1920 m/s and 517 m/s, respectively. The speeds of sound in these two gases at this temperature are 1350 m/s and 350 m/s, respectively.

A question often arises: If molecules move so fast, why does it take as long as a minute or so before you can smell perfume when someone opens a bottle across a room? The answer is that, as we shall discuss in Module 19-5, each perfume molecule may have a high speed but it moves away from the bottle only very slowly because its repeated collisions with other molecules prevent it from moving directly across the room to you.

**Sample Problem 19.03** **Average and rms values**

Here are five numbers: 5, 11, 32, 67, and 89.

(a) What is the average value $n_{avg}$ of these numbers?

*Calculation:* We find this from

$$n_{avg} = \frac{5 + 11 + 32 + 67 + 89}{5} = 40.8. \quad \text{(Answer)}$$

(b) What is the rms value $n_{rms}$ of these numbers?

*Calculation:* We find this from

$$n_{rms} = \sqrt{\frac{5^2 + 11^2 + 32^2 + 67^2 + 89^2}{5}}$$
$$= 52.1. \quad \text{(Answer)}$$

The rms value is greater than the average value because the larger numbers—being squared—are relatively more important in forming the rms value.

# 19-4 TRANSLATIONAL KINETIC ENERGY

## Learning Objectives

*After reading this module, you should be able to . . .*

**19.19** For an ideal gas, relate the average kinetic energy of the molecules to their rms speed.

**19.20** Apply the relationship between the average kinetic energy and the temperature of the gas.

**19.21** Identify that a measurement of a gas temperature is effectively a measurement of the average kinetic energy of the gas molecules.

## Key Ideas

● The average translational kinetic energy per molecule in an ideal gas is

$$K_{avg} = \tfrac{1}{2}mv_{rms}^2.$$

● The average translational kinetic energy is related to the temperature of the gas:

$$K_{avg} = \tfrac{3}{2}kT.$$

## Translational Kinetic Energy

We again consider a single molecule of an ideal gas as it moves around in the box of Fig. 19-4, but we now assume that its speed changes when it collides with other molecules. Its translational kinetic energy at any instant is $\tfrac{1}{2}mv^2$. Its *average* translational kinetic energy over the time that we watch it is

$$K_{avg} = (\tfrac{1}{2}mv^2)_{avg} = \tfrac{1}{2}m(v^2)_{avg} = \tfrac{1}{2}mv_{rms}^2, \qquad (19\text{-}23)$$

in which we make the assumption that the average speed of the molecule during our observation is the same as the average speed of all the molecules at any given time. (Provided the total energy of the gas is not changing and provided we observe our molecule for long enough, this assumption is appropriate.) Substituting for $v_{rms}$ from Eq. 19-22 leads to

$$K_{avg} = (\tfrac{1}{2}m)\frac{3RT}{M}.$$

However, $M/m$, the molar mass divided by the mass of a molecule, is simply Avogadro's number. Thus,

$$K_{avg} = \frac{3RT}{2N_A}.$$

Using Eq. 19-7 ($k = R/N_A$), we can then write

$$K_{avg} = \tfrac{3}{2}kT. \qquad (19\text{-}24)$$

This equation tells us something unexpected:

At a given temperature $T$, all ideal gas molecules—no matter what their mass—have the same average translational kinetic energy—namely, $\tfrac{3}{2}kT$. When we measure the temperature of a gas, we are also measuring the average translational kinetic energy of its molecules.

 **Checkpoint 2**

A gas mixture consists of molecules of types 1, 2, and 3, with molecular masses $m_1 > m_2 > m_3$. Rank the three types according to (a) average kinetic energy and (b) rms speed, greatest first.

# 19-5 MEAN FREE PATH

## Learning Objectives

*After reading this module, you should be able to . . .*

**19.22** Identify what is meant by mean free path.

**19.23** Apply the relationship between the mean free path, the

diameter of the molecules, and the number of molecules per unit volume.

## Key Idea

● The mean free path λ of a gas molecule is its average path length between collisions and is given by

$$\lambda = \frac{1}{\sqrt{2}\pi d^2 \, N/V},$$

where $N/V$ is the number of molecules per unit volume and $d$ is the molecular diameter.

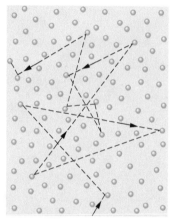

**Figure 19-5** A molecule traveling through a gas, colliding with other gas molecules in its path. Although the other molecules are shown as stationary, they are also moving in a similar fashion.

## Mean Free Path

We continue to examine the motion of molecules in an ideal gas. Figure 19-5 shows the path of a typical molecule as it moves through the gas, changing both speed and direction abruptly as it collides elastically with other molecules. Between collisions, the molecule moves in a straight line at constant speed. Although the figure shows the other molecules as stationary, they are (of course) also moving.

One useful parameter to describe this random motion is the **mean free path** λ of the molecules. As its name implies, λ is the average distance traversed by a molecule between collisions. We expect λ to vary inversely with $N/V$, the number of molecules per unit volume (or density of molecules). The larger $N/V$ is, the more collisions there should be and the smaller the mean free path. We also expect λ to vary inversely with the size of the molecules—with their diameter $d$, say. (If the molecules were points, as we have assumed them to be, they would never collide and the mean free path would be infinite.) Thus, the larger the molecules are, the smaller the mean free path. We can even predict that λ should vary (inversely) as the *square* of the molecular diameter because the cross section of a molecule—not its diameter—determines its effective target area.

The expression for the mean free path does, in fact, turn out to be

$$\lambda = \frac{1}{\sqrt{2}\pi d^2 \, N/V} \quad \text{(mean free path)}. \tag{19-25}$$

To justify Eq. 19-25, we focus attention on a single molecule and assume—as Fig. 19-5 suggests—that our molecule is traveling with a constant speed $v$ and that all the other molecules are at rest. Later, we shall relax this assumption.

We assume further that the molecules are spheres of diameter $d$. A collision will then take place if the centers of two molecules come within a distance $d$ of each other, as in Fig. 19-6a. Another, more helpful way to look at the situation is

**Figure 19-6** (*a*) A collision occurs when the centers of two molecules come within a distance $d$ of each other, $d$ being the molecular diameter. (*b*) An equivalent but more convenient representation is to think of the moving molecule as having a *radius d* and all other molecules as being points. The condition for a collision is unchanged.

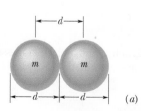

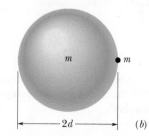

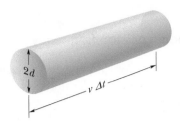

**Figure 19-7** In time $\Delta t$ the moving molecule effectively sweeps out a cylinder of length $v\,\Delta t$ and radius $d$.

to consider our single molecule to have a *radius* of $d$ and all the other molecules to be *points,* as in Fig. 19-6*b*. This does not change our criterion for a collision.

As our single molecule zigzags through the gas, it sweeps out a short cylinder of cross-sectional area $\pi d^2$ between successive collisions. If we watch this molecule for a time interval $\Delta t$, it moves a distance $v\,\Delta t$, where $v$ is its assumed speed. Thus, if we align all the short cylinders swept out in interval $\Delta t$, we form a composite cylinder (Fig. 19-7) of length $v\,\Delta t$ and volume $(\pi d^2)(v\,\Delta t)$. The number of collisions that occur in time $\Delta t$ is then equal to the number of (point) molecules that lie within this cylinder.

Since $N/V$ is the number of molecules per unit volume, the number of molecules in the cylinder is $N/V$ times the volume of the cylinder, or $(N/V)(\pi d^2 v\,\Delta t)$. This is also the number of collisions in time $\Delta t$. The mean free path is the length of the path (and of the cylinder) divided by this number:

$$\lambda = \frac{\text{length of path during } \Delta t}{\text{number of collisions in } \Delta t} \approx \frac{v\,\Delta t}{\pi d^2 v\,\Delta t\,N/V}$$

$$= \frac{1}{\pi d^2\,N/V}. \tag{19-26}$$

This equation is only approximate because it is based on the assumption that all the molecules except one are at rest. In fact, *all* the molecules are moving; when this is taken properly into account, Eq. 19-25 results. Note that it differs from the (approximate) Eq. 19-26 only by a factor of $1/\sqrt{2}$.

The approximation in Eq. 19-26 involves the two $v$ symbols we canceled. The $v$ in the numerator is $v_{avg}$, the mean speed of the molecules *relative to the container.* The $v$ in the denominator is $v_{rel}$, the mean speed of our single molecule *relative to the other molecules,* which are moving. It is this latter average speed that determines the number of collisions. A detailed calculation, taking into account the actual speed distribution of the molecules, gives $v_{rel} = \sqrt{2}\,v_{avg}$ and thus the factor $\sqrt{2}$.

The mean free path of air molecules at sea level is about 0.1 $\mu$m. At an altitude of 100 km, the density of air has dropped to such an extent that the mean free path rises to about 16 cm. At 300 km, the mean free path is about 20 km. A problem faced by those who would study the physics and chemistry of the upper atmosphere in the laboratory is the unavailability of containers large enough to hold gas samples (of Freon, carbon dioxide, and ozone) that simulate upper atmospheric conditions.

### ✓ Checkpoint 3

One mole of gas $A$, with molecular diameter $2d_0$ and average molecular speed $v_0$, is placed inside a certain container. One mole of gas $B$, with molecular diameter $d_0$ and average molecular speed $2v_0$ (the molecules of $B$ are smaller but faster), is placed in an identical container. Which gas has the greater average collision rate within its container?

## Sample Problem 19.04 Mean free path, average speed, collision frequency

**(a)** What is the mean free path $\lambda$ for oxygen molecules at temperature $T = 300$ K and pressure $p = 1.0$ atm? Assume that the molecular diameter is $d = 290$ pm and the gas is ideal.

### KEY IDEA

Each oxygen molecule moves among other *moving* oxygen molecules in a zigzag path due to the resulting collisions. Thus, we use Eq. 19-25 for the mean free path.

*Calculation:* We first need the number of molecules per unit volume, $N/V$. Because we assume the gas is ideal, we can use the ideal gas law of Eq. 19-9 ($pV = NkT$) to write $N/V = p/kT$. Substituting this into Eq. 19-25, we find

$$\lambda = \frac{1}{\sqrt{2}\pi d^2\, N/V} = \frac{kT}{\sqrt{2}\pi d^2 p}$$

$$= \frac{(1.38 \times 10^{-23}\ \text{J/K})(300\ \text{K})}{\sqrt{2}\pi(2.9 \times 10^{-10}\ \text{m})^2(1.01 \times 10^5\ \text{Pa})}$$

$$= 1.1 \times 10^{-7}\ \text{m}. \qquad \text{(Answer)}$$

This is about 380 molecular diameters.

**(b)** Assume the average speed of the oxygen molecules is $v = 450$ m/s. What is the average time $t$ between successive

collisions for any given molecule? At what rate does the molecule collide; that is, what is the frequency $f$ of its collisions?

### KEY IDEAS

(1) Between collisions, the molecule travels, on average, the mean free path $\lambda$ at speed $v$. (2) The average rate or frequency at which the collisions occur is the inverse of the time $t$ between collisions.

*Calculations:* From the first key idea, the average time between collisions is

$$t = \frac{\text{distance}}{\text{speed}} = \frac{\lambda}{v} = \frac{1.1 \times 10^{-7}\ \text{m}}{450\ \text{m/s}}$$

$$= 2.44 \times 10^{-10}\ \text{s} \approx 0.24\ \text{ns}. \qquad \text{(Answer)}$$

This tells us that, on average, any given oxygen molecule has less than a nanosecond between collisions.

From the second key idea, the collision frequency is

$$f = \frac{1}{t} = \frac{1}{2.44 \times 10^{-10}\ \text{s}} = 4.1 \times 10^9\ \text{s}^{-1}. \qquad \text{(Answer)}$$

This tells us that, on average, any given oxygen molecule makes about 4 billion collisions per second.

 Additional examples, video, and practice available at *WileyPLUS*

# 19-6 THE DISTRIBUTION OF MOLECULAR SPEEDS

## Learning Objectives

*After reading this module, you should be able to . . .*

**19.24** Explain how Maxwell's speed distribution law is used to find the fraction of molecules with speeds in a certain speed range.

**19.25** Sketch a graph of Maxwell's speed distribution, showing the probability distribution versus speed and indicating the relative positions of the average speed $v_{avg}$, the most probable speed $v_P$, and the rms speed $v_{rms}$.

**19.26** Explain how Maxwell's speed distribution is used to find the average speed, the rms speed, and the most probable speed.

**19.27** For a given temperature $T$ and molar mass $M$, calculate the average speed $v_{avg}$, the most probable speed $v_P$, and the rms speed $v_{rms}$.

## Key Ideas

● The Maxwell speed distribution $P(v)$ is a function such that $P(v)\,dv$ gives the fraction of molecules with speeds in the interval $dv$ at speed $v$:

$$P(v) = 4\pi\left(\frac{M}{2\pi RT}\right)^{3/2} v^2 e^{-Mv^2/2RT}.$$

● Three measures of the distribution of speeds among the molecules of a gas are

and

$$v_{avg} = \sqrt{\frac{8RT}{\pi M}} \qquad \text{(average speed),}$$

$$v_P = \sqrt{\frac{2RT}{M}} \qquad \text{(most probable speed),}$$

$$v_{rms} = \sqrt{\frac{3RT}{M}} \qquad \text{(rms speed).}$$

## The Distribution of Molecular Speeds

The root-mean-square speed $v_{rms}$ gives us a general idea of molecular speeds in a gas at a given temperature. We often want to know more. For example, what fraction of the molecules have speeds greater than the rms value? What fraction have speeds greater than twice the rms value? To answer such questions, we need to know how the possible values of speed are distributed among the molecules. Figure 19-8a shows this distribution for oxygen molecules at room temperature ($T = 300$ K); Fig. 19-8b compares it with the distribution at $T = 80$ K.

In 1852, Scottish physicist James Clerk Maxwell first solved the problem of finding the speed distribution of gas molecules. His result, known as **Maxwell's speed distribution law,** is

$$P(v) = 4\pi\left(\frac{M}{2\pi RT}\right)^{3/2} v^2 e^{-Mv^2/2RT}. \qquad (19\text{-}27)$$

Here $M$ is the molar mass of the gas, $R$ is the gas constant, $T$ is the gas temperature, and $v$ is the molecular speed. It is this equation that is plotted in Fig. 19-8a, b. The quantity $P(v)$ in Eq. 19-27 and Fig. 19-8 is a *probability distribution function:* For any speed $v$, the product $P(v)\,dv$ (a dimensionless quantity) is the fraction of molecules with speeds in the interval $dv$ centered on speed $v$.

As Fig. 19-8a shows, this fraction is equal to the area of a strip with height $P(v)$ and width $dv$. The total area under the distribution curve corresponds to the fraction of the molecules whose speeds lie between zero and infinity. All molecules fall into this category, so the value of this total area is unity; that is,

$$\int_0^\infty P(v)\,dv = 1. \qquad (19\text{-}28)$$

The fraction (frac) of molecules with speeds in an interval of, say, $v_1$ to $v_2$ is then

$$\text{frac} = \int_{v_1}^{v_2} P(v)\,dv. \qquad (19\text{-}29)$$

### Average, RMS, and Most Probable Speeds

In principle, we can find the **average speed** $v_{avg}$ of the molecules in a gas with the following procedure: We *weight* each value of $v$ in the distribution; that is, we multiply it

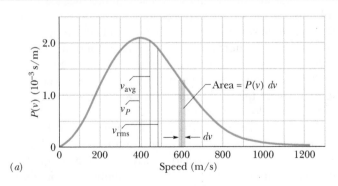

(a)

**Figure 19-8** (a) The Maxwell speed distribution for oxygen molecules at $T = 300$ K. The three characteristic speeds are marked. (b) The curves for 300 K and 80 K. Note that the molecules move more slowly at the lower temperature. Because these are probability distributions, the area under each curve has a numerical value of unity.

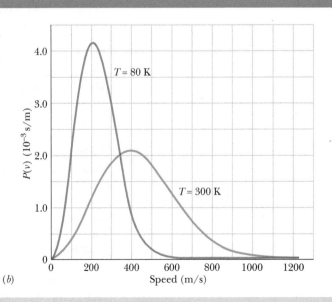

(b)

by the fraction $P(v)\,dv$ of molecules with speeds in a differential interval $dv$ centered on $v$. Then we add up all these values of $v\,P(v)\,dv$. The result is $v_{avg}$. In practice, we do all this by evaluating

$$v_{avg} = \int_0^\infty v\,P(v)\,dv. \tag{19-30}$$

Substituting for $P(v)$ from Eq. 19-27 and using generic integral 20 from the list of integrals in Appendix E, we find

$$v_{avg} = \sqrt{\frac{8RT}{\pi M}} \quad \text{(average speed).} \tag{19-31}$$

Similarly, we can find the average of the square of the speeds $(v^2)_{avg}$ with

$$(v^2)_{avg} = \int_0^\infty v^2\,P(v)\,dv. \tag{19-32}$$

Substituting for $P(v)$ from Eq. 19-27 and using generic integral 16 from the list of integrals in Appendix E, we find

$$(v^2)_{avg} = \frac{3RT}{M}. \tag{19-33}$$

The square root of $(v^2)_{avg}$ is the root-mean-square speed $v_{rms}$. Thus,

$$v_{rms} = \sqrt{\frac{3RT}{M}} \quad \text{(rms speed),} \tag{19-34}$$

which agrees with Eq. 19-22.

The **most probable speed** $v_P$ is the speed at which $P(v)$ is maximum (see Fig. 19-8a). To calculate $v_P$, we set $dP/dv = 0$ (the slope of the curve in Fig. 19-8a is zero at the maximum of the curve) and then solve for $v$. Doing so, we find

$$v_P = \sqrt{\frac{2RT}{M}} \quad \text{(most probable speed).} \tag{19-35}$$

A molecule is more likely to have speed $v_P$ than any other speed, but some molecules will have speeds that are many times $v_P$. These molecules lie in the *high-speed tail* of a distribution curve like that in Fig. 19-8a. Such higher speed molecules make possible both rain and sunshine (without which we could not exist):

**Rain** The speed distribution of water molecules in, say, a pond at summertime temperatures can be represented by a curve similar to that of Fig. 19-8a. Most of the molecules lack the energy to escape from the surface. However, a few of the molecules in the high-speed tail of the curve can do so. It is these water molecules that evaporate, making clouds and rain possible.

As the fast water molecules leave the surface, carrying energy with them, the temperature of the remaining water is maintained by heat transfer from the surroundings. Other fast molecules—produced in particularly favorable collisions—quickly take the place of those that have left, and the speed distribution is maintained.

**Sunshine** Let the distribution function of Eq. 19-27 now refer to protons in the core of the Sun. The Sun's energy is supplied by a nuclear fusion process that starts with the merging of two protons. However, protons repel each other because of their electrical charges, and protons of average speed do not have enough kinetic energy to overcome the repulsion and get close enough to merge. Very fast protons with speeds in the high-speed tail of the distribution curve can do so, however, and for that reason the Sun can shine.

## Sample Problem 19.05 Speed distribution in a gas

In oxygen (molar mass $M = 0.0320$ kg/mol) at room temperature (300 K), what fraction of the molecules have speeds in the interval 599 to 601 m/s?

### KEY IDEAS

1. The speeds of the molecules are distributed over a wide range of values, with the distribution $P(v)$ of Eq. 19-27.
2. The fraction of molecules with speeds in a differential interval $dv$ is $P(v)\,dv$.
3. For a larger interval, the fraction is found by integrating $P(v)$ over the interval.
4. However, the interval $\Delta v = 2$ m/s here is small compared to the speed $v = 600$ m/s on which it is centered.

*Calculations:* Because $\Delta v$ is small, we can avoid the integration by approximating the fraction as

$$\text{frac} = P(v)\,\Delta v = 4\pi \left(\frac{M}{2\pi RT}\right)^{3/2} v^2 e^{-Mv^2/2RT}\,\Delta v.$$

The total area under the plot of $P(v)$ in Fig. 19-8a is the total fraction of molecules (unity), and the area of the thin gold strip (not to scale) is the fraction we seek. Let's evaluate frac in parts:

$$\text{frac} = (4\pi)(A)(v^2)(e^B)(\Delta v), \qquad (19\text{-}36)$$

where

$$A = \left(\frac{M}{2\pi RT}\right)^{3/2} = \left(\frac{0.0320 \text{ kg/mol}}{(2\pi)(8.31 \text{ J/mol}\cdot\text{K})(300 \text{ K})}\right)^{3/2}$$

$$= 2.92 \times 10^{-9} \text{ s}^3/\text{m}^3$$

and $B = -\dfrac{Mv^2}{2RT} = -\dfrac{(0.0320 \text{ kg/mol})(600 \text{ m/s})^2}{(2)(8.31 \text{ J/mol}\cdot\text{K})(300 \text{ K})}$

$$= -2.31.$$

Substituting $A$ and $B$ into Eq. 19-36 yields

$$\text{frac} = (4\pi)(A)(v^2)(e^B)(\Delta v)$$

$$= (4\pi)(2.92 \times 10^{-9} \text{ s}^3/\text{m}^3)(600 \text{ m/s})^2(e^{-2.31})(2 \text{ m/s})$$

$$= 2.62 \times 10^{-3} = 0.262\%. \qquad \text{(Answer)}$$

## Sample Problem 19.06 Average speed, rms speed, most probable speed

The molar mass $M$ of oxygen is 0.0320 kg/mol.

(a) What is the average speed $v_{\text{avg}}$ of oxygen gas molecules at $T = 300$ K?

### KEY IDEA

To find the average speed, we must weight speed $v$ with the distribution function $P(v)$ of Eq. 19-27 and then integrate the resulting expression over the range of possible speeds (from zero to the limit of an infinite speed).

*Calculation:* We end up with Eq. 19-31, which gives us

$$v_{\text{avg}} = \sqrt{\frac{8RT}{\pi M}}$$

$$= \sqrt{\frac{8(8.31 \text{ J/mol}\cdot\text{K})(300 \text{ K})}{\pi(0.0320 \text{ kg/mol})}}$$

$$= 445 \text{ m/s.} \qquad \text{(Answer)}$$

This result is plotted in Fig. 19-8a.

(b) What is the root-mean-square speed $v_{\text{rms}}$ at 300 K?

### KEY IDEA

To find $v_{\text{rms}}$, we must first find $(v^2)_{\text{avg}}$ by weighting $v^2$ with the distribution function $P(v)$ of Eq. 19-27 and then integrating the expression over the range of possible speeds. Then we must take the square root of the result.

*Calculation:* We end up with Eq. 19-34, which gives us

$$v_{\text{rms}} = \sqrt{\frac{3RT}{M}}$$

$$= \sqrt{\frac{3(8.31 \text{ J/mol}\cdot\text{K})(300 \text{ K})}{0.0320 \text{ kg/mol}}}$$

$$= 483 \text{ m/s.} \qquad \text{(Answer)}$$

This result, plotted in Fig. 19-8a, is greater than $v_{\text{avg}}$ because the greater speed values influence the calculation more when we integrate the $v^2$ values than when we integrate the $v$ values.

(c) What is the most probable speed $v_P$ at 300 K?

### KEY IDEA

Speed $v_P$ corresponds to the maximum of the distribution function $P(v)$, which we obtain by setting the derivative $dP/dv = 0$ and solving the result for $v$.

*Calculation:* We end up with Eq. 19-35, which gives us

$$v_P = \sqrt{\frac{2RT}{M}}$$

$$= \sqrt{\frac{2(8.31 \text{ J/mol}\cdot\text{K})(300 \text{ K})}{0.0320 \text{ kg/mol}}}$$

$$= 395 \text{ m/s.} \qquad \text{(Answer)}$$

This result is also plotted in Fig. 19-8a.

 Additional examples, video, and practice available at *WileyPLUS*

# 19-7 THE MOLAR SPECIFIC HEATS OF AN IDEAL GAS

## Learning Objectives

*After reading this module, you should be able to . . .*

**19.28** Identify that the internal energy of an ideal monatomic gas is the sum of the translational kinetic energies of its atoms.

**19.29** Apply the relationship between the internal energy $E_{int}$ of a monatomic ideal gas, the number of moles $n$, and the gas temperature $T$.

**19.30** Distinguish between monatomic, diatomic, and polyatomic ideal gases.

**19.31** For monatomic, diatomic, and polyatomic ideal gases, evaluate the molar specific heats for a constant-volume process and a constant-pressure process.

**19.32** Calculate a molar specific heat at constant pressure $C_p$ by adding $R$ to the molar specific heat at constant volume $C_V$, and explain why (physically) $C_p$ is greater.

**19.33** Identify that the energy transferred to an ideal gas as heat in a constant-volume process goes entirely into the internal energy (the random translational motion) but that

in a constant-pressure process energy also goes into the work done to expand the gas.

**19.34** Identify that for a given change in temperature, the change in the internal energy of an ideal gas is the same for *any* process and is most easily calculated by assuming a constant-volume process.

**19.35** For an ideal gas, apply the relationship between heat $Q$, number of moles $n$, and temperature change $\Delta T$, using the appropriate molar specific heat.

**19.36** Between two isotherms on a $p$-$V$ diagram, sketch a constant-volume process and a constant-pressure process, and for each identify the work done in terms of area on the graph.

**19.37** Calculate the work done by an ideal gas for a constant-pressure process.

**19.38** Identify that work is zero for constant volume.

## Key Ideas

● The molar specific heat $C_V$ of a gas at constant volume is defined as

$$C_V = \frac{Q}{n\,\Delta T} = \frac{\Delta E_{int}}{n\,\Delta T},$$

in which $Q$ is the energy transferred as heat to or from a sample of $n$ moles of the gas, $\Delta T$ is the resulting temperature change of the gas, and $\Delta E_{int}$ is the resulting change in the internal energy of the gas.

● For an ideal monatomic gas,

$$C_V = \tfrac{3}{2}R = 12.5 \text{ J/mol} \cdot \text{K}.$$

● The molar specific heat $C_p$ of a gas at constant pressure is

defined to be

$$C_p = \frac{Q}{n\,\Delta T},$$

in which $Q$, $n$, and $\Delta T$ are defined as above. $C_p$ is also given by

$$C_p = C_V + R.$$

● For $n$ moles of an ideal gas,

$$E_{int} = nC_V T \quad \text{(ideal gas).}$$

● If $n$ moles of a confined ideal gas undergo a temperature change $\Delta T$ due to *any* process, the change in the internal energy of the gas is

$$\Delta E_{int} = nC_V\,\Delta T \quad \text{(ideal gas, any process).}$$

## The Molar Specific Heats of an Ideal Gas

In this module, we want to derive from molecular considerations an expression for the internal energy $E_{int}$ of an ideal gas. In other words, we want an expression for the energy associated with the random motions of the atoms or molecules in the gas. We shall then use that expression to derive the molar specific heats of an ideal gas.

### Internal Energy $E_{int}$

Let us first assume that our ideal gas is a *monatomic gas* (individual atoms rather than molecules), such as helium, neon, or argon. Let us also assume that the internal energy $E_{int}$ is the sum of the translational kinetic energies of the atoms. (Quantum theory disallows rotational kinetic energy for individual atoms.)

The average translational kinetic energy of a single atom depends only on the gas temperature and is given by Eq. 19-24 as $K_{avg} = \tfrac{3}{2}kT$. A sample of $n$ moles of such a gas contains $nN_A$ atoms. The internal energy $E_{int}$ of the sample is then

$$E_{int} = (nN_A)K_{avg} = (nN_A)(\tfrac{3}{2}kT). \quad (19\text{-}37)$$

Using Eq. 19-7 ($k = R/N_A$), we can rewrite this as

$$E_{int} = \tfrac{3}{2}nRT \quad \text{(monatomic ideal gas).} \qquad (19\text{-}38)$$

⭐ The internal energy $E_{int}$ of an ideal gas is a function of the gas temperature *only*; it does not depend on any other variable.

With Eq. 19-38 in hand, we are now able to derive an expression for the molar specific heat of an ideal gas. Actually, we shall derive two expressions. One is for the case in which the volume of the gas remains constant as energy is transferred to or from it as heat. The other is for the case in which the pressure of the gas remains constant as energy is transferred to or from it as heat. The symbols for these two molar specific heats are $C_V$ and $C_p$, respectively. (By convention, the capital letter $C$ is used in both cases, even though $C_V$ and $C_p$ represent types of specific heat and not heat capacities.)

## Molar Specific Heat at Constant Volume

Figure 19-9a shows $n$ moles of an ideal gas at pressure $p$ and temperature $T$, confined to a cylinder of fixed volume $V$. This *initial state i* of the gas is marked on the $p$-$V$ diagram of Fig. 19-9b. Suppose now that you add a small amount of energy to the gas as heat $Q$ by slowly turning up the temperature of the thermal reservoir. The gas temperature rises a small amount to $T + \Delta T$, and its pressure rises to $p + \Delta p$, bringing the gas to *final state f*. In such experiments, we would find that the heat $Q$ is related to the temperature change $\Delta T$ by

$$Q = nC_V \Delta T \quad \text{(constant volume),} \qquad (19\text{-}39)$$

where $C_V$ is a constant called the **molar specific heat at constant volume.** Substituting this expression for $Q$ into the first law of thermodynamics as given by Eq. 18-26 ($\Delta E_{int} = Q - W$) yields

$$\Delta E_{int} = nC_V \Delta T - W. \qquad (19\text{-}40)$$

With the volume held constant, the gas cannot expand and thus cannot do any work. Therefore, $W = 0$, and Eq. 19-40 gives us

$$C_V = \frac{\Delta E_{int}}{n\,\Delta T}. \qquad (19\text{-}41)$$

From Eq. 19-38, the change in internal energy must be

$$\Delta E_{int} = \tfrac{3}{2}nR\,\Delta T. \qquad (19\text{-}42)$$

Substituting this result into Eq. 19-41 yields

$$C_V = \tfrac{3}{2}R = 12.5 \text{ J/mol} \cdot \text{K} \quad \text{(monatomic gas).} \qquad (19\text{-}43)$$

As Table 19-2 shows, this prediction of the kinetic theory (for ideal gases) agrees very well with experiment for real monatomic gases, the case that we have assumed. The (predicted and) experimental values of $C_V$ for *diatomic gases* (which have molecules with two atoms) and *polyatomic gases* (which have molecules with more than two atoms) are greater than those for monatomic gases for reasons that will be suggested in Module 19-8. Here we make the preliminary assumption that the $C_V$ values for diatomic and polyatomic gases are greater than for monatomic gases because the more complex molecules can rotate and thus have rotational kinetic energy. So, when $Q$ is transferred to a diatomic or polyatomic gas, only part of it goes into the translational kinetic energy, increasing the

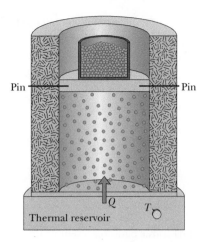

*(a)*

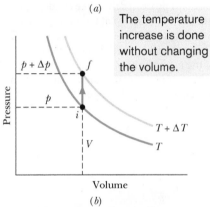

The temperature increase is done without changing the volume.

*(b)*

**Figure 19-9** (*a*) The temperature of an ideal gas is raised from $T$ to $T + \Delta T$ in a constant-volume process. Heat is added, but no work is done. (*b*) The process on a $p$-$V$ diagram.

**Table 19-2** Molar Specific Heats at Constant Volume

| Molecule | | Example | $C_V$ (J/mol·K) |
|---|---|---|---|
| Monatomic | Ideal | | $\tfrac{3}{2}R = 12.5$ |
| | Real | He | 12.5 |
| | | Ar | 12.6 |
| Diatomic | Ideal | | $\tfrac{5}{2}R = 20.8$ |
| | Real | $N_2$ | 20.7 |
| | | $O_2$ | 20.8 |
| Polyatomic | Ideal | | $3R = 24.9$ |
| | Real | $NH_4$ | 29.0 |
| | | $CO_2$ | 29.7 |

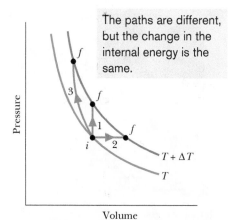

The paths are different, but the change in the internal energy is the same.

**Figure 19-10** Three paths representing three different processes that take an ideal gas from an initial state $i$ at temperature $T$ to some final state $f$ at temperature $T + \Delta T$. The change $\Delta E_{int}$ in the internal energy of the gas is the same for these three processes and for any others that result in the same change of temperature.

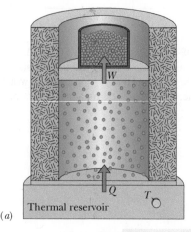

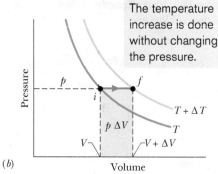

The temperature increase is done without changing the pressure.

*(b)*

**Figure 19-11** (*a*) The temperature of an ideal gas is raised from $T$ to $T + \Delta T$ in a constant-pressure process. Heat is added and work is done in lifting the loaded piston. (*b*) The process on a $p$-$V$ diagram. The work $p\,\Delta V$ is given by the shaded area.

temperature. (For now we neglect the possibility of also putting energy into oscillations of the molecules.)

We can now generalize Eq. 19-38 for the internal energy of any ideal gas by substituting $C_V$ for $\frac{3}{2}R$; we get

$$E_{int} = nC_V T \quad \text{(any ideal gas)}. \tag{19-44}$$

This equation applies not only to an ideal monatomic gas but also to diatomic and polyatomic ideal gases, provided the appropriate value of $C_V$ is used. Just as with Eq. 19-38, we see that the internal energy of a gas depends on the temperature of the gas but not on its pressure or density.

When a confined ideal gas undergoes temperature change $\Delta T$, then from either Eq. 19-41 or Eq. 19-44 the resulting change in its internal energy is

$$\Delta E_{int} = nC_V \Delta T \quad \text{(ideal gas, any process)}. \tag{19-45}$$

This equation tells us:

> A change in the internal energy $E_{int}$ of a confined ideal gas depends on only the change in the temperature, *not* on what type of process produces the change.

As examples, consider the three paths between the two isotherms in the $p$-$V$ diagram of Fig. 19-10. Path 1 represents a constant-volume process. Path 2 represents a constant-pressure process (we examine it next). Path 3 represents a process in which no heat is exchanged with the system's environment (we discuss this in Module 19-9). Although the values of heat $Q$ and work $W$ associated with these three paths differ, as do $p_f$ and $V_f$, the values of $\Delta E_{int}$ associated with the three paths are identical and are all given by Eq. 19-45, because they all involve the same temperature change $\Delta T$. Therefore, no matter what path is actually taken between $T$ and $T + \Delta T$, we can *always* use path 1 and Eq. 19-45 to compute $\Delta E_{int}$ easily.

## Molar Specific Heat at Constant Pressure

We now assume that the temperature of our ideal gas is increased by the same small amount $\Delta T$ as previously but now the necessary energy (heat $Q$) is added with the gas under constant pressure. An experiment for doing this is shown in Fig. 19-11*a*; the $p$-$V$ diagram for the process is plotted in Fig. 19-11*b*. From such experiments we find that the heat $Q$ is related to the temperature change $\Delta T$ by

$$Q = nC_p \Delta T \quad \text{(constant pressure)}, \tag{19-46}$$

where $C_p$ is a constant called the **molar specific heat at constant pressure**. This $C_p$ is *greater* than the molar specific heat at constant volume $C_V$, because energy must now be supplied not only to raise the temperature of the gas but also for the gas to do work — that is, to lift the weighted piston of Fig. 19-11*a*.

To relate molar specific heats $C_p$ and $C_V$, we start with the first law of thermodynamics (Eq. 18-26):

$$\Delta E_{int} = Q - W. \tag{19-47}$$

We next replace each term in Eq. 19-47. For $\Delta E_{int}$, we substitute from Eq. 19-45. For $Q$, we substitute from Eq. 19-46. To replace $W$, we first note that since the pressure remains constant, Eq. 19-16 tells us that $W = p\,\Delta V$. Then we note that, using the ideal gas equation ($pV = nRT$), we can write

$$W = p\,\Delta V = nR\,\Delta T. \tag{19-48}$$

Making these substitutions in Eq. 19-47 and then dividing through by $n\,\Delta T$, we find

$$C_V = C_p - R$$

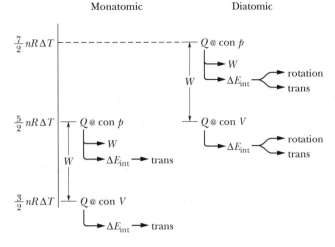

**Figure 19-12** The relative values of $Q$ for a monatomic gas (left side) and a diatomic gas undergoing a constant-volume process (labeled "con $V$") and a constant-pressure process (labeled "con $p$"). The transfer of the energy into work $W$ and internal energy ($\Delta E_{int}$) is noted.

and then

$$C_p = C_V + R. \qquad (19\text{-}49)$$

This prediction of kinetic theory agrees well with experiment, not only for monatomic gases but also for gases in general, as long as their density is low enough so that we may treat them as ideal.

The left side of Fig. 19-12 shows the relative values of $Q$ for a monatomic gas undergoing either a constant-volume process ($Q = \frac{3}{2}nR\,\Delta T$) or a constant-pressure process ($Q = \frac{5}{2}nR\,\Delta T$). Note that for the latter, the value of $Q$ is higher by the amount $W$, the work done by the gas in the expansion. Note also that for the constant-volume process, the energy added as $Q$ goes entirely into the change in internal energy $\Delta E_{int}$ and for the constant-pressure process, the energy added as $Q$ goes into both $\Delta E_{int}$ and the work $W$.

 **Checkpoint 4**

The figure here shows five paths traversed by a gas on a $p$-$V$ diagram. Rank the paths according to the change in internal energy of the gas, greatest first.

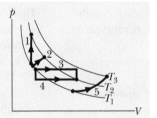

## Sample Problem 19.07    Monatomic gas, heat, internal energy, and work

A bubble of 5.00 mol of helium is submerged at a certain depth in liquid water when the water (and thus the helium) undergoes a temperature increase $\Delta T$ of 20.0 C° at constant pressure. As a result, the bubble expands. The helium is monatomic and ideal.

**(a)** How much energy is added to the helium as heat during the increase and expansion?

### KEY IDEA

Heat $Q$ is related to the temperature change $\Delta T$ by a molar specific heat of the gas.

**Calculations:** Because the pressure $p$ is held constant during the addition of energy, we use the molar specific heat at constant pressure $C_p$ and Eq. 19-46,

$$Q = nC_p\,\Delta T, \qquad (19\text{-}50)$$

to find $Q$. To evaluate $C_p$ we go to Eq. 19-49, which tells us that for any ideal gas, $C_p = C_V + R$. Then from Eq. 19-43, we know that for any *monatomic* gas (like the helium here), $C_V = \frac{3}{2}R$. Thus, Eq. 19-50 gives us

$$Q = n(C_V + R)\,\Delta T = n(\tfrac{3}{2}R + R)\,\Delta T = n(\tfrac{5}{2}R)\,\Delta T$$

$$= (5.00\text{ mol})(2.5)(8.31\text{ J/mol·K})(20.0\text{ C°})$$

$$= 2077.5\text{ J} \approx 2080\text{ J.} \qquad \text{(Answer)}$$

**(b)** What is the change $\Delta E_{int}$ in the internal energy of the helium during the temperature increase?

## KEY IDEA

Because the bubble expands, this is not a constant-volume process. However, the helium is nonetheless confined (to the bubble). Thus, the change $\Delta E_{int}$ is the same as *would occur* in a constant-volume process with the same temperature change $\Delta T$.

*Calculation:* We can now easily find the constant-volume change $\Delta E_{int}$ with Eq. 19-45:

$$\Delta E_{int} = nC_V \Delta T = n(\tfrac{3}{2}R)\,\Delta T$$

$$= (5.00\ \text{mol})(1.5)(8.31\ \text{J/mol·K})(20.0\ \text{C}°)$$

$$= 1246.5\ \text{J} \approx 1250\ \text{J}. \qquad \text{(Answer)}$$

(c) How much work $W$ is done by the helium as it expands against the pressure of the surrounding water during the temperature increase?

## KEY IDEAS

The work done by *any* gas expanding against the pressure from its environment is given by Eq. 19-11, which tells us to in-

tegrate $p\ dV$. When the pressure is constant (as here), we can simplify that to $W = p\,\Delta V$. When the gas is *ideal* (as here), we can use the ideal gas law (Eq. 19-5) to write $p\,\Delta V = nR\,\Delta T$.

*Calculation:* We end up with

$$W = nR\,\Delta T$$

$$= (5.00\ \text{mol})(8.31\ \text{J/mol·K})(20.0\ \text{C}°)$$

$$= 831\ \text{J}. \qquad \text{(Answer)}$$

*Another way:* Because we happen to know $Q$ and $\Delta E_{int}$, we can work this problem another way: We can account for the energy changes of the gas with the first law of thermodynamics, writing

$$W = Q - \Delta E_{int} = 2077.5\ \text{J} - 1246.5\ \text{J}$$

$$= 831\ \text{J}. \qquad \text{(Answer)}$$

*The transfers:* Let's follow the energy. Of the 2077.5 J transferred to the helium as heat $Q$, 831 J goes into the work $W$ required for the expansion and 1246.5 J goes into the internal energy $E_{int}$, which, for a monatomic gas, is entirely the kinetic energy of the atoms in their translational motion. These several results are suggested on the left side of Fig. 19-12.

WILEY **PLUS** Additional examples, video, and practice available at *WileyPLUS*

# 19-8 DEGREES OF FREEDOM AND MOLAR SPECIFIC HEATS

## Learning Objectives

*After reading this module, you should be able to . . .*

**19.39** Identify that a degree of freedom is associated with each way a gas can store energy (translation, rotation, and oscillation).

**19.40** Identify that an energy of $\tfrac{1}{2}kT$ per molecule is associated with each degree of freedom.

**19.41** Identify that a monatomic gas can have an internal energy consisting of only translational motion.

**19.42** Identify that at low temperatures a diatomic gas has energy in only translational motion, at higher temperatures it also has energy in molecular rotation, and at even higher temperatures it can also have energy in molecular oscillations.

**19.43** Calculate the molar specific heat for monatomic and diatomic ideal gases in a constant-volume process and a constant-pressure process.

## Key Ideas

● We find $C_V$ by using the equipartition of energy theorem, which states that every degree of freedom of a molecule (that is, every independent way it can store energy) has associated with it—on average—an energy $\tfrac{1}{2}kT$ per molecule ($= \tfrac{1}{2}RT$ per mole).

● If $f$ is the number of degrees of freedom, then

$E_{int} = (f/2)nRT$ and

$$C_V = \left(\frac{f}{2}\right)R = 4.16f\ \text{J/mol·K}.$$

● For monatomic gases $f = 3$ (three translational degrees); for diatomic gases $f = 5$ (three translational and two rotational degrees).

## Degrees of Freedom and Molar Specific Heats

As Table 19-2 shows, the prediction that $C_V = \tfrac{3}{2}R$ agrees with experiment for monatomic gases but fails for diatomic and polyatomic gases. Let us try to explain the discrepancy by considering the possibility that molecules with more than one atom can store internal energy in forms other than translational kinetic energy.

Figure 19-13 shows common models of helium (a *monatomic* molecule, containing a single atom), oxygen (a *diatomic* molecule, containing two atoms), and

**Table 19-3** Degrees of Freedom for Various Molecules

| Molecule | Example | Degrees of Freedom | | | Predicted Molar Specific Heats | |
|---|---|---|---|---|---|---|
| | | Translational | Rotational | Total ($f$) | $C_V$ (Eq. 19-51) | $C_p = C_V + R$ |
| Monatomic | He | 3 | 0 | 3 | $\frac{3}{2}R$ | $\frac{5}{2}R$ |
| Diatomic | $O_2$ | 3 | 2 | 5 | $\frac{5}{2}R$ | $\frac{7}{2}R$ |
| Polyatomic | $CH_4$ | 3 | 3 | 6 | $3R$ | $4R$ |

methane (a *polyatomic* molecule). From such models, we would assume that all three types of molecules can have translational motions (say, moving left–right and up–down) and rotational motions (spinning about an axis like a top). In addition, we would assume that the diatomic and polyatomic molecules can have oscillatory motions, with the atoms oscillating slightly toward and away from one another, as if attached to opposite ends of a spring.

To keep account of the various ways in which energy can be stored in a gas, James Clerk Maxwell introduced the theorem of the **equipartition of energy:**

> Every kind of molecule has a certain number $f$ of *degrees of freedom*, which are independent ways in which the molecule can store energy. Each such degree of freedom has associated with it—on average—an energy of $\frac{1}{2}kT$ per molecule (or $\frac{1}{2}RT$ per mole).

Let us apply the theorem to the translational and rotational motions of the molecules in Fig. 19-13. (We discuss oscillatory motion below.) For the translational motion, superimpose an *xyz* coordinate system on any gas. The molecules will, in general, have velocity components along all three axes. Thus, gas molecules of all types have three degrees of translational freedom (three ways to move in translation) and, on average, an associated energy of $3(\frac{1}{2}kT)$ per molecule.

For the rotational motion, imagine the origin of our *xyz* coordinate system at the center of each molecule in Fig. 19-13. In a gas, each molecule should be able to rotate with an angular velocity component along each of the three axes, so each gas should have three degrees of rotational freedom and, on average, an additional energy of $3(\frac{1}{2}kT)$ per molecule. *However,* experiment shows this is true only for the polyatomic molecules. According to *quantum theory,* the physics dealing with the allowed motions and energies of molecules and atoms, a monatomic gas molecule does not rotate and so has no rotational energy (a single atom cannot rotate like a top). A diatomic molecule can rotate like a top only about axes perpendicular to the line connecting the atoms (the axes are shown in Fig. 19-13b) and not about that line itself. Therefore, a diatomic molecule can have only two degrees of rotational freedom and a rotational energy of only $2(\frac{1}{2}kT)$ per molecule.

To extend our analysis of molar specific heats ($C_p$ and $C_V$ in Module 19-7) to ideal diatomic and polyatomic gases, it is necessary to retrace the derivations of that analysis in detail. First, we replace Eq. 19-38 ($E_{int} = \frac{3}{2}nRT$) with $E_{int} = (f/2)nRT$, where $f$ is the number of degrees of freedom listed in Table 19-3. Doing so leads to the prediction

$$C_V = \left(\frac{f}{2}\right)R = 4.16f \text{ J/mol}\cdot\text{K}, \qquad (19\text{-}51)$$

which agrees—as it must—with Eq. 19-43 for monatomic gases ($f = 3$). As Table 19-2 shows, this prediction also agrees with experiment for diatomic gases ($f = 5$), but it is too low for polyatomic gases ($f = 6$ for molecules comparable to $CH_4$).

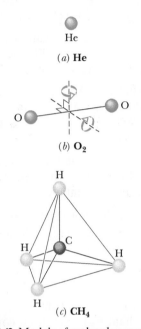

He

(*a*) **He**

(*b*) **O₂**

(*c*) **CH₄**

**Figure 19-13** Models of molecules as used in kinetic theory: (*a*) helium, a typical monatomic molecule; (*b*) oxygen, a typical diatomic molecule; and (*c*) methane, a typical polyatomic molecule. The spheres represent atoms, and the lines between them represent bonds. Two rotation axes are shown for the oxygen molecule.

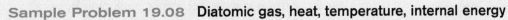

**Sample Problem 19.08** **Diatomic gas, heat, temperature, internal energy**

We transfer 1000 J as heat $Q$ to a diatomic gas, allowing the gas to expand with the pressure held constant. The gas molecules each rotate around an internal axis but do not oscillate. How much of the 1000 J goes into the increase of the gas's internal energy? Of that amount, how much goes into $\Delta K_{tran}$ (the kinetic energy of the translational motion of the molecules) and $\Delta K_{rot}$ (the kinetic energy of their rotational motion)?

## KEY IDEAS

1. The transfer of energy as heat $Q$ to a gas under constant pressure is related to the resulting temperature increase $\Delta T$ via Eq. 19-46 ($Q = nC_p \Delta T$).

2. Because the gas is diatomic with molecules undergoing rotation but not oscillation, the molar specific heat is, from Fig. 19-12 and Table 19-3, $C_p = \frac{7}{2}R$.

3. The increase $\Delta E_{int}$ in the internal energy is the same as would occur with a constant-volume process resulting in the same $\Delta T$. Thus, from Eq. 19-45, $\Delta E_{int} = nC_V \Delta T$. From Fig. 19-12 and Table 19-3, we see that $C_V = \frac{5}{2}R$.

4. For the same $n$ and $\Delta T$, $\Delta E_{int}$ is greater for a diatomic gas than for a monatomic gas because additional energy is required for rotation.

**Increase in $E_{int}$:** Let's first get the temperature change $\Delta T$ due to the transfer of energy as heat. From Eq. 19-46, substituting $\frac{7}{2}R$ for $C_p$, we have

$$\Delta T = \frac{Q}{\frac{7}{2}nR}. \qquad (19\text{-}52)$$

We next find $\Delta E_{int}$ from Eq. 19-45, substituting the molar specific heat $C_V (= \frac{5}{2}R)$ for a constant-volume process and using the same $\Delta T$. Because we are dealing with a diatomic gas, let's call this change $\Delta E_{int,dia}$. Equation 19-45 gives us

$$\Delta E_{int,dia} = nC_V \Delta T = n\frac{5}{2}R\left(\frac{Q}{\frac{7}{2}nR}\right) = \frac{5}{7}Q$$

$$= 0.71428Q = 714.3 \text{ J.} \qquad \text{(Answer)}$$

In words, about 71% of the energy transferred to the gas goes into the internal energy. The rest goes into the work required to increase the volume of the gas, as the gas pushes the walls of its container outward.

**Increases in K:** If we were to increase the temperature of a *monatomic* gas (with the same value of $n$) by the amount given in Eq. 19-52, the internal energy would change by a smaller amount, call it $\Delta E_{int,mon}$, because rotational motion is not involved. To calculate that smaller amount, we still use Eq. 19-45 but now we substitute the value of $C_V$ for a monatomic gas—namely, $C_V = \frac{3}{2}R$. So,

$$\Delta E_{int,mon} = n\frac{3}{2}R \Delta T.$$

Substituting for $\Delta T$ from Eq. 19-52 leads us to

$$\Delta E_{int,mon} = n\frac{3}{2}R\left(\frac{Q}{n\frac{7}{2}R}\right) = \frac{3}{7}Q$$

$$= 0.42857Q = 428.6 \text{ J.}$$

For the monatomic gas, all this energy would go into the kinetic energy of the translational motion of the atoms. The important point here is that for a diatomic gas with the same values of $n$ and $\Delta T$, the same amount of energy goes into the kinetic energy of the translational motion of the molecules. The rest of $\Delta E_{int,dia}$ (that is, the additional 285.7 J) goes into the rotational motion of the molecules. Thus, for the diatomic gas,

$$\Delta K_{trans} = 428.6 \text{ J} \quad \text{and} \quad \Delta K_{rot} = 285.7 \text{ J.} \qquad \text{(Answer)}$$

 **PLUS** Additional examples, video, and practice available at *WileyPLUS*

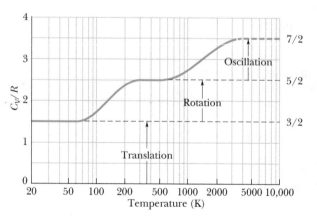

**Figure 19-14** $C_V/R$ versus temperature for (diatomic) hydrogen gas. Because rotational and oscillatory motions begin at certain energies, only translation is possible at very low temperatures. As the temperature increases, rotational motion can begin. At still higher temperatures, oscillatory motion can begin.

## A Hint of Quantum Theory

We can improve the agreement of kinetic theory with experiment by including the oscillations of the atoms in a gas of diatomic or polyatomic molecules. For example, the two atoms in the $O_2$ molecule of Fig. 19-13b can oscillate toward and away from each other, with the interconnecting bond acting like a spring. However, experiment shows that such oscillations occur only at relatively high temperatures of the gas—the motion is "turned on" only when the gas molecules have relatively large energies. Rotational motion is also subject to such "turning on," but at a lower temperature.

Figure 19-14 is of help in seeing this turning on of rotational motion and oscillatory motion. The ratio $C_V/R$ for diatomic hydrogen gas ($H_2$) is plotted there against temperature, with the temperature scale logarithmic to cover several orders of magnitude. Below about 80 K, we find that $C_V/R = 1.5$. This result implies that only the three translational degrees of freedom of hydrogen are involved in the specific heat.

As the temperature increases, the value of $C_V/R$ gradually increases to 2.5, implying that two additional degrees of freedom have become involved. Quantum theory shows that these two degrees of freedom are associated with the rotational motion of the hydrogen molecules and that this motion requires a certain minimum amount of energy. At very low temperatures (below 80 K), the molecules do not have enough energy to rotate. As the temperature increases from 80 K, first a few molecules and then more and more of them obtain enough energy to rotate, and the value of $C_V/R$ increases, until all of the molecules are rotating and $C_V/R = 2.5$.

Similarly, quantum theory shows that oscillatory motion of the molecules requires a certain (higher) minimum amount of energy. This minimum amount is not met until the molecules reach a temperature of about 1000 K, as shown in Fig. 19-14. As the temperature increases beyond 1000 K, more and more molecules have enough energy to oscillate and the value of $C_V/R$ increases, until all of the molecules are oscillating and $C_V/R = 3.5$. (In Fig. 19-14, the plotted curve stops at 3200 K because there the atoms of a hydrogen molecule oscillate so much that they overwhelm their bond, and the molecule then *dissociates* into two separate atoms.)

The turning on of the rotation and vibration of the diatomic and polyatomic molecules is due to the fact that the energies of these motions are quantized, that is, restricted to certain values. There is a lowest allowed value for each type of motion. Unless the thermal agitation of the surrounding molecules provides those lowest amounts, a molecule simply cannot rotate or vibrate.

## 19-9 THE ADIABATIC EXPANSION OF AN IDEAL GAS

### Learning Objectives

*After reading this module, you should be able to . . .*

**19.44** On a $p$-$V$ diagram, sketch an adiabatic expansion (or contraction) and identify that there is no heat exchange $Q$ with the environment.

**19.45** Identify that in an adiabatic expansion, the gas does work on the environment, decreasing the gas's internal energy, and that in an adiabatic contraction, work is done on the gas, increasing the internal energy.

**19.46** In an adiabatic expansion or contraction, relate the initial pressure and volume to the final pressure and volume.

**19.47** In an adiabatic expansion or contraction, relate the initial temperature and volume to the final temperature and volume.

**19.48** Calculate the work done in an adiabatic process by integrating the pressure with respect to volume.

**19.49** Identify that a free expansion of a gas into a vacuum is adiabatic but no work is done and thus, by the first law of thermodynamics, the internal energy and temperature of the gas do not change.

### Key Ideas

● When an ideal gas undergoes a slow adiabatic volume change (a change for which $Q = 0$),

$$pV^\gamma = \text{a constant} \quad \text{(adiabatic process)},$$

in which $\gamma \, (= C_p/C_V)$ is the ratio of molar specific heats for the gas.

● For a free expansion, $pV = $ a constant.

### The Adiabatic Expansion of an Ideal Gas

We saw in Module 17-2 that sound waves are propagated through air and other gases as a series of compressions and expansions; these variations in the transmission medium take place so rapidly that there is no time for energy to be transferred from one part of the medium to another as heat. As we saw in Module 18-5, a process for which $Q = 0$ is an *adiabatic process*. We can ensure that $Q = 0$ either by carrying out the process very quickly (as in sound waves) or by doing it (at any rate) in a well-insulated container.

Figure 19-15$a$ shows our usual insulated cylinder, now containing an ideal gas and resting on an insulating stand. By removing mass from the piston, we can allow the gas to expand adiabatically. As the volume increases, both the pressure and the temperature drop. We shall prove next that the relation between the pressure and the volume during such an adiabatic process is

$$pV^{\gamma} = \text{a constant} \qquad \text{(adiabatic process)}, \qquad (19\text{-}53)$$

in which $\gamma = C_p/C_V$, the ratio of the molar specific heats for the gas. On a $p$-$V$ diagram such as that in Fig. 19-15$b$, the process occurs along a line (called an *adiabat*) that has the equation $p = (\text{a constant})/V^{\gamma}$. Since the gas goes from an initial state $i$ to a final state $f$, we can rewrite Eq. 19-53 as

$$p_iV_i^{\gamma} = p_fV_f^{\gamma} \qquad \text{(adiabatic process)}. \qquad (19\text{-}54)$$

To write an equation for an adiabatic process in terms of $T$ and $V$, we use the ideal gas equation ($pV = nRT$) to eliminate $p$ from Eq. 19-53, finding

$$\left(\frac{nRT}{V}\right)V^{\gamma} = \text{a constant}.$$

Because $n$ and $R$ are constants, we can rewrite this in the alternative form

$$TV^{\gamma-1} = \text{a constant} \qquad \text{(adiabatic process)}, \qquad (19\text{-}55)$$

in which the constant is different from that in Eq. 19-53. When the gas goes from an initial state $i$ to a final state $f$, we can rewrite Eq. 19-55 as

$$T_iV_i^{\gamma-1} = T_fV_f^{\gamma-1} \qquad \text{(adiabatic process)}. \qquad (19\text{-}56)$$

Understanding adiabatic processes allows you to understand why popping the cork on a cold bottle of champagne or the tab on a cold can of soda causes a slight fog to form at the opening of the container. At the top of any unopened carbonated drink sits a gas of carbon dioxide and water vapor. Because the pressure of that gas is much greater than atmospheric pressure, the gas expands out into the atmosphere when the container is opened. Thus, the gas volume increases, but that means the gas must do work pushing against the atmosphere. Because the expansion is rapid, it is adiabatic, and the only source of energy for the work is the internal energy of the gas. Because the internal energy decreases,

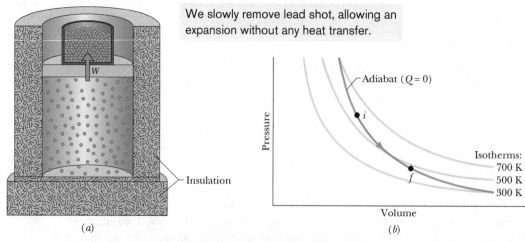

We slowly remove lead shot, allowing an expansion without any heat transfer.

$W$

Insulation

Adiabat ($Q = 0$)

$i$

$f$

Isotherms:
700 K
500 K
300 K

Pressure

Volume

($a$)

($b$)

**Figure 19-15** ($a$) The volume of an ideal gas is increased by removing mass from the piston. The process is adiabatic ($Q = 0$). ($b$) The process proceeds from $i$ to $f$ along an adiabat on a $p$-$V$ diagram.

the temperature of the gas also decreases and so does the number of water molecules that can remain as a vapor. So, lots of the water molecules condense into tiny drops of fog.

### Proof of Eq. 19-53

Suppose that you remove some shot from the piston of Fig. 19-15a, allowing the ideal gas to push the piston and the remaining shot upward and thus to increase the volume by a differential amount $dV$. Since the volume change is tiny, we may assume that the pressure $p$ of the gas on the piston is constant during the change. This assumption allows us to say that the work $dW$ done by the gas during the volume increase is equal to $p\,dV$. From Eq. 18-27, the first law of thermodynamics can then be written as

$$dE_{\text{int}} = Q - p\,dV. \tag{19-57}$$

Since the gas is thermally insulated (and thus the expansion is adiabatic), we substitute 0 for $Q$. Then we use Eq. 19-45 to substitute $nC_V\,dT$ for $dE_{\text{int}}$. With these substitutions, and after some rearranging, we have

$$n\,dT = -\left(\frac{p}{C_V}\right)dV. \tag{19-58}$$

Now from the ideal gas law ($pV = nRT$) we have

$$p\,dV + V\,dp = nR\,dT. \tag{19-59}$$

Replacing $R$ with its equal, $C_p - C_V$, in Eq. 19-59 yields

$$n\,dT = \frac{p\,dV + V\,dp}{C_p - C_V}. \tag{19-60}$$

Equating Eqs. 19-58 and 19-60 and rearranging then give

$$\frac{dp}{p} + \left(\frac{C_p}{C_V}\right)\frac{dV}{V} = 0.$$

Replacing the ratio of the molar specific heats with $\gamma$ and integrating (see integral 5 in Appendix E) yield

$$\ln p + \gamma \ln V = \text{a constant.}$$

Rewriting the left side as $\ln pV^\gamma$ and then taking the antilog of both sides, we find

$$pV^\gamma = \text{a constant.} \tag{19-61}$$

### Free Expansions

Recall from Module 18-5 that a free expansion of a gas is an adiabatic process with *no* work or change in internal energy. Thus, a free expansion differs from the adiabatic process described by Eqs. 19-53 through 19-61, in which work is done and the internal energy changes. Those equations then do *not* apply to a free expansion, even though such an expansion is adiabatic.

Also recall that in a free expansion, a gas is in equilibrium only at its initial and final points; thus, we can plot only those points, but not the expansion itself, on a p-V diagram. In addition, because $\Delta E_{\text{int}} = 0$, the temperature of the final state must be that of the initial state. Thus, the initial and final points on a p-V diagram must be on the same isotherm, and instead of Eq. 19-56 we have

$$T_i = T_f \quad \text{(free expansion).} \tag{19-62}$$

If we next assume that the gas is ideal (so that $pV = nRT$), then because there is no change in temperature, there can be no change in the product $pV$. Thus, instead of Eq. 19-53 a free expansion involves the relation

$$p_iV_i = p_fV_f \quad \text{(free expansion)}. \tag{19-63}$$

---

**Sample Problem 19.09** **Work done by a gas in an adiabatic expansion**

Initially an ideal diatomic gas has pressure $p_i = 2.00 \times 10^5$ Pa and volume $V_i = 4.00 \times 10^{-6}$ m³. How much work $W$ does it do, and what is the change $\Delta E_{int}$ in its internal energy if it expands adiabatically to volume $V_f = 8.00 \times 10^{-6}$ m³? Throughout the process, the molecules have rotation but not oscillation.

**KEY IDEAS**

(1) In an adiabatic expansion, no heat is exchanged between the gas and its environment, and the energy for the work done by the gas comes from the internal energy. (2) The final pressure and volume are related to the initial pressure and volume by Eq. 19-54 ($p_iV_i^{\gamma} = p_fV_f^{\gamma}$). (3) The work done by a gas in any process can be calculated by integrating the pressure with respect to the volume (the work is due to the gas pushing the walls of its container outward).

*Calculations:* We want to calculate the work by filling out this integration,

$$W = \int_{V_i}^{V_f} p \, dV, \tag{19-64}$$

but we first need an expression for the pressure as a function of volume (so that we integrate the expression with respect to volume). So, let's rewrite Eq. 19-54 with indefinite symbols (dropping the subscripts $f$) as

$$p = \frac{1}{V^{\gamma}} p_i V_i^{\gamma} = V^{-\gamma} p_i V_i^{\gamma}. \tag{19-65}$$

The initial quantities are given constants but the pressure $p$ is a function of the variable volume $V$. Substituting this

expression into Eq. 19-64 and integrating lead us to

$$W = \int_{V_i}^{V_f} p \, dV = \int_{V_i}^{V_f} V^{-\gamma} p_i V_i^{\gamma} \, dV$$

$$= p_i V_i^{\gamma} \int_{V_i}^{V_f} V^{-\gamma} dV = \frac{1}{-\gamma + 1} p_i V_i^{\gamma} [V^{-\gamma+1}]_{V_i}^{V_f}$$

$$= \frac{1}{-\gamma + 1} p_i V_i^{\gamma} [V_f^{-\gamma+1} - V_i^{-\gamma+1}]. \tag{19-66}$$

Before we substitute in given data, we must determine the ratio $\gamma$ of molar specific heats for a gas of diatomic molecules with rotation but no oscillation. From Table 19-3 we find

$$\gamma = \frac{C_p}{C_V} = \frac{\frac{7}{2}R}{\frac{5}{2}R} = 1.4. \tag{19-67}$$

We can now write the work done by the gas as the following (with volume in cubic meters and pressure in pascals):

$$W = \frac{1}{-1.4 + 1} (2.00 \times 10^5)(4.00 \times 10^{-6})^{1.4}$$

$$\times [(8.00 \times 10^{-6})^{-1.4+1} - (4.00 \times 10^{-6})^{-1.4+1}]$$

$$= 0.48 \text{ J.} \qquad \text{(Answer)}$$

The first law of thermodynamics (Eq. 18-26) tells us that $\Delta E_{int} = Q - W$. Because $Q = 0$ in the adiabatic expansion, we see that

$$\Delta E_{int} = -0.48 \text{ J.} \qquad \text{(Answer)}$$

With this decrease in internal energy, the gas temperature must also decrease because of the expansion.

---

**Sample Problem 19.10** **Adiabatic expansion, free expansion**

Initially, 1 mol of oxygen (assumed to be an ideal gas) has temperature 310 K and volume 12 L. We will allow it to expand to volume 19 L.

(a) What would be the final temperature if the gas expands adiabatically? Oxygen ($O_2$) is diatomic and here has rotation but not oscillation.

**KEY IDEAS**

**1.** When a gas expands against the pressure of its environment, it must do work.

**2.** When the process is adiabatic (no energy is transferred as heat), then the energy required for the work can come only from the internal energy of the gas.

**3.** Because the internal energy decreases, the temperature $T$ must also decrease.

*Calculations:* We can relate the initial and final temperatures and volumes with Eq. 19-56:

$$T_i V_i^{\gamma-1} = T_f V_f^{\gamma-1}. \tag{19-68}$$

Because the molecules are diatomic and have rotation but not oscillation, we can take the molar specific heats from

Table 19-3. Thus,

$$\gamma = \frac{C_p}{C_V} = \frac{\frac{7}{2}R}{\frac{5}{2}R} = 1.40.$$

Solving Eq. 19-68 for $T_f$ and inserting known data then yield

$$T_f = \frac{T_i V_i^{\gamma-1}}{V_f^{\gamma-1}} = \frac{(310 \text{ K})(12 \text{ L})^{1.40-1}}{(19 \text{ L})^{1.40-1}}$$

$$= (310 \text{ K})(\tfrac{12}{19})^{0.40} = 258 \text{ K}. \qquad \text{(Answer)}$$

(b) What would be the final temperature and pressure if, instead, the gas expands freely to the new volume, from an initial pressure of 2.0 Pa?

## KEY IDEA

The temperature does not change in a free expansion because there is nothing to change the kinetic energy of the molecules.

*Calculation:* Thus, the temperature is

$$T_f = T_i = 310 \text{ K}. \qquad \text{(Answer)}$$

We find the new pressure using Eq. 19-63, which gives us

$$p_f = p_i \frac{V_i}{V_f} = (2.0 \text{ Pa}) \frac{12 \text{ L}}{19 \text{ L}} = 1.3 \text{ Pa}. \qquad \text{(Answer)}$$

**Problem-Solving Tactics**   **A Graphical Summary of Four Gas Processes**

In this chapter we have discussed four special processes that an ideal gas can undergo. An example of each (for a monatomic ideal gas) is shown in Fig. 19-16, and some associated characteristics are given in Table 19-4, including two process names (isobaric and isochoric) that we have not used but that you might see in other courses.

 **Checkpoint 5**

Rank paths 1, 2, and 3 in Fig. 19-16 according to the energy transfer to the gas as heat, greatest first.

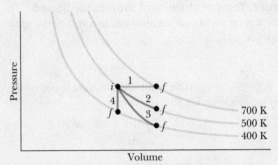

**Figure 19-16** A $p$-$V$ diagram representing four special processes for an ideal monatomic gas.

**Table 19-4** **Four Special Processes**

| Path in Fig. 19-16 | Constant Quantity | Process Type | Some Special Results ($\Delta E_{int} = Q - W$ and $\Delta E_{int} = nC_V\Delta T$ for all paths) |
|---|---|---|---|
| 1 | $p$ | Isobaric | $Q = nC_p\Delta T; W = p\Delta V$ |
| 2 | $T$ | Isothermal | $Q = W = nRT\ln(V_f/V_i); \Delta E_{int} = 0$ |
| 3 | $pV^\gamma, TV^{\gamma-1}$ | Adiabatic | $Q = 0; \quad W = -\Delta E_{int}$ |
| 4 | $V$ | Isochoric | $Q = \Delta E_{int} = nC_V\Delta T; \quad W = 0$ |

 Additional examples, video, and practice available at *WileyPLUS*

# Review & Summary

**Kinetic Theory of Gases**   The *kinetic theory of gases* relates the *macroscopic* properties of gases (for example, pressure and temperature) to the *microscopic* properties of gas molecules (for example, speed and kinetic energy).

**Avogadro's Number**   One mole of a substance contains $N_A$ (*Avogadro's number*) elementary units (usually atoms or molecules), where $N_A$ is found experimentally to be

$$N_A = 6.02 \times 10^{23} \text{ mol}^{-1} \quad \text{(Avogadro's number).} \qquad (19\text{-}1)$$

One molar mass $M$ of any substance is the mass of one mole of the substance. It is related to the mass $m$ of the individual molecules of the substance by

$$M = mN_A. \qquad (19\text{-}4)$$

The number of moles $n$ contained in a sample of mass $M_{sam}$, consisting of $N$ molecules, is given by

$$n = \frac{N}{N_A} = \frac{M_{sam}}{M} = \frac{M_{sam}}{mN_A}. \qquad (19\text{-}2, 19\text{-}3)$$

**Ideal Gas** An *ideal gas* is one for which the pressure $p$, volume $V$, and temperature $T$ are related by

$$pV = nRT \quad \text{(ideal gas law)}. \tag{19-5}$$

Here $n$ is the number of moles of the gas present and $R$ is a constant (8.31 J/mol·K) called the **gas constant**. The ideal gas law can also be written as

$$pV = NkT, \tag{19-9}$$

where the **Boltzmann constant** $k$ is

$$k = \frac{R}{N_A} = 1.38 \times 10^{-23} \text{ J/K}. \tag{19-7}$$

**Work in an Isothermal Volume Change** The work done *by* an ideal gas during an **isothermal** (constant-temperature) change from volume $V_i$ to volume $V_f$ is

$$W = nRT \ln \frac{V_f}{V_i} \quad \text{(ideal gas, isothermal process)}. \tag{19-14}$$

**Pressure, Temperature, and Molecular Speed** The pressure exerted by $n$ moles of an ideal gas, in terms of the speed of its molecules, is

$$p = \frac{nMv_{rms}^2}{3V}, \tag{19-21}$$

where $v_{rms} = \sqrt{(v^2)_{avg}}$ is the **root-mean-square speed** of the molecules of the gas. With Eq. 19-5 this gives

$$v_{rms} = \sqrt{\frac{3RT}{M}}. \tag{19-22}$$

**Temperature and Kinetic Energy** The average translational kinetic energy $K_{avg}$ per molecule of an ideal gas is

$$K_{avg} = \tfrac{3}{2}kT. \tag{19-24}$$

**Mean Free Path** The *mean free path* $\lambda$ of a gas molecule is its average path length between collisions and is given by

$$\lambda = \frac{1}{\sqrt{2}\pi d^2 \, N/V}, \tag{19-25}$$

where $N/V$ is the number of molecules per unit volume and $d$ is the molecular diameter.

**Maxwell Speed Distribution** The *Maxwell speed distribution* $P(v)$ is a function such that $P(v)\,dv$ gives the *fraction* of molecules with speeds in the interval $dv$ at speed $v$:

$$P(v) = 4\pi \left( \frac{M}{2\pi RT} \right)^{3/2} v^2 e^{-Mv^2/2RT}. \tag{19-27}$$

Three measures of the distribution of speeds among the molecules of a gas are

$$v_{avg} = \sqrt{\frac{8RT}{\pi M}} \quad \text{(average speed)}, \tag{19-31}$$

$$v_P = \sqrt{\frac{2RT}{M}} \quad \text{(most probable speed)}, \tag{19-35}$$

and the rms speed defined above in Eq. 19-22.

**Molar Specific Heats** The molar specific heat $C_V$ of a gas at constant volume is defined as

$$C_V = \frac{Q}{n\,\Delta T} = \frac{\Delta E_{int}}{n\,\Delta T}, \tag{19-39, 19-41}$$

in which $Q$ is the energy transferred as heat to or from a sample of $n$ moles of the gas, $\Delta T$ is the resulting temperature change of the gas, and $\Delta E_{int}$ is the resulting change in the internal energy of the gas. For an ideal monatomic gas,

$$C_V = \tfrac{3}{2}R = 12.5 \text{ J/mol·K}. \tag{19-43}$$

The molar specific heat $C_p$ of a gas at constant pressure is defined to be

$$C_p = \frac{Q}{n\,\Delta T}, \tag{19-46}$$

in which $Q$, $n$, and $\Delta T$ are defined as above. $C_p$ is also given by

$$C_p = C_V + R. \tag{19-49}$$

For $n$ moles of an ideal gas,

$$E_{int} = nC_V T \quad \text{(ideal gas)}. \tag{19-44}$$

If $n$ moles of a confined ideal gas undergo a temperature change $\Delta T$ due to *any* process, the change in the internal energy of the gas is

$$\Delta E_{int} = nC_V\,\Delta T \quad \text{(ideal gas, any process)}. \tag{19-45}$$

**Degrees of Freedom and $C_V$** The *equipartition of energy* theorem states that every *degree of freedom* of a molecule has an energy $\tfrac{1}{2}kT$ per molecule $(=\tfrac{1}{2}RT$ per mole$)$. If $f$ is the number of degrees of freedom, then $E_{int} = (f/2)nRT$ and

$$C_V = \left( \frac{f}{2} \right)R = 4.16f \text{ J/mol·K}. \tag{19-51}$$

For monatomic gases $f = 3$ (three translational degrees); for diatomic gases $f = 5$ (three translational and two rotational degrees).

**Adiabatic Process** When an ideal gas undergoes an adiabatic volume change (a change for which $Q = 0$),

$$pV^\gamma = \text{a constant} \quad \text{(adiabatic process)}, \tag{19-53}$$

in which $\gamma\,(= C_p/C_V)$ is the ratio of molar specific heats for the gas. For a free expansion, however, $pV = $ a constant.

# Questions

**1** For four situations for an ideal gas, the table gives the energy transferred to or from the gas as heat $Q$ and either the work $W$ done by the gas or the work $W_{on}$ done on the gas, all in joules. Rank the four situations in terms of the temperature change of the gas, most positive first.

| | a | b | c | d |
|---|---|---|---|---|
| $Q$ | −50 | +35 | −15 | +20 |
| $W$ | −50 | +35 | | |
| $W_{on}$ | | | −40 | +40 |

**2** In the p-V diagram of Fig. 19-17, the gas does 5 J of work when taken along isotherm $ab$ and 4 J when taken along adiabat $bc$. What is the change

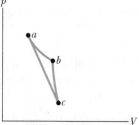

**Figure 19-17** Question 2.

in the internal energy of the gas when it is taken along the straight path from $a$ to $c$?

**3** For a temperature increase of $\Delta T_1$, a certain amount of an ideal gas requires 30 J when heated at constant volume and 50 J when heated at constant pressure. How much work is done by the gas in the second situation?

**4** The dot in Fig. 19-18$a$ represents the initial state of a gas, and the vertical line through the dot divides the $p$-$V$ diagram into regions 1 and 2. For the following processes, determine whether the work $W$ done by the gas is positive, negative, or zero: (a) the gas moves up along the vertical line, (b) it moves down along the vertical line, (c) it moves to anywhere in region 1, and (d) it moves to anywhere in region 2.

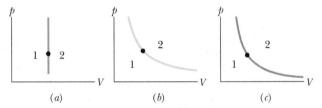

**Figure 19-18** Questions 4, 6, and 8.

**5** A certain amount of energy is to be transferred as heat to 1 mol of a monatomic gas (a) at constant pressure and (b) at constant volume, and to 1 mol of a diatomic gas (c) at constant pressure and (d) at constant volume. Figure 19-19 shows four paths from an initial point to four

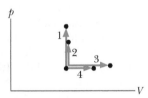

**Figure 19-19** Question 5.

final points on a $p$-$V$ diagram for the two gases. Which path goes with which process? (e) Are the molecules of the diatomic gas rotating?

**6** The dot in Fig. 19-18$b$ represents the initial state of a gas, and the isotherm through the dot divides the $p$-$V$ diagram into regions 1 and 2. For the following processes, determine whether the change $\Delta E_{int}$ in the internal energy of the gas is positive, negative, or zero: (a) the gas moves up along the isotherm, (b) it moves down along the isotherm, (c) it moves to anywhere in region 1, and (d) it moves to anywhere in region 2.

**7** (a) Rank the four paths of Fig. 19-16 according to the work done by the gas, greatest first. (b) Rank paths 1, 2, and 3 according to the change in the internal energy of the gas, most positive first and most negative last.

**8** The dot in Fig. 19-18$c$ represents the initial state of a gas, and the adiabat through the dot divides the $p$-$V$ diagram into regions 1 and 2. For the following processes, determine whether the corresponding heat $Q$ is positive, negative, or zero: (a) the gas moves up along the adiabat, (b) it moves down along the adiabat, (c) it moves to anywhere in region 1, and (d) it moves to anywhere in region 2.

**9** An ideal diatomic gas, with molecular rotation but without any molecular oscillation, loses a certain amount of energy as heat $Q$. Is the resulting decrease in the internal energy of the gas greater if the loss occurs in a constant-volume process or in a constant-pressure process?

**10** Does the temperature of an ideal gas increase, decrease, or stay the same during (a) an isothermal expansion, (b) an expansion at constant pressure, (c) an adiabatic expansion, and (d) an increase in pressure at constant volume?

## Problems

| | |
|---|---|
| **GO** Tutoring problem available (at instructor's discretion) in *WileyPLUS* and WebAssign | |
| **SSM** Worked-out solution available in Student Solutions Manual | **WWW** Worked-out solution is at |
| • – ••• Number of dots indicates level of problem difficulty | **ILW** Interactive solution is at |
| ✈ Additional information available in *The Flying Circus of Physics* and at flyingcircusofphysics.com | http://www.wiley.com/college/halliday |

### Module 19-1 Avogadro's Number

**•1** Find the mass in kilograms of $7.50 \times 10^{24}$ atoms of arsenic, which has a molar mass of 74.9 g/mol.

**•2** Gold has a molar mass of 197 g/mol. (a) How many moles of gold are in a 2.50 g sample of pure gold? (b) How many atoms are in the sample?

### Module 19-2 Ideal Gases

**•3 SSM** Oxygen gas having a volume of 1000 cm$^3$ at 40.0°C and $1.01 \times 10^5$ Pa expands until its volume is 1500 cm$^3$ and its pressure is $1.06 \times 10^5$ Pa. Find (a) the number of moles of oxygen present and (b) the final temperature of the sample.

**•4** A quantity of ideal gas at 10.0°C and 100 kPa occupies a volume of 2.50 m$^3$. (a) How many moles of the gas are present? (b) If the pressure is now raised to 300 kPa and the temperature is raised to 30.0°C, how much volume does the gas occupy? Assume no leaks.

**•5** The best laboratory vacuum has a pressure of about $1.00 \times 10^{-18}$ atm, or $1.01 \times 10^{-13}$ Pa. How many gas molecules are there per cubic centimeter in such a vacuum at 293 K?

**•6** ✈ *Water bottle in a hot car.* In the American Southwest, the temperature in a closed car parked in sunlight during the summer can be high enough to burn flesh. Suppose a bottle of water at a refrigerator temperature of 5.00°C is opened, then closed, and then left in a closed car with an internal temperature of 75.0°C. Neglecting the thermal expansion of the water and the bottle, find the pressure in the air pocket trapped in the bottle. (The pressure can be enough to push the bottle cap past the threads that are intended to keep the bottle closed.)

**•7** Suppose 1.80 mol of an ideal gas is taken from a volume of 3.00 m$^3$ to a volume of 1.50 m$^3$ via an isothermal compression at 30°C. (a) How much energy is transferred as heat during the compression, and (b) is the transfer *to* or *from* the gas?

**•8** Compute (a) the number of moles and (b) the number of molecules in 1.00 cm$^3$ of an ideal gas at a pressure of 100 Pa and a temperature of 220 K.

**•9** An automobile tire has a volume of $1.64 \times 10^{-2}$ m$^3$ and contains air at a gauge pressure (pressure above atmospheric pressure) of 165 kPa when the temperature is 0.00°C. What is the gauge pressure of the air in the tires when its temperature rises to 27.0°C and its volume increases to $1.67 \times 10^{-2}$ m³? Assume atmospheric pressure is $1.01 \times 10^5$ Pa.

•10 A container encloses 2 mol of an ideal gas that has molar mass $M_1$ and 0.5 mol of a second ideal gas that has molar mass $M_2 = 3M_1$. What fraction of the total pressure on the container wall is attributable to the second gas? (The kinetic theory explanation of pressure leads to the experimentally discovered law of partial pressures for a mixture of gases that do not react chemically: *The total pressure exerted by the mixture is equal to the sum of the pressures that the several gases would exert separately if each were to occupy the vessel alone.* The molecule–vessel collisions of one type would not be altered by the presence of another type.)

••11 SSM ILW WWW Air that initially occupies 0.140 m³ at a gauge pressure of 103.0 kPa is expanded isothermally to a pressure of 101.3 kPa and then cooled at constant pressure until it reaches its initial volume. Compute the work done by the air. (Gauge pressure is the difference between the actual pressure and atmospheric pressure.)

••12 ✈ GO *Submarine rescue.* When the U.S. submarine *Squalus* became disabled at a depth of 80 m, a cylindrical chamber was lowered from a ship to rescue the crew. The chamber had a radius of 1.00 m and a height of 4.00 m, was open at the bottom, and held two rescuers. It slid along a guide cable that a diver had attached to a hatch on the submarine. Once the chamber reached the hatch and clamped to the hull, the crew could escape into the chamber. During the descent, air was released from tanks to prevent water from flooding the chamber. Assume that the interior air pressure matched the water pressure at depth $h$ as given by $p_0 + \rho g h$, where $p_0 = 1.000$ atm is the surface pressure and $\rho = 1024$ kg/m³ is the density of seawater. Assume a surface temperature of 20.0°C and a submerged water temperature of −30.0°C. (a) What is the air volume in the chamber at the surface? (b) If air had not been released from the tanks, what would have been the air volume in the chamber at depth $h = 80.0$ m? (c) How many moles of air were needed to be released to maintain the original air volume in the chamber?

••13 GO A sample of an ideal gas is taken through the cyclic process *abca* shown in Fig. 19-20. The scale of the vertical axis is set by $p_b = 7.5$ kPa and $p_{ac} = 2.5$ kPa. At point $a$, $T = 200$ K. (a) How many moles of gas are in the sample? What are (b) the temperature of the gas at point $b$, (c) the temperature of the gas at point $c$, and (d) the net energy added to the gas as heat during the cycle?

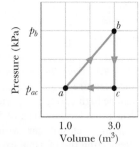

**Figure 19-20** Problem 13.

••14 In the temperature range 310 K to 330 K, the pressure $p$ of a certain nonideal gas is related to volume $V$ and temperature $T$ by

$$p = (24.9 \text{ J/K})\frac{T}{V} - (0.00662 \text{ J/K}^2)\frac{T^2}{V}.$$

How much work is done by the gas if its temperature is raised from 315 K to 325 K while the pressure is held constant?

••15 Suppose 0.825 mol of an ideal gas undergoes an isothermal expansion as energy is added to it as heat $Q$. If Fig. 19-21 shows the final volume $V_f$ versus $Q$, what is the gas temperature? The scale of the vertical axis is set by $V_{fs} = 0.30$ m³, and the scale of the horizontal axis is set by $Q_s = 1200$ J.

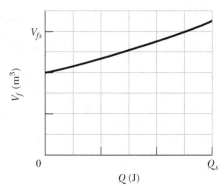

**Figure 19-21** Problem 15.

•••16 An air bubble of volume 20 cm³ is at the bottom of a lake 40 m deep, where the temperature is 4.0°C. The bubble rises to the surface, which is at a temperature of 20°C. Take the temperature of the bubble's air to be the same as that of the surrounding water. Just as the bubble reaches the surface, what is its volume?

•••17 GO Container A in Fig. 19-22 holds an ideal gas at a pressure of $5.0 \times 10^5$ Pa and a temperature of 300 K. It is connected by a thin tube (and a closed valve) to container B, with four times the volume of A. Container B holds the same ideal gas at a pressure of $1.0 \times 10^5$ Pa and a temperature of 400 K. The valve is opened to allow the pressures to equalize, but the temperature of each container is maintained. What then is the pressure?

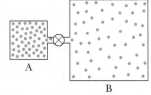

**Figure 19-22** Problem 17.

### Module 19-3 Pressure, Temperature, and RMS Speed

•18 The temperature and pressure in the Sun's atmosphere are $2.00 \times 10^6$ K and 0.0300 Pa. Calculate the rms speed of free electrons (mass $9.11 \times 10^{-31}$ kg) there, assuming they are an ideal gas.

•19 (a) Compute the rms speed of a nitrogen molecule at 20.0°C. The molar mass of nitrogen molecules ($N_2$) is given in Table 19-1. At what temperatures will the rms speed be (b) half that value and (c) twice that value?

•20 Calculate the rms speed of helium atoms at 1000 K. See Appendix F for the molar mass of helium atoms.

•21 SSM The lowest possible temperature in outer space is 2.7 K. What is the rms speed of hydrogen molecules at this temperature? (The molar mass is given in Table 19-1.)

•22 Find the rms speed of argon atoms at 313 K. See Appendix F for the molar mass of argon atoms.

••23 A beam of hydrogen molecules ($H_2$) is directed toward a wall, at an angle of 55° with the normal to the wall. Each molecule in the beam has a speed of 1.0 km/s and a mass of $3.3 \times 10^{-24}$ g. The beam strikes the wall over an area of 2.0 cm², at the rate of $10^{23}$ molecules per second. What is the beam's pressure on the wall?

••24 At 273 K and $1.00 \times 10^{-2}$ atm, the density of a gas is $1.24 \times 10^{-5}$ g/cm³. (a) Find $v_{rms}$ for the gas molecules. (b) Find the molar mass of the gas and (c) identify the gas. See Table 19-1.

### Module 19-4 Translational Kinetic Energy

•25 Determine the average value of the translational kinetic energy of the molecules of an ideal gas at temperatures (a) 0.00°C and (b) 100°C. What is the translational kinetic energy per mole of an ideal gas at (c) 0.00°C and (d) 100°C?

•26   What is the average translational kinetic energy of nitrogen molecules at 1600 K?

••27   Water standing in the open at 32.0°C evaporates because of the escape of some of the surface molecules. The heat of vaporization (539 cal/g) is approximately equal to $\varepsilon n$, where $\varepsilon$ is the average energy of the escaping molecules and $n$ is the number of molecules per gram. (a) Find $\varepsilon$. (b) What is the ratio of $\varepsilon$ to the average kinetic energy of $H_2O$ molecules, assuming the latter is related to temperature in the same way as it is for gases?

### Module 19-5   Mean Free Path

•28   At what frequency would the wavelength of sound in air be equal to the mean free path of oxygen molecules at 1.0 atm pressure and 0.00°C? The molecular diameter is $3.0 \times 10^{-8}$ cm.

•29  SSM   The atmospheric density at an altitude of 2500 km is about 1 molecule/cm³. (a) Assuming the molecular diameter of $2.0 \times 10^{-8}$ cm, find the mean free path predicted by Eq. 19-25. (b) Explain whether the predicted value is meaningful.

•30   The mean free path of nitrogen molecules at 0.0°C and 1.0 atm is $0.80 \times 10^{-5}$ cm. At this temperature and pressure there are $2.7 \times 10^{19}$ molecules/cm³. What is the molecular diameter?

••31   In a certain particle accelerator, protons travel around a circular path of diameter 23.0 m in an evacuated chamber, whose residual gas is at 295 K and $1.00 \times 10^{-6}$ torr pressure. (a) Calculate the number of gas molecules per cubic centimeter at this pressure. (b) What is the mean free path of the gas molecules if the molecular diameter is $2.00 \times 10^{-8}$ cm?

••32   At 20°C and 750 torr pressure, the mean free paths for argon gas (Ar) and nitrogen gas ($N_2$) are $\lambda_{Ar} = 9.9 \times 10^{-6}$ cm and $\lambda_{N_2} = 27.5 \times 10^{-6}$ cm. (a) Find the ratio of the diameter of an Ar atom to that of an $N_2$ molecule. What is the mean free path of argon at (b) 20°C and 150 torr, and (c) −40°C and 750 torr?

### Module 19-6   The Distribution of Molecular Speeds

•33  SSM   The speeds of 10 molecules are 2.0, 3.0, 4.0, . . . , 11 km/s. What are their (a) average speed and (b) rms speed?

•34   The speeds of 22 particles are as follows ($N_i$ represents the number of particles that have speed $v_i$):

| $N_i$ | 2 | 4 | 6 | 8 | 2 |
|---|---|---|---|---|---|
| $v_i$ (cm/s) | 1.0 | 2.0 | 3.0 | 4.0 | 5.0 |

What are (a) $v_{avg}$, (b) $v_{rms}$, and (c) $v_P$?

•35   Ten particles are moving with the following speeds: four at 200 m/s, two at 500 m/s, and four at 600 m/s. Calculate their (a) average and (b) rms speeds. (c) Is $v_{rms} > v_{avg}$?

••36   The most probable speed of the molecules in a gas at temperature $T_2$ is equal to the rms speed of the molecules at temperature $T_1$. Find $T_2/T_1$.

••37  SSM  WWW   Figure 19-23 shows a hypothetical speed distribution for a sample of $N$ gas particles (note that $P(v) = 0$ for speed $v > 2v_0$). What are the values of (a) $av_0$, (b) $v_{avg}/v_0$, and (c) $v_{rms}/v_0$? (d) What fraction of the particles has a speed between $1.5v_0$ and $2.0v_0$?

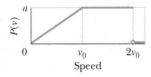

**Figure 19-23**  Problem 37.

••38   Figure 19-24 gives the probability distribution for nitrogen gas. The scale of the horizontal axis is set by $v_s = 1200$ m/s. What are the (a) gas temperature and (b) rms speed of the molecules?

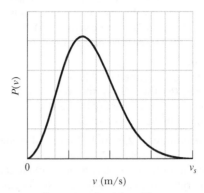

**Figure 19-24**  Problem 38.

••39   At what temperature does the rms speed of (a) $H_2$ (molecular hydrogen) and (b) $O_2$ (molecular oxygen) equal the escape speed from Earth (Table 13-2)? At what temperature does the rms speed of (c) $H_2$ and (d) $O_2$ equal the escape speed from the Moon (where the gravitational acceleration at the surface has magnitude $0.16g$)? Considering the answers to parts (a) and (b), should there be much (e) hydrogen and (f) oxygen high in Earth's upper atmosphere, where the temperature is about 1000 K?

••40   Two containers are at the same temperature. The first contains gas with pressure $p_1$, molecular mass $m_1$, and rms speed $v_{rms1}$. The second contains gas with pressure $2.0p_1$, molecular mass $m_2$, and average speed $v_{avg2} = 2.0v_{rms1}$. Find the mass ratio $m_1/m_2$.

••41   A hydrogen molecule (diameter $1.0 \times 10^{-8}$ cm), traveling at the rms speed, escapes from a 4000 K furnace into a chamber containing *cold* argon atoms (diameter $3.0 \times 10^{-8}$ cm) at a density of $4.0 \times 10^{19}$ atoms/cm³. (a) What is the speed of the hydrogen molecule? (b) If it collides with an argon atom, what is the closest their centers can be, considering each as spherical? (c) What is the initial number of collisions per second experienced by the hydrogen molecule? (*Hint*: Assume that the argon atoms are stationary. Then the mean free path of the hydrogen molecule is given by Eq. 19-26 and not Eq. 19-25.)

### Module 19-7   The Molar Specific Heats of an Ideal Gas

•42   What is the internal energy of 1.0 mol of an ideal monatomic gas at 273 K?

••43  GO   The temperature of 3.00 mol of an ideal diatomic gas is increased by 40.0 C° without the pressure of the gas changing. The molecules in the gas rotate but do not oscillate. (a) How much energy is transferred to the gas as heat? (b) What is the change in the internal energy of the gas? (c) How much work is done by the gas? (d) By how much does the rotational kinetic energy of the gas increase?

••44  GO   One mole of an ideal diatomic gas goes from $a$ to $c$ along the diagonal path in Fig. 19-25. The scale of the vertical axis is set by $p_{ab} = 5.0$ kPa and $p_c = 2.0$ kPa, and the scale of the horizontal axis is set by $V_{bc} = 4.0$ m³ and $V_a = 2.0$ m³. During the transition, (a) what is the change in internal energy of the gas, and (b) how

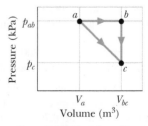

**Figure 19-25**  Problem 44.

much energy is added to the gas as heat? (c) How much heat is required if the gas goes from $a$ to $c$ along the indirect path $abc$?

**••45** **ILW** The mass of a gas molecule can be computed from its specific heat at constant volume $c_V$. (Note that this is not $C_V$.) Take $c_V = 0.075$ cal/g·C° for argon and calculate (a) the mass of an argon atom and (b) the molar mass of argon.

**••46** Under constant pressure, the temperature of 2.00 mol of an ideal monatomic gas is raised 15.0 K. What are (a) the work $W$ done by the gas, (b) the energy transferred as heat $Q$, (c) the change $\Delta E_{int}$ in the internal energy of the gas, and (d) the change $\Delta K$ in the average kinetic energy per atom?

**••47** The temperature of 2.00 mol of an ideal monatomic gas is raised 15.0 K at constant volume. What are (a) the work $W$ done by the gas, (b) the energy transferred as heat $Q$, (c) the change $\Delta E_{int}$ in the internal energy of the gas, and (d) the change $\Delta K$ in the average kinetic energy per atom?

**••48** **GO** When 20.9 J was added as heat to a particular ideal gas, the volume of the gas changed from 50.0 cm³ to 100 cm³ while the pressure remained at 1.00 atm. (a) By how much did the internal energy of the gas change? If the quantity of gas present was $2.00 \times 10^{-3}$ mol, find (b) $C_p$ and (c) $C_V$.

**••49** **SSM** A container holds a mixture of three nonreacting gases: 2.40 mol of gas 1 with $C_{V1} = 12.0$ J/mol·K, 1.50 mol of gas 2 with $C_{V2} = 12.8$ J/mol·K, and 3.20 mol of gas 3 with $C_{V3} = 20.0$ J/mol·K. What is $C_V$ of the mixture?

### Module 19-8 Degrees of Freedom and Molar Specific Heats

**•50** We give 70 J as heat to a diatomic gas, which then expands at constant pressure. The gas molecules rotate but do not oscillate. By how much does the internal energy of the gas increase?

**•51** **ILW** When 1.0 mol of oxygen ($O_2$) gas is heated at constant pressure starting at 0°C, how much energy must be added to the gas as heat to double its volume? (The molecules rotate but do not oscillate.)

**••52** **GO** Suppose 12.0 g of oxygen ($O_2$) gas is heated at constant atmospheric pressure from 25.0°C to 125°C. (a) How many moles of oxygen are present? (See Table 19-1 for the molar mass.) (b) How much energy is transferred to the oxygen as heat? (The molecules rotate but do not oscillate.) (c) What fraction of the heat is used to raise the internal energy of the oxygen?

**••53** **SSM** **WWW** Suppose 4.00 mol of an ideal diatomic gas, with molecular rotation but not oscillation, experienced a temperature increase of 60.0 K under constant-pressure conditions. What are (a) the energy transferred as heat $Q$, (b) the change $\Delta E_{int}$ in internal energy of the gas, (c) the work $W$ done by the gas, and (d) the change $\Delta K$ in the total translational kinetic energy of the gas?

### Module 19-9 The Adiabatic Expansion of an Ideal Gas

**•54** We know that for an adiabatic process $pV^\gamma = $ a constant. Evaluate "a constant" for an adiabatic process involving exactly 2.0 mol of an ideal gas passing through the state having exactly $p = 1.0$ atm and $T = 300$ K. Assume a diatomic gas whose molecules rotate but do not oscillate.

**•55** A certain gas occupies a volume of 4.3 L at a pressure of 1.2 atm and a temperature of 310 K. It is compressed adiabatically to a volume of 0.76 L. Determine (a) the final pressure and (b) the final temperature, assuming the gas to be an ideal gas for which $\gamma = 1.4$.

**•56** Suppose 1.00 L of a gas with $\gamma = 1.30$, initially at 273 K and 1.00 atm, is suddenly compressed adiabatically to half its initial volume. Find its final (a) pressure and (b) temperature. (c) If the gas is then cooled to 273 K at constant pressure, what is its final volume?

**•57** The volume of an ideal gas is adiabatically reduced from 200 L to 74.3 L. The initial pressure and temperature are 1.00 atm and 300 K. The final pressure is 4.00 atm. (a) Is the gas monatomic, diatomic, or polyatomic? (b) What is the final temperature? (c) How many moles are in the gas?

**••58** **GO** *Opening champagne.* In a bottle of champagne, the pocket of gas (primarily carbon dioxide) between the liquid and the cork is at pressure of $p_i = 5.00$ atm. When the cork is pulled from the bottle, the gas undergoes an adiabatic expansion until its pressure matches the ambient air pressure of 1.00 atm. Assume that the ratio of the molar specific heats is $\gamma = \frac{4}{3}$. If the gas has initial temperature $T_i = 5.00°C$, what is its temperature at the end of the adiabatic expansion?

**••59** **GO** Figure 19-26 shows two paths that may be taken by a gas from an initial point $i$ to a final point $f$. Path 1 consists of an isothermal expansion (work is 50 J in magnitude), an adiabatic expansion (work is 40 J in magnitude), an isothermal compression (work is 30 J in magnitude), and then an adiabatic compression (work is 25 J in magnitude). What is the change in the internal energy of the gas if the gas goes from point $i$ to point $f$ along path 2?

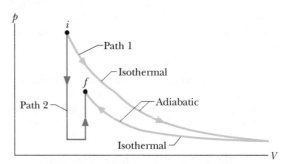

**Figure 19-26** Problem 59.

**••60** **GO** *Adiabatic wind.* The normal airflow over the Rocky Mountains is west to east. The air loses much of its moisture content and is chilled as it climbs the western side of the mountains. When it descends on the eastern side, the increase in pressure toward lower altitudes causes the temperature to increase. The flow, then called a chinook wind, can rapidly raise the air temperature at the base of the mountains. Assume that the air pressure $p$ depends on altitude $y$ according to $p = p_0 \exp(-ay)$, where $p_0 = 1.00$ atm and $a = 1.16 \times 10^{-4}$ m⁻¹. Also assume that the ratio of the molar specific heats is $\gamma = \frac{4}{3}$. A parcel of air with an initial temperature of $-5.00°C$ descends adiabatically from $y_1 = 4267$ m to $y = 1567$ m. What is its temperature at the end of the descent?

**••61** **GO** A gas is to be expanded from initial state $i$ to final state $f$ along either path 1 or path 2 on a $p$-$V$ diagram. Path 1 consists of three steps: an isothermal expansion (work is 40 J in magnitude), an adiabatic expansion (work is 20 J in magnitude), and another isothermal expansion (work is 30 J in magnitude). Path 2 consists of two steps: a pressure reduction at constant volume and an expansion at constant pressure. What is the change in the internal energy of the gas along path 2?

**•••62** **GO** An ideal diatomic gas, with rotation but no oscillation, undergoes an adiabatic compression. Its initial pressure and volume are

1.20 atm and 0.200 m³. Its final pressure is 2.40 atm. How much work is done by the gas?

•••**63**   Figure 19-27 shows a cycle undergone by 1.00 mol of an ideal monatomic gas. The temperatures are $T_1 = 300$ K, $T_2 = 600$ K, and $T_3 = 455$ K. For $1 \rightarrow 2$, what are (a) heat $Q$, (b) the change in internal energy $\Delta E_{int}$, and (c) the work done $W$? For $2 \rightarrow 3$, what are (d) $Q$, (e) $\Delta E_{int}$, and (f) $W$? For $3 \rightarrow 1$, what are (g) $Q$, (h) $\Delta E_{int}$, and (i) $W$? For the full cycle, what are (j) $Q$, (k) $\Delta E_{int}$, and (l) $W$? The initial pressure at point 1 is 1.00 atm ($= 1.013 \times 10^5$ Pa). What are the (m) volume and (n) pressure at point 2 and the (o) volume and (p) pressure at point 3?

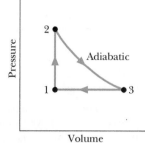

**Figure 19-27**   Problem 63.

## Additional Problems

**64**   Calculate the work done by an external agent during an isothermal compression of 1.00 mol of oxygen from a volume of 22.4 L at 0°C and 1.00 atm to a volume of 16.8 L.

**65**   An ideal gas undergoes an adiabatic compression from $p = 1.0$ atm, $V = 1.0 \times 10^6$ L, $T = 0.0$°C to $p = 1.0 \times 10^5$ atm, $V = 1.0 \times 10^3$ L. (a) Is the gas monatomic, diatomic, or polyatomic? (b) What is its final temperature? (c) How many moles of gas are present? What is the total translational kinetic energy per mole (d) before and (e) after the compression? (f) What is the ratio of the squares of the rms speeds before and after the compression?

**66**   An ideal gas consists of 1.50 mol of diatomic molecules that rotate but do not oscillate. The molecular diameter is 250 pm. The gas is expanded at a constant pressure of $1.50 \times 10^5$ Pa, with a transfer of 200 J as heat. What is the change in the mean free path of the molecules?

**67**   An ideal monatomic gas initially has a temperature of 330 K and a pressure of 6.00 atm. It is to expand from volume 500 cm³ to volume 1500 cm³. If the expansion is isothermal, what are (a) the final pressure and (b) the work done by the gas? If, instead, the expansion is adiabatic, what are (c) the final pressure and (d) the work done by the gas?

**68**   In an interstellar gas cloud at 50.0 K, the pressure is $1.00 \times 10^{-8}$ Pa. Assuming that the molecular diameters of the gases in the cloud are all 20.0 nm, what is their mean free path?

**69** SSM   The envelope and basket of a hot-air balloon have a combined weight of 2.45 kN, and the envelope has a capacity (volume) of $2.18 \times 10^3$ m³. When it is fully inflated, what should be the temperature of the enclosed air to give the balloon a *lifting capacity* (force) of 2.67 kN (in addition to the balloon's weight)? Assume that the surrounding air, at 20.0°C, has a weight per unit volume of 11.9 N/m³ and a molecular mass of 0.028 kg/mol, and is at a pressure of 1.0 atm.

**70**   An ideal gas, at initial temperature $T_1$ and initial volume 2.0 m³, is expanded adiabatically to a volume of 4.0 m³, then expanded isothermally to a volume of 10 m³, and then compressed adiabatically back to $T_1$. What is its final volume?

**71** SSM   The temperature of 2.00 mol of an ideal monatomic gas is raised 15.0 K in an adiabatic process. What are (a) the work $W$ done by the gas, (b) the energy transferred as heat $Q$, (c) the change $\Delta E_{int}$ in internal energy of the gas, and (d) the change $\Delta K$ in the average kinetic energy per atom?

**72**   At what temperature do atoms of helium gas have the same rms speed as molecules of hydrogen gas at 20.0°C? (The molar masses are given in Table 19-1.)

**73** SSM   At what frequency do molecules (diameter 290 pm) collide in (an ideal) oxygen gas ($O_2$) at temperature 400 K and pressure 2.00 atm?

**74**   (a) What is the number of molecules per cubic meter in air at 20°C and at a pressure of 1.0 atm ($= 1.01 \times 10^5$ Pa)? (b) What is the mass of 1.0 m³ of this air? Assume that 75% of the molecules are nitrogen ($N_2$) and 25% are oxygen ($O_2$).

**75**   The temperature of 3.00 mol of a gas with $C_V = 6.00$ cal/mol · K is to be raised 50.0 K. If the process is at *constant volume*, what are (a) the energy transferred as heat $Q$, (b) the work $W$ done by the gas, (c) the change $\Delta E_{int}$ in internal energy of the gas, and (d) the change $\Delta K$ in the total translational kinetic energy? If the process is at *constant pressure*, what are (e) $Q$, (f) $W$, (g) $\Delta E_{int}$, and (h) $\Delta K$? If the process is *adiabatic*, what are (i) $Q$, (j) $W$, (k) $\Delta E_{int}$, and (l) $\Delta K$?

**76**   During a compression at a constant pressure of 250 Pa, the volume of an ideal gas decreases from 0.80 m³ to 0.20 m³. The initial temperature is 360 K, and the gas loses 210 J as heat. What are (a) the change in the internal energy of the gas and (b) the final temperature of the gas?

**77** SSM   Figure 19-28 shows a hypothetical speed distribution for particles of a certain gas: $P(v) = Cv^2$ for $0 < v \le v_0$ and $P(v) = 0$ for $v > v_0$. Find (a) an expression for $C$ in terms of $v_0$, (b) the average speed of the particles, and (c) their rms speed.

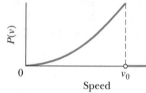

**Figure 19-28**   Problem 77.

**78**   (a) An ideal gas initially at pressure $p_0$ undergoes a free expansion until its volume is 3.00 times its initial volume. What then is the ratio of its pressure to $p_0$? (b) The gas is next slowly and adiabatically compressed back to its original volume. The pressure after compression is $(3.00)^{1/3}p_0$. Is the gas monatomic, diatomic, or polyatomic? (c) What is the ratio of the average kinetic energy per molecule in this final state to that in the initial state?

**79** SSM   An ideal gas undergoes isothermal compression from an initial volume of 4.00 m³ to a final volume of 3.00 m³. There is 3.50 mol of the gas, and its temperature is 10.0°C. (a) How much work is done by the gas? (b) How much energy is transferred as heat between the gas and its environment?

**80**   Oxygen ($O_2$) gas at 273 K and 1.0 atm is confined to a cubical container 10 cm on a side. Calculate $\Delta U_g / K_{avg}$, where $\Delta U_g$ is the change in the gravitational potential energy of an oxygen molecule falling the height of the box and $K_{avg}$ is the molecule's average translational kinetic energy.

**81**   An ideal gas is taken through a complete cycle in three steps: adiabatic expansion with work equal to 125 J, isothermal contraction at 325 K, and increase in pressure at constant volume. (a) Draw a $p$-$V$ diagram for the three steps. (b) How much energy is transferred as heat in step 3, and (c) is it transferred *to* or *from* the gas?

**82**   (a) What is the volume occupied by 1.00 mol of an ideal gas at standard conditions—that is, 1.00 atm ($= 1.01 \times 10^5$ Pa) and 273 K? (b) Show that the number of molecules per cubic centimeter (the *Loschmidt number*) at standard conditions is $2.69 \times 10^9$.

**83** SSM   A sample of ideal gas expands from an initial pressure and volume of 32 atm and 1.0 L to a final volume of 4.0 L. The initial temperature is 300 K. If the gas is monatomic and the expansion isothermal, what are the (a) final pressure $p_f$, (b) final temperature $T_f$, and (c) work $W$ done by the gas? If the gas is monatomic and the expansion adiabatic, what are (d) $p_f$, (e) $T_f$, and (f) $W$? If the gas is diatomic and the expansion adiabatic, what are (g) $p_f$, (h) $T_f$, and (i) $W$?

**84** An ideal gas with 3.00 mol is initially in state 1 with pressure $p_1 = 20.0$ atm and volume $V_1 = 1500$ cm³. First it is taken to state 2 with pressure $p_2 = 1.50p_1$ and volume $V_2 = 2.00V_1$. Then it is taken to state 3 with pressure $p_3 = 2.00p_1$ and volume $V_3 = 0.500V_1$. What is the temperature of the gas in (a) state 1 and (b) state 2? (c) What is the net change in internal energy from state 1 to state 3?

**85** A steel tank contains 300 g of ammonia gas ($NH_3$) at a pressure of $1.35 \times 10^6$ Pa and a temperature of 77°C. (a) What is the volume of the tank in liters? (b) Later the temperature is 22°C and the pressure is $8.7 \times 10^5$ Pa. How many grams of gas have leaked out of the tank?

**86** In an industrial process the volume of 25.0 mol of a monatomic ideal gas is reduced at a uniform rate from 0.616 m³ to 0.308 m³ in 2.00 h while its temperature is increased at a uniform rate from 27.0°C to 450°C. Throughout the process, the gas passes through thermodynamic equilibrium states. What are (a) the cumulative work done on the gas, (b) the cumulative energy absorbed by the gas as heat, and (c) the molar specific heat for the process? (*Hint:* To evaluate the integral for the work, you might use

$$\int \frac{a + bx}{A + Bx}\, dx = \frac{bx}{B} + \frac{aB - bA}{B^2} \ln(A + Bx),$$

an indefinite integral.) Suppose the process is replaced with a two-step process that reaches the same final state. In step 1, the gas volume is reduced at constant temperature, and in step 2 the temperature is increased at constant volume. For this process, what are (d) the cumulative work done on the gas, (e) the cumulative energy absorbed by the gas as heat, and (f) the molar specific heat for the process?

**87** Figure 19-29 shows a cycle consisting of five paths: $AB$ is isothermal at 300 K, $BC$ is adiabatic with work = 5.0 J, $CD$ is at a constant pressure of 5 atm, $DE$ is isothermal, and $EA$ is adiabatic with a change in internal energy of 8.0 J. What is the change in internal energy of the gas along path $CD$?

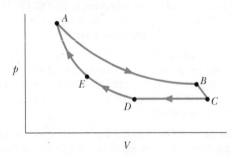

**Figure 19-29** Problem 87.

**88** An ideal gas initially at 300 K is compressed at a constant pressure of 25 N/m² from a volume of 3.0 m³ to a volume of 1.8 m³. In the process, 75 J is lost by the gas as heat. What are (a) the change in internal energy of the gas and (b) the final temperature of the gas?

**89** A pipe of length $L = 25.0$ m that is open at one end contains air at atmospheric pressure. It is thrust vertically into a freshwater lake until the water rises halfway up in the pipe (Fig. 19-30). What is the depth $h$ of the lower end of the pipe? Assume that the temperature is the same everywhere and does not change.

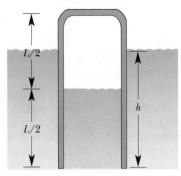

**Figure 19-30** Problem 89.

**90** In a motorcycle engine, a piston is forced down toward the crankshaft when the fuel in the top of the piston's cylinder undergoes combustion. The mixture of gaseous combustion products then expands adiabatically as the piston descends. Find the average power in (a) watts and (b) horsepower that is involved in this expansion when the engine is running at 4000 rpm, assuming that the gauge pressure immediately after combustion is 15 atm, the initial volume is 50 cm³, and the volume of the mixture at the bottom of the stroke is 250 cm³. Assume that the gases are diatomic and that the time involved in the expansion is one-half that of the total cycle.

**91** For adiabatic processes in an ideal gas, show that (a) the bulk modulus is given by

$$B = -V\frac{dp}{dV} = \gamma p,$$

where $\gamma = C_p/C_V$. (See Eq. 17-2.) (b) Then show that the speed of sound in the gas is

$$v_s = \sqrt{\frac{\gamma p}{\rho}} = \sqrt{\frac{\gamma RT}{M}},$$

where $\rho$ is the density, $T$ is the temperature, and $M$ is the molar mass. (See Eq. 17-3.)

**92** Air at 0.000°C and 1.00 atm pressure has a density of $1.29 \times 10^{-3}$ g/cm³, and the speed of sound is 331 m/s at that temperature. Compute the ratio $\gamma$ of the molar specific heats of air. (*Hint:* See Problem 91.)

**93** The speed of sound in different gases at a certain temperature $T$ depends on the molar mass of the gases. Show that

$$\frac{v_1}{v_2} = \sqrt{\frac{M_2}{M_1}},$$

where $v_1$ is the speed of sound in a gas of molar mass $M_1$ and $v_2$ is the speed of sound in a gas of molar mass $M_2$. (*Hint:* See Problem 91.)

**94** From the knowledge that $C_V$, the molar specific heat at constant volume, for a gas in a container is $5.0R$, calculate the ratio of the speed of sound in that gas to the rms speed of the molecules, for gas temperature $T$. (*Hint:* See Problem 91.)

**95** The molar mass of iodine is 127 g/mol. When sound at frequency 1000 Hz is introduced to a tube of iodine gas at 400 K, an internal acoustic standing wave is set up with nodes separated by 9.57 cm. What is $\gamma$ for the gas? (*Hint:* See Problem 91.)

**96** For air near 0°C, by how much does the speed of sound increase for each increase in air temperature by 1 C°? (*Hint:* See Problem 91.)

**97** Two containers are at the same temperature. The gas in the first container is at pressure $p_1$ and has molecules with mass $m_1$ and root-mean-square speed $v_{rms1}$. The gas in the second is at pressure $2p_1$ and has molecules with mass $m_2$ and average speed $v_{avg2} = 2v_{rms1}$. Find the ratio $m_1/m_2$ of the masses of their molecules.

# Entropy and the Second Law of Thermodynamics

## 20-1 ENTROPY

### Learning Objectives

*After reading this module, you should be able to . . .*

**20.01** Identify the second law of thermodynamics: If a process occurs in a closed system, the entropy of the system increases for irreversible processes and remains constant for reversible processes; it never decreases.

**20.02** Identify that entropy is a state function (the value for a particular state of the system does not depend on how that state is reached).

**20.03** Calculate the change in entropy for a process by integrating the inverse of the temperature (in kelvins) with respect to the heat $Q$ transferred during the process.

**20.04** For a phase change with a constant temperature process, apply the relationship between the entropy change $\Delta S$, the total transferred heat $Q$, and the temperature $T$ (in kelvins).

**20.05** For a temperature change $\Delta T$ that is small relative to the temperature $T$, apply the relationship between the entropy change $\Delta S$, the transferred heat $Q$, and the average temperature $T_{avg}$ (in kelvins).

**20.06** For an ideal gas, apply the relationship between the entropy change $\Delta S$ and the initial and final values of the pressure and volume.

**20.07** Identify that if a process is an irreversible one, the integration for the entropy change must be done for a reversible process that takes the system between the same initial and final states as the irreversible process.

**20.08** For stretched rubber, relate the elastic force to the rate at which the rubber's entropy changes with the change in the stretching distance.

### Key Ideas

● An irreversible process is one that cannot be reversed by means of small changes in the environment. The direction in which an irreversible process proceeds is set by the change in entropy $\Delta S$ of the system undergoing the process. Entropy $S$ is a state property (or state function) of the system; that is, it depends only on the state of the system and not on the way in which the system reached that state. The entropy postulate states (in part): If an irreversible process occurs in a closed system, the entropy of the system always increases.

● The entropy change $\Delta S$ for an irreversible process that takes a system from an initial state $i$ to a final state $f$ is exactly equal to the entropy change $\Delta S$ for any reversible process that takes the system between those same two states. We can compute the latter (but not the former) with

$$\Delta S = S_f - S_i = \int_i^f \frac{dQ}{T}.$$

Here $Q$ is the energy transferred as heat to or from the system during the process, and $T$ is the temperature of the system in kelvins during the process.

● For a reversible isothermal process, the expression for an entropy change reduces to

$$\Delta S = S_f - S_i = \frac{Q}{T}.$$

● When the temperature change $\Delta T$ of a system is small relative to the temperature (in kelvins) before and after the process, the entropy change can be approximated as

$$\Delta S = S_f - S_i \approx \frac{Q}{T_{avg}},$$

where $T_{avg}$ is the system's average temperature during the process.

● When an ideal gas changes reversibly from an initial state with temperature $T_i$ and volume $V_i$ to a final state with temperature $T_f$ and volume $V_f$, the change $\Delta S$ in the entropy of the gas is

$$\Delta S = S_f - S_i = nR \ln \frac{V_f}{V_i} + nC_V \ln \frac{T_f}{T_i}.$$

● The second law of thermodynamics, which is an extension of the entropy postulate, states: If a process occurs in a closed system, the entropy of the system increases for irreversible processes and remains constant for reversible processes. It never decreases. In equation form,

$$\Delta S \geq 0.$$

## What Is Physics?

Time has direction, the direction in which we age. We are accustomed to many one-way processes—that is, processes that can occur only in a certain sequence (the right way) and never in the reverse sequence (the wrong way). An egg is dropped onto a floor, a pizza is baked, a car is driven into a lamppost, large waves erode a sandy beach—these one-way processes are **irreversible,** meaning that they cannot be reversed by means of only small changes in their environment.

One goal of physics is to understand why time has direction and why one-way processes are irreversible. Although this physics might seem disconnected from the practical issues of everyday life, it is in fact at the heart of any engine, such as a car engine, because it determines how well an engine can run.

The key to understanding why one-way processes cannot be reversed involves a quantity known as *entropy*.

## Irreversible Processes and Entropy

The one-way character of irreversible processes is so pervasive that we take it for granted. If these processes were to occur *spontaneously* (on their own) in the wrong way, we would be astonished. Yet *none* of these wrong-way events would violate the law of conservation of energy.

For example, if you were to wrap your hands around a cup of hot coffee, you would be astonished if your hands got cooler and the cup got warmer. That is obviously the wrong way for the energy transfer, but the total energy of the closed system (*hands + cup of coffee*) would be the same as the total energy if the process had run in the right way. For another example, if you popped a helium balloon, you would be astonished if, later, all the helium molecules were to gather together in the original shape of the balloon. That is obviously the wrong way for molecules to spread, but the total energy of the closed system (*molecules + room*) would be the same as for the right way.

Thus, changes in energy within a closed system do not set the direction of irreversible processes. Rather, that direction is set by another property that we shall discuss in this chapter—the *change in entropy* $\Delta S$ of the system. The change in entropy of a system is defined later in this module, but we can here state its central property, often called the *entropy postulate:*

> If an irreversible process occurs in a *closed* system, the entropy $S$ of the system always increases; it never decreases.

Entropy differs from energy in that entropy does *not* obey a conservation law. The *energy* of a closed system is conserved; it always remains constant. For irreversible processes, the *entropy* of a closed system always increases. Because of this property, the change in entropy is sometimes called "the arrow of time." For example, we associate the explosion of a popcorn kernel with the forward direction of time and with an increase in entropy. The backward direction of time (a videotape run backwards) would correspond to the exploded popcorn re-forming the original kernel. Because this backward process would result in an entropy decrease, it never happens.

There are two equivalent ways to define the change in entropy of a system: (1) in terms of the system's temperature and the energy the system gains or loses as heat, and (2) by counting the ways in which the atoms or molecules that make up the system can be arranged. We use the first approach in this module and the second in Module 20-4.

# Change in Entropy

Let's approach this definition of *change in entropy* by looking again at a process that we described in Modules 18-5 and 19-9: the free expansion of an ideal gas. Figure 20-1a shows the gas in its initial equilibrium state $i$, confined by a closed stopcock to the left half of a thermally insulated container. If we open the stopcock, the gas rushes to fill the entire container, eventually reaching the final equilibrium state $f$ shown in Fig. 20-1b. This is an irreversible process; all the molecules of the gas will never return to the left half of the container.

The $p$-$V$ plot of the process, in Fig. 20-2, shows the pressure and volume of the gas in its initial state $i$ and final state $f$. Pressure and volume are *state properties*, properties that depend only on the state of the gas and not on how it reached that state. Other state properties are temperature and energy. We now assume that the gas has still another state property—its entropy. Furthermore, we define the **change in entropy** $S_f - S_i$ of a system during a process that takes the system from an initial state $i$ to a final state $f$ as

$$\Delta S = S_f - S_i = \int_i^f \frac{dQ}{T} \quad \text{(change in entropy defined).} \quad (20\text{-}1)$$

Here $Q$ is the energy transferred as heat to or from the system during the process, and $T$ is the temperature of the system in kelvins. Thus, an entropy change depends not only on the energy transferred as heat but also on the temperature at which the transfer takes place. Because $T$ is always positive, the sign of $\Delta S$ is the same as that of $Q$. We see from Eq. 20-1 that the SI unit for entropy and entropy change is the joule per kelvin.

There is a problem, however, in applying Eq. 20-1 to the free expansion of Fig. 20-1. As the gas rushes to fill the entire container, the pressure, temperature, and volume of the gas fluctuate unpredictably. In other words, they do not have a sequence of well-defined equilibrium values during the intermediate stages of the change from initial state $i$ to final state $f$. Thus, we cannot trace a pressure–volume path for the free expansion on the $p$-$V$ plot of Fig. 20-2, and we cannot find a relation between $Q$ and $T$ that allows us to integrate as Eq. 20-1 requires.

However, if entropy is truly a state property, the difference in entropy between states $i$ and $f$ must depend *only on those states* and not at all on the way the system went from one state to the other. Suppose, then, that we replace the irreversible free expansion of Fig. 20-1 with a *reversible* process that connects states $i$ and $f$. With a reversible process we can trace a pressure–volume path on a $p$-$V$ plot, and we can find a relation between $Q$ and $T$ that allows us to use Eq. 20-1 to obtain the entropy change.

We saw in Module 19-9 that the temperature of an ideal gas does not change during a free expansion: $T_i = T_f = T$. Thus, points $i$ and $f$ in Fig. 20-2 must be on the same isotherm. A convenient replacement process is then a reversible isothermal expansion from state $i$ to state $f$, which actually proceeds *along* that isotherm. Furthermore, because $T$ is constant throughout a reversible isothermal expansion, the integral of Eq. 20-1 is greatly simplified.

Figure 20-3 shows how to produce such a reversible isothermal expansion. We confine the gas to an insulated cylinder that rests on a thermal reservoir maintained at the temperature $T$. We begin by placing just enough lead shot on the movable piston so that the pressure and volume of the gas are those of the initial state $i$ of Fig. 20-1a. We then remove shot slowly (piece by piece) until the pressure and volume of the gas are those of the final state $f$ of Fig. 20-1b. The temperature of the gas does not change because the gas remains in thermal contact with the reservoir throughout the process.

The reversible isothermal expansion of Fig. 20-3 is physically quite different from the irreversible free expansion of Fig. 20-1. However, *both processes have the same initial state and the same final state and thus must have the same change in*

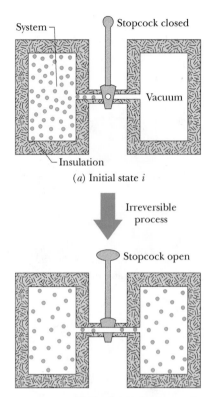

(a) Initial state $i$

Irreversible process

(b) Final state $f$

**Figure 20-1** The free expansion of an ideal gas. (a) The gas is confined to the left half of an insulated container by a closed stopcock. (b) When the stopcock is opened, the gas rushes to fill the entire container. This process is irreversible; that is, it does not occur in reverse, with the gas spontaneously collecting itself in the left half of the container.

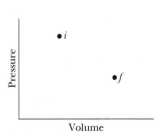

**Figure 20-2** A $p$-$V$ diagram showing the initial state $i$ and the final state $f$ of the free expansion of Fig. 20-1. The intermediate states of the gas cannot be shown because they are not equilibrium states.

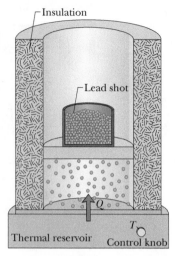

(*a*) Initial state *i*

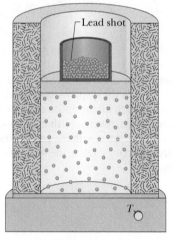

(*b*) Final state *f*

**Figure 20-3** The isothermal expansion of an ideal gas, done in a reversible way. The gas has the same initial state *i* and same final state *f* as in the irreversible process of Figs. 20-1 and 20-2.

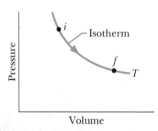

**Figure 20-4** A *p-V* diagram for the reversible isothermal expansion of Fig. 20-3. The intermediate states, which are now equilibrium states, are shown.

*entropy*. Because we removed the lead shot slowly, the intermediate states of the gas are equilibrium states, so we can plot them on a *p-V* diagram (Fig. 20-4).

To apply Eq. 20-1 to the isothermal expansion, we take the constant temperature *T* outside the integral, obtaining

$$\Delta S = S_f - S_i = \frac{1}{T}\int_i^f dQ.$$

Because $\int dQ = Q$, where *Q* is the total energy transferred as heat during the process, we have

$$\Delta S = S_f - S_i = \frac{Q}{T} \qquad \text{(change in entropy, isothermal process).} \qquad (20\text{-}2)$$

To keep the temperature *T* of the gas constant during the isothermal expansion of Fig. 20-3, heat *Q* must have been energy transferred *from* the reservoir *to* the gas. Thus, *Q* is positive and the entropy of the gas *increases* during the isothermal process and during the free expansion of Fig. 20-1.

To summarize:

> To find the entropy change for an irreversible process, replace that process with any reversible process that connects the same initial and final states. Calculate the entropy change for this reversible process with Eq. 20-1.

When the temperature change $\Delta T$ of a system is small relative to the temperature (in kelvins) before and after the process, the entropy change can be approximated as

$$\Delta S = S_f - S_i \approx \frac{Q}{T_{\text{avg}}}, \qquad (20\text{-}3)$$

where $T_{\text{avg}}$ is the average temperature of the system in kelvins during the process.

✓ **Checkpoint 1**

Water is heated on a stove. Rank the entropy changes of the water as its temperature rises (a) from 20°C to 30°C, (b) from 30°C to 35°C, and (c) from 80°C to 85°C, greatest first.

### Entropy as a State Function

We have assumed that entropy, like pressure, energy, and temperature, is a property of the state of a system and is independent of how that state is reached. That entropy is indeed a *state function* (as state properties are usually called) can be deduced only by experiment. However, we can prove it is a state function for the special and important case in which an ideal gas is taken through a reversible process.

To make the process reversible, it is done slowly in a series of small steps, with the gas in an equilibrium state at the end of each step. For each small step, the energy transferred as heat to or from the gas is $dQ$, the work done by the gas is $dW$, and the change in internal energy is $dE_{\text{int}}$. These are related by the first law of thermodynamics in differential form (Eq. 18-27):

$$dE_{\text{int}} = dQ - dW.$$

Because the steps are reversible, with the gas in equilibrium states, we can use Eq. 18-24 to replace $dW$ with $p\, dV$ and Eq. 19-45 to replace $dE_{\text{int}}$ with $nC_V\, dT$. Solving for $dQ$ then leads to

$$dQ = p\, dV + nC_V\, dT.$$

Using the ideal gas law, we replace *p* in this equation with $nRT/V$. Then we divide each term in the resulting equation by *T*, obtaining

$$\frac{dQ}{T} = nR\frac{dV}{V} + nC_V\frac{dT}{T}.$$

Now let us integrate each term of this equation between an arbitrary initial state *i* and an arbitrary final state *f* to get

$$\int_i^f \frac{dQ}{T} = \int_i^f nR\,\frac{dV}{V} + \int_i^f nC_V\,\frac{dT}{T}.$$

The quantity on the left is the entropy change $\Delta S$ ($= S_f - S_i$) defined by Eq. 20-1. Substituting this and integrating the quantities on the right yield

$$\Delta S = S_f - S_i = nR \ln\frac{V_f}{V_i} + nC_V \ln\frac{T_f}{T_i}. \qquad (20\text{-}4)$$

Note that we did not have to specify a particular reversible process when we integrated. Therefore, the integration must hold for all reversible processes that take the gas from state *i* to state *f*. Thus, the change in entropy $\Delta S$ between the initial and final states of an ideal gas depends only on properties of the initial state ($V_i$ and $T_i$) and properties of the final state ($V_f$ and $T_f$); $\Delta S$ does not depend on how the gas changes between the two states.

 **Checkpoint 2**

An ideal gas has temperature $T_1$ at the initial state *i* shown in the *p-V* diagram here. The gas has a higher temperature $T_2$ at final states *a* and *b*, which it can reach along the paths shown. Is the entropy change along the path to state *a* larger than, smaller than, or the same as that along the path to state *b*?

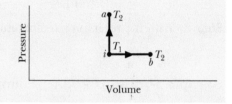

## Sample Problem 20.01    Entropy change of two blocks coming to thermal equilibrium

Figure 20-5*a* shows two identical copper blocks of mass $m = 1.5$ kg: block *L* at temperature $T_{iL} = 60°C$ and block *R* at temperature $T_{iR} = 20°C$. The blocks are in a thermally insulated box and are separated by an insulating shutter. When we lift the shutter, the blocks eventually come to the equilibrium temperature $T_f = 40°C$ (Fig. 20-5*b*). What is the net entropy change of the two-block system during this irreversible process? The specific heat of copper is 386 J/kg·K.

### KEY IDEA

To calculate the entropy change, we must find a reversible process that takes the system from the initial state of Fig. 20-5*a* to the final state of Fig. 20-5*b*. We can calculate the net entropy change $\Delta S_{rev}$ of the reversible process using Eq. 20-1, and then the entropy change for the irreversible process is equal to $\Delta S_{rev}$.

*Calculations:* For the reversible process, we need a thermal reservoir whose temperature can be changed slowly (say, by turning a knob). We then take the blocks through the following two steps, illustrated in Fig. 20-6.

**Step 1:** With the reservoir's temperature set at 60°C, put block *L* on the reservoir. (Since block and reservoir are at the same temperature, they are already in thermal equilib-

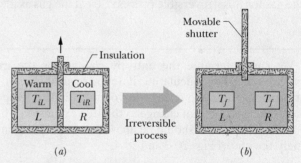

**Figure 20-5** (*a*) In the initial state, two copper blocks *L* and *R*, identical except for their temperatures, are in an insulating box and are separated by an insulating shutter. (*b*) When the shutter is removed, the blocks exchange energy as heat and come to a final state, both with the same temperature $T_f$.

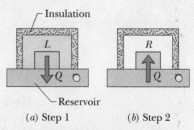

(*a*) Step 1          (*b*) Step 2

**Figure 20-6** The blocks of Fig. 20-5 can proceed from their initial state to their final state in a reversible way if we use a reservoir with a controllable temperature (*a*) to extract heat reversibly from block *L* and (*b*) to add heat reversibly to block *R*.

rium.) Then slowly lower the temperature of the reservoir and the block to 40°C. As the block's temperature changes by each increment $dT$ during this process, energy $dQ$ is transferred as heat *from* the block to the reservoir. Using Eq. 18-14, we can write this transferred energy as $dQ = mc\, dT$, where $c$ is the specific heat of copper. According to Eq. 20-1, the entropy change $\Delta S_L$ of block $L$ during the full temperature change from initial temperature $T_{iL}$ ($= 60°C = 333$ K) to final temperature $T_f$ ($= 40°C = 313$ K) is

$$\Delta S_L = \int_i^f \frac{dQ}{T} = \int_{T_{iL}}^{T_f} \frac{mc\, dT}{T} = mc \int_{T_{iL}}^{T_f} \frac{dT}{T}$$

$$= mc \ln \frac{T_f}{T_{iL}}.$$

Inserting the given data yields

$$\Delta S_L = (1.5 \text{ kg})(386 \text{ J/kg} \cdot \text{K}) \ln \frac{313 \text{ K}}{333 \text{ K}}$$

$$= -35.86 \text{ J/K}.$$

**Step 2:** With the reservoir's temperature now set at 20°C,

put block $R$ on the reservoir. Then slowly raise the temperature of the reservoir and the block to 40°C. With the same reasoning used to find $\Delta S_L$, you can show that the entropy change $\Delta S_R$ of block $R$ during this process is

$$\Delta S_R = (1.5 \text{ kg})(386 \text{ J/kg} \cdot \text{K}) \ln \frac{313 \text{ K}}{293 \text{ K}}$$

$$= +38.23 \text{ J/K}.$$

The net entropy change $\Delta S_{rev}$ of the two-block system undergoing this two-step reversible process is then

$$\Delta S_{rev} = \Delta S_L + \Delta S_R$$

$$= -35.86 \text{ J/K} + 38.23 \text{ J/K} = 2.4 \text{ J/K}.$$

Thus, the net entropy change $\Delta S_{irrev}$ for the two-block system undergoing the actual irreversible process is

$$\Delta S_{irrev} = \Delta S_{rev} = 2.4 \text{ J/K}. \qquad \text{(Answer)}$$

This result is positive, in accordance with the entropy postulate.

---

### Sample Problem 20.02    Entropy change of a free expansion of a gas

Suppose 1.0 mol of nitrogen gas is confined to the left side of the container of Fig. 20-1a. You open the stopcock, and the volume of the gas doubles. What is the entropy change of the gas for this irreversible process? Treat the gas as ideal.

#### KEY IDEAS

(1) We can determine the entropy change for the irreversible process by calculating it for a reversible process that provides the same change in volume. (2) The temperature of the gas does not change in the free expansion. Thus, the reversible process should be an isothermal expansion—namely, the one of Figs. 20-3 and 20-4.

*Calculations:* From Table 19-4, the energy $Q$ added as heat to the gas as it expands isothermally at temperature $T$ from an initial volume $V_i$ to a final volume $V_f$ is

$$Q = nRT \ln \frac{V_f}{V_i},$$

in which $n$ is the number of moles of gas present. From Eq. 20-2 the entropy change for this reversible process in which the temperature is held constant is

$$\Delta S_{rev} = \frac{Q}{T} = \frac{nRT \ln(V_f/V_i)}{T} = nR \ln \frac{V_f}{V_i}.$$

Substituting $n = 1.00$ mol and $V_f/V_i = 2$, we find

$$\Delta S_{rev} = nR \ln \frac{V_f}{V_i} = (1.00 \text{ mol})(8.31 \text{ J/mol} \cdot \text{K})(\ln 2)$$

$$= +5.76 \text{ J/K}.$$

Thus, the entropy change for the free expansion (and for all other processes that connect the initial and final states shown in Fig. 20-2) is

$$\Delta S_{irrev} = \Delta S_{rev} = +5.76 \text{ J/K}. \qquad \text{(Answer)}$$

Because $\Delta S$ is positive, the entropy increases, in accordance with the entropy postulate.

 **PLUS**  Additional examples, video, and practice available at *WileyPLUS*

## The Second Law of Thermodynamics

Here is a puzzle. In the process of going from (a) to (b) in Fig. 20-3, the entropy change of the gas (our system) is positive. However, because the process is reversible, we can also go from (b) to (a) by, say, gradually adding lead shot to the piston, to restore the initial gas volume. To maintain a constant temperature, we need to remove energy as heat, but that means $Q$ is negative and thus the entropy change is also. Doesn't this entropy decrease violate the entropy postulate: entropy always increases? No, because the postulate holds only for irreversible processes in closed systems. Here, the process is *not* irreversible and the system is *not* closed (because of the energy transferred to and from the reservoir as heat).

However, if we include the reservoir, along with the gas, as part of the system, then we do have a closed system. Let's check the change in entropy of the enlarged system *gas + reservoir* for the process that takes it from (*b*) to (*a*) in Fig. 20-3. During this reversible process, energy is transferred as heat from the gas to the reservoir—that is, from one part of the enlarged system to another. Let $|Q|$ represent the absolute value (or magnitude) of this heat. With Eq. 20-2, we can then calculate separately the entropy changes for the gas (which loses $|Q|$) and the reservoir (which gains $|Q|$). We get

$$\Delta S_{gas} = -\frac{|Q|}{T}$$

and

$$\Delta S_{res} = +\frac{|Q|}{T}.$$

The entropy change of the closed system is the sum of these two quantities: 0.

With this result, we can modify the entropy postulate to include both reversible and irreversible processes:

> If a process occurs in a *closed* system, the entropy of the system increases for irreversible processes and remains constant for reversible processes. It never decreases.

Although entropy may decrease in part of a closed system, there will always be an equal or larger entropy increase in another part of the system, so that the entropy of the system as a whole never decreases. This fact is one form of the **second law of thermodynamics** and can be written as

$$\Delta S \geq 0 \qquad \text{(second law of thermodynamics)}, \qquad (20\text{-}5)$$

where the greater-than sign applies to irreversible processes and the equals sign to reversible processes. Equation 20-5 applies only to closed systems.

In the real world almost all processes are irreversible to some extent because of friction, turbulence, and other factors, so the entropy of real closed systems undergoing real processes always increases. Processes in which the system's entropy remains constant are always idealizations.

### Force Due to Entropy

To understand why rubber resists being stretched, let's write the first law of thermodynamics

$$dE = dQ - dW$$

for a rubber band undergoing a small increase in length $dx$ as we stretch it between our hands. The force from the rubber band has magnitude $F$, is directed inward, and does work $dW = -F\,dx$ during length increase $dx$. From Eq. 20-2 ($\Delta S = Q/T$), small changes in $Q$ and $S$ at constant temperature are related by $dS = dQ/T$, or $dQ = T\,dS$. So, now we can rewrite the first law as

$$dE = T\,dS + F\,dx. \qquad (20\text{-}6)$$

To good approximation, the change $dE$ in the internal energy of rubber is 0 if the total stretch of the rubber band is not very much. Substituting 0 for $dE$ in Eq. 20-6 leads us to an expression for the force from the rubber band:

$$F = -T\frac{dS}{dx}. \qquad (20\text{-}7)$$

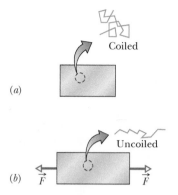

**Figure 20-7** A section of a rubber band (*a*) unstretched and (*b*) stretched, and a polymer within it (*a*) coiled and (*b*) uncoiled.

This tells us that $F$ is proportional to the rate $dS/dx$ at which the rubber band's entropy changes during a small change $dx$ in the rubber band's length. Thus, you can *feel* the effect of entropy on your hands as you stretch a rubber band.

To make sense of the relation between force and entropy, let's consider a simple model of the rubber material. Rubber consists of cross-linked polymer chains (long molecules with cross links) that resemble three-dimensional zig-zags (Fig. 20-7). When the rubber band is at its rest length, the polymers are coiled up in a spaghetti-like arrangement. Because of the large disorder of the molecules, this rest state has a high value of entropy. When we stretch a rubber band, we uncoil many of those polymers, aligning them in the direction of stretch. Because the alignment decreases the disorder, the entropy of the stretched rubber band is less. That is, the change $dS/dx$ in Eq. 20-7 is a negative quantity because the entropy decreases with stretching. Thus, the force on our hands from the rubber band is due to the tendency of the polymers to return to their former disordered state and higher value of entropy.

# 20-2 ENTROPY IN THE REAL WORLD: ENGINES

## Learning Objectives

*After reading this module, you should be able to . . .*

**20.09** Identify that a heat engine is a device that extracts energy from its environment in the form of heat and does useful work and that in an *ideal* heat engine, all processes are reversible, with no wasteful energy transfers.

**20.10** Sketch a $p$-$V$ diagram for the cycle of a Carnot engine, indicating the direction of cycling, the nature of the processes involved, the work done during each process (including algebraic sign), the net work done in the cycle, and the heat transferred during each process (including algebraic sign).

**20.11** Sketch a Carnot cycle on a temperature–entropy diagram, indicating the heat transfers.

**20.12** Determine the net entropy change around a Carnot cycle.

**20.13** Calculate the efficiency $\varepsilon_C$ of a Carnot engine in terms of the heat transfers and also in terms of the temperatures of the reservoirs.

**20.14** Identify that there are no perfect engines in which the energy transferred as heat $Q$ from a high temperature reservoir goes entirely into the work $W$ done by the engine.

**20.15** Sketch a $p$-$V$ diagram for the cycle of a Stirling engine, indicating the direction of cycling, the nature of the processes involved, the work done during each process (including algebraic sign), the net work done in the cycle, and the heat transfers during each process.

## Key Ideas

● An engine is a device that, operating in a cycle, extracts energy as heat $|Q_H|$ from a high-temperature reservoir and does a certain amount of work $|W|$. The efficiency $\varepsilon$ of any engine is defined as

$$\varepsilon = \frac{\text{energy we get}}{\text{energy we pay for}} = \frac{|W|}{|Q_H|}.$$

● In an ideal engine, all processes are reversible and no wasteful energy transfers occur due to, say, friction and turbulence.

● A Carnot engine is an ideal engine that follows the cycle of Fig. 20-9. Its efficiency is

$$\varepsilon_C = 1 - \frac{|Q_L|}{|Q_H|} = 1 - \frac{T_L}{T_H},$$

in which $T_H$ and $T_L$ are the temperatures of the high- and low-temperature reservoirs, respectively. Real engines always have an efficiency lower than that of a Carnot engine. Ideal engines that are not Carnot engines also have efficiencies lower than that of a Carnot engine.

● A perfect engine is an imaginary engine in which energy extracted as heat from the high-temperature reservoir is converted completely to work. Such an engine would violate the second law of thermodynamics, which can be restated as follows: No series of processes is possible whose sole result is the absorption of energy as heat from a thermal reservoir and the complete conversion of this energy to work.

## Entropy in the Real World: Engines

A **heat engine,** or more simply, an **engine,** is a device that extracts energy from its environment in the form of heat and does useful work. At the heart of every engine is a *working substance.* In a steam engine, the working substance is water, in both its vapor and its liquid form. In an automobile engine the working substance is a gasoline–air mixture. If an engine is to do work on a sustained basis, the working substance must operate in a *cycle;* that is, the working substance must pass through a closed series of thermodynamic processes, called *strokes,* returning again and again to each state in its cycle. Let us see what the laws of thermodynamics can tell us about the operation of engines.

### A Carnot Engine

We have seen that we can learn much about real gases by analyzing an ideal gas, which obeys the simple law $pV = nRT$. Although an ideal gas does not exist, any real gas approaches ideal behavior if its density is low enough. Similarly, we can study real engines by analyzing the behavior of an **ideal engine.**

In an ideal engine, all processes are reversible and no wasteful energy transfers occur due to, say, friction and turbulence.

We shall focus on a particular ideal engine called a **Carnot engine** after the French scientist and engineer N. L. Sadi Carnot (pronounced "car-no"), who first proposed the engine's concept in 1824. This ideal engine turns out to be the best (in principle) at using energy as heat to do useful work. Surprisingly, Carnot was able to analyze the performance of this engine before the first law of thermodynamics and the concept of entropy had been discovered.

Figure 20-8 shows schematically the operation of a Carnot engine. During each cycle of the engine, the working substance absorbs energy $|Q_H|$ as heat from a thermal reservoir at constant temperature $T_H$ and discharges energy $|Q_L|$ as heat to a second thermal reservoir at a constant lower temperature $T_L$.

Figure 20-9 shows a *p-V* plot of the *Carnot cycle*—the cycle followed by the working substance. As indicated by the arrows, the cycle is traversed in the clockwise direction. Imagine the working substance to be a gas, confined to an insulating cylinder with a weighted, movable piston. The cylinder may be placed at will on either of the two thermal reservoirs, as in Fig. 20-6, or on an insulating slab. Figure 20-9a shows that, if we place the cylinder in contact with the high-temperature reservoir at temperature $T_H$, heat $|Q_H|$ is transferred *to* the working substance *from* this reservoir as the gas undergoes an isothermal *expansion* from volume $V_a$ to volume $V_b$. Similarly, with the working substance in contact with the low-temperature reservoir at temperature $T_L$, heat $|Q_L|$ is transferred *from*

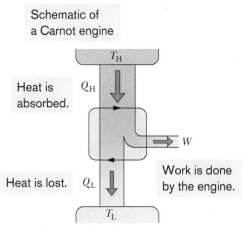

Schematic of a Carnot engine

**Figure 20-8** The elements of a Carnot engine. The two black arrowheads on the central loop suggest the working substance operating in a cycle, as if on a *p-V* plot. Energy $|Q_H|$ is transferred as heat from the high-temperature reservoir at temperature $T_H$ to the working substance. Energy $|Q_L|$ is transferred as heat from the working substance to the low-temperature reservoir at temperature $T_L$. Work $W$ is done by the engine (actually by the working substance) on something in the environment.

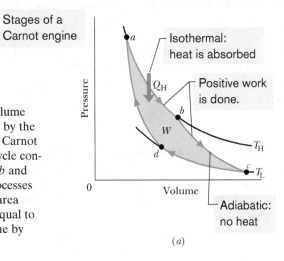

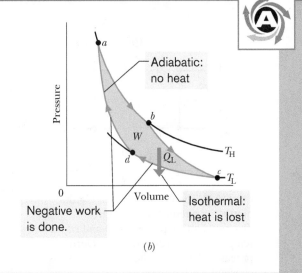

**Figure 20-9** A pressure–volume plot of the cycle followed by the working substance of the Carnot engine in Fig. 20-8. The cycle consists of two isothermal (*ab* and *cd*) and two adiabatic processes (*bc* and *da*). The shaded area enclosed by the cycle is equal to the work $W$ per cycle done by the Carnot engine.

the working substance *to* the low-temperature reservoir as the gas undergoes an isothermal *compression* from volume $V_c$ to volume $V_d$ (Fig. 20-9b).

In the engine of Fig. 20-8, we assume that heat transfers to or from the working substance can take place *only* during the isothermal processes *ab* and *cd* of Fig. 20-9. Therefore, processes *bc* and *da* in that figure, which connect the two isotherms at temperatures $T_H$ and $T_L$, must be (reversible) adiabatic processes; that is, they must be processes in which no energy is transferred as heat. To ensure this, during processes *bc* and *da* the cylinder is placed on an insulating slab as the volume of the working substance is changed.

During the processes *ab* and *bc* of Fig. 20-9a, the working substance is expanding and thus doing positive work as it raises the weighted piston. This work is represented in Fig. 20-9a by the area under curve *abc*. During the processes *cd* and *da* (Fig. 20-9b), the working substance is being compressed, which means that it is doing negative work on its environment or, equivalently, that its environment is doing work on it as the loaded piston descends. This work is represented by the area under curve *cda*. The *net work per cycle*, which is represented by *W* in both Figs. 20-8 and 20-9, is the difference between these two areas and is a positive quantity equal to the area enclosed by cycle *abcda* in Fig. 20-9. This work *W* is performed on some outside object, such as a load to be lifted.

Equation 20-1 ($\Delta S = \int dQ/T$) tells us that any energy transfer as heat must involve a change in entropy. To see this for a Carnot engine, we can plot the Carnot cycle on a temperature–entropy (*T-S*) diagram as in Fig. 20-10. The lettered points *a*, *b*, *c*, and *d* there correspond to the lettered points in the *p-V* diagram in Fig. 20-9. The two horizontal lines in Fig. 20-10 correspond to the two isothermal processes of the cycle. Process *ab* is the isothermal expansion of the cycle. As the working substance (reversibly) absorbs energy $|Q_H|$ as heat at constant temperature $T_H$ during the expansion, its entropy increases. Similarly, during the isothermal compression *cd*, the working substance (reversibly) loses energy $|Q_L|$ as heat at constant temperature $T_L$, and its entropy decreases.

The two vertical lines in Fig. 20-10 correspond to the two adiabatic processes of the Carnot cycle. Because no energy is transferred as heat during the two processes, the entropy of the working substance is constant during them.

**The Work** To calculate the net work done by a Carnot engine during a cycle, let us apply Eq. 18-26, the first law of thermodynamics ($\Delta E_{int} = Q - W$), to the working substance. That substance must return again and again to any arbitrarily selected state in the cycle. Thus, if *X* represents any state property of the working substance, such as pressure, temperature, volume, internal energy, or entropy, we must have $\Delta X = 0$ for every cycle. It follows that $\Delta E_{int} = 0$ for a complete cycle of the working substance. Recalling that *Q* in Eq. 18-26 is the *net* heat transfer per cycle and *W* is the *net* work, we can write the first law of thermodynamics for the Carnot cycle as

$$W = |Q_H| - |Q_L|. \qquad (20\text{-}8)$$

**Entropy Changes** In a Carnot engine, there are *two* (and only two) reversible energy transfers as heat, and thus two changes in the entropy of the working substance—one at temperature $T_H$ and one at $T_L$. The net entropy change per cycle is then

$$\Delta S = \Delta S_H + \Delta S_L = \frac{|Q_H|}{T_H} - \frac{|Q_L|}{T_L}. \qquad (20\text{-}9)$$

Here $\Delta S_H$ is positive because energy $|Q_H|$ is *added to* the working substance as heat (an increase in entropy) and $\Delta S_L$ is negative because energy $|Q_L|$ is *removed from* the working substance as heat (a decrease in entropy). Because entropy is a state function, we must have $\Delta S = 0$ for a complete cycle. Putting $\Delta S = 0$ in Eq. 20-9 requires that

$$\frac{|Q_H|}{T_H} = \frac{|Q_L|}{T_L}. \qquad (20\text{-}10)$$

Note that, because $T_H > T_L$, we must have $|Q_H| > |Q_L|$; that is, more energy is

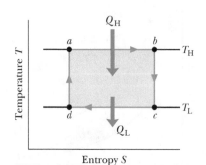

**Figure 20-10** The Carnot cycle of Fig. 20-9 plotted on a temperature–entropy diagram. During processes *ab* and *cd* the temperature remains constant. During processes *bc* and *da* the entropy remains constant.

extracted as heat from the high-temperature reservoir than is delivered to the low-temperature reservoir.

We shall now derive an expression for the efficiency of a Carnot engine.

### Efficiency of a Carnot Engine

The purpose of any engine is to transform as much of the extracted energy $Q_H$ into work as possible. We measure its success in doing so by its **thermal efficiency** $\varepsilon$, defined as the work the engine does per cycle ("energy we get") divided by the energy it absorbs as heat per cycle ("energy we pay for"):

$$\varepsilon = \frac{\text{energy we get}}{\text{energy we pay for}} = \frac{|W|}{|Q_H|} \qquad \text{(efficiency, any engine).} \qquad (20\text{-}11)$$

For any ideal engine we substitute for $W$ from Eq. 20-8 to write Eq. 20-11 as

$$\varepsilon = \frac{|Q_H| - |Q_L|}{Q_H} = 1 - \frac{|Q_L|}{|Q_H|}. \qquad (20\text{-}12)$$

Using Eq. 20-10 for a Carnot engine, we can write this as

$$\varepsilon_C = 1 - \frac{T_L}{T_H} \qquad \text{(efficiency, Carnot engine),} \qquad (20\text{-}13)$$

where the temperatures $T_L$ and $T_H$ are in kelvins. Because $T_L < T_H$, the Carnot engine necessarily has a thermal efficiency less than unity—that is, less than 100%. This is indicated in Fig. 20-8, which shows that only part of the energy extracted as heat from the high-temperature reservoir is available to do work, and the rest is delivered to the low-temperature reservoir. We shall show in Module 20-3 that no real engine can have a thermal efficiency greater than that calculated from Eq. 20-13.

Inventors continually try to improve engine efficiency by reducing the energy $|Q_L|$ that is "thrown away" during each cycle. The inventor's dream is to produce the *perfect engine,* diagrammed in Fig. 20-11, in which $|Q_L|$ is reduced to zero and $|Q_H|$ is converted completely into work. Such an engine on an ocean liner, for example, could extract energy as heat from the water and use it to drive the propellers, with no fuel cost. An automobile fitted with such an engine could extract energy as heat from the surrounding air and use it to drive the car, again with no fuel cost. Alas, a perfect engine is only a dream: Inspection of Eq. 20-13 shows that we can achieve 100% engine efficiency (that is, $\varepsilon = 1$) only if $T_L = 0$ or $T_H \to \infty$, impossible requirements. Instead, experience gives the following alternative version of the second law of thermodynamics, which says in short, *there are no perfect engines:*

> No series of processes is possible whose sole result is the transfer of energy as heat from a thermal reservoir and the complete conversion of this energy to work.

To summarize: The thermal efficiency given by Eq. 20-13 applies only to Carnot engines. Real engines, in which the processes that form the engine cycle are not reversible, have lower efficiencies. If your car were powered by a Carnot engine, it would have an efficiency of about 55% according to Eq. 20-13; its actual efficiency is probably about 25%. A nuclear power plant (Fig. 20-12), taken in its entirety, is an engine. It extracts energy as heat from a reactor core, does work by means of a turbine, and discharges energy as heat to a nearby river. If the power plant operated as a Carnot engine, its efficiency would be about 40%; its actual efficiency is about 30%. In designing engines of any type, there is simply no way to beat the efficiency limitation imposed by Eq. 20-13.

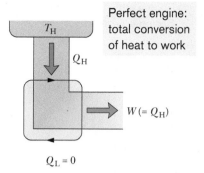

**Figure 20-11** The elements of a perfect engine—that is, one that converts heat $Q_H$ from a high-temperature reservoir directly to work $W$ with 100% efficiency.

© Richard Ustinich

**Figure 20-12** The North Anna nuclear power plant near Charlottesville, Virginia, which generates electric energy at the rate of 900 MW. At the same time, by design, it discards energy into the nearby river at the rate of 2100 MW. This plant and all others like it throw away more energy than they deliver in useful form. They are real counterparts of the ideal engine of Fig. 20-8.

Stages of a
Stirling engine

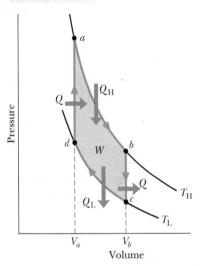

**Figure 20-13** A *p-V* plot for the working substance of an ideal Stirling engine, with the working substance assumed for convenience to be an ideal gas.

### Stirling Engine

Equation 20-13 applies not to all ideal engines but only to those that can be represented as in Fig. 20-9—that is, to Carnot engines. For example, Fig. 20-13 shows the operating cycle of an ideal **Stirling engine.** Comparison with the Carnot cycle of Fig. 20-9 shows that each engine has isothermal heat transfers at temperatures $T_H$ and $T_L$. However, the two isotherms of the Stirling engine cycle are connected, not by adiabatic processes as for the Carnot engine but by constant-volume processes. To increase the temperature of a gas at constant volume reversibly from $T_L$ to $T_H$ (process *da* of Fig. 20-13) requires a transfer of energy as heat to the working substance from a thermal reservoir whose temperature can be varied smoothly between those limits. Also, a reverse transfer is required in process *bc*. Thus, reversible heat transfers (and corresponding entropy changes) occur in all four of the processes that form the cycle of a Stirling engine, not just two processes as in a Carnot engine. Thus, the derivation that led to Eq. 20-13 does not apply to an ideal Stirling engine. More important, the efficiency of an ideal Stirling engine is lower than that of a Carnot engine operating between the same two temperatures. Real Stirling engines have even lower efficiencies.

The Stirling engine was developed in 1816 by Robert Stirling. This engine, long neglected, is now being developed for use in automobiles and spacecraft. A Stirling engine delivering 5000 hp (3.7 MW) has been built. Because they are quiet, Stirling engines are used on some military submarines.

✔️ **Checkpoint 3**

Three Carnot engines operate between reservoir temperatures of (a) 400 and 500 K, (b) 600 and 800 K, and (c) 400 and 600 K. Rank the engines according to their thermal efficiencies, greatest first.

### Sample Problem 20.03  Carnot engine, efficiency, power, entropy changes

Imagine a Carnot engine that operates between the temperatures $T_H = 850$ K and $T_L = 300$ K. The engine performs 1200 J of work each cycle, which takes 0.25 s.

(a) What is the efficiency of this engine?

**KEY IDEA**

The efficiency $\varepsilon$ of a Carnot engine depends only on the ratio $T_L/T_H$ of the temperatures (in kelvins) of the thermal reservoirs to which it is connected.

**Calculation:** Thus, from Eq. 20-13, we have

$$\varepsilon = 1 - \frac{T_L}{T_H} = 1 - \frac{300 \text{ K}}{850 \text{ K}} = 0.647 \approx 65\%. \quad \text{(Answer)}$$

(b) What is the average power of this engine?

**KEY IDEA**

The average power $P$ of an engine is the ratio of the work $W$ it does per cycle to the time $t$ that each cycle takes.

**Calculation:** For this Carnot engine, we find

$$P = \frac{W}{t} = \frac{1200 \text{ J}}{0.25 \text{ s}} = 4800 \text{ W} = 4.8 \text{ kW}. \quad \text{(Answer)}$$

(c) How much energy $|Q_H|$ is extracted as heat from the high-temperature reservoir every cycle?

**KEY IDEA**

The efficiency $\varepsilon$ is the ratio of the work $W$ that is done per cycle to the energy $|Q_H|$ that is extracted as heat from the high-temperature reservoir per cycle ($\varepsilon = W/|Q_H|$).

**Calculation:** Here we have

$$|Q_H| = \frac{W}{\varepsilon} = \frac{1200 \text{ J}}{0.647} = 1855 \text{ J}. \quad \text{(Answer)}$$

(d) How much energy $|Q_L|$ is delivered as heat to the low-temperature reservoir every cycle?

**KEY IDEA**

For a Carnot engine, the work $W$ done per cycle is equal to the difference in the energy transfers as heat: $|Q_H| - |Q_L|$, as in Eq. 20-8.

**Calculation:** Thus, we have

$$|Q_L| = |Q_H| - W$$
$$= 1855 \text{ J} - 1200 \text{ J} = 655 \text{ J}. \quad \text{(Answer)}$$

(e) By how much does the entropy of the working substance change as a result of the energy transferred to it from the high-temperature reservoir? From it to the low-temperature reservoir?

### KEY IDEA

The entropy change $\Delta S$ during a transfer of energy as heat $Q$ at constant temperature $T$ is given by Eq. 20-2 ($\Delta S = Q/T$).

**Calculations:** Thus, for the *positive* transfer of energy $Q_H$ from the high-temperature reservoir at $T_H$, the change in the

entropy of the working substance is

$$\Delta S_H = \frac{Q_H}{T_H} = \frac{1855 \text{ J}}{850 \text{ K}} = +2.18 \text{ J/K}. \quad \text{(Answer)}$$

Similarly, for the *negative* transfer of energy $Q_L$ to the low-temperature reservoir at $T_L$, we have

$$\Delta S_L = \frac{Q_L}{T_L} = \frac{-655 \text{ J}}{300 \text{ K}} = -2.18 \text{ J/K}. \quad \text{(Answer)}$$

Note that the net entropy change of the working substance for one cycle is zero, as we discussed in deriving Eq. 20-10.

### Sample Problem 20.04   Impossibly efficient engine

An inventor claims to have constructed an engine that has an efficiency of 75% when operated between the boiling and freezing points of water. Is this possible?

### KEY IDEA

The efficiency of a real engine must be less than the efficiency of a Carnot engine operating between the same two temperatures.

**Calculation:** From Eq. 20-13, we find that the efficiency of a Carnot engine operating between the boiling and freezing points of water is

$$\varepsilon = 1 - \frac{T_L}{T_H} = 1 - \frac{(0 + 273) \text{ K}}{(100 + 273) \text{ K}} = 0.268 \approx 27\%.$$

Thus, for the given temperatures, the claimed efficiency of 75% for a real engine (with its irreversible processes and wasteful energy transfers) is impossible.

 Additional examples, video, and practice available at *WileyPLUS*

# 20-3 REFRIGERATORS AND REAL ENGINES

## Learning Objectives

*After reading this module, you should be able to . . .*

**20.16** Identify that a refrigerator is a device that uses work to transfer energy from a low-temperature reservoir to a high-temperature reservoir, and that an ideal refrigerator is one that does this with reversible processes and no wasteful losses.

**20.17** Sketch a $p$-$V$ diagram for the cycle of a Carnot refrigerator, indicating the direction of cycling, the nature of the processes involved, the work done during each process (including algebraic sign), the net work done in the cycle,

and the heat transferred during each process (including algebraic sign).

**20.18** Apply the relationship between the coefficient of performance $K$ and the heat exchanges with the reservoirs and the temperatures of the reservoirs.

**20.19** Identify that there is no ideal refrigerator in which all of the energy extracted from the low-temperature reservoir is transferred to the high-temperature reservoir.

**20.20** Identify that the efficiency of a real engine is less than that of the ideal Carnot engine.

## Key Ideas

● A refrigerator is a device that, operating in a cycle, has work $W$ done on it as it extracts energy $|Q_L|$ as heat from a low-temperature reservoir. The coefficient of performance $K$ of a refrigerator is defined as

$$K = \frac{\text{what we want}}{\text{what we pay for}} = \frac{|Q_L|}{|W|}.$$

● A Carnot refrigerator is a Carnot engine operating in reverse. Its coefficient of performance is

$$K_C = \frac{|Q_L|}{|Q_H| - |Q_L|} = \frac{T_L}{T_H - T_L}.$$

● A perfect refrigerator is an entirely imaginary refrigerator in which energy extracted as heat from the low-temperature reservoir is somehow converted completely to heat discharged to the high-temperature reservoir without any need for work.

● A perfect refrigerator would violate the second law of thermodynamics, which can be restated as follows: No series of processes is possible whose sole result is the transfer of energy as heat from a reservoir at a given temperature to a reservoir at a higher temperature (without work being involved).

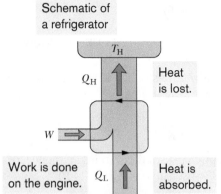

Schematic of a refrigerator

Heat is lost.

Work is done on the engine.

Heat is absorbed.

**Figure 20-14** The elements of a Carnot refrigerator. The two black arrowheads on the central loop suggest the working substance operating in a cycle, as if on a p-V plot. Energy is transferred as heat $Q_L$ to the working substance from the low-temperature reservoir. Energy is transferred as heat $Q_H$ to the high-temperature reservoir from the working substance. Work $W$ is done on the refrigerator (on the working substance) by something in the environment.

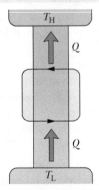

Perfect refrigerator: total transfer of heat from cold to hot without any work

**Figure 20-15** The elements of a perfect refrigerator—that is, one that transfers energy from a low-temperature reservoir to a high-temperature reservoir without any input of work.

# Entropy in the Real World: Refrigerators

A **refrigerator** is a device that uses work in order to transfer energy from a low-temperature reservoir to a high-temperature reservoir as the device continuously repeats a set series of thermodynamic processes. In a household refrigerator, for example, work is done by an electrical compressor to transfer energy from the food storage compartment (a low-temperature reservoir) to the room (a high-temperature reservoir).

Air conditioners and heat pumps are also refrigerators. For an air conditioner, the low-temperature reservoir is the room that is to be cooled and the high-temperature reservoir is the warmer outdoors. A heat pump is an air conditioner that can be operated in reverse to heat a room; the room is the high-temperature reservoir, and heat is transferred to it from the cooler outdoors.

Let us consider an *ideal refrigerator:*

> In an ideal refrigerator, all processes are reversible and no wasteful energy transfers occur as a result of, say, friction and turbulence.

Figure 20-14 shows the basic elements of an ideal refrigerator. Note that its operation is the reverse of how the Carnot engine of Fig. 20-8 operates. In other words, all the energy transfers, as either heat or work, are reversed from those of a Carnot engine. We can call such an ideal refrigerator a **Carnot refrigerator.**

The designer of a refrigerator would like to extract as much energy $|Q_L|$ as possible from the low-temperature reservoir (what we want) for the least amount of work $|W|$ (what we pay for). A measure of the efficiency of a refrigerator, then, is

$$K = \frac{\text{what we want}}{\text{what we pay for}} = \frac{|Q_L|}{|W|} \qquad \begin{array}{c}\text{(coefficient of performance,} \\ \text{any refrigerator),}\end{array} \qquad (20\text{-}14)$$

where $K$ is called the *coefficient of performance.* For any ideal refrigerator, the first law of thermodynamics gives $|W| = |Q_H| - |Q_L|$, where $|Q_H|$ is the magnitude of the energy transferred as heat to the high-temperature reservoir. Equation 20-14 then becomes

$$K = \frac{|Q_L|}{|Q_H| - |Q_L|}. \qquad (20\text{-}15)$$

Because a Carnot refrigerator is a Carnot engine operating in reverse, we can combine Eq. 20-10 with Eq. 20-15; after some algebra we find

$$K_C = \frac{T_L}{T_H - T_L} \qquad \begin{array}{c}\text{(coefficient of performance,} \\ \text{Carnot refrigerator).}\end{array} \qquad (20\text{-}16)$$

For typical room air conditioners, $K \approx 2.5$. For household refrigerators, $K \approx 5$. Perversely, the value of $K$ is higher the closer the temperatures of the two reservoirs are to each other. That is why heat pumps are more effective in temperate climates than in very cold climates.

It would be nice to own a refrigerator that did not require some input of work—that is, one that would run without being plugged in. Figure 20-15 represents another "inventor's dream," a *perfect refrigerator* that transfers energy as heat $Q$ from a cold reservoir to a warm reservoir without the need for work. Because the unit operates in cycles, the entropy of the working substance does not change during a complete cycle. The entropies of the two reservoirs, however, do change: The entropy change for the cold reservoir is $-|Q|/T_L$, and that for the warm reservoir is $+|Q|/T_H$. Thus, the net entropy change for the entire system is

$$\Delta S = -\frac{|Q|}{T_L} + \frac{|Q|}{T_H}.$$

Because $T_H > T_L$, the right side of this equation is negative and thus the net change in entropy per cycle for the closed system *refrigerator + reservoirs* is also negative. Because such a decrease in entropy violates the second law of thermodynamics (Eq. 20-5), a perfect refrigerator does not exist. (If you want your refrigerator to operate, you must plug it in.)

Here, then, is another way to state the second law of thermodynamics:

> No series of processes is possible whose sole result is the transfer of energy as heat from a reservoir at a given temperature to a reservoir at a higher temperature.

In short, *there are no perfect refrigerators.*

 Checkpoint 4

You wish to increase the coefficient of performance of an ideal refrigerator. You can do so by (a) running the cold chamber at a slightly higher temperature, (b) running the cold chamber at a slightly lower temperature, (c) moving the unit to a slightly warmer room, or (d) moving it to a slightly cooler room. The magnitudes of the temperature changes are to be the same in all four cases. List the changes according to the resulting coefficients of performance, greatest first.

## The Efficiencies of Real Engines

Let $\varepsilon_C$ be the efficiency of a Carnot engine operating between two given temperatures. Here we prove that no real engine operating between those temperatures can have an efficiency greater than $\varepsilon_C$. If it could, the engine would violate the second law of thermodynamics.

Let us assume that an inventor, working in her garage, has constructed an engine $X$, which she claims has an efficiency $\varepsilon_X$ that is greater than $\varepsilon_C$:

$$\varepsilon_X > \varepsilon_C \quad \text{(a claim).} \tag{20-17}$$

Let us couple engine $X$ to a Carnot refrigerator, as in Fig. 20-16a. We adjust the strokes of the Carnot refrigerator so that the work it requires per cycle is just equal to that provided by engine $X$. Thus, no (external) work is performed on or by the combination *engine + refrigerator* of Fig. 20-16a, which we take as our system.

If Eq. 20-17 is true, from the definition of efficiency (Eq. 20-11), we must have

$$\frac{|W|}{|Q'_H|} > \frac{|W|}{|Q_H|},$$

where the prime refers to engine $X$ and the right side of the inequality is the efficiency of the Carnot refrigerator when it operates as an engine. This inequality requires that

$$|Q_H| > |Q'_H|. \tag{20-18}$$

**Figure 20-16** (*a*) Engine $X$ drives a Carnot refrigerator. (*b*) If, as claimed, engine $X$ is more efficient than a Carnot engine, then the combination shown in (*a*) is equivalent to the perfect refrigerator shown here. This violates the second law of thermodynamics, so we conclude that engine $X$ *cannot* be more efficient than a Carnot engine.

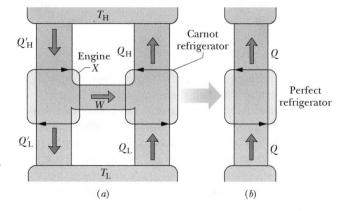

Because the work done by engine $X$ is equal to the work done on the Carnot refrigerator, we have, from the first law of thermodynamics as given by Eq. 20-8,

$$|Q_H| - |Q_L| = |Q'_H| - |Q'_L|,$$

which we can write as

$$|Q_H| - |Q'_H| = |Q_L| - |Q'_L| = Q. \qquad (20\text{-}19)$$

Because of Eq. 20-18, the quantity $Q$ in Eq. 20-19 must be positive.

Comparison of Eq. 20-19 with Fig. 20-16 shows that the net effect of engine $X$ and the Carnot refrigerator working in combination is to transfer energy $Q$ as heat from a low-temperature reservoir to a high-temperature reservoir without the requirement of work. Thus, the combination acts like the perfect refrigerator of Fig. 20-15, whose existence is a violation of the second law of thermodynamics.

Something must be wrong with one or more of our assumptions, and it can only be Eq. 20-17. We conclude that *no real engine can have an efficiency greater than that of a Carnot engine when both engines work between the same two temperatures.* At most, the real engine can have an efficiency equal to that of a Carnot engine. In that case, the real engine *is* a Carnot engine.

# 20-4 A STATISTICAL VIEW OF ENTROPY

## Learning Objectives

*After reading this module, you should be able to . . .*

**20.21** Explain what is meant by the configurations of a system of molecules.

**20.22** Calculate the multiplicity of a given configuration.

**20.23** Identify that all microstates are equally probable but the configurations with more microstates are more probable than the other configurations.

**20.24** Apply Boltzmann's entropy equation to calculate the entropy associated with a multiplicity.

## Key Ideas

● The entropy of a system can be defined in terms of the possible distributions of its molecules. For identical molecules, each possible distribution of molecules is called a microstate of the system. All equivalent microstates are grouped into a configuration of the system. The number of microstates in a configuration is the multiplicity $W$ of the configuration.

● For a system of $N$ molecules that may be distributed between the two halves of a box, the multiplicity is given by

$$W = \frac{N!}{n_1! \, n_2!},$$

in which $n_1$ is the number of molecules in one half of the box and $n_2$ is the number in the other half. A basic assumption of statistical mechanics is that all the microstates are equally probable.

Thus, configurations with a large multiplicity occur most often. When $N$ is very large (say, $N = 10^{22}$ molecules or more), the molecules are nearly always in the configuration in which $n_1 = n_2$.

● The multiplicity $W$ of a configuration of a system and the entropy $S$ of the system in that configuration are related by Boltzmann's entropy equation:

$$S = k \ln W,$$

where $k = 1.38 \times 10^{-23}$ J/K is the Boltzmann constant.

● When $N$ is very large (the usual case), we can approximate $\ln N!$ with Stirling's approximation:

$$\ln N! \approx N(\ln N) - N.$$

## A Statistical View of Entropy

In Chapter 19 we saw that the macroscopic properties of gases can be explained in terms of their microscopic, or molecular, behavior. Such explanations are part of a study called **statistical mechanics.** Here we shall focus our attention on a single problem, one involving the distribution of gas molecules between the two halves of an insulated box. This problem is reasonably simple to analyze, and it allows us to use statistical mechanics to calculate the entropy change for the free expansion of an ideal gas. You will see that statistical mechanics leads to the same entropy change as we would find using thermodynamics.

Figure 20-17 shows a box that contains six identical (and thus indistinguishable) molecules of a gas. At any instant, a given molecule will be in either the left or the right half of the box; because the two halves have equal volumes, the molecule has the same likelihood, or probability, of being in either half.

Table 20-1 shows the seven possible *configurations* of the six molecules, each configuration labeled with a Roman numeral. For example, in configuration I, all six molecules are in the left half of the box ($n_1 = 6$) and none are in the right half ($n_2 = 0$). We see that, in general, a given configuration can be achieved in a number of different ways. We call these different arrangements of the molecules *microstates*. Let us see how to calculate the number of microstates that correspond to a given configuration.

Suppose we have $N$ molecules, distributed with $n_1$ molecules in one half of the box and $n_2$ in the other. (Thus $n_1 + n_2 = N$.) Let us imagine that we distribute the molecules "by hand," one at a time. If $N = 6$, we can select the first molecule in six independent ways; that is, we can pick any one of the six molecules. We can pick the second molecule in five ways, by picking any one of the remaining five molecules; and so on. The total number of ways in which we can select all six molecules is the product of these independent ways, or $6 \times 5 \times 4 \times 3 \times 2 \times 1 = 720$. In mathematical shorthand we write this product as $6! = 720$, where $6!$ is pronounced "six factorial." Your hand calculator can probably calculate factorials. For later use you will need to know that $0! = 1$. (Check this on your calculator.)

However, because the molecules are indistinguishable, these 720 arrangements are not all different. In the case that $n_1 = 4$ and $n_2 = 2$ (which is configuration III in Table 20-1), for example, the order in which you put four molecules in one half of the box does not matter, because after you have put all four in, there is no way that you can tell the order in which you did so. The number of ways in which you can order the four molecules is $4! = 24$. Similarly, the number of ways in which you can order two molecules for the other half of the box is simply $2! = 2$. To get the number of *different* arrangements that lead to the $(4, 2)$ split of configuration III, we must divide 720 by 24 and also by 2. We call the resulting quantity, which is the number of microstates that correspond to a given configuration, the *multiplicity W* of that configuration. Thus, for configuration III,

$$W_{III} = \frac{6!}{4!\,2!} = \frac{720}{24 \times 2} = 15.$$

Thus, Table 20-1 tells us there are 15 independent microstates that correspond to configuration III. Note that, as the table also tells us, the total number of microstates for six molecules distributed over the seven configurations is 64.

Extrapolating from six molecules to the general case of $N$ molecules, we have

$$W = \frac{N!}{n_1!\,n_2!} \qquad \text{(multiplicity of configuration).} \qquad (20\text{-}20)$$

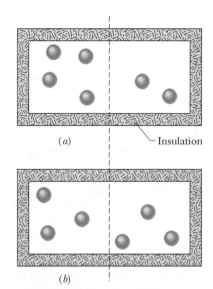

**Figure 20-17** An insulated box contains six gas molecules. Each molecule has the same probability of being in the left half of the box as in the right half. The arrangement in (*a*) corresponds to configuration III in Table 20-1, and that in (*b*) corresponds to configuration IV.

**Table 20-1 Six Molecules in a Box**

| Configuration | | | Multiplicity $W$ | Calculation of $W$ | Entropy $10^{-23}$ J/K |
|---|---|---|---|---|---|
| Label | $n_1$ | $n_2$ | (number of microstates) | (Eq. 20-20) | (Eq. 20-21) |
| I | 6 | 0 | 1 | $6!/(6!\ 0!) = 1$ | 0 |
| II | 5 | 1 | 6 | $6!/(5!\ 1!) = 6$ | 2.47 |
| III | 4 | 2 | 15 | $6!/(4!\ 2!) = 15$ | 3.74 |
| IV | 3 | 3 | 20 | $6!/(3!\ 3!) = 20$ | 4.13 |
| V | 2 | 4 | 15 | $6!/(2!\ 4!) = 15$ | 3.74 |
| VI | 1 | 5 | 6 | $6!/(1!\ 5!) = 6$ | 2.47 |
| VII | 0 | 6 | 1 | $6!/(0!\ 6!) = 1$ | 0 |
| | | | Total = 64 | | |

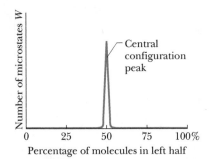

**Figure 20-18** For a *large* number of molecules in a box, a plot of the number of microstates that require various percentages of the molecules to be in the left half of the box. Nearly all the microstates correspond to an approximately equal sharing of the molecules between the two halves of the box; those microstates form the *central configuration peak* on the plot. For $N \approx 10^{22}$, the central configuration peak is much too narrow to be drawn on this plot.

You should verify the multiplicities for all the configurations in Table 20-1.

The basic assumption of statistical mechanics is that

 All microstates are equally probable.

In other words, if we were to take a great many snapshots of the six molecules as they jostle around in the box of Fig. 20-17 and then count the number of times each microstate occurred, we would find that all 64 microstates would occur equally often. Thus the system will spend, on average, the same amount of time in each of the 64 microstates.

Because all microstates are equally probable but different configurations have different numbers of microstates, the configurations are *not* all equally probable. In Table 20-1 configuration IV, with 20 microstates, is the *most probable configuration*, with a probability of 20/64 = 0.313. This result means that the system is in configuration IV 31.3% of the time. Configurations I and VII, in which all the molecules are in one half of the box, are the least probable, each with a probability of 1/64 = 0.016 or 1.6%. It is not surprising that the most probable configuration is the one in which the molecules are evenly divided between the two halves of the box, because that is what we expect at thermal equilibrium. However, it *is* surprising that there is *any* probability, however small, of finding all six molecules clustered in half of the box, with the other half empty.

For large values of $N$ there are extremely large numbers of microstates, but nearly all the microstates belong to the configuration in which the molecules are divided equally between the two halves of the box, as Fig. 20-18 indicates. Even though the measured temperature and pressure of the gas remain constant, the gas is churning away endlessly as its molecules "visit" all probable microstates with equal probability. However, because so few microstates lie outside the very narrow central configuration peak of Fig. 20-18, we might as well assume that the gas molecules are always divided equally between the two halves of the box. As we shall see, this is the configuration with the greatest entropy.

---

## Sample Problem 20.05  Microstates and multiplicity

Suppose that there are 100 indistinguishable molecules in the box of Fig. 20-17. How many microstates are associated with the configuration $n_1 = 50$ and $n_2 = 50$, and with the configuration $n_1 = 100$ and $n_2 = 0$? Interpret the results in terms of the relative probabilities of the two configurations.

### KEY IDEA

The multiplicity $W$ of a configuration of indistinguishable molecules in a closed box is the number of independent microstates with that configuration, as given by Eq. 20-20.

*Calculations:* Thus, for the $(n_1, n_2)$ configuration (50, 50),

$$W = \frac{N!}{n_1! \, n_2!} = \frac{100!}{50! \, 50!}$$

$$= \frac{9.33 \times 10^{157}}{(3.04 \times 10^{64})(3.04 \times 10^{64})}$$

$$= 1.01 \times 10^{29}. \qquad \text{(Answer)}$$

Similarly, for the configuration (100, 0), we have

$$W = \frac{N!}{n_1! \, n_2!} = \frac{100!}{100! \, 0!} = \frac{1}{0!} = \frac{1}{1} = 1. \quad \text{(Answer)}$$

*The meaning:* Thus, a 50–50 distribution is more likely than a 100–0 distribution by the enormous factor of about $1 \times 10^{29}$. If you could count, at one per nanosecond, the number of microstates that correspond to the 50–50 distribution, it would take you about $3 \times 10^{12}$ years, which is about 200 times longer than the age of the universe. Keep in mind that the 100 molecules used in this sample problem is a very small number. Imagine what these calculated probabilities would be like for a mole of molecules, say about $N = 10^{24}$. Thus, you need never worry about suddenly finding all the air molecules clustering in one corner of your room, with you gasping for air in another corner. So, you can breathe easy because of the physics of entropy.

Additional examples, video, and practice available at *WileyPLUS*

### Probability and Entropy

In 1877, Austrian physicist Ludwig Boltzmann (the Boltzmann of Boltzmann's constant $k$) derived a relationship between the entropy $S$ of a configuration of a gas and the multiplicity $W$ of that configuration. That relationship is

$$S = k \ln W \quad \text{(Boltzmann's entropy equation)}. \quad (20\text{-}21)$$

This famous formula is engraved on Boltzmann's tombstone.

It is natural that $S$ and $W$ should be related by a logarithmic function. The total entropy of two systems is the *sum* of their separate entropies. The probability of occurrence of two independent systems is the *product* of their separate probabilities. Because $\ln ab = \ln a + \ln b$, the logarithm seems the logical way to connect these quantities.

Table 20-1 displays the entropies of the configurations of the six-molecule system of Fig. 20-17, computed using Eq. 20-21. Configuration IV, which has the greatest multiplicity, also has the greatest entropy.

When you use Eq. 20-20 to calculate $W$, your calculator may signal "OVER-FLOW" if you try to find the factorial of a number greater than a few hundred. Instead, you can use **Stirling's approximation** for $\ln N!$:

$$\ln N! \approx N(\ln N) - N \quad \text{(Stirling's approximation)}. \quad (20\text{-}22)$$

The Stirling of this approximation was an English mathematician and not the Robert Stirling of engine fame.

### Checkpoint 5

A box contains 1 mol of a gas. Consider two configurations: (a) each half of the box contains half the molecules and (b) each third of the box contains one-third of the molecules. Which configuration has more microstates?

---

## Sample Problem 20.06    Entropy change of free expansion using microstates

In Sample Problem 20.01, we showed that when $n$ moles of an ideal gas doubles its volume in a free expansion, the entropy increase from the initial state $i$ to the final state $f$ is $S_f - S_i = nR \ln 2$. Derive this increase in entropy by using statistical mechanics.

### KEY IDEA

We can relate the entropy $S$ of any given configuration of the molecules in the gas to the multiplicity $W$ of microstates for that configuration, using Eq. 20-21 ($S = k \ln W$).

*Calculations:* We are interested in two configurations: the final configuration $f$ (with the molecules occupying the full volume of their container in Fig. 20-1b) and the initial configuration $i$ (with the molecules occupying the left half of the container). Because the molecules are in a closed container, we can calculate the multiplicity $W$ of their microstates with Eq. 20-20. Here we have $N$ molecules in the $n$ moles of the gas. Initially, with the molecules all in the left

half of the container, their $(n_1, n_2)$ configuration is $(N, 0)$. Then, Eq. 20-20 gives their multiplicity as

$$W_i = \frac{N!}{N!\,0!} = 1.$$

Finally, with the molecules spread through the full volume, their $(n_1, n_2)$ configuration is $(N/2, N/2)$. Then, Eq. 20-20 gives their multiplicity as

$$W_f = \frac{N!}{(N/2)!\,(N/2)!}.$$

From Eq. 20-21, the initial and final entropies are

$$S_i = k \ln W_i = k \ln 1 = 0$$

and

$$S_f = k \ln W_f = k \ln(N!) - 2k \ln[(N/2)!]. \quad (20\text{-}23)$$

In writing Eq. 20-23, we have used the relation

$$\ln \frac{a}{b^2} = \ln a - 2 \ln b.$$

Now, applying Eq. 20-22 to evaluate Eq. 20-23, we find that

$$S_f = k \ln(N!) - 2k \ln[(N/2)!]$$

$$= k[N(\ln N) - N] - 2k[(N/2) \ln(N/2) - (N/2)]$$

$$= k[N(\ln N) - N - N \ln(N/2) + N]$$

$$= k[N(\ln N) - N(\ln N - \ln 2)] = Nk \ln 2. \qquad (20\text{-}24)$$

From Eq. 19-8 we can substitute $nR$ for $Nk$, where $R$ is the universal gas constant. Equation 20-24 then becomes

$$S_f = nR \ln 2.$$

The change in entropy from the initial state to the final is

thus

$$S_f - S_i = nR \ln 2 - 0$$

$$= nR \ln 2, \qquad \text{(Answer)}$$

which is what we set out to show. In the first sample problem of this chapter we calculated this entropy increase for a free expansion with thermodynamics by finding an equivalent reversible process and calculating the entropy change for *that* process in terms of temperature and heat transfer. In this sample problem, we calculate the same increase in entropy with statistical mechanics using the fact that the system consists of molecules. In short, the two, very different approaches give the same answer.

 **WILEY PLUS** Additional examples, video, and practice available at *WileyPLUS*

## Review & Summary

**One-Way Processes** An **irreversible process** is one that cannot be reversed by means of small changes in the environment. The direction in which an irreversible process proceeds is set by the *change in entropy* $\Delta S$ of the system undergoing the process. Entropy $S$ is a *state property* (or *state function*) of the system; that is, it depends only on the state of the system and not on the way in which the system reached that state. The *entropy postulate* states (in part): *If an irreversible process occurs in a closed system, the entropy of the system always increases.*

**Calculating Entropy Change** The **entropy change** $\Delta S$ for an irreversible process that takes a system from an initial state $i$ to a final state $f$ is exactly equal to the entropy change $\Delta S$ for *any reversible process* that takes the system between those same two states. We can compute the latter (but not the former) with

$$\Delta S = S_f - S_i = \int_i^f \frac{dQ}{T}. \qquad (20\text{-}1)$$

Here $Q$ is the energy transferred as heat to or from the system during the process, and $T$ is the temperature of the system in kelvins during the process.

For a reversible isothermal process, Eq. 20-1 reduces to

$$\Delta S = S_f - S_i = \frac{Q}{T}. \qquad (20\text{-}2)$$

When the temperature change $\Delta T$ of a system is small relative to the temperature (in kelvins) before and after the process, the entropy change can be approximated as

$$\Delta S = S_f - S_i \approx \frac{Q}{T_{\text{avg}}}, \qquad (20\text{-}3)$$

where $T_{\text{avg}}$ is the system's average temperature during the process.

When an ideal gas changes reversibly from an initial state with temperature $T_i$ and volume $V_i$ to a final state with temperature $T_f$ and volume $V_f$, the change $\Delta S$ in the entropy of the gas is

$$\Delta S = S_f - S_i = nR \ln \frac{V_f}{V_i} + nC_V \ln \frac{T_f}{T_i}. \qquad (20\text{-}4)$$

**The Second Law of Thermodynamics** This law, which is an extension of the entropy postulate, states: *If a process occurs in a closed system, the entropy of the system increases for irreversible processes and remains constant for reversible processes. It never decreases.* In equation form,

$$\Delta S \geq 0. \qquad (20\text{-}5)$$

**Engines** An **engine** is a device that, operating in a cycle, extracts energy as heat $|Q_H|$ from a high-temperature reservoir and does a certain amount of work $|W|$. The *efficiency* $\varepsilon$ of any engine is defined as

$$\varepsilon = \frac{\text{energy we get}}{\text{energy we pay for}} = \frac{|W|}{|Q_H|}. \qquad (20\text{-}11)$$

In an **ideal engine**, all processes are reversible and no wasteful energy transfers occur due to, say, friction and turbulence. A **Carnot engine** is an ideal engine that follows the cycle of Fig. 20-9. Its efficiency is

$$\varepsilon_C = 1 - \frac{|Q_L|}{|Q_H|} = 1 - \frac{T_L}{T_H}, \qquad (20\text{-}12, 20\text{-}13)$$

in which $T_H$ and $T_L$ are the temperatures of the high- and low-temperature reservoirs, respectively. Real engines always have an efficiency lower than that given by Eq. 20-13. Ideal engines that are not Carnot engines also have lower efficiencies.

A *perfect engine* is an imaginary engine in which energy extracted as heat from the high-temperature reservoir is converted completely to work. Such an engine would violate the second law of thermodynamics, which can be restated as follows: No series of processes is possible whose sole result is the absorption of energy as heat from a thermal reservoir and the complete conversion of this energy to work.

**Refrigerators** A refrigerator is a device that, operating in a cycle, has work $W$ done on it as it extracts energy $|Q_L|$ as heat from a low-temperature reservoir. The coefficient of performance $K$ of a refrigerator is defined as

$$K = \frac{\text{what we want}}{\text{what we pay for}} = \frac{|Q_L|}{|W|}. \qquad (20\text{-}14)$$

A **Carnot refrigerator** is a Carnot engine operating in reverse.

For a Carnot refrigerator, Eq. 20-14 becomes

$$K_C = \frac{|Q_L|}{|Q_H| - |Q_L|} = \frac{T_L}{T_H - T_L}. \qquad \text{(20-15, 20-16)}$$

A *perfect refrigerator* is an imaginary refrigerator in which energy extracted as heat from the low-temperature reservoir is converted completely to heat discharged to the high-temperature reservoir, without any need for work. Such a refrigerator would violate the second law of thermodynamics, which can be restated as follows: No series of processes is possible whose sole result is the transfer of energy as heat from a reservoir at a given temperature to a reservoir at a higher temperature.

**Entropy from a Statistical View**   The entropy of a system can be defined in terms of the possible distributions of its molecules. For identical molecules, each possible distribution of molecules is called a **microstate** of the system. All equivalent microstates are grouped into a **configuration** of the system. The number of microstates in a configuration is the **multiplicity** $W$ of the configuration.

For a system of $N$ molecules that may be distributed between the two halves of a box, the multiplicity is given by

$$W = \frac{N!}{n_1! \, n_2!}, \qquad \text{(20-20)}$$

in which $n_1$ is the number of molecules in one half of the box and $n_2$ is the number in the other half. A basic assumption of **statistical mechanics** is that all the microstates are equally probable. Thus, configurations with a large multiplicity occur most often.

The multiplicity $W$ of a configuration of a system and the entropy $S$ of the system in that configuration are related by Boltzmann's entropy equation:

$$S = k \ln W, \qquad \text{(20-21)}$$

where $k = 1.38 \times 10^{-23}$ J/K is the Boltzmann constant.

# Questions

**1**   Point $i$ in Fig. 20-19 represents the initial state of an ideal gas at temperature $T$. Taking algebraic signs into account, rank the entropy changes that the gas undergoes as it moves, successively and reversibly, from point $i$ to points $a$, $b$, $c$, and $d$, greatest first.

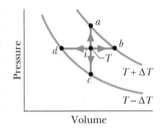

**Figure 20-19**   Question 1.

**2**   In four experiments, blocks $A$ and $B$, starting at different initial temperatures, were brought together in an insulating box and allowed to reach a common final temperature. The entropy changes for the blocks in the four experiments had the following values (in joules per kelvin), but not necessarily in the order given. Determine which values for $A$ go with which values for $B$.

| Block | | Values | | |
|---|---|---|---|---|
| $A$ | 8 | 5 | 3 | 9 |
| $B$ | $-3$ | $-8$ | $-5$ | $-2$ |

**3**   A gas, confined to an insulated cylinder, is compressed adiabatically to half its volume. Does the entropy of the gas increase, decrease, or remain unchanged during this process?

**4**   An ideal monatomic gas at initial temperature $T_0$ (in kelvins) expands from initial volume $V_0$ to volume $2V_0$ by each of the five processes indicated in the $T$-$V$ diagram of Fig. 20-20. In which process is the expan-

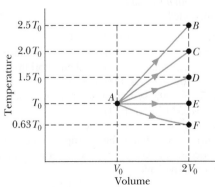

**Figure 20-20**
Question 4.

sion (a) isothermal, (b) isobaric (constant pressure), and (c) adiabatic? Explain your answers. (d) In which processes does the entropy of the gas decrease?

**5**   In four experiments, 2.5 mol of hydrogen gas undergoes reversible isothermal expansions, starting from the same volume but at different temperatures. The corresponding $p$-$V$ plots are shown in Fig. 20-21. Rank the situations according to the change in the entropy of the gas, greatest first.

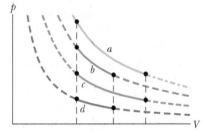

**Figure 20-21**   Question 5.

**6**   A box contains 100 atoms in a configuration that has 50 atoms in each half of the box. Suppose that you could count the different microstates associated with this configuration at the rate of 100 billion states per second, using a supercomputer. Without written calculation, guess how much computing time you would need: a day, a year, or much more than a year.

**7**   Does the entropy per cycle increase, decrease, or remain the same for (a) a Carnot engine, (b) a real engine, and (c) a perfect engine (which is, of course, impossible to build)?

**8**   Three Carnot engines operate between temperature limits of (a) 400 and 500 K, (b) 500 and 600 K, and (c) 400 and 600 K. Each engine extracts the same amount of energy per cycle from the high-temperature reservoir. Rank the magnitudes of the work done by the engines per cycle, greatest first.

**9**   An inventor claims to have invented four engines, each of which operates between constant-temperature reservoirs at 400 and 300 K. Data on each engine, per cycle of operation, are: engine A, $Q_H = 200$ J, $Q_L = -175$ J, and $W = 40$ J; engine B, $Q_H = 500$ J, $Q_L = -200$ J, and $W = 400$ J; engine C, $Q_H = 600$ J, $Q_L = -200$ J, and $W = 400$ J; engine D, $Q_H = 100$ J, $Q_L = -90$ J, and $W = 10$ J. Of the first and second laws of thermodynamics, which (if either) does each engine violate?

**10**   Does the entropy per cycle increase, decrease, or remain the same for (a) a Carnot refrigerator, (b) a real refrigerator, and (c) a perfect refrigerator (which is, of course, impossible to build)?

## Problems

### Module 20-1 Entropy

**•1 SSM** Suppose 4.00 mol of an ideal gas undergoes a reversible isothermal expansion from volume $V_1$ to volume $V_2 = 2.00V_1$ at temperature $T = 400$ K. Find (a) the work done by the gas and (b) the entropy change of the gas. (c) If the expansion is reversible and adiabatic instead of isothermal, what is the entropy change of the gas?

**•2** An ideal gas undergoes a reversible isothermal expansion at 77.0°C, increasing its volume from 1.30 L to 3.40 L. The entropy change of the gas is 22.0 J/K. How many moles of gas are present?

**•3 ILW** A 2.50 mol sample of an ideal gas expands reversibly and isothermally at 360 K until its volume is doubled. What is the increase in entropy of the gas?

**•4** How much energy must be transferred as heat for a reversible isothermal expansion of an ideal gas at 132°C if the entropy of the gas increases by 46.0 J/K?

**•5 ILW** Find (a) the energy absorbed as heat and (b) the change in entropy of a 2.00 kg block of copper whose temperature is increased reversibly from 25.0°C to 100°C. The specific heat of copper is 386 J/kg · K.

**•6** (a) What is the entropy change of a 12.0 g ice cube that melts completely in a bucket of water whose temperature is just above the freezing point of water? (b) What is the entropy change of a 5.00 g spoonful of water that evaporates completely on a hot plate whose temperature is slightly above the boiling point of water?

**••7 ILW** A 50.0 g block of copper whose temperature is 400 K is placed in an insulating box with a 100 g block of lead whose temperature is 200 K. (a) What is the equilibrium temperature of the two-block system? (b) What is the change in the internal energy of the system between the initial state and the equilibrium state? (c) What is the change in the entropy of the system? (See Table 18-3.)

**••8** At very low temperatures, the molar specific heat $C_V$ of many solids is approximately $C_V = AT^3$, where $A$ depends on the particular substance. For aluminum, $A = 3.15 \times 10^{-5}$ J/mol · K⁴. Find the entropy change for 4.00 mol of aluminum when its temperature is raised from 5.00 K to 10.0 K.

**••9** A 10 g ice cube at −10°C is placed in a lake whose temperature is 15°C. Calculate the change in entropy of the cube−lake system as the ice cube comes to thermal equilibrium with the lake. The specific heat of ice is 2220 J/kg · K. (*Hint:* Will the ice cube affect the lake temperature?)

**••10** A 364 g block is put in contact with a thermal reservoir. The block is initially at a lower temperature than the reservoir. Assume that the consequent transfer of energy as heat from the reservoir to the block is reversible. Figure 20-22

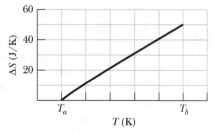

**Figure 20-22** Problem 10.

gives the change in entropy $\Delta S$ of the block until thermal equilibrium is reached. The scale of the horizontal axis is set by $T_a = 280$ K and $T_b = 380$ K. What is the specific heat of the block?

**••11 SSM WWW** In an experiment, 200 g of aluminum (with a specific heat of 900 J/kg · K) at 100°C is mixed with 50.0 g of water at 20.0°C, with the mixture thermally isolated. (a) What is the equilibrium temperature? What are the entropy changes of (b) the aluminum, (c) the water, and (d) the aluminum−water system?

**••12** A gas sample undergoes a reversible isothermal expansion. Figure 20-23 gives the change $\Delta S$ in entropy of the gas versus the final volume $V_f$ of the gas. The scale of the vertical axis is set by $\Delta S_s = 64$ J/K. How many moles are in the sample?

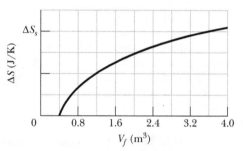

**Figure 20-23** Problem 12.

**••13** In the irreversible process of Fig. 20-5, let the initial temperatures of the identical blocks $L$ and $R$ be 305.5 and 294.5 K, respectively, and let 215 J be the energy that must be transferred between the blocks in order to reach equilibrium. For the reversible processes of Fig. 20-6, what is $\Delta S$ for (a) block $L$, (b) its reservoir, (c) block $R$, (d) its reservoir, (e) the two-block system, and (f) the system of the two blocks and the two reservoirs?

**••14** (a) For 1.0 mol of a monatomic ideal gas taken through the cycle in Fig. 20-24, where $V_1 = 4.00V_0$, what is $W/p_0V_0$ as the gas goes from state $a$ to state $c$ along path $abc$? What is $\Delta E_{int}/p_0V_0$ in going (b) from $b$ to $c$ and (c) through one full cycle? What is $\Delta S$ in going (d) from $b$ to $c$ and (e) through one full cycle?

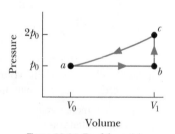

**Figure 20-24** Problem 14.

**••15** A mixture of 1773 g of water and 227 g of ice is in an initial equilibrium state at 0.000°C. The mixture is then, in a reversible process, brought to a second equilibrium state where the water−ice ratio, by mass, is 1.00:1.00 at 0.000°C. (a) Calculate the entropy change of the system during this process. (The heat of fusion for water is 333 kJ/kg.) (b) The system is then returned to the initial equilibrium state in an irreversible process (say, by using a Bunsen burner). Calculate the entropy change of the system during this process. (c) Are your answers consistent with the second law of thermodynamics?

••16 🔵 An 8.0 g ice cube at −10°C is put into a Thermos flask containing 100 cm³ of water at 20°C. By how much has the entropy of the cube–water system changed when equilibrium is reached? The specific heat of ice is 2220 J/kg · K.

••17 In Fig. 20-25, where $V_{23} = 3.00V_1$, n moles of a diatomic ideal gas are taken through the cycle with the molecules rotating but not oscillating. What are (a) $p_2/p_1$, (b) $p_3/p_1$, and (c) $T_3/T_1$? For path 1 → 2, what are (d) $W/nRT_1$, (e) $Q/nRT_1$, (f) $\Delta E_{int}/nRT_1$, and (g) $\Delta S/nR$? For path 2 → 3, what are (h) $W/nRT_1$, (i) $Q/nRT_1$, (j) $\Delta E_{int}/nRT_1$, (k) $\Delta S/nR$? For path 3 → 1, what are (l) $W/nRT_1$, (m) $Q/nRT_1$, (n) $\Delta E_{int}/nRT_1$, and (o) $\Delta S/nR$?

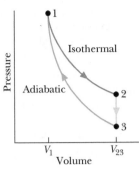

**Figure 20-25** Problem 17.

••18 🔵 A 2.0 mol sample of an ideal monatomic gas undergoes the reversible process shown in Fig. 20-26. The scale of the vertical axis is set by $T_s = 400.0$ K and the scale of the horizontal axis is set by $S_s = 20.0$ J/K. (a) How much energy is absorbed as heat by the gas? (b) What is the change in the internal energy of the gas? (c) How much work is done by the gas?

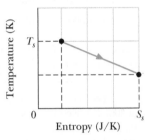

**Figure 20-26** Problem 18.

•••19 Suppose 1.00 mol of a monatomic ideal gas is taken from initial pressure $p_1$ and volume $V_1$ through two steps: (1) an isothermal expansion to volume $2.00V_1$ and (2) a pressure increase to $2.00p_1$ at constant volume. What is $Q/p_1V_1$ for (a) step 1 and (b) step 2? What is $W/p_1V_1$ for (c) step 1 and (d) step 2? For the full process, what are (e) $\Delta E_{int}/p_1V_1$ and (f) $\Delta S$? The gas is returned to its initial state and again taken to the same final state but now through these two steps: (1) an isothermal compression to pressure $2.00p_1$ and (2) a volume increase to $2.00V_1$ at constant pressure. What is $Q/p_1V_1$ for (g) step 1 and (h) step 2? What is $W/p_1V_1$ for (i) step 1 and (j) step 2? For the full process, what are (k) $\Delta E_{int}/p_1V_1$ and (l) $\Delta S$?

•••20 Expand 1.00 mol of an monatomic gas initially at 5.00 kPa and 600 K from initial volume $V_i = 1.00$ m³ to final volume $V_f = 2.00$ m³. At any instant during the expansion, the pressure p and volume V of the gas are related by $p = 5.00 \exp[(V_i − V)/a]$, with p in kilopascals, $V_i$ and V in cubic meters, and a = 1.00 m³. What are the final (a) pressure and (b) temperature of the gas? (c) How much work is done by the gas during the expansion? (d) What is $\Delta S$ for the expansion? (*Hint:* Use two simple reversible processes to find $\Delta S$.)

•••21 🔵 ✈ Energy can be removed from water as heat at and even below the normal freezing point (0.0°C at atmospheric pressure) without causing the water to freeze; the water is then said to be *supercooled*. Suppose a 1.00 g water drop is supercooled until its temperature is that of the surrounding air, which is at −5.00°C. The drop then suddenly and irreversibly freezes, transferring energy to the air as heat. What is the entropy change for the drop? (*Hint:* Use a three-step reversible process as if the water were taken through the normal freezing point.) The specific heat of ice is 2220 J/kg · K.

•••22 🔵 An insulated Thermos contains 130 g of water at 80.0°C. You put in a 12.0 g ice cube at 0°C to form a system of *ice + original water.* (a) What is the equilibrium temperature of the system? What are the entropy changes of the water that was originally the ice cube (b) as it melts and (c) as it warms to the equilibrium temperature? (d) What is the entropy change of the original water as it cools to the equilibrium temperature? (e) What is the net entropy change of the *ice + original water* system as it reaches the equilibrium temperature?

## Module 20-2 **Entropy in the Real World: Engines**

•23 A Carnot engine whose low-temperature reservoir is at 17°C has an efficiency of 40%. By how much should the temperature of the high-temperature reservoir be increased to increase the efficiency to 50%?

•24 A Carnot engine absorbs 52 kJ as heat and exhausts 36 kJ as heat in each cycle. Calculate (a) the engine's efficiency and (b) the work done per cycle in kilojoules.

•25 A Carnot engine has an efficiency of 22.0%. It operates between constant-temperature reservoirs differing in temperature by 75.0 C°. What is the temperature of the (a) lower-temperature and (b) higher-temperature reservoir?

•26 In a hypothetical nuclear fusion reactor, the fuel is deuterium gas at a temperature of $7 \times 10^8$ K. If this gas could be used to operate a Carnot engine with $T_L = 100$°C, what would be the engine's efficiency? Take both temperatures to be exact and report your answer to seven significant figures.

•27 SSM WWW A Carnot engine operates between 235°C and 115°C, absorbing $6.30 \times 10^4$ J per cycle at the higher temperature. (a) What is the efficiency of the engine? (b) How much work per cycle is this engine capable of performing?

••28 In the first stage of a two-stage Carnot engine, energy is absorbed as heat $Q_1$ at temperature $T_1$, work $W_1$ is done, and energy is expelled as heat $Q_2$ at a lower temperature $T_2$. The second stage absorbs that energy as heat $Q_2$, does work $W_2$, and expels energy as heat $Q_3$ at a still lower temperature $T_3$. Prove that the efficiency of the engine is $(T_1 − T_3)/T_1$.

••29 🔵 Figure 20-27 shows a reversible cycle through which 1.00 mol of a monatomic ideal gas is taken. Assume that $p = 2p_0$, $V = 2V_0$, $p_0 = 1.01 \times 10^5$ Pa, and $V_0 = 0.0225$ m³. Calculate (a) the work done during the cycle, (b) the energy added as heat during stroke *abc,* and (c) the efficiency of the cycle. (d) What is the efficiency of a Carnot engine operating between the highest and lowest temperatures that occur in the cycle? (e) Is this greater than or less than the efficiency calculated in (c)?

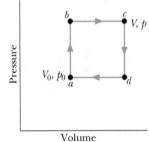

**Figure 20-27** Problem 29.

••30 A 500 W Carnot engine operates between constant-temperature reservoirs at 100°C and 60.0°C. What is the rate at which energy is (a) taken in by the engine as heat and (b) exhausted by the engine as heat?

••31 The efficiency of a particular car engine is 25% when the engine does 8.2 kJ of work per cycle. Assume the process is reversible. What are (a) the energy the engine gains per cycle as heat $Q_{gain}$ from the fuel combustion and (b) the energy the engine loses per cycle as heat $Q_{lost}$? If a tune-up increases the efficiency to 31%, what are (c) $Q_{gain}$ and (d) $Q_{lost}$ at the same work value?

**••32** GO A Carnot engine is set up to produce a certain work $W$ per cycle. In each cycle, energy in the form of heat $Q_H$ is transferred to the working substance of the engine from the higher-temperature thermal reservoir, which is at an adjustable temperature $T_H$. The lower-temperature thermal reservoir is maintained at temperature $T_L = 250$ K. Figure 20-28 gives $Q_H$ for a range of $T_H$. The scale of the vertical axis is set by $Q_{Hs} = 6.0$ kJ. If $T_H$ is set at 550 K, what is $Q_H$?

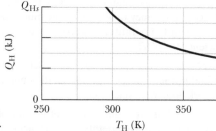

**Figure 20-28** Problem 32.

**••33** SSM ILW Figure 20-29 shows a reversible cycle through which 1.00 mol of a monatomic ideal gas is taken. Volume $V_c = 8.00V_b$. Process $bc$ is an adiabatic expansion, with $p_b = 10.0$ atm and $V_b = 1.00 \times 10^{-3}$ m³. For the cycle, find (a) the energy added to the gas as heat, (b) the energy leaving the gas as heat, (c) the net work done by the gas, and (d) the efficiency of the cycle.

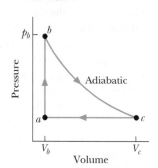

**Figure 20-29** Problem 33.

**••34** GO An ideal gas (1.0 mol) is the working substance in an engine that operates on the cycle shown in Fig. 20-30. Processes $BC$ and $DA$ are reversible and adiabatic. (a) Is the gas monatomic, diatomic, or polyatomic? (b) What is the engine efficiency?

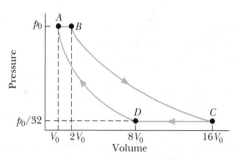

**Figure 20-30** Problem 34.

**•••35** The cycle in Fig. 20-31 represents the operation of a gasoline internal combustion engine. Volume $V_3 = 4.00V_1$. Assume the gasoline–air intake mixture is an ideal gas with $\gamma = 1.30$. What are the ratios (a) $T_2/T_1$, (b) $T_3/T_1$, (c) $T_4/T_1$, (d) $p_3/p_1$, and (e) $p_4/p_1$? (f) What is the engine efficiency?

### Module 20-3  Refrigerators and Real Engines

**•36** How much work must be done by a Carnot refrigerator to transfer 1.0

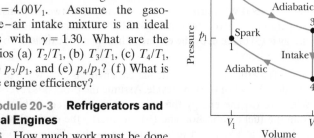

**Figure 20-31** Problem 35.

J as heat (a) from a reservoir at 7.0°C to one at 27°C, (b) from a reservoir at −73°C to one at 27°C, (c) from a reservoir at −173°C to one at 27°C, and (d) from a reservoir at −223°C to one at 27°C?

**•37** SSM A heat pump is used to heat a building. The external temperature is less than the internal temperature. The pump's coefficient of performance is 3.8, and the heat pump delivers 7.54 MJ as heat to the building each hour. If the heat pump is a Carnot engine working in reverse, at what rate must work be done to run it?

**•38** The electric motor of a heat pump transfers energy as heat from the outdoors, which is at −5.0°C, to a room that is at 17°C. If the heat pump were a Carnot heat pump (a Carnot engine working in reverse), how much energy would be transferred as heat to the room for each joule of electric energy consumed?

**•39** SSM A Carnot air conditioner takes energy from the thermal energy of a room at 70°F and transfers it as heat to the outdoors, which is at 96°F. For each joule of electric energy required to operate the air conditioner, how many joules are removed from the room?

**•40** To make ice, a freezer that is a reverse Carnot engine extracts 42 kJ as heat at −15°C during each cycle, with coefficient of performance 5.7. The room temperature is 30.3°C. How much (a) energy per cycle is delivered as heat to the room and (b) work per cycle is required to run the freezer?

**••41** ILW An air conditioner operating between 93°F and 70°F is rated at 4000 Btu/h cooling capacity. Its coefficient of performance is 27% of that of a Carnot refrigerator operating between the same two temperatures. What horsepower is required of the air conditioner motor?

**••42** The motor in a refrigerator has a power of 200 W. If the freezing compartment is at 270 K and the outside air is at 300 K, and assuming the efficiency of a Carnot refrigerator, what is the maximum amount of energy that can be extracted as heat from the freezing compartment in 10.0 min?

**••43** GO Figure 20-32 represents a Carnot engine that works between temperatures $T_1 = 400$ K and $T_2 = 150$ K and drives a Carnot refrigerator that works between temperatures $T_3 = 325$ K and $T_4 = 225$ K. What is the ratio $Q_3/Q_1$?

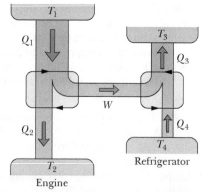

**Figure 20-32** Problem 43.

**••44** (a) During each cycle, a Carnot engine absorbs 750 J as heat from a high-temperature reservoir at 360 K, with the low-temperature reservoir at 280 K. How much work is done per cycle? (b) The engine is then made to work in reverse to function as a Carnot refrigerator between those same two reservoirs. During each cycle, how much work is required to remove 1200 J as heat from the low-temperature reservoir?

### Module 20-4  A Statistical View of Entropy

**•45** Construct a table like Table 20-1 for eight molecules.

**••46** A box contains $N$ identical gas molecules equally divided between its two halves. For $N = 50$, what are (a) the multiplicity $W$ of the central configuration, (b) the total number of microstates, and (c) the percentage of the time the system spends in the central configuration? For $N = 100$, what are (d) $W$ of the central configuration, (e) the total number of microstates, and (f) the percentage of the time the system spends in the central configuration? For $N = 200$, what are (g) $W$ of the central configuration, (h) the total number of microstates, and (i) the percentage of the time the system spends in the central configuration? (j) Does the time spent in the central configuration increase or decrease with an increase in $N$?

•••47 SSM WWW A box contains $N$ gas molecules. Consider the box to be divided into three equal parts. (a) By extension of Eq. 20-20, write a formula for the multiplicity of any given configuration. (b) Consider two configurations: configuration $A$ with equal numbers of molecules in all three thirds of the box, and configuration $B$ with equal numbers of molecules in each half of the box divided into two equal parts rather than three. What is the ratio $W_A/W_B$ of the multiplicity of configuration $A$ to that of configuration $B$? (c) Evaluate $W_A/W_B$ for $N = 100$. (Because 100 is not evenly divisible by 3, put 34 molecules into one of the three box parts of configuration $A$ and 33 in each of the other two parts.)

### Additional Problems

48 Four particles are in the insulated box of Fig. 20-17. What are (a) the least multiplicity, (b) the greatest multiplicity, (c) the least entropy, and (d) the greatest entropy of the four-particle system?

49 A cylindrical copper rod of length 1.50 m and radius 2.00 cm is insulated to prevent heat loss through its curved surface. One end is attached to a thermal reservoir fixed at 300°C; the other is attached to a thermal reservoir fixed at 30.0°C. What is the rate at which entropy increases for the rod–reservoirs system?

50 Suppose 0.550 mol of an ideal gas is isothermally and reversibly expanded in the four situations given below. What is the change in the entropy of the gas for each situation?

| Situation | (a) | (b) | (c) | (d) |
|---|---|---|---|---|
| Temperature (K) | 250 | 350 | 400 | 450 |
| Initial volume (cm³) | 0.200 | 0.200 | 0.300 | 0.300 |
| Final volume (cm³) | 0.800 | 0.800 | 1.20 | 1.20 |

51 SSM As a sample of nitrogen gas ($N_2$) undergoes a temperature increase at constant volume, the distribution of molecular speeds increases. That is, the probability distribution function $P(v)$ for the molecules spreads to higher speed values, as suggested in Fig. 19-8b. One way to report the spread in $P(v)$ is to measure the difference $\Delta v$ between the most probable speed $v_P$ and the rms speed $v_{rms}$. When $P(v)$ spreads to higher speeds, $\Delta v$ increases. Assume that the gas is ideal and the $N_2$ molecules rotate but do not oscillate. For 1.5 mol, an initial temperature of 250 K, and a final temperature of 500 K, what are (a) the initial difference $\Delta v_i$, (b) the final difference $\Delta v_f$, and (c) the entropy change $\Delta S$ for the gas?

52 Suppose 1.0 mol of a monatomic ideal gas initially at 10 L and 300 K is heated at constant volume to 600 K, allowed to expand isothermally to its initial pressure, and finally compressed at constant pressure to its original volume, pressure, and temperature. During the cycle, what are (a) the net energy entering the system (the gas) as heat and (b) the net work done by the gas? (c) What is the efficiency of the cycle?

53 GO Suppose that a deep shaft were drilled in Earth's crust near one of the poles, where the surface temperature is −40°C, to a depth where the temperature is 800°C. (a) What is the theoretical limit to the efficiency of an engine operating between these

temperatures? (b) If all the energy released as heat into the low-temperature reservoir were used to melt ice that was initially at −40°C, at what rate could liquid water at 0°C be produced by a 100 MW power plant (treat it as an engine)? The specific heat of ice is 2220 J/kg·K; water's heat of fusion is 333 kJ/kg. (Note that the engine can operate only between 0°C and 800°C in this case. Energy exhausted at −40°C cannot warm anything above −40°C.)

54 What is the entropy change for 3.20 mol of an ideal monatomic gas undergoing a reversible increase in temperature from 380 K to 425 K at constant volume?

55 A 600 g lump of copper at 80.0°C is placed in 70.0 g of water at 10.0°C in an insulated container. (See Table 18-3 for specific heats.) (a) What is the equilibrium temperature of the copper–water system? What entropy changes do (b) the copper, (c) the water, and (d) the copper–water system undergo in reaching the equilibrium temperature?

56 〜〜 Figure 20-33 gives the force magnitude $F$ versus stretch distance $x$ for a rubber band, with the scale of the $F$ axis set by $F_s = 1.50$ N and the scale of the $x$ axis set by $x_s = 3.50$ cm. The temperature is 2.00°C. When the rubber band is stretched by $x = 1.70$ cm, at what rate does the entropy of the rubber band change during a small additional stretch?

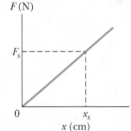

**Figure 20-33**
Problem 56.

57 The temperature of 1.00 mol of a monatomic ideal gas is raised reversibly from 300 K to 400 K, with its volume kept constant. What is the entropy change of the gas?

58 Repeat Problem 57, with the pressure now kept constant.

59 SSM A 0.600 kg sample of water is initially ice at temperature −20°C. What is the sample's entropy change if its temperature is increased to 40°C?

60 A three-step cycle is undergone by 3.4 mol of an ideal diatomic gas: (1) the temperature of the gas is increased from 200 K to 500 K at constant volume; (2) the gas is then isothermally expanded to its original pressure; (3) the gas is then contracted at constant pressure back to its original volume. Throughout the cycle, the molecules rotate but do not oscillate. What is the efficiency of the cycle?

61 An inventor has built an engine X and claims that its efficiency $\varepsilon_X$ is greater than the efficiency $\varepsilon$ of an ideal engine operating between the same two temperatures. Suppose you couple engine X to an ideal refrigerator (Fig. 20-34a) and adjust the cycle

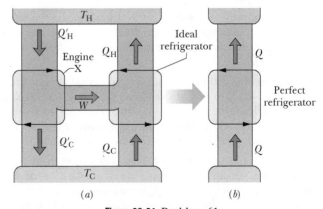

(a)                                              (b)

**Figure 20-34** Problem 61.

of engine X so that the work per cycle it provides equals the work per cycle required by the ideal refrigerator. Treat this combination as a single unit and show that if the inventor's claim were true (if $\varepsilon_X > \varepsilon$), the combined unit would act as a perfect refrigerator (Fig. 20-34b), transferring energy as heat from the low-temperature reservoir to the high-temperature reservoir without the need for work.

**62** Suppose 2.00 mol of a diatomic gas is taken reversibly around the cycle shown in the T-S diagram of Fig. 20-35, where $S_1 = 6.00$ J/K and $S_2 = 8.00$ J/K. The molecules do not rotate or oscillate. What is the energy transferred as heat Q for (a) path $1 \rightarrow 2$, (b) path $2 \rightarrow 3$, and (c) the full cycle? (d) What is the work W for the isothermal process? The volume $V_1$ in state 1 is 0.200 m³. What is the volume in (e) state 2 and (f) state 3?

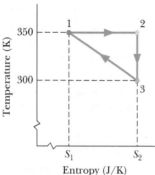
Figure 20-35 Problem 62.

What is the change $\Delta E_{int}$ for (g) path $1 \rightarrow 2$, (h) path $2 \rightarrow 3$, and (i) the full cycle? (*Hint:* (h) can be done with one or two lines of calculation using Module 19-7 or with a page of calculation using Module 19-9.) (j) What is the work W for the adiabatic process?

**63** A three-step cycle is undergone reversibly by 4.00 mol of an ideal gas: (1) an adiabatic expansion that gives the gas 2.00 times its initial volume, (2) a constant-volume process, (3) an isothermal compression back to the initial state of the gas. We do not know whether the gas is monatomic or diatomic; if it is diatomic, we do not know whether the molecules are rotating or oscillating. What are the entropy changes for (a) the cycle, (b) process 1, (c) process 3, and (d) process 2?

**64** (a) A Carnot engine operates between a hot reservoir at 320 K and a cold one at 260 K. If the engine absorbs 500 J as heat per cycle at the hot reservoir, how much work per cycle does it deliver? (b) If the engine working in reverse functions as a refrigerator between the same two reservoirs, how much work per cycle must be supplied to remove 1000 J as heat from the cold reservoir?

**65** A 2.00 mol diatomic gas initially at 300 K undergoes this cycle: It is (1) heated at constant volume to 800 K, (2) then allowed to expand isothermally to its initial pressure, (3) then compressed at constant pressure to its initial state. Assuming the gas molecules neither rotate nor oscillate, find (a) the net energy transferred as heat to the gas, (b) the net work done by the gas, and (c) the efficiency of the cycle.

**66** An ideal refrigerator does 150 J of work to remove 560 J as heat from its cold compartment. (a) What is the refrigerator's coefficient of performance? (b) How much heat per cycle is exhausted to the kitchen?

**67** Suppose that 260 J is conducted from a constant-temperature reservoir at 400 K to one at (a) 100 K, (b) 200 K, (c) 300 K, and (d) 360 K. What is the net change in entropy $\Delta S_{net}$ of the reservoirs in each case? (e) As the temperature difference of the two reservoirs decreases, does $\Delta S_{net}$ increase, decrease, or remain the same?

**68** An apparatus that liquefies helium is in a room maintained at 300 K. If the helium in the apparatus is at 4.0 K, what is the minimum ratio $Q_{to}/Q_{from}$, where $Q_{to}$ is the energy delivered as heat to the room and $Q_{from}$ is the energy removed as heat from the helium?

**69** A brass rod is in thermal contact with a constant-temperature reservoir at 130°C at one end and a constant-temperature reservoir at 24.0°C at the other end. (a) Compute the total change in entropy of the rod–reservoirs system when 5030 J of energy is conducted through the rod, from one reservoir to the other. (b) Does the entropy of the rod change?

**70** A 45.0 g block of tungsten at 30.0°C and a 25.0 g block of silver at −120°C are placed together in an insulated container. (See Table 18-3 for specific heats.) (a) What is the equilibrium temperature? What entropy changes do (b) the tungsten, (c) the silver, and (d) the tungsten–silver system undergo in reaching the equilibrium temperature?

**71** A box contains N molecules. Consider two configurations: configuration A with an equal division of the molecules between the two halves of the box, and configuration B with 60.0% of the molecules in the left half of the box and 40.0% in the right half. For $N = 50$, what are (a) the multiplicity $W_A$ of configuration A, (b) the multiplicity $W_B$ of configuration B, and (c) the ratio $f_{B/A}$ of the time the system spends in configuration B to the time it spends in configuration A? For $N = 100$, what are (d) $W_A$, (e) $W_B$, and (f) $f_{B/A}$? For $N = 200$, what are (g) $W_A$, (h) $W_B$, and (i) $f_{B/A}$? (j) With increasing N, does f increase, decrease, or remain the same?

**72** Calculate the efficiency of a fossil-fuel power plant that consumes 380 metric tons of coal each hour to produce useful work at the rate of 750 MW. The heat of combustion of coal (the heat due to burning it) is 28 MJ/kg.

**73** SSM A Carnot refrigerator extracts 35.0 kJ as heat during each cycle, operating with a coefficient of performance of 4.60. What are (a) the energy per cycle transferred as heat to the room and (b) the work done per cycle?

**74** A Carnot engine whose high-temperature reservoir is at 400 K has an efficiency of 30.0%. By how much should the temperature of the low-temperature reservoir be changed to increase the efficiency to 40.0%?

**75** SSM System A of three particles and system B of five particles are in insulated boxes like that in Fig. 20-17. What is the least multiplicity W of (a) system A and (b) system B? What is the greatest multiplicity W of (c) A and (d) B? What is the greatest entropy of (e) A and (f) B?

**76** Figure 20-36 shows a Carnot cycle on a T-S diagram, with a scale set by $S_s = 0.60$ J/K. For a full cycle, find (a) the net heat transfer and (b) the net work done by the system.

**77** Find the relation between the efficiency of a reversible ideal heat engine and the coefficient of performance of the reversible refrigerator obtained by running the engine backwards.

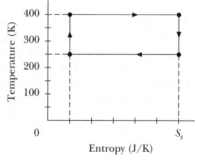
Figure 20-36 Problem 76.

**78** A Carnot engine has a power of 500 W. It operates between heat reservoirs at 100°C and 60.0°C. Calculate (a) the rate of heat input and (b) the rate of exhaust heat output.

**79** In a real refrigerator, the low-temperature coils are at −13°C, and the compressed gas in the condenser is at 26°C. What is the theoretical coefficient of performance?

# THE INTERNATIONAL SYSTEM OF UNITS (SI)*

Table 1 **The SI Base Units**

| Quantity | Name | Symbol | Definition |
|---|---|---|---|
| length | meter | m | "... the length of the path traveled by light in vacuum in 1/299,792,458 of a second." (1983) |
| mass | kilogram | kg | "... this prototype [a certain platinum–iridium cylinder] shall henceforth be considered to be the unit of mass." (1889) |
| time | second | s | "... the duration of 9,192,631,770 periods of the radiation corresponding to the transition between the two hyperfine levels of the ground state of the cesium-133 atom." (1967) |
| electric current | ampere | A | "... that constant current which, if maintained in two straight parallel conductors of infinite length, of negligible circular cross section, and placed 1 meter apart in vacuum, would produce between these conductors a force equal to $2 \times 10^{-7}$ newton per meter of length." (1946) |
| thermodynamic temperature | kelvin | K | "... the fraction 1/273.16 of the thermodynamic temperature of the triple point of water." (1967) |
| amount of substance | mole | mol | "... the amount of substance of a system which contains as many elementary entities as there are atoms in 0.012 kilogram of carbon-12." (1971) |
| luminous intensity | candela | cd | "... the luminous intensity, in a given direction, of a source that emits monochromatic radiation of frequency $540 \times 10^{12}$ hertz and that has a radiant intensity in that direction of 1/683 watt per steradian." (1979) |

*Adapted from "The International System of Units (SI)," National Bureau of Standards Special Publication 330, 1972 edition. The definitions above were adopted by the General Conference of Weights and Measures, an international body, on the dates shown. In this book we do not use the candela.

Table 2 **Some SI Derived Units**

| Quantity | Name of Unit | Symbol | |
|---|---|---|---|
| area | square meter | $m^2$ | |
| volume | cubic meter | $m^3$ | |
| frequency | hertz | Hz | $s^{-1}$ |
| mass density (density) | kilogram per cubic meter | $kg/m^3$ | |
| speed, velocity | meter per second | m/s | |
| | | | |
| angular velocity | radian per second | rad/s | |
| acceleration | meter per second per second | $m/s^2$ | |
| angular acceleration | radian per second per second | $rad/s^2$ | |
| force | newton | N | $kg \cdot m/s^2$ |
| pressure | pascal | Pa | $N/m^2$ |
| | | | |
| work, energy, quantity of heat | joule | J | $N \cdot m$ |
| power | watt | W | J/s |
| quantity of electric charge | coulomb | C | $A \cdot s$ |
| potential difference, electromotive force | volt | V | W/A |
| electric field strength | volt per meter (or newton per coulomb) | V/m | N/C |
| | | | |
| electric resistance | ohm | $\Omega$ | V/A |
| capacitance | farad | F | $A \cdot s/V$ |
| magnetic flux | weber | Wb | $V \cdot s$ |
| inductance | henry | H | $V \cdot s/A$ |
| magnetic flux density | tesla | T | $Wb/m^2$ |
| | | | |
| magnetic field strength | ampere per meter | A/m | |
| entropy | joule per kelvin | J/K | |
| specific heat | joule per kilogram kelvin | $J/(kg \cdot K)$ | |
| thermal conductivity | watt per meter kelvin | $W/(m \cdot K)$ | |
| radiant intensity | watt per steradian | W/sr | |

Table 3 **The SI Supplementary Units**

| Quantity | Name of Unit | Symbol |
|---|---|---|
| plane angle | radian | rad |
| solid angle | steradian | sr |

# SOME FUNDAMENTAL CONSTANTS OF PHYSICS*

| Constant | Symbol | Computational Value | Best (1998) Value | |
|---|---|---|---|---|
| | | | Value[a] | Uncertainty[b] |
| Speed of light in a vacuum | $c$ | $3.00 \times 10^8$ m/s | 2.997 924 58 | exact |
| Elementary charge | $e$ | $1.60 \times 10^{-19}$ C | 1.602 176 487 | 0.025 |
| Gravitational constant | $G$ | $6.67 \times 10^{-11}$ m$^3$/s$^2 \cdot$kg | 6.674 28 | 100 |
| Universal gas constant | $R$ | 8.31 J/mol$\cdot$K | 8.314 472 | 1.7 |
| Avogadro constant | $N_A$ | $6.02 \times 10^{23}$ mol$^{-1}$ | 6.022 141 79 | 0.050 |
| Boltzmann constant | $k$ | $1.38 \times 10^{-23}$ J/K | 1.380 650 4 | 1.7 |
| Stefan–Boltzmann constant | $\sigma$ | $5.67 \times 10^{-8}$ W/m$^2 \cdot$K$^4$ | 5.670 400 | 7.0 |
| Molar volume of ideal gas at STP[d] | $V_m$ | $2.27 \times 10^{-2}$ m$^3$/mol | 2.271 098 1 | 1.7 |
| Permittivity constant | $\epsilon_0$ | $8.85 \times 10^{-12}$ F/m | 8.854 187 817 62 | exact |
| Permeability constant | $\mu_0$ | $1.26 \times 10^{-6}$ H/m | 1.256 637 061 43 | exact |
| Planck constant | $h$ | $6.63 \times 10^{-34}$ J$\cdot$s | 6.626 068 96 | 0.050 |
| Electron mass[c] | $m_e$ | $9.11 \times 10^{-31}$ kg | 9.109 382 15 | 0.050 |
| | | $5.49 \times 10^{-4}$ u | 5.485 799 094 3 | $4.2 \times 10^{-4}$ |
| Proton mass[c] | $m_p$ | $1.67 \times 10^{-27}$ kg | 1.672 621 637 | 0.050 |
| | | 1.0073 u | 1.007 276 466 77 | $1.0 \times 10^{-4}$ |
| Ratio of proton mass to electron mass | $m_p/m_e$ | 1840 | 1836.152 672 47 | $4.3 \times 10^{-4}$ |
| Electron charge-to-mass ratio | $e/m_e$ | $1.76 \times 10^{11}$ C/kg | 1.758 820 150 | 0.025 |
| Neutron mass[c] | $m_n$ | $1.68 \times 10^{-27}$ kg | 1.674 927 211 | 0.050 |
| | | 1.0087 u | 1.008 664 915 97 | $4.3 \times 10^{-4}$ |
| Hydrogen atom mass[c] | $m_{1_H}$ | 1.0078 u | 1.007 825 031 6 | 0.0005 |
| Deuterium atom mass[c] | $m_{2_H}$ | 2.0136 u | 2.013 553 212 724 | $3.9 \times 10^{-5}$ |
| Helium atom mass[c] | $m_{4_{He}}$ | 4.0026 u | 4.002 603 2 | 0.067 |
| Muon mass | $m_\mu$ | $1.88 \times 10^{-28}$ kg | 1.883 531 30 | 0.056 |
| Electron magnetic moment | $\mu_e$ | $9.28 \times 10^{-24}$ J/T | 9.284 763 77 | 0.025 |
| Proton magnetic moment | $\mu_p$ | $1.41 \times 10^{-26}$ J/T | 1.410 606 662 | 0.026 |
| Bohr magneton | $\mu_B$ | $9.27 \times 10^{-24}$ J/T | 9.274 009 15 | 0.025 |
| Nuclear magneton | $\mu_N$ | $5.05 \times 10^{-27}$ J/T | 5.050 783 24 | 0.025 |
| Bohr radius | $a$ | $5.29 \times 10^{-11}$ m | 5.291 772 085 9 | $6.8 \times 10^{-4}$ |
| Rydberg constant | $R$ | $1.10 \times 10^7$ m$^{-1}$ | 1.097 373 156 852 7 | $6.6 \times 10^{-6}$ |
| Electron Compton wavelength | $\lambda_C$ | $2.43 \times 10^{-12}$ m | 2.426 310 217 5 | 0.0014 |

[a]Values given in this column should be given the same unit and power of 10 as the computational value.
[b]Parts per million.
[c]Masses given in u are in unified atomic mass units, where 1 u = 1.660 538 782 $\times 10^{-27}$ kg.
[d]STP means standard temperature and pressure: 0°C and 1.0 atm (0.1 MPa).

*The values in this table were selected from the 1998 CODATA recommended values (www.physics.nist.gov).

# SOME ASTRONOMICAL DATA

## Some Distances from Earth

| | | | |
|---|---|---|---|
| To the Moon* | $3.82 \times 10^8$ m | To the center of our galaxy | $2.2 \times 10^{20}$ m |
| To the Sun* | $1.50 \times 10^{11}$ m | To the Andromeda Galaxy | $2.1 \times 10^{22}$ m |
| To the nearest star (Proxima Centauri) | $4.04 \times 10^{16}$ m | To the edge of the observable universe | $\sim 10^{26}$ m |

*Mean distance.

## The Sun, Earth, and the Moon

| Property | Unit | Sun | Earth | Moon |
|---|---|---|---|---|
| Mass | kg | $1.99 \times 10^{30}$ | $5.98 \times 10^{24}$ | $7.36 \times 10^{22}$ |
| Mean radius | m | $6.96 \times 10^8$ | $6.37 \times 10^6$ | $1.74 \times 10^6$ |
| Mean density | kg/m$^3$ | 1410 | 5520 | 3340 |
| Free-fall acceleration at the surface | m/s$^2$ | 274 | 9.81 | 1.67 |
| Escape velocity | km/s | 618 | 11.2 | 2.38 |
| Period of rotation[a] | — | 37 d at poles[b]   26 d at equator[b] | 23 h 56 min | 27.3 d |
| Radiation power[c] | W | $3.90 \times 10^{26}$ | | |

[a]Measured with respect to the distant stars.
[b]The Sun, a ball of gas, does not rotate as a rigid body.
[c]Just outside Earth's atmosphere solar energy is received, assuming normal incidence, at the rate of 1340 W/m$^2$.

## Some Properties of the Planets

| | Mercury | Venus | Earth | Mars | Jupiter | Saturn | Uranus | Neptune | Pluto[d] |
|---|---|---|---|---|---|---|---|---|---|
| Mean distance from Sun, $10^6$ km | 57.9 | 108 | 150 | 228 | 778 | 1430 | 2870 | 4500 | 5900 |
| Period of revolution, y | 0.241 | 0.615 | 1.00 | 1.88 | 11.9 | 29.5 | 84.0 | 165 | 248 |
| Period of rotation,[a] d | 58.7 | $-243$[b] | 0.997 | 1.03 | 0.409 | 0.426 | $-0.451$[b] | 0.658 | 6.39 |
| Orbital speed, km/s | 47.9 | 35.0 | 29.8 | 24.1 | 13.1 | 9.64 | 6.81 | 5.43 | 4.74 |
| Inclination of axis to orbit | $<28°$ | $\approx 3°$ | 23.4° | 25.0° | 3.08° | 26.7° | 97.9° | 29.6° | 57.5° |
| Inclination of orbit to Earth's orbit | 7.00° | 3.39° | | 1.85° | 1.30° | 2.49° | 0.77° | 1.77° | 17.2° |
| Eccentricity of orbit | 0.206 | 0.0068 | 0.0167 | 0.0934 | 0.0485 | 0.0556 | 0.0472 | 0.0086 | 0.250 |
| Equatorial diameter, km | 4880 | 12 100 | 12 800 | 6790 | 143 000 | 120 000 | 51 800 | 49 500 | 2300 |
| Mass (Earth = 1) | 0.0558 | 0.815 | 1.000 | 0.107 | 318 | 95.1 | 14.5 | 17.2 | 0.002 |
| Density (water = 1) | 5.60 | 5.20 | 5.52 | 3.95 | 1.31 | 0.704 | 1.21 | 1.67 | 2.03 |
| Surface value of $g$,[c] m/s$^2$ | 3.78 | 8.60 | 9.78 | 3.72 | 22.9 | 9.05 | 7.77 | 11.0 | 0.5 |
| Escape velocity,[c] km/s | 4.3 | 10.3 | 11.2 | 5.0 | 59.5 | 35.6 | 21.2 | 23.6 | 1.3 |
| Known satellites | 0 | 0 | 1 | 2 | 67 + ring | 62 + rings | 27 + rings | 13 + rings | 4 |

[a]Measured with respect to the distant stars.
[b]Venus and Uranus rotate opposite their orbital motion.
[c]Gravitational acceleration measured at the planet's equator.
[d]Pluto is now classified as a dwarf planet.

# CONVERSION FACTORS

Conversion factors may be read directly from these tables. For example, 1 degree = $2.778 \times 10^{-3}$ revolutions, so $16.7° = 16.7 \times 2.778 \times 10^{-3}$ rev. The SI units are fully capitalized. Adapted in part from G. Shortley and D. Williams, *Elements of Physics*, 1971, Prentice-Hall, Englewood Cliffs, NJ.

### Plane Angle

| | ° | ' | " | RADIAN | rev |
|---|---|---|---|---|---|
| 1 degree = | 1 | 60 | 3600 | $1.745 \times 10^{-2}$ | $2.778 \times 10^{-3}$ |
| 1 minute = | $1.667 \times 10^{-2}$ | 1 | 60 | $2.909 \times 10^{-4}$ | $4.630 \times 10^{-5}$ |
| 1 second = | $2.778 \times 10^{-4}$ | $1.667 \times 10^{-2}$ | 1 | $4.848 \times 10^{-6}$ | $7.716 \times 10^{-7}$ |
| 1 RADIAN = | 57.30 | 3438 | $2.063 \times 10^{5}$ | 1 | 0.1592 |
| 1 revolution = | 360 | $2.16 \times 10^{4}$ | $1.296 \times 10^{6}$ | 6.283 | 1 |

### Solid Angle

1 sphere = $4\pi$ steradians = 12.57 steradians

### Length

| | cm | METER | km | in. | ft | mi |
|---|---|---|---|---|---|---|
| 1 centimeter = | 1 | $10^{-2}$ | $10^{-5}$ | 0.3937 | $3.281 \times 10^{-2}$ | $6.214 \times 10^{-6}$ |
| 1 METER = | 100 | 1 | $10^{-3}$ | 39.37 | 3.281 | $6.214 \times 10^{-4}$ |
| 1 kilometer = | $10^{5}$ | 1000 | 1 | $3.937 \times 10^{4}$ | 3281 | 0.6214 |
| 1 inch = | 2.540 | $2.540 \times 10^{-2}$ | $2.540 \times 10^{-5}$ | 1 | $8.333 \times 10^{-2}$ | $1.578 \times 10^{-5}$ |
| 1 foot = | 30.48 | 0.3048 | $3.048 \times 10^{-4}$ | 12 | 1 | $1.894 \times 10^{-4}$ |
| 1 mile = | $1.609 \times 10^{5}$ | 1609 | 1.609 | $6.336 \times 10^{4}$ | 5280 | 1 |

1 angström = $10^{-10}$ m
1 nautical mile = 1852 m
  = 1.151 miles = 6076 ft

1 fermi = $10^{-15}$ m
1 light-year = $9.461 \times 10^{12}$ km
1 parsec = $3.084 \times 10^{13}$ km

1 fathom = 6 ft
1 Bohr radius = $5.292 \times 10^{-11}$ m
1 yard = 3 ft

1 rod = 16.5 ft
1 mil = $10^{-3}$ in.
1 nm = $10^{-9}$ m

### Area

| | METER$^2$ | cm$^2$ | ft$^2$ | in.$^2$ |
|---|---|---|---|---|
| 1 SQUARE METER = | 1 | $10^{4}$ | 10.76 | 1550 |
| 1 square centimeter = | $10^{-4}$ | 1 | $1.076 \times 10^{-3}$ | 0.1550 |
| 1 square foot = | $9.290 \times 10^{-2}$ | 929.0 | 1 | 144 |
| 1 square inch = | $6.452 \times 10^{-4}$ | 6.452 | $6.944 \times 10^{-3}$ | 1 |

1 square mile = $2.788 \times 10^7$ ft$^2$ = 640 acres
1 barn = $10^{-28}$ m$^2$

1 acre = 43 560 ft$^2$
1 hectare = $10^4$ m$^2$ = 2.471 acres

## Volume

| | METER$^3$ | cm$^3$ | L | ft$^3$ | in.$^3$ |
|---|---|---|---|---|---|
| 1 CUBIC METER = 1 | | $10^6$ | 1000 | 35.31 | $6.102 \times 10^4$ |
| 1 cubic centimeter = $10^{-6}$ | | 1 | $1.000 \times 10^{-3}$ | $3.531 \times 10^{-5}$ | $6.102 \times 10^{-2}$ |
| 1 liter = $1.000 \times 10^{-3}$ | | 1000 | 1 | $3.531 \times 10^{-2}$ | 61.02 |
| 1 cubic foot = $2.832 \times 10^{-2}$ | | $2.832 \times 10^4$ | 28.32 | 1 | 1728 |
| 1 cubic inch = $1.639 \times 10^{-5}$ | | 16.39 | $1.639 \times 10^{-2}$ | $5.787 \times 10^{-4}$ | 1 |

1 U.S. fluid gallon = 4 U.S. fluid quarts = 8 U.S. pints = 128 U.S. fluid ounces = 231 in.$^3$
1 British imperial gallon = 277.4 in.$^3$ = 1.201 U.S. fluid gallons

## Mass

Quantities in the colored areas are not mass units but are often used as such. For example, when we write 1 kg "=" 2.205 lb, this means that a kilogram is a *mass* that *weighs* 2.205 pounds at a location where g has the standard value of 9.80665 m/s$^2$.

| | g | KILOGRAM | slug | u | oz | lb | ton |
|---|---|---|---|---|---|---|---|
| 1 gram = 1 | | 0.001 | $6.852 \times 10^{-5}$ | $6.022 \times 10^{23}$ | $3.527 \times 10^{-2}$ | $2.205 \times 10^{-3}$ | $1.102 \times 10^{-6}$ |
| 1 KILOGRAM = 1000 | | 1 | $6.852 \times 10^{-2}$ | $6.022 \times 10^{26}$ | 35.27 | 2.205 | $1.102 \times 10^{-3}$ |
| 1 slug = $1.459 \times 10^4$ | | 14.59 | 1 | $8.786 \times 10^{27}$ | 514.8 | 32.17 | $1.609 \times 10^{-2}$ |
| 1 atomic mass unit = $1.661 \times 10^{-24}$ | | $1.661 \times 10^{-27}$ | $1.138 \times 10^{-28}$ | 1 | $5.857 \times 10^{-26}$ | $3.662 \times 10^{-27}$ | $1.830 \times 10^{-30}$ |
| 1 ounce = 28.35 | | $2.835 \times 10^{-2}$ | $1.943 \times 10^{-3}$ | $1.718 \times 10^{25}$ | 1 | $6.250 \times 10^{-2}$ | $3.125 \times 10^{-5}$ |
| 1 pound = 453.6 | | 0.4536 | $3.108 \times 10^{-2}$ | $2.732 \times 10^{26}$ | 16 | 1 | 0.0005 |
| 1 ton = $9.072 \times 10^5$ | | 907.2 | 62.16 | $5.463 \times 10^{29}$ | $3.2 \times 10^4$ | 2000 | 1 |

1 metric ton = 1000 kg

## Density

Quantities in the colored areas are weight densities and, as such, are dimensionally different from mass densities. See the note for the mass table.

| | slug/ft$^3$ | KILOGRAM/ METER$^3$ | g/cm$^3$ | lb/ft$^3$ | lb/in.$^3$ |
|---|---|---|---|---|---|
| 1 slug per foot$^3$ = 1 | | 515.4 | 0.5154 | 32.17 | $1.862 \times 10^{-2}$ |
| 1 KILOGRAM per METER$^3$ = $1.940 \times 10^{-3}$ | | 1 | 0.001 | $6.243 \times 10^{-2}$ | $3.613 \times 10^{-5}$ |
| 1 gram per centimeter$^3$ = 1.940 | | 1000 | 1 | 62.43 | $3.613 \times 10^{-2}$ |
| 1 pound per foot$^3$ = $3.108 \times 10^{-2}$ | | 16.02 | $16.02 \times 10^{-2}$ | 1 | $5.787 \times 10^{-4}$ |
| 1 pound per inch$^3$ = 53.71 | | $2.768 \times 10^4$ | 27.68 | 1728 | 1 |

## Time

| | y | d | h | min | SECOND |
|---|---|---|---|---|---|
| 1 year = 1 | | 365.25 | $8.766 \times 10^3$ | $5.259 \times 10^5$ | $3.156 \times 10^7$ |
| 1 day = $2.738 \times 10^{-3}$ | | 1 | 24 | 1440 | $8.640 \times 10^4$ |
| 1 hour = $1.141 \times 10^{-4}$ | | $4.167 \times 10^{-2}$ | 1 | 60 | 3600 |
| 1 minute = $1.901 \times 10^{-6}$ | | $6.944 \times 10^{-4}$ | $1.667 \times 10^{-2}$ | 1 | 60 |
| 1 SECOND = $3.169 \times 10^{-8}$ | | $1.157 \times 10^{-5}$ | $2.778 \times 10^{-4}$ | $1.667 \times 10^{-2}$ | 1 |

Speed

|  | ft/s | km/h | METER/SECOND | mi/h | cm/s |
|---|---|---|---|---|---|
| 1 foot per second = | 1 | 1.097 | 0.3048 | 0.6818 | 30.48 |
| 1 kilometer per hour = | 0.9113 | 1 | 0.2778 | 0.6214 | 27.78 |
| 1 METER per SECOND = | 3.281 | 3.6 | 1 | 2.237 | 100 |
| 1 mile per hour = | 1.467 | 1.609 | 0.4470 | 1 | 44.70 |
| 1 centimeter per second = | $3.281 \times 10^{-2}$ | $3.6 \times 10^{-2}$ | 0.01 | $2.237 \times 10^{-2}$ | 1 |

1 knot = 1 nautical mi/h = 1.688 ft/s    1 mi/min = 88.00 ft/s = 60.00 mi/h

Force

Force units in the colored areas are now little used. To clarify: 1 gram-force (= 1 gf) is the force of gravity that would act on an object whose mass is 1 gram at a location where $g$ has the standard value of 9.80665 m/s$^2$.

|  | dyne | NEWTON | lb | pdl | gf | kgf |
|---|---|---|---|---|---|---|
| 1 dyne = | 1 | $10^{-5}$ | $2.248 \times 10^{-6}$ | $7.233 \times 10^{-5}$ | $1.020 \times 10^{-3}$ | $1.020 \times 10^{-6}$ |
| 1 NEWTON = | $10^5$ | 1 | 0.2248 | 7.233 | 102.0 | 0.1020 |
| 1 pound = | $4.448 \times 10^5$ | 4.448 | 1 | 32.17 | 453.6 | 0.4536 |
| 1 poundal = | $1.383 \times 10^4$ | 0.1383 | $3.108 \times 10^{-2}$ | 1 | 14.10 | $1.410 \times 10^2$ |
| 1 gram-force = | 980.7 | $9.807 \times 10^{-3}$ | $2.205 \times 10^{-3}$ | $7.093 \times 10^{-2}$ | 1 | 0.001 |
| 1 kilogram-force = | $9.807 \times 10^5$ | 9.807 | 2.205 | 70.93 | 1000 | 1 |

1 ton = 2000 lb

Pressure

|  | atm | dyne/cm$^2$ | inch of water | cm Hg | PASCAL | lb/in.$^2$ | lb/ft$^2$ |
|---|---|---|---|---|---|---|---|
| 1 atmosphere = | 1 | $1.013 \times 10^6$ | 406.8 | 76 | $1.013 \times 10^5$ | 14.70 | 2116 |
| 1 dyne per centimeter$^2$ = | $9.869 \times 10^{-7}$ | 1 | $4.015 \times 10^{-4}$ | $7.501 \times 10^{-5}$ | 0.1 | $1.405 \times 10^{-5}$ | $2.089 \times 10^{-3}$ |
| 1 inch of water$^a$ at 4°C = | $2.458 \times 10^{-3}$ | 2491 | 1 | 0.1868 | 249.1 | $3.613 \times 10^{-2}$ | 5.202 |
| 1 centimeter of mercury$^a$ at 0°C = | $1.316 \times 10^{-2}$ | $1.333 \times 10^4$ | 5.353 | 1 | 1333 | 0.1934 | 27.85 |
| 1 PASCAL = | $9.869 \times 10^{-6}$ | 10 | $4.015 \times 10^{-3}$ | $7.501 \times 10^{-4}$ | 1 | $1.450 \times 10^{-4}$ | $2.089 \times 10^{-2}$ |
| 1 pound per inch$^2$ = | $6.805 \times 10^{-2}$ | $6.895 \times 10^4$ | 27.68 | 5.171 | $6.895 \times 10^3$ | 1 | 144 |
| 1 pound per foot$^2$ = | $4.725 \times 10^{-4}$ | 478.8 | 0.1922 | $3.591 \times 10^{-2}$ | 47.88 | $6.944 \times 10^{-3}$ | 1 |

$^a$Where the acceleration of gravity has the standard value of 9.80665 m/s$^2$.

1 bar = $10^6$ dyne/cm$^2$ = 0.1 MPa        1 millibar = $10^3$ dyne/cm$^2$ = $10^2$ Pa        1 torr = 1 mm Hg

## Energy, Work, Heat

Quantities in the colored areas are not energy units but are included for convenience. They arise from the relativistic mass–energy equivalence formula $E = mc^2$ and represent the energy released if a kilogram or unified atomic mass unit (u) is completely converted to energy (bottom two rows) or the mass that would be completely converted to one unit of energy (rightmost two columns).

| | Btu | erg | ft·lb | hp·h | JOULE | cal | kW·h | eV | MeV | kg | u |
|---|---|---|---|---|---|---|---|---|---|---|---|
| 1 British thermal unit = | 1 | $1.055 \times 10^{10}$ | 777.9 | $3.929 \times 10^{-4}$ | 1055 | 252.0 | $2.930 \times 10^{-4}$ | $6.585 \times 10^{21}$ | $6.585 \times 10^{15}$ | $1.174 \times 10^{-14}$ | $7.070 \times 10^{12}$ |
| 1 erg = | $9.481 \times 10^{-11}$ | 1 | $7.376 \times 10^{-8}$ | $3.725 \times 10^{-14}$ | $10^{-7}$ | $2.389 \times 10^{-8}$ | $2.778 \times 10^{-14}$ | $6.242 \times 10^{11}$ | $6.242 \times 10^{5}$ | $1.113 \times 10^{-24}$ | 670.2 |
| 1 foot-pound = | $1.285 \times 10^{-3}$ | $1.356 \times 10^{7}$ | 1 | $5.051 \times 10^{-7}$ | 1.356 | 0.3238 | $3.766 \times 10^{-7}$ | $8.464 \times 10^{18}$ | $8.464 \times 10^{12}$ | $1.509 \times 10^{-17}$ | $9.037 \times 10^{9}$ |
| 1 horsepower-hour = | 2545 | $2.685 \times 10^{13}$ | $1.980 \times 10^{6}$ | 1 | $2.685 \times 10^{6}$ | $6.413 \times 10^{5}$ | 0.7457 | $1.676 \times 10^{25}$ | $1.676 \times 10^{19}$ | $2.988 \times 10^{-11}$ | $1.799 \times 10^{16}$ |
| 1 JOULE = | $9.481 \times 10^{-4}$ | $10^{7}$ | 0.7376 | $3.725 \times 10^{-7}$ | 1 | 0.2389 | $2.778 \times 10^{-7}$ | $6.242 \times 10^{18}$ | $6.242 \times 10^{12}$ | $1.113 \times 10^{-17}$ | $6.702 \times 10^{9}$ |
| 1 calorie = | $3.968 \times 10^{-3}$ | $4.1868 \times 10^{7}$ | 3.088 | $1.560 \times 10^{-6}$ | 4.1868 | 1 | $1.163 \times 10^{-6}$ | $2.613 \times 10^{19}$ | $2.613 \times 10^{13}$ | $4.660 \times 10^{-17}$ | $2.806 \times 10^{10}$ |
| 1 kilowatt-hour = | 3413 | $3.600 \times 10^{13}$ | $2.655 \times 10^{6}$ | 1.341 | $3.600 \times 10^{6}$ | $8.600 \times 10^{5}$ | 1 | $2.247 \times 10^{25}$ | $2.247 \times 10^{19}$ | $4.007 \times 10^{-11}$ | $2.413 \times 10^{16}$ |
| 1 electron-volt = | $1.519 \times 10^{-22}$ | $1.602 \times 10^{-12}$ | $1.182 \times 10^{-19}$ | $5.967 \times 10^{-26}$ | $1.602 \times 10^{-19}$ | $3.827 \times 10^{-20}$ | $4.450 \times 10^{-26}$ | 1 | $10^{-6}$ | $1.783 \times 10^{-36}$ | $1.074 \times 10^{-9}$ |
| 1 million electron-volts = | $1.519 \times 10^{-16}$ | $1.602 \times 10^{-6}$ | $1.182 \times 10^{-13}$ | $5.967 \times 10^{-20}$ | $1.602 \times 10^{-13}$ | $3.827 \times 10^{-14}$ | $4.450 \times 10^{-20}$ | $10^{6}$ | 1 | $1.783 \times 10^{-30}$ | $1.074 \times 10^{-3}$ |
| 1 kilogram = | $8.521 \times 10^{13}$ | $8.987 \times 10^{23}$ | $6.629 \times 10^{16}$ | $3.348 \times 10^{10}$ | $8.987 \times 10^{16}$ | $2.146 \times 10^{16}$ | $2.497 \times 10^{10}$ | $5.610 \times 10^{35}$ | $5.610 \times 10^{29}$ | 1 | $6.022 \times 10^{26}$ |
| 1 unified atomic mass unit = | $1.415 \times 10^{-13}$ | $1.492 \times 10^{-3}$ | $1.101 \times 10^{-10}$ | $5.559 \times 10^{-17}$ | $1.492 \times 10^{-10}$ | $3.564 \times 10^{-11}$ | $4.146 \times 10^{-17}$ | $9.320 \times 10^{8}$ | 932.0 | $1.661 \times 10^{-27}$ | 1 |

## Power

| | Btu/h | ft·lb/s | hp | cal/s | kW | WATT |
|---|---|---|---|---|---|---|
| 1 British thermal unit per hour = | 1 | 0.2161 | $3.929 \times 10^{-4}$ | $6.998 \times 10^{-2}$ | $2.930 \times 10^{-4}$ | 0.2930 |
| 1 foot-pound per second = | 4.628 | 1 | $1.818 \times 10^{-3}$ | 0.3239 | $1.356 \times 10^{-3}$ | 1.356 |
| 1 horsepower = | 2545 | 550 | 1 | 178.1 | 0.7457 | 745.7 |
| 1 calorie per second = | 14.29 | 3.088 | $5.615 \times 10^{-3}$ | 1 | $4.186 \times 10^{-3}$ | 4.186 |
| 1 kilowatt = | 3413 | 737.6 | 1.341 | 238.9 | 1 | 1000 |
| 1 WATT = | 3.413 | 0.7376 | $1.341 \times 10^{-3}$ | 0.2389 | 0.001 | 1 |

## Magnetic Field

| | gauss | TESLA | milligauss |
|---|---|---|---|
| 1 gauss = | 1 | $10^{-4}$ | 1000 |
| 1 TESLA = | $10^4$ | 1 | $10^7$ |
| 1 milligauss = | 0.001 | $10^{-7}$ | 1 |

1 tesla = 1 weber/meter$^2$

## Magnetic Flux

| | maxwell | WEBER |
|---|---|---|
| 1 maxwell = | 1 | $10^{-8}$ |
| 1 WEBER = | $10^8$ | 1 |

# MATHEMATICAL FORMULAS

## Geometry

Circle of radius $r$: circumference $= 2\pi r$; area $= \pi r^2$.

Sphere of radius $r$: area $= 4\pi r^2$; volume $= \frac{4}{3}\pi r^3$.

Right circular cylinder of radius $r$ and height $h$:
area $= 2\pi r^2 + 2\pi rh$; volume $= \pi r^2 h$.

Triangle of base $a$ and altitude $h$: area $= \frac{1}{2}ah$.

## Quadratic Formula

If $ax^2 + bx + c = 0$, then $x = \dfrac{-b \pm \sqrt{b^2 - 4ac}}{2a}$.

## Trigonometric Functions of Angle $\theta$

$$\sin\theta = \frac{y}{r} \quad \cos\theta = \frac{x}{r}$$

$$\tan\theta = \frac{y}{x} \quad \cot\theta = \frac{x}{y}$$

$$\sec\theta = \frac{r}{x} \quad \csc\theta = \frac{r}{y}$$

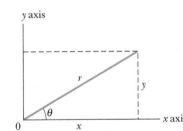

## Pythagorean Theorem

In this right triangle,
$$a^2 + b^2 = c^2$$

## Triangles

Angles are $A$, $B$, $C$

Opposite sides are $a$, $b$, $c$

Angles $A + B + C = 180°$

$$\frac{\sin A}{a} = \frac{\sin B}{b} = \frac{\sin C}{c}$$

$c^2 = a^2 + b^2 - 2ab\cos C$

Exterior angle $D = A + C$

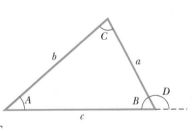

## Mathematical Signs and Symbols

$=$  equals

$\approx$  equals approximately

$\sim$  is the order of magnitude of

$\neq$  is not equal to

$\equiv$  is identical to, is defined as

$>$  is greater than ($\gg$ is much greater than)

$<$  is less than ($\ll$ is much less than)

$\geq$  is greater than or equal to (or, is no less than)

$\leq$  is less than or equal to (or, is no more than)

$\pm$  plus or minus

$\propto$  is proportional to

$\Sigma$  the sum of

$x_{avg}$  the average value of $x$

## Trigonometric Identities

$\sin(90° - \theta) = \cos\theta$

$\cos(90° - \theta) = \sin\theta$

$\sin\theta/\cos\theta = \tan\theta$

$\sin^2\theta + \cos^2\theta = 1$

$\sec^2\theta - \tan^2\theta = 1$

$\csc^2\theta - \cot^2\theta = 1$

$\sin 2\theta = 2\sin\theta\cos\theta$

$\cos 2\theta = \cos^2\theta - \sin^2\theta = 2\cos^2\theta - 1 = 1 - 2\sin^2\theta$

$\sin(\alpha \pm \beta) = \sin\alpha\cos\beta \pm \cos\alpha\sin\beta$

$\cos(\alpha \pm \beta) = \cos\alpha\cos\beta \mp \sin\alpha\sin\beta$

$\tan(\alpha \pm \beta) = \dfrac{\tan\alpha \pm \tan\beta}{1 \mp \tan\alpha\tan\beta}$

$\sin\alpha \pm \sin\beta = 2\sin\frac{1}{2}(\alpha \pm \beta)\cos\frac{1}{2}(\alpha \mp \beta)$

$\cos\alpha + \cos\beta = 2\cos\frac{1}{2}(\alpha + \beta)\cos\frac{1}{2}(\alpha - \beta)$

$\cos\alpha - \cos\beta = -2\sin\frac{1}{2}(\alpha + \beta)\sin\frac{1}{2}(\alpha - \beta)$

## Binomial Theorem

$$(1 + x)^n = 1 + \frac{nx}{1!} + \frac{n(n-1)x^2}{2!} + \cdots \qquad (x^2 < 1)$$

## Exponential Expansion

$$e^x = 1 + x + \frac{x^2}{2!} + \frac{x^3}{3!} + \cdots$$

## Logarithmic Expansion

$$\ln(1 + x) = x - \tfrac{1}{2}x^2 + \tfrac{1}{3}x^3 - \cdots \qquad (|x| < 1)$$

## Trigonometric Expansions ($\theta$ in radians)

$$\sin \theta = \theta - \frac{\theta^3}{3!} + \frac{\theta^5}{5!} - \cdots$$

$$\cos \theta = 1 - \frac{\theta^2}{2!} + \frac{\theta^4}{4!} - \cdots$$

$$\tan \theta = \theta + \frac{\theta^3}{3} + \frac{2\theta^5}{15} + \cdots$$

## Cramer's Rule

Two simultaneous equations in unknowns $x$ and $y$,

$$a_1x + b_1y = c_1 \quad \text{and} \quad a_2x + b_2y = c_2,$$

have the solutions

$$x = \frac{\begin{vmatrix} c_1 & b_1 \\ c_2 & b_2 \end{vmatrix}}{\begin{vmatrix} a_1 & b_1 \\ a_2 & b_2 \end{vmatrix}} = \frac{c_1b_2 - c_2b_1}{a_1b_2 - a_2b_1}$$

and

$$y = \frac{\begin{vmatrix} a_1 & c_1 \\ a_2 & c_2 \end{vmatrix}}{\begin{vmatrix} a_1 & b_1 \\ a_2 & b_2 \end{vmatrix}} = \frac{a_1c_2 - a_2c_1}{a_1b_2 - a_2b_1}.$$

## Products of Vectors

Let $\hat{i}$, $\hat{j}$, and $\hat{k}$ be unit vectors in the $x$, $y$, and $z$ directions. Then

$$\hat{i}\cdot\hat{i} = \hat{j}\cdot\hat{j} = \hat{k}\cdot\hat{k} = 1, \quad \hat{i}\cdot\hat{j} = \hat{j}\cdot\hat{k} = \hat{k}\cdot\hat{i} = 0,$$
$$\hat{i}\times\hat{i} = \hat{j}\times\hat{j} = \hat{k}\times\hat{k} = 0,$$
$$\hat{i}\times\hat{j} = \hat{k}, \quad \hat{j}\times\hat{k} = \hat{i}, \quad \hat{k}\times\hat{i} = \hat{j}$$

Any vector $\vec{a}$ with components $a_x$, $a_y$, and $a_z$ along the $x$, $y$, and $z$ axes can be written as

$$\vec{a} = a_x\hat{i} + a_y\hat{j} + a_z\hat{k}.$$

Let $\vec{a}$, $\vec{b}$, and $\vec{c}$ be arbitrary vectors with magnitudes $a$, $b$, and $c$. Then

$$\vec{a}\times(\vec{b}+\vec{c}) = (\vec{a}\times\vec{b}) + (\vec{a}\times\vec{c})$$
$$(s\vec{a})\times\vec{b} = \vec{a}\times(s\vec{b}) = s(\vec{a}\times\vec{b}) \qquad (s = \text{a scalar}).$$

Let $\theta$ be the smaller of the two angles between $\vec{a}$ and $\vec{b}$. Then

$$\vec{a}\cdot\vec{b} = \vec{b}\cdot\vec{a} = a_xb_x + a_yb_y + a_zb_z = ab\cos\theta$$

$$\vec{a}\times\vec{b} = -\vec{b}\times\vec{a} = \begin{vmatrix} \hat{i} & \hat{j} & \hat{k} \\ a_x & a_y & a_z \\ b_x & b_y & b_z \end{vmatrix}$$

$$= \hat{i}\begin{vmatrix} a_y & a_z \\ b_y & b_z \end{vmatrix} - \hat{j}\begin{vmatrix} a_x & a_z \\ b_x & b_z \end{vmatrix} + \hat{k}\begin{vmatrix} a_x & a_y \\ b_x & b_y \end{vmatrix}$$

$$= (a_yb_z - b_ya_z)\hat{i} + (a_zb_x - b_za_x)\hat{j} + (a_xb_y - b_xa_y)\hat{k}$$

$$|\vec{a}\times\vec{b}| = ab\sin\theta$$

$$\vec{a}\cdot(\vec{b}\times\vec{c}) = \vec{b}\cdot(\vec{c}\times\vec{a}) = \vec{c}\cdot(\vec{a}\times\vec{b})$$

$$\vec{a}\times(\vec{b}\times\vec{c}) = (\vec{a}\cdot\vec{c})\vec{b} - (\vec{a}\cdot\vec{b})\vec{c}$$

## Derivatives and Integrals

In what follows, the letters $u$ and $v$ stand for any functions of $x$, and $a$ and $m$ are constants. To each of the indefinite integrals should be added an arbitrary constant of integration. The *Handbook of Chemistry and Physics* (CRC Press Inc.) gives a more extensive tabulation.

**1.** $\dfrac{dx}{dx} = 1$

**2.** $\dfrac{d}{dx}(au) = a\,\dfrac{du}{dx}$

**3.** $\dfrac{d}{dx}(u + v) = \dfrac{du}{dx} + \dfrac{dv}{dx}$

**4.** $\dfrac{d}{dx}x^m = mx^{m-1}$

**5.** $\dfrac{d}{dx}\ln x = \dfrac{1}{x}$

**6.** $\dfrac{d}{dx}(uv) = u\,\dfrac{dv}{dx} + v\,\dfrac{du}{dx}$

**7.** $\dfrac{d}{dx}e^x = e^x$

**8.** $\dfrac{d}{dx}\sin x = \cos x$

**9.** $\dfrac{d}{dx}\cos x = -\sin x$

**10.** $\dfrac{d}{dx}\tan x = \sec^2 x$

**11.** $\dfrac{d}{dx}\cot x = -\csc^2 x$

**12.** $\dfrac{d}{dx}\sec x = \tan x \sec x$

**13.** $\dfrac{d}{dx}\csc x = -\cot x \csc x$

**14.** $\dfrac{d}{dx}e^u = e^u\,\dfrac{du}{dx}$

**15.** $\dfrac{d}{dx}\sin u = \cos u\,\dfrac{du}{dx}$

**16.** $\dfrac{d}{dx}\cos u = -\sin u\,\dfrac{du}{dx}$

**1.** $\displaystyle\int dx = x$

**2.** $\displaystyle\int au\,dx = a\int u\,dx$

**3.** $\displaystyle\int (u + v)\,dx = \int u\,dx + \int v\,dx$

**4.** $\displaystyle\int x^m\,dx = \dfrac{x^{m+1}}{m + 1}\quad (m \neq -1)$

**5.** $\displaystyle\int \dfrac{dx}{x} = \ln |x|$

**6.** $\displaystyle\int u\,\dfrac{dv}{dx}\,dx = uv - \int v\,\dfrac{du}{dx}\,dx$

**7.** $\displaystyle\int e^x\,dx = e^x$

**8.** $\displaystyle\int \sin x\,dx = -\cos x$

**9.** $\displaystyle\int \cos x\,dx = \sin x$

**10.** $\displaystyle\int \tan x\,dx = \ln |\sec x|$

**11.** $\displaystyle\int \sin^2 x\,dx = \tfrac{1}{2}x - \tfrac{1}{4}\sin 2x$

**12.** $\displaystyle\int e^{-ax}\,dx = -\dfrac{1}{a}e^{-ax}$

**13.** $\displaystyle\int xe^{-ax}\,dx = -\dfrac{1}{a^2}(ax + 1)\,e^{-ax}$

**14.** $\displaystyle\int x^2 e^{-ax}\,dx = -\dfrac{1}{a^3}(a^2x^2 + 2ax + 2)e^{-ax}$

**15.** $\displaystyle\int_0^\infty x^n e^{-ax}\,dx = \dfrac{n!}{a^{n+1}}$

**16.** $\displaystyle\int_0^\infty x^{2n} e^{-ax^2}\,dx = \dfrac{1 \cdot 3 \cdot 5 \, \cdots \, (2n - 1)}{2^{n+1}a^n}\sqrt{\dfrac{\pi}{a}}$

**17.** $\displaystyle\int \dfrac{dx}{\sqrt{x^2 + a^2}} = \ln(x + \sqrt{x^2 + a^2})$

**18.** $\displaystyle\int \dfrac{x\,dx}{(x^2 + a^2)^{3/2}} = -\dfrac{1}{(x^2 + a^2)^{1/2}}$

**19.** $\displaystyle\int \dfrac{dx}{(x^2 + a^2)^{3/2}} = \dfrac{x}{a^2(x^2 + a^2)^{1/2}}$

**20.** $\displaystyle\int_0^\infty x^{2n+1} e^{-ax^2}\,dx = \dfrac{n!}{2a^{n+1}}\quad (a > 0)$

**21.** $\displaystyle\int \dfrac{x\,dx}{x + d} = x - d\ln(x + d)$

# PROPERTIES OF THE ELEMENTS

All physical properties are for a pressure of 1 atm unless otherwise specified.

| Element | Symbol | Atomic Number Z | Molar Mass, g/mol | Density, g/cm³ at 20°C | Melting Point, °C | Boiling Point, °C | Specific Heat, J/(g·°C) at 25°C |
|---|---|---|---|---|---|---|---|
| Actinium | Ac | 89 | (227) | 10.06 | 1323 | (3473) | 0.092 |
| Aluminum | Al | 13 | 26.9815 | 2.699 | 660 | 2450 | 0.900 |
| Americium | Am | 95 | (243) | 13.67 | 1541 | — | — |
| Antimony | Sb | 51 | 121.75 | 6.691 | 630.5 | 1380 | 0.205 |
| Argon | Ar | 18 | 39.948 | $1.6626 \times 10^{-3}$ | −189.4 | −185.8 | 0.523 |
| Arsenic | As | 33 | 74.9216 | 5.78 | 817 (28 atm) | 613 | 0.331 |
| Astatine | At | 85 | (210) | — | (302) | — | — |
| Barium | Ba | 56 | 137.34 | 3.594 | 729 | 1640 | 0.205 |
| Berkelium | Bk | 97 | (247) | 14.79 | — | — | — |
| Beryllium | Be | 4 | 9.0122 | 1.848 | 1287 | 2770 | 1.83 |
| Bismuth | Bi | 83 | 208.980 | 9.747 | 271.37 | 1560 | 0.122 |
| Bohrium | Bh | 107 | 262.12 | — | — | — | — |
| Boron | B | 5 | 10.811 | 2.34 | 2030 | — | 1.11 |
| Bromine | Br | 35 | 79.909 | 3.12 (liquid) | −7.2 | 58 | 0.293 |
| Cadmium | Cd | 48 | 112.40 | 8.65 | 321.03 | 765 | 0.226 |
| Calcium | Ca | 20 | 40.08 | 1.55 | 838 | 1440 | 0.624 |
| Californium | Cf | 98 | (251) | — | — | — | — |
| Carbon | C | 6 | 12.01115 | 2.26 | 3727 | 4830 | 0.691 |
| Cerium | Ce | 58 | 140.12 | 6.768 | 804 | 3470 | 0.188 |
| Cesium | Cs | 55 | 132.905 | 1.873 | 28.40 | 690 | 0.243 |
| Chlorine | Cl | 17 | 35.453 | $3.214 \times 10^{-3}$ (0°C) | −101 | −34.7 | 0.486 |
| Chromium | Cr | 24 | 51.996 | 7.19 | 1857 | 2665 | 0.448 |
| Cobalt | Co | 27 | 58.9332 | 8.85 | 1495 | 2900 | 0.423 |
| Copernicium | Cn | 112 | (285) | — | — | — | — |
| Copper | Cu | 29 | 63.54 | 8.96 | 1083.40 | 2595 | 0.385 |
| Curium | Cm | 96 | (247) | 13.3 | — | — | — |
| Darmstadtium | Ds | 110 | (271) | — | — | — | — |
| Dubnium | Db | 105 | 262.114 | — | — | — | — |
| Dysprosium | Dy | 66 | 162.50 | 8.55 | 1409 | 2330 | 0.172 |
| Einsteinium | Es | 99 | (254) | — | — | — | — |
| Erbium | Er | 68 | 167.26 | 9.15 | 1522 | 2630 | 0.167 |
| Europium | Eu | 63 | 151.96 | 5.243 | 817 | 1490 | 0.163 |
| Fermium | Fm | 100 | (237) | — | — | — | — |
| Flerovium* | Fl | 114 | (289) | — | — | — | — |
| Fluorine | F | 9 | 18.9984 | $1.696 \times 10^{-3}$ (0°C) | −219.6 | −188.2 | 0.753 |
| Francium | Fr | 87 | (223) | — | (27) | — | — |
| Gadolinium | Gd | 64 | 157.25 | 7.90 | 1312 | 2730 | 0.234 |
| Gallium | Ga | 31 | 69.72 | 5.907 | 29.75 | 2237 | 0.377 |
| Germanium | Ge | 32 | 72.59 | 5.323 | 937.25 | 2830 | 0.322 |
| Gold | Au | 79 | 196.967 | 19.32 | 1064.43 | 2970 | 0.131 |

| Element | Symbol | Atomic Number Z | Molar Mass, g/mol | Density, g/cm³ at 20°C | Melting Point, °C | Boiling Point, °C | Specific Heat, J/(g·°C) at 25°C |
|---|---|---|---|---|---|---|---|
| Hafnium | Hf | 72 | 178.49 | 13.31 | 2227 | 5400 | 0.144 |
| Hassium | Hs | 108 | (265) | — | — | — | — |
| Helium | He | 2 | 4.0026 | $0.1664 \times 10^{-3}$ | −269.7 | −268.9 | 5.23 |
| Holmium | Ho | 67 | 164.930 | 8.79 | 1470 | 2330 | 0.165 |
| Hydrogen | H | 1 | 1.00797 | $0.08375 \times 10^{-3}$ | −259.19 | −252.7 | 14.4 |
| Indium | In | 49 | 114.82 | 7.31 | 156.634 | 2000 | 0.233 |
| Iodine | I | 53 | 126.9044 | 4.93 | 113.7 | 183 | 0.218 |
| Iridium | Ir | 77 | 192.2 | 22.5 | 2447 | (5300) | 0.130 |
| Iron | Fe | 26 | 55.847 | 7.874 | 1536.5 | 3000 | 0.447 |
| Krypton | Kr | 36 | 83.80 | $3.488 \times 10^{-3}$ | −157.37 | −152 | 0.247 |
| Lanthanum | La | 57 | 138.91 | 6.189 | 920 | 3470 | 0.195 |
| Lawrencium | Lr | 103 | (257) | — | — | — | — |
| Lead | Pb | 82 | 207.19 | 11.35 | 327.45 | 1725 | 0.129 |
| Lithium | Li | 3 | 6.939 | 0.534 | 180.55 | 1300 | 3.58 |
| Livermorium* | Lv | 116 | (293) | — | — | — | — |
| Lutetium | Lu | 71 | 174.97 | 9.849 | 1663 | 1930 | 0.155 |
| Magnesium | Mg | 12 | 24.312 | 1.738 | 650 | 1107 | 1.03 |
| Manganese | Mn | 25 | 54.9380 | 7.44 | 1244 | 2150 | 0.481 |
| Meitnerium | Mt | 109 | (266) | — | — | — | — |
| Mendelevium | Md | 101 | (256) | — | — | — | — |
| Mercury | Hg | 80 | 200.59 | 13.55 | −38.87 | 357 | 0.138 |
| Molybdenum | Mo | 42 | 95.94 | 10.22 | 2617 | 5560 | 0.251 |
| Neodymium | Nd | 60 | 144.24 | 7.007 | 1016 | 3180 | 0.188 |
| Neon | Ne | 10 | 20.183 | $0.8387 \times 10^{-3}$ | −248.597 | −246.0 | 1.03 |
| Neptunium | Np | 93 | (237) | 20.25 | 637 | — | 1.26 |
| Nickel | Ni | 28 | 58.71 | 8.902 | 1453 | 2730 | 0.444 |
| Niobium | Nb | 41 | 92.906 | 8.57 | 2468 | 4927 | 0.264 |
| Nitrogen | N | 7 | 14.0067 | $1.1649 \times 10^{-3}$ | −210 | −195.8 | 1.03 |
| Nobelium | No | 102 | (255) | — | — | — | — |
| Osmium | Os | 76 | 190.2 | 22.59 | 3027 | 5500 | 0.130 |
| Oxygen | O | 8 | 15.9994 | $1.3318 \times 10^{-3}$ | −218.80 | −183.0 | 0.913 |
| Palladium | Pd | 46 | 106.4 | 12.02 | 1552 | 3980 | 0.243 |
| Phosphorus | P | 15 | 30.9738 | 1.83 | 44.25 | 280 | 0.741 |
| Platinum | Pt | 78 | 195.09 | 21.45 | 1769 | 4530 | 0.134 |
| Plutonium | Pu | 94 | (244) | 19.8 | 640 | 3235 | 0.130 |
| Polonium | Po | 84 | (210) | 9.32 | 254 | — | — |
| Potassium | K | 19 | 39.102 | 0.862 | 63.20 | 760 | 0.758 |
| Praseodymium | Pr | 59 | 140.907 | 6.773 | 931 | 3020 | 0.197 |
| Promethium | Pm | 61 | (145) | 7.22 | (1027) | — | — |
| Protactinium | Pa | 91 | (231) | 15.37 (estimated) | (1230) | — | — |
| Radium | Ra | 88 | (226) | 5.0 | 700 | — | — |
| Radon | Rn | 86 | (222) | $9.96 \times 10^{-3}$ (0°C) | (−71) | −61.8 | 0.092 |
| Rhenium | Re | 75 | 186.2 | 21.02 | 3180 | 5900 | 0.134 |
| Rhodium | Rh | 45 | 102.905 | 12.41 | 1963 | 4500 | 0.243 |
| Roentgenium | Rg | 111 | (280) | — | — | — | — |
| Rubidium | Rb | 37 | 85.47 | 1.532 | 39.49 | 688 | 0.364 |
| Ruthenium | Ru | 44 | 101.107 | 12.37 | 2250 | 4900 | 0.239 |
| Rutherfordium | Rf | 104 | 261.11 | — | — | — | — |

| Element | Symbol | Atomic Number Z | Molar Mass, g/mol | Density, g/cm³ at 20°C | Melting Point, °C | Boiling Point, °C | Specific Heat, J/(g·°C) at 25°C |
|---|---|---|---|---|---|---|---|
| Samarium | Sm | 62 | 150.35 | 7.52 | 1072 | 1630 | 0.197 |
| Scandium | Sc | 21 | 44.956 | 2.99 | 1539 | 2730 | 0.569 |
| Seaborgium | Sg | 106 | 263.118 | — | — | — | — |
| Selenium | Se | 34 | 78.96 | 4.79 | 221 | 685 | 0.318 |
| Silicon | Si | 14 | 28.086 | 2.33 | 1412 | 2680 | 0.712 |
| Silver | Ag | 47 | 107.870 | 10.49 | 960.8 | 2210 | 0.234 |
| Sodium | Na | 11 | 22.9898 | 0.9712 | 97.85 | 892 | 1.23 |
| Strontium | Sr | 38 | 87.62 | 2.54 | 768 | 1380 | 0.737 |
| Sulfur | S | 16 | 32.064 | 2.07 | 119.0 | 444.6 | 0.707 |
| Tantalum | Ta | 73 | 180.948 | 16.6 | 3014 | 5425 | 0.138 |
| Technetium | Tc | 43 | (99) | 11.46 | 2200 | — | 0.209 |
| Tellurium | Te | 52 | 127.60 | 6.24 | 449.5 | 990 | 0.201 |
| Terbium | Tb | 65 | 158.924 | 8.229 | 1357 | 2530 | 0.180 |
| Thallium | Tl | 81 | 204.37 | 11.85 | 304 | 1457 | 0.130 |
| Thorium | Th | 90 | (232) | 11.72 | 1755 | (3850) | 0.117 |
| Thulium | Tm | 69 | 168.934 | 9.32 | 1545 | 1720 | 0.159 |
| Tin | Sn | 50 | 118.69 | 7.2984 | 231.868 | 2270 | 0.226 |
| Titanium | Ti | 22 | 47.90 | 4.54 | 1670 | 3260 | 0.523 |
| Tungsten | W | 74 | 183.85 | 19.3 | 3380 | 5930 | 0.134 |
| Unnamed | Uut | 113 | (284) | — | — | — | — |
| Unnamed | Uup | 115 | (288) | — | — | — | — |
| Unnamed | Uus | 117 | — | — | — | — | — |
| Unnamed | Uuo | 118 | (294) | — | — | — | — |
| Uranium | U | 92 | (238) | 18.95 | 1132 | 3818 | 0.117 |
| Vanadium | V | 23 | 50.942 | 6.11 | 1902 | 3400 | 0.490 |
| Xenon | Xe | 54 | 131.30 | $5.495 \times 10^{-3}$ | −111.79 | −108 | 0.159 |
| Ytterbium | Yb | 70 | 173.04 | 6.965 | 824 | 1530 | 0.155 |
| Yttrium | Y | 39 | 88.905 | 4.469 | 1526 | 3030 | 0.297 |
| Zinc | Zn | 30 | 65.37 | 7.133 | 419.58 | 906 | 0.389 |
| Zirconium | Zr | 40 | 91.22 | 6.506 | 1852 | 3580 | 0.276 |

The values in parentheses in the column of molar masses are the mass numbers of the longest-lived isotopes of those elements that are radioactive. Melting points and boiling points in parentheses are uncertain.

The data for gases are valid only when these are in their usual molecular state, such as $H_2$, He, $O_2$, Ne, etc. The specific heats of the gases are the values at constant pressure.

*Source:* Adapted from J. Emsley, *The Elements,* 3rd ed., 1998, Clarendon Press, Oxford. See also www.webelements.com for the latest values and newest elements.

*The names and symbols for elements 114 (Flerovium, Fl) and 116 (Livermorium, Lv) have been suggested but are not official.

# PERIODIC TABLE OF THE ELEMENTS

Metals
Metalloids
Nonmetals

| | Alkali metals IA | IIA | | | | | | | | | | | | IIIA | IVA | VA | VIA | VIIA | Noble gases 0 |
|---|---|---|---|---|---|---|---|---|---|---|---|---|---|---|---|---|---|---|---|
| 1 | 1 H | | | | | | | | | | | | | | | | | | 2 He |
| 2 | 3 Li | 4 Be | | | | | | | | | | | | 5 B | 6 C | 7 N | 8 O | 9 F | 10 Ne |
| 3 | 11 Na | 12 Mg | IIIB | IVB | VB | VIB | VIIB | VIIIB | | | IB | IIB | | 13 Al | 14 Si | 15 P | 16 S | 17 Cl | 18 Ar |
| 4 | 19 K | 20 Ca | 21 Sc | 22 Ti | 23 V | 24 Cr | 25 Mn | 26 Fe | 27 Co | 28 Ni | 29 Cu | 30 Zn | 31 Ga | 32 Ge | 33 As | 34 Se | 35 Br | 36 Kr |
| 5 | 37 Rb | 38 Sr | 39 Y | 40 Zr | 41 Nb | 42 Mo | 43 Tc | 44 Ru | 45 Rh | 46 Pd | 47 Ag | 48 Cd | 49 In | 50 Sn | 51 Sb | 52 Te | 53 I | 54 Xe |
| 6 | 55 Cs | 56 Ba | 57-71 * | 72 Hf | 73 Ta | 74 W | 75 Re | 76 Os | 77 Ir | 78 Pt | 79 Au | 80 Hg | 81 Tl | 82 Pb | 83 Bi | 84 Po | 85 At | 86 Rn |
| 7 | 87 Fr | 88 Ra | 89-103 † | 104 Rf | 105 Db | 106 Sg | 107 Bh | 108 Hs | 109 Mt | 110 Ds | 111 Rg | 112 Cn | 113 | 114 Fl | 115 | 116 Lv | 117 | 118 |

Transition metals

THE HORIZONTAL PERIODS

Inner transition metals

| Lanthanide series * | 57 La | 58 Ce | 59 Pr | 60 Nd | 61 Pm | 62 Sm | 63 Eu | 64 Gd | 65 Tb | 66 Dy | 67 Ho | 68 Er | 69 Tm | 70 Yb | 71 Lu |
|---|---|---|---|---|---|---|---|---|---|---|---|---|---|---|---|
| Actinide series † | 89 Ac | 90 Th | 91 Pa | 92 U | 93 Np | 94 Pu | 95 Am | 96 Cm | 97 Bk | 98 Cf | 99 Es | 100 Fm | 101 Md | 102 No | 103 Lr |

Evidence for the discovery of elements 113 through 118 has been reported. See www.webelements.com for the latest information and newest elements. The names and symbols for elements 114 and 116 have been suggested but are not official.

# ANSWERS

## To Checkpoints and Odd-Numbered Questions and Problems

### Chapter 1

**P** **1.** (a) $4.00 \times 10^4$ km; (b) $5.10 \times 10^8$ km$^2$; (c) $1.08 \times 10^{12}$ km$^3$
**3.** (a) $10^9$ $\mu$m; (b) $10^{-4}$; (c) $9.1 \times 10^5$ $\mu$m **5.** (a) 160 rods; (b) 40 chains **7.** $1.1 \times 10^3$ acre-feet **9.** $1.9 \times 10^{22}$ cm$^3$ **11.** (a) 1.43; (b) 0.864 **13.** (a) 495 s; (b) 141 s; (c) 198 s; (d) $-245$ s **15.** $1.21 \times 10^{12}$ $\mu$s **17.** C, D, A, B, E; the important criterion is the consistency of the daily variation, not its magnitude **19.** $5.2 \times 10^6$ m **21.** $9.0 \times 10^{49}$ atoms **23.** (a) $1 \times 10^3$ kg; (b) 158 kg/s **25.** $1.9 \times 10^5$ kg **27.** (a) $1.18 \times 10^{-29}$ m$^3$; (b) 0.282 nm **29.** $1.75 \times 10^3$ kg **31.** 1.43 kg/min **33.** (a) 293 U.S. bushels; (b) $3.81 \times 10^3$ U.S. bushels **35.** (a) 22 pecks; (b) 5.5 Imperial bushels; (c) 200 L **37.** $8 \times 10^2$ km **39.** (a) 18.8 gallons; (b) 22.5 gallons **41.** 0.3 cord **43.** 3.8 mg/s **45.** (a) yes; (b) 8.6 universe seconds **47.** 0.12 AU/min **49.** (a) 3.88; (b) 7.65; (c) 156 ken$^3$; (d) $1.19 \times 10^3$ m$^3$ **51.** (a) 3.9 m, 4.8 m; (b) $3.9 \times 10^3$ mm, $4.8 \times 10^3$ mm; (c) 2.2 m$^3$, 4.2 m$^3$ **53.** (a) $4.9 \times 10^{-6}$ pc; (b) $1.6 \times 10^{-5}$ ly **55.** (a) 3 nebuchadnezzars, 1 methuselah; (b) 0.37 standard bottle; (c) 0.26 L **57.** 10.7 habaneros **59.** 700 to 1500 oysters

### Chapter 2

**CP** **1.** b and c **2.** (check the derivative $dx/dt$) (a) 1 and 4; (b) 2 and 3 **3.** (a) plus; (b) minus; (c) minus; (d) plus **4.** 1 and 4 ($a = d^2x/dt^2$ must be constant) **5.** (a) plus (upward displacement on $y$ axis); (b) minus (downward displacement on $y$ axis); (c) $a = -g = -9.8$ m/s$^2$

**Q** **1.** (a) negative; (b) positive; (c) yes; (d) positive; (e) constant **3.** (a) all tie; (b) 4, tie of 1 and 2, then 3 **5.** (a) positive direction; (b) negative direction; (c) 3 and 5; (d) 2 and 6 tie, then 3 and 5 tie, then 1 and 4 tie (zero) **7.** (a) $D$; (b) $E$ **9.** (a) 3, 2, 1; (b) 1, 2, 3; (c) all tie; (d) 1, 2, 3 **11.** 1 and 2 tie, then 3

**P** **1.** 13 m **3.** (a) $+40$ km/h; (b) 40 km/h **5.** (a) 0; (b) $-2$ m; (c) 0; (d) 12 m; (e) $+12$ m; (f) $+7$ m/s **7.** 60 km **9.** 1.4 m **11.** 128 km/h **13.** (a) 73 km/h; (b) 68 km/h; (c) 70 km/h; (d) 0 **15.** (a) $-6$ m/s; (b) $-x$ direction; (c) 6 m/s; (d) decreasing; (e) 2 s; (f) no **17.** (a) 28.5 cm/s; (b) 18.0 cm/s; (c) 40.5 cm/s; (d) 28.1 cm/s; (e) 30.3 cm/s **19.** $-20$ m/s$^2$ **21.** (a) 1.10 m/s; (b) 6.11 mm/s$^2$; (c) 1.47 m/s; (d) 6.11 mm/s$^2$ **23.** $1.62 \times 10^{15}$ m/s$^2$ **25.** (a) 30 s; (b) 300 m **27.** (a) $+1.6$ m/s; (b) $+18$ m/s **29.** (a) 10.6 m; (b) 41.5 s **31.** (a) $3.1 \times 10^6$ s; (b) $4.6 \times 10^{13}$ m **33.** (a) 3.56 m/s$^2$; (b) 8.43 m/s **35.** 0.90 m/s$^2$ **37.** (a) 4.0 m/s$^2$; (b) $+x$ **39.** (a) $-2.5$ m/s$^2$; (b) 1; (d) 0; (e) 2 **41.** 40 m **43.** (a) 0.994 m/s$^2$ **45.** (a) 31 m/s; (b) 6.4 s **47.** (a) 29.4 m/s; (b) 2.45 s **49.** (a) 5.4 s; (b) 41 m/s **51.** (a) 20 m; (b) 59 m **53.** 4.0 m/s **55.** (a) 857 m/s$^2$; (b) up **57.** (a) $1.26 \times 10^3$ m/s$^2$; (b) up **59.** (a) 89 cm; (b) 22 cm **61.** 20.4 m **63.** 2.34 m **65.** (a) 2.25 m/s; (b) 3.90 m/s **67.** 0.56 m/s **69.** 100 m **71.** (a) 2.00 s; (b) 12 cm; (c) $-9.00$ cm/s$^2$; (d) right; (e) left; (f) 3.46 s **73.** (a) 82 m; (b) 19 m/s **75.** (a) 0.74 s; (b) 6.2 m/s$^2$ **77.** (a) 3.1 m/s$^2$; (b) 45 m; (c) 13 s **79.** 17 m/s **81.** $+47$ m/s **83.** (a) 1.23 cm; (b) 4 times; (c) 9 times; (d) 16 times; (e) 25 times **85.** 25 km/h **87.** 1.2 h **89.** $4H$ **91.** (a) 3.2 s; (b) 1.3 s **93.** (a) 8.85 m/s; (b) 1.00 m **95.** (a) 2.0 m/s$^2$; (b) 12 m/s; (c) 45 m **97.** (a) 48.5 m/s; (b) 4.95 s; (c) 34.3 m/s; (d) 3.50 s **99.** 22.0 m/s **101.** (a) $v = (v_0^2 + 2gh)^{0.5}$; (b) $t = [(v_0^2 + 2gh)^{0.5} - v_0]/g$; (c) same as (a); (d) $t = [(v_0^2 + 2gh)^{0.5} + v_0]/g$, greater **103.** 414 ms **105.** 90 m **107.** 0.556 s **109.** (a) 0.28 m/s$^2$; (b) 0.28 m/s$^2$ **111.** (a) 10.2 s;

(b) 10.0 m **113.** (a) 5.44 s; (b) 53.3 m/s; (c) 5.80 m **115.** 2.3 cm/min **117.** 0.15 m/s **119.** (a) 1.0 cm/s; (b) 1.6 cm/s, 1.1 cm/s, 0; (c) $-0.79$ cm/s$^2$; (d) 0, $-0.87$ cm/s$^2$, $-1.2$ cm/s$^2$

### Chapter 3

**CP** **1.** (a) 7 m ($\vec{a}$ and $\vec{b}$ are in same direction); (b) 1 m ($\vec{a}$ and $\vec{b}$ are in opposite directions) **2.** $c, d, f$ (components must be head to tail; $\vec{a}$ must extend from tail of one component to head of the other) **3.** (a) $+, +$; (b) $+, -$; (c) $+, +$ (draw vector from tail of $\vec{d_1}$ to head of $\vec{d_2}$) **4.** (a) 90°; (b) 0° (vectors are parallel—same direction); (c) 180° (vectors are antiparallel—opposite directions) **5.** (a) 0° or 180°; (b) 90°

**Q** **1.** yes, when the vectors are in same direction **3.** Either the sequence $\vec{d_2}, \vec{d_1}$ or the sequence $\vec{d_2}, \vec{d_2}, \vec{d_3}$ **5.** all but ($e$) **7.** (a) yes; (b) yes; (c) no **9.** (a) $+x$ for (1), $+z$ for (2), $+z$ for (3); (b) $-x$ for (1), $-z$ for (2), $-z$ for (3) **11.** $\vec{s}, \vec{p}, \vec{r}$ or $\vec{p}, \vec{s}, \vec{r}$ **13.** Correct: $c, d, f, h$. Incorrect: $a$ (cannot dot a vector with a scalar), $b$ (cannot cross a vector with a scalar), $e, g, i, j$ (cannot add a scalar and a vector).

**P** **1.** (a) $-2.5$ m; (b) $-6.9$ m **3.** (a) 47.2 m; (b) 122° **5.** (a) 156 km; (b) 39.8° west of due north **7.** (a) parallel; (b) antiparallel; (c) perpendicular **9.** (a) $(3.0\,\text{m})\hat{i} - (2.0\,\text{m})\hat{j} + (5.0\,\text{m})\hat{k}$; (b) $(5.0\,\text{m})\hat{i} - (4.0\,\text{m})\hat{j} - (3.0\,\text{m})\hat{k}$; (c) $(-5.0\,\text{m})\hat{i} + (4.0\,\text{m})\hat{j} + (3.0\,\text{m})\hat{k}$ **11.** (a) $(-9.0\,\text{m})\hat{i} + (10\,\text{m})\hat{j}$; (b) 13 m; (c) 132° **13.** 4.74 km **15.** (a) 1.59 m; (b) 12.1 m; (c) 12.2 m; (d) 82.5° **17.** (a) 38 m; (b) $-37.5°$; (c) 130 m; (d) 1.2°; (e) 62 m; (f) 130° **19.** 5.39 m at 21.8° left of forward **21.** (a) $-70.0$ cm; (b) 80.0 cm; (c) 141 cm; (d) $-172°$ **23.** 3.2 **25.** 2.6 km **27.** (a) $8\hat{i} + 16\hat{j}$; (b) $2\hat{i} + 4\hat{j}$ **29.** (a) 7.5 cm; (b) 90°; (c) 8.6 cm; (d) 48° **31.** (a) 9.51 m; (b) 14.1 m; (c) 13.4 m; (d) 10.5 m **33.** (a) 12; (b) $+z$; (c) 12; (d) $-z$; (e) 12; (f) $+z$ **35.** (a) $-18.8$ units; (b) 26.9 units, $+z$ direction **37.** (a) $-21$; (b) $-9$; (c) $5\hat{i} - 11\hat{j} - 9\hat{k}$ **39.** 70.5° **41.** 22° **43.** (a) 3.00 m; (b) 0; (c) 3.46 m; (d) 2.00 m; (e) $-5.00$ m; (f) 8.66 m; (g) $-6.67$; (h) 4.33 **45.** (a) $-83.4$; (b) $(1.14 \times 10^3)\hat{k}$; (c) $1.14 \times 10^3$, $\theta$ not defined, $\phi = 0°$; (d) 90.0°; (e) $-5.14\hat{i} + 6.13\hat{j} + 3.00\hat{k}$; (f) 8.54, $\theta = 130°$, $\phi = 69.4°$ **47.** (a) 140°; (b) 90.0°; (c) 99.1° **49.** (a) 103 km; (b) 60.9° north of due west **51.** (a) 27.8 m; (b) 13.4 m **53.** (a) 30; (b) 52 **55.** (a) $-2.83$ m; (b) $-2.83$ m; (c) 5.00 m; (d) 0; (e) 3.00 m; (f) 5.20 m; (g) 5.17 m; (h) 2.37 m; (i) 5.69 m; (j) 25° north of due east; (k) 5.69 m; (l) 25° south of due west **57.** 4.1 **59.** (a) $(9.19\,\text{m})\hat{i}' + (7.71\,\text{m})\hat{j}'$; (b) $(14.0\,\text{m})\hat{i}' + (3.41\,\text{m})\hat{j}'$ **61.** (a) $11\hat{i} + 5.0\hat{j} - 7.0\hat{k}$; (b) 120°; (c) $-4.9$; (d) 7.3 **63.** (a) 3.0 m$^2$; (b) 52 m$^3$; (c) $(11\,\text{m}^2)\hat{i} + (9.0\,\text{m}^2)\hat{j} + (3.0\,\text{m}^2)\hat{k}$ **65.** (a) $(-40\hat{i} - 20\hat{j} + 25\hat{k})$ m; (b) 45 m **67.** (a) 0; (b) 0; (c) $-1$; (d) west; (e) up; (f) west **69.** (a) 168 cm; (b) 32.5° **71.** (a) 15 m; (b) south; (c) 6.0 m; (d) north **73.** (a) $2\hat{k}$; (b) 26; (c) 46; (d) 5.81 **75.** (a) up; (b) 0; (c) south; (d) 1; (e) 0 **77.** (a) $(1300\,\text{m})\hat{i} + (2200\,\text{m})\hat{j} - (410\,\text{m})\hat{k}$; (b) $2.56 \times 10^3$ m **79.** 8.4

### Chapter 4

**CP** **1.** (draw $\vec{v}$ tangent to path, tail on path) (a) first; (b) third **2.** (take second derivative with respect to time) (1) and (3) $a_x$ and $a_y$ are both constant and thus $\vec{a}$ is constant; (2) and (4) $a_y$ is constant but $a_x$ is not, thus $\vec{a}$ is not **3.** yes **4.** (a) $v_x$ constant; (b) $v_y$ initially positive, decreases to zero, and then becomes progressively more negative; (c) $a_x = 0$ throughout; (d) $a_y = -g$ throughout **5.** (a) $-(4\,\text{m/s})\hat{i}$; (b) $-(8\,\text{m/s}^2)\hat{j}$

**Q** **1.** $a$ and $c$ tie, then $b$ **3.** decreases **5.** $a, b, c$ **7.** (a) 0; (b) 350 km/h; (c) 350 km/h; (d) same (nothing changed about the vertical motion) **9.** (a) all tie; (b) all tie; (c) 3, 2, 1; (d) 3, 2, 1 **11.** 2, then 1 and 4 tie, then 3 **13.** (a) yes; (b) no; (c) yes **15.** (a) decreases; (b) increases **17.** maximum height

**P** **1.** (a) 6.2 m **3.** $(-2.0\,\text{m})\hat{\text{i}} + (6.0\,\text{m})\hat{\text{j}} - (10\,\text{m})\hat{\text{k}}$ **5.** (a) 7.59 km/h; (b) 22.5° east of due north **7.** $(-0.70\,\text{m/s})\hat{\text{i}} + (1.4\,\text{m/s})\hat{\text{j}} - (0.40\,\text{m/s})\hat{\text{k}}$ **9.** (a) 0.83 cm/s; (b) 0°; (c) 0.11 m/s; (d) $-63°$ **11.** (a) $(6.00\,\text{m})\hat{\text{i}} - (106\,\text{m})\hat{\text{j}}$; (b) $(19.0\,\text{m/s})\hat{\text{i}} - (224\,\text{m/s})\hat{\text{j}}$; (c) $(24.0\,\text{m/s}^2)\hat{\text{i}} - (336\,\text{m/s}^2)\hat{\text{j}}$; (d) $-85.2°$ **13.** (a) $(8\,\text{m/s}^2)t\hat{\text{j}} + (1\,\text{m/s})\hat{\text{k}}$; (b) $(8\,\text{m/s}^2)\hat{\text{j}}$ **15.** (a) $(-1.50\,\text{m/s})\hat{\text{j}}$; (b) $(4.50\,\text{m})\hat{\text{i}} - (2.25\,\text{m})\hat{\text{j}}$ **17.** $(32\,\text{m/s})\hat{\text{i}}$ **19.** (a) $(72.0\,\text{m})\hat{\text{i}} + (90.7\,\text{m})\hat{\text{j}}$; (b) 49.5° **21.** (a) 18 cm; (b) 1.9 m **23.** (a) 3.03 s; (b) 758 m; (c) 29.7 m/s **25.** 43.1 m/s (155 km/h) **27.** (a) 10.0 s; (b) 897 m **29.** 78.5° **31.** 3.35 m **33.** (a) 202 m/s; (b) 806 m; (c) 161 m/s; (d) $-171$ m/s **35.** 4.84 cm **37.** (a) 1.60 m; (b) 6.86 m; (c) 2.86 m **39.** (a) 32.3 m; (b) 21.9 m/s; (c) 40.4°; (d) below **41.** 55.5° **43.** (a) 11 m; (b) 23 m; (c) 17 m/s; (d) 63° **45.** (a) ramp; (b) 5.82 m; (c) 31.0° **47.** (a) yes; (b) 2.56 m **49.** (a) 31°; (b) 63° **51.** (a) 2.3°; (b) 1.1 m; (c) 18° **53.** (a) 75.0 m; (b) 31.9 m/s; (c) 66.9°; (d) 25.5 m **55.** the third **57.** (a) 7.32 m; (b) west; (c) north **59.** (a) 12 s; (b) 4.1 m/s²; (c) down; (d) 4.1 m/s²; (e) up **61.** (a) $1.3 \times 10^5$ m/s; (b) $7.9 \times 10^5$ m/s²; (c) increase **63.** 2.92 m **65.** $(3.00\,\text{m/s}^2)\hat{\text{i}} + (6.00\,\text{m/s}^2)\hat{\text{j}}$ **67.** 160 m/s² **69.** (a) 13 m/s²; (b) eastward; (c) 13 m/s²; (d) eastward **71.** 1.67 **73.** (a) $(80\,\text{km/h})\hat{\text{i}} - (60\,\text{km/h})\hat{\text{j}}$; (b) 0°; (c) answers do not change **75.** 32 m/s **77.** 60° **79.** (a) 38 knots; (b) 1.5° east of due north; (c) 4.2 h; (d) 1.5° west of due south **81.** (a) $(-32\,\text{km/h})\hat{\text{i}} - (46\,\text{km/h})\hat{\text{j}}$; (b) $[(2.5\,\text{km}) - (32\,\text{km/h})t]\hat{\text{i}} + [(4.0\,\text{km}) - (46\,\text{km/h})t]\hat{\text{j}}$; (c) 0.084 h; (d) $2 \times 10^2$ m **83.** (a) $-30°$; (b) 69 min; (c) 80 min; (d) 80 min; (e) 0°; (f) 60 min **85.** (a) 2.7 km; (b) 76° clockwise **87.** (a) 44 m; (b) 13 m; (c) 8.9 m **89.** (a) 45 m; (b) 22 m/s **91.** (a) $2.6 \times 10^2$ m/s; (b) 45 s; (c) increase **93.** (a) 63 km; (b) 18° south of due east; (c) 0.70 km/h; (d) 18° south of due east; (e) 1.6 km/h; (f) 1.2 km/h; (g) 33° north of due east **95.** (a) 1.5; (b) (36 m, 54 m) **97.** (a) 62 ms; (b) $4.8 \times 10^2$ m/s **99.** 2.64 m **101.** (a) 2.5 m; (b) 0.82 m; (c) 9.8 m/s²; (d) 9.8 m/s² **103.** (a) 6.79 km/h; (b) 6.96° **105.** (a) 16 m/s; (b) 23°; (c) above; (d) 27 m/s; (e) 57°; (f) below **107.** (a) 4.2 m, 45°; (b) 5.5 m, 68°; (c) 6.0 m, 90°; (d) 4.2 m, 135°; (e) 0.85 m/s, 135°; (f) 0.94 m/s, 90°; (g) 0.94 m/s, 180°; (h) 0.30 m/s², 180°; (i) 0.30 m/s², 270° **109.** (a) $5.4 \times 10^{-13}$ m; (b) decrease **111.** (a) 0.034 m/s²; (b) 84 min **113.** (a) 8.43 m; (b) $-129°$ **115.** (a) 2.00 ns; (b) 2.00 mm; (c) $1.00 \times 10^7$ m/s; (d) $2.00 \times 10^6$ m/s **117.** (a) 24 m/s; (b) 65° **119.** 93° from the car's direction of motion **121.** (a) $4.6 \times 10^{12}$ m; (b) $2.4 \times 10^5$ s **123.** (a) 6.29°; (b) 83.7° **125.** (a) $3 \times 10^1$ m **127.** (a) $(6.0\hat{\text{i}} + 4.2\hat{\text{j}})$ m/s; (b) $(18\hat{\text{i}} + 6.3\hat{\text{j}})$ m **129.** (a) 38 ft/s; (b) 32 ft/s; (c) 9.3 ft **131.** (a) 11 m; (b) 45 m/s **133.** (a) 5.8 m/s; (b) 17 m; (c) 67° **135.** (a) 32.4 m; (b) $-37.7$ m **137.** 88.6 km/h

## Chapter 5

**CP** **1.** $c, d,$ and $e$ ($\vec{F}_1$ and $\vec{F}_2$ must be head to tail, $\vec{F}_{\text{net}}$ must be from tail of one of them to head of the other) **2.** (a) and (b) 2 N, leftward (acceleration is zero in each situation) **3.** (a) equal; (b) greater (acceleration is upward, thus net force on body must be upward) **4.** (a) equal; (b) greater; (c) less **5.** (a) increase; (b) yes; (c) same; (d) yes

**Q** **1.** (a) 2, 3, 4; (b) 1, 3, 4; (c) 1, $+y$; 2, $+x$; 3, fourth quadrant; 4, third quadrant **3.** increase **5.** (a) 2 and 4; (b) 2 and 4 **7.** (a) $M$; (b) $M$; (c) $M$; (d) $2M$; (e) $3M$ **9.** (a) 20 kg; (b) 18 kg; (c) 10 kg; (d) all tie; (e) 3, 2, 1 **11.** (a) increases from initial value $mg$; (b) decreases from $mg$ to zero (after which the block moves up away from the floor)

**P** **1.** 2.9 m/s² **3.** (a) 1.88 N; (b) 0.684 N; (c) $(1.88\,\text{N})\hat{\text{i}} + (0.684\,\text{N})\hat{\text{j}}$ **5.** (a) $(0.86\,\text{m/s}^2)\hat{\text{i}} - (0.16\,\text{m/s}^2)\hat{\text{j}}$; (b) 0.88 m/s²; (c) $-11°$ **7.** (a)

$(-32.0\,\text{N})\hat{\text{i}} - (20.8\,\text{N})\hat{\text{j}}$; (b) 38.2 N; (c) $-147°$ **9.** (a) 8.37 N; (b) $-133°$; (c) $-125°$ **11.** 9.0 m/s² **13.** (a) 4.0 kg; (b) 1.0 kg; (c) 4.0 kg; (d) 1.0 kg **15.** (a) 108 N; (b) 108 N; (c) 108 N **17.** (a) 42 N; (b) 72 N; (c) 4.9 m/s² **19.** $1.2 \times 10^5$ N **21.** (a) 11.7 N; (b) $-59.0°$ **23.** (a) $(285\,\text{N})\hat{\text{i}} + (705\,\text{N})\hat{\text{j}}$; (b) $(285\,\text{N})\hat{\text{i}} - (115\,\text{N})\hat{\text{j}}$; (c) 307 N; (d) $-22.0°$; (e) 3.67 m/s²; (f) $-22.0°$ **25.** (a) 0.022 m/s²; (b) $8.3 \times 10^4$ km; (c) $1.9 \times 10^3$ m/s **27.** 1.5 mm **29.** (a) 494 N; (b) up; (c) 494 N; (d) down **31.** (a) 1.18 m; (b) 0.674 s; (c) 3.50 m/s **33.** $1.8 \times 10^4$ N **35.** (a) 46.7°; (b) 28.0° **37.** (a) 0.62 m/s²; (b) 0.13 m/s²; (c) 2.6 m **39.** (a) $2.2 \times 10^{-3}$ N; (b) $3.7 \times 10^{-3}$ N **41.** (a) 1.4 m/s²; (b) 4.1 m/s **43.** (a) 1.23 N; (b) 2.46 N; (c) 3.69 N; (d) 4.92 N; (e) 6.15 N; (f) 0.250 N **45.** (a) 31.3 kN; (b) 24.3 kN **47.** $6.4 \times 10^3$ N **49.** (a) 2.18 m/s²; (b) 116 N; (c) 21.0 m/s² **51.** (a) 3.6 m/s²; (b) 17 N **53.** (a) 0.970 m/s²; (b) 11.6 N; (c) 34.9 N **55.** (a) 1.1 N **57.** (a) 0.735 m/s²; (b) down; (c) 20.8 N **59.** (a) 4.9 m/s²; (b) 2.0 m/s²; (c) up; (d) 120 N **61.** $2Ma/(a + g)$ **63.** (a) 8.0 m/s; (b) $+x$ **65.** (a) 0.653 m/s³; (b) 0.896 m/s³; (c) 6.50 s **67.** 81.7 N **69.** 2.4 N **71.** 16 N **73.** (a) 2.6 N; (b) 17° **75.** (a) 0; (b) 0.83 m/s²; (c) 0 **77.** (a) 0.74 m/s²; (b) 7.3 m/s² **79.** (a) 11 N; (b) 2.2 kg; (c) 0; (d) 2.2 kg **81.** 195 N **83.** (a) 4.6 m/s²; (b) 2.6 m/s² **85.** (a) rope breaks; (b) 1.6 m/s² **87.** (a) 65 N; (b) 49 N **89.** (a) $4.6 \times 10^3$ N; (b) $5.8 \times 10^3$ N **91.** (a) $1.8 \times 10^2$ N; (b) $6.4 \times 10^2$ N **93.** (a) 44 N; (b) 78 N; (c) 54 N; (d) 152 N **95.** (a) 4 kg; (b) 6.5 m/s²; (c) 13 N **97.** (a) $(1.0\hat{\text{i}} - 2.0\hat{\text{j}})$ N; (b) 2.2 N; (c) $-63°$; (d) 2.2 m/s²; (e) $-63°$

## Chapter 6

**CP** **1.** (a) zero (because there is no attempt at sliding); (b) 5 N; (c) no; (d) yes; (e) 8 N **2.** ($\vec{a}$ is directed toward center of circular path) (a) $\vec{a}$ downward, $\vec{F}_N$ upward; (b) $\vec{a}$ and $\vec{F}_N$ upward; (c) same; (d) greater at lowest point

**Q** **1.** (a) decrease; (b) decrease; (c) increase; (d) increase; (e) increase **3.** (a) same; (b) increases; (c) increases; (d) no **5.** (a) upward; (b) horizontal, toward you; (c) no change; (d) increases; (e) increases **7.** At first, $\vec{f}_s$ is directed up the ramp and its magnitude increases from $mg \sin \theta$ until it reaches $f_{s,\text{max}}$. Thereafter the force is kinetic friction directed up the ramp, with magnitude $f_k$ (a constant value smaller than $f_{s,\text{max}}$). **9.** 4, 3, then 1, 2, and 5 tie **11.** (a) all tie; (b) all tie; (c) 2, 3, 1 **13.** (a) increases; (b) increases; (c) decreases; (d) decreases; (e) decreases

**P** **1.** 36 m **3.** (a) $2.0 \times 10^2$ N; (b) $1.2 \times 10^2$ N **5.** (a) 6.0 N; (b) 3.6 N; (c) 3.1 N **7.** (a) $1.9 \times 10^2$ N; (b) 0.56 m/s² **9.** (a) 11 N; (b) 0.14 m/s² **11.** (a) $3.0 \times 10^2$ N; (b) 1.3 m/s² **13.** (a) $1.3 \times 10^2$ N; (b) no; (c) $1.1 \times 10^2$ N; (d) 46 N; (e) 17 N **15.** 2° **17.** (a) $(17\,\text{N})\hat{\text{i}}$; (b) $(20\,\text{N})\hat{\text{i}}$; (c) $(15\,\text{N})\hat{\text{i}}$ **19.** (a) no; (b) $(-12\,\text{N})\hat{\text{i}} + (5.0\,\text{N})\hat{\text{j}}$ **21.** (a) 19°; (b) 3.3 kN **23.** 0.37 **25.** $1.0 \times 10^2$ N **27.** (a) 0; (b) $(-3.9\,\text{m/s}^2)\hat{\text{i}}$; (c) $(-1.0\,\text{m/s}^2)\hat{\text{i}}$ **29.** (a) 66 N; (b) 2.3 m/s² **31.** (a) 3.5 m/s²; (b) 0.21 N **33.** 9.9 s **35.** $4.9 \times 10^2$ N **37.** (a) $3.2 \times 10^2$ km/h; (b) $6.5 \times 10^2$ km/h; (c) no **39.** 2.3 **41.** 0.60 **43.** 21 m **45.** (a) light; (b) 778 N; (c) 223 N; (d) 1.11 kN **47.** (a) 10 s; (b) $4.9 \times 10^2$ N; (c) $1.1 \times 10^3$ N **49.** $1.37 \times 10^3$ N **51.** 2.2 km **53.** 12° **55.** $2.6 \times 10^3$ N **57.** 1.81 m/s **59.** (a) 8.74 N; (b) 37.9 N; (c) 6.45 m/s; (d) radially inward **61.** (a) 27 N; (b) 3.0 m/s² **63.** (b) 240 N; (c) 0.60 **65.** (a) 69 km/h; (b) 139 km/h; (c) yes **67.** $g(\sin \theta - 2^{0.5}\mu_k \cos \theta)$ **69.** 3.4 m/s² **71.** (a) 35.3 N; (b) 39.7 N; (c) 320 N **73.** (a) 7.5 m/s²; (b) down; (c) 9.5 m/s²; (d) down **75.** (a) $3.0 \times 10^5$ N; (b) 1.2° **77.** 147 m/s **79.** (a) 13 N; (b) 1.6 m/s² **81.** (a) 275 N; (b) 877 N **83.** (a) 84.2 N; (b) 52.8 N; (c) 1.87 m/s² **85.** 3.4% **87.** (a) $3.21 \times 10^3$ N; (b) yes **89.** (a) 222 N; (b) 334 N; (c) 311 N; (d) 311 N; (e) c, d **91.** (a) $v_0^2/(4g \sin \theta)$; (b) no **93.** (a) 0.34; (b) 0.24 **95.** (a) $\mu_k mg/(\sin \theta - \mu_k \cos \theta)$; (b) $\theta_0 = \tan^{-1} \mu_s$ **97.** 0.18 **99.** (a) 56 N; (b) 59 N; (c) $1.1 \times 10^3$ N **101.** 0.76 **103.** (a) bottom of circle; (b) 9.5 m/s **105.** 0.56

**Chapter 7**

**CP** **1.** (a) decrease; (b) same; (c) negative, zero    **2.** (a) positive; (b) negative; (c) zero    **3.** zero

**Q** **1.** all tie    **3.** (a) positive; (b) negative; (c) negative    **5.** $b$ (positive work), $a$ (zero work), $c$ (negative work), $d$ (more negative work)    **7.** all tie    **9.** (a) $A$; (b) $B$    **11.** 2, 3, 1

**P** **1.** (a) $2.9 \times 10^7$ m/s; (b) $2.1 \times 10^{-13}$ J    **3.** (a) $5 \times 10^{14}$ J; (b) 0.1 megaton TNT; (c) 8 bombs    **5.** (a) 2.4 m/s; (b) 4.8 m/s    **7.** 0.96 J    **9.** 20 J    **11.** (a) $62.3°$; (b) $118°$    **13.** (a) $1.7 \times 10^2$ N; (b) $3.4 \times 10^2$ m; (c) $-5.8 \times 10^4$ J; (d) $3.4 \times 10^2$ N; (e) $1.7 \times 10^2$ m; (f) $-5.8 \times 10^4$ J    **15.** (a) 1.50 J; (b) increases    **17.** (a) 12 kJ; (b) $-11$ kJ; (c) 1.1 kJ; (d) 5.4 m/s    **19.** 25 J    **21.** (a) $-3Mgd/4$; (b) $Mgd$; (c) $Mgd/4$; (d) $(gd/2)^{0.5}$    **23.** 4.41 J    **25.** (a) 25.9 kJ; (b) 2.45 N    **27.** (a) 7.2 J; (b) 7.2 J; (c) 0; (d) $-25$ J    **29.** (a) 0.90 J; (b) 2.1 J; (c) 0    **31.** (a) 6.6 m/s; (b) 4.7 m    **33.** (a) 0.12 m; (b) 0.36 J; (c) $-0.36$ J; (d) 0.060 m; (e) 0.090 J    **35.** (a) 0; (b) 0    **37.** (a) 42 J; (b) 30 J; (c) 12 J; (d) 6.5 m/s, $+x$ axis; (e) 5.5 m/s, $+x$ axis; (f) 3.5 m/s, $+x$ axis    **39.** 4.00 N/m    **41.** $5.3 \times 10^2$ J    **43.** (a) 0.83 J; (b) 2.5 J; (c) 4.2 J; (d) 5.0 W    **45.** $4.9 \times 10^2$ W    **47.** (a) $1.0 \times 10^2$ J; (b) 8.4 W    **49.** $7.4 \times 10^2$ W    **51.** (a) 32.0 J; (b) 8.00 W; (c) $78.2°$    **53.** (a) 1.20 J; (b) 1.10 m/s    **55.** (a) $1.8 \times 10^5$ ft·lb; (b) 0.55 hp    **57.** (a) 797 N; (b) 0; (c) $-1.55$ kJ; (d) 0; (e) 1.55 kJ; (f) $F$ varies during displacement    **59.** (a) 11 J; (b) $-21$ J    **61.** $-6$ J    **63.** (a) 314 J; (b) $-155$ J; (c) 0; (d) 158 J    **65.** (a) 98 N; (b) 4.0 cm; (c) 3.9 J; (d) $-3.9$ J    **67.** (a) 23 mm; (b) 45 N    **69.** 165 kW    **71.** $-37$ J    **73.** (a) 13 J; (b) 13 J    **75.** 235 kW    **77.** (a) 6 J; (b) 6.0 J    **79.** (a) 0.6 J; (b) 0; (c) $-0.6$ J    **81.** (a) 3.35 m/s; (b) 22.5 J; (c) 0; (d) 0; (e) 0.212 m    **83.** (a) $-5.20 \times 10^{-2}$ J; (b) $-0.160$ J    **85.** 6.63 m/s

**Chapter 8**

**CP** **1.** no (consider round trip on the small loop)    **2.** 3, 1, 2 (see Eq. 8-6)    **3.** (a) all tie; (b) all tie    **4.** (a) $CD, AB, BC$ (0) (check slope magnitudes); (b) positive direction of $x$    **5.** all tie

**Q** **1.** (a) 3, 2, 1; (b) 1, 2, 3    **3.** (a) 12 J; (b) $-2$ J    **5.** (a) increasing; (b) decreasing; (c) decreasing; (d) constant in $AB$ and $BC$, decreasing in $CD$    **7.** $+30$ J    **9.** 2, 1, 3    **11.** $-40$ J

**P** **1.** 89 N/cm    **3.** (a) 167 J; (b) $-167$ J; (c) 196 J; (d) 29 J; (e) 167 J; (f) $-167$ J; (g) 296 J; (h) 129 J    **5.** (a) 4.31 mJ; (b) $-4.31$ mJ; (c) 4.31 mJ; (d) $-4.31$ mJ; (e) all increase    **7.** (a) 13.1 J; (b) $-13.1$ J; (c) 13.1 J; (d) all increase    **9.** (a) 17.0 m/s; (b) 26.5 m/s; (c) 33.4 m/s; (d) 56.7 m; (e) all the same    **11.** (a) 2.08 m/s; (b) 2.08 m/s; (c) increase    **13.** (a) 0.98 J; (b) $-0.98$ J; (c) 3.1 N/cm    **15.** (a) $2.6 \times 10^2$ m; (b) same; (c) decrease    **17.** (a) 2.5 N; (b) 0.31 N; (c) 30 cm    **19.** (a) 784 N/m; (b) 62.7 J; (c) 62.7 J; (d) 80.0 cm    **21.** (a) 8.35 m/s; (b) 4.33 m/s; (c) 7.45 m/s; (d) both decrease    **23.** (a) 4.85 m/s; (b) 2.42 m/s    **25.** $-3.2 \times 10^2$ J    **27.** (a) no; (b) $9.3 \times 10^2$ N    **29.** (a) 35 cm; (b) 1.7 m/s    **31.** (a) 39.2 J; (b) 39.2 J; (c) 4.00 m    **33.** (a) 2.40 m/s; (b) 4.19 m/s    **35.** (a) 39.6 cm; (b) 3.64 cm    **37.** $-18$ mJ    **39.** (a) 2.1 m/s; (b) 10 N; (c) $+x$ direction; (d) 5.7 m; (e) 30 N; (f) $-x$ direction    **41.** (a) $-3.7$ J; (c) 1.3 m; (d) 9.1 m; (e) 2.2 J; (f) 4.0 m; (g) $(4 - x)e^{-x/4}$; (h) 4.0 m    **43.** (a) 5.6 J; (b) 3.5 J    **45.** (a) 30.1 J; (b) 30.1 J; (c) 0.225    **47.** 0.53 J    **49.** (a) $-2.9$ kJ; (b) $3.9 \times 10^2$ J; (c) $2.1 \times 10^2$ N    **51.** (a) 1.5 MJ; (b) 0.51 MJ; (c) 1.0 MJ; (d) 63 m/s    **53.** (a) 67 J; (b) 67 J; (c) 46 cm    **55.** (a) $-0.90$ J; (b) 0.46 J; (c) 1.0 m/s    **57.** 1.2 m    **59.** (a) 19.4 m; (b) 19.0 m/s    **61.** (a) $1.5 \times 10^{-2}$ N; (b) $(3.8 \times 10^2)g$    **63.** (a) 7.4 m/s; (b) 90 cm; (c) 2.8 m; (d) 15 m    **65.** 20 cm    **67.** (a) 7.0 J; (b) 22 J    **69.** 3.7 J    **71.** 4.33 m/s    **73.** 25 J    **75.** (a) 4.9 m/s; (b) 4.5 N; (c) $71°$; (d) same    **77.** (a) 4.8 N; (b) $+x$ direction; (c) 1.5 m; (d) 13.5 m; (e) 3.5 m/s    **79.** (a) 24 kJ; (b) $4.7 \times 10^2$ N    **81.** (a) 5.00 J; (b) 9.00 J; (c) 11.0 J; (d) 3.00 J; (e) 12.0 J; (f) 2.00 J; (g) 13.0 J; (h) 1.00 J; (i) 13.0 J; (j) 1.00 J; (l) 11.0 J; (m) 10.8 m; (n) It returns to $x = 0$ and stops.    **83.** (a) 6.0 kJ; (b) $6.0 \times 10^2$ W; (c) $3.0 \times 10^2$ W;

(d) $9.0 \times 10^2$ W    **85.** 880 MW    **87.** (a) $v_0 = (2gL)^{0.5}$; (b) $5mg$; (c) $-mgL$; (d) $-2mgL$    **89.** (a) 109 J; (b) 60.3 J; (c) 68.2 J; (d) 41.0 J    **91.** (a) 2.7 J; (b) 1.8 J; (c) 0.39 m    **93.** (a) 10 m; (b) 49 N; (c) 4.1 m; (d) $1.2 \times 10^2$ N    **95.** (a) 5.5 m/s; (b) 5.4 m; (c) same    **97.** 80 mJ    **99.** 24 W    **101.** $-12$ J    **103.** (a) 8.8 m/s; (b) 2.6 kJ; (c) 1.6 kW    **105.** (a) $7.4 \times 10^2$ J; (b) $2.4 \times 10^2$ J    **107.** 15 J    **109.** (a) $2.35 \times 10^3$ J; (b) 352 J    **111.** 738 m    **113.** (a) $-3.8$ kJ; (b) 31 kN    **115.** (a) 300 J; (b) 93.8 J; (c) 6.38 J    **117.** (a) 5.6 J; (b) 12 J; (c) 13 J    **119.** (a) 1.2 J; (b) 11 m/s; (c) no; (d) no    **121.** (a) $2.1 \times 10^6$ kg; (b) $(100 + 1.5t)^{0.5}$ m/s; (c) $(1.5 \times 10^6)/(100 + 1.5t)^{0.5}$ N; (d) 6.7 km    **123.** 54%    **125.** (a) $2.7 \times 10^9$ J; (b) $2.7 \times 10^9$ W; (c) $\$2.4 \times 10^8$    **127.** 5.4 kJ    **129.** $3.1 \times 10^{11}$ W    **131.** because your force on the cabbage (as you lower it) does work    **135.** (a) 8.6 kJ; (b) $8.6 \times 10^2$ W; (c) $4.3 \times 10^2$ W; (d) 1.3 kW

**Chapter 9**

**CP** **1.** (a) origin; (b) fourth quadrant; (c) on $y$ axis below origin; (d) origin; (e) third quadrant; (f) origin    **2.** (a) $-$ (c) at the center of mass, still at the origin (their forces are internal to the system and cannot move the center of mass)    **3.** (Consider slopes and Eq. 9-23.) (a) 1, 3, and then 2 and 4 tie (zero force); (b) 3    **4.** (a) unchanged; (b) unchanged (see Eq. 9-32); (c) decrease (Eq. 9-35)    **5.** (a) zero; (b) positive (initial $p_y$ down $y$; final $p_y$ up $y$); (c) positive direction of $y$    **6.** (No net external force; $\vec{P}$ conserved.) (a) 0; (b) no; (c) $-x$    **7.** (a) 10 kg·m/s; (b) 14 kg·m/s; (c) 6 kg·m/s    **8.** (a) 4 kg·m/s; (b) 8 kg·m/s; (c) 3 J    **9.** (a) 2 kg·m/s (conserve momentum along $x$); (b) 3 kg·m/s (conserve momentum along $y$)

**Q** **1.** (a) 2 N, rightward; (b) 2 N, rightward; (c) greater than 2 N, rightward    **3.** b, c, a    **5.** (a) $x$ yes, $y$ no; (b) $x$ yes, $y$ no; (c) $x$ no, $y$ yes    **7.** (a) $c$, kinetic energy cannot be negative; $d$, total kinetic energy cannot increase; (b) $a$; (c) $b$    **9.** (a) one was stationary; (b) 2; (c) 5; (d) equal (pool player's result)    **11.** (a) $C$; (b) $B$; (c) 3

**P** **1.** (a) $-1.50$ m; (b) $-1.43$ m    **3.** (a) $-6.5$ cm; (b) 8.3 cm; (c) 1.4 cm    **5.** (a) $-0.45$ cm; (b) $-2.0$ cm    **7.** (a) 0; (b) $3.13 \times 10^{-11}$ m    **9.** (a) 28 cm; (b) 2.3 m/s    **11.** $(-4.0 \text{ m})\hat{i} + (4.0 \text{ m})\hat{j}$    **13.** 53 m    **15.** (a) $(2.35\hat{i} - 1.57\hat{j})$ m/s²; (b) $(2.35\hat{i} - 1.57\hat{j})t$ m/s, with $t$ in seconds; (d) straight, at downward angle $34°$    **17.** 4.2 m    **19.** (a) $7.5 \times 10^4$ J; (b) $3.8 \times 10^4$ kg·m/s; (c) $39°$ south of due east    **21.** (a) 5.0 kg·m/s; (b) 10 kg·m/s    **23.** $1.0 \times 10^3$ to $1.2 \times 10^3$ kg·m/s    **25.** (a) 42 N·s; (b) 2.1 kN    **27.** (a) 67 m/s; (b) $-x$; (c) 1.2 kN; (d) $-x$    **29.** 5 N    **31.** (a) $2.39 \times 10^3$ N·s; (b) $4.78 \times 10^5$ N; (c) $1.76 \times 10^3$ N·s; (d) $3.52 \times 10^5$ N    **33.** (a) 5.86 kg·m/s; (b) $59.8°$; (c) 2.93 kN; (d) $59.8°$    **35.** $9.9 \times 10^2$ N    **37.** (a) 9.0 kg·m/s; (b) 3.0 kN; (c) 4.5 kN; (d) 20 m/s    **39.** 3.0 mm/s    **41.** (a) $-(0.15 \text{ m/s})\hat{i}$; (b) 0.18 m    **43.** 55 cm    **45.** (a) $(1.00\hat{i} - 0.167\hat{j})$ km/s; (b) 3.23 MJ    **47.** (a) 14 m/s; (b) $45°$    **49.** $3.1 \times 10^2$ m/s    **51.** (a) 721 m/s; (b) 937 m/s    **53.** (a) 33%; (b) 23%; (c) decreases    **55.** (a) $+2.0$ m/s; (b) $-1.3$ J; (c) $+40$ J; (d) system got energy from some source, such as a small explosion    **57.** (a) 4.4 m/s; (b) 0.80    **59.** 25 cm    **61.** (a) 99 g; (b) 1.9 m/s; (c) 0.93 m/s    **63.** (a) 3.00 m/s; (b) 6.00 m/s    **65.** (a) 1.2 kg; (b) 2.5 m/s    **67.** $-28$ cm    **69.** (a) 0.21 kg; (b) 7.2 m    **71.** (a) $4.15 \times 10^5$ m/s; (b) $4.84 \times 10^5$ m/s    **73.** $120°$    **75.** (a) 433 m/s; (b) 250 m/s    **77.** (a) 46 N; (b) none    **79.** (a) $1.57 \times 10^6$ N; (b) $1.35 \times 10^5$ kg; (c) 2.08 km/s    **81.** (a) 7290 m/s; (b) 8200 m/s; (c) $1.271 \times 10^{10}$ J; (d) $1.275 \times 10^{10}$ J    **83.** (a) 1.92 m; (b) 0.640 m    **85.** (a) 1.78 m/s; (b) less; (c) less; (d) greater    **87.** (a) 3.7 m/s; (b) 1.3 N·s; (c) $1.8 \times 10^2$ N    **89.** (a) $(7.4 \times 10^3 \text{ N·s})\hat{i} - (7.4 \times 10^3 \text{ N·s})\hat{j}$; (b) $(-7.4 \times 10^3 \text{ N·s})\hat{i}$; (c) $2.3 \times 10^3$ N; (d) $2.1 \times 10^4$ N; (e) $-45°$    **91.** $+4.4$ m/s    **93.** $1.18 \times 10^4$ kg    **95.** (a) 1.9 m/s; (b) $-30°$; (c) elastic    **97.** (a) 6.9 m/s; (b) $30°$; (c) 6.9 m/s; (d) $-30°$; (e) 2.0 m/s; (f) $-180°$    **99.** (a) 25 mm; (b) 26 mm; (c) down; (d) $1.6 \times 10^{-2}$ m/s²    **101.** 29 J    **103.** 2.2 kg    **105.** 5.0 kg    **107.** (a) 50 kg/s; (b) $1.6 \times 10^2$ kg/s    **109.** (a) $4.6 \times 10^3$ km; (b) 73%    **111.** 190 m/s

**113.** 28.8 N    **115.** (a) 0.745 mm; (b) 153°; (c) 1.67 mJ    **117.** (a)
$(2.67 \text{ m/s})\hat{i} + (-3.00 \text{ m/s})\hat{j}$; (b) 4.01 m/s; (c) 48.4°    **119.** (a)
$-0.50$ m; (b) $-1.8$ cm; (c) 0.50 m    **121.** 0.22%    **123.** 36.5 km/s
**125.** (a) $(-1.00 \times 10^{-19}\hat{i} + 0.67 \times 10^{-19}\hat{j})$ kg·m/s; (b) $1.19 \times 10^{-12}$ J
**127.** $2.2 \times 10^{-3}$

## Chapter 10

**CP    1.** b and c    **2.** (a) and (d) ($\alpha = d^2\theta/dt^2$ must be a constant)
**3.** (a) yes; (b) no; (c) yes; (d) yes    **4.** all tie    **5.** 1, 2, 4, 3 (see Eq. 10-36)
**6.** (see Eq. 10-40) 1 and 3 tie, 4, then 2 and 5 tie (zero)    **7.** (a)
downward in the figure ($\tau_{net} = 0$); (b) less (consider moment arms)
**Q    1.** (a) $c, a$, then $b$ and $d$ tie; (b) $b$, then $a$ and $c$ tie, then $d$    **3.** all
tie    **5.** (a) decrease; (b) clockwise; (c) counterclockwise    **7.** larger
**9.** $c, a, b$    **11.** less
**P    1.** 14 rev    **3.** (a) 4.0 rad/s; (b) 11.9 rad/s    **5.** 11 rad/s    **7.** (a) 4.0
m/s; (b) no    **9.** (a) 3.00 s; (b) 18.9 rad    **11.** (a) 30 s; (b) $1.8 \times 10^3$ rad
**13.** (a) $3.4 \times 10^2$ s; (b) $-4.5 \times 10^{-3}$ rad/s²; (c) 98 s    **15.** 8.0 s
**17.** (a) 44 rad; (b) 5.5 s; (c) 32 s; (d) $-2.1$ s; (e) 40 s    **19.** (a) $2.50 \times$
$10^{-3}$ rad/s; (b) 20.2 m/s²; (c) 0    **21.** $6.9 \times 10^{-13}$ rad/s    **23.** (a) 20.9
rad/s; (b) 12.5 m/s; (c) 800 rev/min²; (d) 600 rev    **25.** (a) $7.3 \times 10^{-5}$
rad/s; (b) $3.5 \times 10^2$ m/s; (c) $7.3 \times 10^{-5}$ rad/s; (d) $4.6 \times 10^2$ m/s    **27.**
(a) 73 cm/s²; (b) 0.075; (c) 0.11    **29.** (a) $3.8 \times 10^3$ rad/s; (b) $1.9 \times 10^2$
m/s    **31.** (a) 40 s; (b) 2.0 rad/s²    **33.** 12.3 kg·m²    **35.** (a) 1.1 kJ; (b)
9.7 kJ    **37.** 0.097 kg·m²    **39.** (a) 49 MJ; (b) $1.0 \times 10^2$ min    **41.** (a)
0.023 kg·m²; (b) 1.1 mJ    **43.** $4.7 \times 10^{-4}$ kg·m²    **45.** $-3.85$ N·m
**47.** 4.6 N·m    **49.** (a) 28.2 rad/s²; (b) 338 N·m    **51.** (a) 6.00 cm/s²;
(b) 4.87 N; (c) 4.54 N; (d) 1.20 rad/s²; (e) 0.0138 kg·m²    **53.** 0.140 N
**55.** $2.51 \times 10^{-4}$ kg·m²    **57.** (a) $4.2 \times 10^2$ rad/s²; (b) $5.0 \times 10^2$ rad/s
**59.** 396 N·m    **61.** (a) $-19.8$ kJ; (b) 1.32 kW    **63.** 5.42 m/s    **65.** (a)
5.32 m/s²; (b) 8.43 m/s²; (c) 41.8°    **67.** 9.82 rad/s    **69.** $6.16 \times 10^{-5}$
kg·m²    **71.** (a) 31.4 rad/s²; (b) 0.754 m/s²;  (c) 56.1 N;  (d) 55.1 N
**73.** (a) $4.81 \times 10^5$ N; (b) $1.12 \times 10^4$ N·m; (c) $1.25 \times 10^6$ J
**75.** (a) 2.3 rad/s²; (b) 1.4 rad/s²    **77.** (a) $-67$ rev/min²; (b) 8.3 rev
**81.** 3.1 rad/s    **83.** (a) 1.57 m/s²; (b) 4.55 N; (c) 4.94 N    **85.** 30 rev
**87.** 0.054 kg·m²    **89.** $1.4 \times 10^2$ N·m    **91.** (a) 10 J; (b) 0.27 m
**93.** 4.6 rad/s²    **95.** 2.6 J    **97.** (a) $5.92 \times 10^4$ m/s²; (b) $4.39 \times 10^4$ s⁻²
**99.** (a) 0.791 kg·m²; (b) $1.79 \times 10^{-2}$ N·m    **101.** (a) $1.5 \times 10^2$ cm/s;
(b) 15 rad/s; (c) 15 rad/s; (d) 75 cm/s; (e) 3.0 rad/s    **103.** (a) 7.0 kg·m²;
(b) 7.2 m/s; (c) 71°    **105.** (a) 0.32 rad/s; (b) $1.0 \times 10^2$ km/h
**107.** (a) $1.4 \times 10^2$ rad; (b) 14 s

## Chapter 11

**CP    1.** (a) same; (b) less    **2.** less (consider the transfer of energy
from rotational kinetic energy to gravitational potential energy)
**3.** (draw the vectors, use right-hand rule) (a) $\pm z$; (b) $+y$; (c) $-x$
**4.** (see Eq. 11-21) (a) 1 and 3 tie; then 2 and 4 tie, then 5 (zero); (b) 2
and 3    **5.** (see Eqs. 11-23 and 11-16) (a) 3, 1; then 2 and 4 tie (zero);
(b) 3    **6.** (a) all tie (same $\tau$, same $t$, thus same $\Delta L$); (b) sphere, disk,
hoop (reverse order of $I$)    **7.** (a) decreases; (b) same ($\tau_{net} = 0$, so $L$ is
conserved); (c) increases
**Q    1.** $a$, then $b$ and $c$ tie, then $e, d$ (zero)    **3.** (a) spins in place; (b)
rolls toward you; (c) rolls away from you    **5.** (a) 1, 2, 3 (zero); (b) 1
and 2 tie, then 3; (c) 1 and 3 tie, then 2    **7.** (a) same; (b) increase; (c)
decrease; (d) same, decrease, increase    **9.** $D, B$, then $A$ and $C$ tie
**11.** (a) same; (b) same
**P    1.** (a) 0; (b) (22 m/s)$\hat{i}$; (c) $(-22 \text{ m/s})\hat{i}$; (d) 0; (e) $1.5 \times 10^3$ m/s²; (f)
$1.5 \times 10^3$ m/s²; (g) (22 m/s)$\hat{i}$; (h) (44 m/s)$\hat{i}$; (i) 0; (j) 0; (k) $1.5 \times 10^3$ m/s²;
(l) $1.5 \times 10^3$ m/s²    **3.** $-3.15$ J    **5.** 0.020    **7.** (a) 63 rad/s; (b) 4.0 m
**9.** 4.8 m    **11.** (a) $(-4.0 \text{ N})\hat{i}$; (b) 0.60 kg·m²    **13.** 0.50    **15.** (a)
$-(0.11 \text{ m})\omega$; (b) $-2.1$ m/s²; (c) $-47$ rad/s²; (d) 1.2 s; (e) 8.6 m; (f) 6.1
m/s    **17.** (a) 13 cm/s²; (b) 4.4 s; (c) 55 cm/s; (d) 18 mJ; (e) 1.4 J; (f) 27
rev/s    **19.** $(-2.0 \text{ N·m})\hat{i}$    **21.** (a) $(6.0 \text{ N·m})\hat{j} + (8.0 \text{ N·m})\hat{k}$; (b)

$(-22 \text{ N·m})\hat{i}$    **23.** (a) $(-1.5 \text{ N·m})\hat{i} - (4.0 \text{ N·m})\hat{j} - (1.0 \text{ N·m})\hat{k}$;
(b) $(-1.5 \text{ N·m})\hat{i} - (4.0 \text{ N·m})\hat{j} - (1.0 \text{ N·m})\hat{k}$    **25.** (a) $(50 \text{ N·m})\hat{k}$;
(b) 90°    **27.** (a) 0; (b) $(8.0 \text{ N·m})\hat{i} + (8.0 \text{ N·m})\hat{k}$    **29.** (a) 9.8 kg·m²/s;
(b) $+z$ direction    **31.** (a) 0; (b) $-22.6$ kg·m²/s; (c) $-7.84$ N·m;
(d) $-7.84$ N·m    **33.** (a) $(-1.7 \times 10^2$ kg·m²/s)$\hat{k}$; (b) $(+56 \text{ N·m})\hat{k}$;
(c) $(+56$ kg·m²/s²)$\hat{k}$    **35.** (a) $48t\hat{k}$ N·m; (b) increasing    **37.** (a) 4.6 ×
$10^{-3}$ kg·m²; (b) $1.1 \times 10^{-3}$ kg·m²/s; (c) $3.9 \times 10^{-3}$ kg·m²/s
**39.** (a) 1.47 N·m; (b) 20.4 rad; (c) $-29.9$ J; (d) 19.9 W    **41.** (a) 1.6
kg·m²; (b) 4.0 kg·m²/s    **43.** (a) 1.5 m; (b) 0.93 rad/s; (c) 98 J; (d) 8.4
rad/s; (e) $8.8 \times 10^2$ J; (f) internal energy of the skaters    **45.** (a) 3.6
rev/s; (b) 3.0; (c) forces on the bricks from the man transferred energy from the man's internal energy to kinetic energy    **47.** 0.17 rad/s
**49.** (a) 750 rev/min; (b) 450 rev/min; (c) clockwise    **51.** (a) 267 rev/min;
(b) 0.667    **53.** $1.3 \times 10^3$ m/s    **55.** 3.4 rad/s    **57.** (a) 18 rad/s; (b) 0.92
**59.** 11.0 m/s    **61.** 1.5 rad/s    **63.** 0.070 rad/s    **65.** (a) 0.148 rad/s; (b)
0.0123; (c) 181°    **67.** (a) 0.180 m; (b) clockwise    **69.** 0.041 rad/s    **71.**
(a) 1.6 m/s²; (b) 16 rad/s²; (c) $(4.0 \text{ N})\hat{i}$    **73.** (a) 0; (b) 0; (c) $-30t^3\hat{k}$
kg·m²/s; (d) $-90t^2\hat{k}$ N·m; (e) $30t^3\hat{k}$ kg·m²/s; (f) $90t^2\hat{k}$ N·m    **75.** (a)
149 kg·m²; (b) 158 kg·m²/s; (c) 0.744 rad/s    **77.** (a) $6.65 \times 10^{-5}$
kg·m²/s; (b) no; (c) 0; (d) yes    **79.** (a) 0.333; (b) 0.111    **81.** (a) 58.8 J;
(b) 39.2 J    **83.** (a) 61.7 J; (b) 3.43 m; (c) no    **85.** (a) $mvR/(I + MR^2)$;
(b) $mvR^2/(I + MR^2)$

## Chapter 12

**CP    1.** $c, e, f$    **2.** (a) no; (b) at site of $\vec{F}_1$, perpendicular to plane of
figure; (c) 45 N    **3.** $d$
**Q    1.** (a) 1 and 3 tie, then 2; (b) all tie; (c) 1 and 3 tie, then 2 (zero)
**3.** $a$ and $c$ (forces and torques balance)    **5.** (a) 12 kg; (b) 3 kg;
(c) 1 kg    **7.** (a) at $C$ (to eliminate forces there from a torque
equation); (b) plus; (c) minus; (d) equal    **9.** increase    **11.** $A$ and $B$,
then $C$
**P    1.** (a) 1.00 m; (b) 2.00 m; (c) 0.987 m; (d) 1.97 m    **3.** (a) 9.4 N;
(b) 4.4 N    **5.** 7.92 kN    **7.** (a) $2.8 \times 10^2$ N; (b) $8.8 \times 10^2$ N; (c) 71°
**9.** 74.4 g    **11.** (a) 1.2 kN; (b) down; (c) 1.7 kN; (d) up; (e) left;
(f) right    **13.** (a) 2.7 kN; (b) up; (c) 3.6 kN; (d) down    **15.** (a) 5.0 N;
(b) 30 N; (c) 1.3 m    **17.** (a) 0.64 m; (b) increased    **19.** 8.7 N
**21.** (a) 6.63 kN; (b) 5.74 kN; (c) 5.96 kN    **23.** (a) 192 N; (b) 96.1 N;
(c) 55.5 N    **25.** 13.6 N    **27.** (a) 1.9 kN; (b) up; (c) 2.1 kN; (d) down
**29.** (a) $(-80 \text{ N})\hat{i} + (1.3 \times 10^2 \text{ N})\hat{j}$; (b) $(80 \text{ N})\hat{i} + (1.3 \times 10^2 \text{ N})\hat{j}$
**31.** 2.20 m    **33.** (a) 60.0°; (b) 300 N    **35.** (a) 445 N; (b) 0.50; (c) 315 N
**37.** 0.34    **39.** (a) 207 N; (b) 539 N; (c) 315 N    **41.** (a) slides;
(b) 31°; (c) tips; (d) 34°    **43.** (a) $6.5 \times 10^6$ N/m²; (b) $1.1 \times 10^{-5}$ m
**45.** (a) 0.80; (b) 0.20; (c) 0.25    **47.** (a) $1.4 \times 10^9$ N; (b) 75
**49.** (a) 866 N; (b) 143 N; (c) 0.165    **51.** (a) $1.2 \times 10^2$ N; (b) 68 N
**53.** (a) $1.8 \times 10^7$ N; (b) $1.4 \times 10^7$ N; (c) 16    **55.** 0.29    **57.** 76 N
**59.** (a) 8.01 kN; (b) 3.65 kN; (c) 5.66 kN    **61.** 71.7 N    **63.** (a) $L/2$;
(b) $L/4$; (c) $L/6$; (d) $L/8$; (e) $25L/24$    **65.** (a) 88 N; (b) $(30\hat{i} + 97\hat{j})$ N
**67.** $2.4 \times 10^9$ N/m²    **69.** 60°    **71.** (a) $\mu < 0.57$; (b) $\mu > 0.57$
**73.** (a) $(35\hat{i} + 200\hat{j})$ N; (b) $(-45\hat{i} + 200\hat{j})$ N; (c) $1.9 \times 10^2$ N
**75.** (a) $BC, CD, DA$; (b) 535 N; (c) 757 N    **77.** (a) 1.38 kN; (b) 180 N
**79.** (a) $a_1 = L/2, a_2 = 5L/8, h = 9L/8$; (b) $b_1 = 2L/3, b_2 = L/2$,
$h = 7L/6$    **81.** $L/4$    **83.** (a) 106 N; (b) 64.0°    **85.** $1.8 \times 10^2$ N
**87.** (a) $-24.4$ N; (b) 1.60 N; (c) $-3.75°$

## Chapter 13

**CP    1.** all tie    **2.** (a) 1, tie of 2 and 4, then 3; (b) line $d$
**3.** (a) increase; (b) negative    **4.** (a) 2; (b) 1    **5.** (a) path 1
(decreased $E$ (more negative) gives decreased $a$); (b) less
(decreased $a$ gives decreased $T$)
**Q    1.** $3GM^2/d^2$, leftward    **3.** $Gm^2/r^2$, upward    **5.** $b$ and $c$ tie, then $a$
(zero)    **7.** 1, tie of 2 and 4, then 3    **9.** (a) positive $y$; (b) yes, rotates

counterclockwise until it points toward particle $B$ **11.** $b$, $d$, and $f$ all tie, then $e$, $c$, $a$

**P** **1.** $\frac{1}{2}$ **3.** 19 m **5.** 0.8 m **7.** $-5.00d$ **9.** $2.60 \times 10^5$ km
**11.** (a) $M = m$; (b) 0 **13.** $8.31 \times 10^{-9}$ N **15.** (a) $-1.88d$;
(b) $-3.90d$; (c) $0.489d$ **17.** (a) 17 N; (b) 2.4 **19.** $2.6 \times 10^6$ m
**21.** $5 \times 10^{24}$ kg **23.** (a) 7.6 m/s$^2$; (b) 4.2 m/s$^2$ **25.** (a) $(3.0 \times 10^{-7}$ N/kg$)m$; (b) $(3.3 \times 10^{-7}$ N/kg$)m$; (c) $(6.7 \times 10^{-7}$ N/kg $\cdot$ m$)mr$
**27.** (a) 9.83 m/s$^2$; (b) 9.84 m/s$^2$; (c) 9.79 m/s$^2$ **29.** $5.0 \times 10^9$ J
**31.** (a) 0.74; (b) 3.8 m/s$^2$; (c) 5.0 km/s **33.** (a) 0.0451; (b) 28.5
**35.** $-4.82 \times 10^{-13}$ J **37.** (a) 0.50 pJ; (b) $-0.50$ pJ **39.** (a) 1.7 km/s;
(b) $2.5 \times 10^5$ m; (c) 1.4 km/s **41.** (a) 82 km/s; (b) $1.8 \times 10^4$ km/s
**43.** (a) 7.82 km/s; (b) 87.5 min **45.** $6.5 \times 10^{23}$ kg **47.** $5 \times 10^{10}$ stars
**49.** (a) $1.9 \times 10^{13}$ m; (b) $6.4R_P$ **51.** (a) $6.64 \times 10^3$ km; (b) 0.0136
**53.** $5.8 \times 10^6$ m **57.** 0.71 y **59.** $(GM/L)^{0.5}$ **61.** (a) $3.19 \times 10^3$ km;
(b) lifting **63.** (a) 2.8 y; (b) $1.0 \times 10^{-4}$ **65.** (a) $r^{1.5}$; (b) $r^{-1}$; (c) $r^{0.5}$;
(d) $r^{-0.5}$ **67.** (a) 7.5 km/s; (b) 97 min; (c) $4.1 \times 10^2$ km; (d) 7.7 km/s;
(e) 93 min; (f) $3.2 \times 10^{-3}$ N; (g) no; (h) yes **69.** 1.1 s
**71.** (a) $GMmx(x^2 + R^2)^{-3/2}$; (b) $[2GM(R^{-1} - (R^2 + x^2)^{-1/2})]^{1/2}$
**73.** (a) $1.0 \times 10^3$ kg; (b) 1.5 km/s **75.** $3.2 \times 10^{-7}$ N **77.** $.037\hat{\mathrm{j}}$ $\mu$N
**79.** $2\pi r^{1.5} G^{-0.5} (M + m/4)^{-0.5}$ **81.** (a) $2.2 \times 10^{-7}$ rad/s; (b) 89 km/s
**83.** (a) $2.15 \times 10^4$ s; (b) 12.3 km/s; (c) 12.0 km/s; (d) $2.17 \times 10^{11}$ J;
(e) $-4.53 \times 10^{11}$ J; (f) $-2.35 \times 10^{11}$ J; (g) $4.04 \times 10^7$ m; (h) $1.22 \times 10^3$ s; (i) elliptical **85.** $2.5 \times 10^4$ km **87.** (a) $1.4 \times 10^6$ m/s; (b) $3 \times 10^6$ m/s$^2$ **89.** (a) 0; (b) $1.8 \times 10^{32}$ J; (c) $1.8 \times 10^{32}$ J; (d) 0.99 km/s
**91.** (a) $Gm^2/R_i$; (b) $Gm^2/2R_i$; (c) $(Gm/R_i)^{0.5}$; (d) $2(Gm/R_i)^{0.5}$;
(e) $Gm^2/R_i$; (f) $(2Gm/R_i)^{0.5}$; (g) The center-of-mass frame is an inertial frame, and in it the principle of conservation of energy may be written as in Chapter 8; the reference frame attached to body $A$ is noninertial, and the principle cannot be written as in Chapter 8. Answer (d) is correct. **93.** $2.4 \times 10^4$ m/s **95.** $-0.044\hat{\mathrm{j}}$ $\mu$N
**97.** $GM_E m/12R_E$ **99.** $1.51 \times 10^{-12}$ N **101.** $3.4 \times 10^5$ km

## Chapter 14

**CP** **1.** all tie **2.** (a) all tie (the gravitational force on the penguin is the same); (b) $0.95\rho_0, \rho_0, 1.1\rho_0$ **3.** 13 cm$^3$/s, outward
**4.** (a) all tie; (b) 1, then 2 and 3 tie, 4 (wider means slower); (c) 4, 3, 2, 1 (wider and lower mean more pressure)
**Q** **1.** (a) moves downward; (b) moves downward **3.** (a) downward; (b) downward; (c) same **5.** $b$, then $a$ and $d$ tie (zero), then $c$
**7.** (a) 1 and 4; (b) 2; (c) 3 **9.** $B$, $C$, $A$
**P** **1.** 0.074 **3.** $1.1 \times 10^5$ Pa **5.** $2.9 \times 10^4$ N **7.** (b) 26 kN
**9.** (a) $1.0 \times 10^3$ torr; (b) $1.7 \times 10^3$ torr **11.** (a) 94 torr; (b) $4.1 \times 10^2$ torr; (c) $3.1 \times 10^2$ torr **13.** $1.08 \times 10^3$ atm **15.** $-2.6 \times 10^4$ Pa
**17.** $7.2 \times 10^5$ N **19.** $4.69 \times 10^5$ N **21.** 0.635 J **23.** 44 km
**25.** 739.26 torr **27.** (a) 7.9 km; (b) 16 km **29.** 8.50 kg **31.** (a) $6.7 \times 10^2$ kg/m$^3$; (b) $7.4 \times 10^2$ kg/m$^3$ **33.** (a) $2.04 \times 10^{-2}$ m$^3$;
(b) 1.57 kN **35.** five **37.** 57.3 cm **39.** (a) 1.2 kg; (b) $1.3 \times 10^3$ kg/m$^3$ **41.** (a) 0.10; (b) 0.083 **43.** (a) 637.8 cm$^3$; (b) 5.102 m$^3$;
(c) $5.102 \times 10^3$ kg **45.** 0.126 m$^3$ **47.** (a) 1.80 m$^3$; (b) 4.75 m$^3$
**49.** (a) 3.0 m/s; (b) 2.8 m/s **51.** 8.1 m/s **53.** 66 W **55.** $1.4 \times 10^5$ J
**57.** (a) $1.6 \times 10^{-3}$ m$^3$/s; (b) 0.90 m **59.** (a) 2.5 m/s; (b) $2.6 \times 10^5$ Pa
**61.** (a) 3.9 m/s; (b) 88 kPa **63.** $1.1 \times 10^2$ m/s **65.** (b) $2.0 \times 10^{-2}$ m$^3$/s **67.** (a) 74 N; (b) $1.5 \times 10^2$ m$^3$ **69.** (a) 0.0776 m$^3$/s; (b) 69.8 kg/s **71.** (a) 35 cm; (b) 30 cm; (c) 20 cm **73.** 1.5 g/cm$^3$ **75.** $5.11 \times 10^{-7}$ kg **77.** 44.2 g **79.** $6.0 \times 10^2$ kg/m$^3$ **81.** 45.3 cm$^3$
**83.** (a) 3.2 m/s; (b) $9.2 \times 10^4$ Pa; (c) 10.3 m **85.** $1.07 \times 10^3$ g
**87.** 26.3 m$^2$ **89.** (a) $5.66 \times 10^9$ N; (b) 25.4 atm

## Chapter 15

**CP** **1.** (sketch $x$ versus $t$) (a) $-x_m$; (b) $+x_m$; (c) 0 **2.** c ($a$ must have the form of Eq. 15-8) **3.** a ($F$ must have the form of Eq. 15-10)

**4.** (a) 5 J; (b) 2 J; (c) 5 J **5.** all tie (in Eq. 15-29, $m$ is included in $I$)
**6.** 1, 2, 3 (the ratio $m/b$ matters; $k$ does not)
**Q** **1.** a and b **3.** (a) 2; (b) positive; (c) between 0 and $+x_m$
**5.** (a) between $D$ and $E$; (b) between $3\pi/2$ rad and $2\pi$ rad
**7.** (a) all tie; (b) 3, then 1 and 2 tie; (c) 1, 2, 3 (zero); (d) 1, 2, 3 (zero);
(e) 1, 3, 2 **9.** $b$ (infinite period, does not oscillate), $c$, $a$
**11.** (a) greater; (b) same; (c) same; (d) greater; (e) greater
**P** **1.** (a) 0.50 s; (b) 2.0 Hz; (c) 18 cm **3.** 37.8 m/s$^2$ **5.** (a) 1.0 mm;
(b) 0.75 m/s; (c) $5.7 \times 10^2$ m/s$^2$ **7.** (a) 498 Hz; (b) greater
**9.** (a) 3.0 m; (b) $-49$ m/s; (c) $-2.7 \times 10^2$ m/s$^2$; (d) 20 rad; (e) 1.5 Hz;
(f) 0.67 s **11.** 39.6 Hz **13.** (a) 0.500 s; (b) 2.00 Hz; (c) 12.6 rad/s;
(d) 79.0 N/m; (e) 4.40 m/s; (f) 27.6 N **15.** (a) $0.18A$; (b) same direction
**17.** (a) 5.58 Hz; (b) 0.325 kg; (c) 0.400 m **19.** (a) 25 cm; (b) 2.2 Hz
**21.** 54 Hz **23.** 3.1 cm **25.** (a) 0.525 m; (b) 0.686 s
**27.** (a) 0.75; (b) 0.25; (c) $2^{-0.5}x_m$ **29.** 37 mJ **31.** (a) 2.25 Hz;
(b) 125 J; (c) 250 J; (d) 86.6 cm **33.** (a) 1.1 m/s; (b) 3.3 cm
**35.** (a) 3.1 ms; (b) 4.0 m/s; (c) 0.080 J; (d) 80 N; (e) 40 N
**37.** (a) 2.2 Hz; (b) 56 cm/s; (c) 0.10 kg; (d) 20.0 cm **39.** (a) 39.5 rad/s; (b) 34.2 rad/s; (c) 124 rad/s$^2$ **41.** (a) 0.205 kg·m$^2$; (b) 47.7 cm; (c) 1.50 s **43.** (a) 1.64 s; (b) equal **45.** 8.77 s **47.** 0.366 s
**49.** (a) 0.845 rad; (b) 0.0602 rad **51.** (a) 0.53 m; (b) 2.1 s
**53.** 0.0653 s **55.** (a) 2.26 s; (b) increases; (c) same **57.** 6.0%
**59.** (a) 14.3 s; (b) 5.27 **61.** (a) $F_m/b\omega$; (b) $F_m/b$ **63.** 5.0 cm
**65.** (a) $2.8 \times 10^3$ rad/s; (b) 2.1 m/s; (c) 5.7 km/s$^2$ **67.** (a) 1.1 Hz;
(b) 5.0 cm **69.** 7.2 m/s **71.** (a) 7.90 N/m; (b) 1.19 cm; (c) 2.00 Hz
**73.** (a) $1.3 \times 10^2$ N/m; (b) 0.62 s; (c) 1.6 Hz; (d) 5.0 cm; (e) 0.51 m/s
**75.** (a) 16.6 cm; (b) 1.23% **77.** (a) 1.2 J; (b) 50 **79.** 1.53 m
**81.** (a) 0.30 m; (b) 0.28 s; (c) $1.5 \times 10^2$ m/s$^2$; (d) 11 J **83.** (a) 1.23 kN/m; (b) 76.0 N **85.** 1.6 kg **87.** (a) 0.735 kg $\cdot$ m$^2$; (b) 0.0240 N $\cdot$ m; (c) 0.181 rad/s **89.** (a) 3.5 m; (b) 0.75 s **91.** (a) 0.35 Hz; (b) 0.39 Hz;
(c) 0 (no oscillation) **93.** (a) 245 N/m; (b) 0.284 s
**95.** 0.079 kg $\cdot$ m$^2$ **97.** (a) $8.11 \times 10^{-5}$ kg $\cdot$ m$^2$; (b) 3.14 rad/s
**99.** 14.0° **101.** (a) 3.2 Hz; (b) 0.26 m; (c) $x = (0.26$ m$) \cos(20t - \pi/2)$, with $t$ in seconds **103.** (a) 0.44 s; (b) 0.18 m **105.** (a) 0.45 s; (b) 0.10 m above and 0.20 m below; (c) 0.15 m; (d) 2.3 J **107.** $7 \times 10^2$ N/m
**109.** 0.804 m **111.** (a) 0.30 m; (b) 30 m/s$^2$; (c) 0; (d) 4.4 s
**113.** (a) $F/m$; (b) $2F/mL$; (c) 0 **115.** 2.54 m

## Chapter 16

**CP** **1.** a, 2; b, 3; c, 1 (compare with the phase in Eq. 16-2, then see Eq. 16-5) **2.** (a) 2, 3, 1 (see Eq. 16-12); (b) 3, then 1 and 2 tie (find amplitude of $dy/dt$) **3.** (a) same (independent of $f$); (b) decrease ($\lambda = v/f$); (c) increase; (d) increase **4.** 0.20 and 0.80 tie, then 0.60, 0.45 **5.** (a) 1; (b) 3; (c) 2 **6.** (a) 75 Hz; (b) 525 Hz
**Q** **1.** (a) 1, 4, 2, 3; (b) 1, 4, 2, 3 **3.** $a$, upward; $b$, upward; $c$, downward; $d$, downward; $e$, downward; $f$, downward; $g$, upward; $h$, upward
**5.** intermediate (closer to fully destructive) **7.** (a) 0, 0.2 wavelength, 0.5 wavelength (zero); (b) $4P_{\mathrm{avg},1}$ **9.** $d$ **11.** $c$, $a$, $b$
**P** **1.** 1.1 ms **3.** (a) 3.49 m$^{-1}$; (b) 31.5 m/s **5.** (a) 0.680 s; (b) 1.47 Hz; (c) 2.06 m/s **7.** (a) 64 Hz; (b) 1.3 m; (c) 4.0 cm; (d) 5.0 m$^{-1}$;
(e) $4.0 \times 10^2$ s$^{-1}$; (f) $\pi/2$ rad; (g) minus **9.** (a) 3.0 mm; (b) 16 m$^{-1}$;
(c) $2.4 \times 10^2$ s$^{-1}$; (d) minus **11.** (a) negative; (b) 4.0 cm; (c) 0.31 cm$^{-1}$; (d) 0.63 s$^{-1}$; (e) $\pi$ rad; (f) minus; (g) 2.0 cm/s; (h) $-2.5$ cm/s
**13.** (a) 11.7 cm; (b) $\pi$ rad **15.** (a) 0.12 mm; (b) 141 m$^{-1}$; (c) 628 s$^{-1}$;
(d) plus **17.** (a) 15 m/s; (b) 0.036 N **19.** 129 m/s **21.** 2.63 m
**23.** (a) 5.0 cm; (b) 40 cm; (c) 12 m/s; (d) 0.033 s; (e) 9.4 m/s;
(f) 16 m$^{-1}$; (g) $1.9 \times 10^2$ s$^{-1}$; (h) 0.93 rad; (i) plus **27.** 3.2 mm
**29.** 0.20 m/s **31.** $1.41y_m$ **33.** (a) 9.0 mm; (b) 16 m$^{-1}$; (c) $1.1 \times 10^3$ s$^{-1}$; (d) 2.7 rad; (e) plus **35.** 5.0 cm **37.** (a) 3.29 mm; (b) 1.55 rad;
(c) 1.55 rad **39.** 84° **41.** (a) 82.0 m/s; (b) 16.8 m; (c) 4.88 Hz
**43.** (a) 7.91 Hz; (b) 15.8 Hz; (c) 23.7 Hz **45.** (a) 105 Hz; (b) 158 Hz
**47.** 260 Hz **49.** (a) 144 m/s; (b) 60.0 cm; (c) 241 Hz **51.** (a) 0.50 cm;

(b) $3.1\ m^{-1}$; (c) $3.1 \times 10^2\ s^{-1}$; (d) minus **53.** (a) 0.25 cm; (b) $1.2 \times 10^2$ cm/s; (c) 3.0 cm; (d) 0 **55.** 0.25 m **57.** (a) 2.00 Hz; (b) 2.00 m; (c) 4.00 m/s; (d) 50.0 cm; (e) 150 cm; (f) 250 cm; (g) 0; (h) 100 cm; (i) 200 cm
**59.** (a) 324 Hz; (b) eight **61.** 36 N **63.** (a) 75 Hz; (b) 13 ms
**65.** (a) 2.0 mm; (b) 95 Hz; (c) +30 m/s; (d) 31 cm; (e) 1.2 m/s
**67.** (a) 0.31 m; (b) 1.64 rad; (c) 2.2 mm **69.** (a) $0.83y_1$; (b) $37°$
**71.** (a) 3.77 m/s; (b) 12.3 N; (c) 0; (d) 46.4 W; (e) 0; (f) 0; (g) $\pm 0.50$ cm
**73.** 1.2 rad **75.** (a) 300 m/s; (b) no **77.** (a) $[k\,\Delta\ell(\ell + \Delta\ell)/m]^{0.5}$
**79.** (a) 144 m/s; (b) 3.00 m; (c) 1.50 m; (d) 48.0 Hz; (e) 96.0 Hz
**81.** (a) 1.00 cm; (b) $3.46 \times 10^3\ s^{-1}$; (c) $10.5\ m^{-1}$; (d) plus **83.** (a) $2\pi y_m/\lambda$; (b) no **85.** (a) 240 cm; (b) 120 cm; (c) 80 cm **87.** (a) 1.33 m/s; (b) 1.88 m/s; (c) $16.7\ m/s^2$; (d) $23.7\ m/s^2$ **89.** (a) 0.52 m; (b) 40 m/s; (c) 0.40 m **91.** (a) 0.16 m; (b) $2.4 \times 10^2$ N; (c) $y(x, t) =$ (0.16 m) $\sin[(1.57\ m^{-1})x]\sin[(31.4\ s^{-1})t]$ **93.** (c) 2.0 m/s; (d) $-x$
**95.** (a) $\infty$; (b) 1.0; (c) 4.0%

## Chapter 17

**CP** **1.** beginning to decrease (example: mentally move the curves of Fig. 17-6 rightward past the point at $x = 42$ cm) **2.** (a) 1 and 2 tie, then 3 (see Eq. 17-28); (b) 3, then 1 and 2 tie (see Eq. 17-26)
**3.** second (see Eqs. 17-39 and 17-41) **4.** $a$, greater; $b$, less; $c$, can't tell; $d$, can't tell; $e$, greater; $f$, less
**Q** **1.** (a) 0, 0.2 wavelength, 0.5 wavelength (zero); (b) $4P_{avg,1}$
**3.** $C$, then $A$ and $B$ tie **5.** $E, A, D, C, B$ **7.** 1, 4, 3, 2 **9.** 150 Hz and 450 Hz **11.** 505, 507, 508 Hz or 501, 503, 508 Hz
**P** **1.** (a) 79 m; (b) 41 m; (c) 89 m **3.** (a) 2.6 km; (b) $2.0 \times 10^2$
**5.** $1.9 \times 10^3$ km **7.** 40.7 m **9.** 0.23 ms **11.** (a) 76.2 $\mu$m; (b) 0.333 mm
**13.** 960 Hz **15.** (a) $2.3 \times 10^2$ Hz; (b) higher **17.** (a) 143 Hz; (b) 3;
(c) 5; (d) 286 Hz; (e) 2; (f) 3 **19.** (a) 14; (b) 14 **21.** (a) 343 Hz;
(b) 3; (c) 5; (d) 686 Hz; (e) 2; (f) 3 **23.** (a) 0; (b) fully constructive;
(c) increase; (d) 128 m; (e) 63.0 m; (f) 41.2 m **25.** 36.8 nm
**27.** (a) $1.0 \times 10^3$; (b) 32 **29.** 15.0 mW **31.** 2 $\mu$W **33.** 0.76 $\mu$m
**35.** (a) $5.97 \times 10^{-5}\ W/m^2$; (b) 4.48 nW **37.** (a) 0.34 nW; (b) 0.68 nW;
(c) 1.4 nW; (d) 0.88 nW; (e) 0 **39.** (a) 405 m/s; (b) 596 N; (c) 44.0 cm; (d) 37.3 cm **41.** (a) 833 Hz; (b) 0.418 m **43.** (a) 3; (b) 1129 Hz;
(c) 1506 Hz **45.** (a) 2; (b) 1 **47.** 12.4 m **49.** 45.3 N **51.** 2.25 ms
**53.** 0.020 **55.** (a) 526 Hz; (b) 555 Hz **57.** 0 **59.** (a) 1.022 kHz;
(b) 1.045 kHz **61.** 41 kHz **63.** 155 Hz **65.** (a) 2.0 kHz; (b) 2.0
kHz **67.** (a) 485.8 Hz; (b) 500.0 Hz; (c) 486.2 Hz; (d) 500.0 Hz
**69.** (a) $42°$; (b) 11 s **71.** 1 cm **73.** 2.1 m **75.** (a) 39.7 $\mu$W/m$^2$;
(b) 171 nm; (c) 0.893 Pa **77.** 0.25 **79.** (a) 2.10 m; (b) 1.47 m
**81.** (a) 59.7; (b) $2.81 \times 10^{-4}$ **83.** (a) rightward; (b) 0.90 m/s; (c) less
**85.** (a) 11 ms; (b) 3.8 m **87.** (a) $9.7 \times 10^2$ Hz; (b) 1.0 kHz; (c) 60 Hz,
no **89.** (a) 21 nm; (b) 35 cm; (c) 24 nm; (d) 35 cm **91.** (a) 7.70 Hz;
(b) 7.70 Hz **93.** (a) 5.2 kHz; (b) 2 **95.** (a) 10 W; (b) 0.032 W/m$^2$;
(c) 99 dB **97.** (a) 0; (b) 0.572 m; (c) 1.14 m **99.** 171 m **101.** (a)
$3.6 \times 10^2$ m/s; (b) 150 Hz **103.** 400 Hz **105.** (a) 14; (b) 12
**107.** 821 m/s **109.** (a) 39.3 Hz; (b) 118 Hz **111.** $4.8 \times 10^2$ Hz

## Chapter 18

**CP** **1.** (a) all tie; (b) $50°$X, $50°$Y, $50°$W **2.** (a) 2 and 3 tie, then 1, then 4; (b) 3, 2, then 1 and 4 tie (from Eqs. 18-9 and 18-10, assume that change in area is proportional to initial area) **3.** $A$ (see Eq. 18-14) **4.** $c$ and $e$ (maximize area enclosed by a clockwise cycle) **5.** (a) all tie ($\Delta E_{int}$ depends on $i$ and $f$, not on path); (b) 4, 3, 2, 1 (compare areas under curves); (c) 4, 3, 2, 1 (see Eq. 18-26)
**6.** (a) zero (closed cycle); (b) negative ($W_{net}$ is negative; see Eq. 18-26) **7.** $b$ and $d$ tie, then $a$, $c$ ($P_{cond}$ identical; see Eq. 18-32)
**Q** **1.** $c$, then the rest tie **3.** $B$, then $A$ and $C$ tie **5.** (a) $f$, because ice temperature will not rise to freezing point and then drop; (b) $b$ and $c$ at freezing point, $d$ above, $e$ below; (c) in $b$ liquid partly freezes and no ice melts; in $c$ no liquid freezes and no ice melts; in $d$

no liquid freezes and ice fully melts; in $e$ liquid fully freezes and no ice melts **7.** (a) both clockwise; (b) both clockwise **9.** (a) greater;
(b) 1, 2, 3; (c) 1, 3, 2; (d) 1, 2, 3; (e) 2, 3, 1 **11.** $c, b, a$
**P** **1.** 1.366 **3.** 348 K **5.** (a) $320°$F; (b) $-12.3°$F **7.** $-92.1°$X
**9.** 2.731 cm **11.** 49.87 cm$^3$ **13.** 29 cm$^3$ **15.** $360°$C **17.** 0.26 cm$^3$
**19.** 0.13 mm **21.** 7.5 cm **23.** 160 s **25.** 94.6 L **27.** 42.7 kJ
**29.** 33 m$^2$ **31.** 33 g **33.** 3.0 min **35.** 13.5 C° **37.** (a) $5.3°$C; (b) 0;
(c) $0°$C; (d) 60 g **39.** 742 kJ **41.** (a) $0°$C; (b) $2.5°$C **43.** (a) $1.2 \times 10^2$ J; (b) 75 J; (c) 30 J **45.** $-30$ J **47.** (a) 6.0 cal; (b) $-43$ cal;
(c) 40 cal; (d) 18 cal; (e) 18 cal **49.** 60 J **51.** (a) 1.23 kW; (b) 2.28 kW; (c) 1.05 kW **53.** 1.66 kJ/s **55.** (a) 16 J/s; (b) 0.048 g/s
**57.** (a) $1.7 \times 10^4$ W/m$^2$; (b) 18 W/m$^2$ **59.** 0.50 min **61.** 0.40 cm/h
**63.** $-4.2°$C **65.** 1.1 m **67.** 10% **69.** (a) 80 J; (b) 80 J **71.** $4.5 \times 10^2$ J/kg·K **73.** 0.432 cm$^3$ **75.** $3.1 \times 10^2$ J **77.** $79.5°$C **79.** 23 J
**81.** (a) $11p_1V_1$; (b) $6p_1V_1$ **83.** $4.83 \times 10^{-2}$ cm$^3$ **85.** $10.5°$C
**87.** (a) 90 W; (b) $2.3 \times 10^2$ W; (c) $3.3 \times 10^2$ W **89.** (a) $1.87 \times 10^4$;
(b) 10.4 h **91.** 333 J **93.** 8.6 J **95.** (a) $-45$ J; (b) $+45$ J **97.** $4.0 \times 10^3$ min **99.** $-6.1$ nW **101.** 1.17 C° **103.** $8.0 \times 10^{-3}$ m$^2$
**105.** (a) too fast; (b) 0.79 s/h **107.** 1.9

## Chapter 19

**CP** **1.** all but $c$ **2.** (a) all tie; (b) 3, 2, 1 **3.** gas $A$ **4.** 5 (greatest change in $T$), then tie of 1, 2, 3, and 4 **5.** 1, 2, 3 ($Q_3 = 0$, $Q_2$ goes into work $W_2$, but $Q_1$ goes into greater work $W_1$ and increases gas temperature)
**Q** **1.** $d$, then $a$ and $b$ tie, then $c$ **3.** 20 J **5.** (a) 3; (b) 1; (c) 4; (d) 2;
(e) yes **7.** (a) 1, 2, 3, 4; (b) 1, 2, 3 **9.** constant-volume process
**P** **1.** 0.933 kg **3.** (a) 0.0388 mol; (b) $220°$C **5.** 25 molecules/cm$^3$
**7.** (a) $3.14 \times 10^3$ J; (b) from **9.** 186 kPa **11.** 5.60 kJ
**13.** (a) 1.5 mol; (b) $1.8 \times 10^3$ K; (c) $6.0 \times 10^2$ K; (d) 5.0 kJ
**15.** 360 K **17.** $2.0 \times 10^5$ Pa **19.** (a) 511 m/s; (b) $-200°$C; (c) $899°$C
**21.** $1.8 \times 10^2$ m/s **23.** 1.9 kPa **25.** (a) $5.65 \times 10^{-21}$ J; (b) $7.72 \times 10^{-21}$ J; (c) 3.40 kJ; (d) 4.65 kJ **27.** (a) $6.76 \times 10^{-20}$ J; (b) 10.7
**29.** (a) $6 \times 10^9$ km **31.** (a) $3.27 \times 10^{10}$ molecules/cm$^3$; (b) 172 m
**33.** (a) 6.5 km/s; (b) 7.1 km/s **35.** (a) 420 m/s; (b) 458 m/s; (c) yes
**37.** (a) 0.67; (b) 1.2; (c) 1.3; (d) 0.33 **39.** (a) $1.0 \times 10^4$ K; (b) $1.6 \times 10^5$ K; (c) $4.4 \times 10^2$ K; (d) $7.0 \times 10^3$ K; (e) no; (f) yes **41.** (a) 7.0 km/s; (b) $2.0 \times 10^{-8}$ cm; (c) $3.5 \times 10^{10}$ collisions/s **43.** (a) 3.49 kJ;
(b) 2.49 kJ; (c) 997 J; (d) 1.00 kJ **45.** (a) $6.6 \times 10^{-26}$ kg; (b) 40 g/mol **47.** (a) 0; (b) $+374$ J; (c) $+374$ J; (d) $+3.11 \times 10^{-22}$ J
**49.** 15.8 J/mol·K **51.** 8.0 kJ **53.** (a) 6.98 kJ; (b) 4.99 kJ; (c) 1.99 kJ;
(d) 2.99 kJ **55.** (a) 14 atm; (b) $6.2 \times 10^2$ K **57.** (a) diatomic;
(b) 446 K; (c) 8.10 mol **59.** $-15$ J **61.** $-20$ J **63.** (a) 3.74 kJ;
(b) 3.74 kJ; (c) 0; (d) 0; (e) $-1.81$ kJ; (f) 1.81 kJ; (g) $-3.22$ kJ;
(h) $-1.93$ kJ; (i) $-1.29$ kJ; (j) 520 J; (k) 0; (l) 520 J; (m) 0.0246 m$^3$;
(n) 2.00 atm; (o) 0.0373 m$^3$; (p) 1.00 atm **65.** (a) monatomic;
(b) $2.7 \times 10^4$ K; (c) $4.5 \times 10^4$ mol; (d) 3.4 kJ; (e) $3.4 \times 10^2$ kJ;
(f) 0.010 **67.** (a) 2.00 atm; (b) 333 J; (c) 0.961 atm; (d) 236 J
**69.** 349 K **71.** (a) $-374$ J; (b) 0; (c) $+374$ J; (d) $+3.11 \times 10^{-22}$ J
**73.** $7.03 \times 10^9\ s^{-1}$ **75.** (a) 900 cal; (b) 0; (c) 900 cal; (d) 450 cal;
(e) 1200 cal; (f) 300 cal; (g) 900 cal; (h) 450 cal; (i) 0; (j) $-900$ cal;
(k) 900 cal; (l) 450 cal **77.** (a) $3/v_0^3$; (b) $0.750v_0$; (c) $0.775v_0$
**79.** (a) $-2.37$ kJ; (b) 2.37 kJ **81.** (a) 125 J; (c) to **83.** (a) 8.0 atm;
(b) 300 K; (c) 4.4 kJ; (d) 3.2 atm; (e) 120 K; (f) 2.9 kJ; (g) 4.6 atm;
(h) 170 K; (i) 3.4 kJ **85.** (a) 38 L; (b) 71 g **87.** $-3.0$ J **89.** 22.8 m
**95.** 1.40 **97.** 4.71

## Chapter 20

**CP** **1.** a, b, c **2.** smaller ($Q$ is smaller) **3.** c, b, a **4.** a, d, c, b **5.** b
**Q** **1.** $b, a, c, d$ **3.** unchanged **5.** $a$ and $c$ tie, then $b$ and $d$ tie
**7.** (a) same; (b) increase; (c) decrease **9.** A, first; B, first and second; C, second; D, neither

**P** **1.** (a) 9.22 kJ; (b) 23.1 J/K; (c) 0 **3.** 14.4 J/K **5.** (a) 5.79 $\times$ $10^4$ J; (b) 173 J/K **7.** (a) 320 K; (b) 0; (c) +1.72 J/K **9.** +0.76 J/K
**11.** (a) 57.0°C; (b) −22.1 J/K; (c) +24.9 J/K; (d) +2.8 J/K
**13.** (a) −710 mJ/K; (b) +710 mJ/K; (c) +723 mJ/K; (d) −723 mJ/K;
(e) +13 mJ/K; (f) 0 **15.** (a) −943 J/K; (b) +943 J/K; (c) yes
**17.** (a) 0.333; (b) 0.215; (c) 0.644; (d) 1.10; (e) 1.10; (f) 0; (g) 1.10;
(h) 0; (i) −0.889; (j) −0.889; (k) −1.10; (l) −0.889; (m) 0; (n) 0.889;
(o) 0 **19.** (a) 0.693; (b) 4.50; (c) 0.693; (d) 0; (e) 4.50; (f) 23.0 J/K;
(g) −0.693; (h) 7.50; (i) −0.693; (j) 3.00; (k) 4.50; (l) 23.0 J/K
**21.** −1.18 J/K **23.** 97 K **25.** (a) 266 K; (b) 341 K **27.** (a) 23.6%;
(b) 1.49 $\times$ $10^4$ J **29.** (a) 2.27 kJ; (b) 14.8 kJ; (c) 15.4%; (d) 75.0%;
(e) greater **31.** (a) 33 kJ; (b) 25 kJ; (c) 26 kJ; (d) 18 kJ
**33.** (a) 1.47 kJ; (b) 554 J; (c) 918 J; (d) 62.4% **35.** (a) 3.00; (b) 1.98;
(c) 0.660; (d) 0.495; (e) 0.165; (f) 34.0% **37.** 440 W **39.** 20 J
**41.** 0.25 hp **43.** 2.03 **47.** (a) $W = N!/(n_1!\, n_2!\, n_3!)$; (b)
$[(N/2)!\,(N/2)!]/[(N/3)!\,(N/3)!\,(N/3)!]$; (c) 4.2 $\times$ $10^{16}$ **49.** 0.141 J/K·s
**51.** (a) 87 m/s; (b) 1.2 $\times$ $10^2$ m/s; (c) 22 J/K **53.** (a) 78%; (b) 82 kg/s
**55.** (a) 40.9°C; (b) −27.1 J/K; (c) 30.3 J/K; (d) 3.18 J/K **57.** +3.59
J/K **59.** 1.18 $\times$ $10^3$ J/K **63.** (a) 0; (b) 0; (c) −23.0 J/K; (d) 23.0 J/K
**65.** (a) 25.5 kJ; (b) 4.73 kJ; (c) 18.5% **67.** (a) 1.95 J/K; (b) 0.650 J/K;
(c) 0.217 J/K; (d) 0.072 J/K; (e) decrease **69.** (a) 4.45 J/K; (b) no
**71.** (a) 1.26 $\times$ $10^{14}$; (b) 4.71 $\times$ $10^{13}$; (c) 0.37; (d) 1.01 $\times$ $10^{29}$;
(e) 1.37 $\times$ $10^{28}$; (f) 0.14; (g) 9.05 $\times$ $10^{58}$; (h) 1.64 $\times$ $10^{57}$; (i) 0.018;
(j) decrease **73.** (a) 42.6 kJ; (b) 7.61 kJ **75.** (a) 1; (b) 1; (c) 3;
(d) 10; (e) 1.5 $\times$ $10^{-23}$ J/K; (f) 3.2 $\times$ $10^{-23}$ J/K **77.** $e = (1 + K)^{-1}$
**79.** 6.7

## Chapter 21

**CP** **1.** $C$ and $D$ attract; $B$ and $D$ attract **2.** (a) leftward;
(b) leftward; (c) leftward **3.** (a) $a, c, b$; (b) less than **4.** −15$e$
(net charge of −30$e$ is equally shared)
**Q** **1.** 3, 1, 2, 4 (zero) **3.** $a$ and $b$ **5.** $2kq^2/r^2$, up the page
**7.** $b$ and $c$ tie, then $a$ (zero) **9.** (a) same; (b) less than; (c) cancel;
(d) add; (e) adding components; (f) positive direction of $y$;
(g) negative direction of $y$; (h) positive direction of $x$; (i) negative
direction of $x$ **11.** (a) +4$e$; (b) −2$e$ upward; (c) −3$e$ upward;
(d) −12$e$ upward
**P** **1.** 0.500 **3.** 1.39 m **5.** 2.81 N **7.** −4.00 **9.** (a) −1.00 $\mu$C;
(b) 3.00 $\mu$C **11.** (a) 0.17 N; (b) −0.046 N **13.** (a) −14 cm; (b) 0
**15.** (a) 35 N; (b) −10°; (c) −8.4 cm; (d) +2.7 cm **17.** (a) 1.60 N;
(b) 2.77 N **19.** (a) 3.00 cm; (b) 0; (c) −0.444 **21.** 3.8 $\times$ $10^{-8}$ C
**23.** (a) 0; (b) 12 cm; (c) 0; (d) 4.9 $\times$ $10^{-26}$ N **25.** 6.3 $\times$ $10^{11}$
**27.** (a) 3.2 $\times$ $10^{-19}$ C; (b) 2 **29.** (a) −6.05 cm; (b) 6.05 cm
**31.** 122 mA **33.** 1.3 $\times$ $10^7$ C **35.** (a) 0; (b) 1.9 $\times$ $10^{-9}$ N
**37.** (a) $^9$B; (b) $^{13}$N; (c) $^{12}$C **39.** 1.31 $\times$ $10^{-22}$ N **41.** (a) 5.7 $\times$ $10^{13}$ C;
(b) cancels out; (c) 6.0 $\times$ $10^5$ kg **43.** (b) 3.1 cm **45.** 0.19 MC
**47.** −45 $\mu$C **49.** 3.8 N **51.** (a) 2.00 $\times$ $10^{10}$ electrons; (b) 1.33 $\times$ $10^{10}$
electrons **53.** (a) 8.99 $\times$ $10^9$ N; (b) 8.99 kN **55.** (a) 0.5; (b) 0.15;
(c) 0.85 **57.** 1.7 $\times$ $10^8$ N **59.** −1.32 $\times$ $10^{13}$ C **61.** (a) (0.829 N)î;
(b) (−0.621 N)ĵ **63.** 2.2 $\times$ $10^{-6}$ kg **65.** 4.68 $\times$ $10^{-19}$ N
**67.** (a) 2.72$L$; (b) 0 **69.** (a) 5.1 $\times$ $10^2$ N; (b) 7.7 $\times$ $10^{28}$ m/s$^2$
**71.** (a) 0; (b) 3.43 $\times$ $10^9$ m/s$^2$ **73.** (a) 2.19 $\times$ $10^6$ m/s;
(b) 1.09 $\times$ $10^6$ m/s; (c) decrease **75.** 4.16 $\times$ $10^{42}$

## Chapter 22

**CP** **1.** (a) rightward; (b) leftward; (c) leftward; (d) rightward
(p and e have same charge magnitude, and p is farther)
**2.** (a) toward positive $y$; (b) toward positive $x$; (c) toward negative $y$
**3.** (a) leftward; (b) leftward; (c) decrease **4.** (a) all tie; (b) 1 and 3
tie, then 2 and 4 tie
**Q** **1.** $a, b, c$ **3.** (a) yes; (b) toward; (c) no (the field vectors are not
along the same line); (d) cancel; (e) add; (f) adding components;

(g) toward negative $y$ **5.** (a) to their left; (b) no **7.** (a) 4, 3, 1, 2;
(b) 3, then 1 and 4 tie, then 2 **9.** $a, b, c$ **11.** $e, b$, then $a$ and $c$ tie,
then $d$ (zero) **13.** $a, b, c$
**P** **3.** (a) 3.07 $\times$ $10^{21}$ N/C; (b) outward **5.** 56 pC **7.** (1.02 $\times$
$10^5$ N/C)ĵ **9.** (a) 1.38 $\times$ $10^{-10}$ N/C; (b) 180° **11.** −30 cm
**13.** (a) 3.60 $\times$ $10^{-6}$ N/C; (b) 2.55 $\times$ $10^{-6}$ N/C; (c) 3.60 $\times$ $10^{-4}$ N/C;
(d) 7.09 $\times$ $10^{-7}$ N/C; (e) As the proton nears the disk, the forces on
it from electrons e$_s$ more nearly cancel. **15.** (a) 160 N/C; (b) 45°
**17.** (a) −90°; (b) +2.0 $\mu$C; (c) −1.6 $\mu$C **19.** (a) $qd/4\pi\varepsilon_0 r^3$; (b) −90°
**23.** 0.506 **25.** (a) 1.62 $\times$ $10^6$ N/C; (b) −45° **27.** (a) 23.8 N/C;
(b) −90° **29.** 1.57 **31.** (a) −5.19 $\times$ $10^{-14}$ C/m; (b) 1.57 $\times$ $10^{-3}$ N/C;
(c) −180°; (d) 1.52 $\times$ $10^{-8}$ N/C; (e) 1.52 $\times$ $10^{-8}$ N/C **35.** 0.346 m
**37.** 28% **39.** −5$e$ **41.** (a) 1.5 $\times$ $10^3$ N/C; (b) 2.4 $\times$ $10^{-16}$ N; (c) up;
(d) 1.6 $\times$ $10^{-26}$ N; (e) 1.5 $\times$ $10^{10}$ **43.** 3.51 $\times$ $10^{15}$ m/s$^2$
**45.** 6.6 $\times$ $10^{-15}$ N **47.** (a) 1.92 $\times$ $10^{12}$ m/s$^2$; (b) 1.96 $\times$ $10^5$ m/s
**49.** (a) 0.245 N; (b) −11.3°; (c) 108 m; (d) −21.6 m **51.** 2.6 $\times$ $10^{-10}$ N;
(b) 3.1 $\times$ $10^{-8}$ N; (c) moves to stigma **53.** 27 $\mu$m **55.** (a) 2.7 $\times$ $10^6$
m/s; (b) 1.0 kN/C **57.** (a) 9.30 $\times$ $10^{-15}$ C·m; (b) 2.05 $\times$ $10^{-11}$ J
**59.** 1.22 $\times$ $10^{-23}$ J **61.** $(1/2\pi)(pE/I)^{0.5}$ **63.** (a) 8.87 $\times$ $10^{-15}$ N;
(b) 120 **65.** 217° **67.** 61 N/C **69.** (a) 47 N/C; (b) 27 N/C
**71.** 38 N/C **73.** (a) −1.0 cm; (b) 0; (c) 10 pC **75.** +1.00 $\mu$C
**77.** (a) 6.0 mm; (b) 180° **79.** 9:30 **81.** (a) −0.029 C; (b) repulsive
forces would explode the sphere **83.** (a) −1.49 $\times$ $10^{-26}$ J;
(b) (−1.98 $\times$ $10^{-26}$ N·m)k̂; (c) 3.47 $\times$ $10^{-26}$ J **85.** (a) top row: 4, 8,
12; middle row: 5, 10, 14; bottom row: 7, 11, 16; (b) 1.63 $\times$ $10^{-19}$ C
**87.** (a) (−1.80 N/C)î; (b) (43.2 N/C)î; (c) (−6.29 N/C)î

## Chapter 23

**CP** **1.** (a) +$EA$; (b) −$EA$; (c) 0; (d) 0 **2.** (a) 2; (b) 3; (c) 1
**3.** (a) equal; (b) equal; (c) equal **4.** 3 and 4 tie, then 2, 1
**Q** **1.** (a) 8 N·m$^2$/C; (b) 0 **3.** all tie **5.** all tie **7.** $a, c$, then $b$ and $d$
tie (zero) **9.** (a) 2, 1, 3; (b) all tie (+4$q$) **11.** (a) impossible;
(b) −3$q_0$; (c) impossible
**P** **1.** −0.015 N·m$^2$/C **3.** (a) 0; (b) −3.92 N·m$^2$/C; (c) 0; (d) 0
**5.** 3.01 nN·m$^2$/C **7.** 2.0 $\times$ $10^5$ N·m$^2$/C **9.** (a) 8.23 N·m$^2$/C;
(b) 72.9 pC; (c) 8.23 N·m$^2$/C; (d) 72.9 pC **11.** −1.70 nC
**13.** 3.54 $\mu$C **15.** (a) 0; (b) 0.0417 **17.** (a) 37 $\mu$C; (b) 4.1 $\times$ $10^6$ N·m$^2$/C
**19.** (a) 4.5 $\times$ $10^{-7}$ C/m$^2$; (b) 5.1 $\times$ $10^4$ N/C **21.** (a) −3.0 $\times$ $10^{-6}$ C;
(b) +1.3 $\times$ $10^{-5}$ C **23.** (a) 0.32 $\mu$C; (b) 0.14 $\mu$C **25.** 5.0 $\mu$C/m
**27.** 3.8 $\times$ $10^{-8}$ C/m$^2$ **29.** (a) 0.214 N/C; (b) inward; (c) 0.855 N/C;
(d) outward; (e) −3.40 $\times$ $10^{-12}$ C; (f) −3.40 $\times$ $10^{-12}$ C **31.** (a) 2.3 $\times$
$10^6$ N/C; (b) outward; (c) 4.5 $\times$ $10^5$ N/C; (d) inward **33.** (a) 0;
(b) 0; (c) (−7.91 $\times$ $10^{-11}$ N/C)î **35.** −1.5 **37.** (a) 5.3 $\times$ $10^7$ N/C;
(b) 60 N/C **39.** 5.0 nC/m$^2$ **41.** 0.44 mm **43.** (a) 0; (b) 1.31 $\mu$N/C;
(c) 3.08 $\mu$N/C; (d) 3.08 $\mu$N/C **45.** (a) 2.50 $\times$ $10^4$ N/C; (b) 1.35 $\times$
$10^4$ N/C **47.** −7.5 nC **49.** (a) 0; (b) 56.2 mN/C; (c) 112 mN/C;
(d) 49.9 mN/C; (e) 0; (f) 0; (g) −5.00 fC; (h) 0 **51.** 1.79 $\times$ $10^{-11}$ C/m$^2$
**53.** (a) 7.78 fC; (b) 0; (c) 5.58 mN/C; (d) 22.3 mN/C **55.** 6$K\varepsilon_0 r^3$
**57.** (a) 0; (b) 2.88 $\times$ $10^4$ N/C; (c) 200 N/C **59.** (a) 5.4 N/C;
(b) 6.8 N/C **61.** (a) 0; (b) $q_a/4\pi\varepsilon_0 r^2$; (c) $(q_a + q_b)/4\pi\varepsilon_0 r^2$
**63.** −1.04 nC **65.** (a) 0.125; (b) 0.500 **67.** (a) +2.0 nC;
(b) −1.2 nC; (c) +1.2 nC; (d) +0.80 nC **69.** (5.65 $\times$ $10^4$ N/C)ĵ
**71.** (a) −2.53 $\times$ $10^{-2}$ N·m$^2$/C; (b) +2.53 $\times$ $10^{-2}$ N·m$^2$/C
**75.** 3.6 nC **77.** (a) +4.0 $\mu$C; (b) −4.0 $\mu$C **79.** (a) 693 kg/s;
(b) 693 kg/s; (c) 347 kg/s; (d) 347 kg/s; (e) 575 kg/s **81.** (a) 0.25$R$;
(b) 2.0$R$

## Chapter 24

**CP** **1.** (a) negative; (b) positive; (c) increase; (d) higher
**2.** (a) rightward; (b) 1, 2, 3, 5: positive; 4, negative; (c) 3, then 1, 2,
and 5 tie, then 4 **3.** all tie **4.** $a, c$ (zero), $b$ **5.** (a) 2, then 1 and 3
tie; (b) 3; (c) accelerate leftward

**Q**   **1.** $-4q/4\pi\varepsilon_0 d$   **3.** (a) 1 and 2; (b) none; (c) no; (d) 1 and 2, yes; 3 and 4, no   **5.** (a) higher; (b) positive; (c) negative; (d) all tie   **7.** (a) 0; (b) 0; (c) 0; (d) all three quantities still 0   **9.** (a) 3 and 4 tie, then 1 and 2 tie; (b) 1 and 2, increase; 3 and 4, decrease   **11.** $a, b, c$

**P**   **1.** (a) $3.0 \times 10^5$ C; (b) $3.6 \times 10^6$ J   **3.** $2.8 \times 10^5$   **5.** 8.8 mm   **7.** $-32.0$ V   **9.** (a) $1.87 \times 10^{-21}$ J; (b) $-11.7$ mV   **11.** (a) $-0.268$ mV; (b) $-0.681$ mV   **13.** (a) 3.3 nC; (b) 12 nC/m$^2$   **15.** (a) 0.54 mm; (b) 790 V   **17.** 0.562 mV   **19.** (a) 6.0 cm; (b) $-12.0$ cm   **21.** 16.3 $\mu$V   **23.** (a) 24.3 mV; (b) 0   **25.** (a) $-2.30$ V; (b) $-1.78$ V   **27.** 13 kV   **29.** 32.4 mV   **31.** 47.1 $\mu$V   **33.** 18.6 mV   **35.** $(-12$ V/m$)\hat{i} + (12$ V/m$)\hat{j}$   **37.** 150 N/C   **39.** $(-4.0 \times 10^{-16}$ N$)\hat{i} + (1.6 \times 10^{-16}$ N$)\hat{j}$   **41.** (a) 0.90 J; (b) 4.5 J   **43.** $-0.192$ pJ   **45.** 2.5 km/s   **47.** 22 km/s   **49.** 0.32 km/s   **51.** (a) $+6.0 \times 10^4$ V; (b) $-7.8 \times 10^5$ V; (c) 2.5 J; (d) increase; (e) same; (f) same   **53.** (a) 0.225 J; (b) $A$ 45.0 m/s$^2$, $B$ 22.5 m/s$^2$; (c) $A$ 7.75 m/s, $B$ 3.87 m/s   **55.** $1.6 \times 10^{-9}$ m   **57.** (a) 3.0 J; (b) $-8.5$ m   **59.** (a) proton; (b) 65.3 km/s   **61.** (a) 12; (b) 2   **63.** (a) $-1.8 \times 10^2$ V; (b) 2.9 kV; (c) $-8.9$ kV   **65.** $2.5 \times 10^{-8}$ C   **67.** (a) 12 kN/C; (b) 1.8 kV; (c) 5.8 cm   **69.** (a) 64 N/C; (b) 2.9 V; (c) 0   **71.** $p/2\pi\varepsilon_0 r^3$   **73.** (a) $3.6 \times 10^5$ V; (b) no   **75.** $6.4 \times 10^8$ V   **77.** 2.90 kV   **79.** $7.0 \times 10^5$ m/s   **81.** (a) 1.8 cm; (b) $8.4 \times 10^5$ m/s; (c) $2.1 \times 10^{-17}$ N; (d) positive; (e) $1.6 \times 10^{-17}$ N; (f) negative   **83.** (a) $+7.19 \times 10^{-10}$ V; (b) $+2.30 \times 10^{-28}$ J; (c) $+2.43 \times 10^{-29}$ J   **85.** $2.30 \times 10^{-28}$ J   **87.** 2.1 days   **89.** $2.30 \times 10^{-22}$ J   **91.** $1.48 \times 10^7$ m/s   **93.** $-1.92$ MV   **95.** (a) $Q/4\pi\varepsilon_0 r$; (b) $(\rho/3\varepsilon_0)(1.5r_2^2 - 0.50r^2 - r_1^3 r^{-1})$, $\rho = Q/[(4\pi/3)(r_2^3 - r_1^3)]$; (c) $(\rho/2\varepsilon_0)(r_2^2 - r_1^2)$, with $\rho$ as in (b); (d) yes   **97.** (a) 38 s; (b) $2.7 \times 10^2$ days   **101.** (a) 0.484 MeV; (b) 0   **103.** $-1.7$

## Chapter 25

**CP**   **1.** (a) same; (b) same   **2.** (a) decreases; (b) increases; (c) decreases   **3.** (a) $V, q/2$; (b) $V/2$; $q$

**Q**   **1.** $a, 2; b, 1; c, 3$   **3.** (a) no; (b) yes; (c) all tie   **5.** (a) same; (b) same; (c) more; (d) more   **7.** $a$, series; $b$, parallel; $c$, parallel   **9.** (a) increase; (b) same; (c) increase; (d) increase; (e) increase; (f) increase   **11.** parallel, $C_1$ alone, $C_2$ alone, series

**P**   **1.** (a) 3.5 pF; (b) 3.5 pF; (c) 57 V   **3.** (a) 144 pF; (b) 17.3 nC   **5.** 0.280 pF   **7.** $6.79 \times 10^{-4}$ F/m$^2$   **9.** 315 mC   **11.** 3.16 $\mu$F   **13.** 43 pF   **15.** (a) 3.00 $\mu$F; (b) 60.0 $\mu$C; (c) 10.0 V; (d) 30.0 $\mu$C; (e) 10.0 V; (f) 20.0 $\mu$C; (g) 5.00 V; (h) 20.0 $\mu$C   **17.** (a) 789 $\mu$C; (b) 78.9 V   **19.** (a) 4.0 $\mu$F; (b) 2.0 $\mu$F   **21.** (a) 50 V; (b) $5.0 \times 10^{-5}$ C; (c) $1.5 \times 10^{-4}$ C   **23.** (a) $4.5 \times 10^{14}$; (b) $1.5 \times 10^{14}$; (c) $3.0 \times 10^{14}$; (d) $4.5 \times 10^{14}$; (e) up; (f) up   **25.** 3.6 pC   **27.** (a) 9.00 $\mu$C; (b) 16.0 $\mu$C; (c) 9.00 $\mu$C; (d) 16.0 $\mu$C; (e) 8.40 $\mu$C; (f) 16.8 $\mu$C; (g) 10.8 $\mu$C; (h) 14.4 $\mu$C   **29.** 72 F   **31.** 0.27 J   **33.** 0.11 J/m$^3$   **35.** (a) $9.16 \times 10^{-18}$ J/m$^3$; (b) $9.16 \times 10^{-6}$ J/m$^3$; (c) $9.16 \times 10^6$ J/m$^3$; (d) $9.16 \times 10^{18}$ J/m$^3$; (e) $\infty$   **37.** (a) 16.0 V; (b) 45.1 pJ; (c) 120 pJ; (d) 75.2 pJ   **39.** (a) 190 V; (b) 95 mJ   **41.** 81 pF/m   **43.** Pyrex   **45.** 66 $\mu$J   **47.** 0.63 m$^2$   **49.** 17.3 pF   **51.** (a) 10 kV/m; (b) 5.0 nC; (c) 4.1 nC   **53.** (a) 89 pF; (b) 0.12 nF; (c) 11 nC; (d) 11 nC; (e) 10 kV/m; (f) 2.1 kV/m; (g) 88 V; (h) $-0.17$ $\mu$J   **55.** (a) 0.107 nF; (b) 7.79 nC; (c) 7.45 nC   **57.** 45 $\mu$C   **59.** 16 $\mu$C   **61.** (a) 7.20 $\mu$C; (b) 18.0 $\mu$C; (c) Battery supplies charges only to plates to which it is connected; charges on other plates are due to electron transfers between plates, in accord with new distribution of voltages across the capacitors. So the battery does not directly supply charge on capacitor 4.   **63.** (a) 10 $\mu$C; (b) 20 $\mu$C   **65.** 1.06 nC   **67.** (a) 2.40 $\mu$F; (b) 0.480 mC; (c) 80 V; (d) 0.480 mC; (e) 120 V   **69.** 4.9%   **71.** (a) 0.708 pF; (b) 0.600; (c) $1.02 \times 10^{-9}$ J; (d) sucked in   **73.** 5.3 V   **75.** 40 $\mu$F   **77.** (a) 200 kV/m; (b) 200 kV/m; (c) 1.77 $\mu$C/m$^2$; (d) 4.60 $\mu$C/m$^2$; (e) $-2.83$ $\mu$C/m$^2$   **79.** (a) $q^2/2\varepsilon_0 A$

## Chapter 26

**CP**   **1.** 8 A, rightward   **2.** (a)–(c) rightward   **3.** $a$ and $c$ tie, then $b$   **4.** device 2   **5.** (a) and (b) tie, then (d), then (c)

**Q**   **1.** tie of $A$, $B$, and $C$, then tie of $A + B$ and $B + C$, then $A + B + C$   **3.** (a) top-bottom, front-back, left-right; (b) top-bottom, front-back, left-right; (c) top-bottom, front-back, left-right; (d) top-bottom, front-back, left-right   **5.** $a$, $b$, and $c$ all tie, then $d$   **7.** (a) $B, A, C$; (b) $B, A, C$   **9.** (a) $C, B, A$; (b) all tie; (c) $A, B, C$; (d) all tie   **11.** (a) $a$ and $c$ tie, then $b$ (zero); (b) $a, b, c$; (c) $a$ and $b$ tie, then $c$

**P**   **1.** (a) 1.2 kC; (b) $7.5 \times 10^{21}$   **3.** 6.7 $\mu$C/m$^2$   **5.** (a) 6.4 A/m$^2$; (b) north; (c) cross-sectional area   **7.** 0.38 mm   **9.** 18.1 $\mu$A   **11.** (a) 1.33 A; (b) 0.666 A; (c) $J_a$   **13.** 13 min   **15.** 2.4 $\Omega$   **17.** $2.0 \times 10^6$ ($\Omega \cdot$m)$^{-1}$   **19.** $2.0 \times 10^{-8}$ $\Omega \cdot$m   **21.** $(1.8 \times 10^3)$°C   **23.** $8.2 \times 10^{-8}$ $\Omega \cdot$m   **25.** 54 $\Omega$   **27.** 3.0   **29.** $3.35 \times 10^{-7}$ C   **31.** (a) 6.00 mA; (b) $1.59 \times 10^{-8}$ V; (c) 21.2 n$\Omega$   **33.** (a) 38.3 mA; (b) 109 A/m$^2$; (c) 1.28 cm/s; (d) 227 V/m   **35.** 981 k$\Omega$   **39.** 150 s   **41.** (a) 1.0 kW; (b) US$0.25   **43.** 0.135 W   **45.** (a) 10.9 A; (b) 10.6 $\Omega$; (c) 4.50 MJ   **47.** (a) 5.85 m; (b) 10.4 m   **49.** (a) US$4.46; (b) 144 $\Omega$; (c) 0.833 A   **51.** (a) 5.1 V; (b) 10 V; (c) 10 W; (d) 20 W   **53.** (a) 28.8 $\Omega$; (b) $2.60 \times 10^{19}$ s$^{-1}$   **55.** 660 W   **57.** 28.8 kC   **59.** (a) silver; (b) 51.6 n$\Omega$   **61.** (a) $2.3 \times 10^{12}$; (b) $5.0 \times 10^3$; (c) 10 MV   **63.** 2.4 kW   **65.** (a) 1.37; (b) 0.730   **67.** (a) $-8.6$%; (b) smaller   **69.** 146 kJ   **71.** (a) 250°C; (b) yes   **73.** $3.0 \times 10^6$ J/kg   **75.** 560 W   **77.** 0.27 m/s   **79.** (a) 10 A/cm$^2$; (b) eastward   **81.** (a) $9.4 \times 10^{13}$ s$^{-1}$; (b) $2.40 \times 10^2$ W   **83.** 113 min   **85.** (a) 225 $\mu$C; (b) 60.0 $\mu$A; (c) 0.450 mW

## Chapter 27

**CP**   **1.** (a) rightward; (b) all tie; (c) $b$, then $a$ and $c$ tie; (d) $b$, then $a$ and $c$ tie   **2.** (a) all tie; (b) $R_1, R_2, R_3$   **3.** (a) less; (b) greater; (c) equal   **4.** (a) $V/2, i$; (b) $V, i/2$   **5.** (a) 1, 2, 4, 3; (b) 4, tie of 1 and 2, then 3

**Q**   **1.** (a) equal; (b) more   **3.** parallel, $R_2, R_1$, series   **5.** (a) series; (b) parallel; (c) parallel   **7.** (a) less; (b) less; (c) more   **9.** (a) parallel; (b) series   **11.** (a) same; (b) same; (c) less; (d) more   **13.** (a) all tie; (b) 1, 3, 2

**P**   **1.** (a) 0.50 A; (b) 1.0 W; (c) 2.0 W; (d) 6.0 W; (e) 3.0 W; (f) supplied; (g) absorbed   **3.** (a) 14 V; (b) $1.0 \times 10^2$ W; (c) $6.0 \times 10^2$ W; (d) 10 V; (e) $1.0 \times 10^2$ W   **5.** 11 kJ   **7.** (a) 80 J; (b) 67 J; (c) 13 J   **9.** (a) 12.0 eV; (b) 6.53 W   **11.** (a) 50 V; (b) 48 V; (c) negative   **13.** (a) 6.9 km; (b) 20 $\Omega$   **15.** 8.0 $\Omega$   **17.** (a) 0.004 $\Omega$; (b) 1   **19.** (a) 4.00 $\Omega$; (b) parallel   **21.** 5.56 A   **23.** (a) 50 mA; (b) 60 mA; (c) 9.0 V   **25.** $3d$   **27.** $3.6 \times 10^3$ A   **29.** (a) 0.333 A; (b) right; (c) 720 J   **31.** (a) $-11$ V; (b) $-9.0$ V   **33.** 48.3 V   **35.** (a) 5.25 V; (b) 1.50 V; (c) 5.25 V; (d) 6.75 V   **37.** 1.43 $\Omega$   **39.** (a) 0.150 $\Omega$; (b) 240 W   **41.** (a) 0.709 W; (b) 0.050 W; (c) 0.346 W; (d) 1.26 W; (e) $-0.158$ W   **43.** 9   **45.** (a) 0.67 A; (b) down; (c) 0.33 A; (d) up; (e) 0.33 A; (f) up; (g) 3.3 V   **47.** (a) 1.11 A; (b) 0.893 A; (c) 126 m   **49.** (a) 0.45 A   **51.** (a) 55.2 mA; (b) 4.86 V; (c) 88.0 $\Omega$; (d) decrease   **53.** $-3.0$%   **57.** 0.208 ms   **59.** 4.61   **61.** (a) 2.41 $\mu$s; (b) 161 pF   **63.** (a) 1.1 mA; (b) 0.55 mA; (c) 0.55 mA; (d) 0.82 mA; (e) 0.82 mA; (f) 0; (g) $4.0 \times 10^2$ V; (h) $6.0 \times 10^2$ V   **65.** 411 $\mu$A   **67.** 0.72 M$\Omega$   **69.** (a) 0.955 $\mu$C/s; (b) 1.08 $\mu$W; (c) 2.74 $\mu$W; (d) 3.82 $\mu$W   **71.** (a) 3.00 A; (b) 3.75 A; (c) 3.94 A   **73.** (a) $1.32 \times 10^7$ A/m$^2$; (b) 8.90 V; (c) copper; (d) $1.32 \times 10^7$ A/m$^2$; (e) 51.1 V; (f) iron   **75.** (a) 3.0 kV; (b) 10 s; (c) 11 G$\Omega$   **77.** (a) 85.0 $\Omega$; (b) 915 $\Omega$   **81.** 4.0 V   **83.** (a) 24.8 $\Omega$; (b) 14.9 k$\Omega$   **85.** the cable   **87.** $-13$ $\mu$C   **89.** 20 $\Omega$   **91.** (a) 3.00 A; (b) down; (c) 1.60 A; (d) down; (e) supply; (f) 55.2 W; (g) supply; (h) 6.40 W   **93.** (a) 1.0 V; (b) 50 m$\Omega$   **95.** 3   **99.** (a) 1.5 mA; (b) 0; (c) 1.0 mA   **101.** 7.50 V

**103.** (a) 60.0 mA; (b) down; (c) 180 mA; (d) left; (e) 240 mA;
(f) up  **105.** (a) 4.0 A; (b) up; (c) 0.50 A; (d) down; (e) 64 W;
(f) 16 W; (g) supplied; (h) absorbed

## Chapter 28
**CP**  **1.** $a$, $+z$; $b$, $-x$; $c$, $\vec{F}_B = 0$  **2.** (a) 2, then tie of 1 and 3 (zero);
(b) 4  **3.** (a) electron; (b) clockwise  **4.** $-y$  **5.** (a) all tie; (b) 1 and
4 tie, then 2 and 3 tie
**Q**  **1.** (a) no, because $\vec{v}$ and $\vec{F}_B$ must be perpendicular; (b) yes;
(c) no, because $\vec{B}$ and $\vec{F}_B$ must be perpendicular
**3.** (a) $+z$ and $-z$ tie, then $+y$ and $-y$ tie, then $+x$ and $-x$ tie (zero);
(b) $+y$  **5.** (a) $\vec{F}_E$; (b) $\vec{F}_B$  **7.** (a) $\vec{B}_1$; (b) $\vec{B}_1$ into page, $\vec{B}_2$ out of page;
(c) less  **9.** (a) positive; (b) $2 \to 1$ and $2 \to 4$ tie, then $2 \to 3$ (which
is zero)  **11.** (a) negative; (b) equal; (c) equal; (d) half-circle
**P**  **1.** (a) 400 km/s; (b) 835 eV  **3.** (a) $(6.2 \times 10^{-14}\,\text{N})\hat{k}$;
(b) $(-6.2 \times 10^{-14}\,\text{N})\hat{k}$  **5.** $-2.0$ T  **7.** $(-11.4\,\text{V/m})\hat{i} - (6.00\,\text{V/m})\hat{j} + (4.80\,\text{V/m})\hat{k}$  **9.** $-(0.267\,\text{mT})\hat{k}$  **11.** 0.68 MV/m  **13.** 7.4 $\mu$V
**15.** (a) $(-600\,\text{mV/m})\hat{k}$; (b) 1.20 V  **17.** (a) $2.60 \times 10^6$ m/s;
(b) 0.109 $\mu$s; (c) 0.140 MeV; (d) 70.0 kV  **19.** $1.2 \times 10^{-9}$ kg/C
**21.** (a) $2.05 \times 10^7$ m/s; (b) 467 $\mu$T; (c) 13.1 MHz; (d) 76.3 ns
**23.** 21.1 $\mu$T  **25.** (a) 0.978 MHz; (b) 96.4 cm  **27.** (a) 495 mT;
(b) 22.7 mA; (c) 8.17 MJ  **29.** 65.3 km/s  **31.** 5.07 ns
**33.** (a) 0.358 ns; (b) 0.166 mm; (c) 1.51 mm  **35.** (a) 200 eV;
(b) 20.0 keV; (c) 0.499%  **37.** $2.4 \times 10^2$ m  **39.** (a) 28.2 N;
(b) horizontally west  **41.** (a) 467 mA; (b) right  **43.** (a) 0; (b) 0.138 N;
(c) 0.138 N; (d) 0  **45.** $(-2.50\,\text{mN})\hat{j} + (0.750\,\text{mN})\hat{k}$  **47.** (a) 0.10 T;
(b) 31°  **49.** $(-4.3 \times 10^{-3}\,\text{N}\cdot\text{m})\hat{j}$  **51.** 2.45 A  **55.** (a) 2.86 A·m²;
(b) 1.10 A·m²  **57.** (a) 12.7 A; (b) 0.0805 N·m  **59.** (a) 0.30 A·m²;
(b) 0.024 N·m  **61.** (a) $-72.0$ $\mu$J; (b) $(96.0\hat{i} + 48.0\hat{k})$ $\mu$N·m
**63.** (a) $-(9.7 \times 10^{-4}\,\text{N}\cdot\text{m})\hat{i} - (7.2 \times 10^{-4}\,\text{N}\cdot\text{m})\hat{j} + (8.0 \times 10^{-4}\,\text{N}\cdot\text{m})\hat{k}$;
(b) $-6.0 \times 10^{-4}$ J  **65.** (a) 90°; (b) 1; (c) $1.28 \times 10^{-7}$ N·m
**67.** (a) 20 min; (b) $5.9 \times 10^{-2}$ N·m  **69.** 8.2 mm  **71.** 127 u
**73.** (a) $6.3 \times 10^{14}$ m/s²; (b) 3.0 mm  **75.** (a) 1.4; (b) 1.0
**77.** $(-500\,\text{V/m})\hat{j}$  **79.** (a) 0.50; (b) 0.50; (c) 14 cm; (d) 14 cm
**81.** $(0.80\hat{j} - 1.1\hat{k})$ mN  **83.** $-40$ mC  **85.** (a) $(12.8\hat{i} + 6.41\hat{j}) \times 10^{-22}$ N; (b) 90°; (c) 173°  **87.** (a) up the conducting path; (b) rim;
(c) 47.1 V; (d) 47.1 V; (e) 2.36 kW  **89.** $(mV/2ed^2)^{0.5}$  **91.** $n = JB/eE$

## Chapter 29
**CP**  **1.** $b, c, a$  **2.** $d$, tie of $a$ and $c$, then $b$  **3.** $d, a$, tie of $b$ and $c$ (zero)
**Q**  **1.** $c, a, b$  **3.** $c, d$, then $a$ and $b$ tie (zero)  **5.** $a, c, b$
**7.** $c$ and $d$ tie, then $b, a$  **9.** $b, a, d, c$ (zero)  **11.** (a) 1, 3, 2; (b) less
**P**  **1.** (a) 3.3 $\mu$T; (b) yes  **3.** (a) 16 A; (b) east  **5.** (a) 1.0 mT;
(b) out; (c) 0.80 mT; (d) out  **7.** (a) 0.102 $\mu$T; (b) out
**9.** (a) opposite; (b) 30 A  **11.** (a) 4.3 A; (b) out  **13.** 50.3 nT
**15.** (a) 1.7 $\mu$T; (b) into; (c) 6.7 $\mu$T; (d) into  **17.** 132 nT
**19.** 5.0 $\mu$T  **21.** 256 nT  **23.** $(-7.75 \times 10^{-23}\,\text{N})\hat{i}$  **25.** 2.00 rad
**27.** 61.3 mA  **29.** $(80\,\mu\text{T})\hat{j}$  **31.** (a) 20 $\mu$T; (b) into  **33.** $(22.3\,\text{pT})\hat{j}$
**35.** 88.4 pN/m  **37.** $(-125\,\mu\text{N/m})\hat{i} + (41.7\,\mu\text{N/m})\hat{j}$  **39.** 800 nN/m
**41.** $(3.20\,\text{mN})\hat{j}$  **43.** (a) 0; (b) 0.850 mT; (c) 1.70 mT; (d) 0.850 mT
**45.** (a) $-2.5$ $\mu$T·m; (b) 0  **47.** (a) 0; (b) 0.10 $\mu$T; (c) 0.40 $\mu$T
**49.** (a) 533 $\mu$T; (b) 400 $\mu$T  **51.** 0.30 mT  **53.** 0.272 A
**55.** (a) 4.77 cm; (b) 35.5 $\mu$T  **57.** (a) 2.4 A·m²; (b) 46 cm
**59.** 0.47 A·m²  **61.** (a) 79 $\mu$T; (b) $1.1 \times 10^{-6}$ N·m  **63.** (a) $(0.060\,\text{A}\cdot\text{m}^2)\hat{j}$;
(b) $(96\,\text{pT})\hat{j}$  **65.** 1.28 mm  **69.** (a) 15 A; (b) $-z$  **71.** 7.7 mT
**73.** (a) 15.3 $\mu$T  **75.** (a) $(0.24\hat{i})$ nT; (b) 0; (c) $(-43\hat{k})$ pT; (d) $(0.14\hat{k})$
nT  **79.** (a) 4.8 mT; (b) 0.93 mT; (c) 0  **83.** $(-0.20\,\text{mT})\hat{k}$
**87.** (a) $\mu_0 ir/2\pi c^2$; (b) $\mu_0 i/2\pi r$; (c) $\mu_0 i(a^2 - r^2)/2\pi(a^2 - b^2)r$; (d) 0

## Chapter 30
**CP**  **1.** $b$, then $d$ and $e$ tie, and then $a$ and $c$ tie (zero)  **2.** $a$ and $b$
tie, then $c$ (zero)  **3.** $c$ and $d$ tie, then $a$ and $b$ tie  **4.** $b$, out; $c$, out; $d$,

into; $e$, into  **5.** d and e  **6.** (a) 2, 3, 1 (zero); (b) 2, 3, 1
**7.** $a$ and $b$ tie, then $c$
**Q**  **1.** out  **3.** (a) all tie (zero); (b) 2, then 1 and 3 tie (zero)  **5.** $d$
and $c$ tie, then $b, a$  **7.** (a) more; (b) same; (c) same; (d) same (zero)
**9.** (a) all tie (zero); (b) 1 and 2 tie, then 3; (c) all tie (zero)  **11.** $b$
**P**  **1.** 0  **3.** 30 mA  **5.** 0  **7.** (a) 31 mV; (b) left  **9.** 0.198 mV
**11.** (b) 0.796 m²  **13.** 29.5 mC  **15.** (a) 21.7 V; (b) counterclockwise  **17.** (a) $1.26 \times 10^{-4}$ T; (b) 0; (c) $1.26 \times 10^{-4}$ T; (d) yes;
(e) $5.04 \times 10^{-8}$ V  **19.** 5.50 kV  **21.** (a) 40 Hz; (b) 3.2 mV
**23.** (a) $\mu_0 iR^2\pi r^2/2x^3$; (b) $3\mu_0 i\pi R^2 r^2 v/2x^4$; (c) counterclockwise
**25.** (a) 13 $\mu$Wb/m; (b) 17%; (c) 0  **27.** (a) 80 $\mu$V; (b) clockwise
**29.** (a) 48.1 mV; (b) 2.67 mA; (c) 0.129 mW  **31.** 3.68 $\mu$W
**33.** (a) 240 $\mu$V; (b) 0.600 mA; (c) 0.144 $\mu$W; (d) $2.87 \times 10^{-8}$ N;
(e) 0.144 $\mu$W  **35.** (a) 0.60 V; (b) up; (c) 1.5 A; (d) clockwise;
(e) 0.90 W; (f) 0.18 N; (g) 0.90 W  **37.** (a) 71.5 $\mu$V/m; (b) 143 $\mu$V/m
**39.** 0.15 V/m  **41.** (a) 2.45 mWb; (b) 0.645 mH  **43.** 1.81 $\mu$H/m
**45.** (a) decreasing; (b) 0.68 mH  **47.** (b) $L_{eq} = \Sigma L_j$, sum from $j = 1$
to $j = N$  **49.** 59.3 mH  **51.** 46 $\Omega$  **53.** (a) 8.45 ns; (b) 7.37 mA
**55.** 6.91  **57.** (a) 1.5 s  **59.** (a) $i[1 - \exp(-Rt/L)]$; (b) $(L/R) \ln 2$
**61.** (a) 97.9 H; (b) 0.196 mJ  **63.** 25.6 ms  **65.** (a) 18.7 J; (b) 5.10 J;
(c) 13.6 J  **67.** (a) 34.2 J/m³; (b) 49.4 mJ  **69.** $1.5 \times 10^8$ V/m
**71.** (a) 1.0 J/m³; (b) $4.8 \times 10^{-15}$ J/m³  **73.** (a) 1.67 mH; (b) 6.00 mWb
**75.** 13 $\mu$H  **77.** (b) have the turns of the two solenoids wrapped in
opposite directions  **79.** (a) 2.0 A; (b) 0; (c) 2.0 A; (d) 0; (e) 10 V;
(f) 2.0 A/s; (g) 2.0 A; (h) 1.0 A; (i) 3.0 A; (j) 10 V; (k) 0; (l) 0
**81.** (a) 10 $\mu$T; (b) out; (c) 3.3 $\mu$T; (d) out  **83.** 0.520 ms
**85.** (a) $(4.4 \times 10^7\,\text{m/s}^2)\hat{i}$; (b) 0; (c) $(-4.4 \times 10^7\,\text{m/s}^2)\hat{i}$
**87.** (a) 0.40 V; (b) 20 A  **89.** (a) 10 A; (b) $1.0 \times 10^2$ J  **91.** (a) 0;
(b) $8.0 \times 10^2$ A/s; (c) 1.8 mA; (d) $4.4 \times 10^2$ A/s; (e) 4.0 mA; (f) 0
**93.** 1.15 W  **95.** (a) 20 A/s; (b) 0.75 A  **97.** 12 A/s  **99.** $3 \times 10^{36}$ J
**101.** 1.15 $\mu$Wb

## Chapter 31
**CP**  **1.** (a) $T/2$; (b) $T$; (c) $T/2$; (d) $T/4$  **2.** (a) 5 V; (b) 150 $\mu$J
**3.** (a) remains the same; (b) remains the same  **4.** (a) $C, B, A$; (b) 1,
$A$; 2, $B$; 3, $S$; 4, $C$; (c) $A$  **5.** (a) remains the same; (b) increases;
(c) remains the same; (d) decreases  **6.** (a) 1, lags; 2, leads; 3, in
phase; (b) 3 ($\omega_d = \omega$ when $X_L = X_C$)  **7.** (a) increase (circuit is
mainly capacitive; increase $C$ to decrease $X_C$ to be closer to resonance for maximum $P_{avg}$); (b) closer  **8.** (a) greater; (b) step-up
**Q**  **1.** $b, a, c$  **3.** (a) $T/4$; (b) $T/4$; (c) $T/2$; (d) $T/2$  **5.** $c, b, a$  **7.** $a$
inductor; $b$ resistor; $c$ capacitor  **9.** (a) positive; (b) decreased (to
decrease $X_L$ and get closer to resonance); (c) decreased (to increase
$X_C$ and get closer to resonance)  **11.** (a) rightward, increase
($X_L$ increases, closer to resonance); (b) rightward, increase
($X_C$ decreases, closer to resonance); (c) rightward, increase
($\omega_d/\omega$ increases, closer to resonance)  **13.** (a) inductor;
(b) decrease
**P**  **1.** (a) 1.17 $\mu$J; (b) 5.58 mA  **3.** (a) 6.00 $\mu$s; (b) 167 kHz;
(c) 3.00 $\mu$s  **5.** 45.2 mA  **7.** (a) 1.25 kg; (b) 372 N/m;
(c) $1.75 \times 10^{-4}$ m; (d) 3.02 mm/s  **9.** $7.0 \times 10^{-4}$ s  **11.** (a) 6.0;
(b) 36 pF; (c) 0.22 mH  **13.** (a) 0.180 mC; (b) 70.7 $\mu$s; (c) 66.7 W
**15.** (a) 3.0 nC; (b) 1.7 mA; (c) 4.5 nJ  **17.** (a) 275 Hz; (b) 365 mA
**21.** (a) 356 $\mu$s; (b) 2.50 mH; (c) 3.20 mJ  **23.** (a) 1.98 $\mu$J;
(b) 5.56 $\mu$C; (c) 12.6 mA; (d) $-46.9°$; (e) $+46.9°$  **25.** 8.66 m$\Omega$
**29.** (a) 95.5 mA; (b) 11.9 mA  **31.** (a) 0.65 kHz; (b) 24 $\Omega$
**33.** (a) 6.73 ms; (b) 11.2 ms; (c) inductor; (d) 138 mH  **35.** 89 $\Omega$
**37.** 7.61 A  **39.** (a) 267 $\Omega$; (b) $-41.5°$; (c) 135 mA  **41.** (a) 206 $\Omega$;
(b) 13.7°; (c) 175 mA  **43.** (a) 218 $\Omega$; (b) 23.4°; (c) 165 mA
**45.** (a) yes; (b) 1.0 kV  **47.** (a) 224 rad/s; (b) 6.00 A; (c) 219 rad/s;
(d) 228 rad/s; (e) 0.040  **49.** (a) 796 Hz; (b) no change;
(c) decreased; (d) increased  **53.** (a) 12.1 $\Omega$; (b) 1.19 kW

**55.** 1.84 A   **57.** (a) 117 $\mu$F; (b) 0; (c) 90.0 W; (d) 0°; (e) 1; (f) 0;
(g) $-90°$; (h) 0   **59.** (a) 2.59 A; (b) 38.8 V; (c) 159 V; (d) 224 V;
(e) 64.2 V; (f) 75.0 V; (g) 100 W; (h) 0; (i) 0   **61.** (a) 0.743; (b) lead;
(c) capacitive; (d) no; (e) yes; (f) no; (g) yes; (h) 33.4 W
**63.** (a) 2.4 V; (b) 3.2 mA; (c) 0.16 A   **65.** (a) 1.9 V; (b) 5.9 W; (c) 19 V;
(d) $5.9 \times 10^2$ W; (e) 0.19 kV; (f) 59 kW   **67.** (a) 6.73 ms; (b) 2.24 ms;
(c) capacitor; (d) 59.0 $\mu$F   **69.** (a) $-0.405$ rad; (b) 2.76 A;
(c) capacitive   **71.** (a) 64.0 $\Omega$; (b) 50.9 $\Omega$; (c) capacitive
**73.** (a) 2.41 $\mu$H; (b) 21.4 pJ; (c) 82.2 nC   **75.** (a) 39.1 $\Omega$; (b) 21.7 $\Omega$;
(c) capacitive   **79.** (a) 0.577$Q$; (b) 0.152   **81.** (a) 45.0°; (b) 70.7 $\Omega$
**83.** 1.84 kHz   **85.** (a) 0.689 $\mu$H; (b) 17.9 pJ; (c) 0.110 $\mu$C
**87.** (a) 165 $\Omega$; (b) 313 mH; (c) 14.9 $\mu$F   **93.** (a) 36.0 V; (b) 29.9 V;
(c) 11.9 V; (d) $-5.85$ V

## Chapter 32

**CP**   **1.** $d, b, c, a$ (zero)   **2.** $a, c, b, d$ (zero)   **3.** tie of $b, c$, and $d$,
then $a$   **4.** (a) 2; (b) 1   **5.** (a) away; (b) away; (c) less   **6.** (a) toward;
(b) toward; (c) less
**Q**   **1.** 1 $a$, 2 $b$, 3 $c$ and $d$   **3.** a, decreasing; b, decreasing
**5.** supplied   **7.** (a) $a$ and $b$ tie, then $c$, $d$; (b) none (because plate
lacks circular symmetry, $\vec{B}$ not tangent to any circular loop);
(c) none   **9.** (a) 1 up, 2 up, 3 down; (b) 1 down, 2 up, 3 zero
**11.** (a) 1, 3, 2; (b) 2
**P**   **1.** $+3$ Wb   **3.** (a) 47.4 $\mu$Wb; (b) inward   **5.** $2.4 \times 10^{13}$ V/m·s
**7.** (a) $1.18 \times 10^{-19}$ T; (b) $1.06 \times 10^{-19}$ T   **9.** (a) $5.01 \times 10^{-22}$ T;
(b) $4.51 \times 10^{-22}$ T   **11.** (a) 1.9 pT   **13.** $7.5 \times 10^5$ V/s
**17.** (a) 0.324 V/m; (b) $2.87 \times 10^{-16}$A; (c) $2.87 \times 10^{-18}$
**19.** (a) 75.4 nT; (b) 67.9 nT   **21.** (a) 27.9 nT; (b) 15.1 nT
**23.** (a) 2.0 A; (b) $2.3 \times 10^{11}$ V/m·s; (c) 0.50 A; (d) 0.63 $\mu$T·m
**25.** (a) 0.63 $\mu$T; (b) $2.3 \times 10^{12}$ V/m·s   **27.** (a) 0.71 A; (b) 0; (c) 2.8 A
**29.** (a) 7.60 $\mu$A; (b) 859 kV·m/s; (c) 3.39 mm; (d) 5.16 pT   **31.** 55 $\mu$T
**33.** (a) 0; (b) 0; (c) 0; (d) $\pm 3.2 \times 10^{-25}$ J; (e) $-3.2 \times 10^{-34}$ J·s;
(f) $2.8 \times 10^{-23}$ J/T; (g) $-9.7 \times 10^{-25}$ J; (h) $\pm 3.2 \times 10^{-25}$ J
**35.** (a) $-9.3 \times 10^{-24}$ J/T; (b) $1.9 \times 10^{-23}$ J/T   **37.** (b) $+x$;
(c) clockwise; (d) $+x$   **39.** yes   **41.** 20.8 mJ/T   **43.** (b) $K_i/B$;
(c) $-z$; (d) 0.31 kA/m   **47.** (a) $1.8 \times 10^2$ km; (b) $2.3 \times 10^{-5}$
**49.** (a) 3.0 $\mu$T; (b) $5.6 \times 10^{-10}$ eV   **51.** $5.15 \times 10^{-24}$ A·m$^2$
**53.** (a) 0.14 A; (b) 79 $\mu$C   **55.** (a) $6.3 \times 10^8$ A; (b) yes; (c) no
**57.** 0.84 kJ/T   **59.** (a) $(1.2 \times 10^{-13}$ T$)$ exp$[-t/(0.012$ s$)]$;
(b) $5.9 \times 10^{-15}$ T   **63.** (a) 27.5 mm; (b) 110 mm   **65.** 8.0 A
**67.** (a) $-8.8 \times 10^{15}$ V/m·s; (b) $5.9 \times 10^{-7}$ T·m   **69.** (b) sign is
minus; (c) no, because there is compensating positive flux through
open end nearer to magnet   **71.** (b) $-x$; (c) counterclockwise;
(d) $-x$   **73.** (a) 7; (b) 7; (c) $3h/2\pi$; (d) $3eh/4\pi m$; (e) $3.5h/2\pi$;
(f) 8   **75.** (a) 9; (b) $3.71 \times 10^{-23}$ J/T; (c) $+9.27 \times 10^{-24}$ J;
(d) $-9.27 \times 10^{-24}$ J

## Chapter 33

**CP**   **1.** (a) (Use Fig. 33-5.) On right side of rectangle, $\vec{E}$ is in
negative $y$ direction; on left side, $\vec{E} + d\vec{E}$ is greater and in same
direction; (b) $\vec{E}$ is downward. On right side, $\vec{B}$ is in negative $z$
direction; on left side, $\vec{B} + d\vec{B}$ is greater and in same direction.
**2.** positive direction of $x$   **3.** (a) same; (b) decrease   **4.** $a, d, b, c$
(zero)   **5.** $a$
**Q**   **1.** (a) positive direction of $z$; (b) $x$   **3.** (a) same; (b) increase;
(c) decrease   **5.** (a) and (b) $A = 1, n = 4, \theta = 30°$   **7.** $a, b, c$   **9.** $B$
**11.** none
**P**   **1.** 7.49 GHz   **3.** (a) 515 nm; (b) 610 nm; (c) 555 nm;
(d) $5.41 \times 10^{14}$ Hz; (e) $1.85 \times 10^{-15}$ s   **5.** $5.0 \times 10^{-21}$ H   **7.** 1.2 MW/m$^2$
**9.** 0.10 MJ   **11.** (a) 6.7 nT; (b) $y$; (c) negative direction of $y$
**13.** (a) 1.03 kV/m; (b) 3.43 $\mu$T   **15.** (a) 87 mV/m; (b) 0.29 nT;

(c) 6.3 kW   **17.** (a) 6.7 nT; (b) 5.3 mW/m$^2$; (c) 6.7 W   **19.** $1.0 \times 10^7$ Pa
**21.** $5.9 \times 10^{-8}$ Pa   **23.** (a) $4.68 \times 10^{11}$ W; (b) any chance
disturbance could move sphere from directly above source—the
two force vectors no longer along the same axis   **27.** (a) $1.0 \times 10^8$ Hz;
(b) $6.3 \times 10^8$ rad/s; (c) 2.1 m$^{-1}$; (d) 1.0 $\mu$T; (e) $z$; (f) $1.2 \times 10^2$ W/m$^2$;
(g) $8.0 \times 10^{-7}$ N; (h) $4.0 \times 10^{-7}$ Pa   **29.** 1.9 mm/s   **31.** (a) 0.17 $\mu$m;
(b) toward the Sun   **33.** 3.1%   **35.** 4.4 W/m$^2$   **37.** (a) 2 sheets;
(b) 5 sheets   **39.** (a) 1.9 V/m; (b) $1.7 \times 10^{-11}$ Pa   **41.** 20° or 70°
**43.** 0.67   **45.** 1.26   **47.** 1.48   **49.** 180°   **51.** (a) 56.9°; (b) 35.3°
**55.** 1.07 m   **57.** 182 cm   **59.** (a) 48.9°; (b) 29.0°   **61.** (a) 26.8°;
(b) yes   **63.** (a) $(1 + \sin^2 \theta)^{0.5}$; (b) $2^{0.5}$; (c) yes; (d) no   **65.** 23.2°
**67.** (a) 1.39; (b) 28.1°; (c) no   **69.** 49.0°   **71.** (a) 0.50 ms; (b) 8.4 min;
(c) 2.4 h; (d) 5446 B.C.   **73.** (a) (16.7 nT) sin$[(1.00 \times 10^6$ m$^{-1})z +$
$(3.00 \times 10^{14}$ s$^{-1})t]$; (b) 6.28 $\mu$m; (c) 20.9 fs; (d) 33.2 mW/m$^2$; (e) $x$;
(f) infrared   **75.** 1.22   **77.** (c) 137.6°; (d) 139.4°; (e) 1.7°
**81.** (a) $z$ axis; (b) $7.5 \times 10^{14}$ Hz; (c) 1.9 kW/m$^2$   **83.** (a) white;
(b) white dominated by red end; (c) no refracted light
**85.** $1.5 \times 10^{-9}$ m/s$^2$   **87.** (a) 3.5 $\mu$W/m$^2$; (b) 0.78 $\mu$W;
(c) $1.5 \times 10^{-17}$ W/m$^2$; (d) $1.1 \times 10^{-7}$ V/m; (e) 0.25 fT   **89.** (a) 55.8°;
(b) 55.5°   **91.** (a) 83 W/m$^2$; (b) 1.7 MW   **93.** 35°   **97.** $\cos^{-1}(p/50)^{0.5}$
**99.** 8$RI/3c$   **101.** 0.034   **103.** $9.43 \times 10^{-10}$ T   **105.** (a) $-y$; (b) $z$;
(c) 1.91 kW/m$^2$; (d) $E_z = (1.20$ kV/m$)$ sin$[(6.67 \times 10^6$ m$^{-1})y +$
$(2.00 \times 10^{15}$ s$^{-1})t]$; (e) 942 nm; (f) infrared   **107.** (a) 1.60; (b) 58.0°

## Chapter 34

**CP**   **1.** 0.2$d$, 1.8$d$, 2.2$d$   **2.** (a) real; (b) inverted; (c) same
**3.** (a) $e$; (b) virtual, same   **4.** virtual, same as object, diverging
**Q**   **1.** (a) $a$; (b) $c$   **3.** (a) $a$ and $c$; (b) three times; (c) you
**5.** convex   **7.** (a) all but variation 2; (b) 1, 3, 4: right, inverted; 5, 6:
left, same   **9.** $d$ (infinite), tie of $a$ and $b$, then $c$   **11.** (a) $x$; (b) no;
(c) no; (d) the direction you are facing
**P**   **1.** 9.10 m   **3.** 1.11   **5.** 351 cm   **7.** 10.5 cm   **9.** (a) $+24$ cm;
(b) $+36$ cm; (c) $-2.0$; (d) R; (e) I; (f) same   **11.** (a) $-20$ cm;
(b) $-4.4$ cm; (c) $+0.56$; (d) V; (e) NI; (f) opposite   **13.** (a) $+36$ cm;
(b) $-36$ cm; (c) $+3.0$; (d) V; (e) NI; (f) opposite   **15.** (a) $-16$ cm;
(b) $-4.4$ cm; (c) $+0.44$; (d) V; (e) NI; (f) opposite   **17.** (b) plus;
(c) $+40$ cm; (e) $-20$ cm; (f) $+2.0$; (g) V; (h) NI; (i) opposite
**19.** (a) convex; (b) $-20$ cm; (d) $+20$ cm; (f) $+0.50$; (g) V; (h) NI;
(i) opposite   **21.** (a) concave; (c) $+40$ cm; (e) $+60$ cm; (f) $-2.0$;
(g) R; (h) I; (i) same   **23.** (a) convex; (b) minus; (c) $-60$ cm;
(d) $+1.2$ m; (e) $-24$ cm; (g) V; (h) NI; (i) opposite   **25.** (a) concave;
(b) $+8.6$ cm; (c) $+17$ cm; (e) $+12$ cm; (f) minus; (g) R; (i) same
**27.** (a) convex; (c) $-60$ cm; (d) $+30$ cm; (f) $+0.50$; (g) V; (h) NI;
(i) opposite   **29.** (b) $-20$ cm; (c) minus; (d) $+5.0$ cm; (e) minus;
(f) $+0.80$; (g) V; (h) NI; (i) opposite   **31.** (b) 0.56 cm/s; (c) 11 m/s;
(d) 6.7 cm/s   **33.** (c) $-33$ cm; (e) V; (f) same   **35.** (d) $-26$ cm; (e) V;
(f) same   **37.** (c) $+30$ cm; (e) V; (f) same   **39.** (a) 2.00; (b) none
**41.** (a) $+40$ cm; (b) $\infty$   **43.** 5.0 mm   **45.** 1.86 mm   **47.** (a) 45 mm;
(b) 90 mm   **49.** 22 cm   **51.** (a) $-48$ cm; (b) $+4.0$; (c) V; (d) NI;
(e) same   **53.** (a) $-4.8$ cm; (b) $+0.60$; (c) V; (d) NI; (e) same
**55.** (a) $-8.6$ cm; (b) $+0.39$; (c) V; (d) NI; (e) same   **57.** (a) $+36$ cm;
(b) $-0.80$; (c) R; (d) I; (e) opposite   **59.** (a) $+55$ cm; (b) $-0.74$;
(c) R; (d) I; (e) opposite   **61.** (a) $-18$ cm; (b) $+0.76$; (c) V; (d) NI;
(e) same   **63.** (a) $-30$ cm; (b) $+0.86$; (c) V; (d) NI; (e) same
**65.** (a) $-7.5$ cm; (b) $+0.75$; (c) V; (d) NI; (e) same   **67.** (a) $+84$ cm;
(b) $-1.4$; (c) R; (d) I; (e) opposite   **69.** (a) C; (d) $-10$ cm; (e) $+2.0$;
(f) V; (g) NI; (h) same   **71.** (a) D; (b) $-5.3$ cm; (d) $-4.0$ cm; (f) V;
(g) NI; (h) same   **73.** (a) C; (b) $+3.3$ cm; (d) $+5.0$ cm; (f) R; (g) I;
(h) opposite   **75.** (a) D; (b) minus; (d) $-3.3$ cm; (e) $+0.67$; (f) V;
(g) NI   **77.** (a) C; (b) $+80$ cm; (d) $-20$ cm; (f) V; (g) NI; (h) same
**79.** (a) C; (b) plus; (d) $-13$ cm; (e) $+1.7$; (f) V; (g) NI; (h) same

**81.** (a) +24 cm; (b) +6.0; (c) R; (d) NI; (e) opposite
**83.** (a) +3.1 cm; (b) −0.31; (c) R; (d) I; (e) opposite   **85.** (a) −4.6 cm;
(b) +0.69; (c) V; (d) NI; (e) same   **87.** (a) −5.5 cm; (b) +0.12; (c) V;
(d) NI; (e) same   **89.** (a) 13.0 cm; (b) 5.23 cm; (c) −3.25; (d) 3.13;
(e) −10.2   **91.** (a) 2.35 cm; (b) decrease   **93.** (a) 3.5; (b) 2.5
**95.** (a) +8.6 cm; (b) +2.6; (c) R; (d) NI; (e) opposite
**97.** (a) +7.5 cm; (b) −0.75; (c) R; (d) I; (e) opposite   **99.** (a) +24 cm;
(b) −0.58; (c) R; (d) I; (e) opposite   **105.** (a) 3.00 cm; (b) 2.33 cm
**107.** (a) 40 cm; (b) 20 cm; (c) −40 cm; (d) 40 cm   **109.** (a) 20 cm;
(b) 15 cm   **111.** (a) 6.0 mm; (b) 1.6 kW/m$^2$; (c) 4.0 cm   **113.** 100 cm
**115.** 2.2 mm$^2$   **119.** (a) −30 cm; (b) not inverted; (c) virtual; (d) 1.0
**121.** (a) −12 cm   **123.** (a) 80 cm; (b) 0 to 12 cm   **127.** (a) 8.0 cm;
(b) 16 cm; (c) 48 cm   **129.** (a) $\alpha = 0.500$ rad: 7.799 cm; $\alpha = 0.100$ rad:
8.544 cm; $\alpha = 0.0100$ rad: 8.571 cm; mirror equation: 8.571 cm;
(b) $\alpha = 0.500$ rad: −13.56 cm; $\alpha = 0.100$ rad: −12.05 cm; $\alpha = 0.0100$
rad: −12.00 cm; mirror equation: −12.00 cm   **131.** 42 mm
**133.** (b) $P_n$   **135.** (a) $(0.5)(2 - n)r/(n - 1)$; (b) right   **137.** 2.67 cm
**139.** (a) 3.33 cm; (b) left; (c) virtual; (d) not inverted
**141.** (a) $1 + (25 \text{ cm})/f$; (b) $(25 \text{ cm})/f$; (c) 3.5; (d) 2.5

### Chapter 35

**CP** **1.** $b$ (least $n$), $c$, $a$   **2.** (a) top; (b) bright intermediate illumina-
tion (phase difference is 2.1 wavelengths)   **3.** (a) $3\lambda, 3$; (b) $2.5\lambda, 2.5$
**4.** a and d tie (amplitude of resultant wave is $4E_0$), then b and c tie
(amplitude of resultant wave is $2E_0$)   **5.** (a) 1 and 4; (b) 1 and 4
**Q** **1.** (a) decrease; (b) decrease; (c) decrease; (d) blue   **3.** (a) $2d$;
(b) (odd number)$\lambda/2$; (c) $\lambda/4$   **5.** (a) intermediate closer to
maximum, $m = 2$; (b) minimum, $m = 3$; (c) intermediate closer to
maximum, $m = 2$; (d) maximum, $m = 1$   **7.** (a) maximum;
(b) minimum; (c) alternates   **9.** (a) peak; (b) valley   **11.** $c, d$   **13.** $c$
**P** **1.** (a) 155 nm; (b) 310 nm   **3.** (a) 3.60 $\mu$m; (b) intermediate
closer to fully constructive   **5.** $4.55 \times 10^7$ m/s   **7.** 1.56
**9.** (a) 1.55 $\mu$m; (b) 4.65 $\mu$m   **11.** (a) 1.70; (b) 1.70; (c) 1.30;
(d) all tie   **13.** (a) 0.833; (b) intermediate closer to fully
constructive   **15.** 648 nm   **17.** 16   **19.** 2.25 mm   **21.** 72 $\mu$m
**23.** 0   **25.** 7.88 $\mu$m   **27.** 6.64 $\mu$m   **29.** 2.65   **31.** $27 \sin(\omega t + 8.5°)$
**33.** (17.1 $\mu$V/m) $\sin[(2.0 \times 10^{14} \text{ rad/s})t]$   **35.** 120 nm   **37.** 70.0 nm
**39.** (a) 0.117 $\mu$m; (b) 0.352 $\mu$m   **41.** 161 nm   **43.** 560 nm
**45.** 478 nm   **47.** 509 nm   **49.** 273 nm   **51.** 409 nm   **53.** 338 nm
**55.** (a) 552 nm; (b) 442 nm   **57.** 608 nm   **59.** 528 nm   **61.** 455 nm
**63.** 248 nm   **65.** 339 nm   **67.** 329 nm   **69.** 1.89 $\mu$m   **71.** 0.012°
**73.** 140   **75.** $[(m + \frac{1}{2})\lambda R]^{0.5}$, for $m = 0, 1, 2,...$   **77.** 1.00 m
**79.** 588 nm   **81.** 1.00030   **83.** (a) 50.0 nm; (b) 36.2 nm   **85.** 0.23°
**87.** (a) 1500 nm; (b) 2250 nm; (c) 0.80   **89.** $x = (D/2a)(m + 0.5)\lambda$,
for $m = 0, 1, 2,...$   **91.** (a) 22°; (b) refraction reduces $\theta$   **93.** 600
nm   **95.** (a) 1.75 $\mu$m; (b) 4.8 mm   **97.** $I_m \cos^2(2\pi x/\lambda)$   **99.** (a)
42.0 ps; (b) 42.3 ps; (c) 43.2 ps; (d) 41.8 ps; (e) 4   **101.** 33 $\mu$m
**103.** (a) bright; (b) 594 nm; (c) Primary reason: the colored bands
begin to overlap too much to be distinguished. Secondary reason:
the two reflecting surfaces are too separated for the light reflecting from them to be coherent.

### Chapter 36

**CP** **1.** (a) expand; (b) expand   **2.** (a) second side maximum;
(b) 2.5   **3.** (a) red; (b) violet   **4.** diminish   **5.** (a) left; (b) less
**Q** **1.** (a) $m = 5$ minimum; (b) (approximately) maximum between
the $m = 4$ and $m = 5$ minima   **3.** (a) $A, B, C$; (b) $A, B, C$
**5.** (a) 1 and 3 tie, then 2 and 4 tie; (b) 1 and 2 tie, then 3 and 4 tie
**7.** (a) larger; (b) red   **9.** (a) decrease; (b) same; (c) remain in place
**11.** (a) $A$; (b) left; (c) left; (d) right   **13.** (a) 1 and 2 tie, then 3;
(b) yes; (c) no

**P** **1.** (a) 2.5 mm; (b) $2.2 \times 10^{-4}$ rad   **3.** (a) 70 cm; (b) 1.0 mm
**5.** (a) 700 nm; (b) 4; (c) 6   **7.** 60.4 $\mu$m   **9.** 1.77 mm   **11.** 160°
**13.** (a) 0.18°; (b) 0.46 rad; (c) 0.93   **15.** (d) 52.5°; (e) 10.1°; (f) 5.06°
**17.** (b) 0; (c) −0.500; (d) 4.493 rad; (e) 0.930; (f) 7.725 rad; (g) 1.96
**19.** (a) 19 cm; (b) larger   **21.** (a) $1.1 \times 10^4$ km; (b) 11 km
**23.** (a) $1.3 \times 10^{-4}$ rad; (b) 10 km   **25.** 50 m   **27.** $1.6 \times 10^3$ km
**29.** (a) $8.8 \times 10^{-7}$ rad; (b) $8.4 \times 10^7$ km; (c) 0.025 mm   **31.** (a) 0.346°;
(b) 0.97°   **33.** (a) 17.1 m; (b) $1.37 \times 10^{-10}$   **35.** 5   **37.** 3
**39.** (a) 5.0 $\mu$m; (b) 20 $\mu$m   **41.** (a) $7.43 \times 10^{-3}$; (b) between the
$m = 6$ minimum (the seventh one) and the $m = 7$ maximum (the
seventh side maximum); (c) between the $m = 3$ minimum
(the third one) and the $m = 4$ minimum (the fourth one)
**43.** (a) 9; (b) 0.255   **45.** (a) 62.1°; (b) 45.0°; (c) 32.0°   **47.** 3
**49.** (a) 6.0 $\mu$m; (b) 1.5 $\mu$m; (c) 9; (d) 7; (e) 6   **51.** (a) 2.1°; (b) 21°;
(c) 11   **53.** (a) 470 nm; (b) 560 nm   **55.** $3.65 \times 10^3$
**57.** (a) 0.032°/nm; (b) $4.0 \times 10^4$; (c) 0.076°/nm; (d) $8.0 \times 10^4$;
(e) 0.24°/nm; (f) $1.2 \times 10^5$   **59.** 0.15 nm   **61.** (a) 10 $\mu$m; (b) 3.3 mm
**63.** $1.09 \times 10^3$ rulings/mm   **65.** (a) 0.17 nm; (b) 0.13 nm
**67.** (a) 25 pm; (b) 38 pm   **69.** 0.26 nm   **71.** (a) 15.3°; (b) 30.6°;
(c) 3.1°; (d) 37.8°   **73.** (a) $0.7071a_0$; (b) $0.4472a_0$; (c) $0.3162a_0$;
(d) $0.2774a_0$; (e) $0.2425a_0$   **75.** (a) 625 nm; (b) 500 nm; (c) 416 nm
**77.** 3.0 mm   **83.** (a) 13; (b) 6   **85.** 59.5 pm   **87.** 4.9 km   **89.** $1.36 \times 10^4$
**91.** 2   **93.** 4.7 cm   **97.** 36 cm   **99.** (a) fourth; (b) seventh
**103.** (a) 2.4 $\mu$m; (b) 0.80 $\mu$m; (c) 2   **107.** 9

### Chapter 37

**CP** **1.** (a) same (speed of light postulate); (b) no (the start and
end of the flight are spatially separated); (c) no (because his
measurement is not a proper time)   **2.** (a) Eq. 2; (b) +0.90$c$;
(c) 25 ns; (d) −7.0 m   **3.** (a) right; (b) more   **4.** (a) equal; (b) less
**Q** **1.** $c$   **3.** $b$   **5.** (a) $C_1'$; (b) $C_1'$   **7.** (a) 4 s; (b) 3 s; (c) 5 s; (d) 4 s;
(e) 10 s   **9.** (a) a tie of 3, 4, and 6, then a tie of 1, 2, and 5; (b) 1, then
a tie of 2 and 3, then 4, then a tie of 5 and 6; (c) 1, 2, 3, 4, 5, 6; (d) 2
and 4; (e) 1, 2, 5   **11.** (a) 3, tie of 1 and 2, then 4; (b) 4, tie of 1 and 2,
then 3; (c) 1, 4, 2, 3
**P** **1.** 0.990 50   **3.** (a) 0.999 999 50   **5.** 0.446 ps   **7.** $2.68 \times 10^3$ y
**9.** (a) 87.4 m; (b) 394 ns   **11.** 1.32 m   **13.** (a) 26.26 y;
(b) 52.26 y; (c) 3.705 y   **15.** (a) 0.999 999 15; (b) 30 ly
**17.** (a) 138 km; (b) −374 $\mu$s   **19.** (a) 25.8 $\mu$s; (b) small flash
**21.** (a) $\gamma[1.00 \mu s - \beta(400 \text{ m})/(2.998 \times 10^8 \text{ m/s})]$; (d) 0.750;
(e) $0 < \beta < 0.750$; (f) $0.750 < \beta < 1$; (g) no   **23.** (a) 1.25; (b) 0.800 $\mu$s
**25.** (a) 0.480; (b) negative; (c) big flash; (d) 4.39 $\mu$s   **27.** 0.81$c$
**29.** (a) 0.35; (b) 0.62   **31.** 1.2 $\mu$s   **33.** (a) 1.25 y; (b) 1.60 y; (c) 4.00 y
**35.** 22.9 MHz   **37.** 0.13$c$   **39.** (a) 550 nm; (b) yellow
**41.** (a) 196.695; (b) 0.999 987   **43.** (a) 1.0 keV; (b) 1.1 MeV
**45.** 110 km   **47.** $1.01 \times 10^7$ km   **49.** (a) 0.222 cm; (b) 701 ps;
(c) 7.40 ps   **51.** 2.83$mc$   **53.** $\gamma(2\pi m/|q|B)$; (b) no; (c) 4.85 mm;
(d) 15.9 mm; (e) 16.3 ps; (f) 0.334 ns   **55.** (a) 0.707; (b) 1.41;
(c) 0.414   **57.** 18 smu/y   **59.** (a) 2.08 MeV; (b) −1.21 MeV
**61.** (d) 0.801   **63.** (a) $vt \sin \theta$; (b) $t[1 - (v/c) \cos \theta]$; (c) 3.24$c$
**67.** (b) +0.44$c$   **69.** (a) 1.93 m; (b) 6.00 m; (c) 13.6 ns; (d) 13.6 ns;
(e) 0.379 m; (f) 30.5 m; (g) −101 ns; (h) no; (i) 2; (k) no; (l) both
**71.** (a) $5.4 \times 10^4$ km/h; (b) $6.3 \times 10^{-10}$   **73.** 189 MeV
**75.** $8.7 \times 10^{-3}$ ly   **77.** 7   **79.** 2.46 MeV/$c$   **81.** 0.27$c$
**83.** (a) 5.71 GeV; (b) 6.65 GeV; (c) 6.58 GeV/$c$; (d) 3.11 MeV;
(e) 3.62 MeV; (f) 3.59 MeV/$c$   **85.** 0.95$c$   **87.** (a) 256 kV; (b) 0.745$c$
**89.** (a) 0.858$c$; (b) 0.185$c$   **91.** 0.500$c$   **93.** (a) 119 MeV;
(b) 64.0 MeV/$c$; (c) 81.3 MeV; (d) 64.0 MeV/$c$   **95.** 4.00 u, probably
a helium nucleus   **97.** (a) 534; (b) 0.999 998 25; (c) 2.23 T
**99.** (a) 415 nm; (b) blue   **101.** (a) 88 kg; (b) no   **103.** (a) $3 \times 10^{-18}$;
(b) $2 \times 10^{-12}$; (c) $8.2 \times 10^{-8}$; (d) $6.4 \times 10^{-6}$; (e) $1.1 \times 10^{-6}$;
(f) $3.7 \times 10^{-5}$; (g) $9.9 \times 10^{-5}$; (h) 0.10

## Chapter 38

**CP**   **1.** b, a, d, c   **2.** (a) lithium, sodium, potassium, cesium;
(b) all tie   **3.** (a) same; (b)–(d) x rays   **4.** (a) proton; (b) same;
(c) proton   **5.** same

**Q**   **1.** (a) greater; (b) less   **3.** potassium   **5.** only e   **7.** none
**9.** (a) decreases by a factor of $(1/2)^{0.5}$; (b) decreases by a factor of 1/2
**11.** amplitude of reflected wave is less than that of incident wave
**13.** electron, neutron, alpha particle   **15.** all tie

**P**   **1.** (a) 2.1 $\mu$m; (b) infrared   **3.** $1.0 \times 10^{45}$ photons/s   **5.** 2.047 eV
**7.** $1.1 \times 10^{-10}$ W   **9.** (a) $2.96 \times 10^{20}$ photons/s; (b) $4.86 \times 10^{7}$ m;
(c) $5.89 \times 10^{18}$ photons/m$^2 \cdot$ s   **11.** (a) infrared; (b) $1.4 \times 10^{21}$ photons/s
**13.** $4.7 \times 10^{26}$ photons   **15.** 170 nm   **17.** 676 km/s   **19.** 1.3 V;
(b) $6.8 \times 10^{2}$ km/s   **21.** (a) 3.1 keV; (b) 14 keV   **23.** (a) 2.00 eV;
(b) 0; (c) 2.00 V; (d) 295 nm   **25.** (a) 382 nm; (b) 1.82 eV
**27.** (a) 2.73 pm; (b) 6.05 pm   **29.** (a) $8.57 \times 10^{18}$ Hz; (b) $3.55 \times 10^{4}$ eV;
(c) 35.4 keV/c   **31.** 300%   **33.** (a) $-8.1 \times 10^{-9}$%; (b) $-4.9 \times 10^{-4}$%;
(c) $-8.9$%; (d) $-66$%   **35.** (a) 2.43 pm; (b) 1.32 fm; (c) 0.511 MeV;
(d) 939 MeV   **37.** (a) 41.8 keV; (b) 8.2 keV   **39.** 44°   **41.** (a) 2.43 pm;
(b) $4.11 \times 10^{-6}$; (c) $-8.67 \times 10^{-6}$ eV; (d) 2.43 pm; (e) $9.78 \times 10^{-2}$;
(f) $-4.45$ keV   **43.** (a) $2.9 \times 10^{-10}$ m; (b) x ray; (c) $2.9 \times 10^{-8}$ m;
(d) ultraviolet   **45.** (a) 9.35 $\mu$m; (b) $1.47 \times 10^{-5}$ W; (c) $6.93 \times 10^{14}$
photons/s; (d) $2.33 \times 10^{-37}$ W; (e) $5.87 \times 10^{-19}$ photons/s
**47.** 7.75 pm   **49.** (a) $1.9 \times 10^{-21}$ kg $\cdot$ m/s; (b) 346 fm   **51.** 4.3 $\mu$eV
**53.** (a) 1.24 $\mu$m; (b) 1.22 nm; (c) 1.24 fm; (d) 1.24 fm   **55.** (a) 15 keV;
(b) 120 keV   **57.** neutron   **59.** (a) $3.96 \times 10^{6}$ m/s; (b) 81.7 kV
**67.** $2.1 \times 10^{-24}$ kg $\cdot$ m/s   **71.** (a) $1.45 \times 10^{11}$ m$^{-1}$; (b) $7.25 \times 10^{10}$ m$^{-1}$;
(c) 0.111; (d) $5.56 \times 10^{4}$   **73.** 4.81 mA   **75.** (a) $9.02 \times 10^{-6}$;
(b) 3.0 MeV; (c) 3.0 MeV; (d) $7.33 \times 10^{-8}$; (e) 3.0 MeV; (f) 3.0 MeV
**77.** (a) $-20$%; (b) $-10$%; (c) $+15$%   **79.** (a) no; (b) plane wavefronts of infinite extent, perpendicular to x axis   **83.** (a) 38.8 meV;
(b) 146 pm   **85.** (a) $4.14 \times 10^{-15}$ eV $\cdot$ s; (b) 2.31 eV   **89.** (a) no;
(b) 544 nm; (c) green

## Chapter 39

**CP**   **1.** b, a, c   **2.** (a) all tie; (b) $a, b, c$   **3.** a, b, c, d   **4.** $E_{1,1}$ (neither $n_x$
nor $n_y$ can be zero)   **5.** (a) 5; (b) 7
**Q**   **1.** a, c, b   **3.** (a) 18; (b) 17   **5.** equal   **7.** c   **9.** (a) decrease;
(b) increase   **11.** $n = 1, n = 2, n = 3$   **13.** (a) $n = 3$; (b) $n = 1$;
(c) $n = 5$   **15.** b, c, and d
**P**   **1.** 1.41   **3.** 0.65 eV   **5.** 0.85 nm   **7.** 1.9 GeV   **9.** (a) 72.2 eV;
(b) 13.7 nm; (c) 17.2 nm; (d) 68.7 nm; (e) 41.2 nm; (g) 68.7 nm;
(h) 25.8 nm   **11.** (a) 13; (b) 12   **13.** (a) 0.020; (b) 20   **15.** (a) 0.050;
(b) 0.10; (c) 0.0095   **17.** 56 eV   **19.** 109 eV   **23.** 3.21 eV
**25.** $1.4 \times 10^{-3}$   **27.** (a) 8; (b) 0.75; (c) 1.00; (d) 1.25; (e) 3.75; (f) 3.00;
(g) 2.25   **29.** (a) 7; (b) 1.00; (c) 2.00; (d) 3.00; (e) 9.00; (f) 8.00;
(g) 6.00   **31.** 4.0   **33.** (a) 12.1 eV; (b) $6.45 \times 10^{-27}$ kg $\cdot$ m/s; (c) 102
nm   **35.** (a) 291 nm$^{-3}$; (b) 10.2 nm$^{-1}$   **41.** (a) 0.0037; (b) 0.0054
**43.** (a) 13.6 eV; (b) $-27.2$ eV   **45.** (a) $(r^4/8a^5)[\exp(-r/a)] \cos^2 \theta$;
(b) $(r^4/16a^5)[\exp(-r/a)] \sin^2 \theta$   **47.** $4.3 \times 10^{3}$   **49.** (a) 13.6 eV;
(b) 3.40 eV   **51.** 0.68   **59.** (b) $(2\pi/h)[2m(U_0 - E)]^{0.5}$
**61.** (b) meter$^{-2.5}$   **63.** (a) $n$; (b) $2\ell + 1$; (c) $n^2$   **65.** (a) $nh/\pi md^2$;
(b) $n^2 h^2/4\pi^2 md^2$   **67.** (a) $3.9 \times 10^{-22}$ eV; (b) $10^{20}$; (c) $3.0 \times 10^{-18}$ K
**71.** (a) $e^2 r/4\pi\varepsilon_0 a^3$; (b) $e/(4\pi\varepsilon_0 ma_0^3)^{0.5}$   **73.** 18.1, 36.2, 54.3, 66.3,
72.4 $\mu$eV

## Chapter 40

**CP**   **1.** 7   **2.** (a) decrease; (b)–(c) remain the same   **3.** $A, C, B$
**Q**   **1.** (a) 2; (b) 8; (c) 5; (d) 50   **3.** all true   **5.** same number (10)
**7.** 2, $-1, 0,$ and 1   **9.** (a) 2; (b) 3   **11.** (a) $n$; (b) $n$ and $\ell$   **13.** In addition to the quantized energy, a helium atom has kinetic energy; its
total energy can equal 20.66 eV.

**P**   **1.** 24.1°   **3.** (a) $3.65 \times 10^{-34}$ J $\cdot$ s; (b) $3.16 \times 10^{-34}$ J $\cdot$ s   **5.** (a) 3;
(b) 3   **7.** (a) 4; (b) 5; (c) 2   **9.** (a) 3.46; (b) 3.46; (c) 3; (d) 3; (e) $-3$;
(f) 30.0°; (g) 54.7°; (h) 150°   **13.** 72 km/s$^2$   **15.** (a) 54.7°; (b) 125°
**17.** 19 mT   **19.** 5.35 cm   **21.** 44   **23.** 42   **25.** (a) 51; (b) 53; (c) 56
**27.** (a) $(2, 0, 0, +\frac{1}{2}), (2, 0, 0, -\frac{1}{2})$; (b) $(2, 1, 1, +\frac{1}{2}), (2, 1, 1, -\frac{1}{2})$,
$(2, 1, 0, +\frac{1}{2}), (2, 1, 0, -\frac{1}{2}), (2, 1, -1, +\frac{1}{2}), (2, 1, -1, -\frac{1}{2})$   **29.** $g$
**31.** (a) $4p$; (b) 4; (c) $4p$; (d) 5; (e) $4p$; (f) 6   **33.** 12.4 kV   **35.** (a) 35.4 pm;
(b) 56.5 pm; (c) 49.6 pm   **39.** 0.563   **41.** 80.3 pm   **43.** (a) 69.5 kV;
(b) 17.8 pm; (c) 21.3 pm; (d) 18.5 pm   **45.** (a) 49.6 pm; (b) 99.2 pm
**47.** $2.0 \times 10^{16}$ s$^{-1}$   **49.** $2 \times 10^{7}$   **51.** $9.0 \times 10^{-7}$   **53.** $7.3 \times 10^{15}$ s$^{-1}$
**55.** (a) 3.60 mm; (b) $5.24 \times 10^{17}$   **57.** (a) 0; (b) 68 J   **59.** 3.0 eV
**61.** (a) $3.03 \times 10^{5}$; (b) 1.43 GHz; (d) $3.31 \times 10^{-6}$   **63.** 186
**65.** (a) 2.13 meV; (b) 18 T   **69.** (a) no; (b) 140 nm   **71.** $n > 3$;
$\ell = 3; m_\ell = +3, +2, +1, 0, -1, -2, -3; m_s = \pm\frac{1}{2}$   **73.** (a) 6.0;
(b) $3.2 \times 10^{6}$ y   **75.** argon   **79.** $(Ze/4\pi\varepsilon_0)(r^{-2} - rR^{-3})$

## Chapter 41

**CP**   **1.** larger   **2.** a, b, and c
**Q**   **1.** b, c, d (the latter due to thermal expansion)   **3.** 8
**5.** below   **7.** increase   **9.** much less than   **11.** b and d
**P**   **3.** $8.49 \times 10^{28}$ m$^{-3}$   **5.** (b) $6.81 \times 10^{27}$ m$^{-3}$ eV$^{-3/2}$;
(c) $1.52 \times 10^{28}$ m$^{-3}$ eV$^{-1}$   **7.** (a) 0; (b) 0.0955   **9.** (a) $5.86 \times 10^{28}$ m$^{-3}$;
(b) 5.49 eV; (c) $1.39 \times 10^{3}$ km/s; (d) 0.522 nm   **11.** (a) $1.36 \times
10^{28}$ m$^{-3}$ eV$^{-1}$; (b) $1.68 \times 10^{28}$ m$^{-3}$ eV$^{-1}$; (c) $9.01 \times 10^{27}$ m$^{-3}$ eV$^{-1}$;
(d) $9.56 \times 10^{26}$ m$^{-3}$ eV$^{-1}$; (e) $1.71 \times 10^{18}$ m$^{-3}$ eV$^{-1}$   **13.** (a) 6.81 eV;
(b) $1.77 \times 10^{28}$ m$^{-3}$ eV$^{-1}$; (c) $1.59 \times 10^{28}$ m$^{-3}$ eV$^{-1}$
**15.** (a) $2.50 \times 10^{3}$ K; (b) $5.30 \times 10^{3}$ K   **17.** 3   **19.** (a) 1.0; (b) 0.99;
(c) 0.50; (d) 0.014; (e) $2.4 \times 10^{-17}$; (f) $7.0 \times 10^{2}$ K   **21.** (a) 0.0055;
(b) 0.018   **25.** (a) 19.7 kJ; (b) 197 s   **27.** (a) $1.31 \times 10^{29}$ m$^{-3}$;
(b) 9.43 eV; (c) $1.82 \times 10^{3}$ km/s; (d) 0.40 nm   **29.** 57.1 kJ
**31.** (a) 226 nm; (b) ultraviolet   **33.** (a) $1.5 \times 10^{-6}$; (b) $1.5 \times 10^{-6}$
**35.** 0.22 $\mu$g   **37.** (a) $4.79 \times 10^{-10}$; (b) 0.0140; (c) 0.824   **39.** $6.0 \times 10^{5}$
**41.** 4.20 eV   **43.** 13 $\mu$m   **47.** (a) 109.5°; (b) 238 pm
**49.** (b) $1.8 \times 10^{28}$ m$^{-3}$ eV$^{-1}$   **53.** $3.49 \times 10^{3}$ atm

## Chapter 42

**CP**   **1.** $^{90}$As and $^{158}$Nd   **2.** a little more than 75 Bq (elapsed time is
a little less than three half-lives)   **3.** $^{206}$Pb
**Q**   **1.** (a) $^{196}$Pt; (b) no   **3.** yes   **5.** (a) less; (b) greater   **7.** $^{240}$U
**9.** no effect   **11.** yes   **13.** (a) all except $^{198}$Au; (b) $^{132}$Sn and $^{208}$Pb
**15.** d
**P**   **1.** $1.3 \times 10^{-13}$ m   **3.** 46.6 fm   **5.** (a) 0.390 MeV; (b) 4.61 MeV
**7.** (a) $2.3 \times 10^{17}$ kg/m$^3$; (b) $2.3 \times 10^{17}$ kg/m$^3$; (d) $1.0 \times 10^{25}$ C/m$^3$;
(e) $8.8 \times 10^{24}$ C/m$^3$   **9.** (a) 6; (b) 8   **11.** (a) 6.2 fm; (b) yes
**13.** 13 km   **17.** 1.0087 u   **19.** (a) 9.303%; (b) 11.71%
**21.** (b) 7.92 MeV/nucleon   **25.** $5.3 \times 10^{22}$   **27.** (a) 0.250; (b) 0.125
**29.** (a) 64.2 h; (b) 0.125; (c) 0.0749   **31.** (a) $7.5 \times 10^{16}$ s$^{-1}$;
(b) $4.9 \times 10^{16}$ s$^{-1}$   **33.** $1 \times 10^{13}$ atoms   **37.** 265 mg
**39.** (a) $8.88 \times 10^{10}$ s$^{-1}$; (b) $1.19 \times 10^{15}$; (c) 0.111 $\mu$g   **41.** $1.12 \times 10^{11}$ y
**43.** $9.0 \times 10^{8}$ Bq   **45.** (a) $3.2 \times 10^{12}$ Bq; (b) 86 Ci   **47.** (a) $2.0 \times 10^{20}$;
(b) $2.8 \times 10^{9}$ s$^{-1}$   **49.** (a) $1.2 \times 10^{-17}$; (b) 0   **51.** 4.269 MeV
**53.** 1.21 MeV   **55.** 0.783 MeV   **57.** (b) 0.961 MeV   **59.** 78.3 eV
**61.** (a) $1.06 \times 10^{19}$; (b) $0.624 \times 10^{19}$; (c) $1.68 \times 10^{19}$; (d) $2.97 \times 10^{9}$ y
**63.** 1.7 mg   **65.** 1.02 mg   **67.** 2.50 mSv   **69.** (a) $6.3 \times 10^{18}$;
(b) $2.5 \times 10^{11}$; (c) 0.20 J; (d) 2.3 mGy; (e) 30 mSv   **71.** (a) 6.6 MeV;
(b) no   **73.** (a) 25.4 MeV; (b) 12.8 MeV; (c) 25.0 MeV   **75.** $^7$Li
**77.** $3.2 \times 10^{4}$ y   **79.** 730 cm$^2$   **81.** $^{225}$Ac   **83.** 30 MeV   **89.** 27
**91.** (a) 11.906 83 u; (b) 236.2025 u   **93.** 600 keV   **95.** (a) 59.5 d;
(b) 1.18   **97.** (a) $4.8 \times 10^{-18}$ s$^{-1}$; (b) $4.6 \times 10^{9}$ y

## Chapter 43

**CP** **1.** c and d **2.** e

**Q** **1.** (a) 101; (b) 42 **3.** $^{239}$Np **5.** $^{140}$I, $^{105}$Mo, $^{152}$Nd, $^{123}$In, $^{115}$Pd
**7.** increased **9.** less than **11.** still equal to 1

**P** **1.** (a) 16 day$^{-1}$; (b) $4.3 \times 10^8$ **3.** 4.8 MeV **5.** $1.3 \times 10^3$ kg
**7.** $3.1 \times 10^{10}$ s$^{-1}$ **9.** (a) $2.6 \times 10^{24}$; (b) $8.2 \times 10^{13}$ J; (c) $2.6 \times 10^4$ y
**11.** $-23.0$ MeV **13.** (a) 253 MeV; (b) typical fission energy is
200 MeV **15.** (a) 84 kg; (b) $1.7 \times 10^{25}$; (c) $1.3 \times 10^{25}$ **17.** (a) $^{153}$Nd;
(b) 110 MeV; (c) 60 MeV; (d) $1.6 \times 10^7$ m/s; (e) $8.7 \times 10^6$ m/s
**21.** 557 W **23.** 0.99938 **25.** (b) 1.0; (c) 0.89; (d) 0.28; (e) 0.019;
(f) 8 **27.** (a) 75 kW; (b) $5.8 \times 10^3$ kg **29.** $1.7 \times 10^9$ y
**31.** 170 keV **33.** 1.41 MeV **35.** $10^{-12}$ m **37.** (a) $4.3 \times 10^9$ kg/s;
(b) $3.1 \times 10^{-4}$ **41.** $1.6 \times 10^8$ y **43.** (a) 24.9 MeV; (b) 8.65 mega-
tons TNT **45.** (a) $1.8 \times 10^{38}$ s$^{-1}$; (b) $8.2 \times 10^{28}$ s$^{-1}$ **47.** (a) 4.1
eV/atom; (b) 9.0 MJ/kg; (c) $1.5 \times 10^3$ y **49.** 14.4 kW
**51.** $^{238}$U + n $\rightarrow$ $^{239}$U $\rightarrow$ $^{239}$Np + e + $\nu$, $^{239}$Np $\rightarrow$ $^{239}$Pu + e + $\nu$ **55.**
(a) $3.1 \times 10^{31}$ protons/m$^3$; (b) $1.2 \times 10^6$ **57.** (a) 227 J; (b) 49.3 mg;
(c) 22.7 kW

## Chapter 44

**CP** **1.** (a) the muon family; (b) a particle; (c) $L_\mu = +1$
**2.** b and e **3.** c

**Q** **1.** b, c, d **3.** (a) 1; (b) positively charged **5.** a, b, c, d **7.** d
**9.** c **11.** (a) lepton; (b) antiparticle; (c) fermion; (d) yes

**P** **1.** $\pi^- \rightarrow \mu^- + \bar{\nu}$ **3.** 2.4 pm **5.** $2.4 \times 10^{-43}$ **7.** 769 MeV
**9.** 2.7 cm/s **11.** (a) angular momentum, $L_e$; (b) charge, $L_\mu$;
(c) energy, $L_\mu$ **15.** (a) energy; (b) strangeness; (c) charge
**17.** (a) yes; (b)−(d) no **19.** (a) 0; (b) −1; (c) 0 **21.** (a) K$^+$; (b) $\bar{n}$;
(c) K$^0$ **23.** (a) 37.7 MeV; (b) 5.35 MeV; (c) 32.4 MeV **25.** (a) $\bar{u}\bar{u}\bar{d}$;
(b) $\bar{u}\bar{d}\bar{d}$ **27.** s$\bar{d}$ **29.** (a) $\Xi^0$; (b) $\Sigma^-$ **31.** $2.77 \times 10^8$ ly **33.** 668 nm
**35.** $1.4 \times 10^{10}$ ly **37.** (a) 2.6 K; (b) 976 nm **39.** (b) 5.7 H atoms/m$^3$
**41.** $4.57 \times 10^3$ **43.** (a) 121 m/s; (b) 0.00406; (c) 248 y
**47.** $1.08 \times 10^{42}$ J **49.** (a) 0.785$c$; (b) 0.993$c$; (c) C2; (d) C1;
(e) 51 ns; (f) 40 ns **51.** (c) $r\alpha/c + (r\alpha/c)^2 + (r\alpha/c)^3 + \cdots$;
(d) $r\alpha/c$; (e) $\alpha = H$; (f) $6.5 \times 10^8$ ly; (g) $6.9 \times 10^8$ y; (h) $6.5 \times 10^8$ y;
(i) $6.9 \times 10^8$ ly; (j) $1.0 \times 10^9$ ly; (k) $1.1 \times 10^9$ y; (l) $3.9 \times 10^8$ ly
**53.** (a) ss$\bar{d}$; (b) $\bar{s}\bar{s}\bar{d}$

# INDEX

Figures are noted by page numbers in *italics*, tables are indicated by t following the page number.

**A**

$a_g$ (gravitational acceleration), 360, 360t

absolute pressure, 390

absolute zero, 515

absorption:
  of heat, 522–527, *523*
  photon, *see* photon absorption

absorption lines, 1206, *1207*

ac (alternating current), 903, 913

acceleration, 20–30, 283t
  average, 20
  centripetal, 76
  constant, *23,* 23–27, 24t
  free-fall, *27,* 27–28
  graphical integration in motion analysis, *29,* 29–30
  instantaneous, 20–22, *21,* 67–69
  negative, 21–22
  and Newton's first law, 95–98
  Newton's laws applied to, 108–113
  and Newton's second law, 98–101
  principle of equivalence (with gravitation), 374–375
  projectile motion, 70–75
  reference particle, 429
  relating linear to angular, *269,* 269–270
  relative motion in one dimension, 79
  relative motion in two dimensions, 79–80
  rockets, 241–243, *242*
  rolling down ramp, *299,* 299–300
  sign of, 21–22
  simple harmonic motion, 418, *418*
  system of particles, 220–223
  two- and three-dimensional motion, 79–80
  uniform circular motion, *76,* 76–78, *77,* 133
  as vector quantity, 41
  yo-yo, 302

acceleration amplitude, in simple harmonic motion, 418

acceleration vectors, 41

accelerators, 818–819, 1334–1336, *1336*

acceptor atoms, 1264

acre-foot, 9

action at a distance, 630

activity, of radioactive sample, 1287

addition:
  of vectors by components, *46,* 46–47, 49
  of vectors geometrically, *41,* 41–42, *42,* 44

adiabat, 571, *572*

adiabatic expansion, 531–532, *532*
  ideal gas, 571–575, *572*

adiabatic processes:
  first law of thermodynamics for, 531–533, 532t
  summarized, *575,* 575t

adiabatic wind, 580

air:
  bulk modulus, 480–481
  density, 387t
  dielectric properties at 1 atm, 732, 732t
  and drag force, 130–132
  effect on projectile motion, *73,* 73–74
  electric breakdown, 646, *646*
  index of refraction at STP, 992t
  speed of sound in, 480–481, 481t
  terminal speeds in, 131t
  thermal conductivity, 535t
  thin-film interference of water film in, 1067

air conditioners, 596

airplanes:
  projectile dropped from, 74
  turns by top gun pilots, 77–78
  two-dimensional relative motion of, 80–81
  vector components for flight, 44

airspeed, 90

alkali metals, 1235

alpha decay, 1289–1291, *1290*

alpha particles, 621, 705, *1277,* 1277–1279, 1289
  binding energy per nucleon, *1283*
  magic nucleon number, 1299
  radiation dosage, 1296–1297
  in thermonuclear fusion, 1324–1325

alternating current (ac), 903, 913

alternating current circuits, 903–934
  damped oscillations in *RLC,* 910–912, *911*
  forced oscillations, 912–920, *914*
  inductive load, *919*
  *LC* oscillations, 903–910, *904*
  phase and amplitude relationships, 920t
  power in, 927–929
  resistive load, *915*
  series *RLC* circuits, 921–926, *922*
  in transformers, 930–933

alternating current generator, 913–914
  with capacitive load, 916–918, *917*
  with inductive load, 918–919, *919*
  with resistive load, 914–916, *915*

ammeters, 788, *788*

ampere (unit), 614, 746, 843

Ampère, André-Marie, 844

Ampere–Maxwell law, 944–945, 949t

Ampere's law, 844–850
  Amperian loop, *844,* 844–848

amplitude:
  alternating current, 920t
  current, *922,* 922–923, 926
  of emf in ac, 914
  exponentially decaying in *RLC* circuits, 911
  *LC* oscillations, 905
  simple harmonic motion, 416–418, *417*
  waves, 447, *447,* 448, *448*

amplitude ratio, traveling electromagnetic waves, 976

amusement park rides:
  roller coasters, 21
  Rotor, 267–268

analyzer, 988

Andromeda Galaxy, 354–355, *355*

anechoic chamber, 513

angles, 45
  angle between two vectors, 54
  degrees and radian measures, 45
  vector, 43, *43,* 45

angled force, applied to initially stationary block, 128

angle of incidence, 991, *991*

angle of minimum deviation, 1005, 1007

angle of reflection, 991, *991*

angle of refraction, 991, *991*

angular acceleration, 261, 283t
  relating, to linear, *269,* 269–270
  rolling wheel, *299,* 300
  rotation with constant, 266–268

angular amplitude (simple pendulum), 426

angular displacement, *259,* 260, 265

angular frequency:
  circulating charged particle, 815
  damped harmonic oscillator, 430–432
  driving, 914
  *LC* oscillations, 908–909
  natural, 433, 914
  simple harmonic motion, 414–418, *417*

simple pendulum, 426
sound waves, 483
waves, 448

angular magnification:
  compound microscope, 1032
  refracting telescope, 1033
  simple magnifying lens, 1031

angular momentum, 305–318, 312t
  atoms, 1221, *1221*
  conservation of, 312–316, *313, 314*
  defined, *305,* 305–316
  at equilibrium, 328
  intrinsic, 953, 954
  Newton's second law in angular form, 307–308
  nuclear, 1284
  orbital, 954, *955,* 1222–1224, *1223,* 1223t
  rigid body rotating about fixed axis, *311,* 311–312
  sample problems involving, 306, 308–309, 315–316
  spin, 953–954, 1223t, 1224, *1225*
  system of particles, 310–311

angular motion, 259

angular position, *259,* 259–260, 283t
  relating, to linear, 269

angular simple harmonic motion, *423,* 423–424

angular simple harmonic oscillator, *423,* 423–424

angular speed, 261, 262
  relating, to linear, 268–270
  in rolling, 295–297, *296*

angular velocity, 260–264, 283t
  average, 260–261
  instantaneous, 260
  vector nature of, 264–265, *265*

angular wave number, 447, 1171
  sound waves, 483

annihilation:
  electron–positron, 622, *622,* 1338
  particle–antiparticle, 1338
  proton–antiproton, 1339–1340, 1340t

annular cylinder, rotational inertia for, 274t

antenna, 974, *974*

antiderivative, 26

antihydrogen, 1338, 1340, 1340t

antimatter, 1310t, 1338–1339

antineutrino, 1292n

antinodes, *465,* 466, 467–468

antiparticles, 1338–1341, 1359

antiprotons, 1338

antisolar point, 994, *994*

aphelion distance, 371
apparent weight, 104
   in fluids, 396–397
applied force, work and, 688–689
Archimedes' principle, 394–397,
   *395*
areas, law of, *369,* 369–370
area vector, 661
astronomical Doppler effect,
   1135–1136
astronomical unit, 12
atmosphere (atm), 388
atmospheric pressure, 388t
atmospheric sprites, 637–638
atoms, 1186–1187, 1219–1246.
        *See also* electrons; neutrons;
        protons
   Bohr model, *1203,* 1203–1204
   exclusion principle in, 1230
   formation in early universe,
      1360
   and lasers, 1240–1245
   magnetic resonance,
      1229–1230, *1230*
   matter wave interference, *1167,*
      1168
   and multiple electrons in a
      trap, 1230–1234
   and periodic table, 1234–1236
   properties of, 1219–1225
   Stern–Gerlach experiment,
      *1226,* 1226–1228
   x rays and ordering of
      elements, 1236–1240
atoms, elasticity of, 339, *339*
atomic bomb, 1284, 1314–1315,
   1326–1327
atomic clocks, 5–6
atomic clocks, time dilation tests,
   1123–1124
atomic mass, 1280t, 1282–1283
atomic mass units, 7, 1282–1283
atomic number, 621, 1225, 1280
attractive forces, 356, 611
Atwood's machine, 120
aurora, 610
automobile(s). *See also* race cars
   acceleration of motorcycle vs.,
      25–26
   average velocity of truck, 17
   in banked circular turn,
      137–138
   in flat circular turn, 136–137
   magnet applications, 804
   sliding to stop on icy roads,
      129–130
   spark discharge from, 707, *707*
   tire pressure, 388t
average acceleration:
   one-dimensional motion, 20
   two- and three-dimensional
      motion, 67–69
average angular acceleration, 261
average angular velocity, 260–261
average force (of collision), 228
average life, radionuclide,
   1287–1288

average power, 166, 197–198
   engines, 594
   traveling wave on stretched
      string, 455
average speed:
   of gas molecules, 561–563
   one-dimensional motion, 16
average velocity:
   constant acceleration, 24
   one-dimensional motion,
      15–17, *16*
   two- and three-dimensional
      motion, 65
Avogadro's number, 550, 748
axis(--es):
   rotating, of vectors, *47*
   of rotation, 259, *259*
   separation of, in Newton's
      second law, 98–99
   of symmetry, 632

**B**
Babinet's principle, 1109
background noise, 508
ball, motion of, 70–72, *71, 72*
ballet dancing:
   grand jeté, 221–222, *222*
   tour jeté, *314*
ballistic pendulum, 236, *236*
balloons, lifting capacity, 581
Balmer series, 1203, 1206, *1207*
bands, energy bands in crystalline
      solids, 1254, *1254*
band–gap pattern:
   crystalline solid, *1254*
   insulator, *1254*
   metal, *1255*
   semiconductor, *1262*
bar magnets:
   Earth as, 950, *950*
   magnetic dipole moment of
      small, 826, 826t
   magnetic field, 942, *942*
   magnetic field lines, 806–807,
      *807*
barrel units, 11
barrier tunneling, 1176–1179,
   *1177,* 1290–1291
baryons, 1338, 1345–1346
   conservation of baryon
      number, 1345
   and eightfold way, 1347–1348,
      1347t
   and quark model, 1349, 1355
baryonic matter, 1358, 1361, *1361*
baryon number, conservation
      of, 1345
baseball:
   collision of ball with bat, 226,
      *226, 227*
   fly ball, air resistance to, 73, *73,*
      73t
   time of free-fall flight, 28
base quantities, 2
base standards, 2
basic equations for constant
      acceleration, 23–24

basilisk lizards, 249, *249*
basketball free throws, 62
bats, navigation using ultrasonic
      waves, 502
batteries. *See also* electromotive
      force (emf)
   connected to capacitors, *718,*
      718–719, 727–728
   and current, *746,* 746–747
   as emf devices, 772–774
   in multiloop circuits, *781,*
      781–787
   multiple batteries in multiloop
      circuit, 784–785, *785*
   potential difference across,
      777–780, *779*
   and power in circuits, *760,*
      760–761
   in *RC* circuits, 788–792, *789*
   real, 773, *773, 777,* 777–778
   rechargeable, 773–774
   recharging, 779
   in *RL* circuits, 883–886
   in single-loop circuits, 774–775
   work and energy, *773,* 773–774
beam, 976
beam expander, 1044
beam separation, in
      Stern–Gerlach experiment,
      1228
beam splitter, 1071, 1164, *1164*
beats, 496–498, *497*
becquerel, 1287
bends, the, 407, 549
Bernoulli's equation, 401–404
beta decay, 627, 1292–1295, *1293,*
      1351
beta-minus decay, 1292
beta-plus decay, 1292
bi-concave lens, 1044
bi-convex lens, 1044
bicycle wheels:
   rolling, 295–297, *296–297*
   rolling, with friction, *299,*
      299–300
bifurcate (term), 58
Big Bang, 1355–1356, 1358–1361,
   *1359*
billiard balls, Newton's second law
      and motion of, 221
binding energy, *see* nuclear
      binding energy
Biot–Savart law, 837–838, 844,
      852
bivalent atom, 1256
blackbody radiator, 536
black holes, 355
   event horizon, 362
   gravitational lensing caused by,
      375, *376*
   miniature, 379
   supermassive, 355
blocks:
   acceleration of falling, 281
   connected to massless-
      frictionless pulleys, *105,* 106,
      *108,* 108–109
   floating, 397

forces on stationary, 125–126,
   *125–126*
   friction of sliding, 105, *105*
   hanging and sliding, *108,*
      108–109
   Newton's laws applied to, 99,
      108–113
   normal forces, *104,* 104–105
   power used in work on, 168, *168*
   stable static equilibrium,
      328–329, *329,* 332–337
   third-law force pair, *106,*
      106–107
   work done by external force
      with friction, 192–193, *193*
block-spring oscillator, 907–908
block-spring systems:
   damped oscillating systems,
      *430,* 430–431
   and electrical–mechanical
      analogy, 906–907, 906t
   kinetic energy, *159,* 159–162,
      *161*
   oscillating systems, 420–421
   potential energy, 179, *179,*
      182–183
blood pressure, normal systolic,
      387t
blue shift, 1135
bob, of pendulum, 425
body armor, 477–478, *478*
body diagonal, 58–59
body wave, 512
Bohr, Niels, 1193, 1298, 1312
Bohr magneton, 953–955, 1224
Bohr model, of hydrogen, 629,
   *1203,* 1203–1204
Bohr radius, 1204, 1211
boiling point, 526
   for selected substances, 526t
   of water, 518t
Boltzmann, Ludwig, 601, 1243
Boltzmann constant, 551, 1165
Bose, Satyendra Nath, 1337
Bose–Einstein condensate, 1337,
   *1337*
bosons, 1337, *1337*
bottomness, 1346
bottom quark, 1350t, 1351
boundary condition, 1175, 1210
Bragg angle, 1106
Bragg's law, 1106
Brahe, Tycho, 369
branches, circuits, 781
breakdown potential, 732
breakeven, in magnetic confine-
      ment, 1328
Brewster angle, 998, *998*
Brewster's law, 998
bright fringes:
   double-slit interference, 1055,
      *1055,* 1056
   single-slit diffraction, *1083,*
      1083–1085
British thermal unit (Btu),
      524–525
Brookhaven accelerator, 1335

Brout, Robert, 1354
bubble chambers, 622, *622*, 806, *806*
  gamma ray track, 1169, *1169*
  proton–antiproton annihilation event, *1339*, 1339–1340
buildings:
  mile-high, 380
  natural angular frequency, 433
  swaying in wind, 422–424, 468
bulk modulus, 341, 480–481
bungee-cord jumping, 178, *178*
buoyant force, 394–397, *395*

**C**
*c, see* speed of light
Calorie (Cal) (nutritional), 524–525
calorie (cal) (heat), 524–525
cameras, 1030
canal effect, 410
cancer radiation therapy, 1276
capacitance, 717–738
  calculating, 719–723
  of capacitors, 717–718
  of capacitors with dielectrics, 731–734
  and dielectrics/Gauss' law, *735*, 735–737
  and energy stored in electric fields, 728–730
  *LC* oscillations, 903–910
  for parallel and series capacitors, 723–728
  parallel circuits, 783t
  *RC* circuits, 788–792, *789*
  *RLC* circuits, 910–912
  *RLC* series circuits, 921–926
  series circuits, 783t
capacitive reactance, 917
capacitive time constant, for *RC* circuits, *789*, 790
capacitors, *717*, 717–719, *718*. *See also* parallel-plate capacitors
  with ac generator, 916–918, *917*
  capacitance of, 717–718
  charging, 718–719, 727–728, *789*, 789–790, 994
  cylindrical, *721*, 721–722
  with dielectrics, *731*, 731–733
  discharging, 719, *789*, 790–792
  displacement current, *947*, 947–949
  electric field calculation, 720
  energy density, 730
  Faraday's, *731*, 731–732
  induced magnetic field, 944–946
  isolated spherical, 722, 730
  *LC* oscillations, *904*, 905–906
  in parallel, 724, *724*, 726–727, 783t
  and phase/amplitude for ac circuits, 920t
  potential difference calculation, 719–723
  *RC* circuits, 788–792, *789*

in series, 724–727, *725*, 783t, 922, *922*
  series *RLC* circuits, 922
  variable, 742
cars, *see* automobiles
carbon cycle, 1333
carbon¹⁴ dating, 1295
carbon dioxide:
  molar specific heat at constant volume, 565t
  RMS speed at room temperature, 556t
carbon disulfide, index of refraction, 992t
Carnot cycle, 591, *591*, *592*
Carnot engines, 590–593, *591*
  efficiency, 592–593, 597–598
  real vs., 597–598
Carnot refrigerators, 596, 597–598
carrier charge density, 750. *See also* current density
cascade, decay process, 1348–1349
cat, terminal speed of falling, *131*, 131–132
cathode ray tube, *809*, 809–810
cavitation, 508
Celsius temperature scale, *518*, 518–519
center of curvature:
  spherical mirrors, 1015, *1015*
  spherical refracting surfaces, 1020–1021, *1021*
center of gravity, 330–332, *331*
center of mass, 216–219
  and center of gravity, 330–332
  defined, 215
  motion of system's, 220–221
  one-dimensional inelastic collisions, 234–236, *235*
  rolling wheel, 296, *296*
  sample problems involving, 217–218, 223
  solid bodies, 216–219, *219*
  system of particles, *215*, 215–216, 220–223
center of momentum frame, 1151
center of oscillation (physical pendulum), 427
centigrade temperature scale, 518–519
central axis, spherical mirror, 1015, *1016*
central configuration peak, *600*
central diffraction maximum, 1089, *1089*
central interference maximum, 1056
central line, 1099
central maximum, diffraction patterns, 1082, *1082*, 1086–1087
centripetal acceleration, 76
centripetal force, 133–138, *134*
Cerenkov counters, 1366
Ceres, escape speed for, 367t
CERN accelerator, 1335, 1353
  antihydrogen, 1338
  pion beam experiments, 1118

chain-link conversion, of units, 3
chain reaction:
  of elastic collisions, 239–240
  nuclear, 1315
characteristic x-ray spectrum, 1237–1238, *1238*
charge, *see* electric charge
charge carriers, 747
  doped semiconductors, 1263–1265
  silicon vs. copper, 762–763, 762t
charge density. *See also* current density
  carrier, 750
  linear, 638–639, 639t
  surface, 629, 639t
  volume, 626, 628, 639t
charged disk:
  electric field due to, 643–644
  electric potential due to, 700, *700*
charged isolated conductor:
  with cavity, 668, 669
  electric potential, *706*, 706–707
  in external electric field, 707, *707*
  Gauss' law for, 668–670
charge distributions:
  circular arc, 642
  continuous, 638–639, 698–700, *699*, *700*
  ring, 638–640, *639*, 642
  spherically symmetric, 675–677, *676*, 695
  straight line, 642–643
  uniform, *631*, 631–632, *632*, 642–643
charged objects, 631
charged particles, 612
  in cyclotron, 819
  electric field due to, *633*, 633–635
  electric potential due to group of, 695–696, *696*
  electric potential energy of system, 703–705, *704*
  equilibrium of forces on, 618
  helical paths of, *816*, 816–817
  magnetic field due to, 804–805
  motion, in electric field, 647
  net force due to, 616–618
charged rod, electric field of, 641–642
charge number, 1225
charge quantum number, 1341
charging:
  of capacitors, 718–719, 727–728, *789*, 789–790, 944
  electrostatic, 611
charm, 1346
charm quark, 1350t, 1351, 1352
chip (integrated circuits), 1271
chromatic aberration, 1033
chromatic dispersion, *993*, 993–994
circuits, *718*, 719, 771–793, 783t. *See also* alternating current circuits

ammeter and voltmeter for measuring, 788
capacitive load, 916–918, *917*
direct-current (dc), 772
inductive load, 918–919, *919*
integrated, *1270*, 1271
multiloop, 774, *781*, 781–787, *782*
oscillating, 903
parallel capacitors, 724, *724*, 726–727, 783t
parallel resistors, *782*, 782–787, 783t
power in, 760–761
*RC*, 788–792, *789*
resistive load, 914–916, *915*
*RL*, 882–886, *883*, *884*
*RLC*, 910–912, *911*, 921–926, *922*
series capacitors, 724–727, *725*, 783t
series resistors, *776*, 776–777, 783t
single-loop, 771–780, *914*
circuit elements, *718*
circular aperture, diffraction patterns, 1090–1094, *1091*
circular arc, current in, 839–841
circular arc charge distributions, 642
circular orbits, 373–374
clocks:
  event measurement with array of, 1119, *1119*
  time dilation tests, 1123–1124, 1153
closed circuit, 776, *776*
closed cycle processes, first law of thermodynamics for, 532, 532t
closed path, 179–180, *180*
closed-path test, for conservative force, 179–180
closed shell, 1299
closed subshell, 1235
closed surface, electric flux in, 661–664
closed system, 221
  entropy, 589
  linear momentum conservation, 230–231
COBE (Cosmic Background Explorer) satellite, 1360, 1361
coefficient of kinetic friction, 127–130
coefficient of linear expansion, 521, 521t
coefficient of performance (refrigerators), 596
coefficient of static friction, 127–130
coefficient of volume expansion, 521
coherence, 1059–1060
coherence length, 1241
coherent light, 1059, 1241
coils, 823–824. *See also* inductors
  of current loops, 823–824

in ideal transformers, 931, *931*
induced emf, 867–868
magnetic field, 851–854, *852*
mutual induction, 890–892, *891*
self-induction, *881,* 881–882
cold-weld, 126–127, *127*
collective model, of nucleus, 1298
collimated slit, *1226*
collimator, 1100, *1226*
collision(s), 226–229
    elastic in one dimension, *237,* 237–240
    glancing, *240,* 240–241
    impulse of series of, 227–229, *229*
    impulse of single, 226–227, *227*
    inelastic, in one dimension, *234,* 234–236, *235*
    momentum and kinetic energy in, 233
    two-dimensional, *240,* 240–241
color force, 1354–1355
color-neutral quarks, 1354–1355
color-shifting inks, 1048
compass, 950, 964
completely inelastic collisions, *234,* 234–236, *235*
components:
    of light, 993–994
    vector, 42–44, *43, 46,* 46–47, *47,* 49
component notation (vectors), 43
composite slab, conduction through, 535, *535*
compound microscope, 1032, *1032*
compound nucleus, 1298, 1300
compressibility, 342, 388
compressive stress, 340–341
Compton scattering, *1159,* 1159–1162, *1160*
Compton shift, *1159,* 1159–1162
Compton wavelength, 1161
concave lenses, 1044
concave mirrors, 1013, *1016,* 1017–1018
concrete:
    coefficient of linear expansion, 521t
    elastic properties, 341t
condensing, 526
conducting devices, 619, 756–757
conducting path, 612
conducting plates:
    eddy currents, 874
    Gauss' law, *674,* 674–675
conduction, *534,* 535, *535,* 1252–1272
    and electrical properties of metals, 1252–1261
    in *p-n* junctions, 1266–1270
    by semiconductors, 1261–1265
    in transistors, 1270–1271
conduction band, 1262, *1262*
conduction electrons, 612, 746, 752, 1255–1261
conduction rate, 534–535
conductivity, 754, 1257

conductors, 612–613, 746. *See also* electric current
    drift speed in, 749–750, 752
    Hall effect for moving, 812–813
    metallic, 746, 762
    Ohm's law, 756–759
    potential difference across, 812–813
configurations, in statistical mechanics, 599–600
confinement principle, 1187
conical pendulum, 146
conservation of angular momentum, 312–316, *313, 314*
conservation of baryon number, 1345
conservation of electric charge, 621–622
conservation of energy, 149, 195–199, *197*
    in electric field, 688
    mechanical and electric potential energy, 705
    principle of conservation of mechanical energy, 185
    in proton decay, 1348
    sample problems involving, 186–187, 198–199
conservation of lepton number, 1344–1345
conservation of linear momentum, 230–232, 236, 242
conservation of quantum numbers, 1348–1349
conservation of strangeness, 1346
conservative forces, 179–181, *180, 685*
constant acceleration (one-dimensional motion), 23, 23–27, 24t
constant angular acceleration, rotation with, 266–268
constant linear acceleration, 266
constant-pressure molar specific heat, 566–568
constant-pressure processes, *529,* 529–530
    summarized, *575,* 575t
    work done by ideal gases, 554–555
constant-pressure specific heat, 525
constant-temperature processes:
    summarized, *575,* 575t
    work done by ideal gases, 552–553
constant-volume gas thermometer, *516,* 516–517
constant-volume molar specific heat, 565–566
constant-volume processes, *529,* 529–530
    first law of thermodynamics for, 532, 532t
    summarized, *575,* 575t
    work done by ideal gases, 553
constant-volume specific heat, 525
consumption rate, nuclear reactor, 1319–1320

contact potential difference, 1266–1267
continuity, equation of, 398–401, *400*
continuous bodies, 272
continuous charge distribution, 638–639, 698–700, *699, 700*
continuous x-ray spectrum, 1237, *1237*
contracted length, 1126–1128
convection, 537
converging lens, 1023, *1024,* 1025
conversion factors, 3
convex lenses, 1044
convex mirrors, 1013, *1016,* 1017–1018
cooling:
    evaporative, 545
    super-, 605
Coordinated Universal Time (UTC), 6
copper:
    coefficient of linear expansion, 521t
    conduction electrons, 612
    electric properties of silicon vs., 762–763, 762t, 1253t, 1262
    energy levels, 1254, *1254*
    Fermi energy, 1255
    Fermi speed, 1255–1256
    heats of transformation, 526t
    mean free time, 759
    resistivity, 754t, 755, *755,* 1262
    rubbing rod with wool, 612
    temperature coefficient of resistivity, 1262
    unit cell, 1253, *1253*
copper wire:
    as conductor, 612, *612, 746,* 746–747
    drift speed in, 749–750
    magnetic force on current carrying, *820,* 820–822
cord (unit of wood), 11
core (Earth), 380, *380*
    density, 360, *360,* 388t
    pressure, 388t
core (Sun):
    density, 387t
    pressure, 388t
    speed distribution of photons in, 562
corner reflectors, 1046
corn–hog ratio, 12
corona discharge, 707
correspondence principle, 1193
cosine, 45, *45*
cosine-squared rule, for intensity of transmitted polarized light, 987
Cosmic Background Explorer (COBE) satellite, 1360, 1361
cosmic background radiation, 1357–1358, 1360, *1361*
cosmic ray protons, 627
cosmological red shift, 1367–1368
cosmology, 1355–1362

background radiation, 1357–1358
    Big Bang theory, 1358–1361
    dark matter, 1358
    expansion of universe, 1356–1357
coulomb (unit), 614
Coulomb barrier, 1322
coulomb per second, 746
Coulomb's law, 609–622
    conductors and insulators, 612–613
    conservation of charge, 621–622
    electric charge, 610–611
    formulas for, 613–615
    and Gauss' law, 666–667
    quantization of charge, 619–621
    for spherical conductors, 615–619
crimp hold, 348
critical angle, for total internal reflection, 996
crossed magnetic fields:
    and discovery of electrons, 808–810
    Hall effect in, 810–813, *811*
crossed sheets, polarizers, 988, *988*
cross product, 52–55
crust (Earth), *360,* 380, *380,* 387t
crystals:
    matter waves incident after scattering, *1167,* 1168, *1168*
    polycrystalline solids, 963
    x-ray diffraction, *1105,* 1105–1106
crystal defects, 627
crystalline solids:
    electrical properties, 1252–1261, *1253*
    energy bands, 1254, *1254*
crystal planes, 1105, *1105*
curie (unit), 1287
Curie constant, 960
Curie's law, 960
Curie temperature, 962
curled–straight right-hand rule, 838
currency, anti-counterfeiting measures, 1048
current, *see* electric current
current amplitude:
    alternating current, 926
    series *RLC* circuits, *922,* 922–923, 926
current-carrying wire:
    energy dissipation in, 761
    magnetic field due to, *837,* 837–842, *838*
    magnetic field inside long straight, 846, *846*
    magnetic field outside long straight, *845,* 845–846
    magnetic force between parallel, 842–843, *843*
    magnetic force on, *820,* 820–822
current density, *749,* 749–752
current law, Kirchoff's, 781

current-length element, 837, *837*
current loops, 746, *746*
  electrons, *955,* 955–956, *956*
  Faraday's law of induction, 865–866
  Lenz's law for finding direction of current, *868,* 868–871, *869*
  as magnetic dipoles, 851–854, *852*
  solenoids and toroids, 848–851
  torque on, 822–824, *823*
curvature, of space, *375,* 375–376, 1360–1361
cutoff frequency, photoelectric effect, 1156–1157
cutoff wavelength:
  continuous x-ray spectrum, 1237
  photoelectric effect, 1156–1157
cycle:
  engines, 591
  simple harmonic motion, 414
  thermodynamic, *529,* 530, 532
cyclotrons, *818,* 818–819
cylinders:
  of current, 847–848
  rotational inertia, 274t
  tracer study of flow around, *399*
cylindrical capacitor, capacitance of, *721,* 721–722
cylindrical symmetry, Gauss' law, *671,* 671–672

**D**
damped energy, 431
damped oscillations, 430–431, *431,* 910–912
damped simple harmonic motion, *430,* 430–432, *431*
damped simple harmonic oscillator, *430,* 430–432
damping constant, simple harmonic motion, 430–431
damping force, simple harmonic motion, 430–431
dark energy, 1361
dark fringes:
  double-slit interference, 1055, *1055,* 1057
  single-slit diffraction, *1083,* 1083–1085, 1088–1089
dark matter, 1358, 1361, *1361*
daughter nuclei, 622, 1302
day:
  10-hour day, *5*
  variations in length of, *6*
dc (direct current), 772, 913
de Broglie wavelength, 1167, 1171, 1189
decay, *see* radioactive decay
decay constant, 1286
decay rate, 1286–1288
deceleration, 21
decibel, 490–492
decimal places, significant figures with, 4
dees, cyclotron, 818
de-excitation, of electrons, 1190

deformation, 340, *340*
degenerate energy levels, 1200
degrees of freedom, ideal gas molecules, 568–570
density:
  defined, 7
  fluids, 387
  kinetic energy density, 402
  linear, of stretched string, 452, 453
  and liquefaction, 7–8
  nuclear matter, 1285
  occupied states, 1259–1260, *1260*
  selected engineering materials, 341t
  selected materials and objects, 387t
  states, *1257,* 1257–1258
  uniform, for solid bodies, 216–217
density gradient, 1266
depletion zone, *p-n* junction, 1266
derived units, 2
detection, *see* probability of detection
deuterium, *1294*
deuterium–tritium fuel pellets, 1328, *1328*
deuterons, 819, 1327
deuteron–triton reaction, 1327
deviation angle, 1005
diamagnetic material, 957
diamagnetism, 957–958, *958*
diamond:
  as insulator, 1255, 1262
  unit cell, 1253, *1253*
diamond lattice, *1253*
diatomic molecules, 566
  degrees of freedom, 568–570, *569,* 569t
  molar specific heats at constant volume, 565t
  potential energy, 205
dielectrics:
  atomic view, 733–734, *734*
  capacitors with, 731–733
  and Gauss' law, *735,* 735–737
  polarization of light by reflection, 998
dielectric constant, 731–732, 732t
dielectric strength, 731–733, 732t
differential equations, 907
diffraction, 1081–1107. *See also* interference; single-slit diffraction
  circular aperture, 1090–1094, *1091*
  double-slit, 1094–1097, *1095, 1096*
  Fresnel bright spot, *1083*
  intensity in double-slit, *1095,* 1096–1097
  intensity in single-slit, 1086–1090, *1089*
  interference vs., 1097
  neutron, 1168

pinhole, 1082
  and wave theory of light, 1081–1083
  x-ray, 1104–1106, *1105*
  and Young's interference experiment, 1053–1054, *1054*
diffraction factor, 1096
diffraction gratings, *1098,* 1098–1101
  dispersion, 1101–1104
  resolving power, 1102–1104, *1103*
  spacing, 1099–1100
  x rays, 1105
diffraction patterns:
  defined, 1082
  double-slit, 1095–1096, *1096*
  single-slit, 1095–1096, *1096*
diffusion current, *p-n* junctions, 1266
dimensional analysis, 452
dip angle, 141
dip meter, 951
dip north pole, 951
dipole antenna, 974, *974*
dipole axis, 636, 950
dip-slip, 60
direct current (dc), 772, 913
direction:
  of acceleration in one-dimensional motion, 20
  of acceleration in two- and three-dimensional motion, 68
  of angular momentum, 305
  of displacement in one-dimensional motion, 14–15
  of vector components, 43
  of vectors, 41–42, *42*
  of velocity in one-dimensional motion, 16
  of velocity in two- and three-dimensional motion, 66
discharging, 611
  capacitors, 719, *789,* 790–792
  charged objects, 612
disintegration, 1280
disintegration constant, 1286, 1288
disintegration energy, 1290
disks:
  diffraction by circular aperture, 1090–1094, *1091*
  electric field due to charged, 643–644
  electric potential due to charged, 700, *700*
dispersion:
  chromatic, *993,* 993–994
  by diffraction gratings, 1101–1104
displacement:
  damped harmonic oscillator, 430–431, *431*
  electric, 736
  one-dimensional motion, 14–15
  simple harmonic motion, 416, *417, 418*
  traveling waves, 449–450

two- and three-dimensional motion, 63–64, *64*
  as vector quantity, 15, 41, *41*
  waves on vibrating string, 446–448, *447*
displacement amplitude:
  forced oscillations, 433, *433*
  sound waves, *483,* 483–484
displacement current, 946–950, *947*
displacement ton, 11
displacement vector, 15, 41, *41*
dissipated energy, in resistors, 761, 774
distortion parameter, 1314
distribution of molecular speeds, 560–563, *561*
diverging lens, 1023, *1024,* 1025
dominoes, 328, *328*
donor atoms, 1263–1264
doped semiconductors, 762, *1263,* 1263–1265
Doppler effect, 498–502, 1120
  detector moving, source stationary, 500, *500*
  for light, 1134–1137, *1136,* 1357
  source moving, detector stationary, 501, *501*
dose equivalent, radiation, 1297
dot product, 51, *51,* 54, 661
double-slit diffraction, 1094–1097, *1095, 1096*
double-slit interference:
  intensity, 1060–1062, *1061,* 1096
  from matter waves, *1167,* 1167–1168
  single-photon, wide-angle version, 1163–1164, *1164*
  single-photon version, 1162–1164
  Young's experiment, 1053–1058, *1055*
doubly magic nuclides, 1299
down quark, 1349, 1350t, 1351
drag coefficient, 130–131
drag force, 130–132
  damped simple harmonic motion, 430
  mechanical energy not conserved in presence of, 186
  as nonconservative force, 179
drain, FETs, 1270, *1270*
drift current, *p-n* junctions, 1267
drift speed:
  and current density, *749,* 749–750, 752
  Hall effect for determining, 810–813, *811*
driven oscillations, 433, 914, *914*
driving angular frequency, 914
driving frequency, of emf, 914
*d* subshells, 1235, 1236

**E**
E (exponent of 10), 2
Earth, 354–355, 1362. *See also* gravitational force

atmospheric electric field, 717
average density, 387t
density of, as function of
    distance from center, 360
eccentricity of orbit, 369
effective magnetic dipole
    moment, 1225
ellipsoidal shape of, 360
escape speed, 367–368, 367t
gravitation near surface,
    359–362
interior of, 380, 380
Kepler's law of periods, 370t
level of compensation, 408
magnetic dipole moment, 826t
magnetism, 950–951
nonuniform distribution of
    mass, 360, 360
rotation, 360–361, 361
satellite orbits and energy,
    371–373, 372
variation in length of day over
    4-year period, 6
earthquakes:
    building oscillations during, 414
    buildings submerged during, 7
    and liquefaction, 7–8
    natural angular frequency of
        buildings, 433, 433
    S and P waves, 506
Earth's magnetic field, 807, 950,
    950–951
    polarity reversal, 950, 951
    at surface, 806t
eccentricity, of orbits, 369, 369
    and orbital energy, 371–372
    planets of Solar System, 370t
eddy currents, 874
edges, diffraction of light at, 1082
edge effect, 674
effective cross-sectional area, 131
effective magnetic dipole
    moment, 1225
effective phase difference, optical
    interference, 1051
efficiency:
    Carnot engines, 592–593
    real engines, 593, 597–598
    Stirling engines, 594
eightfold way, 1347, 1347–1348,
    1347t
Einstein, Albert, 95, 977, 1117,
    1117, 1120, 1166. See also
    relativity
    Bose–Einstein condensate,
        1337, 1337
    and bosons, 1337
    and lasers, 1242
    view of gravitation, 374,
        374–376
    work on photoelectric effect,
        1156–1158
    work on photons, 1153–1155
Einstein–de Haas experiment,
    1221, 1222
Einstein ring, 376, 376
elastic bodies, 339

elastic collisions:
    defined, 233
    elasticity, 327, 339–342, 340
    in one dimension, with moving
        target, 238–239
    in one dimension, with
        stationary target, 237,
        237–238
    in two dimensions, 240, 240–241
    and wave speed on stretched
        string, 452
elasticity, 338–342
    of atoms and rigid bodies, 339,
        339–340
    and dimensions of solids, 340,
        340
    and equilibrium of indetermi-
        nate structures, 338–339, 339
    hydraulic stress, 341–342, 341t
    sample problem involving, 342
    shearing, 341
    tension and compression,
        340–341, 341
elastic potential energy, 178
    determining, 182–183
    traveling wave on stretched
        string, 454, 454
electrical breakdown, 646, 646
electrically isolated object, 611
electrically neutral objects, 611
electrical–mechanical analogy,
    906–907, 906t
electric charge, 610–611. See also
    circuits
    conservation of, 621–622
    and current, 747–748
    enclosed, 667, 670
    excess, 611
    free, 735
    hypercharge, 1364
    induced, 612–613
    LC oscillations, 904, 908
    lines of, 638–643, 639, 699,
        699–700
    measures of, 639t
    negative, 611, 611
    net, 611
    neutralization of, 611
    positive, 611, 734
    quantization of, 619–621
    in RLC circuits, 911, 912
    sharing of, 619
    in single-loop circuits, 772
electric circuits, see circuits
electric current, 745–752, 746, 747
    in alternating current, 913–914
    for capacitive load, 918
    current density, 748–752, 749
    decay, 885
    direction in circuits, 747,
        747–748
    induced, 864–865, 870–874
    for inductive load, 920
    LC oscillations, 904, 908–910
    magnetic field due to, 837,
        837–842, 838
    in multiloop circuits, 781–782

power in, 760–761
    for resistive load, 916
    in single-loop circuits, 774,
        774–775
    time-varying, in RC circuits, 790
electric dipole, 825
    in electric field, 647–650
    electric field due to, 635–638,
        636
    electric potential due to,
        697–698, 698
    induced, 698
    potential energy of, 648
electric dipole antenna, 974,
    974–975
electric dipole moment, 637, 648
    dielectrics, 733–734
    induced, 698
    permanent, 698
electric displacement, 736
electric field, 630–651, 804
    calculating from potential,
        701, 701–702
    calculating potential from,
        691, 691–693
    capacitors, 720
    crossed fields, 810–813, 811
    as displacement current,
        948–949
    due to charged disk, 643–644,
        700, 700
    due to charged particle, 633,
        633–635
    due to electric dipole, 635–638,
        636
    due to line of charge, 638–643,
        639
    electric dipole in, 647–650
    energy stored in capacitor,
        728–730
    equipotential surfaces, 690,
        690–691, 691
    external, 669–670, 707, 707
    field lines in, 631–632
    and Gauss' law, 666–667, 844,
        942, 949t
    Hall effect, 810–813, 811, 820
    induced, 874–879, 875, 977,
        977–978
    net, 634–635
    nonuniform, 632, 663–664
    point charge in, 645–647
    polarized light, 907, 988
    potential energy in, 687–689,
        730
    rms of, 982–983
    in spherical metal shell, 670
    system of charged particles in,
        703–705, 704
    traveling electromagnetic
        waves in, 974–977, 975, 976
    uniform, 632, 660–662, 692
    as vector field, 631
    work done by, 686–689
electric field lines, 631, 631–632,
    632
electric fish, 786–787

electric flux, 659–664
    in closed surface, 661–664
    and Gauss' law, 659–664
    and induction, 872
    net, 661–662
    through Gaussian surfaces, 660,
        660–664, 661
    in uniform electric fields,
        660–662
electric force, 803
electric generator, 772
electric motor, 822–824, 823, 950
electric potential, 685–708
    calculating field from, 701,
        701–702
    charged isolated conductor,
        706, 706–707
    defined, 686
    due to charged particles, 694,
        694–696, 695
    due to continuous charge
        distribution, 698–700, 699,
        700
    due to electric dipole, 697–698,
        698
    from electric fields, 691–693
    and electric potential energy,
        686, 686–689, 689
    equipotential surfaces, 690–691,
        691
    and induced electric field,
        877–878
    in LC oscillator, 909–910
    potential energy of charged
        particle system, 703–705, 704
    and power/emf, 779
    and self-induction, 882
electric potential energy:
    and electric potential, 686,
        686–689, 689
    for system of charged particles,
        703–705, 704
electric quadrupole, 654
electric spark, 646, 646
    airborne dust explosions set off
        by, 729–730
    dangers of, 707, 707
    and pit stop fuel dispenser fire,
        792, 792
electric wave component, of
    electromagnetic waves,
    975–976, 976
electromagnets, 804, 804, 806t
electromagnetic energy, 909.
    See also electromagnetic
    waves
electromagnetic force, 1338,
    1352–1353
electromagnetic oscillations, 904
    damped, in RLC circuits,
        910–912
    defined, 904
    forced, 912–920, 914
    LC oscillations, 903–910
electromagnetic radiation, 974
electromagnetic spectrum, 973,
    973–974

electromagnetic waves, 445, 972–999. *See also* reflection; refraction
  energy transport and Poynting vector, 980–983, *982*
  Maxwell's rainbow, 973–974
  polarization, *907,* 985–990, *986, 988,* 997–998
  radiation pressure, 983–985
  reflection of, 990–998, *998*
  refraction of, 990–996
  traveling, 974–980, *976, 977*
electromagnetism, 836, 950, 1334
electromotive force (emf), 772–774. *See also* emf devices
  in alternating current, 913–914
  defined, 772, 876–877
  and energy and work, *773,* 773–774
  induced, 865, 867–868, 870–871
  potential and power in circuits, 779
  self-induced, 881
electrons, 612, 1335
  accelerator studies, 818
  in alternating current, 913
  barrier tunneling, 1176–1179, *1177*
  in Bohr model, *1203,* 1203–1204
  bubble chamber tracks, 622, *622, 806*
  charge, 620, 620t
  Compton scattering, 1159–1162, *1160*
  conduction, 1255–1261
  discovery by Thomson, 808–810, *809,* 1276
  energy of, 1142, 1186–1191
  excitation of, 1189, *1189,* 1255
  as fermions, 1336
  in hydrogen atom, 1212
  kinetic energy of, 1118, *1118*
  as leptons, 1338, 1344, 1344t
  magnetic dipole moment, 826, 826t
  and magnetism, 952–957
  majority carrier in *n*-type semiconductors, 1264, 1264t
  matter waves, 1166–1170, *1167, 1168*
  matter waves of, 1166–1170, *1167, 1168,* 1173, 1186
  momentum, *954*
  momentum of, 953–955, *955,* 1142
  orbits of, *955,* 955–956, *956*
  from proton–antiproton annihilation, 1340t
  in *p*-type semiconductors, 1264, 1264t
  radial probability density of, 1211–1212
  radiation dosage, 1296–1297
  speed of, 1118, *1118*
  spin, 1336–1337, *1337*
  spin-flip, 966

  in superconductors, 763
  valence, 1187, 1235, 1256
  wave functions of trapped, 1191–1195
electron capture, 1292n
electron diffraction, 1168
electron microscope, 1183
electron neutrinos, 1343–1344, 1344t
electron–positron annihilation, 622, *622,* 1338
electron spin, 1336–1337, *1337*
electron traps:
  finite well, *1195,* 1195–1197
  hydrogen atoms as, 1202
  multiple electrons in rectangular, 1230–1234
  nanocrystallites, 1197–1198, *1198*
  one-dimensional, 1187–1199
  quantum corrals, 1199, *1199*
  quantum dots, 1187, *1198,* 1198–1199
  two- and three-dimensional, *1200,* 1200–1201
  wave functions, 1191–1195, *1192*
electron-volt, 689, 1258
electroplaques, *786,* 786–787
electrostatic equilibrium, 668
electrostatic force, 611, 631
  and Coulomb's law, *613,* 613–619
  electric field due to point charge, *633,* 633–635
  point charge in electric field, 645–647
  work done by, 686, 688–689
electrostatic stress, 744
electroweak force, 1353
elementary charge, 620, 645–646
elementary particles, 1334–1354
  bosons, 1337, *1337*
  conservation of strangeness, 1346–1347
  eightfold way patterns, 1347–1348
  fermions, 1336, *1337*
  general properties, 1334–1343
  hadrons, 1338, 1345–1346
  leptons, 1338, 1343–1345
  messenger particles, 1352–1354
  quarks, 1349–1352
elevator cab, velocity and acceleration of, 18–19
elliptical orbits, 373–374
emf, *see* electromotive force
emf devices, 772, 773. *See also* batteries
  internal dissipation rate, 779
  real and ideal, 773, *773*
emf rule, 775
emission. *See also* photon emission
  from hydrogen atom, 1212
  spontaneous, *1242,* 1242–1243
  stimulated, 1242–1243

emission lines, *1098,* 1098–1099, 1206
emissivity, 536, 1166
enclosed charge, 667, 670
endothermic reactions, 1343
energy. *See also* kinetic energy; potential energy; work
  for capacitor with dielectric, 733
  conservation of, 149, 195–199, *197,* 705
  in current-carrying wire, 761
  damped, 431
  defined, 149
  of electric dipole in electric field, 650
  in electric field, 728–730
  and induction, 873
  and magnetic dipole moment, 825, 954
  in magnetic field, 887–888
  and relativity, 1138–1143
  in *RLC* circuits, 911
  scalar nature of, 41
  in simple harmonic motion, 421–423, *422*
  as state property, 585
  in transformers, 932
  transport, by electromagnetic waves, 980–983, *982*
  of trapped electrons, 1186–1191
  traveling wave on stretched string, *454,* 454–455
energy bands, 1254, *1254*
energy density, 730, 889–890
energy density, kinetic, 402
energy gap, 1254, *1254*
energy levels:
  excitation and de-excitation, 1189–1190
  full, empty, and partially occupied, 1231
  hydrogen, *1204,* 1206, *1207*
  in infinite potential well, 1190–1191, 1201, 1232–1234
  multiple electron traps, 1231–1233
  nuclear, 1284
  in single electron traps, *1188, 1189*
  of trapped electrons, 1187–1191
energy-level diagrams, 1189, *1189,* 1232, *1232*
energy method, of calculating current in single-loop circuits, 774
engines:
  Carnot, 590–593, *591,* 597–598
  efficiency, *591,* 592–593, *596, 597,* 597–598
  ideal, 591–592
  perfect, 593, *593*
  Stirling, 594, *594*
Englert, François, 1354
entoptic halos, 1108, 1110
entropy, 583–603
  change in, 584–588

engines, 590–595
  force due to, 589–590
  and irreversible processes, 584
  and probability, 601–602
  refrigerators, 595–598, *596*
  sample problems involving, 587–588, 594–595, 600–602
  and second law of thermodynamics, 588–590
  as state function, 585, 586–587
  statistical mechanics view of, 598–602
entropy changes, 584–588
  Carnot engines, 592–593
  Stirling engines, 594
entropy postulate, 584
envelope, in diffraction intensity, *1095*
equation of continuity, 398–401, *400*
equations of motion:
  constant acceleration, 24, 24t
  constant linear vs. angular acceleration, 266t
  free-fall, 27–28
equilibrium, 99, 327–342, 1308
  and center of gravity, 330–332, *331*
  electrostatic, 668
  of forces on particles, 618
  and Hall effect, 811
  of indeterminate structures, 338–339, *339*
  protons, 618
  requirements of, 329–330
  sample problems involving, 332–337, 526–527
  secular, 1304
  static, 327–329, *328, 329*
  thermal, 515
equilibrium charge, capacitors in *RC* circuits, 789
equilibrium points, in potential energy curves, 189–190
equilibrium position, simple pendulum, 425
equilibrium separation, atoms in diatomic molecules, 205
equipartition of energy, 569
equipotential surfaces, *690,* 690–691
equivalence, principle of, 374–375
equivalent capacitance:
  in parallel capacitors, 724, *724,* 726–727, 783t
  in series capacitors, 724–727, 783t
equivalent resistance:
  in parallel resistors, *782,* 782–787, 783t
  in series resistors, 777, 783t
escape speed, 367–368, 367t, 704, 713
evaporative cooling, 545
events, 1117
  Lorentz factor, 1122–1123, *1123,* 1138

Lorentz transformation, 1129–1133
measuring, 1118–1119, *1119*
relativity of length, 1125–1128, *1126*, 1132–1133
relativity of simultaneity, *1120*, 1120–1121, 1131
relativity of time, *1121*, 1121–1125, 1131
relativity of velocity, *1133*, 1133–1134
event horizon, 362
excess charge, 611
exchange coupling, 962
excitation, of electrons, 1189, *1189*, 1255
excitation energy, 1217
excited states, 1189, *1189*
expansion, of universe, 1356–1357
exploding bodies, Newton's second law and motion of, 221
explosions:
one-dimensional, 231, *231*
two-dimensional, 232, *232*
extended objects, 108
drawing rays to locate, 1026, *1026*
in plane mirrors, *1012*, 1012–1013
external agents, applied force from, 688
external electric field:
Gaussian surfaces, 669–670
isolated conductor in, 707, *707*
external forces, 99
collisions and internal energy transfers, 196–197
system of particles, 220–223
work done with friction, 192–194
work done without friction, 192
external magnetic field:
and diamagnetism, 958
and ferromagnetism, 957
and paramagnetism, 957, 959, 960
external torque, 310–311, 313, 314
eye, *see* human eye
eyepiece:
compound microscope, 1032, *1032*
refracting telescope, 1033

**F**
face-centered cubic, *1253*
Fahrenheit temperature scale, *518*, 518–519
falling body, terminal speed of, 130–132, *131*
farad, 718
Faraday, Michael, 610, 631, 731–732, 865, 880
Faraday's experiments, 865–866
and Lenz's law, *868*, 868–871, *869*
Maxwell's equation form, 949t
mutual induction, 891

reformulation, 876–877
self-induction, *881*, 881–882
Faraday's law of induction, 865–866, 943, 978
faults, rock, 60
femtometer, 1282
fermi (unit), 1282
Fermi, Enrico, 1310, 1320, 1336
Fermi–Dirac statistics, 1258
Fermi energy, 1255, 1257–1259, 1261
Fermilab accelerator, 1335, 1352
Fermi level, 1255
fermions, 1336, *1337*
Fermi speed, 1255–1256
ferromagnetic materials, 957, 996
ferromagnetism, 957, 961–964, *962*. *See also* iron
FET (field-effect-transistor), *1270*, 1270–1271
field declination, 951
field-effect-transistor (FET), *1270*, 1270–1271
field inclination, 951
field of view:
refracting telescope, 1033
spherical mirror, 1015
final state, 528, *529*, 565
finite well electron traps, *1195*, 1195–1197
fires, pit stop fuel dispenser, 792, *792*
first law of thermodynamics, 528–533
equation and rules, 531
heat, work, and energy of a system, 528–530, 533
sample problem involving, 533
special cases of, 532–533, 532t
first-order line, 1099
first reflection point, 1006
fish, electric, 786–787
fission, nuclear, 1309–1316
fission rate, nuclear reactor, 1319–1320
fixed axis, 259, *311*, 311–312
floaters, 1082
floating, 395, *395*
flow, 398–400, *399, 400*, 402
flow calorimeter, 547
fluids, 130, 386–405
apparent weight in, 396–397
Archimedes' principle, 394–397, *395*
Bernoulli's equation, 401–404
defined, 386–387
density, 387
equation of continuity, 398–401, *400*
motion of ideal, *398*, 398–399
Pascal's principle, *393*, 393–394
pressure, 387–388
pressure measurement, *392*, 392–393
at rest, 388–391, *389*
sample problems involving, 388, 391, 397, 401, 403–404

fluid streamlines, 399–400, *400*
flux. *See also* electric flux
magnetic, 866–867, 880, 942
volume, 660
focal length:
compound microscope, 1032, *1032*
refracting telescope, 1033, *1033*
simple magnifying lens, *1031*, 1031–1032
spherical mirrors, *1015*, 1015–1016
thin lenses, *1024*, 1024–1025
focal plane, 1057
focal point:
compound microscope, 1032, *1032*
objects outside, 1017
real, 1016, *1016*
refracting telescope, 1033, *1033*
simple magnifying lens, *1031*, 1031–1032
spherical mirrors, *1015*, 1015–1016
thin lenses, *1024*, 1024–1025
two-lens system, *1027*, 1027–1028
virtual, 1016, *1016*
force(s), 312t. *See also specific forces, e.g.:* gravitational force
attractive, 356
buoyant, 394–397, *395*
centripetal, 133–138, *134*
conservative, 179–181, *180*
in crossed magnetic fields, 809–810
defined, 94
and diamagnetism, 958
due to entropy, 589–590
electric field vs., 631
equilibrium, 99
equilibrium of, on particles, 618
external vs. internal, 99
and linear momentum, 224–225
lines of, 631
and motion, 14
net, 99, 616–618
and Newton's first law, 96–98
Newton's laws applied to, 108–113
and Newton's second law, 98–101
and Newton's third law, 106–107
nonconservative, 179
normal, *104*, 104–105
path independence of conservative, 179–181, *180*
principle of superposition for, 96
and radiation pressure, 984
resultant, 99
of rolling, *299*, 299–301
superposition principle for, 615
tension, *105*, 105–106
unit of, *96*, 96–97
as vector quantities, 96
and weight, 103–104

force constant, 159
forced oscillations, 432–433, *433*, 912–920, *914*
forced oscillators, 432–433, *433*
force law, for simple harmonic motion, 419
forward-bias connection, junction rectifiers, 1267–1268, *1268*
fractional efficiency, 1182
Franklin, Benjamin, 611, 619, 621
Fraunhofer lines, 1250–1251
free-body diagrams, 99–101, *100*, 108–113
free charge, 735
free electrons, 746
free-electron model, 758, 1255
free expansion:
first law of thermodynamics for, 532, 532t
ideal gases, 573–575, *585*, 585–588, *586*
free-fall acceleration *(g)*, 27, 27–28, 427
free-fall flight, 28
free oscillations, 432–433, 914
free particle:
Heisenberg's uncertainty principle for, *1172*, 1172–1174
matter waves for, 1187
free space, 974
freeze-frames, 414, *415*, 416
freezing point, 518t, 525
freight ton, 11
frequency. *See also* angular frequency
of circulating charged particles, 814–817
cutoff, 1156–1157
of cyclotrons, 818–819
driving, 914
and index of refraction, 1050
of photons, 1154
proper, 1135
simple harmonic motion, 414–417, *417*
sound waves, 483
waves, 448
and wavelength, 446–449
wave on stretched string, 453
Fresnel bright spot, 1082–1083, *1083*
friction, 105, *105*, 124–130, *125–126*
cold-weld, 126–127, *127*
as nonconservative force (kinetic friction), 179
properties of, 127
and rolling, 299, *299*
sample problems involving, 128–130, 132
types of, 125, 126
work done by external force with, *192*, 192–194, *193*
frictionless surface, 95, 105
fringing, 674
*f* subshells, 1235
fuel charge, nuclear reactor, 1320–1321

fuel rods, 1317, 1320–1321
fulcrum, 345
full electron levels, 1231
fully charged capacitor, 719
fully constructive interference, 460, *460*, 461t, *465*, 486
fully destructive interference, 460, *460*, 461t, *465*, 486–487
fundamental mode, 468, 494
fused quartz:
  coefficient of linear expansion for, 521t
  index of refraction, 992t
  index of refraction as function of wavelength, *993*
  resistivity, 754t
fusion, 1140, 1284, 1322–1329
  controlled, 1326–1329
  laser, 1328–1329
  most probable speed in, 1322, 1333
  process of, 1322–1323
  in Sun and stars, 1322, *1324*, 1324–1326

**G**

$g$ (free-fall acceleration), *27*, 27–28
  measuring, with physical pendulum, 427
$G$ (gravitational constant), 355
$g$ units (acceleration), 21
galaxies, 354
  Doppler shift, 1135–1136, 1148, *1148*
  formation in early universe, 1360
  gravitational lensing caused by, 375, *376*
  matter and antimatter in, 1338–1339
  recession of, and expansion of universe, 1356
  superluminal jets, 1149
Galilean transformation equations, 1129
*Galileo*, 382
gamma rays, 622, 806, 974
  bubble chamber track, 1169, *1169*
  radiation dosage, 1297
  ultimate speed, 1118
gamma-ray photons, 1324, 1338
gas constant, 551
gases, 549. *See also* ideal gases; kinetic theory of gases
  compressibility, 387
  confined to cylinder with movable piston, 528–530, *529*
  density of selected, 387t
  as fluids, 387
  polyatomic, 565
  specific heats of selected, 525t
  speed of sound in, 481t
  thermal conductivity of selected, 535t
gas state, 526
gauge pressure, 390

gauss (unit), 806
Gauss, Carl Friedrich, 660
Gaussian form, of thin-lens formula, 1043
Gaussian surfaces:
  capacitors, 719–723
  defined, 660
  electric field flux through, *660*, 660–664, *661*
  external electric field, *669*, 669–670
  and Gauss' law for magnetic fields, 942
Gauss' law, 659–677
  charged isolated conductor, 668–670
  and Coulomb's law, 666–667
  cylindrical symmetry, *671*, 671–672
  dielectrics, *735*, 735–737
  for electric fields, 942, 949t
  and electric flux, 659–664
  formulas, 664–665
  for magnetic fields, 941–943, *942*, 949t
  and Maxwell's equation, 949t
  planar symmetry, *673*, 673–675, *674*
  spherical symmetry, 675–677, *676*
Geiger counter, 1276
general theory of relativity, 374–376, 1117, 1123–1124
generator. *See also* alternating current generator
  electric, 772
  homopolar, 835
geomagnetic pole, 807, 950
geometric addition of vectors, *41*, 41–42, *42*, 44
geometrical optics, 991, 1054, 1082
geosynchronous orbit, 382
glass:
  coefficient of linear expansion, 521t
  index of refraction, 992t
  as insulator, 612
  polarization of light by reflection, 998
  rubbing rod with silk, 610, *610*, 621
  shattering by sound waves, *490*
Global Positioning System (GPS), 1, 1117
$g$-LOC ($g$-induced loss of consciousness), 77, 408
gluons, 818, 1350, 1354
gold, 1239
  alpha particle scattering, 1277–1279
  impact with alpha particle, 705
  isotopes, 1280
GPS (Global Positioning System), 1, 1117
grand jeté, 221–222, *222*
grand unification theories (GUTs), 1355

graphs, average velocity on, *15*, 16, *16*
graphical integration:
  of force in collision, 227, *227*
  for one-dimensional motion, *29*, 29–30
  work calculated by, 164–166
grating spectroscope, *1100*, 1100–1101
gravitation, 354–377
  and Big Bang, 1360
  defined, 355
  Einstein's view of, 374–376, *376*
  gravitational acceleration ($a^g$), 360
  inside Earth, 362–364
  near Earth's surface, 359–362, *360*
  Newton's law of, 355–356, 369
  potential energy of, 364–368
  sample problems involving, 358, 362, 368, 373–374
  variation with altitude, 360t
gravitational constant *(G)*, 355
gravitational force, 102–103, 621, 1338
  center of gravity, 330–332, *331*
  and Newton's law of gravitation, 355–356, *356*
  pendulums, 425, *425*
  and potential energy, 366–367
  and principle of superposition, 357–359
  work done by, 155–158, *156*
gravitational lensing, 376, *376*
gravitational potential energy, 178, 364–368, *365*
  determining, 182
  and escape speed, 367–368
  and gravitational force, 366–367
graviton, 376
gray (unit), 1296
grounding, 612
ground speed, 90
ground state, *1189*, 1189–1190
  wave function of hydrogen, 1208–1210, *1209*
  zero-point energy, 1193–1194
gry (unit), 8
$g$ subshells, 1235
gyroscope precession, *317*, 317–318

**H**

hadrons, 1338, 1345–1346
half-life, 1281, 1287, 1295, 1335
half-width of diffraction grating lines, *1098*, 1099–1100
Hall effect, 810–813, *811*, 820
Hall potential difference, 811
halogens, 1236
halo nuclides, 1282
hang, in basketball, 86–87
hanging blocks, *108*, 108–109
hard reflection, of traveling waves at boundary, 467

harmonic motion, 414
harmonic number, 468, 492–496
harmonic series, 468
hearing threshold, 490t
heat, 520–538, 594–595
  absorption of, 522–527
  defined, 523
  first law of thermodynamics, 528–533
  path-dependent quantity, 531
  sample problems involving, 526–527, 533, 537–538
  signs for, 523–524
  and temperature, *523*, 523–524, 526–527
  thermal expansion, *520*
  and thermal expansion, 520–522
  transfer of, 534–538
  and work, 528–530
heat capacity, 524
heat engines, 590–595
heat of fusion, 526, 526t
heats of transformation, 525–527, 526t
heat of vaporization, 526, 526t
heat pumps, 596
heat transfer, 534–538
hectare, 11
hedge maze, searching through, 48–49
height, of potential energy step, 1174
Heisenberg's uncertainty principle, *1172*, 1172–1174
helical paths, charged particles, *816*, 816–817
helium burning, in fusion, 1325
helium–neon gas laser, *1243*, 1243–1245
henry (unit), 880
hertz, 414
Higgs, Peter, 1354
Higgs boson, 1354
Higgs field, 1354
holes, 1238, 1262
  majority carrier in *p*-type semiconductors, 1264, 1264t
  minority carrier in *n*-type semiconductors, 1264, 1264t
holograms, 1241
home-base level, for spectral series, 1206
homopolar generator, 835
Hooke, Robert, 159
Hooke's law, 159–160, 188
hoop, rotational inertia for, 274t
horizontal motion, in projectile motion, 72, *73*
horizontal range, in projectile motion, *71*, 73
horsepower (hp), 167
hot chocolate effect, 506
$h$ subshells, 1235
Hubble constant, 1356
Hubble's law, 1356–1357

human body:
  as conductor, 612
  physiological emf devices, 772
human eye, 1031
  floaters, 1082
  image production, 1012
  and resolvability in vision,
    1092, 1093
  sensitivity to different
    wavelengths, *973,* 974
human wave, 472
Huygens, Christian, 1048
Huygens' principle, *1048,* 1048–1049
Huygens' wavelets, 1083
hydraulic compression, 341
hydraulic engineering, 386
hydraulic jack, 394
hydraulic lever, *393,* 393–394
hydraulic stress, 341–342, 341t
hydrogen, 1201–1212
  Bohr model, *1203,* 1203–1204
  as electron trap, 1202
  emission lines, *1100,* 1100–1101
  formation in early universe,
    1360
  in fusion, 1140, 1322–1329
  heats of transformation, 526t
  quantum numbers, 1206–1208,
    1208t
  RMS speed at room
    temperature, 556t
  and Schrödinger's equation,
    1205–1212
  spectrum of, 1206
  speed of sound in, 481t
  thermal conductivity, 535t
  wave function of ground state,
    1208–1210, *1209*
hydrogen bomb (thermonuclear
  bomb), 1326–1327
hydrostatic pressures, 388–391
hypercharge, 1364
hysteresis, *963,* 963–964

**I**
icicles, 546
ideal emf devices, 773
ideal engines, 591–592
ideal fluids, *398,* 398–399
ideal gases, 550–554
  adiabatic expansion, 571–575,
    *572*
  average speed of molecules,
    561–563
  free expansion, *585,* 585–588,
    *586*
  ideal gas law, 551–552
  internal energy, 564–568
  mean free path, *558,* 558–560
  molar specific heats, 564–568
  most probable speed of
    molecules, 562
  RMS speed, 554–556, *555,* 556t
  sample problems involving,
    553–554, 556, 560, 563,
    567–570, 574–575

  translational kinetic energy, 557
  work done by, 552–554
ideal gas law, 551–552, *552*
ideal gas temperature, 517
ideal inductor, 882
ideal refrigerators, 596
ideal solenoid, 849–850
ideal spring, 160
ideal toroids, 850
ideal transformers, *931,* 931–932
ignition, in magnetic confinement,
  1328
images, 1010–1036
  extended objects, 1026, *1026*
  locating by drawing rays, 1026,
    *1026*
  from plane mirrors, 1010–1014,
    *1012*
  from spherical mirrors,
    1014–1020, *1015, 1016, 1033,*
    1033–1034
  from spherical refracting sur-
    faces, 1020–1022, *1021,* 1034,
    *1034*
  from thin lenses, 1023–1030,
    *1025, 1026,* 1034–1036, *1035*
  types of, 1010–1011
image distances, 1012
impedance, 923, 926, 932
impedance matching, in
  transformers, 932
impulse, 227
  series of collisions, 227–228, *228*
  single collision, *226,* 226–227
incident ray, 991, *991*
incoherent light, 1059
incompressible flow, 398
indefinite integral, 26
independent particle model, of
  nucleus, 1298–1299
indeterminate structures,
  equilibrium of, 338–339, *339*
index of refraction:
  and chromatic dispersion,
    993–994
  common materials, 992t
  defined, 992, 1049
  and wavelength, 1050–1052
induced charge, 612–613
induced current, 864–865
induced electric dipole moment,
  698
induced electric fields, 874–879,
  *875, 977,* 977–978
induced emf, 865, 867–868, 870–873
induced magnetic fields, 943–946,
  *944*
  displacement current, *947,* 948
  finding, 948
  from traveling electromagnetic
    waves, *979,* 979–980
inductance, 879–880
  *LC* oscillations, 903–910
  *RLC* circuits, 910–912
  *RL* circuits, 882–886
  series *RLC* circuits, 921–926
  solenoids, 880, *881*

induction:
  of electric fields, 874–879
  and energy density of magnetic
    fields, 889–890
  and energy stored in magnetic
    fields, 887–888
  and energy transfers, 871–874,
    *872*
  Faraday's and Lenz's laws,
    864–871, 978
  in inductors, 879–880
  Maxwell's law, 944, 979
  mutual, 890–892, *891*
  and *RL* circuits, 882–886
  self-, *881,* 881–882, 890
inductive reactance, 919
inductive time constant, 884–885
inductors, 879–880
  with ac generator, *918,* 918–919,
    *919*
  phase and amplitude relation-
    ships for ac circuits, 920t
  *RL* circuits, 882–886
  series *RLC* circuits, 922
inelastic collisions:
  defined, 233
  in one dimension, *234,*
    234–236, *235*
  in two dimensions, 240–241
inertial confinement, 1328
inertial reference frames,
  86–87, 1117
inexact differentials, 531
infinitely deep potential energy
  well, 1188, *1189*
infinite potential well, *1189*
  detection probability in,
    1192–1194
  energy levels in, 1190–1191,
    1201, 1232–1234
  wave function normalization in,
    1194–1195
inflation, of early universe, 1359
initial conditions, 420
initial state, 528, *529,* 565
in phase:
  ac circuits, 920t
  resistive load, 915
  sound waves, 486
  thin-film interference, 1064
  traveling electromagnetic
    waves, 974
  waves, 459, 461
instantaneous acceleration:
  one-dimensional motion,
    20–22, *21*
  two- and three-dimensional
    motion, 67–69
instantaneous angular
  acceleration, 261
instantaneous angular
  velocity, 260
instantaneous power, 167, 198
instantaneous velocity:
  one-dimensional motion, 18–19
  two- and three-dimensional
    motion, 65

insulators, 612–613, 762
  electrical properties, *1254,*
    1254–1255
  resistivities of selected, 754t
  unit cell, 1253
integrated circuits, 1271
intensity:
  defined, 981
  diffraction gratings, 1098–1099
  double-slit diffraction, *1095,*
    1096–1097
  double-slit interference,
    1060–1062, *1061,* 1096
  electromagnetic waves, *982,*
    982–983
  single-slit diffraction,
    1086–1090, *1087, 1089*
  of sound waves, 488–492, *489*
  of transmitted polarized light,
    987–990, *988*
interference, *459,* 459–461, *460,*
  1047–1072. *See also*
  diffraction
  combining more than two
    waves, 1062
  diffraction vs., 1095–1097
  double-slit from matter waves,
    *1167,* 1167–1168
  double-slit from single photons,
    *1162,* 1162–1164
  fully constructive, 460, *460,*
    461t, *465,* 486
  fully destructive, 460, *460,* 461t,
    *465,* 486–487
  intensity in double-slit,
    1059–1063, *1061*
  intermediate, 460, *460,* 461t, 487
  and rainbows, 1051–1052, *1052*
  sound waves, 485–488, *486*
  thin films, *1064,* 1064–1071
  and wave theory of light,
    1047–1052
  Young's double-slit experiment,
    1053–1058, *1055*
interference factor, 1096
interference fringes, *1055,*
  1055–1056
interference pattern, 1055,
  *1055,* 1057
interfering waves, 459
interferometer, 1070–1071
intermediate interference, 460,
  *460,* 461t, 487
internal energy, 514
  and conservation of total
    energy, 195
  and external forces, 196–197
  and first law of
    thermodynamics, 531
  of ideal gas by kinetic theory,
    564–568
internal forces, 99, 220–223
internal resistance:
  ammeters, 788
  circuits, 776, *776*
  emf devices, 779–780
internal torque, 310

International Bureau of Weights and Standards, 3, 6–7
International System of Units, 2–3, 2t
interplanar spacing, 1106
intrinsic angular momentum, 953, 954
inverse cosine, 45, *45*
inverse sine, 45, *45*
inverse tangent, 45, *45*
inverse trigonometric functions, 45, *45*
inverted images, *1016,* 1017
ionization energy, *1220,* 1221
ionized atoms, 1206
ion tail, 1002
iron, 1236
  Curie temperature, 962
  ferromagnetic material, 957, 962
  quantum corral, 1199, *1199*
  radius of nucleus, 620–621
  resistivity, 754t
iron filings:
  bar magnet's effect on, 942, *942*
  current-carrying wire's effect on, *838*
irreversible processes, 584, 588–590
irrotational flow, 398, 402
island of stability, 1281
isobars, 1281
isobaric processes summarized, *575,* 575t
isochoric processes summarized, *575,* 575t
isolated spherical capacitors, 722, 730
isolated system, 184–185
  conservation of total energy, 196
  linear momentum conservation, 230–231
isospin, 1364
isotherm, 552, *552*
isothermal compression, 552, 591, *591*
isothermal expansion, 552
  Carnot engine, 591, *591*
  entropy change, 585–586, *586*
isothermal processes, *575,* 575t
isotopes, 1280
isotopic abundance, 1280n.a
isotropic materials, 754
isotropic point source, 982
isotropic sound source, 489

**J**
joint, in rock layers, 141
Josephson junction, 1178
joule (J), 150, 524
jump seat, 443
junctions, circuits, 781. *See also* *p-n* junctions
junction diodes, 762
junction lasers, 1269, *1269*
junction plane, 1266, *1266*

junction rectifiers, 1267–1268, *1268*
junction rule, Kirchoff's, 781
Jupiter, escape speed for, 367t

**K**
kaons, 1124–1125, 1335
  and eightfold way, 1347t
  and strangeness, 1346
kelvins, 515, 516, 518, 521
Kelvin temperature scale, *515,* 516–517, *518*
Kepler, Johannes, 369
Kepler's first law (law of orbits), 369, *369*
Kepler's second law (law of areas), *369,* 369–370
Kepler's third law (law of periods), *370,* 370–371, 370t
kilogram, *6,* 6–7
kilowatt-hour, 167
kinematics, 14
kinetic energy, 283t
  in collisions, 233
  and conservation of mechanical energy, 184–187
  and conservation of total energy, 195–199
  defined, 150
  and momentum, 1141, 1142
  in pion decay, 1342
  and relativity, 1140–1141, *1141*
  of rolling, *297,* 298–301
  of rotation, 271–273, *272*
  sample problems involving, 150, 161–162, 277
  satellites in orbit, 371–372, *372*
  simple harmonic motion, 422, *422*
  traveling wave on stretched string, *454,* 454–455
  and work, *152,* 152–155
  yo-yo, 302
kinetic energy density, of fluids, 402
kinetic energy function, 188
kinetic frictional force, 126–127, *127*
  as nonconservative force, 179
  rolling wheel, 299
kinetic theory of gases, 549–576
  adiabatic expansion of ideal gases, 571–575, *572*
  average speed of molecules, 561–563
  and Avogadro's number, 550
  distribution of molecular speeds, 560–563, *561*
  ideal gases, 550–554
  mean free path, *558,* 558–560
  molar specific heat, 564–571
  most probable speed of molecules, 562
  pressure, temperature, and RMS speed, 554–556
  and quantum theory, 569, 570–571
  RMS speed, 554–556, 556t
  translational kinetic energy, 557

Kirchoff's current law, 781
Kirchoff's junction rule, 781
Kirchoff's loop rule, 775
Kirchoff's voltage law, 775
*K* shell, 1238, *1238*
Kundt's method, 513

**L**
lagging, in ac circuits, 920, 920t
lagging waves, 461
lambda particles, eightfold way and, 1347t
lambda-zero particle, 1348
laminar flow, 398
Large Magellanic Cloud, 1293
lasers, 1240–1245
  coherence, 1060
  helium–neon gas, *1243,* 1243–1245
  junction, 1269, *1269*
  operation, *1242,* 1242–1245
  radiation pressure, 985
  surgery applications, 1241, *1241*
laser fusion, 1328–1329
lasing, 1244
lateral magnification:
  compound microscope, 1032
  spherical mirrors, 1017
  two-lens system, 1027–1030
lateral manipulation, using STM, 1178
lattice, 339, *339,* 1253, *1253*
law of areas (Kepler's second law), *369,* 369–370
law of Biot and Savart, 837–838, 844, 852
law of conservation of angular momentum, *312,* 312–316
law of conservation of electric charge, 621–622
law of conservation of energy, 195–199, *197*
law of conservation of linear momentum, 230
law of orbits (Kepler's first law), 369, *369*
law of periods (Kepler's third law), 370, *370,* 370t
laws of physics, 47
law of reflection, 991
law of refraction, 992, *1048,* 1048–1052
Lawson's criteria, 1327, 1328–1329
*LC* oscillations, 903–910
  and electrical–mechanical analogy, 906–907, 906t
  qualitative aspects, *904,* 904–906
  quantitative aspects, 907–910
*LC* oscillators, 906–910, 906t
  electrical–mechanical analogy, 906–907
  quantitative treatment of, 907–910
  radio wave creation, *974,* 974–977

lead:
  coefficient of linear expansion, 521t
  heats of transformation, 526t
  specific heats, 525t
  thermal conductivity, 535t
leading, in ac circuits, 920, 920t
leading waves, 461
LEDs (light-emitting diodes), 1268–1270, *1269*
Leidenfrost effect, 545
length:
  coherence, 1241
  consequences of Lorentz transformation equations, 1131–1132
  length contraction, 1126–1128
  proper, 1126
  relativity of, 1125–1128, *1126,* 1131–1132
  rest, 1126
  of selected objects, 4t
  units of, 3–4
  in wavelengths of light, 1071
lens, 1023. *See also* thin lenses
  bi-concave, 1044
  bi-convex, 1044
  converging, 1023, *1024,* 1025
  diffraction by, 1091
  diverging, 1023, *1024,* 1025
  magnifying, *1031,* 1031–1032
  meniscus concave, 1044
  meniscus convex, 1044
  plane-concave, 1044
  plane-convex, 1044
  simple magnifying, *1031,* 1031–1032
  symmetric, 1025–1026
  thin-film interference of coating on, 1068
lens maker's equation, 1024
Lenz's law, *868,* 868–871, *869,* 881
leptons, 1338, 1343–1345, 1344t
  conservation of lepton number, 1344–1345
  formation in early universe, 1359
lepton number, 1344–1345
lifetime:
  compound nucleus, 1300
  radionuclide, 1287–1288
  subatomic particles, 1123
lifting capacity, balloons, 581
light, 445, 977. *See also* diffraction; interference; photons; reflection; refraction
  absorption and emission by atoms, 1221
  coherent, 1059, 1241
  components of, 993–994
  Doppler effect, 499
  in early universe, 1359–1360
  Huygens' principle, *1048,* 1048–1049
  incoherent, 1059

law of reflection, 991
law of refraction, 992, *1048,*
　1048–1052
monochromatic, 993, 995–996,
　1241
polarized light, *907, 986,*
　*986–989, 988*
as probability wave, 1162–1164
speed of, 445
travel through media of
　different indices of
　refraction, 1050, *1050*
unpolarized light, 986, *986*
visible, 974, 1118
as wave, 1047–1052, *1048*
wave theory of, 1047–1052,
　1081–1083
white, *993, 993–994, 994,* 1085
light-emitting diodes (LEDs),
　1268–1270, *1269*
light-gathering power refracting
　telescope, 1033
lightning, 610, 717
　in creation of lodestones, 964
　upward streamers, 672, *672*
light quantum, 1154–1155
light wave, 977, 982–983
light-year, 12
line(s):
　diffraction gratings, 1099–1100
　spectral, 1206
　as unit, 8
linear charge density, 638–639, 639t
linear density, of stretched string,
　452, 453
linear expansion, 521, *521*
linear momentum, 224–225, 312t
　completely inelastic collisions
　　in one dimension, 234–236
　conservation of, 230–232, 242
　elastic collisions in one
　　dimension, with moving
　　target, 238–239
　elastic collisions in one
　　dimension, with stationary
　　target, 237–238
　elastic collisions in two
　　dimensions, 240–241
　at equilibrium, 328
　and impulse of series of
　　collisions, 227–228
　and impulse of single collision,
　　226–227
　inelastic collisions in one
　　dimension, *234,* 234–236, *235*
　inelastic collisions in two
　　dimensions, 240–241
　of photons, *1159,* 1159–1162,
　　*1160*
　sample problems involving, 229,
　　231–232, 236, 239–240, 243
　system of particles, 225
linear momentum-impulse
　theorem, 227
linear motion, 259
linear oscillator, *419,* 419–421

linear simple harmonic oscillators,
　*419,* 419–421
line integral, 692
line of action, of torque, 278, *278*
lines of charge:
　electric field due to, 638–643,
　　*639*
　electric potential due to, *699,*
　　699–700
lines of force, 631
line of symmetry, center of mass
　of solid bodies with, 217
line shapes, diffraction grating,
　1103
liquefaction, of ground during
　earthquakes, 7–8
liquids:
　compressibility, 341, 387
　density of selected, 387t
　as fluids, 386–387
　heat absorption, 524–527
　speed of sound in, 481t
　thermal expansion, 520–522
liquid state, 525–526
Local Group, 354
Local Supercluster, 354
lodestones, 950, 964
longitudinal magnification, 1045
longitudinal motion, 446
longitudinal waves, 446, *446*
long jump, conservation of
　angular momentum in,
　314, *314*
loop model, for electron orbits,
　*955,* 955–956, *956*
loop rule, 775, 781, *883,* 883–884
Lorentz factor, 1122–1123, *1123,*
　1138
Lorentz transformation,
　1129–1133
　consequences of, 1131–1133
　pairs of events, 1130t
Loschmidt number, 581
loudness, 489
*L* shell, 1238, *1238*
Lyman series, 1206, *1207,* 1212

## M

Mach cone, 503, *503*
Mach cone angle, 503, *503*
Mach number, 503
magic electron numbers, 1299
magnets, 610, 803–808, *804, 807,*
　950–952
　applications, 803–804
　bar, 806–807, *807,* 826, 826t,
　　942, *942,* 950, *950*
　electromagnets, 804, *804,* 806t
　north pole, 807, *807, 942*
　permanent, 804
magnetically hard material,
　966, 996
magnetically soft material,
　966, 996
magnetic confinement, 1327
magnetic dipoles, 807, 824–826,
　*825, 942, 942*

magnetic dipole moment,
　824–826, *825,* 1221, 1222,
　*1222. See also* orbital mag-
　netic dipole moment; spin
　magnetic dipole moment
　of compass needle, 964
　diamagnetic materials, 957–958
　effective, 1225
　ferromagnetic materials,
　　957, 962
　paramagnetic materials,
　　957, 959
magnetic domains, 962–964, *963*
magnetic energy, 887–888
magnetic energy density, 889–890
magnetic field, 803–827, 836–854.
　　*See also* Earth's magnetic
　　field
　Ampere's law, *844,* 844–848
　circulating charged particle,
　　814–817, *815, 816*
　crossed fields and electrons,
　　808–810, *811*
　current-carrying coils as
　　magnetic dipoles, 851–854
　cyclotrons and synchrotrons,
　　*818,* 818–819
　defined, 804–808, *805*
　dipole moment, 824–826
　displacement current,
　　946–950, *947*
　due to current, 836–842
　Earth, *950,* 950–951
　energy density of, 889–890
　energy stored in, 887–888
　external, 957–960
　and Faraday's law of induction,
　　865–866
　force on current-carrying wires,
　　820–822
　Gauss' law for, 941–943, 949t
　Hall effect, *810,* 810–813
　induced, 943–946, *944*
　induced electric field from,
　　878–879
　induced emf in, 870–871
　and Lenz' law, *868,* 868–871, *869*
　parallel currents, 842–843, *843*
　producing, 804
　rms of, 982–983
　selected objects and situations,
　　806t
　solenoids and toroids, 848–851
　torque on current loops,
　　822–824, *823*
　traveling electromagnetic
　　waves, 974–977, *975, 976*
magnetic field lines, 806–807, *807,*
　838–839
magnetic flux, 866–867, 880, 942
magnetic force, 610, 805
　circulating charged particle,
　　814–817, *815, 816*
　current-carrying wire, *820,*
　　820–822, 842–843, *843*
　particle in magnetic field, *805,*
　　805–806

magnetic materials, 941, 956–957
magnetic monopole, 804, 942
magnetic potential energy,
　887–888
magnetic resonance, 1229–1230,
　*1230*
magnetic resonance imaging
　(MRI), 941, *941*
magnetic wave component, of
　electromagnetic waves,
　975–976, *976*
magnetism, 941–965. *See also*
　Earth's magnetic field
　of atoms, 1221, *1221*
　diamagnetism, 957–958, *958*
　and displacement current,
　　946–950
　of electrons, 952–957
　ferromagnetism, 957, 961–964,
　　*962*
　Gauss' law for magnetic fields,
　　941–943
　induced magnetic fields,
　　943–946
　magnets, 950–952
　Mid-Atlantic Ridge, 951, *951*
　paramagnetism, 957, 959,
　　959–961
magnetization:
　ferromagnetic materials, *962*
　paramagnetic materials,
　　*959,* 960
magnetization curves:
　ferromagnetic materials, *962*
　hysteresis, 963, *963*
　paramagnetic materials, 960
magnetizing current, transformers,
　931
magnetometers, 951
magnification:
　angular, 1031–1033
　lateral, 1017, 1027–1030, 1032
　longitudinal, 1045
magnifying lens, simple, *1031,*
　1031–1032
magnitude:
　of acceleration, in one-
　　dimensional motion, 20
　of acceleration, in two- and
　　three-dimensional motion, 68
　of angular momentum, 305–306
　of displacement in one-dimen-
　　sional motion, 14–15
　estimating order of, 4–5
　of free-fall acceleration, 27
　of vectors, 41–42, *42*
　of velocity, in one-dimensional
　　motion, 15
　of velocity, in two- and three-
　　dimensional motion, 68
magnitude-angle notation (vec-
　tors), 43
magnitude ratio, traveling electro-
　magnetic waves, 976
majority carriers, 1264, *1266,*
　1266–1267
mantle (Earth), *360,* 380, *380*

mass, 283t
  defined, 97–98
  sample problems involving, 243
  scalar nature of, 41, 98
  of selected objects, 7t
  units of, 6–8
  and wave speed on stretched
    string, 452
  weight vs., 104
mass dampers, 422
mass energy, 1138–1139, 1139t
mass excess, 1283
mass flow rate, 400
massless cord, 105, *105*
massless-frictionless pulleys, *105,*
  106, *108,* 108–109
massless spring, 160
mass number, 621–622, 1280,
  1280t
mass spectrometer, 817, *817*
matter:
  antimatter, 1310t, 1338–1339
  baryonic, 1358, *1361*
  dark, 1358, 1361, *1361*
  energy released by 1 kg, 1310t
  magnetism of, *see* magnetism
  nonbaryonic, 1361, *1361*
  nuclear, 1285
  particle nature of, *1168,*
    1168–1169, *1169*
  wave nature of, 1166–1170
matter waves, 445, 1166–1179,
  1186–1213
  barrier tunneling by, 1176–1179
  of electrons, 1166–1170, *1167,*
    *1169,* 1173, 1186
  of electrons in finite wells,
    1195–1197, *1196*
  energies of trapped electrons,
    1186–1191
  and Heisenberg uncertainty
    principle, 1172–1174
  hydrogen atom models,
    1201–1212
  reflection from a potential step,
    1174–1176
  Schrödinger's equation for,
    1170–1172
  two- and three-dimensional
    electron traps, 1197–1201
  wave functions of trapped
    electrons, 1191–1195
matter wave interference, 1168
maxima:
  diffraction patterns, 1082, *1082*
  double-slit interference, *1055,*
    1055–1057, 1060–1061
  single-slit diffraction,
    1083–1085, 1088, 1090
  thin-film interference, 1066
Maxwell, James Clerk, 561, 569,
  610, 844, 944, 973–974, 984,
  1048, 1353
Maxwellian electromagnetism,
  1334
Maxwell's equations, 941,
  949t, 1171

Maxwell's law of induction, 944,
  979
Maxwell's rainbow, 973–974
Maxwell's speed distribution law,
  *561,* 561–563
mean free distance, 759
mean free path, of gases, *558,*
  558–560
mean free time, 759
mean life, radioactive decay, 1287,
  1335
measurement, 1–8
  of angles, 45
  conversion factors, 3
  International System of
    Units, 2–3
  of length, 3–4
  of mass, 6–8
  of pressure, *392,* 392–393
  sample problems involving,
    4–5, 7–8
  significant figures and decimal
    places, 4
  standards for, 1–2
  of time, 5–6
mechanical energy:
  conservation of, 184–187
  and conservation of total
    energy, 195
  damped harmonic oscillator,
    430–431
  and electric potential energy,
    705
  satellites in orbit, 371–372, *372*
  simple harmonic motion,
    421–422, *422*
mechanical waves, 445. *See also*
  wave(s)
medium, 977
megaphones, 1082
melting point, 525, 526t
meniscus concave lens, 1044
meniscus convex lens, 1044
mercury barometer, 388, 392, *392*
mercury thermometer, 520
mesons, 1338, 1345–1346
  and eightfold way, 1347–1348,
    1347t
  and quark model, 1349–1351,
    1355
  underlying structure suggested,
    1348
messenger particles, 1352–1354
metals:
  coefficients of linear expansion,
    521t
  density of occupied states,
    1259–1260, *1260*
  density of states, *1257,*
    1257–1258
  elastic properties of selected,
    341t
  electrical properties, 1252–1261
  lattice, 339, *339*
  occupancy probability, *1258,*
    1258–1259
  resistivities of selected, 754t

speed of sound in, 481t
  thermal conductivity of
    selected, 535t
  unit cell, 1253
metallic conductors, 746, 762
metal-oxide-semiconductor-field
  effect-transistor (MOSFET),
  *1270,* 1270–1271
metastable states, 1242
meter (m), 1–4
metric system, 2
Michelson's interferometer,
  1070–1071, *1071*
microfarad, 718
micron, 8
microscopes, 1030, 1032, *1032*
microscopic clocks, time dilation
  tests, 1123
microstates, in statistical mechanics, 599–600
microwaves, 445, 499, 649
Mid-Atlantic Ridge, magnetism,
  951, *951*
Milky Way galaxy, 354, 355
Millikan oil-drop experiment,
  *645,* 645–646
millimeter of mercury (mm Hg),
  388
miniature black holes, 379
minima:
  circular aperture diffraction,
    1091, *1091*
  diffraction patterns, 1082, *1082*
  double-slit interference, 1055,
    *1055,* 1056, 1060–1061
  single-slit diffraction,
    1083–1088, *1087*
  thin-film interference, 1067
minority carriers, 1264, 1267
mirage, 1011, *1011*
mirrors, 1012
  in Michelson's interferometer,
    1071
  plane, 1010–1014, *1012*
  spherical, *1015,* 1015–1021,
    *1016, 1033,* 1033–1034
moderators, for nuclear reactors,
  1317
modulus of elasticity, 340
Mohole, 380
molar mass, 550
molar specific heat, 525, 564–571
  at constant pressure, 566–567,
    *566–567*
  at constant volume, *565,*
    565–566, 565t, *567*
  and degrees of freedom,
    568–570, 569t
  of ideal gas, 564–568
  and rotational/oscillatory
    motion, *570,* 570–571
  of selected materials, 525t
molecular mass, 550
molecular speeds, Maxwell's
  distribution of, 560–563, *561*
molecules, 1220
moment arm, 278, *278*

moment of inertia, 272
momentum, 224–225. *See also*
  angular momentum; linear
  momentum
  center of momentum frame,
    1151
  and kinetic energy, 1141, 1142
  in pion decay, 1342
  in proton decay, 1348
  and relativity, 1138
  and uncertainty principle,
    1173–1174
monatomic molecules, 564,
  568–570, *569, 569t*
monochromatic light, 993
  lasers, 1241
  reflection and refraction of,
    995–996
monovalent atom, 1256
Moon, 354, 355
  escape speed, 367t
  potential effect on humans,
    378–379
  radioactive dating of rocks,
    1296
more capacitive than inductive
  circuit, 924
more inductive than capacitive
  circuit, 924
Moseley plot, *1238,* 1239–1240
MOSFET (metal-oxide-semicon-
  ductor-field-effect transistor), *1270,* 1270–1271
most probable configuration, *600*
most probable speed in fusion,
  562, 1322, 1333
motion:
  graphical integration, *29,* 29–30
  one-dimensional, *see* one-
    dimensional motion
  oscillatory and rotational, *570,*
    570–571
  projectile, *70,* 70–75
  properties of, 14
  relative in one dimension, *78,*
    78–79
  relative in two dimensions, *80,*
    80–81
  of system's center of mass,
    220–221
  three-dimensional, *see*
    three-dimensional motion
  two-dimensional, *see*
    two-dimensional motion
motorcycle, acceleration of, 25–26
mountain pull, 380
MRI (magnetic resonance
  imaging), 941, *941*
*M* shell, 1238, *1238*
multiloop circuits, *781,*
  781–787, *782*
  current in, 781–782
  resistances in parallel, *782,*
    782–787
multimeter, 788
multiplication factor, nuclear
  reactors, 1318

multiplication of vectors, 50–55
  multiplying a vector by a
    scalar, 50
  multiplying two vectors, 50–55
  scalar product of, *51,* 51–52
  vector product of, 50, 52–55, *53*
multiplicity, of configurations in
  statistical mechanics, 599
muons, 1123–1124, 1335, 1343,
  1344t
  decay, 1341–1342
  from proton–antiproton
    annihilation, 1340, 1340t
muon neutrinos, 1343, 1344t
musical sounds, 492–496, *493, 495*
mutual induction, 890–892, *891*
mysterious sliding stones, 140

**N**
nanotechnology, 1187
National Institute of Standards
  and Technology (NIST), 6
natural angular frequency, 433,
  914
nautical mile, 11
NAVSTAR satellites, 1117
*n* channel, in MOSFET, 1270
near point, 1031, *1031*
negative charge, 611, *611*
negative charge carriers, 747, 750
negative direction, 14, *14*
negative lift, in race cars, *136,*
  136–137
negative terminal, batteries, *718,*
  718–719, 773
negative work, 530
net current, 845, 850
net electric charge, 611
net electric field, 634–635
net electric flux, 661–662
net electric potential, 692
net force, 99, 616–618
net torque, 278, 310–311, 823
net wave, 458, 495
net work, 153, 592
neutral equilibrium (potential
  energy curves), 190
neutralization, of charge, 611
neutral pion, 1118
neutrinos, 1292
  and beta decay, 1292, 1293
  and conservation of lepton
    number, 1344–1345
  in fusion, 1325
  as leptons, 1338
  as nonbaryonic dark matter,
    1358
  from proton–antiproton annihi-
    lation, 1340t
neutrons, 612, 1335
  accelerator studies, 818
  balance in nuclear reactors,
    *1317,* 1317–1318
  charge, 620, 620t
  control in nuclear reactors,
    *1317,* 1317–1320
  discovery of, 1353

and eightfold way, 1347t
as fermions, 1336
formation in early universe,
  1359
as hadrons, 1338
magnetic dipole moment, 826
and mass number, 621–622
as matter wave, 1168
spin angular momentum, 953
thermal, 1311–1315, 1317
neutron capture, 1300
neutron diffraction, 1168
neutron excess, 1281
neutron number, 1280, 1280t
neutron rich fragments, 1312
neutron stars, 88, 380
  density of core, 387t
  escape speed, 367t
  magnetic field at surface of,
    806t
newton (N), 96
Newton, Isaac, 95, 355, 369, 1082
Newtonian form, of thin-lens formula, 1043
Newtonian mechanics, 95, 1171,
  1334
Newtonian physics, 1187
newton per coulomb, 631
Newton's first law, 95–98
Newton's law of gravitation,
  355–356, 369
Newton's laws, 95, 108–113
Newton's second law, 98–101
  angular form, 307–308
  and Bohr model of hydrogen,
    1203–1204
  for rotation, 279–281
  sample problems involving,
    100–101, 108–113, 223,
    280–281
  system of particles, 220–223,
    *221*
  in terms of momentum,
    224–225
  translational vs. rotational
    forms, 283t, 312t
  units in, 99t
Newton's third law, 106–107
NIST (National Institute of
  Standards and Technology), 6
NMR (nuclear magnetic resonance), 1229–1230
NMR spectrum, 1229–1230, *1230*
noble gases, 1235, 1299
nodes, *465,* 466, 467–468
noise, background, 508
nonbaryonic dark matter, 1358
nonbaryonic matter, 1361, *1361*
nonconductors, 612
  electric field near parallel,
    674–675
  Gauss' law for, 673, *673*
nonconservative forces, 179
noninertial frame, 97
nonlaminar flow, 398
nonpolar dielectrics, 734
nonpolar molecules, 698

nonquantized portion, of energy
  level diagram, 1196, *1196*
nonsteady flow, 398
nonuniform electric field, 632,
  663–664
nonuniform magnetic field, *955,*
  956, *956*
nonviscous flow, 398
normal (optics), 991, *991*
normal force, *104,* 104–105
normalizing, wave function,
  1193–1195
normal vector, for a coil of
  current loop, 824
north magnetic pole, 950
north pole, magnets, 807, *807,* 942,
  *942*
*n*-type semiconductors, *1263,*
  1263–1264. *See also p-n*
  junctions
nuclear angular momentum,
  1284
nuclear binding energy, 1217,
  *1283,* 1283–1284, 1312, 1313
  per nucleon, 1283, *1283,* 1285,
    1312
  selected nuclides, 1280t
nuclear energy, 1284, 1309–1329
  fission, 1309–1316
  in nuclear reactors, 1316–1321
  thermonuclear fusion,
    1322–1329
nuclear fission, 1284, 1309–1316,
  *1313*
nuclear force, 1284
nuclear fusion, *see* thermonuclear
  fusion
nuclear magnetic moment, 1284
nuclear magnetic resonance
  (NMR), 1229–1230
nuclear physics, 1276–1301
  alpha decay, 1289–1291
  beta decay, 1292–1295
  discovery of nucleus, 1276–1279
  nuclear models, 1297–1300
  nuclear properties, 1279–1287
  radiation dosage, 1296–1297
  radioactive dating, 1295–1296
  radioactive decay, 1286–1289
nuclear power plant, *1318*
nuclear radii, 1282
nuclear reactions, 1139
nuclear reactors, 1316–1321
nuclear spin, 1284
nuclear weapons, 1284
nucleons, 1280, 1338
  binding energy per, 1283, *1283,*
    1285, 1312
  magic nucleon numbers, 1299
nucleus, 612
  discovery of, 1276–1279
  models, 1297–1300, *1298*
  mutual electric repulsion in,
    620–621
  properties of, 1279–1287
  radioactive decay, 621–622,
    1335–1336

nuclides, 1279, 1280t. *See also*
  radioactive decay
  halo, 1282
  magic nucleon numbers, 1299
  organizing, 1280–1281, *1281*
  transuranic, 1319
  valley of, 1294, *1294*
nuclidic chart, 1280–1281, *1281,*
  1293–1294, *1294*
number density:
  of charge carriers, 811–812,
    1253t, 1262
  of conduction electrons, 1256

**O**
objects:
  charged objects, 631
  electrically isolated, 611
  electrically neutral, 611
  extended, *1012,* 1012–1013,
    1026, *1026*
object distance, 1012
objective:
  compound microscope, 1032,
    *1032*
  refracting telescope, 1033, *1033*
occupancy probability, *1258,*
  1258–1259
occupied levels, 1231
occupied state density, 1259–1260,
  *1260*
ohm (unit), 753, 754
ohmic losses, 930
ohmmeter, 754
ohm-meter, 754
Ohm's law, 756–759, *757, 758*
oil slick, interference patterns
  from, 1064
one-dimensional elastic collisions,
  *237,* 237–240
one-dimensional electron traps:
  infinite potential well, 1188
  multiple electrons in, 1231
  single electron, 1187–1199
one-dimensional explosions, 231,
  *231*
one-dimensional inelastic
  collisions, *234,* 234–236, *235*
one-dimensional motion, 13–32
  acceleration, 20–30
  average velocity and speed,
    15–17
  constant acceleration, 23–27
  defined, 13
  free-fall acceleration, 27–28
  graphical integration for,
    29–30
  instantaneous acceleration,
    20–22
  instantaneous velocity and
    speed, 18–19
  position and displacement,
    14–15
  properties of, 14
  relative, *78,* 78–79
  sample problems involving,
    17–19, 22, 25–26, 28, 30, 79

Schrödinger's equation for, 1170–1172
one-dimensional variable force, 162–163, *163*
one-half rule, for intensity of transmitted polarized light, 987
one-way processes, 584
open ends (sound waves), 493–495
open-tube manometer, *392,* 392–393
optics, 973
optical fibers, 997, 1241, 1269
optical instruments, 1030–1036
optical interference, 1047. *See also* interference
orbit(s):
    circular vs. elliptical, 373–374
    eccentricity of, 369, 370t, 371–372
    geosynchronous, 382
    law of, 369, *369*
    sample problems involving, 373–374
    of satellites, 371–373, *372*
    semimajor axis of, 369, *369*
    of stars, 382
orbital angular momentum, 954, *955,* 1222–1224, *1223,* 1223t
orbital energy, 1204–1205
orbital magnetic dipole moment, 954, 1223–1224
    diamagnetic materials, 957–958
    ferromagnetic materials, 957
    paramagnetic materials, 959
orbital magnetic quantum number, 954–955, 1208, 1208t, 1223t
orbital quantum number, 1208, 1208t, 1223t, 1254
orbital radius, 1203–1204
order numbers, diffraction gratings, 1099
order of magnitude, 4–5
organizing tables, for images in mirrors, 1018, 1018t
orienteering, 44
origin, 14
oscillation(s), 413–434. *See also* electromagnetic oscillations; simple harmonic motion (SHM)
    of angular simple harmonic oscillator, *423,* 423–424
    damped, 430–431, *431*
    damped simple harmonic motion, 430–432
    energy in simple harmonic motion, 421–423
    forced, 432–433, *433*
    free, 432–433
    and molar specific heat, *570,* 570–571
    of pendulums, 424–428
    simple harmonic motion, 413–421
    simple harmonic motion and uniform circular motion, 428–429

oscillation mode, 467–468
out of phase:
    ac circuits, 920t
    capacitive load, 917–918
    inductive load, 919
    sound waves, 486
    thin-film interference, 1066
    waves, 459
overpressure, 393
oxygen, *569*
    distribution of molecular speeds at 300 K, *561*
    heats of transformation, 526t
    molar specific heat and degrees of freedom, 569t
    molar specific heat at constant volume, 565t
    paramagnetism of liquid, *959*
    RMS speed at room temperature, 556t

**P**
pair production, 622
pancake collapse, of tall building, 253
parallel-axis theorem, for calculating rotational inertia, 273–275, *274*
parallel circuits:
    capacitors, 724, *724,* 726–727, 783t
    resistors, *782,* 782–787, 783t
    summary of relations, 783t
parallel components, of unpolarized light, 998
parallel currents, magnetic field between two, 842–843, *843*
parallel-plate capacitors, 718, *718*
    capacitance, 720–721
    with dielectrics, 733–734, *734, 735*
    displacement current, *947,* 947–949
    energy density, 730
    induced magnetic field, 943–946
paramagnetic material, 957
paramagnetism, 957, *959,* 959–961
parent nucleus, 622
partial derivatives, 484, 978
partially occupied levels, 1231
partially polarized light, *907,* 986
particles, 14, 620. *See also specific types, e.g.:* alpha particles
particle accelerators, 818–819, 1334–1335, *1336*
particle–antiparticle annihilation, 1338
particle detectors, 1335, *1336*
particle nature of matter, *1168,* 1168–1169, *1169*
particle systems. *See also* collisions
    angular momentum, 310–311
    center of mass, 214–219, *215, 219*
    electric potential energy of, 703–705, *704*
    linear momentum, 225
    Newton's second law for, 220–223, *221*

pascal (Pa), 388, 480, 985
Pascal's principle, *393,* 393–394
Paschen series, 1206, *1207*
patch elements, 661
path-dependent quantities, 530
path-independent quantities, 688
    conservative forces, 179–181, *180*
    gravitational potential energy, 366
path length difference:
    double-slit interference, *1055,* 1055–1056, 1061–1063
    and index of refraction, 1051
    single-slit diffraction, *1083,* 1083–1084, *1084,* 1086
    sound waves, 486
    thin-film interference, 1065–1066
Pauli exclusion principle, 1230
    and bosons, 1337
    and energy levels in crystalline solids, 1254
    and fermions, 1337
    and Fermi speed, 1255–1256
    nucleons, 1298–1299
    and periodic table, 1235
pendulum(s), 424–428
    as angular simple harmonic oscillator, *423,* 423–424
    ballistic, 236, *236*
    bob of, 425
    conical, 146
    conservation of mechanical energy, *185,* 185–186
    physical, *426,* 426–428, *427*
    simple, *425,* 425–426
    torsion, *423, 423*
    underwater swinging (damped), 430
perfect engines, 593, *593*
perfect refrigerators, 596, *596*
perihelion distance, 371
period(s):
    law of, 370, *370,* 370t
    of revolution, 76
    simple harmonic motion, 414, *417, 418*
    sound waves, 483
    waves, 448, *448*
periodic motion, 414
periodic table, 1154, 1221
    building, 1234–1236
    x rays and ordering of elements, 1236–1240
permanent electric dipole moment, 698
permanent magnets, 804
permeability constant, 837
permittivity constant, 614–615
perpendicular components, of unpolarized light, 998
phase:
    simple harmonic motion, 416, *417*
    waves, 447, *447*
phase angle:
    alternating current, 920t

simple harmonic motion, 416, *417*
phase change, 525–526
phase constant:
    alternating current, 920t, 926
    series *RLC* circuits, 923–924, *924,* 926
    simple harmonic motion, 416, *417*
    waves, *448,* 448–449
phase difference:
    double-slit interference, 1055, 1060, 1061–1063
    Michelson's interferometer, 1071
    optical interference, 1050–1052
    and resulting interference type, 461t
    single-slit diffraction, 1086
    sound waves, 486
    thin-film interference, 1066
    waves, 459–460
phase shifts, reflection, 1065, *1065*
phase-shifted sound waves, 487
phase-shifted waves, 459–460
phasors, 462–464, *463*
    capacitive load, *917,* 917–918
    double-slit interference, 1061–1063
    inductive load, 919
    resistive load, 915–916
    series *RLC* circuits, *924*
    single-slit diffraction, 1086–1090, *1087, 1089*
phasor diagram, 462–463
phosphorus, doping silicon with, 1265
photodiode, 1269
photoelectric current, 1156
photoelectric effect, *1057,* 1155–1158
photoelectric equation, 1157–1158
photoelectrons, 1156
photomultiplier tube, 1164
photons, 1153–1155
    as bosons, 1337
    in early universe, 1359
    gamma-ray, 1324, 1338
    and light as probability wave, 1162–1164
    momentum, *1159,* 1159–1162, *1160*
    and photoelectric effect, 1155–1158
    as quantum of light, 1153–1155
    in quantum physics, 1164–1166
    virtual, 1353
photon absorption, 1154, 1155, 1221
    absorption lines, 1206, *1207*
    energy changes in hydrogen atom, 1205
    energy for electrons from, 1189–1190
    lasers, 1242
photon emission, 1154, 1221
    emission lines, 1206, *1207*

energy changes in hydrogen atom, 1205
    energy from electrons for, 1190
    lasers, *1242,* 1242–1243
    stimulated emission, 1242, 1243
physics, laws of, 47
physical pendulum, 426–428, *427*
picofarad, 718
piezoelectricity, 1178
pinhole diffraction, 1082
pions, 1118, 1335
    decay, 1341, 1342
    and eightfold way, 1347t
    as hadrons, 1338
    as mesons, 1338
    from proton–antiproton annihilation, 1339–1343, 1340t
    reaction with protons, 1342–1343
pipes, resonance between, 495–496
pitch, 387
pitot tube, 410–411
planar symmetry, Gauss' law, *673,* 673–675, *674*
planar waves, 480
Planck, Max, 1165–1166
Planck constant, 1154
plane-concave lens, 1044
plane-convex lens, 1044
plane mirrors, 1010–1014, *1012*
plane of incidence, 991
plane of oscillation, polarized light, 986, *986*
plane of symmetry, center of mass of solid bodies with, 217
plane-polarized waves, 985–986
plane waves, 974
plastics:
    electric field of plastic rod, 641–642
    as insulators, 612
plates, capacitor, *718,* 718–719
plate tectonics, 13
plum pudding model, of atom, 1277
*p-n* junctions, *1266,* 1266–1270
    junction lasers, 1269, *1269*
    junction rectifiers, 1267–1268, *1268*
    light-emitting diodes (LEDs), 1268–1270, *1269*
*pn* junction diode, 757, 762
point (unit), 8
point charges. *See also* charged particles
    Coulomb's law, *613,* 613–619
    in electric field, 633–635, 645–647
    electric potential due to, *694,* 694–695, *695*
pointillism, 1092–1093
point image, 1012–1013
point of symmetry, center of mass of solid bodies with, 217
point source, 480
    isotropic, 489, 982
    light, 982, 1012

polar dielectrics, 733–734
polarity:
    of applied potential difference, 756–757
    of Earth's magnetic field, reversals in, *950,* 951
polarization, *907,* 985–990, *986, 988*
    intensity of transmitted polarized light, 987–990
    and polarized light, *986,* 986–987
    by reflection, 997–998, *998*
polarized light, *907, 986,* 986–989, *988*
polarized waves, *907,* 985–990, *986*
polarizer, 988
polarizing direction, 986–987, *987*
polarizing sheets, *907, 988,* 988–990
polarizing sunglasses, 998
polar molecules, 698
Polaroid filters, 986
pole faces, horseshoe magnet, 807
polyatomic gases, 565
polyatomic molecules, 566
    degrees of freedom, 568–570, *569,* 569t
    molar specific heats at constant volume, 565t
polycrystalline solids, 963
population inversion, in lasers, 1243–1245, 1269
porcelain, dielectric properties, 733
position, 283t
    one-dimensional motion, *14,* 14–15, *15*
    reference particle, 429
    relating linear to angular, 269
    simple harmonic motion, *417*
    two- and three-dimensional motion, *63,* 63–64, *64*
    uncertainty of particle, 1173–1174
position vector, 63, *63*
positive charge, 611, *734*
positive charge carriers, 747
    drift speed, 750
    emf devices, 773
positive direction, 14, *14*
positive ions, 612
positive kaons, 1124–1125
positive terminal, batteries, *718,* 718–719, 773
positrons:
    antihydrogen, 1338
    bubble chamber tracks, *622,* 806
    electron–positron annihilation, 622, *622,* 1338
    in fusion, 1322–1323
potassium, radioactivity of, 1289
potential, *see* electric potential
potential barrier, 1176–1179, *1177,* 1290–1291, 1314
potential difference, 779
    across moving conductors, 812–813
    across real battery, 778–780

for capacitive load, 918
    capacitors, 719–723, *720*
    capacitors in parallel, 724, *724,* 726–727
    capacitors in series, 724–727, *725*
    Hall, 811
    for inductive load, 920
    *LC* oscillations, 904
    and Ohm's law, 756–757
    for resistive load, 916
    resistors in parallel, 782–787
    resistors in series, *776,* 776–777, 784–787
    *RL* circuits, 882–886, *883*
    single-loop circuits, *774,* 774–775
    between two points in circuit, *777,* 777–780, *779*
potential energy, 177–183
    and conservation of mechanical energy, *184,* 184–187, *185*
    and conservation of total energy, 195–196
    defined, 177
    determining, 181–183
    electric, *686,* 686–689, *689,* 703–705, *704*
    of electric dipoles, 648
    in electric field, 689, 730
    magnetic, 887–888
    sample problems involving, 181, 183, 190–191, 194
    satellites in orbit, 371–372, *372*
    simple harmonic motion, 421–422, *422*
    and work, *178,* 178–181, *179*
    yo-yo, 301–302
potential energy barrier, 1176–1179, *1177*
potential energy curves, 187–191, *189*
potential energy function, 188–190, *189*
potential energy step, reflection from, 1174–1176, *1175*
potential method, of calculating current in single-loop circuits, 774–775
potential well, 190
potentiometer, *732*
pounds per square inch (psi), 388
power, 166–168, *167,* 197–198, 283t
    in alternating current circuits, 927–929
    average, 166
    defined, 166
    in direct current circuits, 760–761
    of electric current, 760–761
    and emf in circuits, 779
    radiated, 1166
    resolving, 1033, *1033,* 1102–1104, *1103,* 1183

in *RLC* circuit, 929, 933
    in rotation, 283
    sample problem involving, 168
    traveling wave on stretched string, *454,* 454–455
power factor, 927, 929
power lines, transformers for, 930
power transmission systems, 745, 930–931
Poynting vector, 980–983, *982*
precession, of gyroscope, *317,* 317–318
precession rate, of gyroscope, 318
prefixes, for SI units, 2t
pressure:
    fluids, 387–388
    and ideal gas law, 550–554
    measuring, *392,* 392–393
    radiation, 983–985
    and RMS speed of ideal gas, 554–556
    scalar nature of, 41
    as state property, 585
    triple point of water, 516
    work done by ideal gas at constant, 553
pressure amplitude (sound waves), 483, *484*
pressure field, 631
pressure sensor, *387*
pressurized-water nuclear reactor, 1318, *1318*
primary coil, transformer, 931
primary loop, pressurized-water reactor, *1318,* 1318–1319
primary rainbows, 994, 1007, 1052, *1052*
primary winding, transformer, 931
principal quantum number, 1208, 1208t, 1223t, 1254
principle of conservation of mechanical energy, 185
principle of energy conservation, 149
principle of equivalence, 374–375
principle of superposition, 96, 615
    for gravitation, 357–359
    for waves, 458, *458*
prisms, 994, *994,* 1005
probability, entropy and, 601–602
probability density, 1171–1172
    barrier tunneling, 1177
    finding, 1172
    trapped electrons, *1192,* 1192–1194
probability distribution function, 561
probability of detection:
    hydrogen electron, 1209, 1212
    trapped electrons, 1192–1194
probability wave:
    light as, 1162–1164
    matter wave as, 1167
projectile(s):
    defined, 70
    dropped from airplane, 74

elastic collisions in one dimension, with moving target, 238–239
elastic collisions in one dimension, with stationary target, 237–238
inelastic collisions in one dimension, 234
launched from water slide, 75
series of collisions, *228*
single collision, 226–227
projectile motion, *70,* 70–75
effects of air on, 73, *73*
trajectory of, 73, *73*
vertical and horizontal components of, 70–73, *71–73*
proper frequency, 1135
proper length, 1126
proper period, 1137
proper time, 1122
proper wavelength, 1135
protons, 612, 1335
accelerator studies, 818
and atomic number, 621
as baryons, 1338
charge, 620, 620t
decay of, 1348
in equilibrium, 618
as fermions, 1336
formation in early universe, 1359
in fusion, 1322–1329
as hadrons, 1338
magnetic dipole moment, 826, 826t
mass energy, 1139t
and mass number, 621–622
as matter wave, 1168, 1187
reaction with pions, 1342–1343
spin angular momentum, 953
ultrarelativistic, 1142–1143
proton number, 1280, 1280t
proton-proton (p-p) cycle, *1324,* 1324–1326
proton synchrotrons, 819
*p* subshells, 1235
*p*-type semiconductors, 1264, *1264*
pulleys, massless-frictionless, *105,* 106, *108,* 108–109
pulsars, secondary time standard based on, 9
pulse, wave, *445,* 446
P waves, 506

**Q**
QCD (quantum chromodynamics), 1354
QED (quantum electrodynamics), 954, 1352
quadrupole moment, 654
quanta, 1154
quantization, 629, 1154, 1187
electric charge, 619–621
energy of trapped electrons, 1187–1191
orbital angular momentum, 954
of orbital energy, 1204–1205
spin angular momentum, 953

quantum, 1154
quantum chromodynamics (QCD), 1354
quantum corrals, 1199, *1199*
quantum dots, 1187, *1198,* 1198–1199
quantum electrodynamics (QED), 954, 1352
quantum jump, 1189
quantum mechanics, 95, 1154
quantum numbers, 1188, 1223t
charge, 1341
conservation of, 1348–1349
for hydrogen, 1206–1208
orbital, 1208, 1208t, 1223t, 1254
orbital magnetic, 954–955, 1208, 1208t, 1223t
and Pauli exclusion principle, 1230
and periodic table, 1234–1236
principal, 1208, 1208t, 1223t, 1254
spin, 1223t, 1225, 1335–1336
spin magnetic, 953, 1223t, 1224, 1335–1336
quantum physics. *See also* electron traps; Pauli exclusion principle; photons; Schrödinger's equation
barrier tunneling, 1176–1179, *1177*
and basic properties of atoms, 1220–1222
confinement principle, 1187
correspondence principle, 1193
defined, 1154
Heisenberg's uncertainty principle, *1172,* 1172–1174
hydrogen wave function, 1208–1210
matter waves, 1187
nucleus, 1276
occupancy probability, *1258,* 1258–1259
particles, 1335
photons in, 1164–1166
and solid-state electronic devices, 1253
quantum states, 1187, 1221
degenerate, 1200
density of, *1257,* 1257–1258
density of occupied, 1259–1260, *1260*
hydrogen with n = 2, *1210,* 1210–1211
quantum theory, 569, 570–571, 1154, 1187
quantum transition, 1189
quantum tunneling, 1176–1179, *1177*
quarks, 818, 1349–1352, *1350,* 1350t
charge, 620
formation in early universe, 1359
quark family, 1350t
quark flavors, 1350, 1353–1354

quasars, *376,* 1356
quicksand, 412
Q value, 1140, 1291, 1294–1295, 1316, 1324–1325

**R**
*R*-value, 534–535
race cars:
collision with wall, 229, *229*
fuel dispenser fires, 792, *792*
negative lift in Grand Prix cars, *136,* 136–137
rad (unit), 1296–1297
radar waves, 445
radial component:
of linear acceleration, 270
of torque, 278
radial probability density, *1209,* 1211–1212
radians, 45, 260
radiated power, 1166
radiated waves, 974
radiation:
in cancer therapy, 1276
cosmic background, 1357–1358, 1360, *1361*
dose equivalent, 1297
electromagnetic, 974
reflected, 984
short wave, 974
ultraviolet, 950
radiation dosage, 1296–1297
radiation heat transfer, 536–538
radiation pressure, 983–985
radioactive dating, *1295,* 1295–1296
radioactive decay, 621–622, 1286–1289, 1335–1336
alpha decay, 1289–1291, *1290*
beta decay, 1292–1295, *1293,* 1351
muons, 1123
and nuclidic chart, 1293–1294, *1294*
process, 1286–1288
radioactive elements, 1277
radioactive wastes, *1318,* 1319
radioactivity, of potassium, 1289
radionuclides, 1280
radio waves, 445, 499, 974
radius of curvature:
spherical mirrors, *1015,* 1015–1016, *1016*
spherical refracting surfaces, 1020–1021, *1021*
radon, 1276
rail gun, 843, *843*
rainbows, *994,* 994–995
Maxwell's, 973–974
and optical interference, 1051–1052, *1052*
primary, 994, 1007, 1052, *1052*
secondary, 994, *994,* 1007, 1052
tertiary, 1007
ramp, rolling down, *299,* 299–300
randomly polarized light, 986, *986*

range, in projectile motion, 73, *73*
rare earth elements, 957, 1239
rattlesnake, thermal radiation sensors, 537
rays, 480, *480*
incident, 991, *991*
locating direct images with, *1018,* 1018–1019
locating indirect object images with, 1026, *1026*
reflected, 991, *991*
refracted, 991, *991*
ray diagrams, *1018,* 1018–1019
Rayleigh's criterion, *1091,* 1091–1094
RBE (relative biology effectiveness factor), 1297
*RC* circuits, 788–792, *789*
capacitor charging, *789,* 789–790
capacitor discharging, *789,* 790–792
real batteries, 773, *773, 777,* 777–778
real emf devices, 773, *773*
real engines, efficiency of, 593, 597–598
real fluids, 398
real focal point, 1016, *1016*
real images, 1011
spherical mirrors, 1017
spherical refracting surfaces, 1020–1021, *1021*
thin lenses, 1025, *1025*
real solenoids, *849*
recessional speed, of universe, 1357
rechargeable batteries, 773–774
recharging batteries, 779
red giant, 1325
red shift, 1135, 1367–1368
reference circle, 429
reference configuration, for potential energy, 182
reference frames, 78–79
inertial, 86–87
noninertial, 97
reference line, 259, *259*
reference particle, 429
reference point, for potential energy, 182
reflected light, 991
reflected radiation, 984
reflected ray, 991, *991*
reflecting planes, *1105,* 1105–1106
reflection, 990–998, *991*
first and second reflection points, 1006
law of, 991
polarization by, 997–998, *998*
from potential energy step, 1174–1176, *1175*
from a potential step, 1174–1176
of standing waves at boundary, 466–467, *467*
total internal, 996–997, *997*

reflection coefficient, 1176
reflection phase shifts, 1065, *1065*
reflectors, corner, 1046
refracted light, 991
refracted ray, 991, *991*
refracting telescope, 1032–1033, *1033*
refraction, 990–996, *991. See also* index of refraction
    angle of, 991, *991*
    and chromatic dispersion, *993,* 993–994
    law of, *992, 1048,* 1048–1052
refrigerators, 595–598, *596*
relative biology effectiveness (RBE) factor, 1297
relative motion:
    in one dimension, *78,* 78–79
    in two dimensions, *80,* 80–81
relative speed, 242
relativistic particles, 1124–1125
relativity, 1116–1144, 1153, 1334
    Doppler effect for light, 1134–1137, *1136*
    and energy, 1138–1143
    general theory of, 374–376, 1117, 1123–1124
    of length, 1125–1128, *1126,* 1131–1132
    Lorentz transformation, 1129–1133
    measuring events, 1118–1119, *1119*
    and momentum, 1138
    postulates, 1117–1118
    simultaneity of, *1120,* 1120–1121, 1131
    special theory of, 95, 977, 1117
    of time, *1121,* 1121–1125, 1131
    of velocities, *1133,* 1133–1134
relaxed state, of spring, *159,* 159–160
released energy, from fusion reaction, 1140
rem (unit), 1297
repulsion, in nucleus, 620–621
repulsive force, 610
resistance, 752–763
    alternating current, 920t
    Ohm's law, 756–759, *757*
    parallel circuits, *782,* 782–787
    and power in electric current, 760–761
    *RC* circuits, 788–792
    and resistivity, 752–756, *754*
    *RLC* circuits, 910–912, 921–926
    *RL* circuits, 882–886
    in semiconductors, 762–763
    series circuits, *776,* 776–777, 921–926
    superconductors, 763
resistance rule, 775
resistivity, 754, 1253
    calculating resistance from, *754,* 754–755
    Ohm's law, 756–759

selected materials at room temperature, 754t
semiconductors, 1262
silicon vs. copper, 762–763, 762t, 1253t
resistors, *753,* 753–754
    with ac generator, *914,* 914–916
    in multiloop circuits, 781–787, *782, 785*
    Ohm's law, 756–759, *757*
    in parallel, *782,* 782–787
    phase and amplitude in ac circuits, 920t
    power dissipation in ac circuits, 927
    and power in circuits, 760–761
    *RC* circuits, 788–792, *789*
    *RLC* circuits, 922
    *RL* circuits, 882–886, *883*
    in series, *776,* 776–777, 922
    single-loop circuits, *774,* 774–775
    work, energy, and emf, *773,* 773–774
resolvability, *1091,* 1091–1093
resolving power:
    diffraction grating, 1102–1104, *1103*
    microscope, 1183
    refracting telescope, 1033, *1033*
resolving vectors, 43
resonance:
    forced oscillations, 433
    magnetic, 1229–1230, *1230*
    magnetic resonance imaging, 941, *941*
    nuclear magnetic, 1229–1230
    between pipes, 495–496
    series *RLC* circuits, 924–926, *925*
    and standing waves, *467,* 467–470, *468*
resonance capture, of neutrons in nuclear reactors, 1317
resonance condition cyclotrons, 818
resonance curves, series *RLC* circuits, *925,* 925–926
resonance peak, 433, 1230
resonant frequencies, *467,* 467–468, 493, 494
response time, nuclear reactor control rods, 1318
rest, fluids at, 388–391, *389*
rest energy, 1139
rest frame, 1123
rest length, 1126
restoring torque, 425–426
resultant, of vector addition, 41
resultant force, 99
resultant torque, 278
resultant wave, 458, *458*
reverse saturation current, junction rectifiers, 1274
reversible processes, 585–588
right-handed coordinate system, 46, *46*

right-hand rule, 264–265, *265*
    Ampere's law, 843, *845*
    angular quantities, 264–265, *265*
    displacement current, *947*
    induced current, 868, *869*
    Lenz's law, 868, *868*
    magnetic dipole moment, 825, *825*
    magnetic field due to current, 838, *838*
    magnetic force, 805, 805–806
    magnetism, 843
    vector products, 52, *53,* 54, 842
rigid bodies:
    angular momentum of rotation about fixed axis, *311,* 311–312
    defined, 259
    elasticity of real, 339–340
ring charge distributions, 638–640, *639,* 642
Ritz combination principle, 1218
*RLC* circuits, 910–912, *911*
    resonance curves, *925,* 925–926
    series, 921–926, *922*
    transient current series, 923
*RL* circuits, 882–886, *883, 884*
RMS, *see* root-mean-square
RMS current:
    in ac circuits, 927–928
    in transformers, 933
rock climbing:
    crimp hold, 348, *348*
    energy conservation in descent using rings, 196, *196*
    energy expended against gravitational force climbing Mount Everest, 211
    friction coefficients between shoes and rock, 127
    lie-back climb along fissure, 347, *347*
rockets, 241–243, *242*
roller coasters, maximum acceleration of, 21
rolling, 295–302
    down ramp, *299,* 299–301
    forces of, *299,* 299–301
    friction during, *299,* 299
    kinetic energy of, *297,* 298–301
    as pure rotation, *296,* 296–297
    sample problem involving, 301
    as translation and rotation combined, 295–297, *297*
    yo-yo, 301–302, *302*
room temperature, 515
root-mean-square (RMS):
    and distribution of molecular speeds, 562
    of electric/magnetic fields, 982–983
    for selected substances, 556t
    speed, of ideal gas, 554–556, *555*
rotation, 257–287
    angular momentum of rigid body rotating about fixed axis, *311,* 311–312

conservation of angular momentum, *313,* 313–315, *314, 315*
    constant angular acceleration, 266–268
    kinetic energy of, 271–273, *272*
    and molar specific heat, *570,* 570–571
    Newton's second law for, 279–281
    relating linear and angular variables, 268–271, *269*
    in rolling, 295–297, *296*
    sample problems involving, 262–264, 267–268, 270–271, 275–277, 280–281, 284
rotational equilibrium, 329
rotational inertia, 272, 273–277, 283t
rotational kinetic energy, 271–272
    of rolling, 299
    and work, 282–284
    yo-yo, 301–302
rotational symmetry, *632, 633*
rotational variables, 259–265, 312t
rotation axis, 259, *259*
Rotor (amusement park ride), 267–268
Rowland ring, 962, *962*
rubber band, entropy change on stretching, 589–590
rulers, 2
rulings, diffraction grating, 1098
Rutherford, Ernest, 1276–1277
Rutherford scattering, 1278–1279
Rydberg constant, 1205

**S**
Sagittarius A*, 355
satellites:
    energy of, in orbit, 371–373
    geosynchronous orbit, 382
    gravitational potential energy, 365
    Kepler's laws, 368–371
    orbits and energy, *372*
scalars:
    multiplying vectors by, 50
    vectors vs., 40–41
scalar components, 46
scalar fields, 631
scalar product, *51,* 51–52
scanning tunneling microscope (STM), 1178, *1178,* 1199, *1199*
scattering:
    Compton, *1159,* 1159–1162, *1160*
    of polarized light, 988
    Rutherford, 1278–1279
    x rays, 1105, *1105*
schematic diagrams, 718
Schrödinger's equation, 1170–1172
    for electron in finite well, 1195
    for electron in infinite well, 1192

for electron in rectangular box, 1200
for electron in rectangular corral, 1200
and hydrogen, 1205–1212
for hydrogen ground state, 1208–1210, *1209*
for multicomponent atoms, 1234
probability density from, 1172
scientific notation, 2–3
Scoville heat unit, 12
screen, in Young's experiment, 1057
seat of emf, 772
secondary coil, transformer, 931
secondary loop, pressurized water reactor, *1318*, 1319
secondary maxima, diffraction patterns, 1082, *1082*
secondary rainbows, 994, *994*, 1007, 1052
secondary standards, 3
secondary winding, transformer, 931
second law of thermodynamics, 588–590
second minima:
    and interference patterns, 1057
    for single-slit diffraction, 1084, 1087–1088
second-order bright fringes, 1056–1057
second-order dark fringes, 1057
second-order line, 1099
second reflection point, 1006
second side maxima, interference patterns of, 1056–1057
secular equilibrium, 1304
seismic waves, 445, 512
self-induced emf, 881, *881*
self-induction, *881*, 881–882, 890
semi-classical angle, 1223
semiconducting devices, 762
semiconductors, 612, 1261–1265. *See also p-n* junctions; transistors
    doped, *1263*, 1263–1265
    electrical properties, 1262, *1262*
    LEDs, 1268–1270, *1269*
    nanocrystallites, 1198, *1198*
    *n*-type, *1263*, 1263–1264. *See also p-n* junctions
    *p*-type, 1264, *1264*
    resistance in, 762–763
    resistivities of, 754t
    unit cell, 1253
semimajor axis, of orbits, 369, *369*, 370t
series, of spectral lines, 1206
series circuits:
    capacitors, 724–727, *725*, 783t
    *RC*, 788–792, *789*
    resistors, *776*, 776–777, 783t
    *RLC*, *911*, 921–926, *922*
    summary of relations, 783t
series limit, 1206, *1207*

shake (unit), 11
shearing stress, 340, *340*
shear modulus, 341
shells, 1211, 1225
    and characteristic x-ray spectrum, 1237–1238
    and electrostatic force, 615
    and energy levels in crystalline solids, 1254
    and periodic table, 1234–1236
shell theorem, 356
SHM, *see* simple harmonic motion
shock waves, 33, 503, *503*
short wave radiation, 974
side maxima:
    diffraction patterns, 1082, *1082*
    interference patterns, 1056–1057
sievert (unit), 1297
sigma particles, 1335, 1346, 1347t
sign:
    acceleration, 21–22
    displacement, 14–15
    heat, 523
    velocity, 21–22, 29
    work, 153
significant figures, 4
silicon:
    doping of, 1265
    electric properties of copper vs., 762–763, 762t, 1253t, 1262
    in MOSFETs, 1270
    properties of *n*- vs. *p*-doped, 1264t
    resistivity of, 754t
    as semiconductor, 612, 762–763, 1262
    unit cell, 1253, *1253*
silk, rubbing glass rod with, 610, *610*, 621
simple harmonic motion (SHM), 413–434, *415*, *417*
    acceleration, 418, *418*, 420
    angular, *423*, 423–424
    damped, *430*, 430–432, *431*
    energy in, 421–423, *422*
    force law for, 419
    freeze-frames of, 414–416, *415*
    pendulums, 424–428, *425*, *426*
    quantities for, *416*, 416–417
    sample problems involving, 420–424, 427–428, 432
    and uniform circular motion, 428–429, *428–429*
    velocity, *417*, 417–418, *418*, 421
    waves produced by, 445–446
simple harmonic oscillators:
    angular, *423*, 423–424
    linear, *419*, 419–421
simple magnifying lens, *1031*, 1031–1032
simple pendulum, *425*, 425–426
simultaneity:
    and Lorentz transformation equations, 1131
    relativity of, *1120*, 1120–1121

sine, 45, *45*
single-component forces, 96
single-loop circuits, 771–780, *914*
    charges in, 772
    current in, *774*, 774–775
    internal resistance, 776, *776*
    potential difference between two points, *777*, 777–780, *779*
    with resistances in series, *776*, 776–777
    work, energy, and emf, *773*, 773–774
single-slit diffraction, 1081–1090
    intensity in, 1086–1090, *1087*, *1089*
    minima for, *1083*, 1083–1085, *1084*
    and wave theory of light, 1081–1083
    Young's interference experiment, 1053–1054, *1055*
sinusoidal waves, 446–448, *447*, *448*
siphons, 412
Sirius B, escape speed for, 367t
SI units, 2–3
skateboarding, motion analyzed, *73*
slab (rotational inertia), 274t
sliding block, *108*, 108–109
sliding friction, 126, *127*
slope, of line, 15–16, *16*
Snell's law, 992, 1048–1049
snorkeling, 407
soap bubbles, interference patterns from, 1064, 1067, *1067*
sodium, 1235
sodium chloride, 1236
    index of refraction, 992t
    x-ray diffraction, 1105, *1105*
sodium doublet, 1250
sodium vapor lamp, 1155
soft reflection, of traveling waves at boundary, 467
solar system, 1361
solar wind, 1002
solenoids, 848–851, *849*
    induced emf, 867–868
    inductance, 880
    magnetic energy density, 889
    magnetic field, 848–851, *849*
    real, *849*
solids:
    compressibility, 342
    crystalline, 1252–1261, *1253*, *1254*
    elasticity and dimensions of, 340, *340*
    heat absorption, 524–527
    polycrystalline, 963
    specific heats of selected, 525t
    speed of sound in, 481t
    thermal conductivity of selected, 535t
    thermal expansion, 520–522, *521*
solid bodies:
    center of mass, 216–219
    Newton's second law, 221

solid state, 525
solid-state electronic devices, 1253
sonar, 480
sonic boom, 503
sound intensity, 488–492, *489*
sound levels, 488–492, 490t
sound waves, 445–446, 479–504
    beats, 496–498, *497*
    defined, 479–480
    Doppler effect, 498–502
    intensity and sound level, 488–492, *489*, 490t
    interference, 485–488, *486*
    sample problems involving, 485, 487–488, 491–492, 495–496, 498, 502
    sources of musical, 492–496, *493*, *495*
    speed of, 480–482, 481t
    supersonic speed, *503*, 503–504
    traveling waves, 482–485, *483*
south pole, magnet's, 807, *807*, 942, *942*
space charge, 1266
space curvature, *375*, 375–376
space time, 375, 1153, 1359
spacetime coordinates, 1118–1119
spark, *see* electric spark
special theory of relativity, 95, 977, 1117
specific heat, 524–525, 525t. *See also* molar specific heat
speckle, 1059
spectral radiancy, 1165–1166
spectroscope, grating, *1100*, 1100–1101
spectrum, 1206
speed:
    average in one-dimensional motion, 16
    drift, *749*, 749–750, 752, 810–813, *811*
    escape, 704, 713
    Fermi, 1255–1256
    most probable, 1322, 1333
    one-dimensional motion, 18
    recessional, of universe, 1357
    relating linear to angular, 269
    relative, 242
    in rolling, 296–297, *297*
    waves, *see* wave speed
speed amplifier, 254
speed deamplifier, 254
speed of light, 445, 977, 1117–1118
speed of light postulate, 1117–1118
speed of sound, 480–482
    and RMS speed in gas, 556
    in various media, 481t
speed parameter, in time dilation, 1122–1123, *1123*
spherical aberrations, 1033
spherical capacitors, 722, 730
spherical conductors, Coulomb's law for, 615–619
spherically symmetric charge distribution, 675–677, *676*, 695

spherical mirrors, *1015, 1016*
   focal points, 1015–1016, *1016*
   images from, 1014–1020, *1015,*
     *1016, 1033,* 1033–1034
spherical refracting surfaces,
   1020–1022, *1021,* 1034, *1034*
spherical shell:
   Coulomb's law for, 615–619
   electric field and enclosed
     charge, 670
   rotational inertia of, 274t
spherical symmetry, Gauss' law,
   675–677, *676*
spherical waves, 480
spin, 1223t, 1336–1337
   electron, 1336–1337, *1337*
   isospin, 1364
   nuclear, 1284
   nuclides, 1280t, 1284
spin angular momentum, 953–954,
   1223t, 1224, *1225*
spin-down electron state, 953,
   1224, 1229, *1229*
spin-flipping, 966, 1229, *1230*
spin magnetic dipole moment,
   953–954, *954,* 1225, *1225*
   diamagnetic materials, 957
   ferromagnetic materials, 957
   paramagnetic materials, 957, 959
spin magnetic quantum number,
   953, 1223t, 1224, 1335–1336
spin quantum number, 1223t,
   1225, 1335–1336
spin-up electron state, 953, 1224,
   1229, *1229*
spontaneous emission, *1242,*
   1242–1243
spontaneous otoacoustic
   emission, 508
spring constant, 159
spring force, 159–161
   as conservative force, 179, *179*
   work done by, *159,* 159–162
spring scale, *103,* 103–104
sprites, *637,* 637–638
*s* subshells, 1235
stable equilibrium potential
   energy curves, 190
stable static equilibrium, *328,*
   328–329, *329*
stainless steel, thermal
   conductivity of, 535t
standards, 1–2
standard kilogram, *6,* 6–7
standard meter bar, 3
Standard Model, of elementary
   particles, 1336
standing waves, 465–470, *466, 467,*
   1187
   reflections at boundary,
     466–467, *467*
   and resonance, *467,* 467–470,
     *468*
   transverse and longitudinal
     waves on, *445,* 446, *446*
   wave equation, 456–457
   wave speed on, 452–453, *453*

stars, 1153
   Doppler shift, 1135–1136
   formation in early universe,
     1360
   fusion in, 1284, 1322, *1324,*
     1324–1326
   matter and antimatter in,
     1338–1339
   neutron, 806t
   orbiting, 382
   rotational speed as function of
     distance from galactic center,
     1358, *1358*
state, 525
state function, entropy as,
   586–587
state properties, 585–586
static equilibrium, 327–329, *328,*
   *329*
   fluids, *389,* 390
   indeterminate structures,
     338–339, *339*
   requirements of, 329–330
   sample problems involving,
     332–337
static frictional force, *125–126,*
   125–127, 299
statistical mechanics, 598–602
steady flow, 398
steady-state current, 746, 923
Stefan–Boltzmann constant, 536,
   1166
step-down transformer, 931
step-up transformer, 931
Stern–Gerlach experiment, *1226,*
   1226–1228
stick-and-slip, 127
stimulated emission, 1242–1243
Stirling engines, 594, *594*
Stirling's approximation, 601
STM, *see* scanning tunneling
   microscope
stopping potential, photoelectric
   effect, *1057,* 1156, 1157
straight line charge distributions,
   642–643
strain, 339–342, *340*
strain gage, 341, *341*
strangeness, conservation of,
   1346–1357
strange particles, 1346
strange quark, 1349, 1350, 1350t
streamlines:
   in electric fields, 749
   in fluid flow, 399, *400*
strength:
   ultimate, 340, *340,* 341t
   yield, 340, *340,* 341t
stress, 340, *340*
   compressive, 340–341
   electrostatic, 744
   hydraulic, 341–342, 341t
   shearing, 340, *340*
   tensile, 340, *340*
stress-strain curves, 340, *340*
stress-strain test specimen, *340*
stretched strings, 480

energy and power of traveling
   wave on, *454,* 454–455
   harmonics, 469–470
   resonance, *467,* 467–470
strike-slip, 60
string theory, 1354
string waves, 451–455
strokes, 591
strong force, 123, 1284, 1338
   conservation of strangeness, 1346
   messenger particle, 1353–1354
strong interaction, 1340–1341
strong nuclear force, 621
subcritical state, nuclear reactors,
   1318
submarines:
   rescue from, 578
   sonar, 480
subshells, 1211, 1223t, 1225
   and energy levels in crystalline
     solids, 1254
   and periodic table, 1234–1236
substrate, MOSFET, 1270
subtraction:
   of vectors by components, 49
   of vectors geometrically, 42, *42*
Sun, 1361
   convection cells in, 536
   density at center of, 387t
   escape speed, 367t
   fusion in, 1284, 1322, *1324,*
     1324–1326
   monitoring charged particles
     from, 745
   neutrinos from, 1293
   period of revolution about
     galactic center, 382
   pressure at center of, 388t
   randomly polarized light, 986
   speed distribution of photons
     in core, 562
sunglasses, polarizing, 998
sunjamming, 118
sunlight, coherence of, 1059
superconductivity, 763
superconductors, 612, 763
supercooling, 605
supercritical state, nuclear
   reactors, 1318
supermassive black holes, 355
supernovas, 88, 367t, *1325,*
   1325–1326, 1361
supernova SN1987a, *1325*
supernumeraries, 1052, *1052*
superposition, principle of, *see*
   principle of superposition
supersonic speed, *503,* 503–504
surface charge density, 629, 639t
surface wave, 512
S waves, 506
symmetric lenses, 1025–1026
symmetry:
   axis of, 632
   center of mass of bodies
     with, 217
   cylindrical, Gauss' law, *671,*
     671–672

   importance in physics, 659
   of messenger particles, 1354
   planar, Gauss' law, *673,*
     673–675, *674*
   rotational, *632, 633*
   spherical, Gauss' law, 675–677,
     *676*
system, 99, 523. *See also* particle
   systems
systolic blood pressure, normal,
   387t

**T**
tangent, 45, *45*
tangential component:
   of linear acceleration, 269–270
   of torque, 278
target:
   collisions in two dimensions,
     *240,* 240–241
   elastic collisions in one dim-
     ension, with moving, 238–239
   elastic collisions in one
     dimension, with stationary,
     *237,* 237–238
   inelastic collisions in one
     dimension, 234
   series of collisions, 228, *228*
   single collision, 226–227
tattoo inks, magnetic particles in,
   941, *941*
tau neutrinos, 1344, 1344t
tau particles, 1344, 1344t
teapot effect, 406
telescopes, 1030, 1032–1033, *1033*
television, 803–804, 950
television waves, 445
temperature, 514–519
   defined, 515
   for fusion, 1323
   and heat, *523,* 523–524, 526–527
   and ideal gas law, 550–554
   measuring, 516–517
   and RMS speed of ideal gas,
     554–556
   sample problems involving, 519,
     522
   scalar nature of, 41
   selected values, 518t
   as state property, 585
   work done by ideal gas at
     constant, *552,* 552–553
   and zeroth law of thermody-
     namics, 515–516, *516*
temperature coefficient of
   resistivity, 755, 1253
   selected materials, 754t
   semiconductors, 1262
   silicon vs. copper, 762t, 1253t
temperature field, 631
temperature scales:
   Celsius, 518–519
   compared, *518*
   Fahreheit, 518–519
   Kelvin, *515,* 516–517
temporal separation, of events,
   1121

10-hour day, *5*
tensile stress, 340, *340*
tension force, *105*, 105–106
 and elasticity, 340–341
 and wave speed on stretched
  string, 453
terminals, battery, 718–719, 773
terminal speed, 130–132, *131*
tertiary rainbows, 1007
tesla (unit), 806
test charge, 631, *631, 632*
Tevatron, 1352
theories of everything (TOE), 1354
thermal agitation:
 of ferromagnetic materials, 962
 of paramagnetic materials,
  959–960
thermal capture, of neutrons, 1317
thermal conduction, 535, *535*
thermal conductivity, 535, 535t
thermal conductor, 535
thermal efficiency:
 Carnot engines, 592–593
 Stirling engines, 594
thermal energy, 179, 195, 514, 873
thermal equilibrium, 515
thermal expansion, *520*, 520–522
thermal insulator, 535
thermal neutrons, 1311–1315, 1317
thermal radiation, 536–538
thermal reservoir, 528, *529*
thermal resistance to conduction,
 535
thermodynamics:
 defined, 514
 first law, 528–533
 second law, 588–590
 zeroth law, 515–516, *516*
thermodynamic cycles, *529*, 530,
 532
thermodynamic processes,
 528–531, *529, 575*
thermometers, 515
 constant-volume gas, *516*,
  516–517
 liquid-in-glass, 520
thermonuclear bomb, 1326–1327
thermonuclear fusion, 1140, 1284,
 1322–1329
 controlled, 1326–1329
 process of, 1322–1323
 in Sun and stars, 1322, *1324*,
  1324–1326
thermopiles, 772
thermoscope, 515, *515*
thin films, interference, *1064*,
 1064–1071
thin lenses, 1023–1030
 formulas, 1024
 images from, 1023–1030, *1025*,
  *1026*, 1034–1036, *1035*
 two-lens systems, *1027*,
  1027–1029
thin-lens approximation, 1035–1036
third-law force pair, 106, 356
three-dimensional electron traps,
 *1200*, 1200–1201

three-dimensional motion:
 acceleration, 66, *66*
 position and displacement,
  63, *63*
 velocity, 64–66, *65, 66*
three-dimensional space, center of
 mass in, 216
thrust, 242
thunderstorm sprites, *637*, 637–638
time:
 directional nature of, 584
 for free-fall flight, 28
 intervals of selected events, 5t
 proper, 1122
 between relativistic events,
  *1121*, 1121–1125
 relativity of, *1121*, 1121–1125,
  1131
 sample problems involving, 7–8
 scalar nature of, 41
 space, 1153, 1359
 units of, 5–6
time constants:
 inductive, 884–885
 for *LC* oscillations, 904
 for *RC* circuits, *789*, 790
 for *RL* circuits, 884–885
time dilation, 1122
 and length contraction,
  1127–1128
 and Lorentz transformation,
  1131
 tests of, 1123–1125
time intervals, 5, 5t
time signals, 6
TOE (theories of everything),
 1354
tokamak, 1327
ton, 11
top gun pilots, turns by, 77–78
top quark, 1350t, 1351, 1352
toroids, 850, *850*
torque, 277–281, 302–304, 312t
 and angular momentum of system of particles, 310–311
 and conservation of angular
  momentum, 313
 for current loop, 822–824, *823*
 of electric dipole in electric
  field, 650
 and gyroscope precession, 317,
  *317*
 internal and external, 310–311
 and magnetic dipole moment,
  825
 net, 278, 310–311
 Newton's second law in angular
  form, 307
 particle about fixed point, *303*,
  303–304
 restoring, 425–426
 rolling down ramp, 299–300
 sample problems involving, 304,
  308–309
 and time derivative of angular
  momentum, 308–309
torr, 388

torsion constant, 423
torsion pendulum, 423, *423*
total energy, relativity of,
 1139–1140
total internal reflection, 996–997,
 *997*
tour jeté, *314*, 314–315
Tower of Pisa, 337
tracer, for following fluid flow,
 398–399, *399*
trajectory, in projectile motion, 73
transfer:
 collisions and internal energy
  transfers, 196–197
 heat, 534–538
transformers, 930–933
 energy transmission requirements, 930–931
 ideal, *931*, 931–932
 impedance matching, 932
 in *LC* oscillators, *974*
transient current series *RLC*
 circuits, 923
transistors, 762, 1270–1271
 FET, *1270*, 1270–1271
 MOSFET, *1270*, 1270–1271
transition elements, paramagnetism of, 957
translation, 258, 295–297, *296*
translational equilibrium, 329
translational kinetic energy:
 ideal gases, 557
 of rolling, 298
 yo-yo, 301–302
translational variables, 312t
transmission coefficient, 1176,
 1177
transparent materials, 991
 in Michelson's interferometer,
  1071
 thin-film interference in,
  1068–1070, *1069*
transuranic nuclides, 1319
transverse Doppler effect, *1136*,
 1136–1137
transverse motion, 446
transverse waves, *445*, 445–446,
 450–451, 975
travel distance, for relativistic
 particle, 1124–1125
traveling waves, 446, 1187
 electromagnetic, 974–980, *976*,
  *977*
 energy and power, *454*,
  454–455
 hard vs. soft reflection of, at
  boundary, 467
 sound, 482–485, *483*
 speed, *449*, 449–451
 wave function, 1170–1172
travel time, 1119, 1142–1143
triangular prisms, 994, *994*
trigonometric functions, 45, *45*
triple point cell, 516, *516*
triple point of water, 516–517
tritium, *1294*, 1327, 1328
triton, 1327

tube length, compound microscope, 1032
tunneling, barrier, 1176–1179,
 *1177*, 1290–1291
turbulent flow, 398
turns:
 in coils, 823–824
 in solenoids, 848–849
turning points, in potential energy
 curves, 188–189, *189*
turns ratio, transformer, 932, 933
two-dimensional collisions, *240*,
 240–241
two-dimensional electron traps,
 *1200*, 1200–1201
two-dimensional explosions,
 232, *232*
two-dimensional motion:
 acceleration, 67–69, *68*
 position and displacement,
  63–64, *64*
 projectile motion, 70–75
 relative, *80*, 80–81
 sample problems involving,
  63–64, 67, 69, 74–78, 80–81,
  229
 uniform circular motion, 76–78
 velocity, 64–67

**U**
ultimate strength, 340, *340*, 341t
ultrarelativistic proton, 1142–1143
ultrasound (ultrasound imaging),
 480, *480*
 bat navigation using, 502
 blood flow speed measurement
  using, 511
ultraviolet light, 445
ultraviolet radiation, 950
uncertainty principle, *1172*,
 1172–1174
underwater illusion, 506
uniform charge distributions:
 electric field lines, *631*, 631–632,
  *632*
 types of, 642–643
uniform circular motion, 76–78
 centripetal force in, 133–138,
  *134*
 sample problems involving,
  135–138
 and simple harmonic motion,
  428–429, *428–429*
 velocity and acceleration for,
  *76, 77*
uniform electric fields, 632
 electric potential of, 692
 flux in, 660–662
units, 1–2
 changing, 3
 heat, 524
 length, 3–4
 mass, 6–8
 time, 5–6
unit cell, *1105*
 determining, with x-ray
  diffraction, 1106

metals, insulators, and semiconductors, 1253, *1253*
United States Naval Observatory time signals, 6
unit vectors, 46, *46*, 49, 54–55
universe:
    Big Bang, 1358–1361, *1359*
    color-coded image of universe at 379 000 yrs old, 1360, *1360*
    cosmic background radiation, 1357–1358, *1361*
    dark energy, 1361
    dark matter, 1358
    estimated age, 1356
    expansion of, 1356–1357
    temperature of early, 515
unoccupied levels, 1231, 1255, 1299
unpolarized light, 986, *986*
unstable equilibrium, 190
unstable static equilibrium, 328–329
up quark, 1349, 1350t, 1351
uranium, 387t
    enrichment of, 1317
    mass energy of, 1139t
uranium$^{228}$:
    alpha decay, 1289–1290
    half-life, 1290, 1291t
uranium$^{235}$:
    enriching fuel, 1317
    fission, 1311–1315, *1313*
    fissionability, 1314–1316, 1314t, 1321
    in natural nuclear reactor, 1320–1321
uranium$^{236}$, 1312, 1314t
uranium$^{238}$, 621–622, 1286
    alpha decay, 1289–1291, *1290*
    binding energy per nucleon, *1283*
    fissionability, 1314–1315, 1314t, 1321
    half-life, 1291, 1291t
uranium$^{239}$, 1314t
UTC (Coordinated Universal Time), 6

**V**

vacant levels, 1255
valence band, 1262, *1262*, 1263
valence electrons, 1187, 1235, 1256
valence number, 1263
valley of nuclides, 1294, *1294*
vaporization, 526
vapor state, 526
variable capacitor, 742
variable force:
    work done by general variable, 162–166, *163*
    work done by spring force, *159*, 160–162
variable-mass systems, rockets, 241–243, *242*
vector(s), 40–55, 631
    adding, by components, 46–47, 49
    adding, geometrically, *41*, 41–42, *42*, 44

area, 661
    for a coil of current loop, 824
    coupled, 1221
    and laws of physics, 47
    multiplying, 50–55, *51*, *53*
    Poynting, 980–983, *982*
    problem-solving with, 45
    resolving, 43
    sample problems involving, 44–45, 48–49, 54–55
    scalars vs., 40–41
    unit, 46, *46*, 49, 54–55
    velocity, 41
vector angles, 43, *43*, 45
vector-capable calculator, 46
vector components, 42–44, *43*
    addition, 46–49
    rotating axes of vectors and, *47*
vector equation, 41
vector fields, 631
vector product, 50, 52–55, *53*
vector quantities, 15, 41, 96
vector sum (resultant), *41*, 41–42
velocity, 283t
    angular, 260–265, *265*, 283t
    average, 15–17, *16*, 24, 65
    graphical integration in motion analysis, 29, *29*
    instantaneous, 18–19
    line of sight, 382
    and Newton's first law, 95–98
    and Newton's second law, 98–101
    one-dimensional motion, 15–19
    projectile motion, 70–75
    reference particle, 429
    relative motion in one dimension, 78–79
    relative motion in two dimensions, 80–81
    relativity of, *1133*, 1133–1134
    rockets, 241–243
    sign of, 21–22
    simple harmonic motion, *417*, 417–418, *418*, 421
    two- and three-dimensional motion, 64–67, *65–67*
    uniform circular motion, *76*, 76–78, *77*
    as vector quantity, 41
velocity amplitude:
    forced oscillations, 433, *433*
    simple harmonic motion, 418
velocity vectors, 41
venturi meter, 411
vertical circular loop, 135
vertical motion, in projectile motion, 72–73, *73*
virtual focal point, 1016, *1016*
virtual images:
    defined, 1011
    spherical mirrors, 1017
    spherical refracting surfaces, 1020–1021, *1021*
    thin lenses, 1025, *1025*
virtual photons, 1353
viscous drag force, 398

visible light, 445, 974, 1118
vision, resolvability in, 1092–1093
volcanic bombs, 90
volt, 687, 689
voltage. *See also* potential difference
    ac circuits, 920t
    transformers, 931–932
voltage law, Kirchoff's, 775
volt-ampere, 761
voltmeters, 788, *788*
volume:
    and ideal gas law, 550–554
    as state property, 585
    work done by ideal gas at constant, 553
volume charge density, 626, 628, 639t
volume expansion, 521–522
volume flow rate, 400, 660–661
volume flux, 660
volume probability density, *1209*, *1210*, 1211

**W**

water:
    boiling/freezing points of, in Celsius and Fahrenheit, 518t
    bulk modulus, 341, 481
    as conductor, 612
    density, 387t
    dielectric properties, 732t, 733–734
    diffraction of waves, *1053*
    as electric dipole, 648, *648*
    heats of transformation, 525–526, 526t
    index of refraction, 992t
    as insulator, 612
    in microwave cooking, 649
    as moderator for nuclear reactors, 1317
    polarization of light by reflection in, 998
    RMS speed at room temperature, 556t
    specific heats, 525t
    speed of sound in, 481, 481t
    thermal properties, 521
    thin-film interference of, 1067
    triple point, 516
water waves, 445
watt (W), 2, 167
Watt, James, 167
wave(s), 444–470. *See also* electromagnetic waves; matter waves
    amplitude, 447, *447*, 448
    lagging vs. leading, 461
    light as, 1047–1052
    net, 458, 495
    phasors, 462–464, *463*
    principle of superposition for, 458, *458*
    probability, 1162–1164, 1167
    resultant, 458, *458*

sample problems involving, 450–452, 455, 461, 464, 469–470
    seismic, 512
    shock, 33, 503, *503*
    sinusoidal, 446–448, *447*
    sound, *see* sound waves
    speed of traveling waves, 449–451
    standing, *see* standing waves
    on stretched string, 452
    string, 451–455
    transverse and longitudinal, *445*, 445–446, *446*, 450–451
    traveling, *see* traveling waves
    types of, 445
    wavelength and frequency of, 446–449
wave equation, 456–457
wave forms, *445*, 446
wavefronts, 480, *480*, 966
wave function, 1170–1172. *See also* Schrödinger's equation
    hydrogen ground state, 1208–1210, *1209*
    normalizing, 1193–1195
    of trapped electrons, 1191–1195, *1192*
wave interference, *459*, 459–461, *460*, 485–488, *486*
wavelength, 447, *447*
    Compton, 1161
    cutoff, 1156–1157, 1237
    de Broglie, 1167, 1171, 1189
    determining, with diffraction grating, 1099
    and frequency, 446–449
    of hydrogen atom, 1203
    and index of refraction, 1050–1052
    proper, 1135
    sound waves, 483
wavelength Doppler shift, 1136
wave shape, 446
wave speed, *449*, 449–453
    sound waves, 483
    on stretched string, 452–453, *453*
    traveling waves, *449*, 449–451
wave theory of light, 1047–1052, 1081–1083
wave trains, 1241
weak force, 1338, 1353
weak interaction, 1341
weber (unit), 866
weight, 103–104
    apparent, 104, 396–397
    mass vs., 104
weightlessness, 134
whiplash injury, 30
white dwarfs, 367t, 387t
white light:
    chromatic dispersion, *993*, 993–994, *994*
    single-slit diffraction pattern, 1085

Wien's law, 1166
Wilkinson Microwave Anisotropy
    Probe (WMAP), 1360
windings, solenoid, 848–849
window glass, thermal
    conductivity of, 535t
Wintergreen LifeSaver, blue
    flashes from, 613
WMAP (Wilkinson Microwave
    Anisotropy Probe), 1360
W messenger particle, 1353
work, 283t
    and applied force, 688
    for capacitor with dielectric,
        733
    Carnot engines, 592
    and conservation of mechanical
        energy, 184–187
    and conservation of total
        energy, 195–199, *197*
    defined, 151
    done by electric field, 688–689
    done by electrostatic force,
        688–689
    done by external force with
        friction, 192–194

done by external force without
    friction, 192
done by gravitational force,
    155–158, *156*
done by ideal gas, 552–554
done by spring force, *159,*
    159–162
done by variable force,
    162–166, *163*
done in lifting and lowering
    objects, *156,* 156–158
done on system by external
    force, 191–194, *193*
and energy/emf, 773–774
first law of thermodynamics,
    531–533
and heat, 524, 528–530
and induction, 872, 873
and kinetic energy, *152,*
    152–155, 1141
and magnetic dipole moment,
    825–826
negative, 530
net, 153, 592
path-dependent quantity,
    530

path independence of
    conservative forces,
    179–181, *180*
and photoelectric effect, 1158
and potential energy, *178,*
    178–181, *179*
and power, 166–168, *167*
and rotational kinetic energy,
    282–284
sample problems involving,
    154–155, 157–158, 161–162,
    164–166, 533
signs for, 153
work function, 1157
working substance, 590–591
work-kinetic energy theorem,
    153–155, 164–166, 283t

**X**

*x* component, of vectors, 42–43, *43*
xenon, decay chain, 1311
xi-minus particle, 1347t,
    1348–1349, 1352
x rays, 445, 974
    characteristic x-ray spectrum,
        1237–1238, *1238*

continuous x-ray spectrum,
    1237, *1237*
and ordering of elements,
    1236–1240
radiation dosage, 1296
x-ray diffraction, 1104–1106, *1105*

**Y**

*y* component, of vectors, 42–43, *43*
yield strength, 340, *340,* 341t
Young's double-slit interference
    experiment, 1054–1058, *1055*
    single-photon version, *1162,*
        1162–1164
    wide-angle version, 1163–1164,
        *1164*
Young's modulus, 341, 341t
yo-yo, 301–302, *302*

**Z**

zero angular position, 259
zero-point energy, 1193–1194
zeroth law of thermodynamics,
    515–516, *516*
zeroth-order line, 1099
Z messenger particle, 1353

# SOME PHYSICAL CONSTANTS*

| | | |
|---|---|---|
| Speed of light | $c$ | $2.998 \times 10^8$ m/s |
| Gravitational constant | $G$ | $6.673 \times 10^{-11}$ N · m$^2$/kg$^2$ |
| Avogadro constant | $N_A$ | $6.022 \times 10^{23}$ mol$^{-1}$ |
| Universal gas constant | $R$ | $8.314$ J/mol · K |
| Mass–energy relation | $c^2$ | $8.988 \times 10^{16}$ J/kg |
| | | $931.49$ MeV/u |
| Permittivity constant | $\varepsilon_0$ | $8.854 \times 10^{-12}$ F/m |
| Permeability constant | $\mu_0$ | $1.257 \times 10^{-6}$ H/m |
| Planck constant | $h$ | $6.626 \times 10^{-34}$ J · s |
| | | $4.136 \times 10^{-15}$ eV · s |
| Boltzmann constant | $k$ | $1.381 \times 10^{-23}$ J/K |
| | | $8.617 \times 10^{-5}$ eV/K |
| Elementary charge | $e$ | $1.602 \times 10^{-19}$ C |
| Electron mass | $m_e$ | $9.109 \times 10^{-31}$ kg |
| Proton mass | $m_p$ | $1.673 \times 10^{-27}$ kg |
| Neutron mass | $m_n$ | $1.675 \times 10^{-27}$ kg |
| Deuteron mass | $m_d$ | $3.344 \times 10^{-27}$ kg |
| Bohr radius | $a$ | $5.292 \times 10^{-11}$ m |
| Bohr magneton | $\mu_B$ | $9.274 \times 10^{-24}$ J/T |
| | | $5.788 \times 10^{-5}$ eV/T |
| Rydberg constant | $R$ | $1.097\,373 \times 10^7$ m$^{-1}$ |

*For a more complete list, showing also the best experimental values, see Appendix B.

# THE GREEK ALPHABET

| | | | | | | | | | |
|---|---|---|---|---|---|---|---|---|---|
| Alpha | A | $\alpha$ | Iota | I | $\iota$ | Rho | P | $\rho$ |
| Beta | B | $\beta$ | Kappa | K | $\kappa$ | Sigma | $\Sigma$ | $\sigma$ |
| Gamma | $\Gamma$ | $\gamma$ | Lambda | $\Lambda$ | $\lambda$ | Tau | T | $\tau$ |
| Delta | $\Delta$ | $\delta$ | Mu | M | $\mu$ | Upsilon | Y | $\upsilon$ |
| Epsilon | E | $\epsilon$ | Nu | N | $\nu$ | Phi | $\Phi$ | $\phi, \varphi$ |
| Zeta | Z | $\zeta$ | Xi | $\Xi$ | $\xi$ | Chi | X | $\chi$ |
| Eta | H | $\eta$ | Omicron | O | $o$ | Psi | $\Psi$ | $\psi$ |
| Theta | $\Theta$ | $\theta$ | Pi | $\Pi$ | $\pi$ | Omega | $\Omega$ | $\omega$ |

# SOME CONVERSION FACTORS*

**Mass and Density**

$1 \text{ kg} = 1000 \text{ g} = 6.02 \times 10^{26} \text{ u}$

$1 \text{ slug} = 14.59 \text{ kg}$

$1 \text{ u} = 1.661 \times 10^{-27} \text{ kg}$

$1 \text{ kg/m}^3 = 10^{-3} \text{ g/cm}^3$

**Length and Volume**

$1 \text{ m} = 100 \text{ cm} = 39.4 \text{ in.} = 3.28 \text{ ft}$

$1 \text{ mi} = 1.61 \text{ km} = 5280 \text{ ft}$

$1 \text{ in.} = 2.54 \text{ cm}$

$1 \text{ nm} = 10^{-9} \text{ m} = 10 \text{ Å}$

$1 \text{ pm} = 10^{-12} \text{ m} = 1000 \text{ fm}$

$1 \text{ light-year} = 9.461 \times 10^{15} \text{ m}$

$1 \text{ m}^3 = 1000 \text{ L} = 35.3 \text{ ft}^3 = 264 \text{ gal}$

**Time**

$1 \text{ d} = 86\,400 \text{ s}$

$1 \text{ y} = 365\tfrac{1}{4} \text{ d} = 3.16 \times 10^7 \text{ s}$

**Angular Measure**

$1 \text{ rad} = 57.3° = 0.159 \text{ rev}$

$\pi \text{ rad} = 180° = \tfrac{1}{2} \text{ rev}$

**Speed**

$1 \text{ m/s} = 3.28 \text{ ft/s} = 2.24 \text{ mi/h}$

$1 \text{ km/h} = 0.621 \text{ mi/h} = 0.278 \text{ m/s}$

**Force and Pressure**

$1 \text{ N} = 10^5 \text{ dyne} = 0.225 \text{ lb}$

$1 \text{ lb} = 4.45 \text{ N}$

$1 \text{ ton} = 2000 \text{ lb}$

$1 \text{ Pa} = 1 \text{ N/m}^2 = 10 \text{ dyne/cm}^2$

$\qquad = 1.45 \times 10^{-4} \text{ lb/in.}^2$

$1 \text{ atm} = 1.01 \times 10^5 \text{ Pa} = 14.7 \text{ lb/in.}^2$

$\qquad = 76.0 \text{ cm Hg}$

**Energy and Power**

$1 \text{ J} = 10^7 \text{ erg} = 0.2389 \text{ cal} = 0.738 \text{ ft} \cdot \text{lb}$

$1 \text{ kW} \cdot \text{h} = 3.6 \times 10^6 \text{ J}$

$1 \text{ cal} = 4.1868 \text{ J}$

$1 \text{ eV} = 1.602 \times 10^{-19} \text{ J}$

$1 \text{ horsepower} = 746 \text{ W} = 550 \text{ ft} \cdot \text{lb/s}$

**Magnetism**

$1 \text{ T} = 1 \text{ Wb/m}^2 = 10^4 \text{ gauss}$

*See Appendix D for a more complete list.

# Notes

# Notes

# Notes

# Notes

# Notes

# Notes

# Notes

# Notes

# Notes

# Notes

# Notes

# Notes

# Notes